Springer
Berlin
Heidelberg
New York
Barcelona
Budapest
Hong Kong
London
Milan
Paris
Tokyo

R. Brun   L. Z. Dumitrescu (Eds.)

Shock Waves @ Marseille II

Physico-Chemical Processes and Nonequilibrium Flow

Proceedings
of the 19th International Symposium
on Shock Waves
Held at Marseille, France, 26-30 July 1993

With 416 Figures

Springer

Professor Dr. Raymond Brun

Professor Dr. Lucien Z. Dumitrescu

Université de Provence, Centre Saint-Jérôme
IUSTI-MHEQ, Case 321, F-13397 Marseille Cedex 20, France

The Preface and lists of the Symposium Committees and of the Sponsoring Organisations are printed in Volume I.

ISBN 3-540-57711-4 Springer-Verlag Berlin Heidelberg New York
ISBN 0-387-57711-4 Springer-Verlag New York Berlin Heidelberg

Set (Volumes I-IV):
ISBN 3-540-57713-0 Springer-Verlag Berlin Heidelberg New York

CIP data applied for

Typesetting: Camera ready by editors
SPIN: 10132532 55/3144 - 5 4 3 2 1 0 - Printed on acid-free paper

Contents - Volume II*

Survey Paper

Plenary Lectures

Part 1: Chemical Kinetics

* The Contents of the other volumes are given at the end of the book.

Part 2: Combustion Kinetics

Part 3: Non-Equilibrium Flow

Survey Paper

Plenary Lectures

Volume II: Physico-Chemical Processes and Non-Equilibrium Flows - An Introductory Survey

John H. Kiefer
Department of Chemical Engineering, University of Illinois at Chicago, Chicago, Illinois 60607, USA

This volume contains 72 papers on physico-chemical processes and non-equilibrium flows presented at the 19th ISSW, Marseille, July 26-30, 1993. These papers are divided herein into groups concerned with chemical kinetics (22), combustion kinetics (12), non-equilibrium flows (24), and plasmas/astrophysics (14). This introduction attempts to place all this work in the context of current knowledge and practice in the application of shock waves to the observation of chemical and physical processes. As such, it begins with a brief consideration of this context, the history and current status of the field, and then considers the present papers in light of this context.

There are but three papers dealing with liquid-gas systems, the remaining papers presented herein confined to shock waves in gases, the great majority using the shock tube in one of its various guises, depending on application. This writer's experience is also with shock-tube applications, so the following discussion will be limited to shock-tube studies on gases.

Shock wave heating has long been used for the study of thermal physico-chemical phenomena in gases. Virtually instantaneous heating, freedom from wall effects, and an almost unlimited usable range of temperature and pressure have kept shock waves the preferred method for the investigation of high-temperature phenomena.

Shock tubes have been used for physico-chemical measurements since the 1950's. The field took off during the 60's, largely motivated by interest in hypersonic flight through the various space programs. The early work and the fundamentals of shock tube function and application are well covered in the three dedicated early volumes by Bradley (1962), Greene and Toennies (1964), and Gaydon and Hurle (1963). The volume edited by A. Lifshitz (1981) contains an extensive discussion of various diagnostic techniques as well as listing applications. Early reviews of shock tube chemistry were given by Bauer (1965) and Belford and Strehlow (1969). More recent are the reviews of Tsang and Lifshitz (1990) and Michael and Lim (1993). These latest compilations show that chemical applications of the shock tube continue to proliferate. The use of the shock tube on physico-chemical phenomena thus remains a thriving area of research. This is evidenced most currently and directly by the 72 papers of this volume.

The shock tube has dominated the fields of high-temperature chemistry and physics since its introduction. There have been serious attempts to dethrone it, for example with high-temperature fast-flow reactors (Fontijn and Zellner 1983), laser heating (Smith 1984), or flash photolysis (Brand et al. 1990). Although these methods have been of great value in themselves, they have in no way displaced the shock tube. The fast-flow and laser heating techniques cannot reach temperatures much above 1500K; here the shock tube reigns supreme. High-energy photolysis, e.g., using excimer lasers, can also identify very endothermic dissociation and isomerization channels, but it is often difficult to make the connection to thermal processes.

Of course, all is not perfect. As with any reactor, the shock tube has its faults and limitations. It is first a one-shot experiment, difficult to reproduce and allowing no direct signal averaging. The heating time is short (ms) and fast diagnostics with their attendant lower signal to noise are often necessary. The high spatial resolution needed to retain temporal resolution in such a fast-flow device increases the difficulty of obtaining adequate signal in optical diagnostics. These last problems are in large part responsible for the growing use of laser diagnostics in shock-tube research.

Shock Waves @ Marseille II
Editors: R. Brun, L. Z. Dumitrescu

The one serious inherent source of problems in shock-tube applications is the cold wall boundary layer which grows rapidly behind the incident wave and whose effects are exacerbated in the reflected wave. In some early single-pulse shock-tube pyrolysis experiments this was responsible for truly major errors in reported rate constants. It appears that unreacted parent gas retained in the boundary layer led to grossly underestimated conversions in experiments with (otherwise) large conversions (Skinner 1977).

At one time it was thought the boundary layer would always severely limit the use of reflected shock techniques, like the single-pulse chemical shock tube, in chemical kinetics (Belford and Strehlow 1969). However, it is now clear that this concern was much overstated. Reliable and accurate rate constants are now routinely extracted from optical measurements in reflected waves (see Lifshitz 1981, Tsang and Lifshitz 1990, Michael and Lim 1993). The putative problems with the single-pulse technique are routinely and neatly avoided (Lifshitz, Frenklach, and Burcat 1976), or circumvented (Tsang 1981). Corrections for non-ideal behavior in incident (Fujii et al. 1979) and reflected (Michael and Sutherland 1986) waves are well-established and in general use. Of course it is essential that the need for such corrections be recognized when appropriate. Then, used properly, the shock tube remains the optimum device for the investigation of high-temperature phenomena.

What follows is an introduction to the papers, grouped more or less by technique, and certainly not in order of importance. The author is a physical chemist, so this consideration regretably but unavoidably emphasizes the chemical kinetics/combustion kinetics portions of the program. The papers on non-equilibrium flows and plasmas are not, therefore, given the kind of informed consideration they deserve.

The plenary lecture by R. K. Hanson sets the stage for many of the included papers using laser diagnostics and/or laser flash photolysis for reactant formation. Hanson lists some advantages of the use of tuned, narrow line-width CW lasers for species detection in absorption: e.g., greater sensitivity (S/N) and selectivity, which allow increased reactant dilution, reducing temperature changes and secondary reactions. With their excellent collimation, lasers are also capable of much improved spatial/temporal resolution. He also discusses the advantages of the combination of laser photolysis and the shock tube, which permits the study of radical reactions over a larger range of temperature, and reduces some of the complications involved in pyrolytic radical production.

The above laser techniques represent the state of the art in shock-tube diagnostics, and it is gratifying to see a good number of papers using them in a wide range of application. Here we have Mertens et al. using both laser absorption of OH and laser photolytic generation of H atoms to study the reactions $H+N_2O \rightarrow N_2+OH$. Braun-Unkhoff and Frank, and Woolridge, Hanson, and Bowman also used laser absorption of OH to study CH_3+O_2 reactions and the pyrolysis of HNO_3, respectively. Markus and Roth observe both CH and NO in their mutual reaction; He et al. follow both CO and NO with a CO laser; and Thoma and Hindelang observed O_3 absorption at 257.25 nm with a frequency-doubled argon ion laser. Laser photolysis was also used by Tsuchiya et al. to form O-atoms.

Other laser diagnostics include a laser-schlieren study of CH_3 recombination by Su and Teitelbaum, and the elegant and potentially very useful introduction of laser-induced predissociation fluorescence, which avoids the quenching problems of LIF in shock waves, by Sutton et al. Perhaps this approach could be used in the shock tube by combining it with the axial imaging technique described by Gray (1990) in which excitation is provided by a short-pulse laser directed axially, the resulting LIF imaged into a one-dimensional array detector viewing from the side. This technique then offers both spatial (temporal) resolution and increased sensitivity.

Other recognized shock-tube diagnostic techniques are well represented, often by their leading practitioners. Atomic resonance absorption spectrometry (ARAS) was used by Mick and Roth (O, Si, N), Natarajan and Roth (H) Woiki and Roth ($S(^3P)$ and $S(^1D)$, Wintergerst and Frank

(H, I), Tsuchiga et al. (O), Miyoshi, et al. (O) and Saito et al. (Cl atom). This technique has well-established virtues of specificity and sensitivity which are exploited here for a remarkable range of atomic species. The detection of two electronic states of S-atom by Woiki and Roth is a particularly striking extension of the method.

The single-pulse, chemical shock tube is represented by its two leading practitioners: W. Tsang and A. Lifshitz. Lifshitz continues his extensive and invaluable investigations of the pyrolysis of large organics with a study of three oxirans. Tsang and Cui continue their elegant use of hexamethylethane for the pyrolytic production of H-atoms, here investigating reactions with acetone and dimethylether. It is worth noting that others, Wintergerst and Frank (H from C_2H_5I) and Fujii et al. (C_6H_5 from C_6H_5NO) have also found good use for pyrolytic radical formation.

R. D. Kern and coworkers have continued their important exploration of the reactions of C_3 hydrocarbons using the difficult but irreplacable technique of time-of-flight mass spectrometry.

The measurement of ignition delays, a unique contribution of the shock tube to the understanding/modeling of combustion, continues to be well-represented. Here we have Burcat and Drinyaninov (n-pentane/O_2) Djebaili et al. (H_2/air/ or H_2O), Fieweger et al. (several prototype hydrocarbon fuels/air), and Lissianski, Gardner, and Zamansky, who considered the application of a global mechanism to modeling CH_4/O_2 delays. These ignition delay measurements provide one of the most important experimental criteria in modern combustion modeling. Other papers explore the usefulness of the shock tube in the study of soot formation (Yao et al.) and even as a means of chemical production (Mattik et al.).

The section on non-equilibrium flows is dominated by applications of free-piston shock tubes/shock tunnels for the production of high-enthalpy flows, as introduced in the plenary paper by Eitelberg. Tunnel and shock flows have been observed or modeled in CO_2 (Morgan et al., Wen and Hornung), N_2 (Wen and Hornung, Sharma, Boyce et al.), air (Herman and Brenner; Eitelberg; Koreeda et al.). Vibrational relaxation receives further attention in the Raman study of N_2 expansions by Gillespie and Sharma. Numerous theory/modeling studies of high-enthalpy flows are presented, including two papers (Baillion and Taquin, Mogstad) on radiation generated on entry to the Titan atmosphere.

The papers on plasma applications include an interesting analysis of impurity effects on shock ionization of krypton (Klages and Demmig), which points up the continuing danger of impurity effects in shock tube studies. They also include a number of theoretical studies of plasma dynamics (Tirskiy, Hemmati, Babaeva et al. , Figura, Barmin and Pushkar, and Pushkar and Grib). Experimental studies are quite varied: Sarnet et al. have studied shocks generated in laser-produced phasmas; Uteza et al., the effect of shock waves on excimer laser beam quality; Sentis et al. VUV generated plasma shocks; Chéron et al., describe a N_2 plasma set; Alyapin et al. a nanosecond discharge plasma shock generator; and finally, Koudriavtsev, Maziar, and Sukhov describe a neutral atom beam source using an electromagnetic shock tube.

In summary, this symposium exhibits the best the shock-tube community has to offer. The papers are expert, innovative, and have important application. The field of shock-waves in chemistry and physics is evidently active, vital, and forward-looking.

References

Bauer SH (1965) Shock waves. Ann. Rev. Phys. Chem. 16:245

Belford RL, Strehlow RA (1969) Shock tube technique in chemical kinetics. Ann. Rev. Phys. Chem. 20:247

Bradley JN (1962) Shock waves in chemistry and physics. Wiley, New York

Brand U, Hippler H, Lindemann L, Troe J (1990) C-C and C-H bond splits of laser-excited aromatic molecules. 1. Specific and thermally averaged rate constants. J. Phys. Chem. 94:6305

Fontijn A, Zellner R (1983) Reactions of small transient species, kinetics and energetics. Academic Press, London

Fujii N, Koshi M, Ando H, Asaba T (1979) Evaluation of boundary layer effects in shock-tube studies of chemical kinetics. Intl. J. Chem. Kinet. 11:285

Gaydon AG, Hurle IR (1963) The shock tube in high temperature chemical physics. Reinhold, New York

Gray JA (1990) A shock tube with tunable, pulsed laser-based diagnostics. Rev. Sci. Instrum. 61:1825

Greene EF, Toennies JP (1964) Chemical reactions in shock waves. Academic Press, New York

Lifshitz A (ed) (1981) Shock waves in chemistry. Marcel Dekker, New York

Lifshitz A, Frenklach M, Burcat A (1981) Pyrolysis of allene and propyne behind reflected shocks. J. Phys. Chem. 80:245

Michael JV, Lim KP (1993) Shock tube techniques in chemical kinetics. Ann. Rev. Phys. Chem. 44:429

Michael JV, Sutherland JW (1986) The thermodynamic state of the hot gas behind reflected shock waves: implications to chemical kinetics. Intl. J. Chem. Kinet. 18:409

Skinner GB (1977) The importance of wall cooling in single-pulse shock tube studies. Intl. J. Chem. Kinet. 9:863

Smith GP (1984) Laser pyrolysis techniques: application to catalysis, combustion diagnostics, and kinetics. Proc. Soc. Photo. Intr. Eng; 11:458

Tsang W (1981) Comparative rate single-pulse shock tube studies of the thermal stability of polyatomic molecules. Chapter 2, p. 59 of Lifshitz (1981)

Tsang W, Lifshitz A(1990) Shock tube techniques in chemical kinetics. Ann. Rev. Phys. Chem. 41:559

The Role of Lasers in Shock Tube Studies of Chemical Kinetics

R.K. Hanson
High Temperature Gasdynamics Laboratory, Stanford University, Stanford, CA 94305 USA

Abstract. The role of lasers in shock tube studies of chemical kinetics has grown remarkably since the first application about 20 years ago, leading to dramatic improvements in experimental capability for measuring reaction rate coefficients at high temperatures. In this paper we discuss two of the most prominent laser methods, namely laser absorption diagnostics for sensitive and accurate species detection, and laser photolysis of shock-heated gases for direct preparation of molecular and atomic radical reactants.

Key words: Laser Absorption, Photolysis, Kinetics

1. Background

Shock tubes have provided the bulk of modern chemical kinetic data for gaseous reactions at high temperatures, particularly above the temperature limit of heated steady-flow reactors (about 1500 K). Although spectroscopic diagnostics have played a key role in such work, from the time of the first shock tube kinetics study about forty years ago, the introduction of laser-based absorption diagnostics about twenty years ago has had a major impact on the quality of kinetics data obtainable in shock tube experiments. In particular, the increased sensitivity of laser absorption has allowed use of more dilute reaction mixtures, leading to less interference from secondary reactions in inferring the reaction rate of interest; and the increased measurement accuracy with laser absorption has led to a significant reduction in the scatter of kinetic data. Within the past six years or so, there has been a further significant development involving lasers in shock tube experimentation, namely the introduction of pulsed laser photolysis to provide an effectively instantaneous source of reactive radical species, thereby facilitating the direct study of elementary radical-radical and radical-molecule reactions in the high temperature environment provided by shock-wave heating. This new approach to high temperature reaction kinetics seeks to marry the advantages of shock wave heating in preparing gases at accurately known temperatures, and without diffusion effects, with the advantages of laser photolysis for direct preparation of reactant species. A further merit of laser photolysis methods is that they allow extension of shock tube methods to lower temperatures than normally possible with pyrolytic methods of reactant preparation. In this paper we will discuss both of these modern shock tube methods, namely sensitive cw laser absorption spectroscopy and the laser-photolysis shock tube.

2. CW dye laser absorption

The development of laser absorption as a shock tube diagnostic for reaction kinetics began nearly 20 years ago (e.g., see Hanson et al. 1976). In that early work, the laser source was a discretely-tunable cw CO electric-discharge infrared laser, and fortuitous coincidences between CO laser lines and NO absorption lines formed the basis of the measurement. Subsequently other coincidences were identified to enable quantitative detection of H_2O (Salimian and Hanson 1983) and N_2O using the same laser source tuned to other emission transitions; detection of excited-state populations of CO was also demonstrated (Louge and Hanson 1984) in our laboratory and subsequently utilized in a number of studies at the Naval Research Laboratory by Lin and co-workers. The

Shock Waves @ Marseille II
Editors: R. Brun, L. Z. Dumitrescu

critical advantages provided by laser absorption in these various studies were the improved sensitivity and accuracy of the species concentration measurements relative to conventional broadband absorption and emission measurements.

An important further improvement in the laser absorption approach was provided in 1983 (e.g., see Hanson et al. 1983) when a cw UV dye laser was first coupled to a shock tube to enable measurements of the OH radical. This work was significant for two reasons: (1) it demonstrated feasibility for accessing the larger absorption coefficients which apply in the UV (electronic rather than vibrational transitions), thereby lowering the species detection limits considerably; and (2) it confirmed the ability to monitor OH, which is a critical species for a number of important combustion reactions. Even in this initial demonstration, the use of laser absorption enabled substantial improvements in OH detection accuracy and sensitivity over previous diagnostic methods. These diagnostics improvements have already resulted in a substantial number of improved rate coefficient determinations.

More recently, extensions of this same general approach have been carried out to enable cw dye laser absorption measurements of CH (at 431 nm) (Dean and Hanson 1989), NCO (440 nm), NH_2 (597 nm), NH (336 nm), NO (226 nm), CH_3 (216 nm) (Davidson et al. 1993), and CN (at 388 nm) (Wooldridge et al. 1993). Unfortunately, each of these measurements requires a different dye laser arrangement, either in terms of frequency-doubling crystal, laser dye, or laser optics, and hence it is not always easy to change from detecting one species to another. Also, it is necessary to perform supporting spectroscopic calculations, and supporting laboratory validations, for each species to provide the absorption coefficient as a function of temperature at the chosen transitions (wavelengths). Progress in developing and validating these new species diagnostics thus requires a substantial effort, but the long-term advantages of the improved methods in determining reaction rate coefficients clearly justify the effort involved.

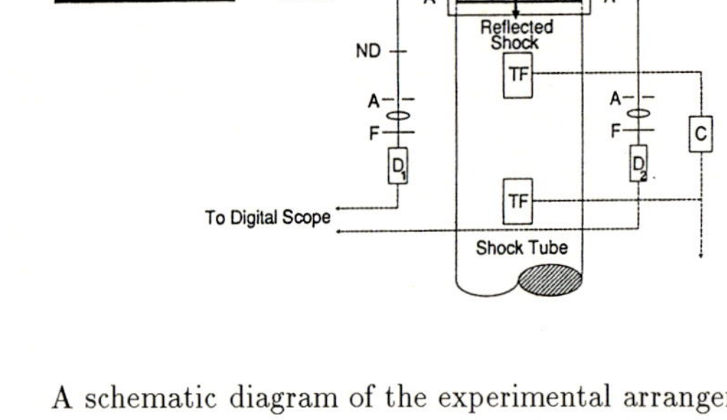

Fig. 1. Experimental arrangement for cw dye laser absorption measurements in a shock tube

A schematic diagram of the experimental arrangement employed for cw dye laser absorption is shown in Fig.1. An argon-ion laser is used to pump a single-frequency ring dye laser which emits 0.1-0.5 W at visible wavelengths. For several species the laser light is frequency-doubled internal to the ring cavity in a non-linear crystal, yielding a final output power of 1-50 mW. The spectrally narrow (1 MHz) laser light is single-passed through the shock tube about 1.9 cm from the end wall, and low levels of absorption are monitored using a two-beam differencing scheme.

The nominal laser wavelength is monitored with a wavemeter, but the final wavelength selection is typically made by peaking the absorption signal of a small fraction of the beam passed through either a flat flame burner, a static absorption cell, or a low pressure electrical discharge tube. Further details of the experimental arrangement and strategies are given in recent papers (e.g., see Davidson et al.1993); here we will focus on the detection sensitivity which can be achieved with this measurement technique.

The single-frequency laser absorption is governed by Beer's law evaluated at the center of a Voigt-broadened line in the electronic absorption spectrum of the species under study. The relevant equation for the fractional transmission of the laser light is

$$T_V = (I/I_0)_\nu = \exp(-\beta_\nu P X_i L)$$

where β_ν is the spectral absorption coefficient ($cm^{-1} \cdot atm^{1}$), P is the static pressure of the gas mixture (atm), L is the optical pathlength (one or two passes of the tube diam, 14.3 or 15 cm), and X_i is the mole fraction of the absorbing species.The absorption coefficient is simply the product of an integrated line strength ($cm^{-2}.atm^{-1}$) and the line-center value of the lineshape function (cm); the former can be calculated directly from the electronic oscillator strength of the molecule and the latter can be calculated using a Voigt lineshape function. Measurements of the collision-broadening coefficient, needed for the Voigt calculation, are made separately. The data reduction process is quite simple and involves converting each local value of the transmission directly to a value for the mole fraction of the species using Beer's law and the known values of L, P and β_ν. The value of P is found from the initial pressure and standard shock wave calculations in terms of the measured shock speed; the same calculations yield the post-shock temperature needed to specify the absorption coefficient.

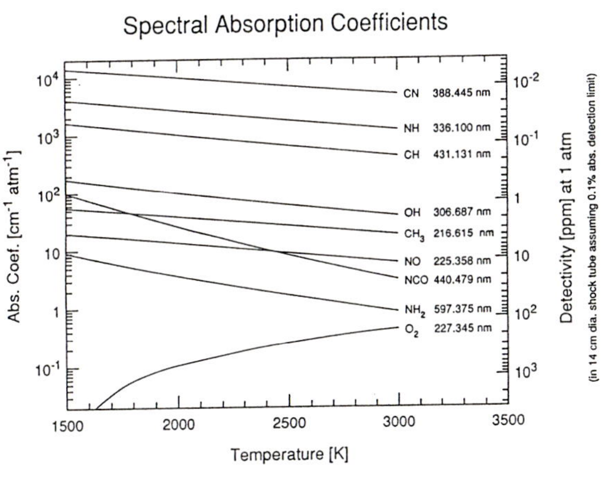

Fig. 2. Spectral absorption coefficients as a function of temperature for several radical species; the detection limit (ppm) for $P = 1$ atm, $L = 14.3$ cm, is shown on the right-hand ordinate

Fig.2 presents calculated values of the absorption coefficient for several species of interest. These coefficients were evaluated at line center (or at the peak absorption in the case of overlapping lines) using a Voigt profile and an assumed pressure of 1 atm. Note that larger absorption coefficients lead to better species detection limits, and hence diatomic species can generally be monitored at lower levels than triatomic species. In most cases, our detection limit is about 0.1 % absorption (SNR = 1), and hence it is easy to estimate the lower level of detection for a given species once the appropriate absorption coefficient has been calculated. This detection limit for single-pass absorption ($L = 14.3$ cm, $P = 1$ atm) is given on the right-hand ordinate of Fig. 2.

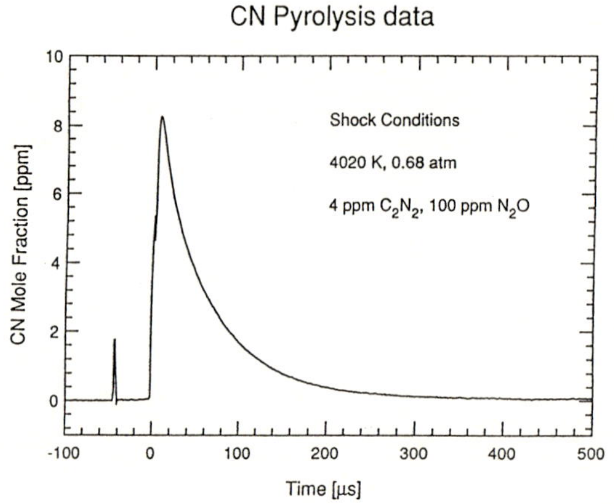

Fig. 3. CN laser absorption data for an experiment to measure the rate of O+CN→N+CO

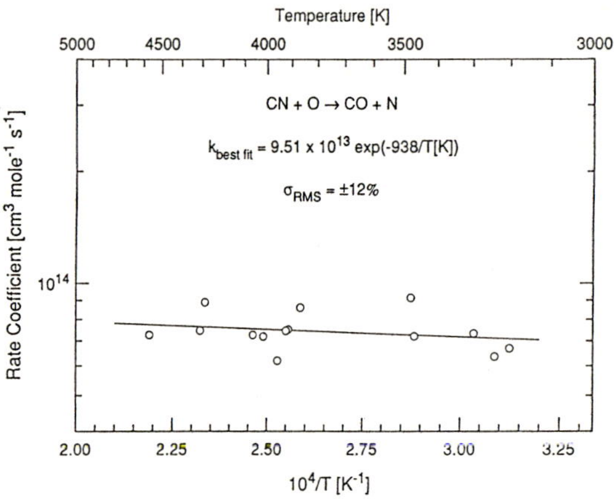

Fig. 4. Arrhenius diagram for the reaction of O+CN→N+CO

By the way of example, the detection limit for OH near 2000 K is below 1 ppm, while that for CN is about 0.01 ppm! These detection limits illustrate clearly the high sensitivity of the laser-absorption method. Such high sensitivity allows study of very dilute systems which are relatively free of the complications which arise when several reactions are active simultaneously.

An example data trace for CN absorption in a reflected-shock C_2N_2-N_2O-Ar mixture at 4020 K, 0.68 atm is shown in Fig.3. These data were acquired in a study of the rate coefficient for O+CN→N+CO (Davidson et al.1991). At these high temperatures, the N_2O decomposes instantaneously to N_2 and O-atoms; the decomposition of cyanogen (C_2+N_2 → 2CN) is not quite as fast and is responsible for the finite time to reach the peak CN concentration. (The known absorption coefficient for CN (Davidson et al. 1991, and Wooldridge et al. 1993) was used to convert the fractional absorption to mole fraction). The subsequent decay of CN is due entirely to the reaction of interest, and the high quality of the data, even at the ppm levels of CN utilized, allow quite accurate determination of the rate coefficient in each experiment. A summary of rate coefficient results is shown in Fig.4. The important point is that the rms deviation of these data is quite small, about 12%, relative to that achievable with non-laser-based spectroscopic methods.

3. Laser photolysis shock tube

The second experimental technique to be covered in this paper is a relatively new type of chemical kinetics shock tube. The objective is to provide improved capability for generating reactive radical species in shock tube experiments. This goal is motivated by the strong current interest in obtaining rate coefficient data for elementary reactions involving these species (e.g., radical-molecule and radical-radical reactions). Previously, the shock tube method employed to generate radical species involved use of a precursor molecule which decomposed by thermal reaction to produce the radical of interest. Unfortunately, this process often is coupled to many other simultaneous pyrolysis reactions, with the result that interference reactions often obscure the reaction of interest. In the new approach, an intense pulse of UV laser light is used to photolyze the precursor molecules, thereby instantly converting a fraction of these molecules to the radical of interest. In general, this is done in the reflected shock region of a shock tube, following shock heating, so that the subsequent reactions of the radical species occur at the high temperature of interest.

The method under development is similar in many respects to that used by reaction kineticists working with low temperature static cells. The major difference is that the shock tube provides a transient high temperature environment, and because the experiment is a single-shot event the diagnostics employed cannot be signal-averaged as is often done in low temperature work. Although our work utilizes laser photolysis in a shock tube, it should be noted that previous workers in the shock tube community employed flashlamps to photolyze molecules in shock-heated gases. Our hope is that the greater monochromaticity and higher spectral intensities available with laser sources will provide much more efficient photolysis and less ambiguity in the identity and quantum states of the photofragments produced.

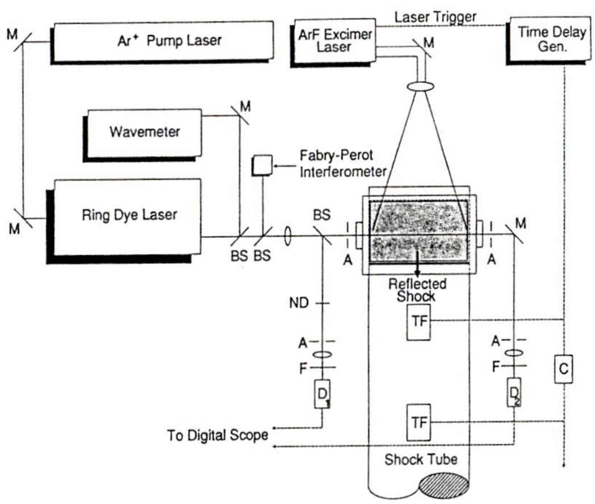

Fig. 5. Experimental arrangement for the laser photolysis shock tube in the axial illumination mode

The experimental arrangement employed in our current laser photolysis shock tube is shown schematically in Fig.5. The photolysis source is usually an argon fluoride excimer laser which provides up to 100 mJ of 193 nm radiation in a 25 nsec pulse, although excimer light at 157 nm and 248 nm has also been employed when needed. The radiation is formed into a rectangular beam (about 6 mm in height) and transmitted through a rectangular quartz window in the shock tube end wall so as to nearly fill the entire width of the tube (14.3 cm) at the location of the cw probe laser beam (typically 1.9 cm from the end wall). Typical fluence levels are 5 to 10 mJ/cm^2

in this illumination mode, and up to 100 mJ/cm^2 when the laser light is introduced transverse to the shock tube axis and coaxially with the probe beam.

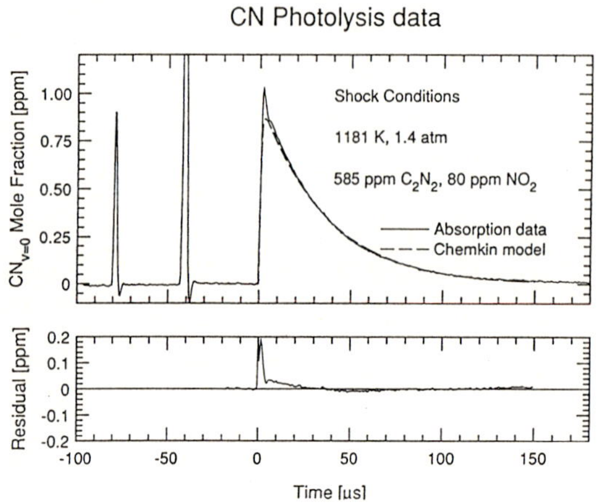

Fig. 6. CN laser absorption data in a laser photolysis experiment to measure the rate of $CN+NO_2 \rightarrow NCO+NO$

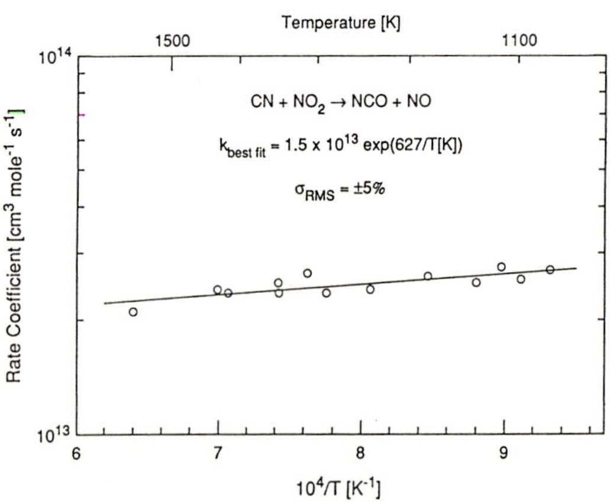

Fig. 7. Arrhenius diagram for the reaction $CN+NO_2 \rightarrow NCO+NO$

Thus far we have investigated several candidate source compounds for preparation of atomic and molecular radicals: NH_3 as a source of H, NH and NH_2; H_2O as a source of H and OH; O_2 as a source of O; HNCO as a source of NH and NCO; NO as a source of N and O; C_3O_2 as a source of C; and C_2N_2 as a source of CN. Critical elements in planning laser-photolysis shock tube experiments are knowledge of the absorption coefficient of the dissociating species and the quantum yield (i.e., the fraction of absorbed photons which yield dissociated products), since these parameters together with the known fluence allow calculation of the reactant concentrations which will be produced. In most cases the absorption coefficient data are not available at high temperatures and must be measured prior to reaction kinetics experiments, although information on absorption coefficients may often be found for room temperature gases to select optimum excitation wavelengths and to estimate the high temperature absorption.

An example data trace for the CN time history following pulsed illumination of a shock-heated C_2N_2-NO_2-Ar mixture (585 ppm cyanogen and 80 ppm nitrogen dioxide) is shown in Fig.6. In this case the 193 nm laser light photolyzes a small fraction of the C_2N_2, producing about 1 ppm of CN, with little photolysis of the NO_2. The CN then reacts with NO_2, allowing a direct (i.e., interference-free) determination of the rate coefficient for $CN+NO_2 \rightarrow NCO+NO$; this is a reaction relevant to propellant chemistry. The spikes in the signal at -80 and -40 microseconds are due to laser schlieren effects as the incident and reflected shock waves passed by the probe beam location. Note the high quality of the absorption data possible even at sub-ppm levels of CN; this is due to both the favorable oscillator strength of CN and the brightness of the laser source. A calculated CN time history is also shown in Fig.6, together with a plot of the residual difference between the calculated and measured CN profiles. Although not displayed here, we also routinely employ computer-based sensitivity analyses to verify that data have minimum influence from interference reactions.

The high precision of the rate coefficient data which can be achieved is illustrated in Fig.7 for the same reaction. As indicated in the figure, the data are well fit with a simple straight line, and the rms deviation of the data is below 5%. The improved precision possible with laser-photolysis, laser-absorption methods is important, because in the past shock tube experiments have sometimes been viewed with suspicion by physical chemists.

Another advantage of the laser-photolysis shock tube is evident in this work (Figs.6 and 7), namely that the method allows measurement of reaction rates at lower temperatures than possible when pyrolysis is used to produce the reactants. In pyrolysis experiments, the lower temperature bound of reaction studies is usually set by the requirement for thermal reactions to produce sufficient radicals in the sub-millisecond test times available in a shock tube. With photolysis, however, the laser produces the radicals instantaneously, and the relatively fast rates of radical reactions can then be measured at temperatures which overlap those achievable in flow reactors.

4. Conclusions

In this paper we have demonstrated the power of laser methods for shock tube experimentation of reaction kinetics, both for controlled production of reactant species and for their detection. These methods offer prospects for substantial improvements in the measurement of rate coefficients of elementary reactions in high temperature gases.

Acknowledgements

This work was supported by the Air Force Office of Scientific Research, Aerospace Sciences Directorate, and the Office of Naval Research.

References

Davidson DF, Chang AY, Hanson RK (1988) Laser photolysis shock tube for combustion kinetics studies. Twenty-Second Combustion Symposium (International) on Combustion, The Combustion Institute, pp 1877-1885

Davidson DF, Chang AY, DiRosa MD, Hanson RK (1991) Shock tube measurements of the reactions of CN with O and O_2. Intl. J. of Chem. Kinetics 23:1035-1050

Davidson DF, Chang AY, DiRosa MD, Hanson RK (1993) A CW laser absorption diagnostic for methyl radicals. J. Quant. Spectrosc. Radiat. Trans. 49:559-571

Dean AJ, Hanson RK (1989) Development of a laser absorption diagnostic for shock tube studies of CH. J. Quant. Spectrosc. Radiat. Trans. 42:375-384

Hanson RK, Monat JP, Kruger CH (1976) Absorption of CO laser radiation by NO. J. Quant. Spectrosc. Radiat. Trans. 16:705-713

Hanson RK, Salimian S, Kychakoff G, Booman RA (1983) Shock tube absorption measurements of OH using a remotely located dye laser. Appl. Opt. 22:641-643

Louge MY, Hanson RK (1984) Shock tube study of cyanogen oxidation kinetics. Intl. J. Chem. Kin. 16:231

Salimian S, Hanson RK (1983) Absorption measurements of H_2O at high temperatures using a CO laser. J. Quant. Spectrosc. Radiat. Trans. 30:1

Wooldridge ST, Hanson RK, Bowman CT (1993) Development of a CW laser absorption diagnostic for CN in shock tube flows. J. Quant. Spectrosc. Radiat. Trans. 50:19-34

Experimental Verification of Real Gas Effects in High-Enthalpy Flows

G. Eitelberg
Institute for Experimental Fluid Mechanics, DLR, Germany

Abstract. This paper deals with the possibilities to study the effects of oxygen and nitrogen dissociation in experiments where this reaction occurs over a sufficiently short distance in stagnating flows. The scope of this paper does not cover the real gas effects of combustion in high enthalpy flows. First, the capabilities of the various types of experimental facilities for generating high enthalpy flows will be discussed. Samples of aerothermodynamic tests performed in these facilities will be discussed. Some of these experiments are reconfirmations of experiments T4 where the difference lies in the new instrumentation applied to the old flow problems, others are new studies of external aerothermodynamics like the experiments on shock/shock interactions. The significance of the dissociation effects in the above experiments is pointed out.

Key words: Real gas effects, High enthalpy, Wind tunnels

1. Introduction

In various spaceflight applications, be it winged or ballistic entry into planetary atmospheres, be it powered ascent to orbital flight or exploratory missions that traverse planetary atmospheres, the aerodynamic design of a vehicle's performance has to consider real gas effects. In particular the desire to build vehicles capable of flying to low earth orbit in single stage using air breathing engines and utilising aerodynamic lift challenges our understanding of real gas effects occurring both in aerodynamics and combustion. Another aspect of the same problem is the consideration of the gas composition and temperature when designing the thermal protection system (TPS) of a vehicle. Further interest in real gas effects is stimulated by the wish to understand the flight phenomena occurring prior to a meteorite striking the earth. The dominant real gas effects in high enthalpy flows are the effects caused by high energy molecular collisions. High pressure, compressibility, effects leading to the change in the polytropic index of a gas occur in the simulation facilities for those effects also, but they will not be considered when the simulation of flow phenomena or free flight is discussed. In most aerodynamic applications it is sufficient to describe the air, which is the gas mixture mainly considered in this paper, as a thermally and calorically perfect gas. When the model is applied to hypervelocity flows, then the gas temperatures can be significantly overpredicted and densities in shock layers underpredicted. A perfect gas going through a strong shock at near (earth) orbital velocity $u = 7000$ m/s would reach a temperature of 25×10^3 K. Fortunately, real gas effects reduce that temperature to the order of 10^4 K, (Hornung 1991).

2. Discussion of the effects to be simulated

In high enthalpy flows, the main real gas effects to be considered are those of: 1)vibrational excitation; 2)dissociation and recombination and; 3)electronic excitation and ionisation.

It is customary to define the starting of high enthalpy flows of air at a total temperature of approximately 3000 K, i.e. at a temperature above which oxygen molecule dissociation becomes significant under atmospheric pressures. This corresponds to a free flight velocity of $u = 2800$ m/s or specific stagnation enthalpy of $h_s = 4$ MJ/kg.

Shock Waves @ Marseille II
Editors: R. Brun, L. Z. Dumitrescu

When the diatomic constituents of air, O_2 and N_2 are considered, it is helpful to look at their dissociation energies h_{O_2} and h_{N_2}. They are given as: $h_{O_2} = 17$ MJ/kg, $h_{N_2} = 34$MJ/kg. In order to be able to concentrate on the simulation of the real gas effects, we first have to discuss their phenomenological importance and the cases where these effects manifest themselves.

The phenomena of dissociation and vibrational excitation occur as a result of two-body collisions. The production of vibrationally excited molecules and the production of atoms from diatomic molecules are therefore proportional to the number density of the original species. The recombination reaction, however, requires the presence of a third collision partner, and is therefore proportional to the square of the density of the original species. When we further assume that the vibrational excitation is in equilibrium with the rotational and translational excitation and consider a monomolecular gas only, then, for the approximation of an ideal dissociating gas, the thermal and caloric equations of state (Hornung 1991, Vincenti and Kruger 1965):

$$p = (k/2m)\rho T(1 + \alpha) \tag{1}$$

$$h = (k/2m)\left[(4 + \alpha)T + \alpha\Theta_d\right] \tag{2}$$

have to be complemented by an equation for the rate of change of the degree of dissociation α:

$$d\alpha/dt = C\rho T^{\eta}\left\{(1-\alpha)e^{-\Theta_d/T} - \rho/\rho_d\alpha^2\right\}, \tag{3}$$

where: ρ = density; α = degree of dissociation; Θ_d = characteristic dissociation temperature ($\Theta_d = 113000$ K for N_2); ρ_d = characteristic dissociation density ($\rho_d = 130000$ kg.m^{-3} for N_2); T = temperature; m = mass of molecule; k = Boltzmann's constant; $\eta = -2.5$; $C = 2.7 \times 10^{21}$ m.^{3}kg.$^{-1}$s.$^{-1}$(K)$^{-2.5}$.

In the above equation (3) the first term describes the dissociation phenomenon and the second term describes the recombination phenomenon. In real gas flows, where both the forward and backward reactions are in equilibrium, the degree of dissociation is described by a simple algebraic equation when the expression in the square brackets is set to equal zero. The assumption made about the vibrational equilibrium is a good one in stagnating dissociative flows, since the dissociation behind a shock removes preferentially molecules from high vibrational levels (e.g. Park 1990). In air, the presence of the second species, oxygen, affects the path to equiibrium, but not the thermodynamic state of equilibrium reached behind a strong shock in high enthalpy flow. The electronic excitation becomes significant at $h_s > 35$ MJ/kg and will not be further considered here.

From Eq.3 it is clear that the rate of dissociation is proportional to the density of the gas and the rate of recombination is proportional to the square of the density of the gas. It is possible to form a characteristic length describing the distance required by the gas to reach the equilibrium between the forward and backward reactions. In fact it is possible to form two characteristic lengths: X_d for the dissociation reaction and X_r for the recombination reaction; the latter being orders of magnitude larger than the former. It is now possible to classify the flowfields of a real gas, using the foregoing discussion:

- Equilibrium flow: the characteristic relaxation length is small compared with the geometric scales in the flowfield.
- Frozen flow: the relaxation length is large compared with the geometry scales.
- Non-equilibrium flow: the case between the above two limits.

This classification is illustrated for the earth reentry corridor of a winged vehicle and some wind tunnels in Fig.1, where the specific enthalpy of flight/simulation is normalised with the dissociation energy of nitrogen molecules and the dissociation length is normalised with the vehicle/model size L (Eitelberg et al. 1992).

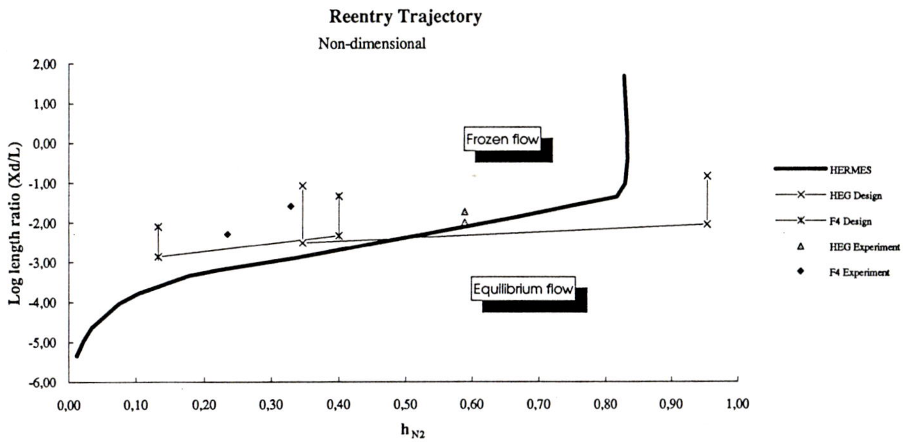

Fig. 1. Atmospheric reentry corridor and facility capabilities

The equilibrium/non-equilibrium effects can only be simulated in a wind tunnel, where the specific enthalpy of the gas is sufficiently high to excite the dissociation reaction. For fluid dynamics simulation it is obviously not sufficient to simulate the high Mach Number alone, and it is not sufficient to simulate the high enthalpy alone. When the enthalpy is high, but the relaxation length is long, the real gas effects in the flow cannot be resolved. In the following, the facilities which can simulate the real gas effects are discussed.

3. Facilities

For the study of real gas effects, clean conditions for high enthalpies are required. High enthalpy flows are achieved by heating the test gas prior to expanding it through the wind tunnel nozzle. It is possible to heat the gases by adiabatic compression, electrical plasma arcs or by shock waves. The latter two methods are predominantly used for generating high enthalpy flow. Further, a combination of adiabatic compression with compression shocks is used in the hypersonic gun tunnels, but these do not achieve a reservoir enthalpy sufficiently high to generate significant real gas effects.

3.1. Facilities heating the test gas by using electrical power

For the study of the thermal protection systems of space vehicles, where the correct simulation of the reaction lengths is not of relevance, high enthalpy flows are produced by using electrically created plasma arcs. The arc heated wind tunnels are continuous wind tunnels where stagnation enthalpies of up to $h_s = 40$ MJ/kg have been reported. With the application of special technologies the specific stagnation enthalpy can be reached in excess of $h_s > 100$ MJ/kg (e.g. Auweter-Kurtz 1992), albeit at low densities so that the reaction length cannot be correctly simulated. The arc heated continuous wind tunnels, however, reproduce the appropriate pressures.

A special version of an electric arc heated facility is the "Hot Shot". The "Hot Shot" is an impulsive wind tunnel where the test gas, contained in a finite volume, is heated by passing an extremely strong current through it until the desired reservoir condition is reached. When the gas, which is initially heated in a non-uniform manner, has reached the design conditions (pressure and temperature), then a plug is removed from the nozzle throat and the flow starts. The only major "Hot Shot" currently operational is the F4 at ONERA, France (Chanetz et al. 1992). In a "Hot Shot" the reservoir conditions change in a monotonic way after the removal of the plug from the nozzle throat. This is similar to operating a shock tunnel in an undertailored manner. When the F4 is operated under perfect gas conditions, this continuous drop in the pressure and reservoir temperature is of no significant consequence, as long as the Mach numbers are staying constant. When, however, high enthalpy effects are to be studied, then this decrease of enthalpy during a test run has to be carefully accounted for.

The F4 is capable of operating in both perfect gas and real gas regimes. The calibrated region is indicated in Fig.1.

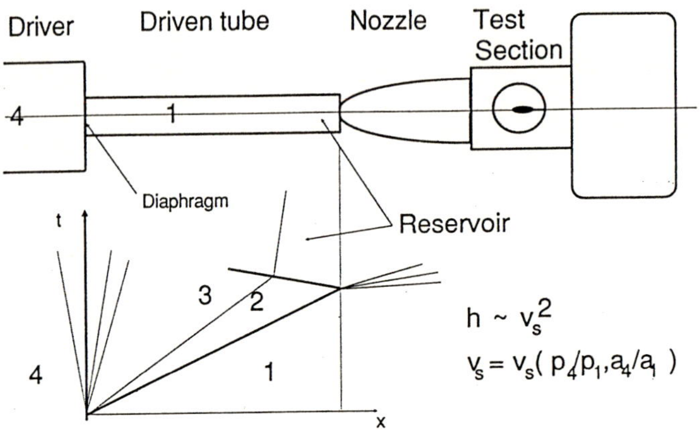

Fig. 2. Shock tunnel with an arbitrary driver and the corresponding wave diagram

3.2. Shock tunnels and expansion tubes

The study of real gas effects alone, when not the flow but the reactions are of interest, is frequently performed in shock tubes. These tools are particularly suited for the provision of controlled conditions at high enthalpies. It is a logical step, to use the same tools for generating high enthalpy flow.

A shock tube consists of a high pressure driver, a driven section and to make it into a shock tunnel, a nozzle is added to the downstream end of the driven tube, Fig.2. In this configuration, the volume of the tube between the downstream end of the driven tube and the reflected shock is used as the reservoir for the blowdown wind tunnel. Since the shock heating process is very fast, and the volume of available test gas small leading to very short test times the shock tunnels are not usually called blowdown tunnels but (im)pulse facilities instead. When the shock is not being reflected at the end of the driven tube but further accelerated into the next section, the resulting facility is called an expansion tube. The expansion tubes can produce extremely high enthalpies at moderate relaxation lengths (Erdos et al. 1990). The enthalpy levels are achieved at the cost of test time when compared with the reflected shock tunnels.

Since the specific enthalpy h_s achievable in a reflected mode is proportional to the square of the incident shock velocity v_s, $h_s \sim v_s^2$, Mach numbers of the incident shock into test gas air or N_2, M_{S_i}, required for producing high enthalpy flows, have to be $M_{S_i} > 6$. $M_{S_i} = 6.16$ is the theoretical limit obtainable in simple air to air shock tubes, when the driver and driven gas are of identical temperatures.

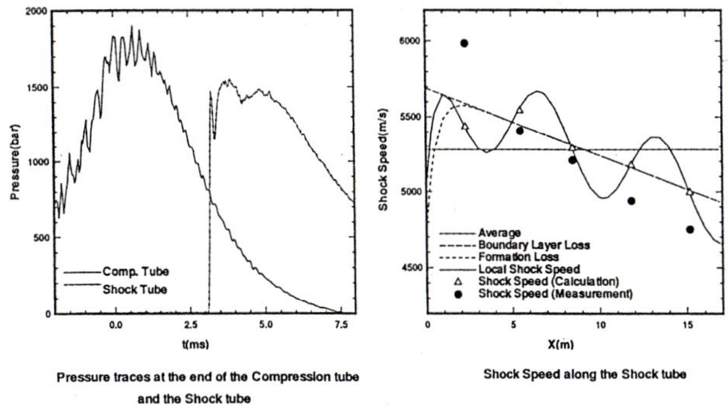

Fig. 3. HEG Shot #54: driver pressure, reservoir pressure and shock speed

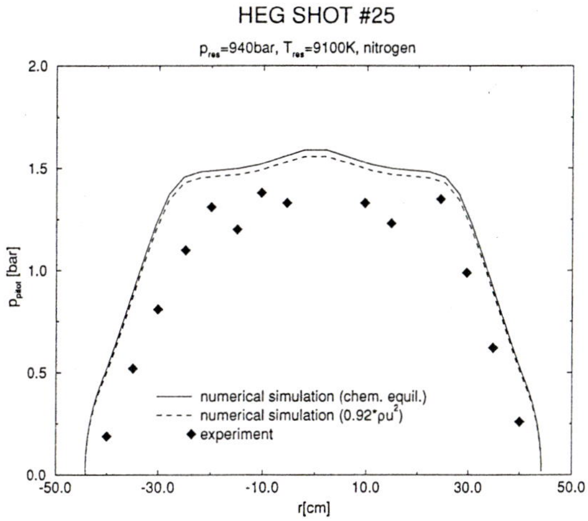

Fig. 4. Pitot profile at the HEG nozzle exit plane, reservoir pressure $p_0 \approx 100$ MPa

In order to achieve the required high incident Mach numbers in the range $6 < M_{S_i} < 20$ for high enthalpy testing, a number of techniques can be used. The equation for the incident shock (Becker 1965),

$$\frac{p_4}{p_1} = \frac{2\gamma_1 M_{S_i}^2 - (\gamma_1 - 1)}{\gamma_1 + 1} \left[1 - \frac{\gamma_4 - 1}{\gamma 1 + 1} \frac{a_1}{a_4} \left(M_{S_i} - \frac{1}{M_{S_i}} \right) \right]^{-\frac{2\gamma_4}{\gamma_4 - 1}} \tag{4}$$

where p_4 = driver pressure at burst, p_1 = driven tube charge pressure, γ_4 = ratio of specific heats of the driver gas, γ_1 = ratio of specific heats of driven gas, a_4 = speed of sound of driver gas, a_1 = speed of sound of driver gas, indicates that M_{S_i} can be increased not only by increasing the pressure ratio from the driver to the driven tube but also the ratio of the speeds of sound. In heated driver tunnels like the CALSPAN tunnel (Witliff 1987) or the tunnel TH2 in Aachen (Grönig 1991), the temperature (and the speed of sound) of the driver gas (helium) is increased by heating the driver tube by electrical resistance heaters from outside. Stagnation enthalpies up to 8 MJ/kg are achievable with the heated driver technique.

Another way to heat up the driver gas in order to obtain high incident Mach numbers is to heat the driver gas by a combustion process. In the combustion driven shock tunnel the driver gas He is enriched by a combustible mixture of oxygen O_2 and hydrogen H_2. The mixture is ignited by an electrical spark and the combustion heats the driver gas. The reservoir temperatures achieved are reported up to 8000 K, corresponding to reservoir enthalpies of up to 15 MJ/kg. The speed of sound achieved in the combustion heated drivers is limited by the formation of water vapour as the combustion product (Lukasiewicz 1973). Notable examples of the combustion driven shot tunnels are at NASA Ames Research Centre (Deiwert 1992) and at the Boeing company tunnel.

The highest driver temperatures achievable under current technology are obtained by using a (near) adiabatic and isentropic compression of the driver gas helium by a heavy free flying piston accelerated into the driver tube by compressed air (Stalker 1972). Here test conditions with h_s in excess of 30 MJ/kg have been reported. In the new tunnels T5 (Hornung 1992) and the HEG (Eitelberg et al. 1992) the standard operating conditions do not exceed $h_s = 20$ MJ/kg, because of the requirement to operate at tailored conditions. The shock tunnel T5 has been operated both in order to study the real gas effects in external flow as well as in supersonic air breathing combustion. Out of the consideration of the requirements of combustion experiments, the T5 nozzle expansion is such that the Mach numbers are at the lower end of the hypersonic approximation ($M = 5$) and the free stream densities very high. While this provides for a high value of the binary scaling parameter ρL, in some applications it may produce relaxation lengths which cannot be resolved with the help of existing instrumentation.

The HEG and the the shock tunnel TCM2 at the Université de Provence, Marseille were designed for external aerodynamics, with the ultimate goal of testing in the framework of the European HERMES space plane programme in mind. Therefore the HEG was designed to produce M numbers in the range of $7 < M < 10$, depending on the reservoir enthalpy. For this purpose the nozzle area ratio is $A^*/A = 1600$. At perfect gas conditions this area ratio would produce Mach numbers around 20.

The high enthalpy flows, coupled with the capability to produce high density (and high values of the binary scaling parameterρL) for any significant times (in case of the HEG for approximately 2 to 3 ms in the nozzle reservoir) set also the operational limits for the high enthalpy shock tunnels. The calibration procedures for the latest tunnels HEG, T5 and the G-Range at the AEDC have all been held up by the need to replace the ablating steel tube walls with some other material which better withstands the thermal loading.

The limited testing with the HEG so far has shown that the prediction methods used for the design of the HEG can be made to work reliably. The main deviation from prediction of the nozzle reservoir conditions came from the requirement to operate in the tailored interface mode, i.e. provide an approximately constant pressure in the nozzle reservoir. This condition has been reached by experimentation and shown to work up to reservoir pressures of $p_0 \approx 150$ MPa at the specific enthalpy of $h_s \approx 25$ MJ/kg, Fig.3, and an empirical correlation for the tailoring of the HEG conditions has been established. This empirical correlation accounts for the loss mechanisms in the driven section of the shock tube and the transition from the driver to the driven section. Although driver pressures in excess of 200 MPa have been reached, standard operating conditions at lower diaphragm burst pressures ($p_r = 50,\ 100$ MPa) and enthalpies (10 MJ/kg$< h_s <$25 MJ/kg) are used, due to the high mechanical and thermal loads on the facility at the higher pressure and enthalpy levels.

Computer codes, validated with the data from the existing facilities (Maus et al. 1992), used in the design of a new facility at AEDC, Tullahoma, are capable of excellent prediction of the HEG operating conditions (McIntyre et al. 1993). The newer, more sophisticated codes, show also some fine detail in the behaviour of the free piston driven compression cycle, which can safely be ignored in the design of experiments but may help in understanding the limitations of the shock tunnels. A pressure wave travelling in the driver gas, caused by the fast piston stroke, has been shown to exist by the numerical codes (e.g Burtschell et al. 1991) and has been confirmed by experiments in the TCM2 driver (Labracherie et al. 1993). In the HEG this has been shown to influence the shock speed in the driven tube: Fig.3.

Once the tailored operating condition has been established, the nozzle performance predictions show good agreement with the calibration data ("good" in the sense of an impulse facility relying on spontaneous bursting of diaphragms may be different from the "good" that aerodynamicists have come to expect from conventional wind tunnels). The Navier-Stokes calculations (Hannemann et al. 1993) performed for the HEG nominal $p_r = 100$ MPa burst condition tend to overpredict the Pitot pressure p_{t2}, Fig.4, at the exit plane of the nozzle. This may have many causes, including the uncertainty of the chemical and thermal reaction rate data and the uncertainty of the state of the boundary layer in the nozzle. The vibrational relaxation may well be frozen in the free stream of a high Mach number nozzle in the case of pure nitrogen testing. This has not been considered in the current comparison. When allowing for enhanced deexcitation through the presence of oxygen atoms, the vibrational temperature may well equilibrate with the translational temperature (Beck et al. 1991).

The pollution levels in the high enthalpy conditions have to be considered when preparing and evaluating experiments. So far there is only a limited amount of quantified work reported in the high enthalpy conditions (Palma et al. 1993, Beck et al. 1993).

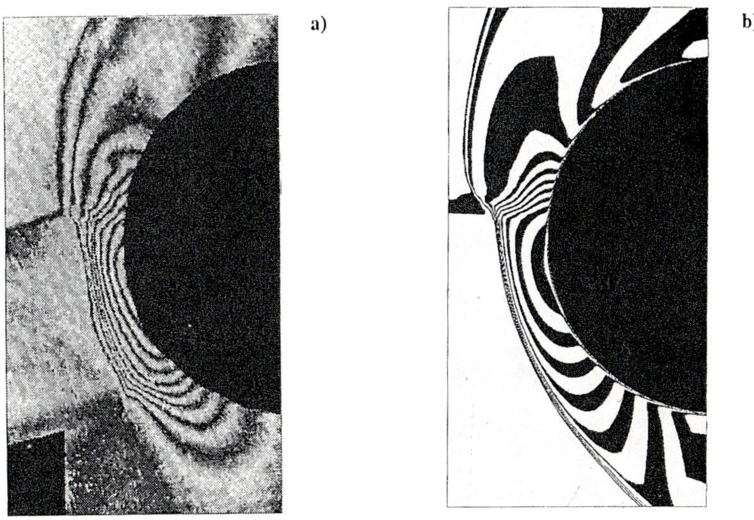

Fig. 5. Shock/shock interaction in dissociating nitrogen, $p_0 = 45$ MPa, $h_s \approx 20$ MJ/kg. a) experiment, b) calculation by Brück et al. (1993)

4. Experiments

The overriding motivation for experiments in the real gas domain is to provide further understanding of the phenomenology of the the flow and to provide input for validation of the numerical codes. Therefore many of the experiments are performed with very simple geometries placed in the flowfield. The geometries are selected for their capability to produce measurable and significant real gas effects. Some of the early experiments were performed at the ANU shock tunnel T3 on a cylinder in nitrogen flow (Hornung 1972). These experiments still serve as a benchmark for numerical codes: Candler (1989), Shapar et al. (1992), Grasso and Belluci (1992). In this sort of flowfield the relaxation process can be observed and quantified with optical methods.The shock stand-off distance is a strong function of the degree of dissociation produced behind the shock, and therefore also a sensitive indicator of the level of enthalpy the wind tunnel is actually being operated at. The dissociation quenching by the expansion over the cylinder is another phenomenon that the numerical methods can be checked against.

Innumerable further calculations are performed simulating stagnation-point flows in air. One of the results of this activity is the need to further refine the instrumentation around the wind tunnels, since different models predict different chemical and temperature distributions in the shock layer. Some of these differences could be resolved by the use of high spatial and spectral resolution instrumentation (Beck and Hannemann 1993). These numerical simulations also point to the difficulties of transfering experimental results from wind tunnels to flight. In order to correctly simulate the dissociative flow phenomena, the density in the flow has to be increased inversely to the model scale. This leads to higher stagnation area density in the shock layer in the experiment and as a result the flow equilibrates at higher temperatures than it does in the free flight. As a consequence the wind tunnels overpredict the thermal loading.

The motivation to provide data for code validation also drives the experiments on biconic shapes, rearward facing steps, spheres and sphere-cone combinations (Gai et al. 1984 and 1989). Of course, all these configurations can be explained as some generic parts of a winged space flight vehicle. The effects of the shift in centre of pressure on delta wings and flat plates at angle of attack due to relaxation effects have been studied in T3 and T4 (Stalker 1989).

The next step in the development of the experiments in high enthalpy hypersonic flows is the study of the interaction of two (or more) flowfields. Two recent examples are the studies of shock/shock interactions by Sanderson and Sturtevant (1993) at moderate reservoir enthalpies, $5\ MJ/kg < h_s < 12\ MJ/kg$, and by Kortz et al. (1993) at high enthalpies, $h_s \approx 20 MJ/kg$. The latter experiments were accompanied by numerical calculations by Brück et al. (1993), simulating the phenomenon and confirming the finding that the relaxation occuring in the flow field behind the shocks can significantly alter the behaviour of the flow, Fig.5.

There have been a series of high enthalpy real gas experimental and numerical studies performed, which have been motivated by the direct application to some extraterrestrial flight project; these studies are both post-dictive and predictive. Studies of the Shuttle Orbiter are an ongoing topic, since these studies provide a good data base for validation of numerical design methods and can be checked against available flight data. As an example the studies performed in the free piston shock tunnel T4 in Brisbane (Krek and Stalker 1992) can be mentioned.

A somewhat simpler geometry, but also motivated by real flight requirements is the geometry of the NASA Aeroassisted Flight Experiment AFE (Deiwert 1992), This geometry has been extensively studied in cold hypersonic tunnels and comparative measurements have been performed in the free piston driven shock tunnel T3 (Gai and Mudford 1992). In these studies the real gas effects are evident in the shock standoff distance and somewhat less in the heat transfer rates to the windward surface. No studies of the wake region, where the dissociation process will be rapidly quenched by the expansion and strong thermal nonequilibrium can occur, are yet available.

5. Summary

The study of chemical and thermal non-equilibrium real gas effects in high enthalpy flows is a challenge for experimental facility and instrumentation developers. A number of new wind tunnels capable of simulating some parameters of the reentry flight conditions for reasonable model sizes have been recently commissioned and are producing results. For the transfer of the results from the wind tunnel to free-flight conditions, feedback from the numerical work is required, in order to adequately account for those parameters that cannot be simulated. In general, it should be obvious that whatever the shortcomings of all the high enthalpy experimental capabilities, their intelligent use can provide valuable data for the process of designing the next generation of space transport vehicles.

Acknowledgement

I wish to thank all colleagues who have been associated with obtaining results in the HEG. This includes also our external partners from Siemens and Fluidyne. The financial support of ESA through CNES and the State of Lower Saxony is gratefully acknowledged.

References

Auweter-Kurtz M, Habinger H, Laure S, Messerschmid E, Röck W, Tubanos N (1991) The IRS plasma wind tunnels for the investigation of thermal protection materials for reentry vehicles. Proc. 1st European Symposium: Aerothermodynamics for space vehicles, ESTEC, Noordwijk, The Netherlands

Beck WH, Eitelberg G, McIntyre TJ, Baird JP, Lacey J, Simon M (1991) The high enthalpy shock tunnel in Göttingen (HEG). In: Takayama K (ed) Proc. 18th ISSW, Sendai, Japan pp 677-682

Beck WH, Hanneman K (1993) High enthalpy wind tunnel instrumentation. CNES-Report, Activity ES 7 (1992)

Beck WH, Müller M, Wollenhaupt M (1993) Application of spectroscopic diagnostic techniques to studies on HEG: preparatory LIF work and emission spectroscopy results. Proc. 15th ICIASF, Saint-Louis, France

Becker E (1965) Gasdynamik. Teubner, Stuttgart

Brück S, Brenner G, Kortz S (1993) Numerical simulations of shock-shock interactions in nonequilibrium flow. DLR-IB 221-93 A 09, DLR Germay

Burtschell Y, Brun R, Zeitoun D (1991) Two dimensional numerical simulation of the Marseille University free piston shock tunnel - TCM2. In: Takayama K (ed) Proc. 18th ISSW, Sendai, Japan pp 583-590

Candler G (1989) On the computation of shock shapes in nonequilibrium hypersonic flows. AIAA Paper 89-0312

Chanetz B, Coët M-C, Nieout D, Pot T, Brossand P, François G, Masson A (1992) Nouveaux moyens d'essais hypersoniques dévelppés à L'ONERA. AGARD Conf. Proc. 514, Theoretical and experimental methods in hypersonic flows

Deiwert SG (1992) Issues and approaches to develop validated analysis tools for hypersonic flows: one perspective. AGARD Conf. Proc. 514, Theoretical and experimental methods in hypersonic flows

Eitelberg G, McIntyre TJ, Beck WH, Lacey J (1992) The high entahlpy shock tunnel in Göttingen. AIAA Paper 92-39 42

Erdes J, Tamaguo J, Bakos R (1989) Hypervelocity real gas capabilites of GASL's HYPULSE facility. In: Rogers RC (ed) Workshop on the application of pulse facilites to hypervelocity combustion simulation, NASP Workshop Publication 1008, NASA-LRC

Gai SL, Mudford NT (1992) Heat flux and shock shape measurements on a aeroassist flight experiment model in a high enthalpy free piston shock tunnel. AIAA Paper 92-3407

Gai SL, Reynolds NT, Ross C, Baird JP (1989) Measurements of heat transfer in separated high enthalpy dissociated laminar hypersonic flow behind a step. J. Fluid Mech. 199: 541-561

Gai SL, Sandeman RJ, Lyons P, Kilpin D (1984) Shock shape over a sphere cone in hypersonic high enthalpy flow. AIAA Journal 22:7

Grasso F, Bellucci V (1992) Thermal and chemical nonequilibrium hypersonic flow computations. AGARD Conf. Proc. 514. Theoretical and experimental methods in hypersonic flows

Grönig H (1991) Shock tube application: High enthalpy European wind tunnels. In: Takayama K (ed) Proc. 18th ISSW, Sendai, Japan pp 3-16

Hannemann K (1990) Design of a axisymmetric contoured mozzle for the HEG. DLR-FB 90-04

Hannemann K, Brenner G, Brück S (1993) Numerical simulation of reacting flows related to the HEG. These Proceedings

Hornung HG (1972) Non-Equilibrium dissociating introgen flow over shperes and circular cylinders. J. Fluid Mech. 53: 149

Hornung HG (1991) Class notes: Ae 234. Lecture Notes, Course given at CALCIT, Calif. Inst. of Technology, Pasadena

Hornung HG (1992) Performance data of the new free-piston shock tunnel at GALCIT. AIAA Paper 92-3943

Kortz S, McIntyre TJ, Eitelberg G (1993) Experimental investigations of shock-shock interactions in the high enthalpy shock tunnel Göttingen (HEG). These Proceedings

Krek RM, Stalker RJ (1992) Experiments on Space Shuttle Orbiter models in a free piston shock tunnel. Aeronaut. Journal, Aug.-Sept., pp 249-259

Labracherie L, Dumitrescu MP, Burschell Y, Houas L (1993) On the compression processs in a free piston shock tunnel. Shock Waves 3.1: 19-23

Lukasiewicz J (1973) Experimental methods of hypersonics. Gas dynamics series, Vol. 3. Marcel Dekker, NY

Maus JR, Laster ML, Hornung HG (1992) The G-Range impulse facility: a high-performance free piston shock tunnel. AIAA Paper 92-3946

McIntyre TJ, Maus JR, Laster ML, Eitelberg G (1993) Comparison of the flow in the high enthalpy shock tunnel in Göttingen with numerical simulations. These Proceedings

Palma PC, Houwing AFP, Sandeman RJ (1993) Absolute intensity measurements of impurity emissions in a shock tunnel and their consequences for laser-induced fluorescence experiments. Shock Waves, 3.1: 49-53

Park C (1990) Nonequilibrium hypersonic aerothermodynamics. Wiley, NY

Sanderson SR, Sturtevant B (1993) Shock wave interactions in hypervelocity flow. These Proceedings

Shahpar S, Kennaugh A, Hall IM, Poll A (1992) Comparison of computation and experiment for a high enthalpy, heated-driver, shock-tube environment. AGARD Conf. Proc. 514, Theoretical and exprimental methods in hypersonic flows

Stalker RJ (1972) Develpment of a hypervelocity wind tunnel. J. Roy. Aeronaut. Soc. 76:376-384

Vincenti WG, Kruger CH jr (1965) Introduction to physical gas dynamics. Wiley, NY

Witliff CE (1987) A survey of existing hypersonic ground test facilities - North America. AGARD Conf. Proc. 428, Aerodynamics of Hypersonic Lifting Vehicles

Part 1: Chemical Kinetics

Vibrational Temperature Measurements in a Shock Layer Using Laser Induced Predissociation Fluorescence

D.J. Sutton*, A.F.P. Houwing, P.C. Palma and R.J. Sandeman
Department of Physics and Theoretical Physics, Faculty of Science, The Australian National University, Canberra, A.C.T. 0200 Australia

Abstract. The feasibility of applying laser induced predissociation fluorescence (LIPF) to the measurement of vibrational temperatures in a shock layer flow around a cylinder in a pulsed supersonic free stream is examined. Single shot spatially and spectrally resolved LIPF measurements are presented and a comparison with theoretical predictions is made.

Key words: Laser-induced predissociation fluorescence, Vibrational temperature, Shock layer

1. Introduction

The measurement of temperature and species concentration in the shock layer on a blunt body in a shock layer flow is a useful way to test theoretical models that describe nonequilibrium reacting flows, such as those associated with the flight of hypersonic vehicles. The measurement of these parameters using laser induced fluorescence (LIF) is often complicated by the problem of collisional quenching (Seitzman and Hanson 1992). Collisional quenching is especially important in the application of LIF to pulsed high enthalpy flows because of the high collision rates in such environments. The development of LIPF (Kelly and Hudson 1985), where the effect of quenching is avoided by exciting the molecule to a fast predissociating state, from which the fluorescence is emitted only during the predissociation lifetime largely overcomes this problem. Provided the predissociation lifetime is sufficiently short, the probability of deexcitation collisions is extremely small, and hence there will be insignificant quenching even at high pressures. This technique has been shown (Andresen et al. 1988) to be applicable to O_2, OH and H_2O. Compared to standard LIF, however, LIPF signals are very weak and hence the presence of strong emission lines from flow impurities (Palma et al. 1993), has the potential to make LIPF studies difficult in shock tunnel flows. The current study was designed to determine whether the LIPF signals generated near a bluff body in the supersonic stream of a free-piston shock tunnel are sufficiently large to be detectable over the background luminosity to allow the measurement of the temperature in the shock layer.

2. Theoretical considerations

The qualitative features of the LIPF scheme are illustrated in Fig.1, which shows a number of rovibrational levels defined by rotational and vibrational quantum numbers N and v respectively. The laser is tuned to the wavenumber $\omega_{i'i''}$ of a transition which promotes a molecule from a rovibrational level $|i'>$ in a lower electronic state to a rovibrational level $|i''>$ in an upper electronic state. This upper level subsequently decays by the emission of a photon of wavenumber $\omega_{i'i'''}$ to another rovibrational level $|i'''>$ in the lower electronic state. By pumping two levels simultaneously with the same laser, it is possible to induce two fluorescence schemes. The vibrational temperature T_V may be determined by measuring the ratio ϕ_1/ϕ_2 of fluorescence from the two different schemes and is given by Sutton et al. (1993):

*Current address : Defence Science Technology Organisation, P.O. Box 1500, Salisbury, S.A. 5108 Australia

Shock Waves @ Marseille II
Editors: R. Brun, L. Z. Dumitrescu

$$T_V = \frac{(G_{2''} - G_{1''})\,\hbar c}{k \ln\left(\frac{g_{2''}\phi_1 \tau_{P_2} E_2 q_{2''2'} q_{2'''2'} S_{2''2'} S_{2'''2'}}{g_{1''}\phi_2 \tau_{P_1} E_1 q_{1''1'} q_{1'''1'} S_{1''1'} S_{1'''1'}}\right)} \tag{1}$$

where $G_{i''}$ are the vibrational energy terms; $g_{i''}$ are the degeneracy of the rotational energy levels; ϕ_i are the fluorescent yields; τ_{P_i} are the predissociation lifetimes; E_i are the laser energies coincident with the absorption lines; $q_{i'''i'}$ and $q_{i''i'}$ are the Franck-Condon factors; $S_{i'''i'}$ and $S_{i''i'}$ are the Hönl-London factors; $\hbar$ is Planck's constant; c is the speed of light and k is Boltzmann's constant; the subscripts $i = 1$ or 2 indicate the fluorescence scheme; while the single, double and triple primes refer to the levels illustrated in Fig.1.

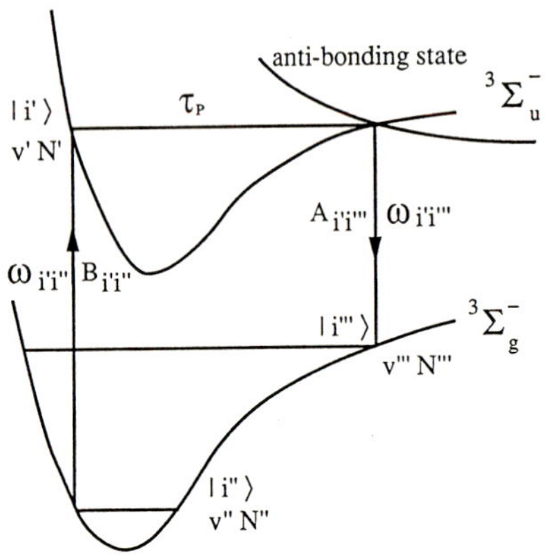

Fig. 1. Schematic of LIPF scheme

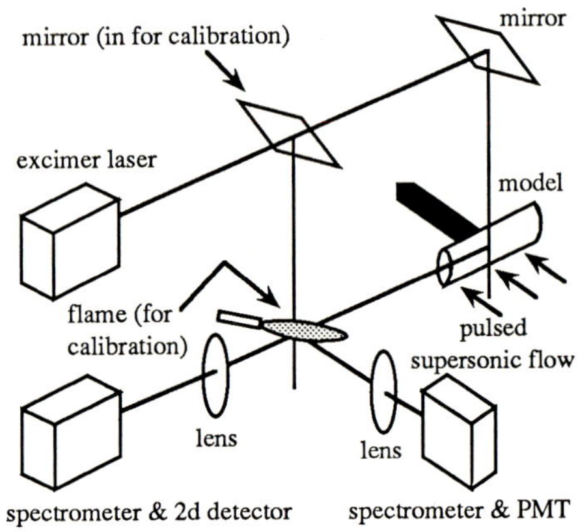

Fig. 2. Experimental arrangement

3. Experimental arrangement

The optical arrangement is illustrated in Fig.2. A pulsed ArF excimer laser operating in a tunable oscillator-amplifier configuration was used to produce a narrowband pulse (0.02 nm FWHM), directed vertically downwards into the shock layer flow around a 25.4 mm diameter cylinder. The laser beam was rectangular in cross section and unfocussed. The region pumped by the laser in front of the cylinder was a vertical rectangular prism, with a width of 20 mm along the axis of the cylinder and a thickness of 5 mm measured perpendicular to this axis. The cylinder was mounted with its axis perpendicular to the direction of the incident flow. The LIPF signals were imaged into a diffraction grating spectrometer (dispersion 0.7 nm/mm). The imaging system allowed spatial resolution along the length of the entrance slit, and had a magnification of 0.2. With this arrangement, it was possible to determine the variation of LIPF signals, and hence vibrational temperatures, along a vertical line in the flow. The signals dispersed by the spectrometer were imaged onto a two dimensional detector, with an overall gain of approximately 10^6. To tune the laser wavelength, the following method was adopted: An additional mirror was placed in the path of the laser beam to direct it into the flame from a hydrogen/oxygen burner. The fluorescence signals from the flame were then imaged onto the entrance slit of another spectrometer, set to transmit the $(10,8)R(17)$, $(10,8)P(19)$, $(16,9)R(19)$ and $(16,9)P(21)$ lines of the $B^3\Sigma_u^- \rightarrow X^3\Sigma_g^-$ system of O_2 at 232.0 nm. The fluorescence signal transmitted by the spectrometer was detected by a photomultiplier and output to an oscilloscope and a boxcar integrator. The tuning mirror on the oscillator of the excimer laser was scanned at a constant rate so that the laser scanned through its tuning range (193.0 to 193.5 nm). The excitation spectrum thus produced was compared with the excitation spectra from earlier work (Kim et al. 1991) to identify the pumping transitions. At the same time other fluorescence lines [$(16,12)R(19)$; $(16,12)P(21)$; $(10,11)R(17)$ and $(10,11)P(19)$] in the vicinity of 256 nm were detected by the 2D detector, with the image displayed in real time on a computer monitor. The tuning mirror was then repositioned by the stepper motor to reproduce the peak on the excitation spectrum corresponding to the desired pumping transitions.

The experiments were performed on the Australian National University's T2 free piston shock tunnel (Stalker 1967) which produced the supersonic air flow over the model illustrated in Fig.2. The flow conditions were calculated using the algorithm described by Vardavas (1984). For the flow through the nozzle, a chemical reaction rate model was used, with allowances made for vibrational freezing, using the algorithm described by Rein (1989). The test conditions calculated in this manner and used in the current experiments are summarised in Table 1. In order to synchronise the firing of the laser pulse and the gating of the 2D detector to coincide with the steady flow period of the pulsed flow produced by the shock tunnel, it was necessary for a PC controlled sequence to be employed. Two pressure transducers situated at different positions along the shock tube were used to determine the transit time (and hence the speed) of the primary shock wave. The output of the second of these transducers was used to initiate the PC controlled sequence, which sent pulses to trigger the laser and gate the camera. The signals from the transducers were displayed on an oscilloscope to facilitate the measurement of shock speeds and pressures. A synchronisation pulse from the laser was displayed on a second oscilloscope together with the signal from the nozzle reservoir transducer. This enabled the determination of the time delay between shock reflection at the nozzle entrance and the firing of the laser. The correct choice for this delay is based upon an understanding of the nozzle starting processes (Smith 1966) and is chosen so that flow conditions over the cylindrical model are reasonably steady at the time the laser and detector are triggered.

4. Results

The LIPF signals recorded by the 2D detector were transferred to a workstation for image processing. The first step of this image processing was the removal of the background signal generated by the thermal noise of the camera. Following this, a number of dispersed emission spectra were extracted from the data for different positions above the stagnation streamline to determine the intensity of the LIPF lines, background radiation and shot noise. The LIPF lines were identified by comparison with LIPF lines produced in a flame. In this way, the LIPF lines were determined to be the $(16,12)R(19)$; $(16,12)P(21)$; $(10,11)R(17)$ and $(10,11)P(19)$ emission lines. The dispersed fluorescence spectra contain a significant amount of background radiation, which was determined to be laser-induced, though did not appear to be significantly dependent on the laser wavelength, as it was still present when the laser was detuned from the absorption frequency. The variation of the peak intensity of one of the LIPF lines with distance through the shock layer is displayed in Fig.3. It is apparent that there is a pronounced reduction in the overall intensity of the LIPF signals with penetration through the flow. All the LIPF lines display similar behaviour, as does the laser induced background emission, suggesting that the cause is a reduction in the intensity of the laser with distance into the shock layer.

A significant level of shot noise was also present in the data. The relative level of this noise was reduced by spatially averaging the data, at the expense of spatial resolution. Such spatially averaged spectra are presented in Figs.4a and 4b, together with fitted curves to the background and the LIPF lines. The curve fits to the lines are Gaussian shapes, each of identical width, superimposed on the background plateau. The fitted width is consistent with that of the instrument function of the spectrograph and detector system. The integrated line intensity was calculated as

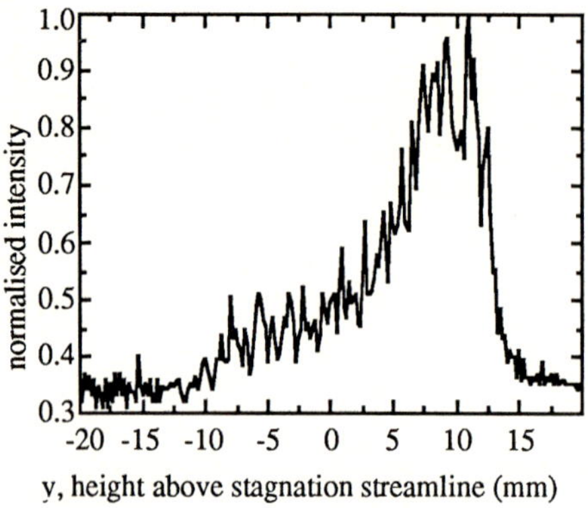

Fig. 3. Variation of LIPF with position

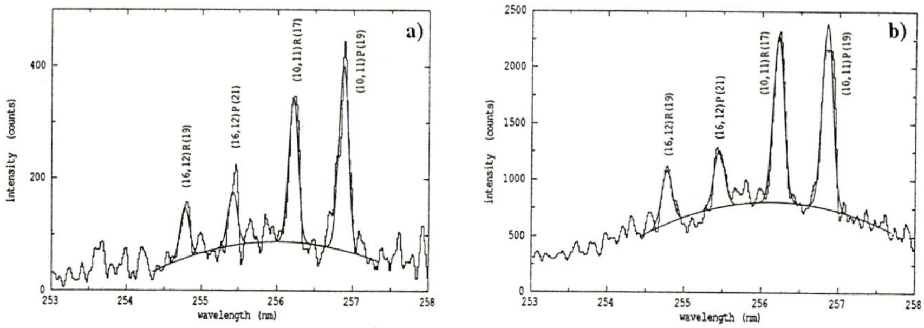

Fig. 4. Emission spectrum for regions (a) and (b)

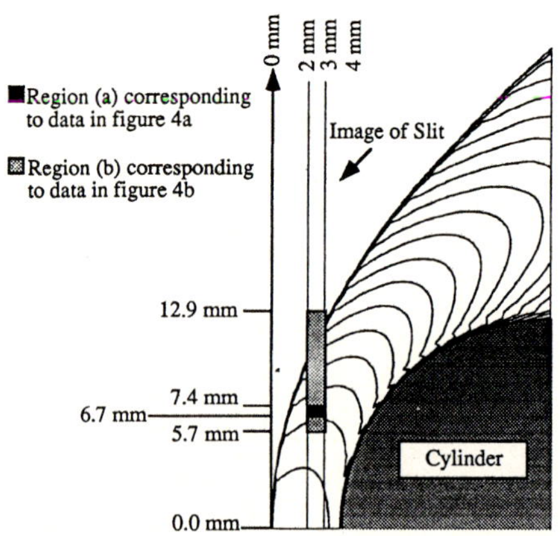

Fig. 5. Model, shock and regions interrogated

the area between the fitted Gaussian profile and the fit to the background. The ratio of the photon counts ϕ_1/ϕ_2 required for use in Eq.1 is determined by dividing the integrated line intensity from one of the (16,12) lines by the integrated line intensity from one of the (10,11) lines. The positions and sizes of the two regions (a) and (b), corresponding to the spectra displayed in Figs.4a and 4b, are indicated by the rectangles displayed in Fig.5. The errors in the measured values of ϕ_1/ϕ_2 are estimated by determining the discrepancy between the fitted curves and the data points and the estimating the uncertainty in the background level. Because of the different number of samples used in the averaging of the data from the two regions, the errors in the measured values of ϕ_1/ϕ_2

for region (b) are much less than those for region (a). By determining the convolution of the laser profile with the excitation transitions it is found that E_1/E_2 (the ratio of laser energies used to pump the two LIPF systems) is equal to 2.5 ± 0.2.

The vibrational temperatures determined from the data given in Figs.4a and 4b are presented in Table 2. The predissociation linewidths, Hönl London factors and Franck Condon factors used in the data reduction are obtained from Lewis et al (1986), Tatum and Watson (1971) and Harris et al. (1969), respectively. The percentage errors associated with the spectroscopic data from these sources are much less than those associated with the values of ϕ_1/ϕ_2 and E_1/E_2. It can be seen from the values in this table that, for region (a), there is a significant discrepancy between the value of T_V determined from the P lines and that determined from the R lines. This is a result of the difficulty of deconvoluting the LIPF signals from the laser induced background. This background contains a significant amount of spectral structure. However, an examination of different spectra at different positions in the flow shows that this structure does not contain a strong spatial correlation. Hence by spatially averaging the data over a larger region, such as region (b) in Fig.5, the background is smoothed and can be easily subtracted from the data. When this is done, the discrepancy between the values of T_V determined from the different lines is removed.

The nozzle flow calculation indicates that the rotational and vibrational temperatures in the freestream are significantly different (T_r = 325 K, T_V = 1040 K). Because of the combination of low rotational temperature and flow density in the freestream, the population of the two levels (N'' = 17, v'' = 2) and (N'' = 19, v'' = 3) of the $^3\Sigma_g^-$ electronic state are too low to be detectable. Behind the bow shock, the collision rate is much larger than in the freestream and equilibrium between rotation and vibration is quickly established (i.e. $T_r = T_V$ behind the shock). The shape of the shock wave and the temperature distribution inside were determined by a CFD Code (Mundt 1992) which accounted for nonequilibrium chemistry. This code yielded theoretical values for the spatially averaged temperatures, giving 3250 K for region (a) and 3100 K for region (b), which are in good agreement with the experimentally measured values.

5. Conclusion

The current work has shown the feasability of employing the LIPF method to measure vibrational temperatures in the shock layer. While the presence of a laser induced background has led to noise problems, which cause large errors in the temperature calculation, it has been possible to reduce this noise through spatially averaging the data. There appears to be significant absorption of the laser in the shock layer, resulting in a reduction in the LIPF signals. As the laser beam was unfocussed, its intensity in the shock layer was significantly lower than the intensity that can be achieved with a focussed beam, hence, saturation effects are not expected to be important.

Acknowledgements

We express our thanks to Mr Paul Walsh for his assistance with the design of the experiment and his constructive technical advice. We are grateful to Mr David Cooper and Dr Stan Newman for their work in the design and construction of the timing electronics. We are indebted to Dr Aidan Byrne for his assistance in the data reduction and to Dr Christian Mundt for the CFD calculations. This work has been financially supported by the Australian Research Council and NASA contract number NAGW 1467.

Table 1. Theoretically calculated shock tunnel conditions

Speed of primary shock	1.86 km/s	Nozzle reservoir temperature	3480 K
Nozzle reservoir pressure	3.14×10^7 Pa	Free stream temperature T_r	325 K
Free stream pressure	3.00×10^3 Pa	T_V	1040 K
Free stream Mach number	7.8		

Table 2. Experimentally measured vibrational temperatures

Region	ϕ_1/ϕ_2 (P lines)	T_V (P lines)	ϕ_1/ϕ_2 (R lines)	T_V (R lines)	T_V (CFD)
(a)	0.29±0.05	2682±500 K	0.35±0.05	3473±500 K	3250 K
(b)	0.31±0.02	2890±300 K	0.31±0.02	2960±300 K	3100 K

References

Andresen P, Bath A, Gröger W, Lülf HW, Meijer G, ter Meulen JJ (1988) Laser induced fluorescence with tunable excimer lasers as a possible method for instantaneous temperature field measurements at high pressures: checks with an atmospheric flame. Applied Optics 27: 365

Hanson RK, Seitzman JM, Paul PH (1990) Planar laser-fluorescence imaging of combustion gases. Applied Physics B 50: 441

Harris R, Blackledge M, Generosa J (1969) Rydberg-Klein-Rees (RKR) Franck-Condon factors for the O_2 Schumann-Runge system including high vibrational quantum numbers. J. Molec. Spectrosc. 30: 506

Hiller B, and Hanson RK (1988) Simultaneous planar measurements of velocity and pressure fields in gas flows using laser-induced fluorescence. Applied Optics 27: 33

Kelly PB, Hudson RS (1985) Resonance fluorescence and resonance Raman spectra of molecular oxygen. Chem. Phys. Lett. 114: 45

Kim G-S, Hitchcock, LM, Rothe EW, Reck GP (1991) Identification and imaging of hot O_2 ($v'' = 2, 3$, or 4) in hydrogen flames using 193 nm- and 210 nm-range light. Applied Physics B 53: 180

Lewis BR, Berzins L, Carver JH, Gibson ST (1986) Rotational variation of predissociation linewidth in the Schumann-Runge Bands of $^{16}O_2$. J. Quant. Spectrosc. Radiat. Transfer 36: 187

Mundt Ch (1992) Calculation of hypersonic, viscous non-equilibrium flows around reenrty bodies usinf a coupled boundary layer/Euler method. AIAA Paper 92 - 2856

Palma PC, Houwing AFP, Sandeman RJ (1993) Absolute intensity measurements of impurity emissions in a shock tunnel and their consequences for laser induced fluorescence experiments. Shock Waves 3:49

Rein M (1989) Surf : A program for calculating inviscid supersonic reacting flows in nozzles. GALCIT FM 89-1

Seitzman JM, Hanson RK (1992) Quantitative fluorescence imaging: A comparison of linear, predissociative and saturated pumping techniques. AIAA Paper 92 - 0879

Smith CE (1966) The starting process in a hypersonic nozzle. J. Fluid Mech. 24: 625

Stalker RJ (1967) A study of the free-piston shock tunnel. AIAA J. 5: 2160

Sutton DJ, Houwing AFP, Palma PC, Sandeman RJ (1993) Vibrational temperature measurements in a shock layer using laer induced predissociation fluorescence. Shock Waves 3:141

Tatum JB, Watson JKG (1971) Rotational line strengths in $^3\Sigma^{\pm} \rightarrow ^3\Sigma^{\pm}$ transitions with intermediate coupling. Can. J. Phys. 49: 2693

Vardavas IM (1984) Modelling reactive gas flows within shock tunnels. Austr. J. Phys. 37: 157

Semiconductor Laser Diagnostics of Kinetic and Population Temperatures in High-Enthalpy Flows

H. Andrew Chang, Douglas S. Baer and Ronald K. Hanson
Department of Mechanical Engineering, Stanford University, Stanford, CA 94305, USA

Abstract. Semiconductor (GaAlAs) diode lasers have been used to probe the $3s^5S_2^o \rightarrow 3p^5P_3$ (7772 Å) excited state transition of atomic oxygen in a pressure-driven shock tube. Lineshape analyses of the measured spectral profiles enable the simultaneous and independent determination of kinetic and population temperatures. Measurements were made in the reflected-shock region with an initial gas mixture of 1.5% O_2 and 98.5% Ar. The experimentally determined temperatures ranged from 6900 to 11400 K, and the pressure range was 0.27 - 1.37 atm. The measured kinetic and population temperatures were found to be in good agreement with each other.

Key words: Semiconductor laser, Absorption, Lineshape, O-atom

1. Introduction

Effective diagnostic techniques for shock tube studies must be non-intrusive and have sufficiently fast time response. The inherently high bandwidth of semiconductors enables CW, narrow-linewidth diode lasers to be rapidly tuned, and thus provides the opportunity to make high spectral resolution measurements in short test times. In addition, the entire laser system is relatively compact, electrically efficient, and economical compared to other laser systems.

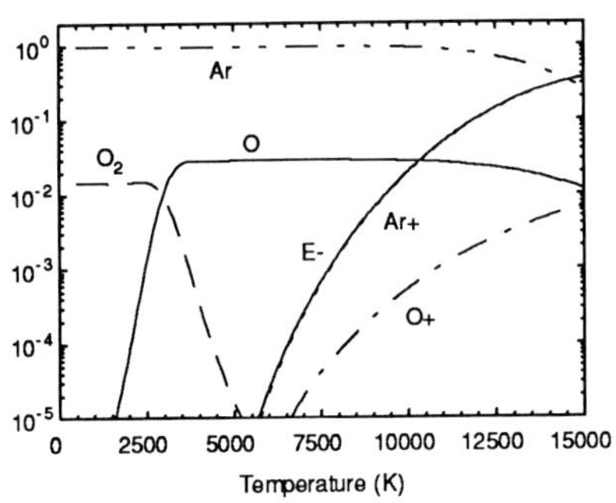

Fig. 1. Equilibrium mole fractions of species in test gas mixture of 98.5% Ar and 1.5% 0_2 (P=1.0 atm)

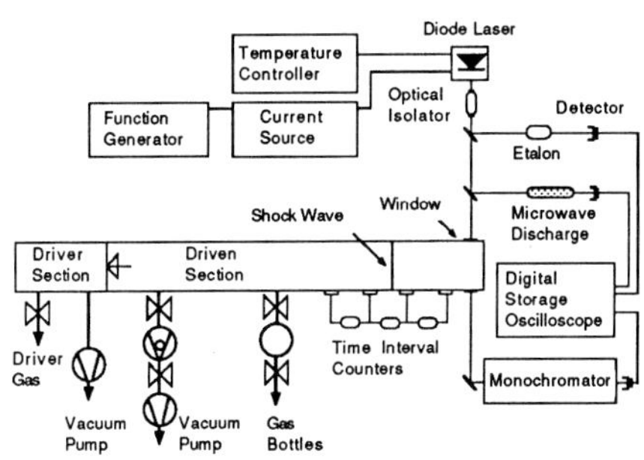

Fig. 2. Schematic diagram of the shock tube experimental setup

Previous studies of high-enthalpy shock-heated gases using tunable near-IR semiconductor lasers have not utilized the rapid-tuning capabilities of these lasers nor their ability to obtain spectrally resolved lineshapes (McIntyre et al. 1988). In the present investigation, a narrow-bandwidth semiconductor (GaAlAs) laser was scanned to record spectrally resolved absorption lineshapes of atomic oxygen corresponding to the $3s^5S_2^o \rightarrow 3p^5P_3$ (7772 Å) excited state transition. Measurements were made in the reflected-shock region of a pressure-driven shock tube to simulate the high-temperature environment found in high-enthalpy hypersonic flows. The initial gas mixture consisted of 1.5% O_2 and 98.5% Ar. The equilibrium mole fractions of the species present in the test gas mixture are shown in Fig.1. For the temperature range studied in this experiment, virtually all of the O_2 dissociates into atomic oxygen. The kinetic and population temperatures were determined from Voigt profile fits of the lineshapes and the frequency-integrated line intensities, respectively. Theoretical calculations of kinetic temperatures were made using measured

Shock Waves @ Marseille II
Editors: R. Brun, L. Z. Dumitrescu

shock velocities and normal shock wave relations taking into account dissociation and ionization chemistry.

2. Experimental details

A schematic of the experimental setup is shown in Fig.2. The pressure-driven stainless steel shock tube consisted of a 2.3 m driver section and a 10.5 m driven section of 15.24 cm ID. The measurements were made 19.6 mm from the end wall behind reflected shock waves.

The commercially available diode laser system (ILX Lightwave) used in the experiment consisted of a low-noise current source (LDX-3620), a thermoelectric temperature controller (LDT-5910B), and a laser diode mount (LDM-4412). The semiconductor diode laser (Sharp LTO-21MD) was repetitively scanned with a 6 kHz triangular waveform to obtain multiple absorption lineshapes for a given shock condition. These conditions provide a 54 GHz (1.8 cm^{-1}) scan centered about the oxygen atom absorption line (7772 Å). The laser power was approximately 5 mW.

The output laser beam was directed through an optical isolator (Optics for Research NIR-5) to minimize unwanted retro-reflections into the laser cavity. The laser beam was then split to enable simultaneous measurements of transmission through the shock tube and a low-pressure discharge tube which served as an absolute wavelength reference for the excited oxygen atom transition probed. A fraction of the beam was also directed through a fixed confocal etalon (FSR = 2 GHz) to provide a means for converting data from the time domain to the frequency domain. The transmission beam through the shock tube was directed through a monochromator to filter out emission from the high-temperature gas. The transmission intensities were converted to voltage signals by using silicon photodiode detectors with a frequency response (-3 dB) of 500 kHz.

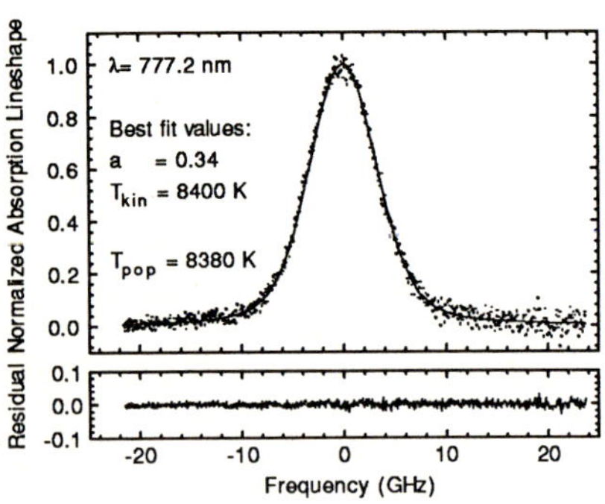

Fig. 3. Example of a reduced absorbtion lineshape for a single scan of the oxygen $3s^5S_2^o \rightarrow 3p^5P_3$ transition. The residual is the difference between the data and the best-fit Voigt profile

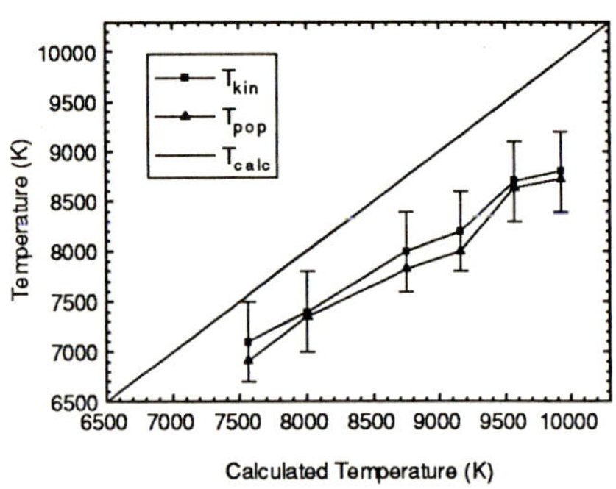

Fig. 4. Comparison of calculated, kinetic, and population temperatures in the reflected-shock region for different shock conditions at approximately 0.25 msec after the reflected shock

3. Data analysis

Spectral lineshapes were obtained from the O-atom absorption data by applying the Beer-Lambert relation for a probe beam passing through a linearly absorbing medium

$$\frac{I(\nu)}{I_0} = \exp(-k(\nu)L) \tag{1}$$

where $I(\nu)$ is the transmission intensity, I_0 is the incident probe beam intensity, $k(\nu)$ is the spectral absorption coefficient, and L the absorption pathlength (shock tube diameter).

Each measured lineshape was then least-squares-fit to a Voigt profile using two parameters: the Voigt a parameter and the kinetic temperature, T_{kin}, which describes the Maxwellian velocity distribution of a gaseous species. A Voigt profile describes the convolution of the Gaussian and Lorentzian profiles assuming that the effects of Doppler and collisional broadening are decoupled. The Voigt a parameter is defined as

$$a \equiv \sqrt{\ln 2}\frac{\Delta\nu_C}{\Delta\nu_D} \tag{2}$$

where $\Delta\nu_D$ is the Doppler-broadened Gaussian linewidth (FWHM) and $\Delta\nu_C$ is the Lorentzian or collisional linewidth (FWHM). Minor corrections for the baseline were sometimes necessary so that the residuals approached zero in the wings, as expected theoretically (Baer et al. 1991). The results of a typical lineshape analysis are shown in Fig.3. The top frame shows the normalized measured lineshape and the best-fit Voigt profile for a single scan, and the bottom frame shows the residual difference between the best-fit and the data.

The population temperature, T_{pop}, may be determined from the integrated absorption coefficient assuming Boltzmann equilibrium between the ground state and the lower laser-coupled level. The frequency integral of the spectral absorption coefficient yields the integrated absorption from which n_i, the lower state number density of the probed species, can be calculated (Baer et al. 1991). For a system in Boltzmann equilibrium, T_{pop} and n_i are related through the expression

$$\frac{n_i}{n_a} = \frac{g_i}{Z_a}\exp\left(\frac{-E_i}{kT_{pop}}\right) \tag{3}$$

where g_i and E_i are the lower state degeneracy and excitation energy; Z_a and n_a are the atomic partition function and total number density of the probed species, respectively. Thus, the population temperature describes the state specific population of a species relative to its ground state. For gases in local thermodynamic equilibrium (LTE), $T_{pop} = T_{kin}$.

Fig.4 compares the kinetic and population temperatures determined from lineshape analyses with calculated temperatures. The experimentally determined temperatures are from a single scan of the O-atom absorption lineshape for different shock conditions at approximately 0.25 msec after the reflected shock. Equilibrium temperatures in the reflected-shock region were first calculated using shock speeds measured from thin film gauges mounted in the shock tube, and a computer code utilizing one-dimensional normal shock relations with a high-temperature thermodynamic database (Gupta et al. 1990). Enthalpy and pressure values calculated from the reflected shock conditions were held constant in determining the changes in temperature with time due to dissociation and ionization chemistry (Kee et al. 1989; Gupta et al. 1990). The changes in the different temperatures with time after the reflected shock for a typical experiment are shown in Fig.5. The measured kinetic and population temperatures show good agreement with each other to within experimental error. However, the calculated equilibrium temperatures are consistently higher than the kinetic temperatures. This may be due to either non-ideal shock wave behavior at the low initial test pressures used or to radiative losses from the high-temperature gases.

Sensitivity analysis was carried out to determine the accuracy and reliability of the Voigt-fit procedure. By determining Δ, the frequency-integrated squared-difference between the fit and the data, the existence and uniqueness of a set of best-fit Voigt parameters can be verified (Baer et al. 1991). The value of a was optimized for a given value of T, and the procedure was repeated over a range of temperatures until the minimum Δ value, Δ_{min}, corresponding to the best-fit Voigt parameters was determined. A Voigt-fit sensitivity analysis may be displayed on a graph with the variation from Δ_{min} (% variation) plotted as a function of T, where % variation is defined as $(\Delta - \Delta_{min})/\Delta_{min} \times 100\%$. The plot is shown in Fig.6. The sets of a and T values which yield Δ values of 1% higher than Δ_{min} was used to estimate the relative accuracy of the best-fit

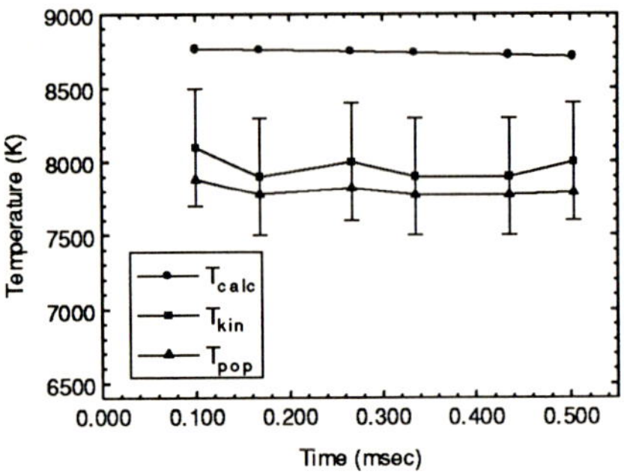

Fig. 5. Variation of kinetic, population and calculated temperatures with time after the reflected shock. Kinetic and population temperatures were determined from Voigt fits of the oxygen (7772 Å) absorption lineshape. The initial pressure of the test gas was 2.5 torr, and the shock speed was 2.07 mm/μsec

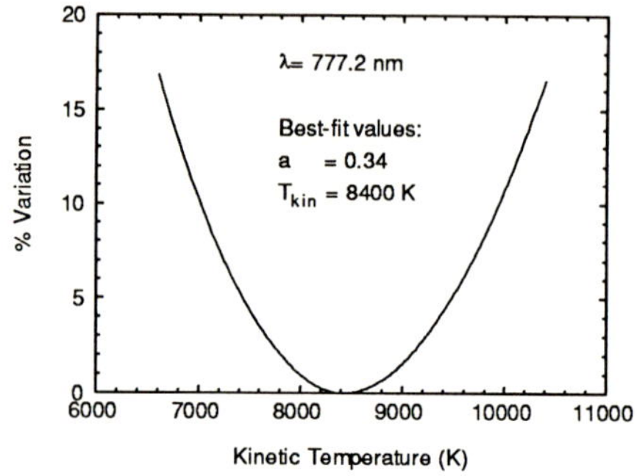

Fig. 6. Voigt-fit sensivity analysis applied to measurement of the oxygen (7772 Å) absorption lineshape. The % variation is defined as $[(\Delta - \Delta_{min})/\Delta_{min} \times 100\%]$, where Δ is the frequency-integrated squared-difference between the data and the Voigt profile fit

parameters for a particular lineshape. The estimated uncertainties in the Voigt-fit parameters for the oxygen absorption line were $\Delta T_{kin} = \pm 400$ K and $\Delta a = \pm 0.03$.

4. Conclusions

These results represent the first application of semiconductor (GaAlAs) lasers for the measurement of fully resolved atomic oxygen absorption spectra in high-enthalpy flowfields. The line-of-sight method can be applied in a variety of high-enthalpy flowfields to simultaneously and independently determine kinetic and population temperatures. Such measurements could prove useful in studying non-equilibrium processes known to be important in modeling hypersonic, high-enthalpy flows. In addition, the method may also allow the determination of velocity in high-speed flows by comparing Doppler-shifted lineshapes recorded with different laser propagation vectors with respect to the flow direction.

Acknowledgment

This research was supported by the U.S. Air Force Office of Scientific Research, Aerospace Sciences Directorate, with Dr. J. Tishkoff as Technical Monitor.

References

Baer S, Hanson RK (1991) Tunable diode laser absorption diagnostics for atmospheric pressure plasmas. J. Quant. Spectrosc. Radiat. Transfer 47:455-475

Chang HA, Baer DS, Hanson RK (1993) Semiconductor laser diagnostics of atomic oxygen for hypersonic flowfield measurements. AIAA 31st Aerospaces Sciences Meeting, Reno NV, AIAA Paper 93-0628

Gupta RN, Yos JM, Thompson RA, Lee KP (1990) A review of reaction rates and thermodynamic and transport properties for an 11-Species air model for chemical and thermal nonequilibrium calculations to 30000 K. NASA Ref. Pub. 1232

Kee RJ, Rupley FM, Miller JA (1989) Chemkin-II: a FORTRAN chemical kinetics package for the analysis of gas-phase chemical kinetics. Sandia National Laboratory, Livermore CA

McIntyre TJ, Houwing AFP, Bachor HA, Sandeman RJ (1988) Absorption measurements of the $1s_5$ metastable state population behind unstable waves in ionizing argon. In: Grönig H (ed) Proc. 16th Intl. Symp. on Shock Tubes and Waves, Aachen, p 161

A Laser Photolysis Shock Tube Study of the Reaction of OH with NH_3

John D. Mertens*, Margaret S. Wooldridge† and Ronald K. Hanson†
*Department of Engineering, Trinity College, Hartford, CT 06106 USA
†High Temperature Gasdynamics Laboratory, Department of Mechanical Engineering, Stanford University, Stanford, CA 94305 USA

Abstract. The reaction of OH with NH_3 has been studied in reflected shock wave experiments using laser photolysis of $NH_3/N_2O/Ar$ mixtures. Quantitative time-histories of the $OH(X^2\Pi_i)$ radical were measured behind the shock waves using cw, narrow-linewidth laser absorption at 307 nm. OH was generated using post-shock laser photolysis of ammonia followed by the reaction of atomic hydrogen with N_2O:

$$\begin{aligned} NH_3 + \hbar\nu \ (193\text{nm}) &\longrightarrow NH_2 + H \\ N_2O + H &\longrightarrow N_2 + OH \end{aligned} \qquad (1)$$

Ammonia photolysis yields were confirmed by quantitative measurements of the NH_2 radical using cw, narrow-linewidth laser absorption at 597 nm. Initial mixture concentrations and test conditions were chosen such that photolysis and pyrolysis of N_2O and pyrolysis of NH_3 do not play a significant role in OH reaction kinetics. Following the production of OH by reaction (1), OH-removal is dominated by reaction (2):

$$NH_3 + OH \longrightarrow NH_2 + H_2O \qquad (2)$$

This makes it possible to determine the second-order rate coefficient of reaction (2) by adjusting the value of k_2 in a detailed reaction mechanism until calculated OH concentrations fit the measured OH profiles. A least-squares two-parameter fit of the results is given by:

$$k_2 = 1.1 \times 10^{14} \exp(-4600/T, \mathrm{K}) \ \ \mathrm{cm^3 mole^{-1} s^{-1}} (f = 0.65, F = 1.45) \quad T = 1240 - 1480 \ \mathrm{K},$$

where f and F are the minimum and maximum rate coefficient factors. The activation energy of the above expression is a parameter determined by the least-squares fitting procedure, and is subject to large uncertainties.

This result is in excellent agreement with the expression recommended by Cohen and Westberg (1991) in their review of reaction (2):

$$k_2 = 5.0 \times 10^7 T^{1.6} \exp(-480/T, \mathrm{K}) \ \ \mathrm{cm^3 mole^{-1} s^{-1}} \quad T = 225 - 3000 \ \mathrm{K}.$$

Key words: Shock Tube, Reaction Kinetics, Ammonia, Hydroxyl, Laser absorption

1. Introduction

The reaction of the hydroxyl radical with ammonia plays an important role in the oxidation of ammonia and the Thermal DeNOx process. While there have been many determinations of the rate coefficient of this reaction over a wide temperature range, there is significant disagreement among the measurements above 1000 K. A recent review by Cohen and Westberg (1991) concluded that this reaction exhibits significant non-Arrhenius behaviour. The aim of the present study is to measure the rate coefficient of this reaction at temperatures above 1000 K to help refine current detailed kinetics models.

In the present investigation, quantitative measurements of OH using cw, narrow-linewidth laser absorption at 307 nm were used in reflected shock wave experiments to study the reaction

Shock Waves @ Marseille II
Editors: R. Brun, L. Z. Dumitrescu

of OH with NH_3. OH was generated using post-shock laser photolysis of NH_3 in the presence of N_2O:

$$\begin{aligned} NH_3 + \hbar\nu\ (193\text{nm}) &\longrightarrow NH_2 + H \\ N_2O + H &\longrightarrow N_2 + OH \end{aligned} \tag{1}$$

Ammonia photolysis yields were calculated using the results of Davidson et al. (1989) and confirmed with quantitative measurements of the NH_2 radical using cw, narrow-linewidth laser absorption at 597 nm (Mertens et al. 1991b). Initial mixture concentrations and test conditions were chosen such that photolysis and pyrolysis of N_2O and pyrolysis of NH_3 do not play a significant role in OH reaction kinetics. Upper limits on N_2O photolysis yields were determined using OH measurements in photolyzed $N_2O/H_2/Ar$ mixtures.

Following the production of OH by reaction (1), OH-removal is dominated by reaction (2):

$$NH_3 + OH \longrightarrow NH_2 + H_2O \tag{2}$$

This makes it possible to determine the second-order rate coefficient of reaction (2) by adjusting the value of k_2 in a detailed reaction mechanism until calculated OH concentrations fit the measured OH profiles.

Table 1. Conditions and results for determination of k_2 : $NH_3 + OH \rightarrow NH_2 + H_2O$

Temperature (K)	Total pressure (atm)	Initial mole fractions		Phot.yield (H/NH_2 ppm)	$k_2 \times 10^{-12}$ ($cm^3 mole^{-1} s-1$)
		NH_3	N_2O		
1254	1.364	0.500%	5.137%	11.0	3.0
1333	1.364	0.500%	5.137%	10.0	3.1
1391	1.343	0.500%	5.137%	9.5	3.9
1405	1.301	0.500%	5.137%	9.0	4.2
1257	1.376	0.998%	5.043%	21.5	2.95
1325	1.384	0.998%	5.043%	19.0	3.7
1327	1.313	0.998%	5.043%	20.0	2.9
1373	1.371	0.998%	5.043%	20.0	3.55
1415	1.397	0.998%	5.043%	18.0	4.2
1440	1.369	0.998%	5.043%	17.0	3.8
1482	1.255	0.998%	5.043%	17.0	4.8
1242	1.362	1.001%	4.988%	23.0	2.5
1330	1.270	1.001%	4.988%	21.0	3.6
1365	1.448	1.001%	4.988%	20.0	4.0
1257	0.856	1.996%	5.262%	39.5	2.4
1426	0.843	1.996%	5.262%	32.5	4.6

2. Experimental details

The experimental apparatus comprises a shock tube (8 meter-long) and gas handling facility previously described by Mertens et al. (1991a), an excimer laser photolysis system previously described by Kohse-Höinghaus et al. (1989), and a narrow-line laser absorption diagnostic system. All measurements were performed behind reflected shock waves.

The narrow-line laser absorption diagnostic system used a Spectra-Physics 380 ring dye laser with Rhodamine 590 dye pumped with 3-4 W (all lines) from a Spectra-Physics 164 argon ion laser. For OH measurements, the ring dye laser was intracavity doubled using a temperature-tuned AD*A crystal (Spectra-Physics 398A) as described by Rea et al. (1984). OH was monitored in the electronic ground state using the $R_1(5)$ line of the $A^2\Sigma^+ \leftarrow X^2\Pi_i$ system at 306.687 nm (vac). Approximately 5 mW of single-mode 307 nm radiation was obtained. NH_2 was monitored in the electronic ground state using the $A^2A_1 \leftarrow X^2B_1$ $(090 \leftarrow 000)\Sigma^P$ $Q_{1,N}(7)$ transition at 597.375 nm as described by Kohse-Höinghaus et al. (1989). Approximately 30 mW of single-mode 597 nm radiation was obtained. The optical arrangement for the OH and NH_2 diagnostics has been shown previously (Mertens et al. 1991a).

A shock tube passivation technique described by Kohse-Höinghaus et al. (1989) was employed to insure that the NH_3 mole fraction was well established for each experiment. NH_3, N_2O, and argon were taken directly from commercial cylinders (Liquid Carbonic) with the following purities: $NH_3 \geq 99.9\%$, $N_2O \geq 99.99\%$, Argon$\geq 99.998\%$. Helium was used as the driver gas.

3. Results

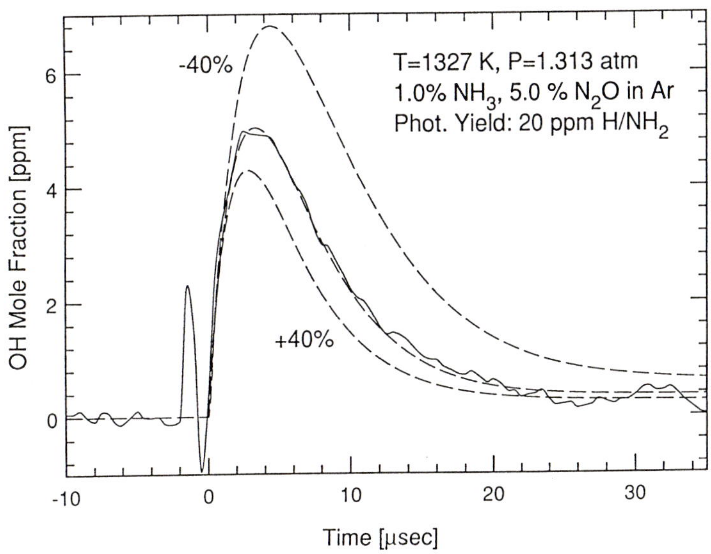

Fig. 1. Determination of k_2 from a fit of a measured OH profile. Solid line: measured profile. Dashed lines: calculated profiles using $k_2 = 2.9 \times 10^{12} cm^3 mole^{-1} s^{-1}$ with $\pm 40\%$ variation in k_2

The test conditions and results of the experiments used to determine k_2 are given in Table 1. All of the measured OH profiles were modeled with a reaction mechanism developed by Mertens et al. (1991a) using the Chemkin (Kee et al. 1980) and Senkin (Lutz et al. 1988) computer programs and the Sandia National Laboratories thermodynamic data base (Kee et al. 1991). In addition to calculating species concentration profiles, the computer programs were used to calculate reaction contribution factors and sensitivity coefficients.

The rate coefficient of reaction (2) was determined by adjusting k_2 in the complete reaction mechanism until calculated OH concentration profiles fit the peak measured OH concentrations and subsequent OH-removal profiles. Fig.1 shows a typical fit to a measured OH profile, as well as the effect of a $\pm 40\%$ adjustment in k_2. Fig.2 is a sensitivity coefficient plot for OH for the

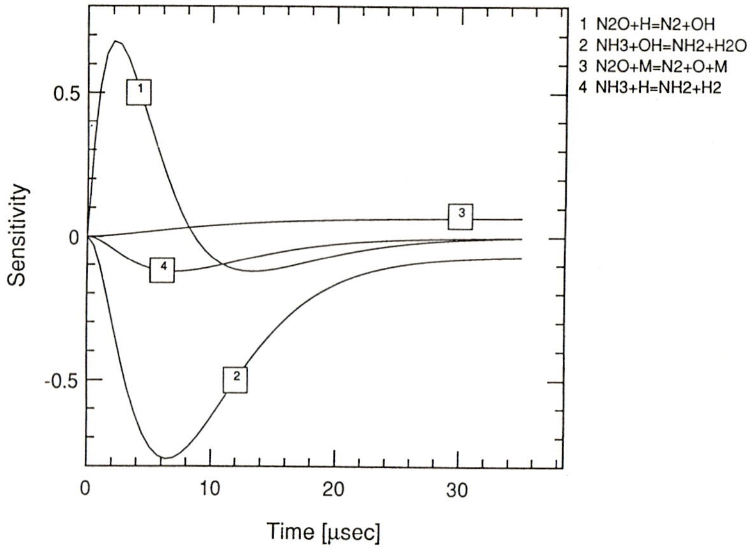

Fig. 2. OH sensitivity coefficient plot for the conditions of Fig.1

same experiment that shows that reactions (1) and (2) are the dominant OH reactions. The rate coefficient of reaction (1) has been well established by many investigators (Marshall et al. 1987). A least-squares two-parameter fit of all of the results of the current study is given by:

$$k_2 = 1.1 \times 10^{14} \exp(-4600/T,\ \mathrm{K})\ \mathrm{cm^3 mole^{-1} s^{-1}} \quad (f = 0.65, F = 1.45) \quad T = 1240 - 1480\ \mathrm{K},$$

where f and F are the minimum and maximum rate coefficient factors (Hanson and Salimian 1985). The activation energy of the above expression is a parameter determined by the least-squares fitting procedure, and is subject to large uncertainties. The minimum and maximum rate constant factors (f, F) were calculated using dependent and independent error propagation analysis (Meyer 1975). The major sources of uncertainty in k_2 are the uncertainties in k_1, the determination of the ammonia photolysis yields, and the scatter of the measurements of k_2. However, the uncertainties in the rate coefficients of all of the reactions in the complete mechanism were included in the uncertainty analysis.

4. Conclusion

Fig.3 is an Arrhenius plot which shows the results of the current study and the results of six other high-temperature studies discussed by Cohen and Westberg (1991) in their review of reaction (2). Fig.3 also shows that the results of the current study is in excellent agreement with the expression recommended by Cohen and Westberg (1991):

$$k_2 = 5.0 \times 10^7 T^{1.6} \exp(-480/T, \mathrm{K}) \quad \mathrm{cm^3 mole^{-1} s^{-1}} \quad T = 225 - 3000\ \mathrm{K}.$$

Cohen and Westberg (1991) performed transition-state theory calculations that resulted in a non-Arrhenius expression which provides good agreement with low temperature measurements of k_2 and the higher values of the high-temperature measurements of k_2. The results of the present study support the conclusions of Cohen and Westberg (1991) and the use of the above expression for k_2 in current detailed kinetics models.

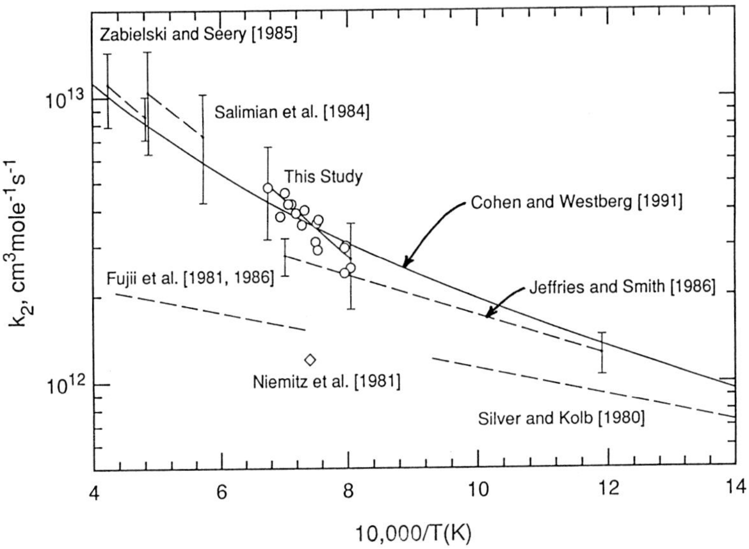

Fig. 3. Arrhenius diagram for reaction (2). Circles: this study. Solid line: Cohen and Westberg (1991). Dashed lines: six high temperature studies discussed by Cohen and Westberg (1991)

Acknowledgments

This work was supported by the Office of Naval Research, with Dr. Richard Miller as technical monitor, and the Environmental Protection Agency.

References

Cohen N, Westberg KR (1991) Chemical kinetic data sheets for high-temperature chemical reactions. J. Phys. Chem. Ref. Data. 20:1211-1311

Davidson DF, Chang AY, Kohse-Höinghaus K, Hanson RK (1989) High temperature absorption coefficients of O_2, NH_3, and H_2O for broadband ArF excimer laser radiation. J. Quant. Spectrosc. Radiat. Transfer 42:267-278

Hanson RK, Salimian S (1984) Survey of rate constants in the N/H/O system. Ch. 6 Combustion Chemistry, Gardiner WC (ed) Springer-Verlag New York:361-421

Kee RJ, Miller JA, Jefferson TH (1980) Chemkin: A general-purpose problem-independent, transportable, Fortran chemical kinetics code package, Sandia National Laboratory Report No. SAND80-8003

Kee RJ, Rupley FM, Miller JA (1991) The Chemkin thermodynamic data base, Sandia National Laboratory Report No. SAND87-8215B

Kohse-Höinghaus K, Davidson DF, Chang AY, Hanson RK (1989) Quantitative NH_2 shock tube laser-absorption experiments. J. Quant. Spectrosc. Radiat. Transfer 42:1-17

Lutz AE, Kee RJ, Miller JA (1988) Senkin: A Fortran program for predicting homogeneous gas phase chemical kinetics with sensitivity analysis. Sandia National Laboratory Report No. SAND87-8248

Marshall P, Fontijn A, Melius CF (1987) High-temperature photochemistry and BAC-MP4 studies of the reaction between ground-state H atoms and N_2O. J. Chem. Phys. 86:5540-5549

Mertens JD, Chang AY, Hanson RK, Bowman CT (1991a) A shock tube study of the reactions of NH with NO, O_2, and O. Intl. J. Chem. Kinetics 23:173-196

Mertens JD, Kohse-Höinghaus K, Hanson RK, Bowman CT (1991b) A shock tube study of H + $HNCO \rightarrow NH_2 + CO$. Intl. J. Chem. Kinetics 23:655-668

Meyer SL (1975) Data analysis for science and engineering. John Wiley and Sons, Inc. New York

Rea EC, Salimian S, Hanson RK (1984) Rapid-tuning frequency-doubled ring dye laser for high resolution absorption spectroscopy in shock-heated gases. Appl. Opt. 23:1691-1694

Pressure Influence on the Rate of NF_3 Decomposition Behind Shock Waves

N.N. Koudriavtsev, A.M. Sukhov and D.P. Shamshev
Moscow Institute of Physics and Technology, Dolgoprudny, Moscow Region 141700, Russia

Abstract. The rate constant of the NF_3 dissociation reaction in a helium mixture in the low-pressure limit $P < 1$ atm. was directly measured in a shock tube by a UV absorption method. The rate constant may be expressed as $K = 10^{14.38\pm0.07} \times \exp\{-[36600 \pm 2100(\text{cal/mol})]/RT\}$, cm^3/mol.s in the temperature range $T = 1050 - 1600$ K. Levels of the NF_3 stationary dissociation fraction were also measured as a function of the initial temperature of the mixture behind the reflected shock wave. It increases monotonically with temperature and reaches 50% of the value at 1600 K, which corresponds to high NF_2 concentration up to 2×10^{17} cm^{-3}. Rates of NF_3 thermal decomposition in the temperature range 1950-2050 K and pressure range 1.2-4.2 atm were also measured. The bimolecular rate constant of the NF_3 decomposition reaction depends strongly upon pressure. An empirical formula was obtained from the experimental data which correctly expresses the dependence of the measured rate constant of decomposition upon the temperature and pressure of the reacting mixture behind the shock wave.

Key words: Rate constant, Thermal decomposition, Shock tube

1. Introduction

Unimolecular reaction kinetics is a fundamental branch of modern chemical physics and is also of essential practical interest (Kuznetsov 1982). Experimental and theoretical investigations carried out so far were mainly concerned with dissociation and isomerization reactions of complex organic molecules and decomposition reactions of diatomic molecules. Decomposition reactions of simple polyatomic molecules containing 3-5 atoms were studied in only a few works (Kuznetsov 1982, Benson and O'Neel 1970).

The kinetics of NF_3 thermal decomposition were explored previously (McFadden et al. 1973, Dorko et al. 1975). Experimental data of Dorko et al. (1975), McFadden et al. (1973), Evans et al. (1976) and Breshears and Bird (1978) were treated in terms of the two-reaction mechanism:

$$NF_3 + M \longrightarrow NF_2 + F + M, \tag{1}$$

$$NF_2 + F + M \longrightarrow NF_3 + M. \tag{2}$$

Values of the rate constants given by different workers are not in satisfactory agreement. Values of the bimolecular rate constant for process (1) in the low pressure limit at 1500 K vary from 3.5×10^6 (Dorko et al 1975) to 4.5×10^9 cm^3/mol.s (Evans et al. 1976). In addition to this, there is no reliable information concerning NF_3 thermal decomposition kinetics in the intermediate pressure range 1-10 atm. Direct rate measurements of NF_3 thermal decomposition at intermediate pressures have been performed in the present work.

2. Experimental apparatus

NF_3 decomposition was studied in a helium mixture behind reflected shock waves in a shock tube. The NF_2 radical concentration was monitored by UV absorption at $\lambda = 260$ nm. Shock waves were generated in a stainless steel shock tube (70 mm ID). The gas temperature behind the reflected shock wave was calculated from the velocity of the incident shock wave. The spectrophotometric detection system consisted of a deuterium discharge lamp, quartz lenses to form the light beam, a 0.25 m grating monochromator and a photomultiplier. High NF_3 dilution with He was used in our experiments.

Shock Waves @ Marseille II
Editors: R. Brun, L. Z. Dumitrescu

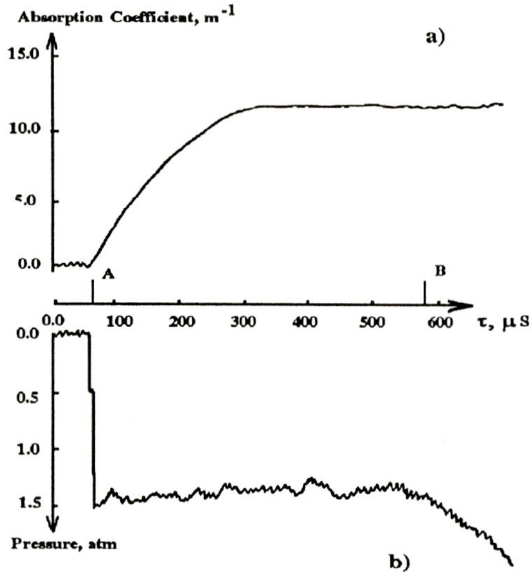

Fig. 1. Typical oscillograms of absorption and pressure behind the reflected shock wave; (a) absorption, (b) pressure

3. Results

3.1. Low pressure limit

Typical oscillograms of absorption and pressure behind the reflected shock waves are represented in Fig.1 ((a) absorption, (b) pressure)). The time interval A-B on the oscillograms is considered to be a period of stationary conditions behind the reflected shock wave. The presence of a stationary plateau on the pressure oscillogram is connected with the isobaric progress of the chemical reactions.

The absorption oscillograms during the initial period of time have a linear region. Then the rate of increase of absorption becomes smaller, before reaching a stationary level. The duration of existence of stationary conditions decreased from 780 to 450 μs and the duration of the linear region on the absorption oscillograms decreased from 80 to 10 μs with increasing temperature from 1050 to 1600 K. UV absorption in this system results from the NF_2 radical absorption band with its maximum near 260 nm (Kuznetsova et al 1964).

The slope of the linear region on the absorption oscillogram is connected with the rate of reaction (1) during the initial period. According to McFadden (1973) both the decrease in the NF_2 formation rate and the stabilization of its concentration are due to the influence of secondary reactions. Thus, the values of the rate constants were measured in the temperature range from 1050 to 1600 K and pressure range from 0.7 to 1.3 atm. A plot of $\log K$ versus $10^4/T$ is given in Fig.2. The values of the rate constant at different pressures and the same temperature coincide within experimental error. From this it follows that reaction (1) proceeds as a second-order reaction in these conditions. The temperature dependence of the rate constant may be written as

$$K = 10^{14.38\pm0.07} \times \exp \frac{36600 \pm 2100\text{cal/mol}}{RT}, \qquad \text{cm}^3/\text{mol.s} \qquad \text{for } T = 1050 - 1600 \text{ K.} \quad (3)$$

Comparison of the rate constant with the results of McFadden et al. (1973), Dorko et al. (1975), Evans et al. (1976), Breshears et al. (1978) shows discrepancies both in the preexponential factor and the activation energy. However, NF_3 was highly diluted with an inert gas (argon) in these investigations. Furthermore, the authors of these works used simplified kinetic models for NF_3

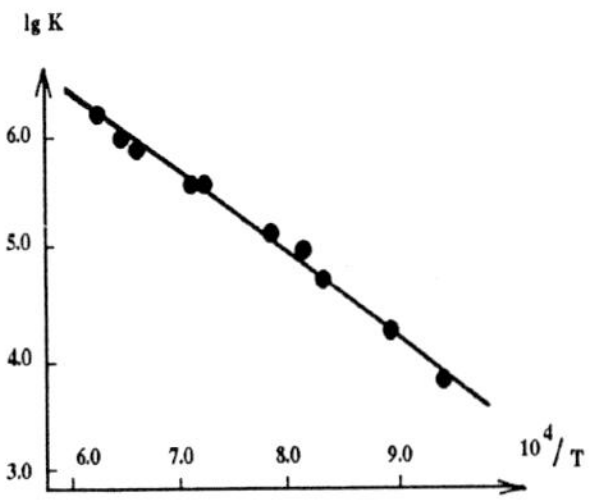

Fig. 2. Experimental values of $\log K$ vs. $10^4/T$. K is the rate constant of the dissociation reaction (1), T is the temperature behind the reflected shock wave. The best fit straight line corresponds to an Arrhenius dependence represented by Eq.3

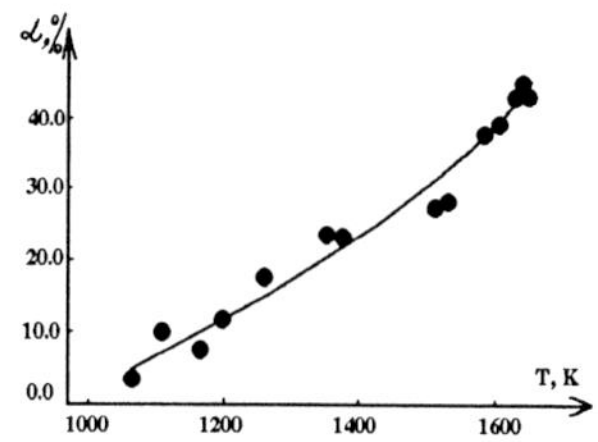

Fig. 3. Experimental values of the NF_3 stationary dissociation fraction vs. T, the temperature behind the reflected shock wave

decomposition in treating their oscillograms. In the present work direct measurements of the dissociation reaction rate (1) were actually performed.

The levels of stationary NF_2 UV absorption were also measured in our experiments. Experimental data are represented as a plot of the stationary dissociation fraction versus the initial temperature behind the reflected shock wave in Fig.3. The dissociation fraction increases monotonically to the 50% level with temperature increasing to 1600 K. This level corresponds to a high concentration of NF_2 radicals of 2×10^{17} cm^{-3}. One should mention that such an NF_2 concentration is supposed to be sufficient for the experimental realization of chemically pumped visible NF-IF laser models (Basov et al. 1987). The use of NF_3 dissociation as a source of fluorine atoms and NF_2 radicals allows to avoid the use of the unstable and expensive reagent tetrafluorohydrazine N_2F_4.

3.2. Intermediate pressure range

In this case the duration of existence of stationary conditions decreased from 480 to 350 μs and the duration of the linear region on the absorption oscillograms decreased from 20 to 8 μs with increasing temperature from 1450 to 2050 K.

The rate of the decomposition reaction (1) was measured directly in the temperature interval from 1450 to 2050 K and pressure interval from 1.2 to 4.2 atm. These experimental rate values cannot be described by the usual simple formula $W = K(T) \times P^n$, where $K(T)$ is the rate constant and n is the order of the reaction. The experimental rate values at different pressures and temperatures are represented in Fig.4 as plots of $\xi N^2/W$ values versus N, where $\xi = 0.09$ is the initial molar fraction of NF_3 in the mixture, and N is total concentration of the mixture. $\xi N^2/W$ values increase linearly with increasing N in the explored interval of total concentrations $0.8 \times 10^{-5} < N < 3.2 \times 10^{-5}$ mol/cm^3 at each fixed value of the temperature. The slopes of the appropriate best straight lines decrease monotonically with temperature growth, while their crossing points with the ordinate axis increase monotonically. Temperature dependencies of these values are represented in Figs.5a and 5b, as the Arrhenius coordinates $\log K$ versus $10^4/T$. Appropriate relationships are linear and so may be represented in the form of Arrhenius exponents. The facts mentioned allow us to describe the dependence of the experimental values of reaction rate upon temperature and total concentration of the mixture at intermediate pressures by the following empirical formula:

$$W(M,T) = \frac{M_2\xi \cdot N^2}{[K_2(T)/K_1(T)] - 1}, \qquad (4)$$

where $K_1 = (2.9 \pm 0.6) \times 10^5 \ \exp[(-18600\,\text{cal/mol})/RT]\ \text{s}^{-1}$,
$K_2 = (2.3 \pm 0.7) \times 10^{10} \ \exp[(-15400\,\text{cal/mol})/RT]$, $P = 1.2 - 4.2$ atm, $T = 1450 - 2050$ K.

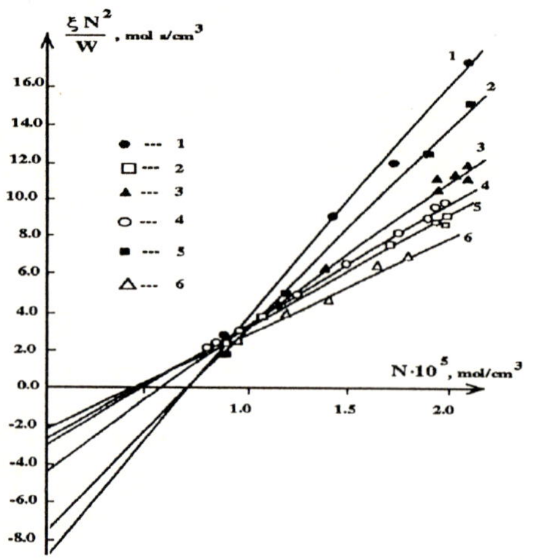

Fig. 4. Experimental decomposition rate values W in the form $\xi N^2/W$ vs. N at different pressures and temperatures in the reacting mixture: 1 - $T = 1510$ K; 2 - $T = 1830$ K; 3 - $T = 1680$ K; 4 - $T = 1880$ K; 5 - $T = 1710$ K; 6 - $T = 1980$ K

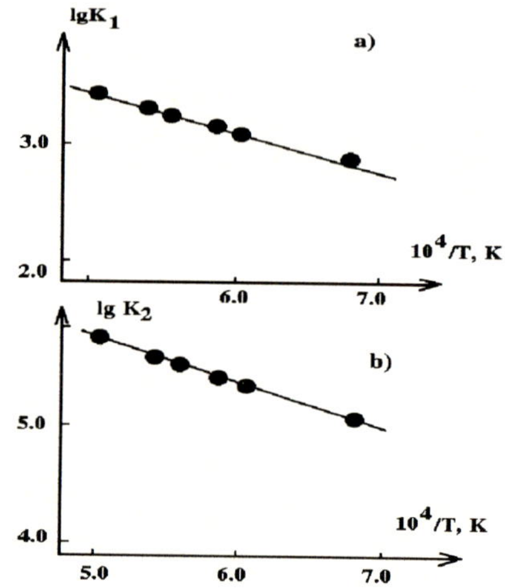

Fig. 5. Temperature dependencies of functions $K_1(T)$ and K_2 in the Arrhenius coordinates $\log K$ vs. $10^4/T$:(a) - K_1, (b) - $K_2 2$

The appropriate bimolecular rate constant of the decomposition reaction (1):

$$K_{bim} = \frac{W(N,T)}{\xi N^2} = \frac{K_2(T)}{[K_2(T)/K_1(T)] - 1} \tag{5}$$

is obviously a function of both the temperature and pressure of the reacting mixture. $K_1(T)$ and $K_2(T)$ are not considered to be rate constants of any elementary chemical reactions. In former works concerning NF_3 decomposition (McFadden et al. 1973, Dorko et al. 1975, Evans et al. 1976 and Breshears et al. 1978) the rate constant was measured in the limits of high and low pressures only. At pressures of the order of 1 atm (total concentration $N \approx (0.5 - 0.7) \times 10^{-5}$ mol/cm^3) values of K_{bim} given by Eq.(5) are in satisfactory agreement with our above results of rate constant measurements in the low pressure limit.

The results reported herein are of interest for several branches of physico-chemical exploration in which NF_3 thermal decomposition is used as a source of NF_2 radicals and fluorine atoms, such as generation of inert gas excimers (Stater 1987), visible and IR chemical lasers (Basov et al. 1987 and Bashkin et al. 1982), etc.

References

Bashkin AS, Igoshin VI, Oraevsky AN, Shcheglov VA (1982) Chemical Lasers. Nauka, Moscow
Basov NG, Gavrikov VF, Shcheglov VA (1987) Kvantovaya Electron. 14:1754
Benson SV, O'Neel HE (1970) Kinetic data on gas phase unimolecular reactions. Wiley, NY
Breshears WD, Bird PF (1978) J. Chem. Phys. 68: 2996
Dorko EA, Grimm UW, Scheller K, Mueller GW (1975) J. Chem. Phys. 63: 3596
Evans PJ, Tschuikow-Roux E (1976) J. Chem. Phys. 65: 4202
Kuznetsov NM (1982) Kinetics of monomolecular reactions. Nauka, Moscow (in Russian)
Kuznetsova IA, Kuzjakov YuYA, Tatevsky VM (1964) Optika i Spectroskopiya 16: 542
McFadden KO, Tschuikow-Roux E (1973) J. Phys. Chem. 77: 1475
Stater R (1987) Appl. Phys. B 42:17

A Shock Tube Study of the Oxidation of Silicon Atoms by NO and CO_2

Hans Jürgen Mick and Paul Roth
Institut für Verbrennung und Gasdynamik, Universität Duisburg, 47048 Duisburg, Germany

Abstract. In the present investigation the very fast thermal dissociation of silane at temperatures above 1800 K was used as a Si atom source to study its reactions with NO and CO_2. In the first part experiments in mixtures of 0.5 to 10 ppm SiH_4 and 75 to 300 ppm NO diluted in argon were performed. Absorption profiles of the atomic species Si, N, and O were measured in the temperature range 1600 K $\leq T \leq$ 3200 K at pressures 0.4 bar $\leq p \leq$ 1.5 bar. The measured Si atom absorption profiles were analysed by assuming a first-order type reaction. From the present measurements the rate coefficient of the reaction

$$\mathrm{Si} + \mathrm{NO} \overset{k_3}{\rightleftharpoons} \mathrm{SiO} + \mathrm{N} \qquad \text{(R3)}$$

$$k_3 = 3.2 \times 10^{13} \exp(-1775K/T) \quad \pm 30\% \qquad \mathrm{cm^3 mol^{-1} s^{-1}}$$

was determined. In addition, N and O atom concentration profiles were observed confirming the correctness of reaction channel (R3).

In the second part of the paper Si atom measurements were performed in mixtures containing 0.5 to 0.75 ppm SiH_4 and 200 ppm CO_2 diluted in argon. The experimental conditions were 2000 K $\leq T \leq$ 3200 K and 0.4 bar $\leq p \leq$ 1.5 bar. Again the first-order assumption was used to determine the rate coefficient of the reaction

$$\mathrm{Si} + \mathrm{CO_2} \overset{k_6}{\rightleftharpoons} \mathrm{SiO} + \mathrm{CO} \qquad \text{(R6)}$$

$$k_6 = 6.0 \times 10^{14} \exp(-9420K/T) \quad \pm 30\% \qquad \mathrm{cm^3 mol^{-1} s^{-1}}$$

from Si atom measurements.

Key words: Oxidation, Silicon atoms

1. Introduction

In spite of the major role of silane chemistry in chemical vapour deposition for semiconductor manufacturing, the thermal decomposition and oxidation of SiH_4 was rarely studied. This paper deals with the oxidation of silicon atoms by nitric oxide and carbon dioxide at high temperatures, where no literature values were found. Husain and Norris (1978a) studied this reactions at room temperature. They generated Si atoms by pulsed irradiation of $SiCl_4$ and detected the absorption of Si atoms at 251.6 nm.

In the present investigation initial SiH_4/NO/Ar and SiH_4/CO_2/Ar mixtures were used to study the respective reactions of Si atoms with NO and CO_2 by detecting Si, N, and O atom absorption profiles. Previous shock tube experiments on the dissociation of silane revealed that highly diluted silane is a very effective source for silicon atoms at temperatures above 1800 K, see Mick et al. (1993a). Under these conditions, SiH_4 decomposes very rapidly to form Si atoms via SiH_2 as intermediate species.

Shock Waves @ Marseille II
Editors: R. Brun, L. Z. Dumitrescu

2. Experimental

The experiments were conducted behind reflected shock waves in a stainless-steel shock tube of 79 mm inner diameter. The tube is constructed as a heatable ultra-high vacuum apparatus and can be evacuated through a special end-plate valve using a fore pump, a turbo-molecular pump, and a liquid nitrogen-cooled titanium sublimation pump, down to pressures of about 1×10^{-8} mbar. Typical leak-plus-outgassing rates were 5×10^{-6} mbar/min. A more detailed description of the apparatus and the experimental procedure is given in Roth and Just (1977).

The gases used were carefully mixed in a stainless steel cylinder by the partial pressure method. The gas mixing setup can be heated and evacuated like the shock tube. Due to the high sensitivity of the ARAS detection technique the test gases were of ultra-high purity: Ar (99.9999 %), SiH_4 (99.995 %), NO (99.5 %), CO_2 (99.995 %). Initial mixtures of 0.5 to 10 ppm SiH_4 and 75 to 300 ppm NO or 200 ppm CO_2 diluted in Ar were used. Prior to each run the tube was evacuated to pressures of 5×10^{-7} mbar or lower. Diaphragms made of aluminum, 30 or 50 μm thick, were ruptured by increasing the driver gas (H_2) pressure. Temperatures and pressures behind the reflected shock waves were calculated from the incident shock speed, which was measured by four thin film gauges, placed along the test section at known intervals.

The optical detection technique for measuring Si, N, and O atom absorptions is a line emission - line absorption method. In the case of N and O atoms the arrangement consists of a microwave-excited discharge lamp, the absorption zone in the shock tube separated on both sides by two thin LiF windows, a 1 m McPherson vacuum UV monochromator, and a special solar blind photomultiplier. The spectral lines NI (λ = 119.9 nm) and OI (λ = 130.5 nm) were excited in gas mixtures of He containing 1% N_2 or O_2 passing through the lamp at pressures of 6 mbar. In the case of Si atoms the detection system consists of a pulsed hollow-cathode lamp, the shock tube absorption zone, a 0.25 m Jarrel-Ash monochromator, and a photomultiplier. The Si triplet transition $[4s(^3P_J) \leftarrow 3p^2(^3P_J)]$ at $\lambda = 251.6$ nm is known to be the most sensitive line for atomic absorption spectroscopy (Husain and Norris 1978b; Dean and Rains 1975).

The spectral shapes of the lines emitted by the discharge as well as by the hollow-cathode lamp are not known in detail. Hence a simple relation between the directly measured absorption and the atom concentration does not exist. Therefore a series of carefully controlled shock wave calibrations have been done to relate the measured absorptions to the corresponding Si, N, and O atom concentrations. Results of the N and O atom calibration procedures are given by Thielen and Roth (1984). The Si atom calibration was carried out by Mick et al. (1993a,b).

3. Results and discussion

3.1. Si+NO reaction

The first group of experiments was performed in the SiH_4/NO/Ar system. The reaction of Si atoms with NO was investigated by Si, N, and O atom measurements in the temperature range 1660 K $\leq T \leq$ 3360 K at pressures between 0.4 and 1.5 bar. Initial mixtures of 0.5 to 10 ppm SiH_4 and 75 to 300 ppm NO in Ar were used. A typical individual example of Si spectral absorption is illustrated in Fig.1 showing the measured light extinction versus reaction time. Since no interference absorption by SiH_4, H_2, and NO was observed, the extinction signal is equal to the absorption by Si atoms. The reaction time zero indicates the arrival of the reflected shock wave in the measurement plane. The Si atom absorption profile increases very rapidly after the arrival of the reflected shock wave and decreases at longer reaction times down to the zero level. To the contrary, experiments with SiH_4/Ar mixtures without NO performed under the same conditions also show a very rapid absorption increase behind the reflected shock wave, but the signals remain constant for the remaining reaction times. From these experiments SiH_4 is known to be a

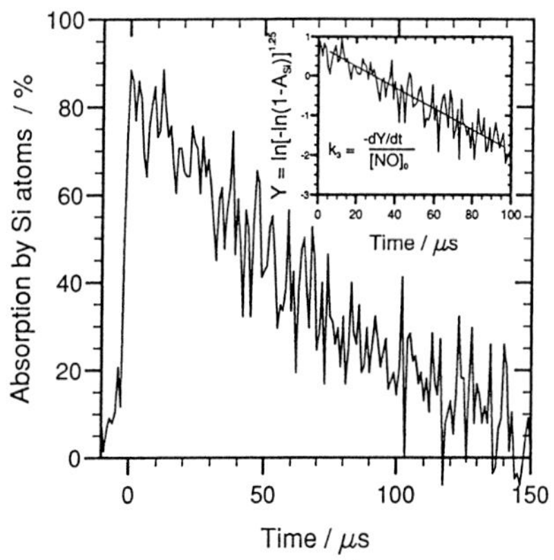

Fig. 1. Measured Si atom absorption. Initial conditions: 0.75 ppm SiH_4 and 300 ppm NO in Ar, $T = 2955$ K, $p = 1.1$ bar.-Insert: First-order representation of the absorption profile.

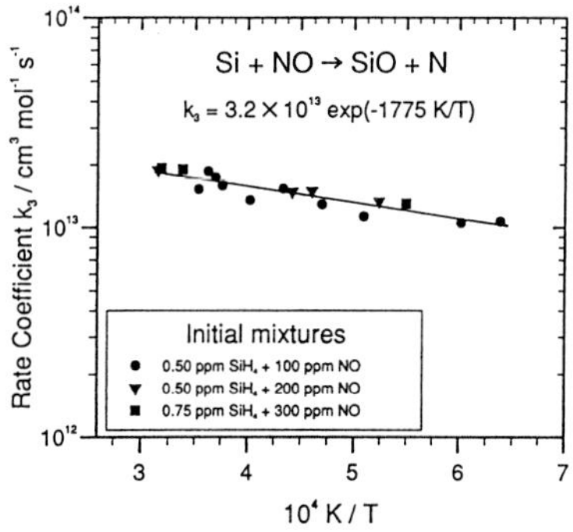

Fig. 2. Arrhenius diagram of rate coefficient k_3

very effective Si atom source at temperatures above 1800 K, see Mick et al. (1993a). The strong absorption decrease observed during the present experiments must be caused by the reaction of Si with NO.

The evaluation of the absorption signal of Fig.1 and all other signals was done by assuming a first-order-type reaction between Si and NO:

$$Si + NO \xrightarrow{k_3} \text{products} \tag{3}$$

which is justified in the present case by the high excess of NO. In the insertion of Fig.1, the data evaluation based on the first-order method is demonstrated. In that case the value $Y = \ln(-\ln(1 - A_{Si}))^{1/0.8}$, with A_{Si} being the measured time-dependent spectral absorption of Si, is shown as a function of time. It can obviously be fitted by a straight line, from which dY/dt was determined. This method of data interpretation is independent on the detailed knowledge of a calibration relation. All k_3 values obtained from the different experiments are summarized in Fig.2. The mean value can be approximated by the expression:

$$k_3 = 3.2 \times 10^{13} \exp(-1775K/T) \qquad \text{cm}^3\text{mol}^{-1}\text{s}^{-1},$$

which is shown as a solid line in Fig.2. The standard deviation due to experimental scatter is about ±10%. The combined uncertainty from the calibration procedure and the experimental scatter is about ±30%. More details of data interpretation are given by Mick et al. (1993b). The comparison of the rate coefficient determined in this study seems inconsistent with the only available literature value of Husain and Norris (1978a) obtained at room temperature. The linear extrapolation of the present rate coefficients to their room temperature is 1000 times smaller. This indicates a strong non-Arrhenius behaviour.

Table 1. Simplified reaction mechanism of the SiH_4/NO system highly diluted in Ar.
$k_i = A \times T^n \times \exp(-E_a/T)$ cm^3 mol^{-1} s^{-1}

No	Reaction	Rate Coefficient A	n	E_A	Reference
1	SiH_4+Ar⇌SiH_2+H_2+Ar	9.9×10^{15}		24000	Mick et al.(1993a)
2	SiH_2+Ar⇌Si+H_2+Ar	9.1×10^{13}		15100	Mick et al.(1993a)
3	Si+NO⇌SiO+N	3.2×10^{13}		1775	see text
4	N+NO⇌N_2+O	2.0×10^{13}			see text
5	O+H_2 ⇌OH+H	3.7×10^{6}	2.17	4080	Natarajan and Roth (1987)

Up to now it was not necessary to consider the possible reaction products of reaction (R3). It is assumed that the reaction can proceed via a two-channel mechanism:

$$\mathrm{Si} + \mathrm{NO} \overset{k_3}{\rightleftharpoons} \mathrm{SiO} + \mathrm{N}, \qquad \Delta^R h = -168.6 \text{ kJ/mol} \qquad \text{(R3)}$$

$$\overset{k_{3'}}{\rightleftharpoons} \mathrm{SiN} + \mathrm{O}, \qquad \Delta^R h = +80.7 \text{ kJ/mol} \qquad \text{(R3')}$$

leading to N and/or O atom formation. To distinguish between both, additional N and O atom measurements were performed. A typical N atom concentration profile is illustrated in Fig.3. The signal increases during the first 100 μs and decreases at longer reaction times. This behaviour can well be described by the reaction mechanism of Table 1, where reaction (R3) is responsible for the formation and the reaction

$$\mathrm{N} + \mathrm{NO} \overset{k_4}{\rightleftharpoons} \mathrm{N_2} + \mathrm{O} \qquad \text{(R4)}$$

for the consumption of N atoms. Computer simulations based on this mechanism are also shown in Fig.3. Best fits were obtained by varying the rate coefficient k_4, see solid smooth line. Variations of k_4 by factors 2 and 0.5 indicating the sensitivity of the rate coefficient to the measured N atom concentration are also included in Fig.3. Other reactions like NO+O, NO+NO, or the NO decomposition, which were not considered in the reaction mechanisms of Table 1 are of less influence under the present conditions.The rate coefficients k_4 obtained from all N atom measurements can be summarized by a temperature-independent mean value of

$$k_4 = 2.0 \times 10^{13} \qquad \mathrm{cm^3 mol^{-1} s^{-1}}.$$

The combined uncertainty from the calibration procedure and the experimental scatter is about ±50%. This reaction was also part of numerous investigations studied over a large temperature interval, summarized by Westley et al.(1992).

The literature rate coefficients scattered between 1×10^{13} and 3.7×10^{13} cm^3 mol^{-1} s^{-1}. Most authors agree that the activation energy of this reaction is close to zero. The value determined in this study confirms a temperature-independent rate coefficient. The present investigation illustrates a new and effective thermal source for N atoms via the NO+Si reaction allowing to study the reaction of N with NO at relatively low temperatures.

O atoms which are either a result of the N+NO reaction or from reaction (R3') were also measured during the present investigation. A typical O atom result is shown in Fig.4. The signal increases with an S-shape behaviour during the first 300 μs and decreases slowly for longer reaction times. The solid smooth line in Fig.4 corresponds to a computer simulation based on the reaction mechanism of Table 1. It can be seen that the agreement between measured and calculated O atom profiles is quite good and no further adjustment is necessary. We conclude that the combined measurements of Si, N, and O atoms can well be represented by the mechanism of Table 1 and gives a sufficiently complete description of the SiH_4/NO system. It shows that reaction (R3) is

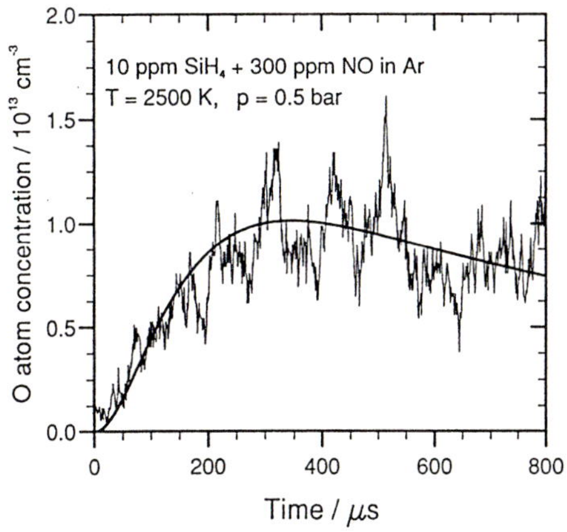

Fig. 3. Measured and calculated N atom concentration profiles

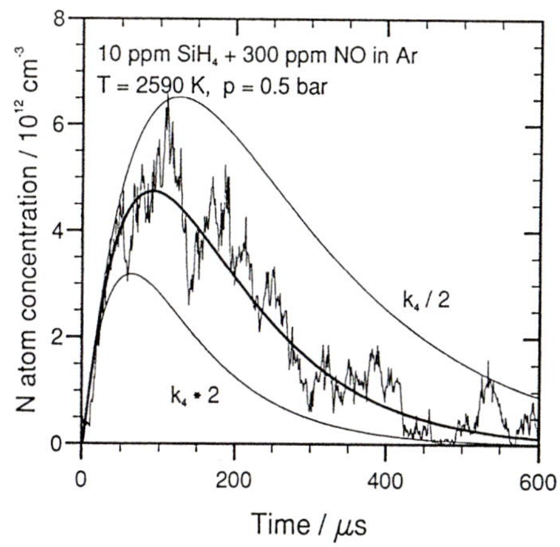

Fig. 4. Measured and calculated O atom concentration profiles

the dominant channel for the oxidation of Si under our conditions and that the alternative path (R3') is not necessary to be considered in the present case.

3.2. Si+CO_2 reaction

The reaction of Si atoms with CO_2 was studied by Si atom concentration measurements in shock tube experiments at conditions 2000 K $\leq T \leq$ 3200 K and 0.4 bar $\leq p \leq$ 1.5 bar.Initial mixtures of 0.5 to 0.75 ppm SiH_4 and 200 ppm CO_2 diluted in Ar were used to investigate the reaction system. The measured light extinction of the hollow cathode Si emission was identified to be caused by Si atoms and interference absorption by other species was not observed.

A typical example is illustrated in Fig.5, showing the Si atom concentration versus reaction time. The time zero indicates the arrival of the reflected shock wave in the measurement plane. The Si atoms were again generated very rapidly from the SiH_4 decomposition and decreased to zero in about 110 μs. It seems clear that the observed Si atom consumption is caused by the reaction with CO_2. The evaluation of the absorption signals was again done by assuming first-order behaviour of the reaction

$$\mathrm{Si} + \mathrm{CO_2} \overset{k_6}{\rightleftharpoons} \mathrm{SiO} + \mathrm{CO}, \qquad \Delta^R h = -619.2\ \mathrm{kJ/mol} \qquad \text{(R6)}$$

to be the only Si consuming reaction.The rate coefficients k_6 obtained from all Si atom profiles are illustrated in Fig 6. The points scatter around a line, which can best be represented by the Arrhenius expression:

$$k_6 = 6.0 \times 10^{14} \exp(-9420\ K/T) \qquad \pm 30\% \qquad \mathrm{cm^3 mol^{-1} s^{-1}}$$

From energetic reasons all other product channels were excluded.

Reaction (R6) was also part of the study of Husain and Norris (1978a). They found a value of $k_6 = 6.6 \times 10^{12}$ cm^3 mol^{-1} s^{-1} at room temperature. Extrapolation of the present value to their room temperature results shows significant differences in the rate coefficients.

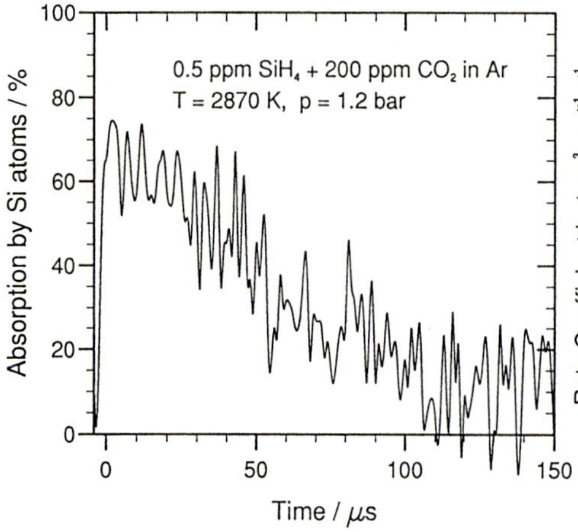

Fig. 5. Measured Si atom absorption profile

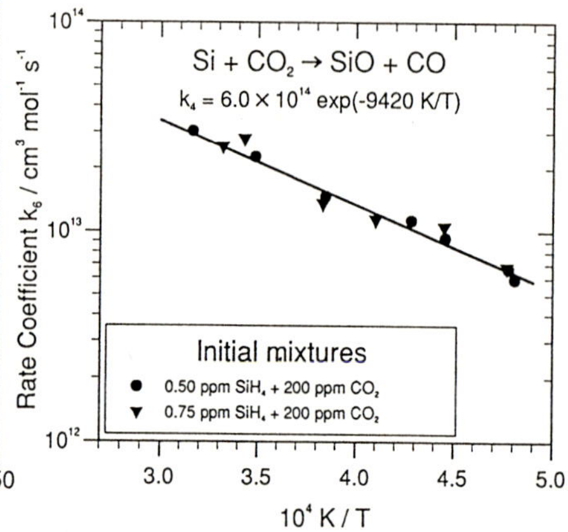

Fig. 6. Arrhenius diagram of rate coefficient k_6

Acknowledgement

The authors thank Mrs. C. Kmiecik for her help in conducting the experiments. The financial support of the Deutsche Forschungsgemeinschaft is gratefully acknowledged.

References

Dean JA, Rains TC (1975) Flame emission and atomic absorption spectrometry Vol. 3: Elements and matrices. Marcel Dekker, Inc., New York

Husain D, Norris PE (1978a) Kinetic study of reactions of ground state Silicon atoms, Si[$3p^2(^3P_j)$], by atomic absorption spectroscopy. J. Chem. Soc. Faraday II 74:106

Husain D, Norris PE (1978b) Kinetic study of ground state Silicon atoms, Si[$3p^2(^3P)$] by atomic absorption spectroscopy. J. Chem. Soc. Faraday II 74:93

Mick HJ, Smirnov VN, Roth P (1993a) ARAS measurements on the thermal decomposition of silane. Ber. Bunsenges. Phys. Chem. 97(6):793

Mick HJ, Matsui H, Roth P (1993b) High temperature kinetics of Si atom oxidation by NO based on Si,N, and O atom measurements. J. Phys. Chem. 97(26):6839

Natarajan K, Roth P (1987) High temperature rate coefficient for the reaction of O(^3P) with H_2 obtained by the resonance absorption of O and H atoms. Combustion and Flame 70:267

Roth P, Just Th (1977) Atomabsorptionsmessungen zur Kinetik der Reaktion CH_4+O→CH_3+OH im Temperaturbereich 1500 K$\leq T \leq$ 2250 K. Ber. Bunsenges. Phys. Chem. 81(6):572

Thielen K, Roth P (1984) Resonance absorption measurements of NO and O atoms in high temperature NO dissociation and formation kinetics. Symp. (Intl.) Comb., [Proc.] 20:685

Westley F, Herron JT, Hampson RF, Mallard WG (1992) NIST chemical kinetics database. Version 4.0, Gaithersburg

On the Formation of $S(^3P)$ and $S(^1D)$ During the Thermal Decomposition of COS Behind Shock Waves

D. Woiki and P. Roth
Institut für Verbrennung und Gasdynamik, Universität Duisburg, 47048 Duisburg, Germany

Abstract. The formation of ground state $S(^3P)$ and electronically excited $S(^1D)$ atoms was studied during the thermal decomposition of carbonyl sulfide behind reflected shock waves by using atomic resonance absorption spectroscopy (ARAS). The experiments were performed in highly diluted mixtures of Ar with relative concentrations of COS between 4 to 100 ppm in the temperature range 1830 K $\leq T \leq$ 3020 K at pressures around 1 bar. The measured slopes of the linear $S(^3P)$ and $S(^1D)$ concentrations divided by the respective concentration of the initial reactants were found to obey the Arrhenius laws:

$$\left.\frac{d[S(^3P)]/dt}{[COS]\,[Ar]}\right|_{t\approx 0} = 2.9 \times 10^{14} \times \exp\left(-\frac{63360 \text{ cal/mol}}{RT}\right) \text{cm}^3\text{mol}^{-1}\text{s}^{-1} \tag{1}$$

$$\left.\frac{d[S(^1D)]/dt}{[COS]\,[Ar]}\right|_{t\approx 0} = 4.8 \times 10^{12} \times \exp\left(-\frac{71800 \text{ cal/mol}}{RT}\right) \text{cm}^3\text{mol}^{-1}\text{s}^{-1} \tag{2}$$

These results indicate a fast thermalisation of the partial equilibrium between the $S(^3P)$ and $S(^1D)$ states during the non-equilibrium dissociation of COS.

Key words: Thermal decomposition of COS

1. Introduction

The thermal decomposition of triatomic molecules like N_2O, CO_2, CS_2, and COS in the ground state $X^1\Sigma$ is of particular interest because it is associated with electronic excitation processes. The energetically more favorable direct formation of ground state products is connected with a spin-forbidden transition from a singlet state to a triplet state of the decomposing molecules. The spin-allowed dissociation forming singlet excited state atoms has to be regarded as a second possible reaction channel. Beside the electronical ground state of COS a bond or an unbound triplet excited state ($^3\Sigma$ or $^3\Pi$) are possible. The exact shape of the lowest triplet potential surface of COS is unknown. Because of the different excitation states of the atomic reaction product S, both possible reaction channels lead to different values of the reaction enthalpy ΔH_0^0. A comparison of these values with the apparent activation energy of the dissociation reaction provides information about the most favorable reaction pathway. This advance is limited by a lack of knowledge of the proper shape of the potential surfaces for the decomposing molecule.

In two former studies on the thermal decomposition of COS by Schecker and Wagner (1969) and by Hay and Belford (1967) the existence of a bond triplet excited state of the COS molecule was discussed to be the most probable one. Schecker and Wagner (1969) conclude from their activation energy of E_{a0} = 61.0 kcal/mol for the low pressure range and $E_{a\infty}$ = 68.0 kcal/mol for the high pressure range that dissociation of COS proceeds via the low energy path forming directly $S(^3P)$ atoms.

From a more simplified point of view the thermal decomposition of COS can be regarded as a two-channel process which can be described by the two elementary reactions:

$$COS(^1\Sigma) + M \overset{k_{1a}}{\rightleftharpoons} CO(^1\Sigma^+) + S(^3P) + M \qquad \Delta H_0^0 = 71.6 \text{ kcal/mol} \tag{R1a}$$

$$\overset{k_{1b}}{\rightleftharpoons} CO(^1\Sigma^+) + S(^1D) + M \qquad \Delta H_0^0 = 97.9 \text{ kcal/mol} \tag{R1b}$$

Shock Waves @ Marseille II
Editors: R. Brun, L. Z. Dumitrescu

Beside this possible direct formation of $S(^3P)$ and $S(^1D)$ atoms the quenching and excitation reaction

$$S(^1D) + M \overset{k_2}{\rightleftharpoons} S(^3P) + M \qquad \Delta H_0^0 = 26.3 \text{ kcal/mol} \tag{R2}$$

contributes to the balance between the different excitation states of S. The equilibrium constant of reaction (R2) is given by the Boltzmann distribution of the electronic excitation states to be:

$$\left.\frac{[S(^3P)]}{[S(^1D)]}\right|_{equ.} = \frac{g(^3P)}{g(^1D)} \times \exp\left(+\frac{26308\,\text{cal/mol}}{RT}\right) \tag{3}$$

where $g(^3P) = 9$ and $g(^1D) = 5$ are the statistical weights of the triplet and singlet states. It is the aim of the present work, to measure the formation of both $S(^3P)$ and $S(^1D)$ behind shock waves under nearly the same conditions using initial mixtures of a few ppm COS diluted in argon.

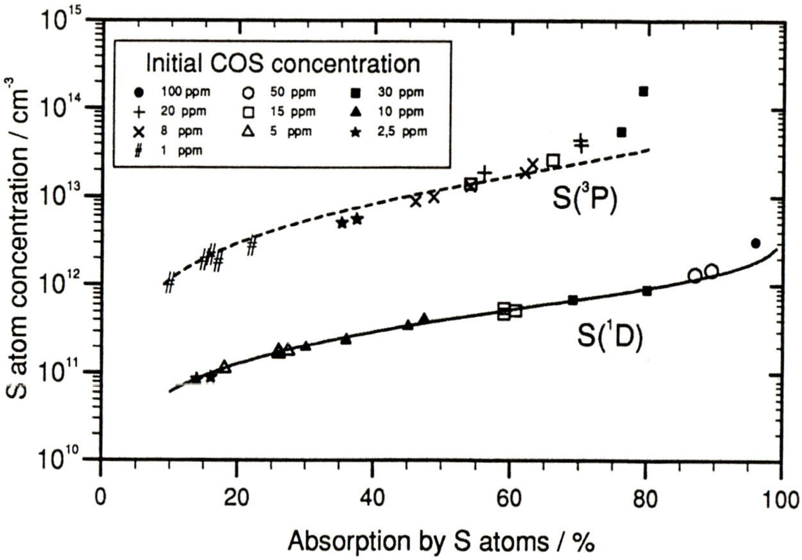

Fig. 1. Concentration of $S(^3P)$ and $S(^1D)$ versus spectral absorption

2. Experimental and calibration procedure

Gas mixtures of Ar containing 4 to 100 ppm COS were heated by reflected shock waves in an UHV shock tube of 79 mm internal diameter having a running section of 6.0 m in length, previously described elsewhere (Roth and Just 1977; Thielen and Roth 1987). The formation of S atoms in the electronical ground state $S(^3P)$ and in the electronically excited state $S(^1D)$ was monitored by Atomic Resonance Absorption Spectroscopy (ARAS). The spectral resonance lines of the S atoms were excited by microwave discharge in mixtures of He with 1% H_2S and separated by a 1 m McPherson vacuum UV monochromator. The time resolution of the detection system is less than 10 μs for a signal-to-noise ratio of about 10. The ARAS measurement technique applied in this study is based on the transitions $^3D^0 - {}^3P(4s' - 3p)$ at $\lambda = 147.4$ nm for the $S(^3P)$ atoms and $^1D^0 - ^1D(4s' - 3p)$ at $\lambda = 166.7$ nm for the $S(^1D)$ atoms. To obtain a relation between the measured absorption and corresponding $S(^3P)$ or $S(^1D)$ atom concentrations, calibration experiments with mixtures of 1 - 100 ppm COS diluted in Ar have been performed at temperatures between 3000 K and 6500 K. According to equilibrium calculations the yield of S atoms resulting from the COS decomposition was assumed to be approximately one in these reaction systems. In all calibration experiments the absorption signal increased step-like to a stationary value after the reflected shock wave. The measured absorption was interpreted based on the known total concentration of S atoms and the equilibrium distribution between the two electronic states, see Eq.(1). The results of all calibration experiments are summarized in Fig.1. The data points scatter around two lines which represent the Lambert-Beer law for the respective absorption measurements. The useful detection ranges were $1.5 \times 10^{12} cm^3 mol^{-1} s^{-1} \leq [S(^3P)] \leq 4.0 \times 10^{13} cm^3 mol^{-1} s^{-1}$ and $8.0 \times 10^{10} cm^3 mol^{-1} s^{-1} \leq [S(^1D)] \leq 2.0 \times 10^{12} cm^3 mol^{-1} s^{-1}$.

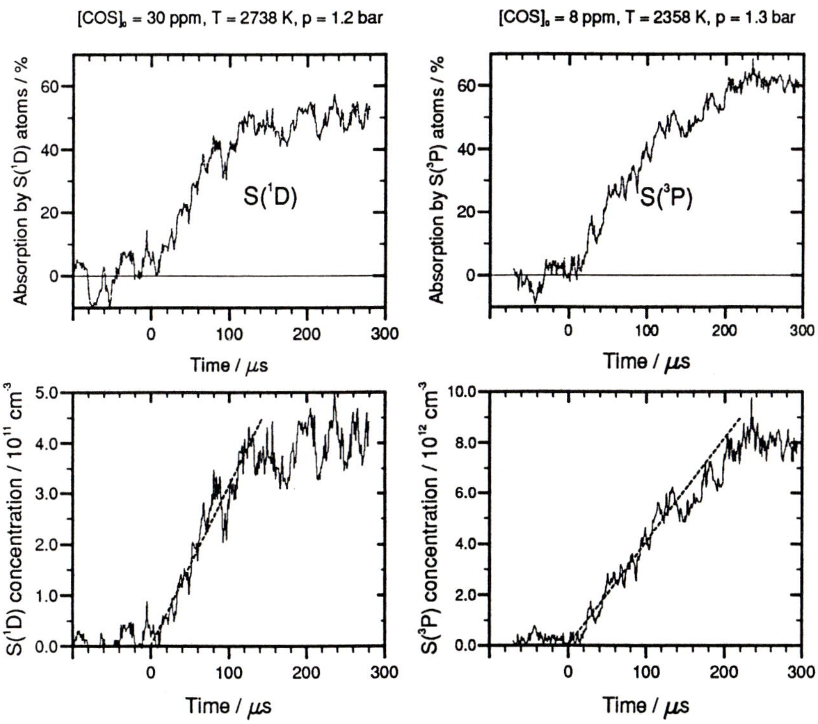

Fig. 2. Measured absorption and corresponding concentration profiles of $S(^3P)$ and (^1D) atoms during the thermal decomposition of COS

Because of the high sensitivity of the ARAS detection technique the absorption of the respective S line intensities by the initial reactant COS was carefully considered. This was done both at room temperature and in a series of shock tube experiments performed at temperatures, for which the COS dissociation could be neglegted. From the absorption signals behind the incident and the reflected shock waves mean values of the absorption cross sections of COS were determined to be $\sigma_{S(^3P)}(COS) = 2.5 \times 10^{-17} \times \exp(1000K/T)\,cm^2$ and $\sigma_{S(^1D)}(COS) = 2.5 \times 10^{-17}\,cm^2$. Based on this values the interference absorption of COS on the absorption profiles was calculated to be: $A_{COS} \leq 0.08$ which was subtracted from the measured absorption profiles of the individual experiments. No detectable interference absorption of the reaction product CO of both line intensity was found.

3. Results and discussion

Two groups of experiments on the thermal decomposition of COS were performed monitoring the absorption of the ground state $S(^3P)$ or of the excited state $S(^1D)$ of the S atoms. In the first group measurement of the $S(^3P)$ formation was carried out on gas mixtures of 4 - 100 ppm COS in Ar in the temperature range 1830 K $\leq T \leq$ 2991 K at pressures around 1 bar. In the second group the $S(^1D)$ formation was measured in gas mixtures of 8 - 100 ppm COS in Ar in the temperature range 2130 K $\leq T \leq$ 3020 K at similar pressures as in the first group.

All absorption profiles belonging to both groups show an increase of the absorption signal immediately after the reflected shock wave has passed the measurement plane. Typical examples of the measured $S(^3P)$ and (^1D) absorption signals are given in the upper part of Fig.2 The transformation into concentrations show an almost linear increase of the $S(^3P)$ and the $S(^1D)$ concentrations at early reaction times, see lower part of Fig.2. A first data reduction was done by dividing the slope of the S concentration profile at early reaction times by the initial concentrations of COS and Ar obtaining the apparent rate coeffcients of the $S(^3P)$ or $S(^1D)$ formation, respectively. The individual values of $k_{1a,app.}$ and $k_{1b,app.}$ are plotted in the Arrhenius diagram of Fig.3. The data values scatter around two straight lines which can be represented by the expressions:

$$k_{1a,app.} = \frac{(d[S(^3P)]/dt)_{t\approx 0}}{[COS]_0[Ar]} = 2.9 \times 10^{14} \times \exp\Big(-\frac{63360\ \text{cal/mol}}{RT}\Big)\,cm^3mol^{-1}s^{-1} \quad (4)$$

$$k_{1b,app.} = \frac{(d[S(^1D)]/dt)_{t\approx 0}}{[COS]_0[Ar]} = 4.8 \times 10^{12} \times \exp\Big(-\frac{71800\ \text{cal/mol}}{RT}\Big)\,cm^3mol^{-1}s^{-1} \quad (5)$$

Neither the $S(^3P)$ nor the $S(^1D)$ concentration profiles show any measurable time delay for the appearance of the particular atomic species, which is restricted by the experimental time resolution. Therefore no a priori decision which of the mentioned processes is predominant can be made. The ratio of both apparent rate coefficients $k_{1a,app.}/k_{1b,app.}$ is close to the equilibrium distribution of $S(^3P)$ and $S(^1D)$, see Eq.(1). This indicates that the partial equilibrium between $S(^3P)$ and $S(^1D)$ is fulfilled during the chemical nonequilibrium process of the S atom formation. The resulting temperature dependence of the ratio $k_{1a,app.}/k_{1b,app.}$ differs from the expected equilibrium value, which might be caused by possible systematic experimental errors.

For further data interpretation computer simulations were made taking into account only either reaction (R1a) or (R1b) as initial reaction step followed by reaction (R2). The rate coefficient of the quenching reaction (R2) was taken from the room temperature data of Donovan et al.(1970) to be $k_2 \geq 1.0 \times 10^{12} cm^3mol^{-1}s^{-1}$. The reverse rate coefficient was calculated from the equilibrium constant of Eq. (1). Starting with the reactions (R1a) and (R2) with $k_{1a} = k_{1a,app.}$ and $k_2 = 1.0 \times 10^{12} cm^3mol^{-1}s^{-1}$, both, the measured $S(^1D)$ and $S(^3P)$ atom concentration profiles could well be verified within the experimental error limits. A sensitivity analysis of this model shows that reaction (R2) is sufficiently fast to provide a prompt thermalisation between the $S(^1D)$ and $S(^3P)$ states within the time scale of the experiments. Regarding reaction (R1b) as the initial decomposition step, the calculated $S(^1D)$ and $S(^3P)$ concentrations were significantly lower than the measured ones when starting the simulations with $k_{1b} = k_{1b,app.}$. A complete fitting of the measured time profiles of both sulfur atom states is again possible by increasing k_{1b} to the value

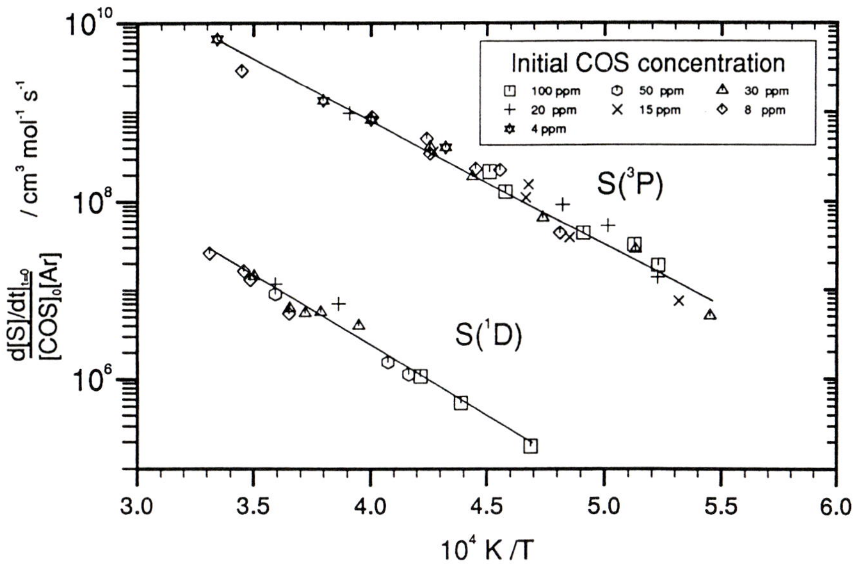

Fig. 3. Arrhenius plot of the values $(d[\mathrm{S}]/dt)_0/([\mathrm{COS}]_0[\mathrm{Ar}])$, with [S] equal to $S(^3P)$ or $S(^1D)$

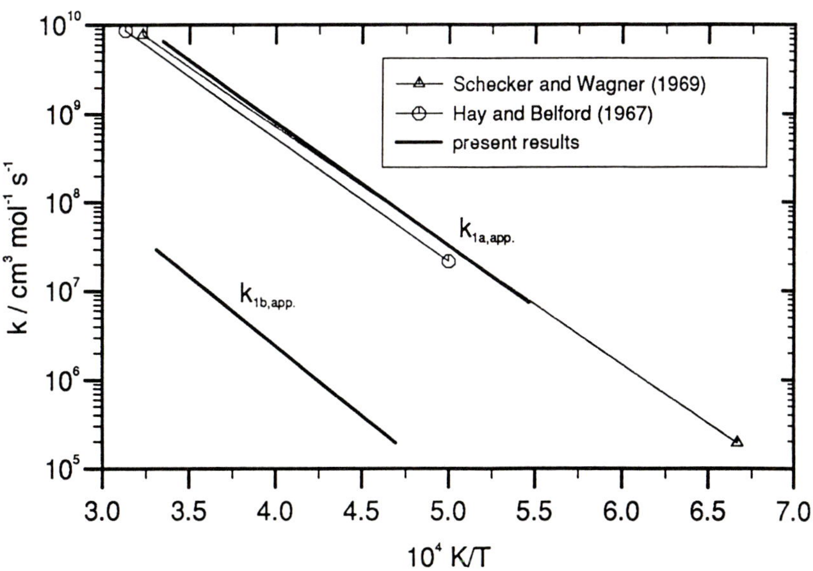

Fig. 4. Comparison of the obtained $k_{1a,app.}$ and $k_{1b,app.}$ with literature data

of $k_{1b} = k_{1a,app.}$ and k_2 to a value of $k_2 \geq 5.0 \times 10^{12} \mathrm{cm^3 mol^{-1} s^{-1}}$. This data interpretation indicates that our shock tube experiments on $S(^3P)$ and $S(^1D)$ formation do not allow to decide whether reaction (R1a) or (R1b) dominates the thermal decomposition of COS. Assuming a decomposition process taking place under quasi partial equilibrium conditions of the electronical states of S our $k_{1a,app.}$ value has to be identified with the rate coefficient for the decomposition of COS.

A comparison of the rate coefficients determined in the present study with the literature values is given in the Arrhenius plot of Fig.4. The activation energy obtained by Hay and Belford (1967) is identical to the own result of $k_{1a,app.}$, the absolute value is a factor of 1.5 lower. The agreement of the present rate coefficient $k_{1a,app.}$ to the low pressure value of Schecker and Wagner (1969) is nearly perfect with a minor difference in the activation temperature of about 4%.

Acknowledgement

The financial support of the *Deutsche Forschungsgemeinschaft* is gratefully acknowledged.

References

Donovan RJ, Kirch LJ, Husain D (1970) Collisional deactivation of the electronically excited atoms $S(3^1D_2)$ and $S(3^1S_0)$, by the noble gases. Trans. Faraday Soc. 66:774

Hay AJ, Belford RL (1967) High-temperature gas-kinetic study of Carbonyl Sulfide pyrolysis performed with a shock tube and quadrupole mass filter. J. Chem. Physics 47:3944

Roth P, Just Th (1977) Absorptionsmessungen zur Kinetik der Reaktion $CH_4 \rightarrow CH_3 + OH$ im Temperaturbereich $1500 \leq T \leq 2250$ K. Ber. Bunsenges. Phys. Chem. 81:572

Schecker HG, Wagner HGg (1969) On the thermal decomposition of COS. Intl. J. Chem. Kinet. 1:541

Thielen K, Roth P (1987) Resonance absorption measurements of N, O, and H atoms in shock heated $HCN/O_2/Ar$ mixtures. Comb. Flame 69:141

Experiments on Shock-Heated Ozone Dissociation in Oxygen/Air Using UV Laser Spectroscopy

M.L. Thoma and F.J. Hindelang
Fakultät für Luft-und Raumfahrttechnik, Institut für Strömungsmechanik, Universität der Bundeswehr München, D-85577 Neubiberg, Germany

Abstract. Measurements on the rate of thermal ozone decomposition $O_3 + M \overset{k_1}{\rightleftharpoons} O_2 + O + M$ in mixtures with pure oxygen and air respectively were performed within the temperature range 1100 - 3000 K. The initial O_3 mole fractions varied between 0.8 and 4.4%. The results were corrected to equivalent ozone (i.e. $M=O_3$) using Johnston's recommended relative efficiencies (O_2: 0.44, N_2: 0.39). No further significant dependencies from the type of bath gas were observed. The experimental equipment (frequency doubled, power stabilized argon ion laser with $\lambda = 257.25$ nm, high speed UV-photodiode) allowed monitoring of the time-dependent O_3 concentration. UV absorption signals of high quality could be detected. Real-gas effects, such as vibrational relaxation and the influence of several chemical reactions, were considered. Our findings indicate a preexponential temperature dependence of k_1, which becomes evident even in the temperature range 1400 - 2000 K. Here the evaluation requires very accurate signals and the use of numerical methods, as the reaction $O_3 + O \rightleftharpoons 2O_2$ (k_2) cannot be neglected. Up to now, no measurements of k_1 within this range have been presented in the respective review articles. We compared our measurements to those of previous investigations (Jones and Davidson 1962, Michael 1971, Center and Kung 1975) and to recommended expressions for k_1. At high temperatures (T $\geq$ 1600 K) our results correspond to the findings of Center and Kung (1975), well described by Heimerl and Coffee's k_1 data (1979), whereas at lower temperatures ($T \leq 1250$ K) they agree with Johnston's k_1 data (1968). Based on our conclusions, an expression is suggested (300 - 3000 K) taking three parameters into account: $k_1 = 2.07 \cdot 10^{-5} \cdot T^{-1.25} \cdot e^{(-12262/T)}\, \mathrm{cm^3 s^{-1}}$ (HRRK theory, extrapolated according to Jones and Davidson 1962). Another result of our experiments is the temperature dependence of the ozone absorption cross-section at 257.25 nm: $\alpha(T) = 14.33 - 7.45 \cdot e^{(-1300/T)} \mathrm{m^2/g}$ (T = 200 - 3000 K). This two-parameter fit uses the Boltzmann factor proposed by Steinfeld et al. (1987) and corresponds to the low- and room-temperature values of Molina (1986).

Key words: High temperature reaction rate, Decomposition, Ozone

1. Introduction

The effects of shock waves on atmospheric ozone became a subject of research in order to estimate one part of the pollution associated with hypersonic vehicles. Besides, the thermal ozone decomposition is used as an important source of $O(^3P)$ atoms for chemical kinetics studies. From the literature two values are known for the reaction rate k_1:

$$k_1 = 1.65 \cdot 10^{-9} \cdot e^{(-11441/T)}\, \mathrm{cm^3 s^{-1}} \quad (M = O_3,\ T = 200 - 1000 \text{ K, Johnston 1968}) \tag{1}$$

The high-temperature values used in Johnston's evaluation of k_1 were taken from Jones and Davidson (1962). These authors measured thermal ozone decomposition with Ar and N_2 as bath gases (689-863 K). Their experiments corresponded to earlier low-temperature measurements of Benson and Axworthy (1957), (T= 343-383 K, M=O_2, N_2, CO_2, He):

$$k_1 = 0.72 \cdot 10^{-9} \cdot e^{(-11179/T)}\, \mathrm{cm^3 s^{-1}} \quad (M = O_3, T = 300 - 3000 \text{ K, Heimer and Coffee 1979}) \tag{2}$$

Shock Waves @ Marseille II
Editors: R. Brun, L. Z. Dumitrescu

In 1971 Michael's results (988-1400 K, M=Kr) indicated a deviation of k_1 towards lower values with increasing temperatures. In 1975 Center and Kung showed with their measurements (1000-3000 K, M=Ar) that k_1 deviates - also towards lower values - by a factor of 2-3 at high temperatures ($T \approx 3000$ K) from Johnston's Arrhenius form. Their results showed a similar behaviour to our findings. Unfortunately, their measurements were not presented in such a way as to make a comparison possible. In addition to the previous results Heimerl and Coffee (1979) used the findings of these two studies on high-temperature ozone decomposition to evaluate their rate expression in an extended temperature range (300 - 3000 K). From Center and Kung's investigation (1975) only the data which are above 2000 K were taken into account, as supplied by R.T.V. Kung (non published table) to Heimerl and Coffee.

Thus, the temperature range from 1400 - 2000 K is not taken into account in their fit function. The reason is presumably the following: Involved with the evaluation of k_1 are the respective values of k_2, whereby a few authors conclude that the ratio k_2/k_1 increases with temperature against Johnston's Arrhenius fit [Center and Kung (1975): at most by a factor of 2 at 2000 K; Kiefer and Lutz (1967): a factor of 5 at 3000 K; Wray (1963):~ factor of 10]. Up to now it is not clear, how much k_2 deviates from the Arrhenius form within the related temperature range. In the intermediate temperature range (1400 - 2000 K) the ozone concentration depends also significantly on k_2. Here very accurate signals and the use of numerical methods for evaluating k_1 are required as the rate expressions for k_2, known from the literature, scatter considerably. As Heimerl and Coffee's k_1 deviates at temperatures below 1000 K from the measurements of Jones and Davidson, but corresponds at temperatures higher than 2000 K well to Center and Kung's data, we think that there is a need for new measurements and we present some in this publication.

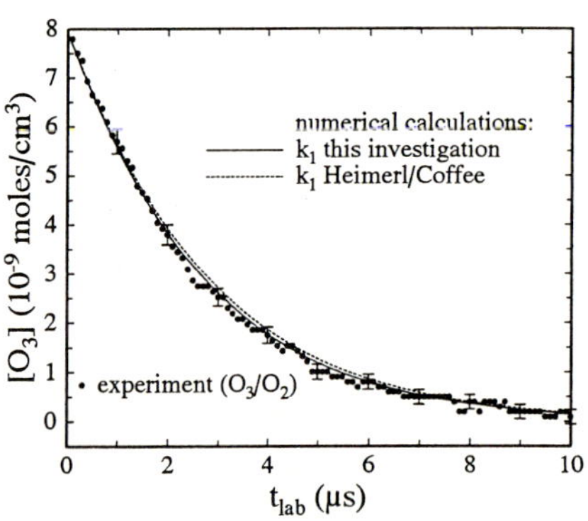

Fig. 1. Experimental conditions: 3.8% O_3, 96.2% O_2, $p_1 = 1.00$ hPa, $v_s = 1659.3$ ms^{-1}, and $T_2 = 1750$ K

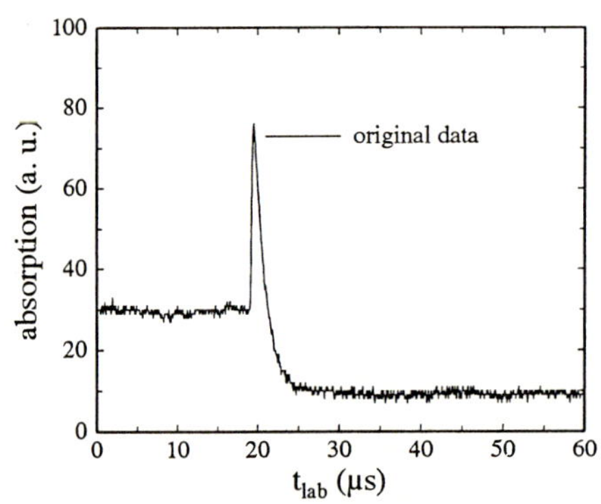

Fig. 2. Sample oscillogram of the experiment (O_3/O_2 mixture) shown in Fig.1.

2. Experimental

A pressure-driven UHV shock tube (stainless steel) of about 20 m length and an inner diameter of 10 cm was used for the experiments, which were performed behind incident shock waves. Capacitance pressure gauges (MKS Baratron 127A), made of ozone-resistant Inconel, allowed precise (~1%) determination of very low initial pressures. Ozone was produced by a commercial ozonizer (Anseros Ozomat COM) using pure oxygen (Linde 5.0). Mixing with pure nitrogen (Linde

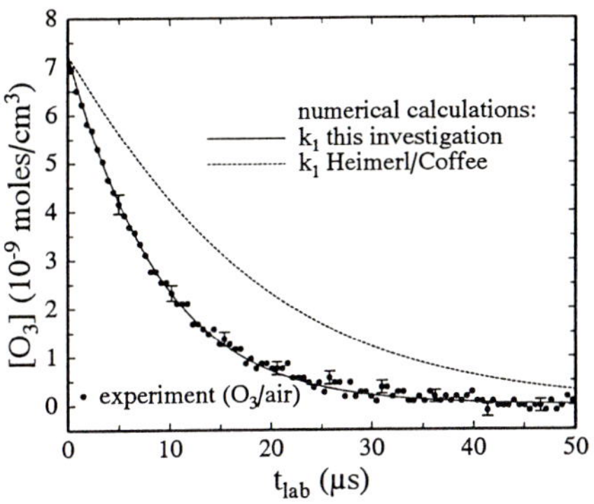

Fig. 3. Experimental conditions: 0.8% O_3, 21.2% O_2, 78.0% N_2, p_1 = 4.96 hPa, v_s = 1354.8 m.s^{-1}, T_2 = 1163 K

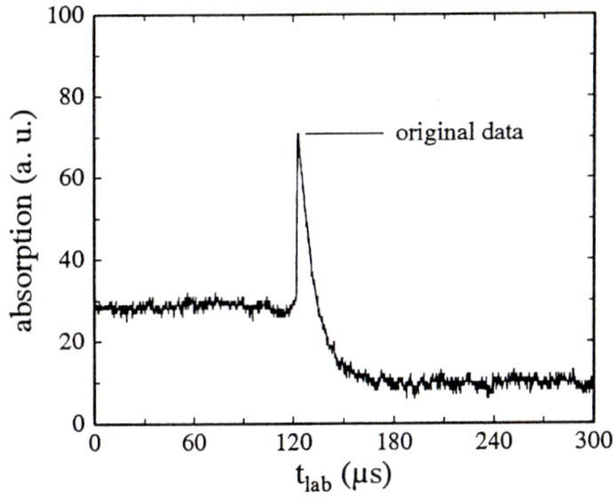

Fig. 4. Sample oscillogram of the experiment (O_3/air mixture) shown in Fig.3

5.0) was carried out in a vessel. This is, to our knowledge, the first time a frequency-doubled argon ion laser (Spectra Physics 2065) was used to examine ozone in a shock tube. Power stabilization was necessary to keep absorption signals stable. The laser beam (approximate diameter $\sim$ 1.5 mm) was focused on the center of the shock tube in order to improve time resolution. Additional slits to minimize the spatial beam extension had several disadvantages like diffraction and beam deflection (Schlieren effects) and were not used. The focused laser beam (f=500 mm) passes the shock tube via two CaF_2 windows. It illuminates the photodiode (Hamamatsu S1722-02) after passing an interference filter (Schott UV-KMD 12-1). The signals were transformed by a 100 MHz low noise preamplifier (HMS-571) to a digital oscilloscope (LeCroy 9400A) and then transferred to a UNIX workstation. The use of the photodiode instead of a photomultiplier improved the achievable signal-to-noise ratio considerably. Thus, very low initial pressures, down to 0.22 mbar, were possible. Further experimental details are described by von Vogt and Habekost (1993).

3. Results and discussion

Typical examples of experimental (UV laser absorption) and calculated ozone decomposition are shown in Figs.1 - 4. In Fig.1 (T_2 = 1750 K) Heimerl and Coffee's rate corresponds with the experiment while in Fig.3 (T_2 = 1163 K) no agreement can be observed. The measurements were compared to theoretical predictions making use of the CHEMKIN chemical kinetics code package (Kee and al. 1989, Mitchell and Kee 1982). The source codes were modified by the authors, so it was possible to consider effects of vibrational relaxation (only necessary in our O_3/O_2 experiments, since the vibrational relaxation in air takes much longer than our ozone decomposition time). The vibrational relaxation times $\tau_{vib}^{O_2}$ and$\tau_{vib}^{N_2}$ were taken from Anderson (1989). Corrections concerning $\tau_{vib}^{O_2}$ due to the findings of Kiefer and Lutz (1967) were not necessary. Vibrational relaxation of ozone itself was neglected (frozen conditions) as the small O_3 concentration (which becomes even smaller by its decomposition) only weakly contributes to the vibrational energy of the whole mixture. Boundary layer corrections (in most of the experiments of minor importance) according to the theory of Mirels were also taken into account. Changes of the absorption cross section due to temperature changes could be neglected during the measuring period. This statement can be quantitatively justified considering the calculated temperature changes behind the shock front and the measured temperature dependence $\alpha(T)$ (see Fig.6).

The complete reaction mechanism consists of the reactions with k_1 and k_2 (see Abstract) and $O + O + M \rightleftharpoons O_2 + M$ (k_3, taken from Baulch and al. 1976). In the experiments with O_3/air,

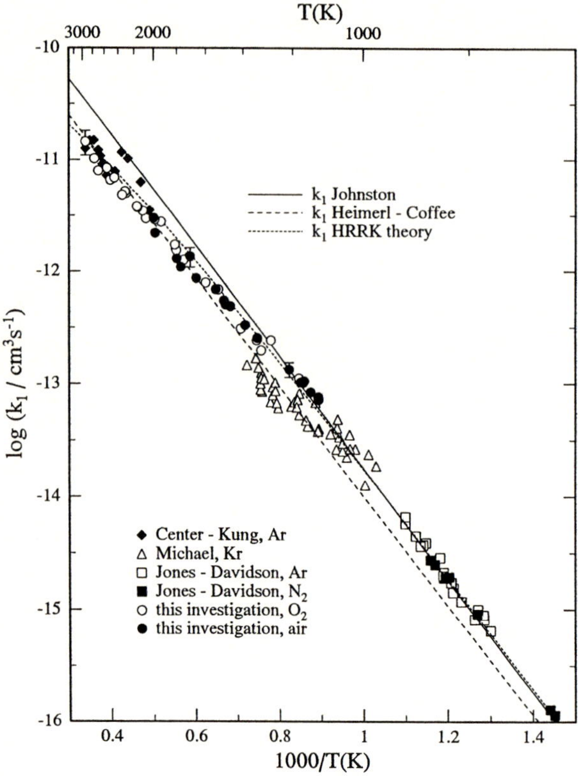

Fig. 5. Arrhenius plot for the thermal decomposition of ozone. Typical error bars vary from 15 % (at ~1200 K) to25 % (at ~3000 K). They increase with temperature due to the decreasing signal-to-noise ratio The respective type of bath gas is given after the authors' names in the legend. Relative efficiencies of the bath gases, used for data reduction: O_3:1.00, Ar:0.25, Kr:0.20, N_2:0.39, O_2:0.44

16 additional reactions were considered, none of them showed - in most of the experiments - significant importance regarding the results. Reverse reactions were involved using the equilibrium constants from the CHEMKIN thermodynamic data base (1987). In our calculations the values of both k_1 and k_2 were varied until accurate agreement - as close to our signals as possible - was achieved (see Figs.1 and 3.).The values obtained for k_1 are shown in Fig.5. The values of k_2 showed much more scattering than those of k_1: in many cases the expressions of Baulch (1976) or Shaw (1977) for k_2 were suited to describe the experimental results. At higher temperatures k_2 tends to increase even higher than stated with Shaw's three- parameter fit function. In this temperature range the k_2 values reach or even surpass Hampson's (1973) rate, which lies up to a factor of ~ 2.5 above Baulch's values. Based on these results and on the rate of k_1 (HRRK-theory) of Fig.5 we may conclude for the fraction k_2/k_1:

a) at T=2000 K: an increase by a factor of 2 - 4 compared to the extrapolated low-temperature ratio of Baulch's k_2 and Johnston's k_1 was found.

b) at T=3000 K: an increase by at least a factor of 4 (lower limit) was found. The upper limit may be much higher, but cannot be determined as our data become insensitive to that rate-ratio at these high temperatures.

Table 1. Observed values of k_1 (cm^3/s) in O_3/O_2 mixtures, corrected to equivalent ozone

T(K)	$\log k_1$	T(K)	$\log k_1$	T(K)	$\log k_1$
1180	-12.956	1811	-11.800	2345	-11.315
1283	-12.617	1818	-11.762	2451	-11.161
1323	-12.708	1934	-11.556	2510	-11.181
1342	-12.616	1990	-11.537	2562	-11.076
1412	-12.511	2076	-11.526	2700	-11.099
1527	-12.163	2109	-11.457	2769	-10.990
1602	-12.104	2162	-11.426	2942	-10.837
1710	-11.868	2170	-11.418		
1750	-11.899	2293	-11.298		
1806	-11.813	2309	-11.283		

Table 2. Values of k_1 in O_3/Air mixtures

T(K)	$\log k_1$	T(K)	$\log k_1$
1119	-13.146	1489	-12.295
1120	-13.119	1497	-12.262
1143	-13.076	1542	-12.161
1163	-12.978	1663	-12.062
1166	-12.991	1704	-11.868
1174	-12.987	1772	-11.962
1214	-12.875	1804	-11.887
1339	-12.594	1987	-11.656
1392	-12.480	2001	-11.522
1463	-12.312		

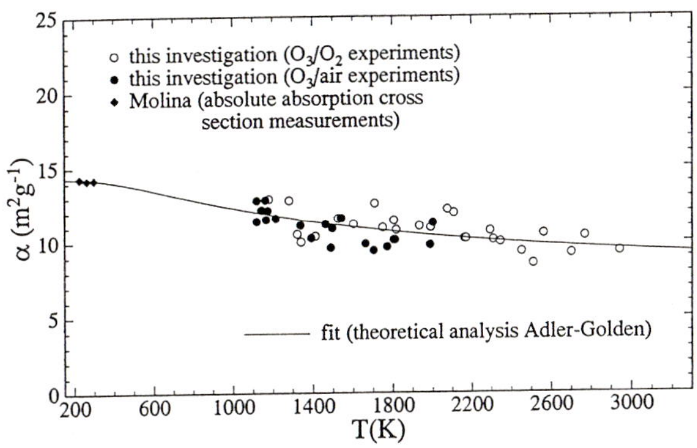

Fig. 6. Measured temperature dependence $\alpha(T)$. We compared our values to the low- and room-temperature measurements of Molina (1986). The evaluated least square fit-function (two parameters) is described in the text

Both conclusions should be compared with the results of other investigators, quoted in the introduction. The results on k_1 of Fig.5 are listed in Tables 1 and 2 and discussed below: Johnston's k_1 gave satisfying results up to 1250 K, which is even above the stated temperature range. Calculations using Heimerl and Coffee's k_1 corresponded approximately with our experimental data above temperatures of $\sim$ 1600 K. The intermediate temperature range can be described by the expression: $k_1 = 2.07 \cdot 10^{-5} \cdot T^{-1.25} \cdot e^{(-12262/T)}$ $cm^3\ s^{-1}$. This is the rate given by Jones and Davidson (1962), corrected to equivalent ozone. They applied the Hinshelwood-Rice-Ramsperger-Kassel (HRRK) theory to their measurements in O_3/N_2. In Fig.5 we can see that this function agrees - over the whole presented temperature range - quite well with the data of Center and Kung, Jones and Davidson, and those of this publication. Due to this suggested expression, k_1 decreases at 2000 K by a factor of $\sim$1.6 and at 3000 K by a factor of $\sim$2.3 with respect to Johnston's extrapolation. These findings agree with Center and Kung's (1975) conclusions (decrease of factor 2-3 stated). Only Michael's values show a diverging behaviour. He used the extrapolated values of Benson and Axworthy (1965) for k_2 in his evaluation of k_1. These values of k_2 do not correspond to our findings. Another result is shown in Fig.6, namely the temperature dependence of the absorption coefficient α at a wavelength of 257.25 nm. It can be summarized in the fit function:

$\alpha(T) = 14.33 - 7.45 \cdot e^{(-1300/T)} m^2/g$ (T = 200 - 3000 K). This fit follows a theoretical analysis of the Hartley band temperature dependence performed by Steinfeld et al. (1987) and agrees well - within the whole temperature range (200 - 3000 K) - with the experimental data shown.

Acknowledgement

The authors gratefully acknowledge the valuable assitance of Lt.z. S.G. von Vogt and OLt. H. Habekost during the measurements and the computational evaluation.

References

Anderson JD Jr. (1989) Hypersonic and high temperature gas dynamics. McGraw-Hill p.489

Baulch DL, Drysdale DD, Duxbury J, Grant SJ (1976) Evaluated kinetic data for high temperature reactions. Butterworths, London, 3

Benson SW, Axworthy AE (1957) Mechanism of the gas phase thermal decomposition of ozone. J. Chem. Phys. 26(6):1718

Benson SW, Axworthy AE (1965) Reconsideration of the rate constants from the thermal decomposition of ozone. J. Chem. Phys. 42:2614

Center RE, Kung RTV (1975) Shock tube study of the thermal decomposition of O_3 from 1000 to 3000 K. J. Chem. Phys. 62(3):802

Hampson RF (1973) Survey of photochemical and rate data for twenty-eight reactions of interest in atmospheric chemistry. J. Phys. Chem. Ref. Data 2:267

Heimerl JM, Coffee TP (1979) The unimolecular ozone decomposition reaction. Combust. Flame 35:117

Johnston HS (1968) Gas phase reaction kinetics of neutral oxygen species. Natl. Stand. Ref. Data Ser. NSRDS-NBS 20

Jones WM, Davidson N (1962) The thermal decomposition of ozone in a shock tube. J. Am. Chem. Soc. 84:2868

Kee RJ, Rupley FM, Miller JA (1987) The Chemkin thermodynamic data base. SAND87-8215B

Kee RJ, Rupley FM, Miller JA (1989) CHEMKIN II: A Fortran chemical kinetics package for the analysis of gas-phase chemical kinetics. SAND89-8009

Kiefer JH, Lutz RW (1967) The effect of oxygen atoms on the vibrational relaxation of oxygen. Symp.(Intl.) Combust., [Proc], 11th:67

Michael JV (1971) Thermal decomposition of ozone. J. Chem. Phys. 54:4455

Mitchell RE, Kee RJ (1982) A general-purpose computer code for predicting chemical behaviour behind incident and reflected shocks. SAND82-8205

Molina LT, Molina MJ (1986) Absolute absorption cross sections of ozone in the 185- to 350-nm wavelength range. J. Geophys. Res. 91(14):501

Shaw R (1977) Estimation of rate constants as a function of temperature for the reactions W + XYZ = WX + YZ, where W, X, Y, and Z are H or O atoms. Intl. J. Chem. Kin. 9:929

Steinfeld JI, Adler-Golden SM, Gallagher JW (1987) Critical survey of data on the spectroscopy and kinetics of ozone in the mesosphere and thermosphere. J. Phys. Chem. Ref. Data 16(4):911

von Vogt G, Habekost H (1993) Diploma thesis. Institut für Strömungsmechanik, Fakultät für Luft- und Raumfahrttechnik, Universität der Bundeswehr München, D-85577 Neubiberg, Germany

Wray KL (1963) Shock-tube study of the recombination of O atoms by Ar catalysts at high temperatures. J. Chem. Phys. 38(7):1518

High-Temperature Reactions of Ozone with Bromine Behind Shock Waves. Rate Constant for the Reaction $BrO+BrO \longrightarrow 2Br+O_2$

Kazuo Takahashi, Junichi Honda, Tadaaki Inomata and Hiroshi Jinno
Department of Chemistry, Faculty of Science and Technology, Sophia University, 7-1 Kioi-cho, Chiyoda-ku, Tokyo 102, Japan

Abstract. A mixture of ozone and bromine, diluted by argon, was heated by reflected shock waves in a shock tube to the temperature range of 600-1100 K and the pressure range of 3.5-5.5 atm. From the concentration profiles of the BrO radical, the rate constant k_1 for the reaction $BrO+BrO \rightarrow 2Br+O_2$ was determined as

$$k_1/\text{cm}^3\,\text{molecule}^{-1}\,\text{s}^{-1} = (1.9 \pm 0.1) \times 10^{-12} \exp[+(42 \pm 13)/T].$$

A model calculation was carried out for the reactions of ozone with bromine. The results showed that bromine catalytically decomposed ozone in the chain cycle formed by the reactions $BrO+BrO \rightarrow 2Br+O_2$ and $O_3+Br \rightarrow BrO+O_2$, but that this acceleration effect of bromine fell with increasing temperature.

Key words: High-temperature reactions, Ozone, Bromine, Rate constants

1. Introduction

The kinetics of reaction involving halogens and oxygen have been studied fairly extensively in the past. The interest in the halogen-oxygen system has been recently renewed with the concern over the possible role of halogens in the chemistry of the upper atmosphere. A very large amount of chlorine is artificially released from chlorofluorocarbons used as refrigerants and cleaners, and injected into the atmosphere. Although the amount of bromine is not as large as that of chlorine, it is also released from bromofluorocarbons used as extinguishants and so on. Once released into the atmosphere, chlorine and bromine could react via several processes ultimately to remove atmospheric ozone. Because of the seriousness of the possible consequences, more research is needed on the chemical kinetics involving halogens and oxygen. One of the questions remaining unanswered is that of the reaction rates of the halogen-oxygen system at the temperatures prevailing in the upper atmosphere, i.e., around 200 K. The kinetic studies at around 200 K are insufficient, because experiments are hampered by the fact that chlorine, bromine, and their compounds tend to condense at such low temperatures. To circumvent this problem, the reaction rates can be measured at high temperatures. Knowledge of the kinetic data at room temperature and at high temperatures would be helpful in estimating the kinetic data at low temperature.

In the present study, a mixture of ozone and bromine was heated by reflected shock waves in a shock tube. The rate constant for the disproportionation reaction of bromine monoxide radicals, i.e.

$$BrO + BrO \rightarrow 2Br + O_2, \qquad (1)$$

which remains highly uncertain, was determined. Furthermore, in order to clarify the addition effect of bromine on the ozone decomposition at elevated temperatures, a model calculation was carried out.

Shock Waves @ Marseille II
Editors: R. Brun, L. Z. Dumitrescu

2. Experimental

The experiments were carried out behind reflected shock waves in a 6.2 cm i.d. stainless steel shock tube. The driver and test sections were 1.7 and 3.8 m long, respectively, so that dwell times above 1.5 ms could be obtained in all of the experiments. Other apparatus for measurements of shock velocity have already been described (Inomata et al. 1985). The concentrations of two reactants, i.e. ozone and bromine, and BrO radical were measured using time-resolved electric absorption spectrophotometry. A xenon analytical lamp (Toshiba, 500 W) was used as a light source. The wavelengths of monitored transmission light were 254, 415, and 339 nm for O_3 (Hartley band), Br_2 ($^3\Pi - ^1\Sigma$), and BrO ($A^2\Pi - X^2\Pi$), respectively. The compositions of the test gas mixtures used are shown in Table 1. The measurements were performed over the temperature range of 600–1100 K and the pressure range of 3.5–5.5 atm.

Table 1. Test gas mixtures

No.	O_3%	Br_2%	Ar%
1	1.00	–	99.00
2	0.25	–	99.75
3	0.10	–	99.90
4	2.00	0.30	97.70
5	1.00	0.20	98.80

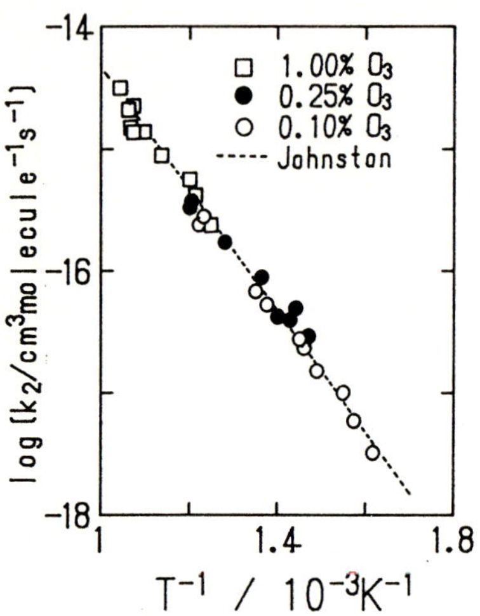

Fig.1. Arrhenius plot of k_2

3. Results and discussion

3.1. Rate constant for the reaction $O_3+Ar \rightarrow O+O_2+Ar$

In a series of reactions occurring in the ozone-bromine mixtures, the thermal decomposition reaction of ozone (2) is very important, because it acts as an initiation reaction:

$$O_3 + Ar \longrightarrow O + O_2 + Ar \tag{2}$$

Preliminarily, the rate constant k_2 for reaction (2) was determined by monitoring the O_3 absorption at 254 nm in the O_3-Ar mixtures (Mixtures 1-3 in Table 1). Following reaction (2), reaction (3) also occurs in these mixtures:

$$O_3 + O \longrightarrow O_2 + O_2 \tag{3}$$

With the steady state assumption for the oxygen atoms, the rate constant k_2 can be directly derived from

$$\ln[\ln(I_0/I_t)] = -2k_2[Ar]t + \text{Const.} \tag{E1}$$

where I_0 is the incident light intensity and I_t refers the transmission light intensity at time t.

The Arrhenius plot of k_2 is shown in Fig.1. All data obtained for the three mixtures follow a common straight line. The expression for the temperature dependence of k_2 was given as follows:

$$k_2/\text{cm}^3.\text{molecule}^{-1}.\text{s}^{-1} = (2.6 \pm 0.5) \times 10^{-10} \exp[-(1.11 \pm 0.02) \times 10^4/T] \quad (600 - 1100\ \text{K})(E2)$$

This result is in good agreement with the form recommended by Johnston (1968), as described with a broken line in Fig.1.

3.2. Rate constant for the reaction $BrO+BrO \rightarrow 2Br+O_2$

In the ozone-bromine mixtures, the dominant reactions at elevated temperatures are shown in Table 2. Although many other reactions can occur simultaneously with reactions (1)-(7), there are several reasons why such additional reactions can be neglected in the present test conditions. For example, the reaction $BrO+O_3 \rightarrow Br+2O_2$ is two orders of magnitude slower than reactions (6) and (7), due to the smaller rate constant.

Table 2. Reaction mechanism used in the calculation

Reaction	log A	n	E/R	Reference
(1) $BrO+BrO \rightarrow 2Br+O_2$	[Eq.E3	in	text]	This study
(2) $O_3+Ar \rightarrow O+O_2+Ar$	[Eq.E2	in	text]	This study
(3) $O_3+O \rightarrow 2O_2$	-10.70	0.00	2410	Baulch et al. (1980)
(4) $Br_2+Ar \rightarrow 2Br+Ar$	- 5.62	-0.50	23200	Westbrook (1983)
(5) $Br_2+O \rightarrow BrO+Br$	-10.85	0.00	0	Baulch et al. (1980)
(6) $O_3+Br \rightarrow BrO+O_2$	-10.77	0.00	800	Baulch et al. (1980)
(7) $BrO+O \rightarrow Br+O_2$	-10.52	0.00	0	Baulch et al. (1980)

Rate constants in the form: $k = AT^n \exp(-E/RT)$, in cm^3, molecule, s and K units

To determine the rate constant k_1, the concentration profiles of the BrO radical were calculated by numerically integrating the rate equations with k_1 as a variable parameter. The calculated curves were then compared with the observed curve. Through a process of trial and error, the calculated curve that most closely matched the observed one was selected, and k_1 was determined. The kinetic data on reactions (3)-(7) have been reported by numerous workers. In this calculation, the values recommended in the reviews (Baulch et al. 1980; Westbrook 1983) were used as the rate constants $k_3 - k_7$. For the rate constant k_2, the Arrhenius form given by Eq.(E1) was used.

Before determining k_1 from the BrO concentration profiles, a sensitivity analysis was carried out. For the rate constant k_1, the kinetic data estimated by NASA (DeMore et al. 1983) were used in the analysis. The result showed that reaction (1) is most sensitive to the BrO concentration in reactions (1)-(7). This proves that the determination of k_1 from the BrO concentration is reasonable.

Fig.2 shows an example of the BrO concentration profiles obtained from the absorption at 339 nm. The three smooth lines in the figure are the BrO concentration curves calculated assuming k_1. Both profiles are normalized by the maximum concentration. By this comparison between the observation and calculation, k_1 could be determined with an accuracy of ±20 %. The Arrhenius plot of k_1 is shown in Fig.3. By coupling this result with $k_1 = 2.2 \times 10^{-12}/\text{cm}^3.\text{molecule}^{-1}.\text{s}^{-1}$ at 298 K, which is recommended by CODATA (Baulch et al. 1984), the following form was obtained:

$$k_1/\text{cm}^3.\text{molecule}^{-1}.\text{s}^{-1} = (1.9 \pm 0.1) \times 10^{-12} \exp[+(42 \pm 13)/T] \quad (298 - 1100\ \text{K}) \qquad (3)$$

This result supports the kinetic data estimated by NASA (DeMore et al. 1983: broken line in Fig.3), but not the data reported by Clyne and Cruse (1970: dotted and dashed line).

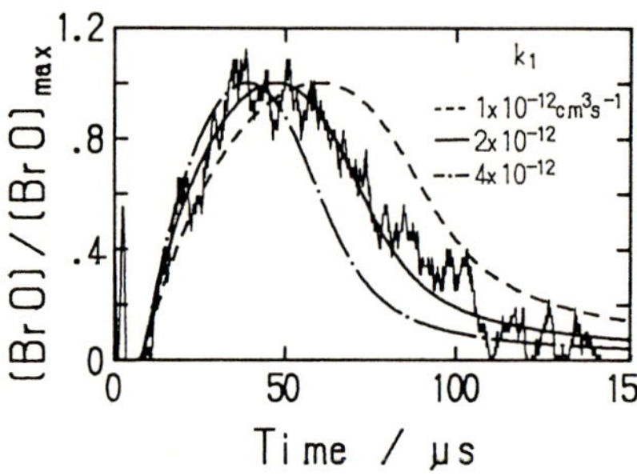

Fig. 2. Comparison of relative BrO concentration profiles as observed and as calculated: 1% O_3-0.2% Br_2 in Ar, T=755 K, P=4.54 atm.

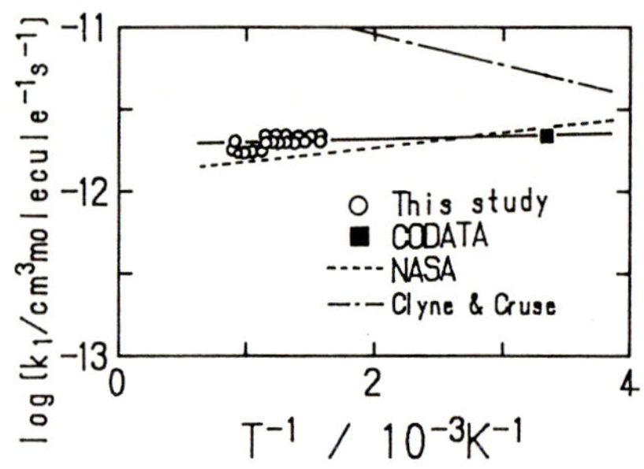

Fig. 3. Arrhenius plot of k_1

3.3. Addition effect of bromine on ozone decomposition

Halogens and halogenated compounds accelerate the ozone decomposition. To clarify the chemical effect of bromine on the ozone decomposition at elevated temperatures, a model calculation was carried out. The reaction mechanism and kinetic data shown in Table 2 were used in the calculation. As the rate constants for reactions (1) and (2), the Arrhenius forms determined in this study Eqs. (E2) and (E3) were employed. Fig.4 shows the reaction rate profiles at the early stage of reaction time, calculated at 755 K in the O_3-Br_2-Ar mixture. In this system, the thermal decomposition of ozone, namely reaction (2), acts as an initiation reaction, producing oxygen atoms. Sequentially, the oxygen atoms react with ozone or bromine molecules through reactions (3) or (5). In the present conditions, reaction (5) is faster than reaction (3). Once the bromine atoms and BrO radicals are produced through reaction (5), ozone is rapidly decomposed in the chain cycle formed by reactions (1) and (6). This process for the ozone decomposition is summarized in Fig.5.

Fig.6 shows the variations of the calculated half-life ($\tau_{0.5}$) of ozone against the initial concentration ratio of bromine to ozone. As described with the solid line in the figure, even a little amount of bromine makes ozone decompose. For example, when the ratio $[Br_2]_0/[O_3]_0$ is 0.1, $\tau_{0.5}$ reduces to a half. The broken line drawn in Fig.6 shows the half-life of ozone calculated by putting k_1 to zero. When reaction (1) is deleted and the chain cycle is broken, bromine does not affect the ozone decomposition. This result means that the acceleration effect of bromine on the ozone decomposition is caused by the chain cycle consisting of reactions (1) and (6).

Fig.7 shows the comparison of the addition effects of bromine on the ozone decomposition between high temperature (755 K, solid line) and room temperature (300 K, broken line). The ordinate means the time until ozone decays to 90 % of the initial concentration, normalized by the time without bromine ($\tau_{0.9}/\tau_{0.9(none)}$). The acceleration effect of bromine at high temperature is smaller than that at room temperature. The ozone decomposition process shown in Fig.5 is divided into the following two paths. One is the self-decomposition of ozone formed by reactions (2) and (3) (Path I). The other is the ozone destruction caused by the reactions with bromine-containing species, consisting of reactions (1) and (6) (Path II). The overall activation energy of Path I is much higher than that of Path II, because the rate-determining step in Path I is reaction (2), of which the activation energy is very high (Table 2). If the decomposition rates of ozone through Paths I and II are expressed as R_I and R_{II}, respectively, the ratio R_{II}/R_I decreases with increasing temperature. This consideration means that the acceleration effect of bromine on the ozone decomposition is less at elevated temperatures.

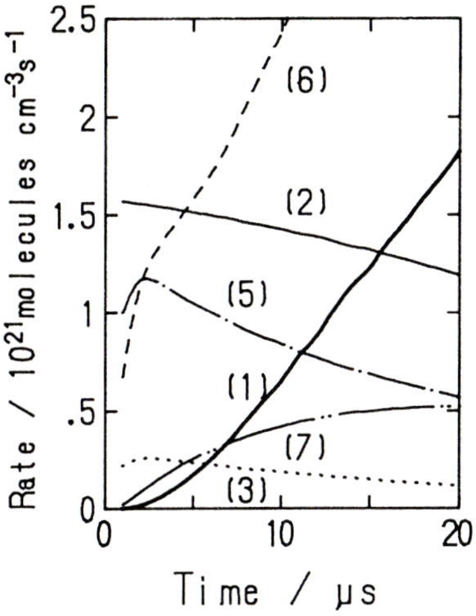

Fig. 4. Calculated reaction rates at early stage of reaction time: 1%O_3-0.2%Br_2 in Ar, T=755 K, P=4.54 atm

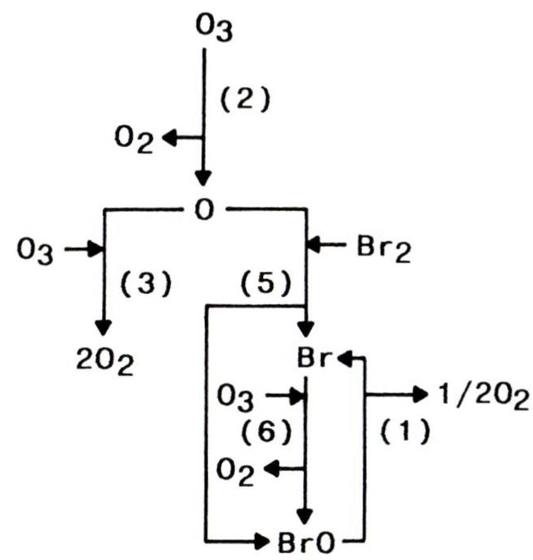

Fig. 5. Ozone decomposition process

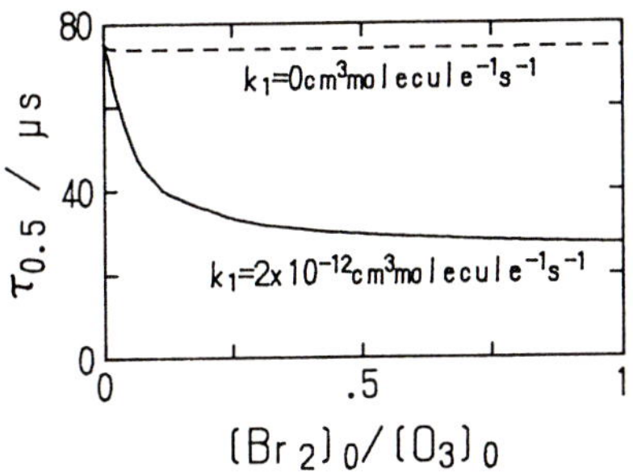

Fig. 6. Calculated half-lives of ozone ($\tau_{0.5}$): T=755 K, P=4.54 atm

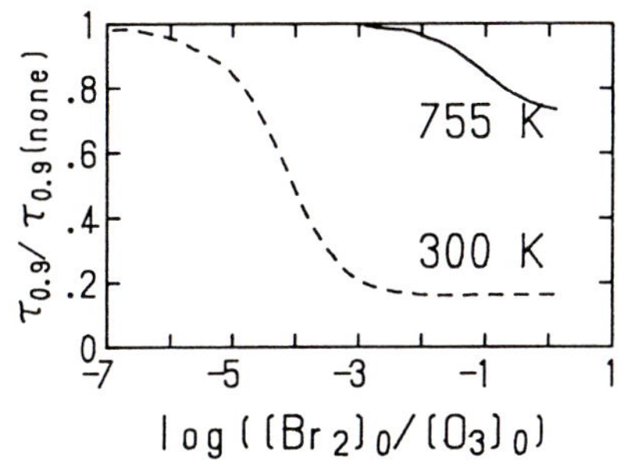

Fig. 7. Comparison of calculated relative nine tenth lives of ozone ($\tau_{0.9}/\tau_{0.9(\mathrm{none})}$) between high and room temperature: c = 7.33 x 10^{-5}mol.cm^{-3}

Acknowledgements

The authors wish to express their gratitude to Dr. Frank Scott Howell, S. J., for correcting the English manuscript.

References

Baulch DL, Cox RA, Hampson Jr RF, Kerr JA, Troe J, Watson RT (1980) Evaluated kinetic and photochemical data for atmospheric chemistry. J. Phys. Chem. Ref. Data 9:295

Baulch DL, Cox RA, Hampson, Jr RF, Kerr JA, Troe J, Watson RT(1984) Evaluated kinetic and photochemical data for atmospheric chemistry. J. Phys. Chem. Ref. Data 13:1259

Clyne MAA, Cruse HW (1970) Rates of elementary reaction involving the BrO ($X^2\Pi$) and IO ($X^2\Pi$) radicals: Part 1. formation and decay of the BrO radical. Trans. Faraday Soc. 66:2214

DeMore WB, Watson RT, Golden DM, Hampson RF, Kurylo MJ, Howard CJ, Molina MJ, Ravishankara AR (1983) Chemical kinetics and photochemical data for use in stratospheric modeling: evaluation number 6. NASA Panel for Data Evaluation, JPL Publ. 82-57

Inomata T, Okazaki S, Moriwaki T (1985) Effect of bromomethane on the ignition in methane-oxygen-argon mixtures behind reflected shock waves. Combust & Flame 62:183

Johnston HS (1968) Natl. Stand. Ref. Data Ser. No.20

Westbrook CK (1983) Numerical modeling of flame inhibition by CF_3Br. Combust. Sci. Technol. 34:201

High-Temperature Reaction of $O(^3P)+H_2S$

Kentaro Tsuchiya*, Hiroyuki Matsui†, Masaaki Oya* and Gabrielle Dupré‡
†Department of Reaction Chemistry, The University of Tokyo, Hongo, Bunkyo-ku, Tokyo 113, Japan
*On leave from National Institute for Resources and Environment, Onagawa, Tsukuba, Ibaraki 305, Japan
‡Centre National de la Recherche scientifique, Orléans, CEDEX 2, France

Abstract. The reaction of atomic oxygen ($3P$) with hydrogen sulfide was investigated by the shock tube - laser photolysis method at high temperatures (1100-2000 K), where time dependences of O and H atoms were monitored with atomic resonance absorption spectrometry (ARAS). O atoms were produced by the photolysis of SO_2 by an KrF excimer laser behind reflected shock waves. The overall rate constant for the reaction

$$O + H_2S \longrightarrow \text{products} \tag{1}$$

was determined from the decay rate of the absorption of O atoms as $k_1 = (2.8 \pm 0.3) \times 10^{-10} \exp\left[-(32 \pm 2)\text{kJmol}^{-1}/RT\right] \text{cm}^3\text{molecule}^{-1}\text{s}^{-1}$. The branching fraction for the substitution channel

$$O + H_2S \longrightarrow H + HSO \tag{1c}$$

was determined from the time dependence of the H atoms, that is, $k_{1c}/k_1 = 0.24 \pm 0.06$ at 1520-1820 K. The rate constant k_1 and the branching fraction k_{1c}/k_1 are also theoretically discussed based on the conventional transition state theory with the Wigner's tunneling correction for the potential energies.

Key words: Reaction rate, branching fraction, Oxgen atom, Hydrogen sulfide

1. Introduction

The reaction mechanism of $O(^3P)$ atoms with hydrogen sulfide is a key process of the initiation stage of oxidation and combustion of hydrogen sulfide (Frenklach et al. 1981). This reaction has been studied extensively (Singleton et al. 1982 and references therein), but the reaction rate constants have been measured only at low temperatures below 500 K. Most of the rate constants at temperatures important in a combustion system and the detailed reaction mechanisms are still unknown. It has been often speculated that the reaction proceeds via the competition of direct abstraction (Reaction 1a) and addition path which is followed by decomposition of the excited adduct (Reaction 1b):

$$O(^3P) + H_2S \longrightarrow OH + HS \tag{1a}$$

$$\longrightarrow (H_2SO)^* \longrightarrow H + HSO \tag{1b}$$

Crossed molecular beam studies (Balucani et al. 1991), however, suggested that HSO was a direct product of the titled reaction. Recently, Yokoyama (1992) has performed ab initio quantum mechanical calculation of the potential energy surfaces for this reaction. The calculation indicates that instead of addition channel (1b), the reaction proceeds via substitution (1c)

$$O(^3P) + H_2S \longrightarrow H + HSO \tag{1c}$$

i.e., no stable reaction intermediate H_2SO is found along the reaction coordinate. This channel seems to compete with the direct abstraction channel (1a) with almost the same barrier height for the transition states. In the present study, the overall rate constant k_1 and the branching fraction

Shock Waves @ Marseille II
Editors: R. Brun, L. Z. Dumitrescu

k_{1c}/k_1 were measured directly by monitoring the concentration of $O(^3P)$ and $H(^2S)$ atoms, using the ARAS technique at high temperatures.

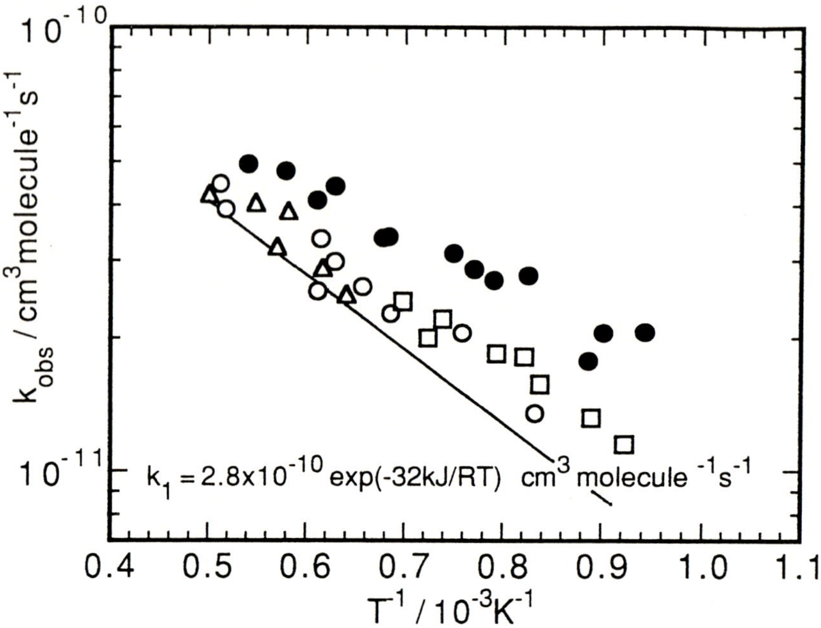

Fig. 1. Summary of the experimental result on the overall rate constant for the reaction $O+H_2S \rightarrow$ products (1) measured via KrF: [o : 43. 5ppm H_2S/200 ppm SO_2; Δ: 22.8 ppm H_2S/106 ppm SO_2; □ :22.9 ppm H_2S/358 ppm SO_2] and ArF: [•: 32.3 ppm H_2S/99.2 ppm SO_2] laser photolysis behind reflected shock waves

2. Experimental

The present experiments were conducted in a piston-actuated shock tube of 5 cm diameter and 4 m long. $O(^3P)$ atoms are prepared behind reflected shock waves by the photolysis of SO_2 using mainly a KrF (248 nm) excimer laser. An ArF excimer laser photolysis (193 nm) was also tried: in this case, due to the efficient photodecomposition of H_2S, higher initial concentration of HS is produced. From the difference of the decay rates of O atoms between ArF and KrF laser photolysis, the effect of the side reaction, O+HS was examined. $O(^3P)$ and $H(^2S)$ atoms were detected by using atomic resonant absorption spectrometry (ARAS).

Experiments are carried out on H_2S/SO_2 mixtures highly diluted in Ar (H_2S/SO_2=23-44/106-358 ppm). The initial concentration of O atoms was always kept so that pseudo-first-order kinetic condition for the decay of O atoms, i.e., $(O)_0/(H_2S) \gg 1$, was maintained. The temperature range covered was 1100-2000 K. As the reaction channel of producing H atoms was found to be minor, ARAS signals for H Lyman- α (121.6 nm) had to be averaged over 4-6 times in order to improve the signal to noise ratio at the same shock wave condition.

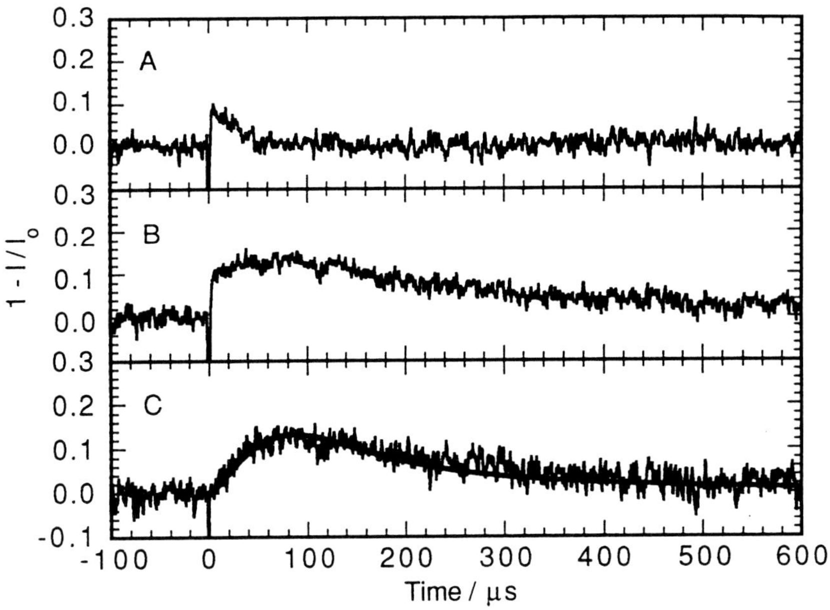

Fig. 2. An example of the time dependence of H atom absorption at 121.6 nm (Lyman-α), A: without addition of SO_2, B: with addition of 200 ppm SO_2 to 23 ppm H_2S, C: net ARAS signal of H atoms, where the contribution of the initial production of H atoms is eliminated, i.e., [B - A]. The signals are averaged over for 6 shots at the same shock wave condition: [23 ppm H_2S+197 ppm SO_2 in Ar, T = 1517±20 K, ρ = 1.31 x 10^{19}molecule/cm^3]

3. Results and discussion

3.1. The overall rate constant for the $H_2S+O(^3P)$ reaction

After a 100 μs delay from the arrival of the reflected shock wave at the observation port, the KrF excimer laser was fired and the first-order rate constant for the $O(^3P)$ atom decay was determined by a least-squares fit of the decay profile of O atoms. The Arrhenius plot of the measured rate constant for Reaction 1 is shown in Fig.1. In order to examine the effects of the consecutive reactions, the decay rate for $O(^3P)$ was measured in several gas mixtures, i.e., initial concentrations are $[H_2S]/[SO_2]$ = 23/106, 23/358, and 44/200 ppm. The effect of the initial concentration of H_2S is not clear in the KrF laser photolysis. However, it is found that the rate constants are 30 - 50% smaller than those of ArF laser photolysis, as shown in the same figure. Initial concentrations of $O(^3P)$ are estimated to be $(0.8\text{-}2.2)\times10^{13}/\text{cm}^3$ both in the KrF and ArF laser photolysis experiments: in contrast, the initial amounts of H and HS produced in the 193 nm photolysis is $(4\text{-}6)\times10^{12}/\text{cm}^3$ which is about 10 times larger than those produced in the 248 nm photolysis. Thus, the effect of the initial concentration of HS on the disappearance rate constants of O atoms is tentatively assigned as due to the side reaction of O+HS, i.e.,

$$O + HS \longrightarrow \text{products} \qquad (2).$$

The rate constants for Reactions 1 and 2 are determined by adjusting k_1 and k_2 in the reaction mechanism in Table 1 until the observed relation between the O atom decay rate and the initial concentration of HS is satisfied. The experimental results are expressed in the Arrhenius form as,

$$k_1 = (2.8 \pm 0.3) \times 10^{-10} \exp\left[-(32 \pm 2)\text{kJmol}^{-1}/RT\right] \text{cm}^3\text{molecule}^{-1}\text{s}^{-1}$$

over the temperatures of 1100-2000 K. This result is the first direct measurement of the rate constants for Reaction 1 at elevated temperatures above 1000 K. Also, the rate constant for Reaction 2 is estimated to be

$$k_2 = (2.8 \pm 0.5) \times 10^{-10} \mathrm{cm^3 molecule^{-1} s^{-1}}.$$

3.2. Measurement of the branching fraction of the hydrogen atom channel

The relative sensitivities of ARAS signals for O and H atoms were calibrated by conducting ArF laser photolysis in $SO_2/H_2/Ar$ mixtures. The initial O atoms prepared in the photolysis were rapidly converted to twice as many H atoms through the reactions

$$O + H_2 \longrightarrow OH + H,$$
$$OH + H_2 \longrightarrow H_2O + H.$$

The absorption profiles of H and O atoms were repeatedly monitored at 1530, 1630, and 1820 K. Examples of the time profile of H atoms with and without coexistence of O atoms are shown in Fig.2. Although the absorption coefficient of H_2S at 248 nm is smaller, H_2S was still photodecomposed so that H atoms were produced as is shown in Fig.2(A), where only H_2S was photolyzed (no SO_2). Without oxygen atoms, the concentration of H atoms produced in the photolysis of H_2S decays exponentially due to the rapid consumption by the reaction

$$H + H_2S \longrightarrow H_2 + HS \qquad (3).$$

In contrast, a gradual increase of the concentration of H atoms (following the initial production by photolysis) was observed when O atoms were present: the time dependence of H atoms is shown in Fig.2(B). The contribution of the initial production by the photolysis and the consecutive decay for hydrogen atom shown in this figure was canceled simply by subtraction of the ARAS signals for H atom in the H_2S/Ar mixtures from those of $H_2S/SO_2/Ar$ mixtures after averaging over for several shots at fixed shock wave conditions. An example of such data processing is shown in Fig.2(C). As shown in this figure, the increase of H atoms at the initial stage is clearly demonstrated: this indicates that H atoms are formed directly by the reaction of O atoms with H_2S (1). By taking account of the ab initio CI calculation (Yokoyama 1992), Reaction 1c is assumed to be the initial source for emitting H atoms in this study. This process is followed by the successive reactions listed in Table 1. Some of the successive reactions are found to have sensitivities to the initial rise of H atoms, but these contributions are not significant. The observed decay part is mainly caused by the reaction of $H+H_2S$ (3) whose rate constant has been recently revised (Yoshimura 1992). An example of the kinetic simulation compared with observed profile of H atoms is shown by the solid curve in Fig.2(C). The branching fraction for Reaction 1c is determined to be $k_{1c}/k_1 = 0.24 \pm 0.06$ over the temperatures T = 1530, 1630, and 1820K: no obvious temperature dependence for the branching fraction is detected. As has been suggested in the previous studies at room temperature, the fraction for producing HSO+H in Reaction 1 is confirmed to be minor. Although the reaction rate constants and the mechanism are not fully understood for the consecutive reactions for the oxidation process of H_2S, the sensitivity analysis indicate that the time dependence of H atoms is sensitive only to the reactions (1 - 3), and the branching fraction decided in this study seems meaningful.

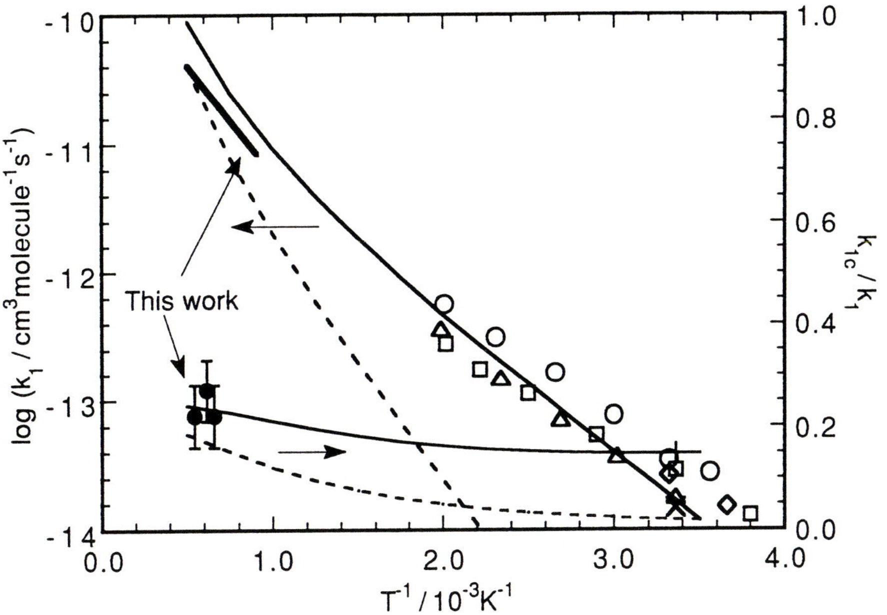

Fig. 3. Comparison of the TST calculation with experimental results on the overall rate constant k_1. [o: Slagle et al. (1978), □: Whytok et al. (1976), △: Singleton et al. (1979), (+): Liuti et al. (1966), x: Cuppit et al. (1970), ◇: Hollinden et al. (1970), Solid curves: k_1 and k_{1c}/k_1 by TST calculation with E_0(1a,1c) = 5.6, 5.2 kcal/mol, Dotted curves: k_1 and k_{1c}/k_1 by TST calculation with E_0(1a,1c) = 8.4, 9.3 kcal/mol

Table 1. Elementary reactions and the Arrhenius rate parameters in the O+H_2S study system. [$k = A\exp(-E/RT)$; units in molecule, cm, and s]
*the rate parameters employed by Frenklach et al. (1981)

	Elementary reaction			A	E(kJ/mol)	Reference
1a.	O+H_2S	→	HS+OH	2.13×10^{-10}	32.0	This study
1c.	O+H_2S	→	HSO+H	0.67×10^{-10}	32.0	This study
2.	O+HS	→	SO+H	2.80×10^{-10}	0.0	This study
3.	H_2S+H	→	H_2+HS	3.20×10^{-10}	20.7	Yoshimura (1992)
4.	H_2S+OH	→	HS+H_2O	5.30×10^{-12}	0.0	*
5.	HSO+M	→	H+SO+M	1.40×10^{-08}	220.0	This study
6.	SO+OH	→	SO_2+H	8.60×10^{-11}	0.0	*
7.	HS+HS	→	H_2S+S	1.30×10^{-11}	0.0	*
8.	HS+H	→	H_2+S	3.30×10^{-11}	0.0	*
9.	HS+S	→	H+S_2	4.50×10^{-11}	0.0	*

3.3. Theoretical calculation of k_1 and k_{1c}/k_1

Yokoyama (1992) has performed ab initio CI calculations for the reaction of $O(^3P)$+H_2S (UHF/6-31G** and larger basis sets). The barrier heights (E_0) with ZPE correction for Reactions 1a and 1c are calculated to be 8.4 and 9.3 kcal/mol, respectively. The rate constants for Reactions 1a and 1c are evaluated on the basis of a conventional transition state theory with Wigner tunneling correction, where magnitudes of E_0 are adjusted so that good agreement is achieved between the calculation and experiment. The calculated rate constant k_1 and the branching fraction k_{1c}/k_1 are compared with the experimental data in Fig.3. In this figure, the energies of the transition states for the channels (1a) and (1c) are assumed as 5.6 and 5.2 kcal/mol (including ZPE correction), respectively. Combination of the present and previous experimental data indicates the nonlinear Arrhenius temperature dependence for Reaction 1 as can be seen in Fig.3. Such a skewed Arrhenius plot is probably due to the tunneling effect at low temperatures, as well as the contribution of the vibrational partition functions of the transition state at high temperature range. The branching fraction determined in this study is found to be consistent with TST calculations, and also consistent with the conclusion given by Singleton et al.(1982), who suggested that $k_{1c}/k_1 < 0.2$ at room temperature.

References

Balucani N, Beneventi L, Casavecchia P, Stranges D, Volpi GG (1991) The effect of reagent electronic energy on the dynamics of chemical reactions: A high-resolution crossed beam study of $O(^3P,^1D)+H_2S$. J. Chem. Phys. 94:8611

Frenklach M, Lee JH, White JN, Gardiner Jr. WC (1981) Oxidation of hydrogen sulfide. Combustion and Flame 41:1

Singleton DL, Paraskevopoulos G, Irwin RS (1982) Mechanism of the $O(^3P)+H_2S$ reaction. Abstraction or addition? J. Phys. Chem. 86:2605

Yokoyama K (1992) private communication

Yoshimura M, Koshi M, Matsui H (1992) Non-Arrhenius temperature dependence of the rate constant for the $H+H_2S$ reaction. Chem. Phys. Lett. 89:199

Direct Measurement of the Reaction $H+CO_2$ at Elevated Temperatures

K. Wintergerst and P. Frank
DLR-Institut für Physikalische Chemie der Verbrennung 70569 Stuttgart, Germany

Abstract. The reaction CO_2+H has been investigated at elevated temperatures in the postshock region behind reflected shocks. The thermal decomposition of very low concentrations (0.2-3 ppm) of C_2H_5I served as H-atom source. Atomic resonance absorption spectrometry (ARAS) was used to record simultaneously H-atom and I-atom profiles. The experiments cover the temperature range 1450 to 1820 K, at total pressures from 1.5 to 7.8 bar. The values of the deduced rate coefficient (k in $cm^3\ mol^{-1}\ sec^{-1}$, uncertainties on one standard deviation level) is:

$$k_1 = (1.0 \pm 0.05) \times 10^{14} \exp(-(12000 \pm 50)/T) \qquad (R1)$$

Key words: Shock tubes, Chemical kinetics

1. Introduction

The reverse of the reaction of carbon dioxide with H-atoms, $CO+OH \longrightarrow CO_2+H$ (R1), which is responsible for the oxidation of CO via OH-radicals, is one of the most important reactions in flame chemistry (Warnatz 1984). Although a large number of individual studies have been carried out in the high temperature regime, there is no sufficient consensus because of the relatively high spread of the data (Baulch et al. 1976; Tsang and Hampson 1986). Under the assumption of detailed chemical equilibrium and the very precisely known thermochemical data (Tsang and Hampson 1986) the evaluation of a rate coefficient for reaction (R1) will also give a reliable value for CO+OH. Therefore the reaction of CO_2 with H-atoms was investigated under shock tube conditions. In this work, k-values were measured directly by using a shock tube technique. The thermal decomposition of C_2H_5I provides H-atoms down to temperatures of about 1150 K. Therefore C_2H_5I was used as the H-atom source.

2. Experimental

Details of the shock tube as well as the optical setup have been described in more detail elsewhere (Frank and Just 1985; Braun-Unkhoff et al. 1990). Only a short summary will be given: The measurements were performed behind the reflected shock front close to the end flange (5 mm) of a shock tube (ID: 72 mm). Atomic resonance absorption spectrometry (ARAS) was used to monitor time dependent H and I concentrations. As light sources for the absorption measurements, microwave-excited discharge lamps which contain H_2 or CH_3I diluted in He, were used. H and I concentrations can be measured simultaneously by using an oxygen spectral filter (for H-atoms at 121.5 nm) and a 1-m vacuum-UV monochromator (for I-atoms at 164.2 nm). The transmitted intensity signals were recorded by a 2-channel digital storage oscilloscope and transmitted to a personal computer. A program computes an averaged signal and transforms the intensity with the aid of calibration curves into particle concentration. Because of the high sensitivity of the optical measurement technique ([H]> 10^{11} cm^{-3}, [I]> 10^{12} cm^{-3}) very small initial concentrations of the reactants could be used. The background absorption of carbon dioxide at both wavelengths were measured in a separate series of experiments. For CO_2 a linear increase with temperature was determined for the absorption cross section at the L_α-line. The absorption cross sections for carbon dioxide were evaluated to be

Shock Waves @ Marseille II
Editors: R. Brun, L. Z. Dumitrescu

$$\sigma_{121.5nm} = (3.75 \times 10^{-3} \times T - 3.45) \times 10^{-18}\ \mathrm{cm}^2$$

$$\sigma_{164.2nm} = 5.8 \times 10^{-18}\ \mathrm{cm}^2.$$

The errors of the absorption cross section measurements are less than ±12%. The initial concentrations were 1250 to 5000 ppm for CO_2 and 0.5 to 3 ppm for C_2H_5I, both diluted with argon. The test gases were of high purity (Ar=99.9999 %, CO_2=99.9993 %, C_2H_5I=99.99 %).

3. Evaluation and results

The H- and I-concentration profiles reported here are based on calibration measurements. Because of the slow thermal dissociation of H_2 at temperatures below 2100 K, the calibration for the H-atoms in the temperature range of 1650 to 2100 K was performed by using the reaction of O-atoms (from N_2O-dissociaton) with H_2 to produce H-atoms (Frank and Just 1985). For sensitive iodine-atom detection I-ARAS was exploited. To achieve an absolute calibration for iodine-atom densities, known amounts of methyl iodide, analyzed by a gas chromatograph with ECD, were shock heated. The absorption of iodine atoms, produced during the thermal dissociation of CH_3I, was recorded. For temperatures above 1350 K the dissociation is sufficiently fast that constant absorption levels are attained during the observation time interval of about 800 μsec (Braun-Unkhoff et al. 1990). No temperature dependence for the [I]-calibration in the temperature range 1350 - 1900 K was observed. The source of hydrogen atoms exploited in the present experiments, is the thermal decomposition of ethyl iodide, followed by the fast decay of ethyl radicals at temperatures above 1100 K:

$$\begin{aligned} C_2H_5I &\longrightarrow C_2H_5 + I \\ C_2H_5 &\longrightarrow C_2H_4 + H. \end{aligned} \qquad (RI)$$

The investigation of the C_2H_5I decay, which has been executed in a separate series of experiments, showed a constant ratio of about 1.0 ± 0.1 between the measured hydrogen- and iodine-atom concentration.

Experiments with N_2O/H_2-concentrations of 100/2000 ppm were carried out to allow for establishing calibration curves down to temperatures of about 1250 K. From the data of Frank and Just (1985) an error of ±20% was estimated for extrapolating the rate constant of the N_2O decay to 1250 K. This results in a change of the absorption values for the calibration curve of less than 5 %. No temperature dependence of the H-calibration curves was found.

Therefore the experimentally derived factor of unity for the H-/I-atom concentration ratio allows C_2H_5I to be exploited as a quantitative H-atom source (see Fig.1), if the iodine-atom concentration is monitored during the reaction time (Fig.2).

In each series of experiments with CO_2, the concentrations of carbon dioxide did not decrease noticeably during the observation time. As there are - besides the H-atoms - no other species contributing to the time-dependent part of the overall absorption, extrapolation to the starting point of the reaction is straightforward. This absorption value gives, after correcting for the background absorption of CO_2, a precise measure of the initial H-atom concentration. The concentration values of hydrogen atoms determined by this method agreed very well with the H-concentration values from the in-situ iodine-ARAS measurements. Therefore we decided to exploit this method of producing H-atoms by a thermal process for our experiments.

The experiments for evaluating a rate coefficient expression for the reaction

$$CO_2 + H \longrightarrow CO + OH \qquad (R1)$$

were carried out at temperatures between 1450 to 1820 K. A typical experimental profile of the total absorption is shown in Fig.3.

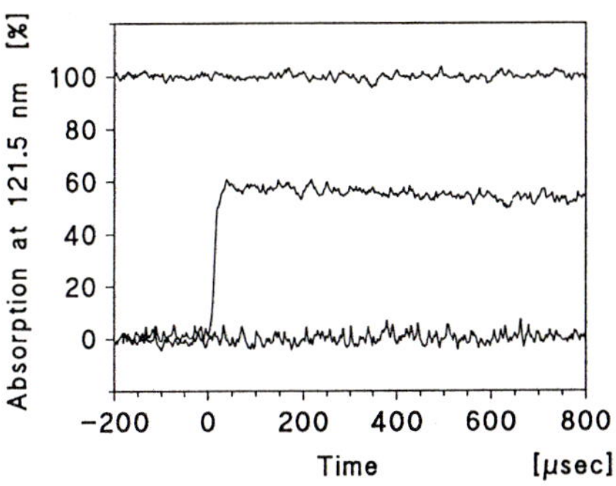

Fig. 1. H-ARAS profile from thermal C_2H_5I- decay: $X_0(C_2H_5I)$=1.5 ppm, T_5 = 1354 K, p_5 = 2.12 bar

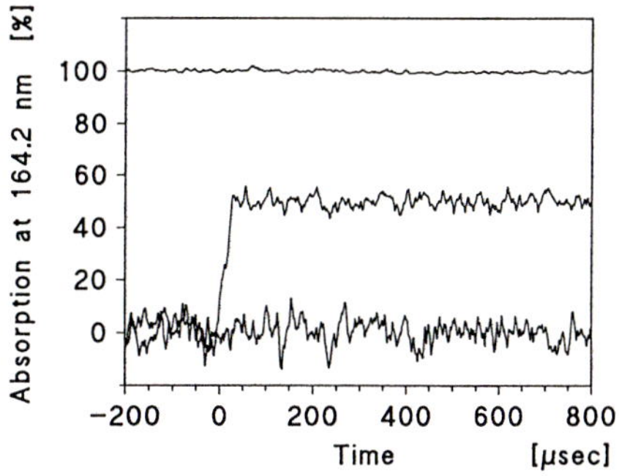

Fig. 2. I-ARAS profile (same experimental data as in Fig.1)

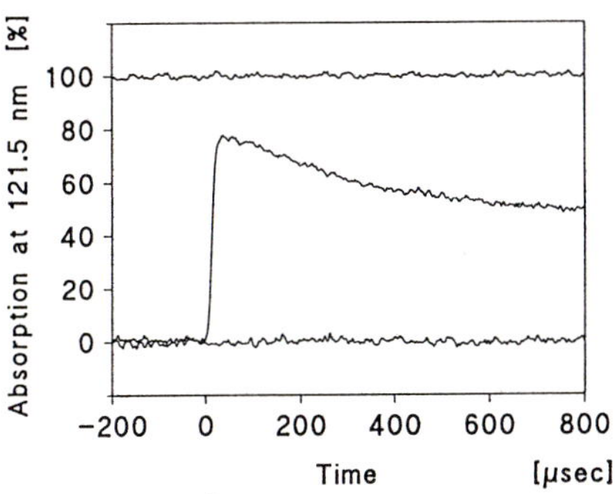

Fig. 3. Total absorption profiles for: $X_0(CO_2)$=2500 ppm, X_0(H)=1.1 ppm T_5 = 1835 K, p_5 = 2.30 bar

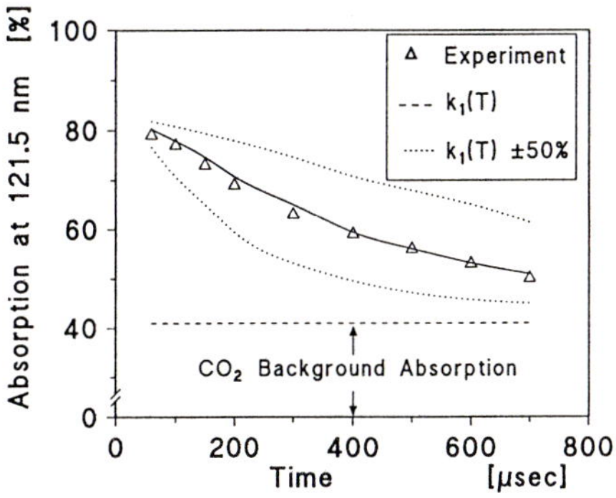

Fig. 4. k_1-sensitivity of the measured total absorption profile: variation of k_1 by ±50% (same experimental data as in fig.3)

Under the given experimental conditions, the signals correspond to the superposition of the - within the observation time interval of less than 1 msec - practically constant CO_2 background absorption and the decreasing H-ARAS signals. Sensitivity analysis shows, that the measured absorption profiles can be completely simulated by using only reaction R1. As an example, Fig.4 shows the comparison between measured and calculated profile. It also indicates the sensitivity of the profile with respect of varying k_1 by ±50%. Due to the very low initial concentrations of H-atoms and the excess carbon dioxide concentrations ((CO_2)/(H)=600-1500), first-order behaviour was observed, at least during the first 300 μsec, over the whole temperature range of investigation. Separate runs of experiments with only pure argon indicated that for temperatures below 1900 K no measureable H-atom absorption resulting from impurity background could be detected. Fig.7 shows the Arrhenius plot for reaction (R1) and rate coefficient data from the literature (Baulch et al. 1992; Just and Stepanek 1969).

4. Discussion

4.1. C_2H_5I decay

This is to our knowledge the first study under shock tube conditions in which hydrogen atoms are instantaneously produced by a very fast thermal process, namely

$$\begin{aligned} C_2H_5I &\longrightarrow C_2H_5 + I \\ C_2H_5 &\longrightarrow C_2H_4 + H, \end{aligned} \qquad (RI)$$

followed by the immediate reaction with carbon dioxide. Computer simulations and experimental results show that ethene ($X_0 \leq 3$ ppm), which is also formed as a product, is not an additional H-atom source under the chosen experimental conditions. Bimolecular reactions of ethylene with carbon dioxide and other products are completely negligible under the experimental conditions of this study. Actually, the high temperature decomposition mechanism of ethyl iodide is not completely clarified. One could assume that a second reaction channel is responsible for the depletion of the H-concentration levels. However, the thermochemically most plausible reaction pathway (Yuang and Conway 1965; Benson 1976)

$$\begin{aligned} C_2H_5I &\longrightarrow C_2H_4 + HI \\ HI &\longrightarrow H + I \end{aligned} \qquad (RII)$$

cannot directly influence the experimental results, because the thermal dissociation of the hydrogen iodide is expected to proceed relatively slow under the experimental conditions (Baulch et al. 1981). However, the reaction of hydrogen atoms with hydrogen iodide : H+HI$\longrightarrow$ H_2+I, which proceeds very fast ($\sim 4 \times 10^{13}$ cm^3 mol^{-1} s^{-1} (Baulch et al. 1981), could influence the apparent H-concentration levels.

Therefore a series of model calculations were carried out, varying the product concentration ratio of $r = RII/RI$ from 0.1 to 0.5. These calculated H and I profiles are shown Figs.5 and 6, respectively. In all cases with $r \geq 0.1$, a pronounced deviation from constant concentration levels of hydrogen- and iodine-atoms is observed, in contrast to the experimental profiles (Fig.2). These results give further evidence that the second channel RII plays only a minor role in the ethyl iodide decomposition under the present experimental conditions. For the CO_2+H investigations we can also show with a crude estimation, that the second channel cannot influence the evaluation of k_1 noticeably:

for $\frac{d[H]}{dt} = -k_1 \bullet [CO_2] \bullet [H] - k_{HI+H} \bullet [HI] \bullet [H]$

with $[H]/[CO_2] \simeq 1/1000$, $k_1/k_{HI+H} \simeq 1/1000$ and $[HI]/[H] \leq 1/10$ (value for r from above),

we calculate $k_1 \bullet [CO_2] \bullet [H] \;/\; (k_{HI+H} \bullet [HI] \bullet [H]) \geq 10$.

This estimation indicates that under the "worst case" assumption with respect to the C_2H_5I-decomposition mechanism, a second C_2H_5I-decomposition channel could alter the evaluated value for k_1 only by less than 10 %.

Therefore this method allows to study the process of H-addition to carbon dioxide under kinetically isolated conditions at elevated temperatures.

4.2. CO_2+H

There have been not many studies in the high temperature regime for reaction $R1$. The results of the present work are compared in Fig.7 with those of a few previous studies. The study of Just and Stepanek (1969) (full-dashed line) has been carried out with mixtures of molecular hydrogen and carbon dioxide in shock tubes at temperatures above 2000 K.

The extrapolation of their data to the lower temperature regime of the present work gives values which are lower by about 25 %. The recommended rate coefficient expression of Baulch et

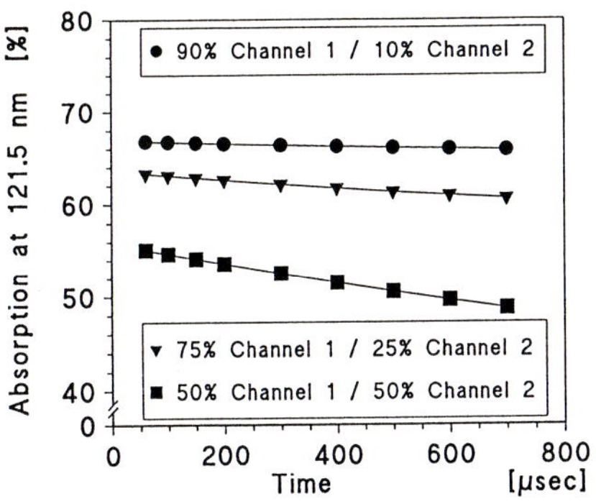

Fig. 5. Calculated H-profiles for different branching ratios. $X_0(C_2H_5I)$=1.6 ppm, T_5=1590 K, p_5=2.5 bar

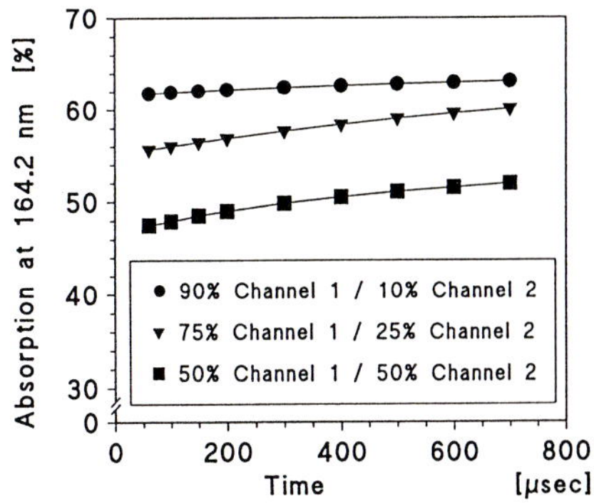

Fig. 6. Calculated I-profiles for different branching ratios. (same experimental data as in Fig.5)

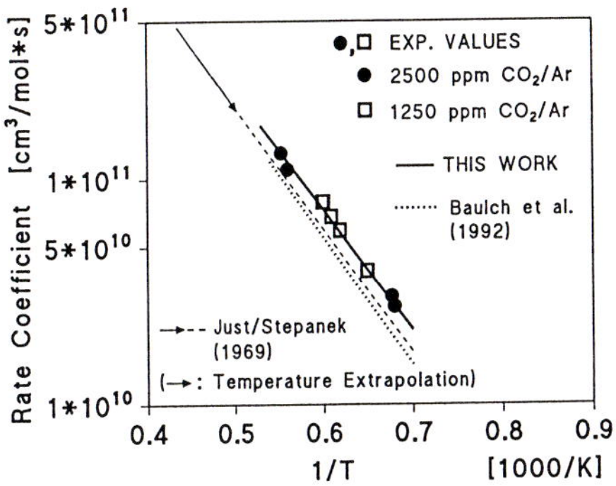

Fig. 7. Arrhenius-plot of CO_2+H ($R1$)

al. (1992), based upon CO+OH ($R1$) and K_c, which agrees fairly well with that of Baulch et al. (1976) and Tsang and Hampson (1986), is also included. If we allow an error of $\Delta G_T \simeq 2$ kJ/mol for the thermodynamic data, the uncertainty in calculating K_c is less than 20 %. The comparison shows, that our data are larger by factors of 1.3-1.5. In a recent study, which was carried out by Yu et al. (1990) on measuring OH-radicals in shock heated H_2-O_2-CO mixtures, the rate constant for the reverse reaction was deduced. Their data agree very well with the recommended value of Baulch et al. (1992). All these data allow a rate coefficient for $R1$ to be calculated with values lying between the rate data from Just and Stepanek (1969) and those of the present study.

Acknowledgements

The authors wish to thank Prof.Th.Just for his interest in this study and for helpful discussions. We thank our colleague Dr.M.Braun-Unkhoff for the oral presentation of the paper at the 19th ISSW. We also like to thank N.Ackermann and M.Kallfass for technical assistance. We are also very grateful to Dr.C.Wahl and M.Kapernaum for analyzing ethyl iodide.

References

Baulch DL, Drysdale DD, Duxbury J, Grant SJ (1976) Evaluated kinetic data for high temperature reactions. Vol.3, Butterworths, London Boston

Baulch DL, Duxbury J, Grant J, Montague DC (1981) J. Phys. Chem. Ref. Data 10, Suppl.1,1-1

Baulch DL, Cobos A, Cox RA, Frank P, Just Th, Kerr JA, Pilling M, Troe J, Walker RW, Warnatz J (1992) Evaluated kinetic data for combustion modeling. J. Phys. Chem. Ref. Data 21 No.3:566

Benson SW (1976) Thermochemical kinetics. J.Wiley, New York

Braun-Unkhoff M, Frank P, Just Th (1990) High temperature reactions of benzyl radicals. Ber. Bunsenges. Phys. Chem. 94:1417

Frank P, Just Th (1985) High temperature reaction rate for $H+O_2=OH+O$ and $OH+H_2=H_2O+H$. Ber. Bunsenges. Phys. Chem. 89:181

Just Th, Stepanek S (1969) Precise measurement of rate constant for the reaction $H+CO_2 \longrightarrow OH+CO$ in the temperature range between 2000 and 2700 K behind reflected shocks using an infrared technique. In: Glass II (ed) Proc. 7th Intl. Shock Tube Symposium, Toronto, pp.626-643

Tsang W, Hampson RF (1986) Chemical kinetic data base for methane combustion. J. Phys. Chem. Ref. Data 15:1087

Warnatz J (1984) Rate coefficients in the C/H/O system. In: Gardiner WC Jr.(ed) Combustion chemistry. Springer, New York Berlin Heidelberg Tokyo, pp 197-360

Yu CL, Wang C, Frenklach M (1990) Measurement of the rate coefficient of reaction $CO+OH \longrightarrow CO_2+H$. Presented at Eastern States Meeting of the Combustion Institute, pp.1-4

Yuang J, Conway DC (1965) Pyrolysis of Ethyl Iodide by the toluene-carrier flow technique. J. Phys. Chem. 43:1296

A Shock Tube Study of Nitric Acid Decomposition

Margaret S. Wooldridge, Ronald K. Hanson and Craig T. Bowman
High Temperature Gasdynamics Laboratory, Department of Mechanical Engineering, Stanford University, Stanford, California, 94305, USA

Abstract. The rate coefficient for the reaction

$$HNO_3 + M \longrightarrow OH + NO_2 + M \qquad (1)$$

has been determined in mixtures of nitric acid (HNO_3) and argon in incident shock wave experiments. Quantitative OH time-histories were obtained by cw narrow-linewidth uv laser absorption of the $R_1(5)$ line of the $A^2\,\Sigma^+ \leftarrow X^2\Pi_i(0,0)$ transition at 32606.56 cm^{-1} (vacuum). The experiments were conducted over the temperature range 990 - 1400 K and the pressure range 0.18 - 0.44 atm. The second-order rate coefficient was determined to be

$$k_1 = 1.01 \times 10^{16}\ \exp(-17280/T[\mathrm{K}])[\mathrm{cm}^3\mathrm{mol}^{-1}\mathrm{s}^{-1}] \qquad [3]$$

with overall uncertainties of +19%, -12% at high temperatures and +40%, -38% at low temperatures.

Key words: Reaction rate, HNO_3 pyrolysis, Shock tube experiment

1. Introduction

The rapid thermal decomposition of nitric acid offers a potentially attractive source of hydroxyl radicals for use in shock tube studies of combustion chemistry. As part of an investigation of this concept, we have determined the rate of the bimolecular reaction

$$HNO_3 + M \longrightarrow OH + NO_2 + M \qquad (1)$$

over a range of conditions by measuring OH time-histories behind incident shock waves. The OH was monitored by cw, narrow-linewidth uv laser absorption of the $R_1(5)$ line of the $A^2\,\Sigma^+ \leftarrow X^2\Pi_i(0,0)$ OH transition at 32606.56 cm^{-1} (vacuum). The direct, quantitative measurement of OH allowed an accurate determination of the rate coefficient for reaction (1).

2. Experimental details

All experiments were conducted behind incident shock waves in a 15.24 cm diameter, stainless steel, pressure-driven shock tube with a 2.3 m long driver section and a 10.5 m long driven section. Prior to each experiment, the driven section was evacuated to ca. 5×10^{-6} torr; the subsequent leak-plus-outgassing rate was ca. 20×10^{-6} torr/minute. Shock speeds were measured using four platinum thin-film gauges equipped with signal amplifiers that triggered three time-interval counters. Shock wave attenuation was typically 1-2% per meter. Helium was used as the driver gas for all experiments.

The gases employed were Matheson Purity argon (99.9995%) and commercial grade helium (Liquid Carbonic, 99.9% minimum purity). The nitric acid was purchased in 90% aqueous solution and subsequently dehydrated using a technique described by Stern et al. (1960). All test gas mixtures were prepared in a 2.3 liter Teflon-coated mixing tank with a magnetic stirrer. A passivation process was utilized to control wall-adsorption effects in the shock tube; details are given in Wooldridge et al. (1993).

Shock Waves @ Marseille II
Editors: R. Brun, L. Z. Dumitrescu

The diagnostic system consisted of a Spectra-Physics 380A cw ring dye laser operated on Rhodamine 590 laser dye and pumped by a Spectra-Physics 164 argon ion laser (4 W, all lines visible). The visible wavelength output of the ring dye laser cavity was monitored by a Burleigh WA-10 wavemeter, and the laser mode quality was monitored by a Spectra-Physics 470 scanning interferometer (2 GHz free spectral range). The fundamental laser frequency was intracavity doubled to the uv by a temperature-controlled AD^*A crystal. Wavelength control was improved by conducting a manual scan across the absorption lineshape in a reference flame of methane and air. The wavelength was centered at the peak absorption of the $R_1(5)$ line prior to each experiment. The beam intensity was measured before and after passing through the shock tube via fast uv photodiode detectors (EG&G UV100BQ, electronic 3-dB bandwidth of 350 kHz). The incident (I_0) and differenced ($I_0 - I$) laser intensities were recorded on a Nicolet 2090 digital oscilloscope. The noise on the differenced signal was typically 0.1% of the incident signal, corresponding to a detectivity limit of approx. 1 ppm at $T = 1200$ K, $P = 0.2$ atm and a path length of 15.24 cm.

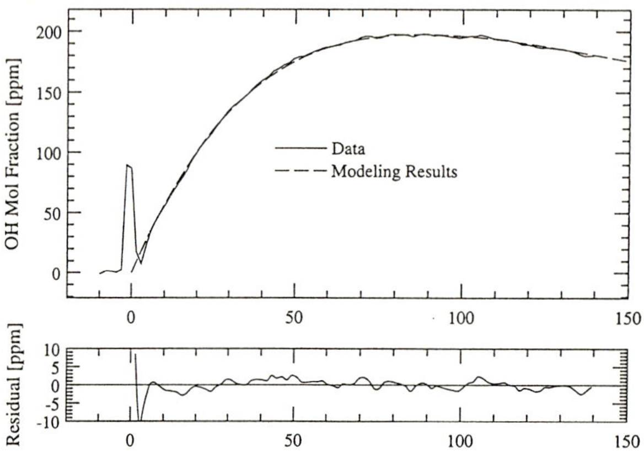

Fig. 1. Comparison of experimental and Chemkin-calculated OH profiles using best-fit $k_1 = 5.25 \times 10^9$ cm³ mol^{-1} s^{-1} and $HNO_3^\circ = 283$ ppm (incident shock, $T = 1198$ K, $P = 0.421$ atm)

3. Results

Fig.1 shows a typical OH mol fraction profile. The experimental conditions corresponding to Fig.1 are $T = 1198$ K, $P = 0.421$ atm and $HNO_3^\circ = 283$ ppm, where HNO_3° is the initial concentration of nitric acid in the shock tube at the time of the experiment. The absorption data were converted to mol fraction data by applying Beer's law for a narrowline radiation source:

$$I/I_0 = \exp\left(-k_v\ P\ \chi_{OH}\ L\right) \qquad [1]$$

Here, I and I_0 are the transmitted and incident radiation intensities, respectively, k_v is the spectral absorption coefficient [cm^{-1} atm^{-1}], P is the total pressure [atm], χ_{OH} is the mole fraction of OH, and L is the path length through the absorbing medium [cm]. The absorption coefficient k_v was determined using the technique described in Wooldridge et al. (1993). The uncertainty in k_v for the conditions of this investigation is ±5%, set primarily by uncertainty in the oscillator strength for this transition.

Table 1. Reaction mechanismm; rate coefficients in the form $k = AT^b \exp(-E_A/RT)$ (units are: mol, cc, s, K and cal.)

	reaction	A	b	E_A	Reference
1	$HNO_3+M=OH+NO_2+M$	1.01×10^{16}	0.0	34355.	This work
2	$OH+OH=H_2O+O$	9.40×10^{-5}	4.77	-9081.	Wooldridge et al.(1993)
3	$HNO_3+OH=H_2O+NO_3$	9.50×10^{10}	0.0	0.	Glänzer & Troe (1974)
4	$OH+HO_2=H_2O+O_2$	7.50×10^{12}	0.0	0.	Miller & Bowman (1989)
5	$HO_2+NO=NO_2+OH$	2.11×10^{12}	0.0	-479.	Miller and Bowman (1989)

The mole fraction data were converted from laboratory time (t_1) to particle time (t_p) using the relation $t_p = (\rho_2/\rho_1)t_1$, where ρ_1 is the initial density of the test gas mixture and ρ_2 is the density of the test gas mixture behind the incident shock wave.

All experiments were modeled using the Chemkin-II chemical kinetics simulation package (Kee et al. 1989) and the Chemkin thermodynamic data base (Kee et al. 1990). The reaction mechanism employed is given in Table 1. This five-reaction mechanism was developed from a more comprehensive 82 reaction model using sensitivity and contribution factor analyses to eliminate reactions that were unimportant for the conditions of the present study.

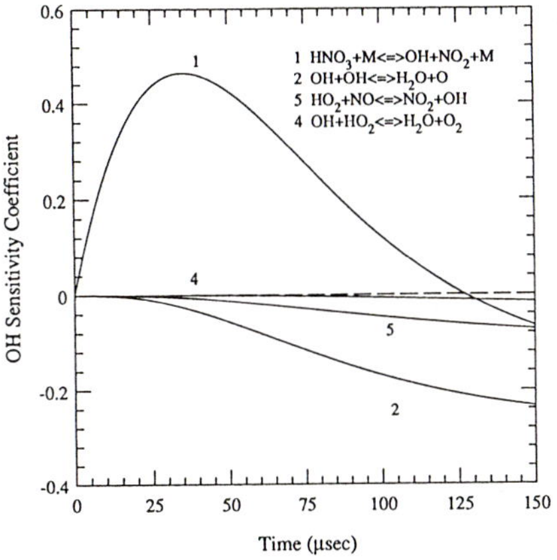

Fig. 2. OH sensivity plot for the conditions of Fig.1

The data were analyzed by adjusting the rate coefficient of reaction (1) and the initial amount of HNO_3 present in the model, and then comparing the calculated OH profiles with the experimental traces. The determination of the best fit to the experimental profile was aided by the residual trace shown at the bottom of Fig.1. The residual was generated by subtracting the calculated profile from the experimental data trace. The optimum values for k_1 and HNO_3^o were obtained by minimizing this residual. Fig.2 shows the OH sensitivity analysis for the conditions of Fig.1, as generated by the Senkin modeling package (Lutz et al. 1988). The sensitivity coefficient for a particular species α_{ij} is defined as the partial derivative of the species mol fraction χ_j with respect to the rate constant parameter A_i, normalized by the maximum species mol fraction χ_j^{max} and the rate constant parameter A_i:

$$\alpha_{ij} = (A_i/\chi_j^{max})(\partial\chi_j/\partial A_i) \quad [2]$$

The sensitivity analysis indicates that reaction (1) is the dominant reaction affecting the OH profile for times up to ca. 80 ms, for this experiment.

Table 2. Experimental conditions and results

T_2 K	P_2 atm	HNO_3^o ppm	k_1 $cm^3\ mol^{-1}\ s^{-1}$
986	0.248	270	2.60×10^{8}
1050	0.217	640	7.20×10^{8}
1069	0.215	374	9.00×10^{8}
1103	0.205	583	1.80×10^{9}
1105	0.365	590	1.50×10^{9}
1118	0.205	717	1.80×10^{9}
1183	0.202	714	5.15×10^{9}
1198	0.421	283	5.25×10^{9}
1228	0.219	562	7.30×10^{9}
1248	0.182	640	1.05×10^{10}
1256	0.183	709	9.90×10^{9}
1300	0.439	342	1.70×10^{10}
1398	0.305	545	4.50×10^{10}

Table 2 summarizes the shock conditions and values determined for k_1. The experiments span a temperature range of 990 - 1400 K and a pressure range of 0.18 - 0.44 atm. The data are plotted in Fig.3 and fit by the following expression for the rate coefficient of the bimolecular reaction:

$$k_1 = 1.01 \times 10^{16} \exp(-17280/T[\mathrm{K}])[\mathrm{cm}^3\mathrm{mol}^-1\mathrm{s}^-1] \quad [3]$$

Overall uncertainties of +19% -12% at high temperatures and +40% -38% at low-temperatures were determined using the process outlined in Wooldridge et al. (1994). The uncertainty at high temperatures is due to a convolution of the size of the laser beam, the speed of the post-shock gases and the electronic response time of the data acquisition system. The uncertainty at low temperatures is primarily due to uncertainty in HNO_3^o.

4 . Discussion

The current work agrees well with the results of other experimental investigations of reaction (1): Harrison et al. (1962), Gershenzon et al. (1973), and Basevich and Kogarko (1979). However, the activation energy of the current work, (E_a = 34.3 kcal/mol) is lower than that quoted by Glänzer and Troe (1974) for the low-pressure limit (E_a = 40.0 kcal/mol).

Following the technique outlined in Troe (1977), the results have been analyzed in terms of the theoretical expression

$$k_o = [M]\beta_c\ Z_{LJ}\ \rho_{vib,h}\ (E_o)\ (k_BT/Q_{vib})\ F_{anh}\ F_{rot}\ F_{rot\ int}\ F_E\ \exp(-E_o/k_BT) \quad [4]$$

where β_c is the collision efficiency, Z_{LJ} is the Lennard-Jones collision frequency, E_0 is the critical energy for dissociation (47.7 kcal/mol), $\rho_{vib,h}(E_o)$ is the harmonic oscillator density of states, Q_{vib} is the vibrational partition function, F_{anh} is the anharmonicity correction, F_{rot} is the external rotational contribution, $F_{rot\ int}$ is the internal rotor correction and F_E is the energy dependence of the density of states. As given in Glänzer and Troe (1974), the barrier to the torsional motion V_o is 7.8 kcal/mol and the HNO_3 vibrational frequencies ν_i are: 3550, 1708.2, 1330.7, 1324.9,

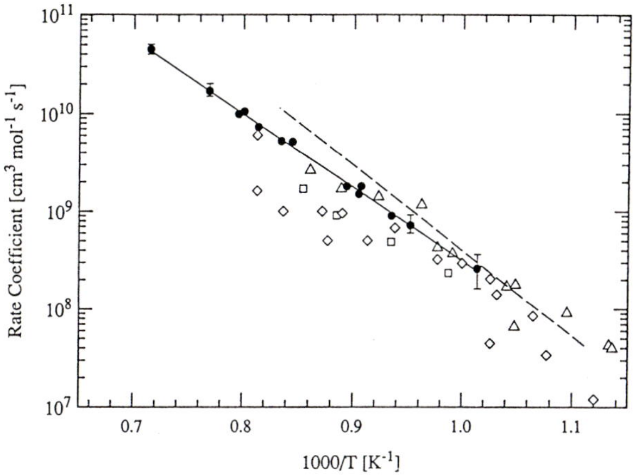

Fig. 3. Arrhenius plot of rate coefficients for reaction (1).The solid line is a least-squares fit to th current data. The vertical error bars are the overall uncertainties determined for the present work. • present work; △ Harrison, Johnston and Hardwick (1962); □ Gershenzon, Dement'ev and Nalbandyan (1973); ◇ Basevich and Kogarko (1979); – – low-pressure limit, Glänzer and Troe (1974)

878.6, 762.2, 646.6, 579, 455.8 cm^{-1}. A Lennard-Jones collision diameter (σ_{HNO_3}) of 4.24 Å and a Lennard-Jones well depth (ε_{HNO_3}) of 390 K are used for this analysis (Troe 1977). Treating the torsional motion as a free internal rotation, we find $\beta_c = 0.13$ for $T = 990$ K, and $\beta_c = 0.08$ at $T = 1400$ K. The value of β_c is related to an average energy transferred per collision $< \Delta E >$, via

$$\beta_c/(1 - \beta_c^{1/2}) = - < \Delta E > /(F_E k_B T) \qquad [5]$$

We find $- < \Delta E >= 530$ cal/mol for $T = 990$ K and $- < \Delta E >= 440$ cal/mol for $T = 1400$ K. These results agree well with those of Glänzer and Troe (1974).

Acknowledgment

This work was supported by the U. S. Environmental Protection Agency and the National Science Foundation.

References

Basevich VY, Kogarko SM (1979) The mechanism of methane combustion. 5. The reactions of methane with nitric acid. Bull. Acad. Sciences USSR 27:1988-1992

Gershenzon YM, Dement'ev AP, Nalbandyan AB (1973) A study of the rate of the thermal dissociation of nitric acid by gaseous esr spectroscopy. Dokl. Phys. Chem. 210:403-406

Gershenzon YM, Dement'ev AP, Nalbandyan AB (1979) A study of the heterogeneous activation and decomposition of HNO_3 molecules by the method of accumulating radicals in the resonator cavity of an epr spectrometer. Kinet. Catal. 20:461-465

Glänzer K, Troe J (1974) Thermal decomposition of nitrocompounds in shock waves. IV: Decomposition of nitric acid. Ber. Bunsenges. Physik. Chem. 78:71-76

Harrison H, Johnston HS, Hardwick ER (1962) Kinetics of the thermal decomposition of nitric acid vapor. J. Am. Chem. Soc. 84:2478-2482

Kee RJ, Rupley FM, Miller JA (1989) Chemkin-II: A Fortran chemical kinetics package for the analysis of gas-phase chemical kinetics. Sandia National Laboratories, Livermore California

Kee RJ, Rupley FM, Miller JA (1990) The Chemkin thermodynamic data base. Sandia National Laboratories, Livermore California

Lutz AE, Kee RJ, Miller JA (1988) Senkin: A Fortran program for predicting homogeneous gas phase chemical kinetics with sensitivity analysis. Sandia National Laboratories, Livermore California

Miller JA, Bowman CT (1989) Mechanism and modeling of nitrogen chemistry in combustion. Prog. Energy Combust. Sci. 15:287-338

Stern SA, Mullhaupt JT, Kay WB (1960) The physicochemical properties of pure nitric acid. Chem. Reviews 60:185-207

Troe J (1977) Theory of thermal unimolecular reactions at low pressures. II. Strong collision rate constants. Applications. J. Chem. Phys. 66: 4758-4775

Wooldridge MS, Hanson RK, Bowman CT (1994) 1993) A shock tube study of the $OH+OH \rightarrow H_2O+O$ reaction. Intl. J. Chem. Kinetics 26

The Reaction of CN with NO at High Temperatures in Shock Waves

Y. He*, **C.H. Wu**†, **M.C. Lin*** **and C.F. Melius**‡
*Department of Chemistry, Emory University, Atlanta, GA 30322, USA
†Naval Research Laboratory, Washington, DC 20375, USA
‡Combustion Research Facility, Sandia National Laboratories, Livermore, CA 94551, USA

Abstract. The reaction of CN with NO has been studied between 2200 and 2810 K with a shock tube equipped with two stabilized cw CO lasers using BrCN as the CN radical source. The rates of CO formation and NO removal, measured simultaneously with the lasers, were kinetically modeled employing a detailed reaction mechanism constructed on the basis of the result of our quantum chemical (BAC-MP4) calculations. The kinetic modeling, aided by sensitivity analysis, suggests that the CN+NO reaction occurs primarily by the abstraction channel, producing NCO+N (1), via a triplet NCON intermediate. The rate constant for this reaction can be effectively represented by $k_1 = 5.5 \times 10^{12} \exp(-15,410/T)$ cm^3/mole.s. This is consistent with the result of the quantum calculations which, however, also indicates the presence of two other minor product channels: CO+N_2 (2) and NCN+O (3). The latter is enthalpically limited, whereas the former is entropically controlled because of the tight cyclic transition state, -CNNO-‡.

Key words: CN+NO reaction, High temperature, BAC-MP4 calculations

1. Introduction

The reaction of CN radicals with NO is very important to the combustion of nitramines (Ermolin et al. 1986; Melius 1988). At low temperatures (T < 1000 K), the reaction of CN with NO was found to occur exclusively by the recombination process, CN+NO+M→NCNO+M (Wang et al. 1989). The four centered, exothermic reaction, CN+NO→N_2+CO (ΔH = -640 kJ/mole), which has been invoked frequently in the literature, was found to be inconsequential up to 740 K.

In the two studies by Colket (1984) and Natarajan and Roth (1986), using molecular resonance absorption to measure CN disappearance rates and atomic resonance absorption to monitor O- and N- atom production rates, respectively, these authors concluded that the CN+NO reaction occurred by two almost equally important product channels: CN+NO→NCO+N (1) and CO+N_2 (2) with the rate constants, $k_1 = 1.0 \times 10^{14} \exp(-21,200/T)$ and $k_2 = 2.5 \times 10^{10}$cm^3/mole.s. This result suggests that the energy barrier for reaction (2) is negligible, whereas that of the reverse of reaction (1), an atom-radical reaction, is as large as 110 kJ/mole under the conditions employed.

In order to elucidate the mechanism of this reaction, we have recently carried out a series of experiments using BrCN as the source of the CN radical for the temperature range 2000-2700 K by simultaneously measuring CO production and NO decay rates. In addition, we have performed a comprehensive set of quantum-chemical calculations using the BAC-MP4 (bond-additivity-corrected-Møller Plesset 4th order perturbation) technique developed by Melius and Binkley (1986) to determine transition state structures and activation energies of various accessible reaction paths. On the basis of these results, we were able to establish a reasonable reaction mechanism by which the measured kinetic data could be more realistically modeled than before.

2. Experimental technique

The reaction was investigated with a 6.35-cm. i.d. stainless steel shock tube, originally constructed by Shaub, Lin and coworkers (Hsu et al. 1982) at the U.S. Naval Research Laboratory. The shock tube was equipped with two lock-in stabilized cw CO lasers for simultaneous detection of CO and NO (Hsu and Lin 1984; Lin et al. 1990). CO was detected by resonant absorption with the $2 \rightarrow 1P(10)$ line of the CO laser, while NO was monitored by near-resonant absorption with the $7 \rightarrow 6P(13)$ line. These measurements were carried out with two liquid-N_2 cooled InSb detectors using fast response amplifiers. The measured absorption coefficients, $\alpha = \ln(I/I_0)/L$ (where $L =$ 6.35 cm), were directly converted into absolute concentrations with a computer using previous calibration data (Hsu et al. 1982, 1984).

Incident shocks were employed to heat the reaction mixtures with the following compositions: BrCN:NO:Ar = 1:1.94:201.53 and 1:0.625:110.55. Reaction temperature was varied from 2200 to 2810 K. The initial temperature, pressure and density of the shocked mixtures were computed with the NASA/Lewis equilibrium program (Hsu et al. 1982), whereas the kinetics and the temperature and concentration changes were calculated by CHEMKIN (Kee et al. 1989) and SENKIN (Lutz et al. 1988) programs.

BrCN (97% purity) and NO (99% purity) used in the present study were obtained from Aldrich Chemical Co., whereas the Ar diluent (99.995%) was that of Matheson Gas Products. BrCN was purified by vacuum distillation near 230 K; NO was purified by passing it through a silica gel trap maintained at dry-ice temperature and condensing it in a storage bulb with liquid N_2. Ar was used for preparation of the aforementioned mixtures without further purification.

3. Results and discussion

Two dozen sets of experiments have been performed with the two mixtures at temperatures between 2200 and 2810 K. Both CO and NO concentration profiles were measured simultaneously as mentioned before. A typical set of these profiles is presented in Fig.1 in comparison with the kinetically modeled values.

Kinetic modeling was carried out with the mechanism summarized in Table 1. The mechanism was built in part upon those employed by Colket (1984) and Natarajan and Roth (1986) and in part upon the result of our BAC-MP4 quantum calculations (He et al. 1993) which suggests that the CN+NO reaction occurs by the following three lower energy paths via different intermediates:

$$\mathrm{CN + NO \longrightarrow {}^3NCON \longrightarrow NCO + N} \quad (1)$$
$$\mathrm{\longrightarrow {}^1CNNO \longrightarrow CO + N_2} \quad (2)$$
$$\mathrm{\longrightarrow {}^3NCNO \longrightarrow NCN + O} \quad (3)$$

The superscript to each intermediate given above represents its spin multiplicity. Inasmuch as the rate-limiting transition states of these three reaction channels are calculable by the BAC-MP4 method, the rate constants for these reactions can be computed by the transition state theory. Since reactions (2) and (3) contribute much less to the production of CO as well as the removal of NO, the calculated values of k_2 and k_3 were kept constant, whereas that of the dominant channel (1) was varied to match the rate of CO production which could be more reliably measured due to the resonant nature of the absorption process.

Table 1. Mechanisms for the CN+NO reaction*

No.	Reaction	A	B	E/R
1	$CN+NO=NCO+N$	5.5 E+12	0.0	15,410
2	$CN+NO=N_2+CO$	3.9 E+11	0.0	14,000
3	$CN+NO=NCN+O$	1.8 E+13	0.0	19,220
4	$O+CN=CO+N$	3.0 E+13	0.0	0
5	$N+NO=N_2+O$	4.3 E+13	0.0	790
6	$C+N_2=N+CN$	6.3 E+13	0.0	23,150
7	$C+NO=CO+N$	2.8 E+13	0.0	0
8	$CN+NCN=N+C_2N_2$	2.0 E+13	0.0	0
9	$CN+O=C+NO$	2.0 E+13	0.0	0
10	$N+NCN=N_2+CN$	1.0 E+13	0.0	0
11	$N+NCO=N_2+C$	2.0 E+13	0.0	0
12	$O+NCO=NO+CO$	2.0 E+13	0.0	0
13	$BrCN+Ar=Br+CN+Ar$	0.72 E+15	0.0	37,800
14	$Br+BrCN=CN+Br_2$	1.0 E+13	0.5	16,000
15	$CN+BrCN=Br+C_2N_2$	4.0 E+09	1.2	950
16	$O+BrCN=Br+NCO$	1.0 E+13	0.0	4,000
17	$O+Br_2=BrO+Br$	8.4 E+12	0.0	0
18	$Br_2+M=Br+Br+M$	2.35 E+14	0.0	21,630
19	$CN+NO+M=NCNO+M$	1.0 E+14	-0.5	0
20	$CN+NCO=NCN+CO$	2.0 E+13	0.0	0
21	$NCN+Ar=N+CN+Ar$	5.1 E+15	0.0	53,300
22	$C_2N_2+Ar=CN+CN+Ar$	1.1 E+34	-4.3	61,900
23	$O+C_2N_2=CN+NCO$	1.3 E+14	0.0	7,180
24	$NCO+Ar=N+CO+Ar$	3.1 E+16	-0.5	23,900
25	$O+N_2O=NO+NO$	2.9 E+13	0.0	11,650
26	$O+N_2O=N_2+O_2$	1.4 E+12	0.0	5,450
27	$N_2O+Ar=N_2+O+Ar$	6.9 E+23	-2.5	32,700
28	$CN+N_2O=NCN+NO$	3.9 E+03	2.6	1,860
29	$N+N+Ar=N_2+Ar$	3.0 E+14	0.0	-5,030
30	$NO+Ar=N+O+Ar$	4.0 E+20	-1.5	75,500
31	$O+O+Ar=O_2+Ar$	1.9 E+13	0.0	-905
32	$O+CO+Ar=CO_2+Ar$	6.2 E+14	0.0	1,500
33	$CN+O_2=NCO+O$	1.05 E+13	0.0	0
34	$N+CO_2=NO+CO$	1.9 E+11	0.0	1,710
35	$NCN+NCO=CN+N_2+CO$	1.0 E+13	0.0	0
36	$NCO+NO=N_2O+CO$	6.42 E+16	-1.34	360
37	$NCO+NO=N_2+CO_2$	5.80 E+18	-1.97	560
38	$NO_2+Ar=NO+O+Ar$	6.8 E+14	0.0	26,400
39	$N+O_2=NO+O$	6.4 E+09	1.0	3,170
40	$C+NCN=CN+CN$	1.0 E+13	0.0	0
41	$C+O_2=CO+O$	2.0 E+13	0.0	0
42	$C+NCO=CN+CO$	1.0 E+14	0.0	0
43	$C+CO=C_2+O$	1.0 E+12	0.5	58,000
44	$C+CN=C_2+N$	2.5 E+11	0.5	19,300
45	$CO+Ar=O+C+Ar$	8.8 E+29	-3.5	128,840
46	$CN+Ar=C+N+Ar$	2.5 E+14	0.0	70,960
47	$C_2O+Ar=C+CO+Ar$	1.0 E+16	-0.5	26,700
48	$N+C_2O=CN+CO$	2.0 E+13	0.0	0
49	$O+C_2O=CO+CO$	2.0 E+13	0.0	0

*The rate constants are given by $k = AT^B \exp(-E/RT)$ in units of cm^3, mole and second

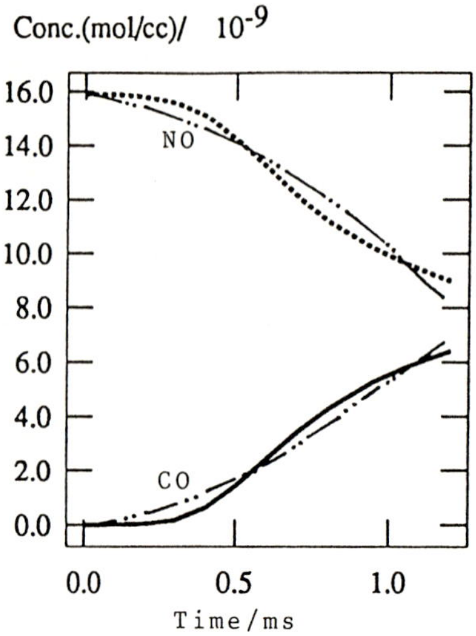

Fig. 1. Measured and calculated concentration profiles of CO and NO for the 1:2/BrCN:NO mixture at 2636 K, 0.354 atm. The calculated values are given by the dash-dotted curves

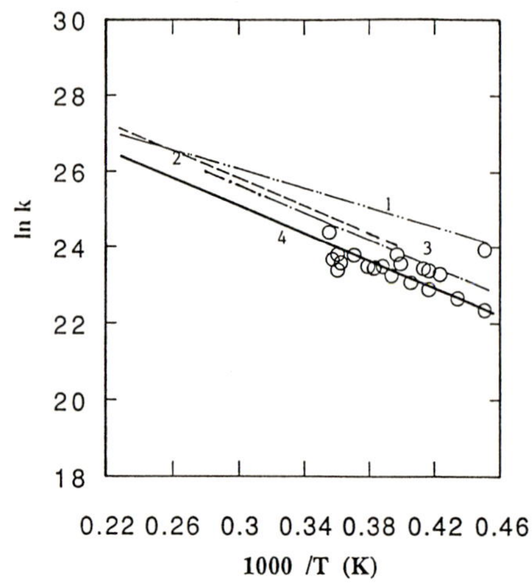

Fig. 2. Arrhenius plots of k_1 obtained from various studies. Points are modeled values from this study. Line (1) - TST/BAC-MP4 result (He et al. 1993); line (2) - Colket (1984), line (3) - Natarajan and Roth (1986); line (4) - He et al. (1993) obtained by the remodeling of Colket and Roth's data with the present mechanism

The kinetically modeled values of k_1 are presented in Fig. 2 and compared with the published data by Colket and by Natarajan and Roth. Also included in the figure are the result of the transition state calculation using the structure, activation energy and vibrational frequencies of the $^3NCON^\ddagger$ computed by the BAC-MP4 method. Both the theoretical result (line 1) and the results of Colket (line 2) and Roth (line 3), modeled with their simpler mechanisms, are higher than the present values given by the open circles. On the other hand, the remodeling of both Colket and Roth's data with the present mechanism gave rise to line (4) (He et al. 1993), which is consistent with our present result within the scatter:

$$k_2 = 5.5 \times 10^{12} \exp(-15,410/T)\, \mathrm{cm^3/mole.s}.$$

The modeled activation energy, 128 kJ/mole, is consistent with the calculated value, 121 kJ/mole, in the same temperature range using the transition-state theory with the BAC-MP4 energy barrier of 78 kJ/mole at 0 K.

The result of the present modeling, based in large part on the fundamental mechanism of the CN + NO reaction provided by the quantum chemical calculation suggests that the reaction occurs by the three channels as given above with the relative importance, $k_1 > k_3 > k_2$. Contrary to the earlier studies of Colket (1984) and Natarajan and Roth (1986), which indicated that $k_1 \approx k_2$ with $E_1 = 176$ and $E_2 \approx 0$ kJ/mole, the present study rules out the importance of reaction (2) because of its tight transition state and high activation energy, $E_2 = 116$ kJ/mole.

On account of the presence of N, O, NCO and NCN species and the high temperature involved, the overall reaction mechanism becomes quite complex as shown in Table 1. Despite the complexity, the dominant processes controlling the formation of CO and the removal of NO are:

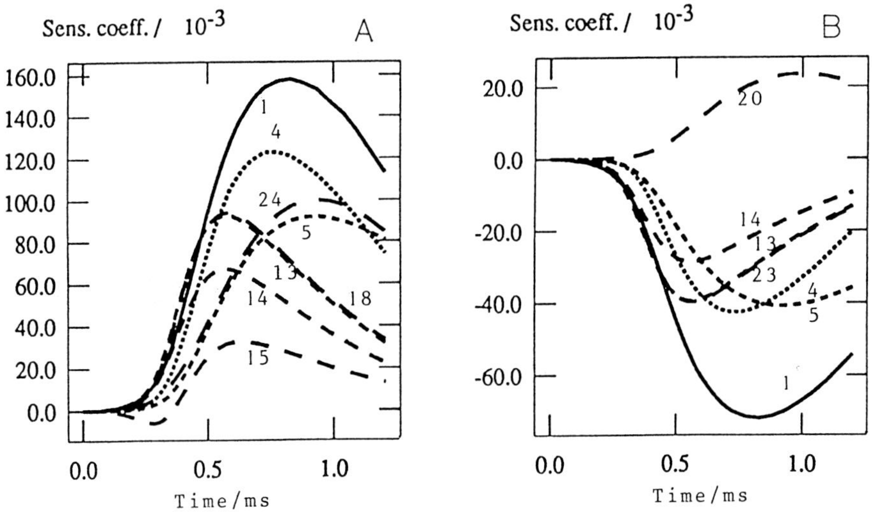

Fig. 3. Sensitivity coefficients for the key processes affecting the concentrations of CO (A) and NO (B)

$$CN + NO = NCO + N$$
$$O + CN = N + CO$$
$$N + NO = O + N_2$$
$$NCO + M = N + CO + M$$

with some minor contributions from the reactions involving BrCN. This conclusion is clearly illustrated by the results of a sensitivity analysis in Fig.3 using the SENKIN program (Lutz et al. 1988).

4. Conclusion

We have studied the reaction of CN with NO behind incident shock waves using the CO laser resonance absorption method. The measured concentration profiles of CO and NO could be kinetically modeled using a detailed reaction mechanism constructed on the basis of the results of BAC-MP4 quantum calculations. The quantum result suggests that the CN+NO reaction takes place primarily by the three lower energy paths producing NCO+N (1), CO+ N_2 (2) and NCN+O (3), with (1) being the dominant channel and (2) the least important one.

Acknowledgment

The authors gratefully acknowledge the support of this work from the Office of Naval Research and the Office of Naval Technology (to CHW).

References

Colket MB (1984) Cyanogen pyrolysis and CN+NO reaction behind incident shock waves. Intl. J. Chem. Kinet. 16:353-369

Ermolin NE, Korobeinichev OP, Kuibida LV, Fomin VM (1985) Study of the kinetics and mechanism of chemical reactions in hexogen flames. Fiz. Goren. i Vzry. 22:54-64

He Y, Lin MC, Melius CF (1993) Kinetic modeling of the CN+NO reaction aided by BAC-MP4 calculations (unpublished)

Hsu DSY, Lin MC (1984) Infrared laser probing of combustion and energetic molecule decomposition reactions in shock waves. SPIE 482:79-88

Hsu DSY, Shaub WM, Blackburn M, Lin MC (1982) Thermal decomposition of formic acid at high temperatures in shock waves. 19th Symp. (Intl.) on Combust. The Combustion Institute, pp. 89-94

Kee RJ, Rupley FM, Miller JA (1989) CHEMKIN-II: A Fortran chemical kinetic package for the analysis of gas-phase chemical kinetics. SNL Report SAND 89-8009

Lin CY, Wang HT, Lin MC, Melius CF (1990) A shock tube study of the CH_2O+NO_2 reaction at high temperatures. Intl. J. Chem. Kinet. 22:455-82

Lutz AE, Kee RJ, Miller JA (1988). SENKIN: A Fortran program for predicting homogenous gas phase chemical kinetics with sensitivity analysis. SNL Report SAND 87-8248

Melius CF (1988) The gas-phase flame chemistry of nitramine combustion. Proc. 25th JANNAF Combustion Meeting. The Combustion Institute, vol. II, pp. 155-62

Melius CF, Binkley JS (1986) Thermochemistry of the decomposition of nitramines in the gas phase. 21st Symp. (Intl.) on Combust. The Combustion Institute pp. 1953-63

Natarajan K, Roth P (1986) A shock tube study of CN radical reactions with H_2 and NO verified by H, N and O atom measurements. 21st Symp. (Intl.) on Combust., The Combustion Institute, pp. 729-37

Wang NS, Yang DL, Lin MC (1989) Kinetics of CN radical reactions with NO_x between 297 and 740 K. Chem. Phys. Lett. 163:480-4

A Shock Tube Study of the Reaction CH+NO⇌Products Using a Perturbation Method

Michael W. Markus and Paul Roth
Institut für Verbrennung und Gasdynamik, Universität Duisburg, D-47048 Duisburg, Germany

Abstract. A ring dye laser spectrometer was employed for in-situ measurements of CH concentrations in the reaction zone behind reflected shock waves. The time dependent absorption in the Q-branch of the $A\,^2\Delta - X\,^2\Pi$ band of CH at 431.1311 nm caused by the formation and consumption of CH radicals during the shock induced pyrolysis of a few ppm ethane in argon was recorded. The CH concentration could directly be calculated from the measured absorption by using spectroscopic data. The addition of some amount of NO to the initial hydrocarbon/argon mixtures leads to a significant reduction of the CH concentration during the reaction time mostly due to the direct reaction of CH radicals with the added molecules. The perturbed and unperturbed profiles were compared with those calculated by computer simulations based on a kinetic model. By a fitting procedure a rate coefficient for the most important perturbation reaction:

$$\mathrm{CH} + \mathrm{NO} \overset{k_1}{\rightleftharpoons} \text{products}, \qquad k_1 = 8.10 \times 10^{13}\mathrm{cm}^3\,\mathrm{mol}^{-1}\,\mathrm{s}^{-1}$$

could be determined. The given value which is valid in the temperature range between 2500 K and 3700 K at pressures about 1 bar confirms the few high temperature data available in literature.

Key words: Hydrocarbon, Nitrogen oxyde, Reaction products.

1. Introduction

The reactions of CH radicals with molecules like N_2, NO, NO_2 are believed to play a major role in the complex formation mechanism of various N-containing air pollutants formed in technical combustion processes. Unfortunately, the knowledge of the kinetic parameter of these reactions at high temperatures is still unsatisfactory. In the present study the exothermic reaction:

$$\mathrm{CH} + \mathrm{NO} \overset{k_1}{\rightleftharpoons} \text{products} \qquad (R1)$$

was studied behind reflected shock waves at temperatures between 2500 K and 3700 K by measuring CH concentations with a ring dye laser absorption diagnostic. Lichtin et al. (1984) determined the reaction channel:

$$\mathrm{CH} + \mathrm{NO} \rightleftharpoons \mathrm{NH}(A^3\Pi_i) + \mathrm{CO}; \qquad \Delta H_{298} = -15\ \mathrm{kcal/mol} \qquad (R1a)$$

as most important at room temperature by measuring the NH chemiluminescence. The rate coefficient of $k_{1a} = 1.5 \times 10^{14}\ \mathrm{cm}^3\,\mathrm{mol}^{-1}\,\mathrm{s}^{-1}$ given for this channel is comparable with earlier room-temperature measurements of the overall rate coefficient of ($R1$) (Butler et al. 1981, Wargal et al. 1982). A quite similar temperature independent rate coefficient of $k_1 = 1.1 \times 10^{14}\ \mathrm{cm}^3\,\mathrm{mol}^{-1}\,\mathrm{s}^{-1}$ was found by Berman et al. (1982) at temperatures between 300 K and 675 K.

For the high temperature region only two studies were found in the literature. Caselton (1990) measured the removal of photolytically generated CH by NO by measuring the CH laser-induced fluorescence in the $A - X$ band and determined a rate coefficient of $k_1 = 9.0 \times 10^{13}\ \mathrm{cm}^3\,\mathrm{mol}^{-1}\,\mathrm{s}^{-1}$ in the temperature interval between 300 K and 1500 K. Dean et al. (1991) reported a value of $k_1 = 1.0 \times 10^{14}\ \mathrm{cm}^3\,\mathrm{mol}^{-1}\,\mathrm{s}^{-1}$ for temperatures between 2570 K and 3790 K using a CH ring dye laser absorption diagnostic. For the conditions discussed in their study the reaction channel:

Shock Waves @ Marseille II
Editors: R. Brun, L. Z. Dumitrescu

$$CH + NO \rightleftharpoons HCN + O; \qquad \Delta H_{298} = -72 \text{ kcal/mol} \qquad (R1b)$$

was found to be the most probable one taking into account several measured product species. As a result of all these studies reaction ($R1$) seems to have a temperature-independent rate coefficient of about $k_1 = 1.0 \times 10^{14}$ cm^3 mol^{-1} s^{-1} in the temperature range 300 K $\leq T \leq$ 3800 K. Although the experiments presented here were measured in nearly the same temperature range by using a comparable experimental setup as Dean et al. (1991) we decided to present our data because of two reasons: to verify the only available high-temperature kinetic data of reaction ($R1$) and to study the reasons for a surprisingly high difference in the CH yield obtained from very similar experiments performed in Stanford and in our laboratory, see Dean and Hanson (1992) and Markus et al. (1992). Therefore the present results can improve both the knowledge of reaction ($R1$) at temperatures $T \geq 2500$ K and can confirm the CH ring dye laser diagnostic including ethane pyrolysis as the CH source.

2. Experimental

The experiments were performed in a conventional UHV-shock tube made of stainless steel. A detailed description of this apparatus was given by Markus et al. (1992). Gas mixtures of a few ppm ethane plus some hundred ppm NO diluted in argon were prepared in a stainless steel vessel. The gases were supplied by Messer-Griesheim, Germany. They were certified to be of the following purities: Ar= 99.9999%, C_2H_6 = 99.95%, NO= 99.5%. Except of NO which was cleaned using an adsorption trap filled with activated carbon, the gases were used without further purification. The initial mixtures of 20 ppm C_2H_6 plus several hundred ppm of the perturbing species diluted in argon were shock-heated to temperatures between 2500 K and 3700 K at pressures of about 1 bar.

A narrow-bandwidth ring dye laser absorption diagnostic was used to detected small CH concentrations with a detection limit of about 10^{12} molecules/cm^3 and a time resolution of about 2 μs. For details of the diagnostics and the ring laser setup see Dean and Hanson (1989) and Markus and Roth (1991). By applying a spectroscopic model the absorption data traces obtained could directly be calculated in concentration-over-time profiles using Beer's law.

3. Results and discussion

- The CH source

From previous studies C_2H_6 is known as a good precursor for CH. In Fig.1 two typical CH concentration profiles obtained from the pyrolysis of 20 ppm ethane are shown for different experimental conditions. Obviously, the time history of the data traces is strongly temperature dependent. Higher temperatures lead to higher peak concentrations at shorter reaction times and vice versa. Such CH concentration profiles were obtained both by Dean and Hanson (1992) as well as by Markus et al. (1992). They could be described well by computer simulations based on nearly the same reaction mechanism. In both studies the reaction:

$$CH_3 + M \overset{k_2}{\rightleftharpoons} CH + H_2 + M \qquad (R2)$$

was discussed to be most sensitive to the CH formation at early reaction times. Surprisingly, both groups found significantly different rate coefficients for this reaction in nearly the same temperature range:

$$k_2 = 5.0 \times 10^{15}\, exp(-42800\ K/T)\ \text{cm}^3\,\text{mol}^{-1}\,\text{s}^{-1} \qquad \text{(Dean and Hanson)}$$

$$k_{2,app.} = 6.9 \times 10^{14}\, exp(-41500\ K/T)\ \text{cm}^3\,\text{mol}^{-1}\,\text{s}^{-1} \qquad \text{(Markus et al.)}$$

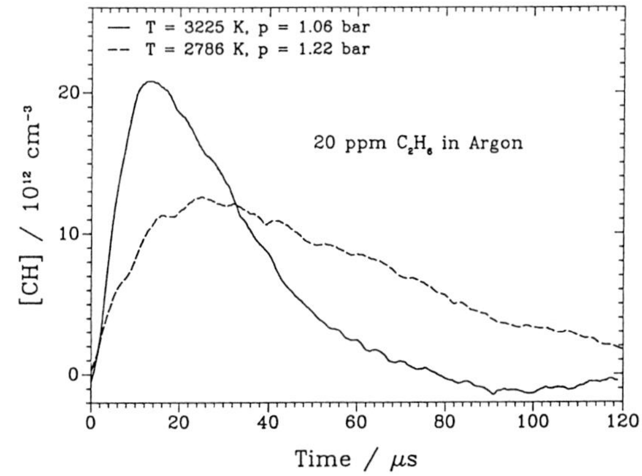

Fig. 1. Typical CH concentration profiles measured during the pyrolysis of 20 ppm ethane/argon

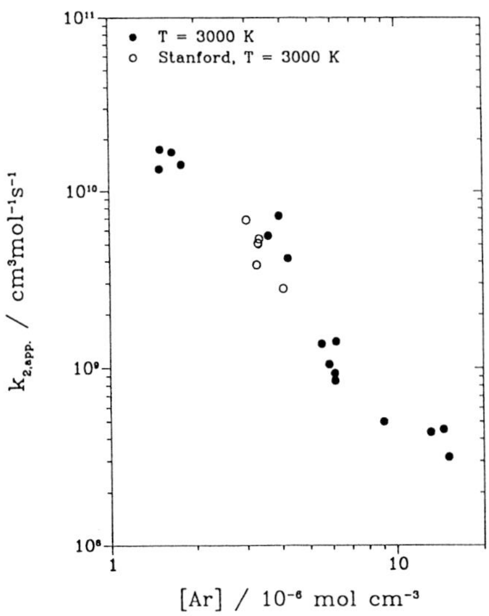

Fig. 2. Pressure dependence of the CH formation rate coefficient in ethane pyrolysis

as a consequence of different observed CH formation rates:

$$(d[\mathrm{CH}]/dt)_{t\approx 0} = k_{2,app.}\ 2[\mathrm{C_2H_6}]_0[\mathrm{Ar}]$$

at similar temperatures and only slightly different pressures. This effect could not be explained by uncertainties in the spectroscopic calibration, especially in the model for pressure broadening.

Because no methodic or experimental reasons were found for the discrepancy in the rate coefficient, we started to study the influence of pressure on the CH formation during C_2H_6 pyrolysis systematically. The CH formation rate coefficient $k_{2,app.}$ was determined from experiments performed at nearly the same temperature but at different post-shock pressures. The results obtained are summarized in Fig.2 as a function of the concentration of the collision partner argon. A strong pressure dependence is obvious indicating an interesting "inverse" trend compared to the typical fall-off behaviour of unimolecular thermal decomposition reactions.

The open symbols represent the CH formation rate coefficients determined from the few raw data traces which we got from Dean (1988). In this figure both experimental results seem to fit quite well to each other. Unfortunately, the pressure-dependent CH formation from high-temperature ethane pyrolysis cannot be represented by the simplified reaction mechanisms used by both groups. Therefore we fitted the measured CH concentration profiles during the high-temperature pyrolysis of C_2H_6 by introducing an individual pressure-dependent rate coefficient for the reaction ($R2$).

- CH+NO perturbation

The perturbation experiments were performed in mixtures containing 20 ppm C_2H_6 and 150, 300, 305, or 500 ppm NO diluted in Ar at temperatures between 2500 K and 3700 K and at pressures around 1 bar. Fig.3 shows the influence of NO on the ethane pyrolysis system.

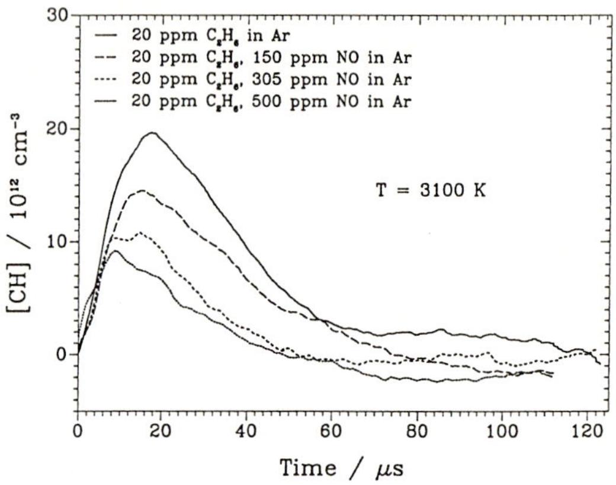

Fig. 3. CH concentration profiles measured in 20 ppm ethane/argon plus various amounts of NO

With increasing amount of NO the CH peak concentration decreases while the typical shape of the profiles is nearly conserved. This behaviour was found in all NO experiments.

To analyse the experimental findings we started from a reaction mechanism containing the simplified C_2H_6 pyrolysis system enlarged by additional perturbation reactions of NO with CH or its precursors CH_3 and CH_2. A sensitivity analysis of the reaction system using the SENKIN program code (Lutz et al. 1988) revealed that only the direct reaction of CH with NO, i.e. ($R1$), has a significant influence on the calculated CH concentration profiles under the given experimental conditions. The influence of the other possible perturbation reactions was very weak allowing to reduce the perturbation mechanism to only reaction ($R1$). As a further result, the practice of fitting the very early part of the CH concentration profiles by varying k_2 was verified because of the different time constants of these two reactions. Therefore best fit values for ($R1$) also include an adjustment of the rate coefficient k_2.

In Fig.4 some examples of measured and calculated CH concentration time-histories are shown for various reaction conditions. It was easily possible to fit the observed profiles over the whole temperature and NO concentration range. The individually determined best fit values for k_1 are summarized in Fig.5. The mean value of:

$$k_1 = 8.10 \times 10^{13}\ \mathrm{cm^3\,mol^{-1}\,s^{-1}}$$

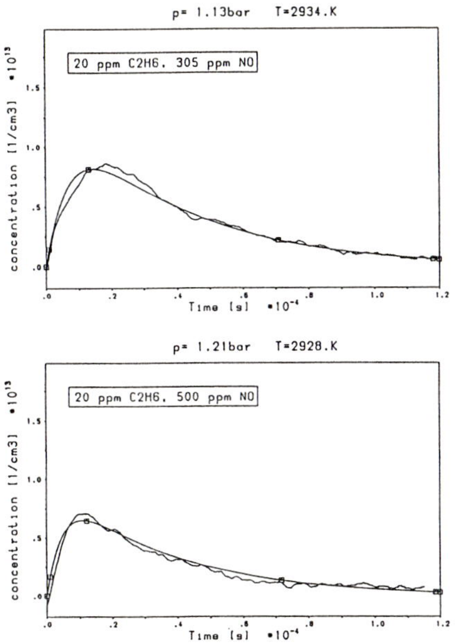

Fig. 4. Examples of calculated CH concentration profiles in comparison to measured ones for different initial reaction conditions

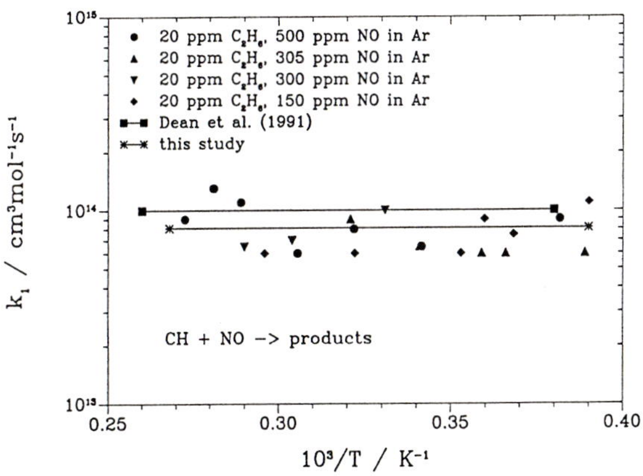

Fig. 5. Arrhenius plot of the individual best fit values determined for k_1 in comparison with literature values

is in excellent agreement with those reported by Dean et al. (1991) and Casleton (1990). Although the CH formation process in the C_2H_6 pyrolysis is not completely understood, bimolecular CH reactions can be analysed in perturbation studies starting from the ethane pyrolysis system. Other CH reactions with CO, CO_2, and O_2 were recently studied in our laboratory and are published elsewhere (Markus and Roth 1993).

Acknowledgements

The authors thank Mr. L. Jerig for his help in conducting the experiments. The financial support of the Deutsche Forschungsgemeinschaft is gratefully acknowledged.

References

Berman MR, Fleming JW, Harvey AB, Lin MC (1982) Temperature dependence of CH radical reactions with O_2, NO, CO and CO_2. 19th Symp. (Intl.) on Combustion, p. 73, The Combustion Institute, Pittsburgh, PA

Butler JE, Fleming JW, Goss LP, Lin MC (1981) Kinetics of CH radical reactions with selected molecules at room temperature. Chem. Phys. 56:355

Casleton KH (1990) unpublished data, in Dean AJ (1990) Thesis, Stanford University, Stanford, CA, USA

Dean AJ (1988) private communication

Dean AJ, Hanson RK (1989) Development of a laser absorption diagnosic for shock tube studies of CH. J. Quant. Spectrosc. Radiat. Transfer 42:5:375

Dean AJ (1990) A shock tube study of NO_X reactions involving CH and C-atoms using laser and atomic resonance absorption diagnostics. Thesis, Stanford University, Stanford, CA, USA

Dean AJ, Hanson RK, Bowman CT (1991) A shock tube study of reactions of C atoms and CH with NO including product channel measurements. J. Phys. Chem. 95:3180

Dean AJ, Hanson RK (1992) CH and C-atom time histories in dilute hydrocarbon pyrolysis: measurements and kinetics calculations. Intl. J. Chem. Kinetics 24:517

Lichtin DA, Berman MR, Lin MC (1984) NH($A\,^3\Pi \rightarrow X\,^3\Sigma^-$) Chemiluminescence from the CH($X\,^2\Pi$)+NO reaction. Chem. Phys. Lett. 108:18

Lutz AE, Kee RJ, Miller JA (1988) SENKIN: A FORTRAN program for predicting homogeneous gas phase chemical kinetics with sensitivity analysis. Sandia National Laboratories Report SAND87-8248

Markus MW, Roth P (1991) Ring dye laser-based absorption measurements of CH radicals behind shock waves. In: Takayama K (ed) Proc. 18th Intl. Symp. Shock Waves, p. 1011

Markus MW, Woiki D, Roth P (1992) Two-channel thermal decomposition of CH_3. 24th Symp. (Intl.) on Combustion, p. 581, The Combustion Institute, Pittsburgh, PA

Markus MW, Roth P (1993) Experiments on some high temperature CH reactions. Proc. Anglo-German Combustion Symp. British Section of the Combustion Institute, Cambridge, GB, p. 204

Wagal SS, Carrington T, Filseth SV, Sadowski C M (1982) Absolute rate constants for the reaction of CH($X\,^2\Pi$) with NO, N_2O, NO_2, and N_2 at room temperature. Chem. Phys. 69:61

Kinetics of the Decay of CH_3 Radicals in Shock Waves

Joseph Zhichun Su and Heshel Teitelbaum
Dept. of Chemistry, University of Ottawa, Ottawa, Ontario, Canada K1N 6N5

Abstract. Mixtures containing up to 13% CH_3 radicals diluted in Ar and N_2 were produced by the near-instantaneous decomposition of azomethane behind incident shock waves at 1360 to 4070 K and at concentrations of 1.0 to 1.5 × 10^{-6} mol cm^{-3}. The laser-Schlieren technique was used to monitor the rate of heat evolution and to determine precisely the rate coefficients for $2CH_3 \rightarrow C_2H_6$ and for $2CH_3 \rightarrow C_2H_4 + H_2$.

Key words: Kinetics, Methyl, Recombination, High-temperature, Laser Schlieren

1. Introduction

CH_3 radicals are prevalent during the combustion of most hydrocarbons. Despite the fact that elementary reactions of CH_3 have been studied for more than 40 years, there is a surprising amount of uncertainty and even controversy regarding the magnitude of the rate constants involved, especially at high temperatures (Warnatz 1984). Various approaches have been tried. One of the most successful methods is to shock heat a precursor, such as CH_4 highly diluted in Ar, and to monitor the product formation as a function of time by a sensitive technique such as ARAS (Just 1981, Frank et al. 1988). Another successful technique, which is especially useful for reactions occuring on the submicrosecond time scale, is the laser Schlieren method. It has been applied to the pyrolysis of C_2H_6 (Kiefer and Budach 1984) and to CH_4 (Kiefer and Kumaran 1993). However, these techniques suffer from complications due to the secondary nature of the reactions of CH_3. The present study applies the laser Schlieren technique to mixtures containing pure CH_3 diluted in Ar and N_2. Consequently, we are able to observe the primary reactions of CH_3. They leave a distinctive signature characteristic of exothermic reactions, permitting unambiguous identification of the reactions.

2. Procedure

The laser-Schlieren method has been described previously by its originators (Kiefer and Lutz 1966) and by us (Dove and Teitelbaum 1974). Essentially, a narrow 5 mW He-Ne laser beam is deflected from its path perpendicular to the flow of the shock wave, and thus off the centre of a split Si-photodiode by the time-dependent refractive index gradients generated by the chemical reaction. Electronic signals amplified by the detector are stored by a transient waveform recorder, and subsequently transfered onto disks for computer processing. We used two different shock tubes, one of which has been described previously (Dove and Teitelbaum 1974). In that apparatus shock velocities are measured with a 6-channel, quartz-crystal controlled time-interval meter working on the signals generated by 4 gold resistance gauges spanning the laser Schlieren station, and whose positions are accurately known to within 0.1 mm . The second (new) shock tube's test-section was constructed of stainless steel plates welded together under ultra-high vacuum with an electron-beam weld. This created a 9.3 cm×8.4 cm perfectly rectangular inner cross-section tube into which perfectly parallel pairs of observation windows could be attached absolutely flush with the inner walls of the shock tube. Here the velocity of the shock wave was determined to within 0.1% by measuring the time intervals between 4 successive electronic pulses generated by the passage of the shock wave across 4 additional laser-Schlieren stations spanning the main observation port. These signals were sequentially fed into and could be subsequently read directly from a digital oscilloscope. Both shock tubes gave rate measurements which were indistinguishable from each other. Shock waves were generated by pressure-bursting aluminum diaphragms with helium. The test gas consisted of mixtures of 0.6% to 7.7% azomethane diluted in Ar. Azomethane was prepared for us by Cambridge Isotope Laboratories. Mass spectrometric analysis indicated 99% purity, with the main impurity being water, which was removed by trap-to-trap distillation. Pyrolysis of azomethane above 1400 K behind the shock wave resulted in quantitative amounts of CH_3 and N_2 within 10^{-7} s or less (Glänzer et al. 1976), thus creating instant mixtures containing as much as 13% methyl radicals in an inert gas.We covered the range 1360 - 4070 K. The chemical reactions initiated by the methyl radicals were monitored 6.13 m downstream of the diaphragm.

3. Observations

Fig.1 shows an example of a Schlieren signal for a shock wave in 6.3% azomethane. The temperature and pressure behind the incident shock were 1655 K and 166 Torr. After passage of the shock front (indicated by the negative and positive spikes), there follows a prominent negative signal which decays on the microsecond time scale. This is the signature for a strongly exothermic reaction. In experiments at higher temperatures, up to 2400 K or more, we observed the same shape of signal with the decay time still on the microsecond time scale. However, at higher pressures (see Fig.2) the signal becomes progressively more positive. The pressure threshold for positive signals decreases at higher temperatures, until at 3000 K the signal is dominated by the positive component (see Fig.3) even at relatively low pressures. As we diluted the azomethane more and more at intermediate temperatures we observed that not only did the signal decrease in magnitude, as expected, but the positive component shifted to longer and longer times, eventually separating completely from the negative component (see Fig.4). This indicates quite clearly that the positive component is caused by secondary reactions rather than by a primary dissociation of CH_3,

$$CH_3 + Ar \longrightarrow CH_2 + H + Ar \quad \Delta H^0_{298} = 109.63\,\text{kcal.mol}^{-1} \tag{1}$$

Indeed, recent calculations by us (Su and Teitelbaum 1994) and the numerical simulations described below show that the rate of reaction (1) is much too small to contribute significantly to the observed signals even at temperatures as high as 4000 K. Rather, it is the following set of bimolecular reactions which are responsible for the exothermic component,

$$2CH_3 \longrightarrow \begin{cases} C_2H_6 & \Delta H^0_{298} = -89.68\,\text{kcal.mol}^{-1} \quad (2) \\ C_2H_5 + H & \Delta H^0_{298} = +11.19\,\text{kcal.mol}^{-1} \quad (3) \\ C_2H_4 + H_2 & \Delta H^0_{298} = -57.10\,\text{kcal.mol}^{-1} \quad (4) \\ CH_4 + CH_2 & \Delta H^0_{298} = +\ 4.81\,\text{kcal.mol}^{-1} \quad (5) \end{cases}$$

and the subsequent dissociation reactions which are responsible for the endothermic signal:

$$C_2H_4 \longrightarrow C_2H_3 + H \qquad \Delta H^0_{298} = 107.96\,\text{kcal.mol}^{-1} \tag{6}$$

$$C_2H_5 \longrightarrow C_2H_4 + H \qquad \Delta H^0_{298} = 35.91\,\text{kcal.mol}^{-1} \tag{7}$$

4. Data reduction

The power of the laser-Schlieren technique lies in its ability to probe the primary reaction of a complicated mechanism (Dove et al. 1975). Consequently, our first step in analyzing the negative signals is to ignore the endothermic steps (6) and (7). As has been shown by Dove et al. 1975), and as will be shown below, the observed signal in a laser Schlieren experiment is proportional to the sum of the elementary rate coefficients, k_i, weighted by their heats of reaction. By virtue

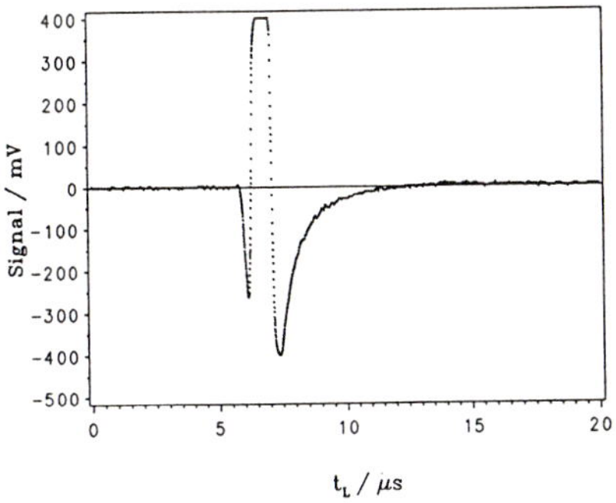

Fig. 1. Schlieren signal for 6.3% azomethane, $T_2 = 1655$ K, $P_2 = 166$ torr

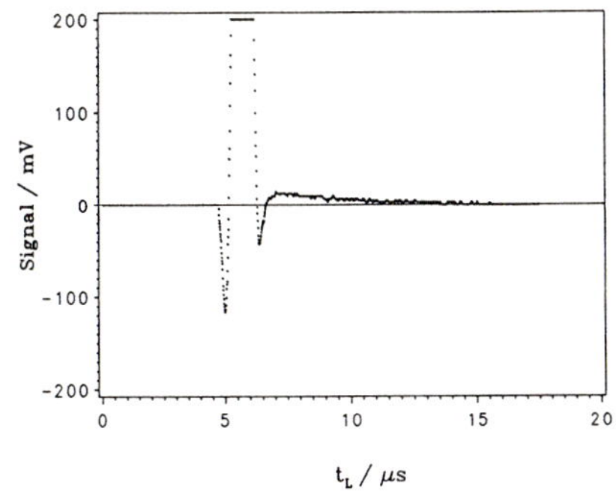

Fig. 2. Schlieren signal for 7.7% azomethane, $T_2 = 2025$ K, $P_2 = 451$ torr

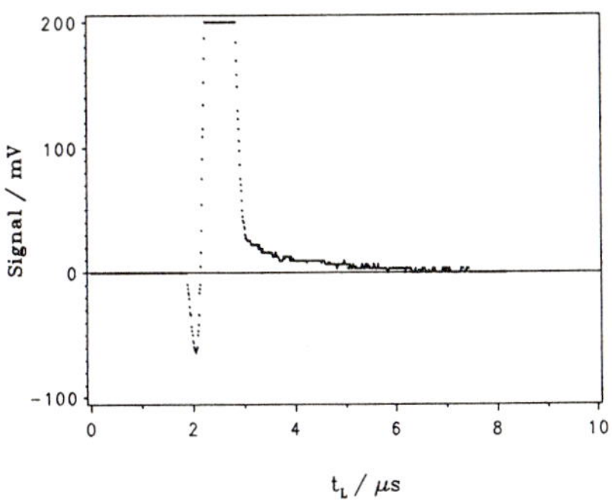

Fig. 3. Schlieren signal for 6.5% azomethane, $T_2 = 3099$ K, $P_2 = 312$ torr

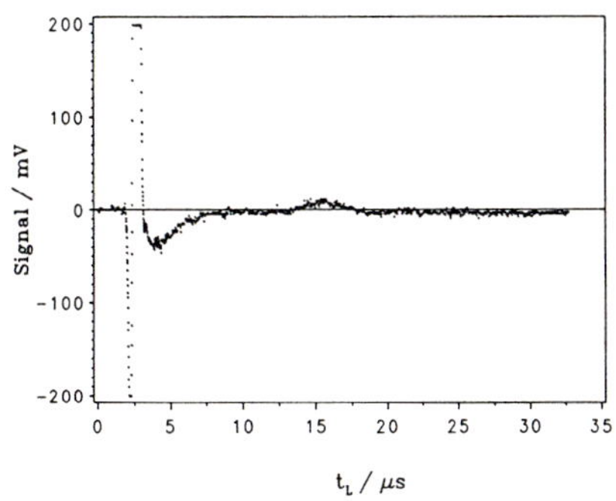

Fig. 4. Schlieren signal for 1.9% azomethane, $T_2 = 2051$ K, $P_2 = 139$ torr

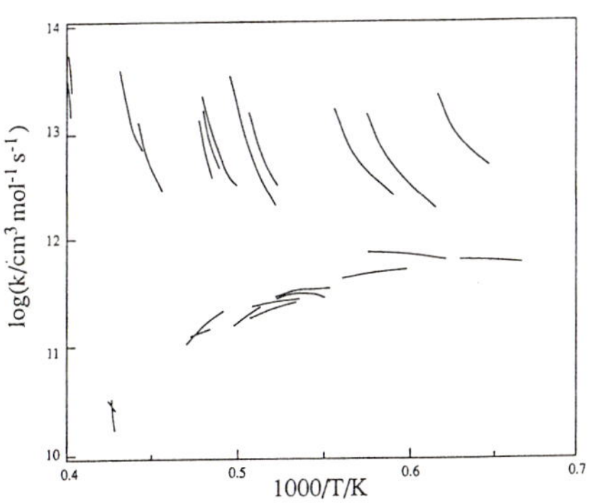

Fig. 5. Rate coefficients for the disappearance of CH_3 as determined by a point-by-point analysis. Upper curves assume products are $C_2H_4+H_2$. Lower curve assume products are C_2H_6.

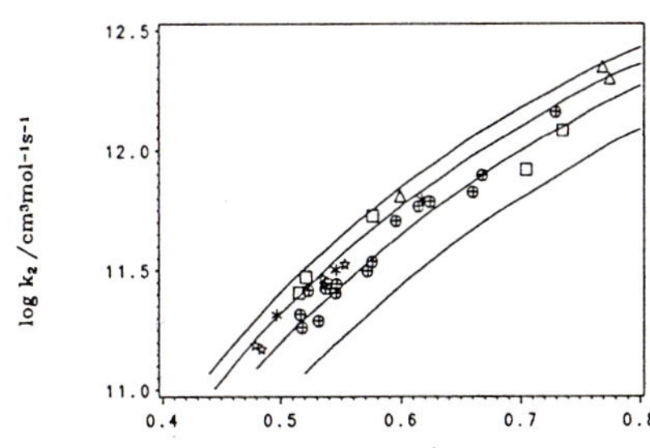

Fig. 6. Rate coefficients for C_2H_6 production from fig.5. □ 1.75%, 1.90%; ⋆ 3.2%; ∗ 4.25%; ⊕ 6.3%; △7.0%; [M]= 1.0 – 1.5 x 10^{-6}mol.cm^{-3}. Fall-off curves are calculated from Hippler et al.(1984), Wagner et al. (1988) and Baulch et al. 1992), where [M]= 0.5, 1.0, 1.5 and 2.0 x 10^{-6} mol. cm^{-3} from bottom to top

of this property a thermoneutral reaction will not contribute to the overall signal. Such is very nearly the case for reaction (5) relative to the other reactions. We can therefore analyze the exothermic signal very simply and analytically point-by-point if we consider merely reactions (2-4). The rate coefficients, k_3 and k_4, for reactions (3) and (4) are not very well known. For example, the results of Frank and Braun-Unkhoff (1988) for k_3 differ from those of Kiefer and Budach (1984) by a factor of 30, while the results of Kern et al. (1988) for k_4 differ from those of Hidaka et al. (1990) by more than a factor of 200. Even if we take the most conservative choice, following the recommendations of Warnatz (1984), we still conclude that k_4 is larger than k_3 at all temperatures. In addition, the rate of reaction (3) is weighted by a much smaller heat of reaction. Thus our simple analysis reduces to considering only reactions (2) and (4).

k_2 decreases with increasing temperature, while k_4 increases with temperature. Somewhere between 1800 and 2200 K, C_2H_6 and C_2H_4 formation compete for domination. Note that it is exactly in this temperature range that k_2 is least well known. It is also the region where thermodynamics and kinetics battle for control of the fate of CH_3: In this temperature range the equilibrium of reaction (2) is far to the left, whereas that of reaction (4) is far to the right. Thus, thermodynamically, C_2H_4 formation is preferable to C_2H_6 formation. Yet, k_4 is smaller than k_2 below 2000 K. It may be that in this range C_2H_4 is produced by other processes, e.g.

$$CH_3 + C_2H_6 \longrightarrow CH_4 + C_2H_5 \qquad \Delta H^0_{298} = -3.95\,\text{kcal.mol}^{-1} \tag{8}$$

followed by the thermal decomposition of C_2H_5, i.e. reaction (7). However, reaction (8) did not actually affect our observations because it is nearly thermoneutral.

The first step in the analysis is to convert the observed signal to density gradient. The signal is proportional to the laser beam deflection, Δ, with the constant of proportionality deduced from a calibration using a rotating mirror. Δ in return is related to the density gradient, $d\rho/dy$, by

$$\Delta = Dl\frac{d(\kappa\rho)}{dy} \tag{9}$$

where l is the width of the shock tube, D is the distance between the shock tube axis and the detector, and κ is the Gladstone-Dale constant for the gas mixture (Dove and Teitelbaum 1974). The latter is a mass fraction average over the κ's of the individual component species. The latter can be estimated for the unstable species according to the usual bond additivity rules (Weast 1974). κ is dominated by the contribution of Ar. The contribution of the disappearing CH_3 diminishes with time but is compensated by the contribution of reaction products. Thus, κ is essentially constant. The relationship between the density gradient and the reaction rate is found from the solution of the hydrodynamic equations of flow coupled to the rate equations:

$$\rho u = \rho_0 u_0 \tag{10}$$

$$p + \rho u^2 = p_0 + \rho_0 u_0^2 \tag{11}$$

$$h + u^2/2 = h_0 + u_0^2/2 \tag{12}$$

$$p = \rho RT/\mu \tag{13}$$

where $h\mu = \sum_n [(H - H^0_{298})_n + \Delta H^0_{f,n}(298)]$. $X_n$is the mixture enthalpy, μ is the effective molecular weight, ρ is the gas density, u is the particle velocity, p is the pressure, T is the temperature, while X_n, $(H_0 - H^0_{298})_n$,and $\Delta H^0_{f,n}(298)$, are the mole fraction, the molar enthalpy, and the heat of formation of species n respectively. The latter two thermodynamic quantities are obtainable directly from the JANAF Thermochemical Tables (Stull and Prophet 1971).

At low enough temperatures, reaction (2) alone is responsible for the signal. Defining α as the degree of reaction at time t_g, ie $(1 - 2\alpha)$ = the amount of CH_3 remaining, compared to the initial amount of CH_3, then the rate of reaction can be cast into the form

$$d\alpha/dt_g = k_2[(1 - 2\alpha)^2 X_0\rho/\mu_0' - \alpha/K_c] \tag{14}$$

where X_0, the initial mole-fraction of CH_3 in the post-shock gas, $= 2X_A/(1+2X_A)$, with X_A being the mole fraction of azomethane in the preshock gas. μ_0' is the initial effective molecular weight of the preshock gas. $\mu = \mu_0'/(1 - X_0\alpha)$ is the effective molecular weight at time t_g. t_g is the real time that the gas has been processed at the high temperature, in distinction to the lab time, t_l, which corresponds to the distance to the shock front. $dy/dt_l = u_0 =$ the shock velocity, while $dt_g/dt_l = \rho/\rho_0$.We can treat ρ as a known quantity, since it can be determined directly from the integral of the Schlieren trace, using the first spike minimum as time zero. Hence we have 5 unknowns (u, p, T, α, and k_2) and 5 equations. We can solve these equations (10-14) in terms of the observable $d\rho/dt_l$ and/or ρ . All 5 depend on time, even k_2 through its dependence on T. In fact, we shall use this property as a diagnostic tool below. The solution set is

$$u = u_0\rho_0/\rho \tag{15}$$

$$p = p_0 + \rho_0 u_0^2(1 - \rho_0/\rho) \tag{16}$$

$$T = (\rho_0/\rho)[T_0 + \mu_0 u_0^2(1 - \rho_0/\rho)/R]/[1 + 2X_A(1-\alpha)] \tag{17}$$

$$\alpha = \{\mu_0 u_0^2[1-(\rho_0/\rho)^2]/2X_A - \Delta H_1 - (H_{\mathrm{Azo}} - H^0_{\mathrm{Azo}}) - (H_{Ar} - H^0_{Ar})(1-X_A)/X_A\}/2\Delta H_2 \tag{18}$$

$$\frac{d\rho}{dt_g} = \frac{2X_A\rho k_2[\frac{2X_A}{\mu_0}\rho(1-2\alpha)^2 - \alpha/K_c]\{\tilde{C}T + \Delta H_2[1+2X_A(1-\alpha)]\}}{\mu_0 u_0^2(\rho_0/\rho)^2\{[1+2X_A(1-\alpha)] - \tilde{C}/R\} + \tilde{C}T[1+2X_A(1-\alpha)]} \tag{19}$$

where ΔH_1 is the heat of dissociation of azomethane. Note that we have included the reverse dissociation reaction for cases close to equilibrium, and we have expressed the reverse rate coefficient in terms of the forward one through the equilibrium constant, K_c. $\tilde{C}$ is the effective heat capacity of the gas mixture at time t_g. Eq.19 shows that the observable is directly proportional to the product of k_2 and the effective heat of reaction, ΔH_2, modulated by a term reflecting the decreasing amount of reagents. At each point in time, one determines T and α by combining Eqs.17 and 18. k_2 can then be deduced by one of 3 methods: a) Invert Eq.19 directly. (This exploits the proportionality of the observable to k_2). Alternatively, b) one can focus on the time scale of the decay, ie on the slope of the observable rather than bank on absolute calibration factors or worry about accumulating numerical errors arising from estimates of T, α, and ρ, based on the graphical integration of the observable. Yet another choice is to c) solve Eq.14 for α in terms of the observable, $d\rho/dt = (d\rho/d\alpha)(d\alpha/dt)$, and use it instead of eq.18.

5. Results

We have tried all three methods, and have found that the second one would be the most reliable if we knew the activation energy in advance. The results in Fig.5 stem from the use of method a), where we have drawn an Arrhenius plot of k_2 for selected experiments. Note that each experiment gives $k_2(T(t))$, ie an Arrhenius plot of its own. We then repeated the analysis assuming that reaction (4) alone is responsible for the signal. The solution for k_4, in analogy to Eq.19 now takes the form

$$\frac{d\rho}{dt_g} = \frac{4X_A^2\rho^2 k_4[(1-2\alpha)^2 - \alpha^2/K_c]\Delta H_4(1+2X_A)/\mu_0}{\mu_0 u_0^2(\rho_0/\rho)^2\{(1+2X_A) - \tilde{C}/R\} + \tilde{C}T(1+2X_A)} \tag{20}$$

Again we drew the results as Arrhenius plots for selected experiments. For internal consistency the rate coefficients must overlap if the temperature ranges overlap, ie. each experiment's Arrhenius plot must be a continuation of another one. Clearly, k_2 shows the required internal consistency at temperatures less than 2300 K; whereas the assumption that k_4 is responsible for the signal leads to a contradiction at such low temperatures. On the other hand, above 2300 K the signal *is* sensitive to k_4.

Fig.6 shows the results for k_2 classified according to the original mixture composition. We saw no evident dependence on azomethane composition. This indicates that CH_3 is not too much more efficient than Ar as an energy transfer partner with excited C_2H_6. Also shown in Fig.6 are the theoretical fall-off curves from Hippler et al. (1984), Wagner et al. (1988) and Baulch et al. (1992), which are in very good agreement with the present results. The slight spread in pressures may also contribute to the apparent scatter of the results. The present results support those of Moller et al. (1986), which indicate a steeply declining rate coefficient as the temperature increases. The values of k_4 determined from the left branch of Fig.5 appear to support a nonlinear extrapolation of the results of Hidaka et al.(1990). However, in view of the complications arising from the participation of endothermic processes, as noted above, it seems wiser not to rely on the point-by-point analysis for k_4. Rather, we have embarked on a numerical simulation of our signals using a 44-step mechanism. The simulation was carried out using Sandia's CHEMKIN-II chemical kinetics integrator (Kee et al.1992). The details of the analysis are given elsewhere (Su and Teitelbaum (1994). It turns out that the signals are sensitive to the value of k_4 over virtually the entire range, 1700 - 4000 K. Fixing the values of k_2 according to Fig.6, and treating k_3 as known (Frank and Braun-Unkhoff 1988) (because of a much smaller sensitivity to it), we were able to extract values of k_4 for each experiment and apply a linear least-squares Arrhenius fit to them. We found

$$k_4(\text{cm}^3\ \text{mol}^{-1}\ \text{s}^{-1}) = (1.10 \pm 0.15) \times 10^{14} \exp[-(9026 \pm 285)/T]$$

References

Baulch DL, Cobos CJ, Cox RA, Esser C, Frank P, Just Th, Kerr JA, Pilling MJ, Troe J, Walker W, Warnatz J (1992) J. Phys. Chem. Ref. Data 21: 411
Dove JE, Teitelbaum H (1974) Chem. Phys. 6: 431
Dove JE, Nip WS, Teitelbaum H (1975) Proc. 15th Intl. Symp. on Combustion, The Combustion Institute, Pittsburgh, p. 903
Frank P, Braun-Unkhoff M (1988) in: Grönig H (ed.) Proc. 16th Intl. Symposium on Shock Tubes and Waves, VCH Verlag, Weinheim, p. 379
Glanzer K, Quack M, Troe J (1976) Chem. Phys. Lett. 39: 304
Hidaka Y, Nakamura T, Tanaka H, Inami K, Kawano H (1990) Intl. J. Chem. Kinet. 22: 701
Hippler H, Luther K, Ravishankara AR, Troe J(1984), Z. Phys. Chemie N. F. 142: 1
Just Th (1981) in: Treanor CE, Hall JG (eds.) 13th Intl. Symp. on Shock Tubes and Waves, SUNY Press, Albany,p.59
Kee RJ, Rupley FM, Miller JA (1992) Chemkin-II, Sandia Report SAND89-8009B-UC-706 ; also SAND82-8205-UC-401 (1992); and SAND87-8248-UC-401 (1991)
Kern RD, Singh HJ, Wu CH (1988) Intl. J. Chem. Kinet. 20: 731
Kiefer JH, Lutz RW (1966) J. Chem. Phys. 44: 658
Kiefer JH , Budach KA (1984) Intl. J. Chem. Kinet. 16: 679
Kiefer JH, Kumaran SS (1993) J. Phys. Chem. 97: 414
Moller W, Mozzhukhin E, Wagner HGg (1986) Ber. Bunsenges. Phys. Chem. 90: 854
Stull D, Prophet H (1971) JANAF Thermochemical Tables, 2nd. ed. NBS, Washington
Su JZ, Teitelbaum H (1994) Intl. J. Chem. Kinetics to be published
Su JZ, Teitelbaum H (1994) J. Phys. Chem. to be published
Wagner AF, Wardlaw DM (1988) J. Phys. Chem. 92: 2462
Warnatz J (1984) In: Gardiner Jr WC (ed) Combustion Chemistry. Springer, NY, p. 197
Weast RC (ed) (1974) Handbook of Chemistry and Physics, CRC Press, Cleveland, p.E222

Thermal Reactions of Oxiran, Methyloxiran and Dimethyloxiran. Single-Pulse Shock Tube Investigation

Assa Lifshitz
Department of Physical Chemistry, The Hebrew University, Jerusalem 91904, Israel

Abstract. The present investigation deals with the differences and similarities in the energetics and the kinetics of the thermal reactions of oxiran (epoxyethane), methyloxiran and 1,2-dimethyloxiran. Experimental results on the isomerization of methyloxiran and 1,2- dimethyloxiran obtained in a single pulse shock tube are presented and compared with data obtained at low temperature. The plethora of decomposition and isomerization products observed in each compound is shown. A reaction scheme for methyloxiran is constructed and a comparison between experimental and calculated mole percents of products are shown.

Key words: Oxirans, Decomposition, Isomerization

1. Introduction

Oxiran, methyloxiran and 1,2-dimethyloxiran are members of the epoxy family of molecules. It is an unstable three member ring structure containing oxygen which tends to break down when subjected to high temperatures. The opening of the ring is initiated by the rupture of the C-O bond in the molecule. The rearrangements that follow the bond rupture produce isomers which are considerably more stable than the epoxy molecule itself. In view of large differences in the heats of formation of the epoxy molecules and its isomers, the latter are produced with a large excess of thermal excitation. This excess energy is equal to $E_{act} + \Delta H_r$ where E_{act} is the activation energy of the C-O bond cleavage-rearrangement process and ΔH_r is the exothermicity of the isomerization reaction. The thermally excited products can either lose their energy by collision to produce the stable isomers or, alternatively, split up with the production of unstable intermediates that can initiate free radical reactions. Post-shock mixtures of heated epoxy molecules contain, therefore, both isomerization and decomposition products.

Ethylene oxide (C_2H_4O), the smallest molecule in the line of epoxides, isomerizes to acetaldehyde (Newfeld and Blades 1963; Lifshitz and Ben-Hamou 1983). Being a small symmetrical molecule, is has only one isomerization channel. There is a difference of $\sim$ 27 kcal/mol (Stull, Westrum and Sinke 1969) between the heats of formation of ethylene oxide and acetaldehyde. With an activation energy of $\sim$ 57 kcal/mol in the ethylene oxide $\rightarrow$ acetaldehyde isomerization and an exothermicity of some 27 kcal/mol an acetaldehyde with more than 84 kcal/mole of thermal energy above the ground state is generated in the process. This excess energy is enough for the molecule to split up into $CH_3^\bullet$ and $HCO^\bullet$ before losing its energy by collision. Therefore, in addition to acetaldehyde, compounds such as methane, ethane, carbon monoxide and others which result from consecutive reactions of these free radicals are also observed as products in the reaction (Lifshitz and Ben-Hamou 1983).

Similar to the thermochemistry of ethylene oxide decomposition, the thermally excited isomers which are produced in the reactions of methyloxiran, the second member in the series of epoxides, also possess enough energy to decompose to unstable fragments (Hardwick 1968; Blades 1968; Flowers 1977; Zhiwei Huixing and Degang 1990). Thus propanal can decompose to $C_2H_5^\bullet$ and $HCO^\bullet$, methyl-vinyl ether to $CH_2CHO^\bullet$ and $CH_3^\bullet$ etc. These fragments initiate free radical reactions which produce methane, ethane and other products.

The energy diagram of the decomposition of methyloxiran is shown in Fig.1. The diagram is based on an average activation energy of 59($\pm$2) kcal/mol which was found experimentally for production of acetone, propanal, methyl-vinyl ether and allyl alcohol (Flowers 1977). It also

Shock Waves @ Marseille II
Editors: R. Brun, L. Z. Dumitrescu

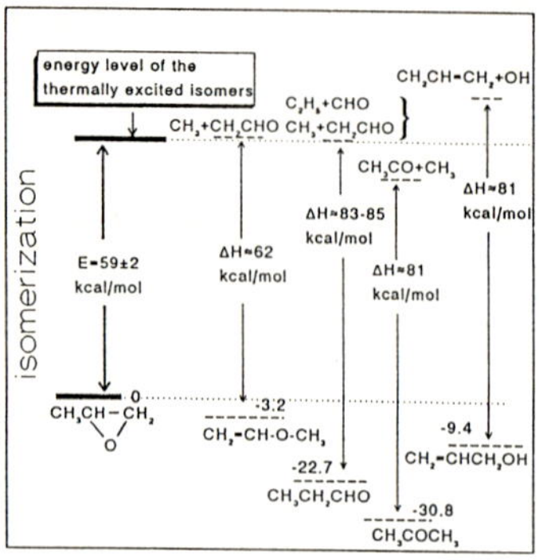

Fig. 1. Energy diagram of methyloxiran decomposition

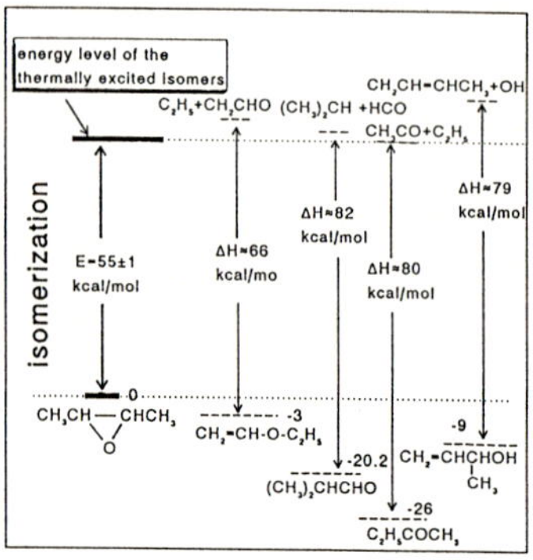

Fig. 2. Dimethyloxiran decomposition

shows the thermochemistry of the unstable intermediates that can be generated by the decomposition of these thermally exited species. As can be seen, the isomers are produced with sufficient thermal energy to allow decomposition. Propanal has two decomposition channels with almost identical thermochemistry. One additional channel in methyl-vinyl ether: CH_3-O-CH=CH_2 → $CH_3O^\bullet$ + CH_2=$CH^\bullet$ is ~ 94 kcal/mol endothermic and is thus inaccessible. Allyl alcohol has two decomposition channels: CH_2=$CHCH_2OH$ → CH_2=$CHCH_2^\bullet$ + $OH^\bullet$ (shown in Fig.1) and CH_2=$CHCH_2OH$ → CH_2=$CH^\bullet$ + $CH_2OH^\bullet$. Both are too endothermic and are thus inaccessible. Fig.2 shows a similar energy diagram for 1,2-dimethyloxiran. In addition to the cis → trans isomerization, it produces a number of stable isomers; methyl- ethyl ketone, isobulyl aldehyde, 2-butene-3-ol and ethyl-vinyl ether (Flowers and Parker 1971). As can be seen in Fig.2 methyl-ethyl ketone is produced with enough energy to decompose from the thermally excited state to $C_2H_5^\bullet$ and $CH_3CO^\bullet$. These species initiate free radical reactions and yield many decomposition products with ethylene being the major product.

This article reviews some recent work done in the shock tube laboratory at the Hebrew University of Jerusalem and presents new results on the isomerization and decomposition of two of the epoxy molecules using the single pulse shock tube technique. Determining the mechanism of the overall reaction and the construction of reaction schemes for the product distribution is assisted by isotope labeling and free radical scavengers.

2. Experimental

2.1. Apparatus

The isomerizations and decompositions in the epoxy molecules were studied behind reflected shocks in a 52 mm i.d. single pulse shock tube. Its driven section was 4 m long, and its driver had a variable length up to a maximum of 2.7 m. It could be varied in small steps in order to tune for the best cooling conditions. A 36-L dump tank was connected to the driven section at 45° angle toward the driver near the diaphragm holder, in order to prevent reflection of transmitted shock waves and repetitive heating of the reaction mixture. Prior to performing an experiment the tube was pumped down to better than 5×10^{-5} torr and was then filled with 80-120 torr of the reaction mixture. After firing the shock, gas samples were collected from the driven section and were analyzed by gas chromatography using FID and NPD. A detailed description of the tube and its mode of operation has been given in previous publications (Lifshitz, Moran and Bidani 1987).

2.2. Analysis

Several reaction mixtures containing 0.1-1% of the epoxy molecule in question and approximately 0.1% cyclopropanecarbonitrile (cpcn), which served as an internal standard for the temperature calculations, were prepared and stored in 12-L glass bulbs which were then filled with argon to 1 atm. Both the bulbs and the vacuum line were pumped down to better than 2×10^{-5} torr before the preparation of the mixtures.

The gas chromatographic analyses of the post-shock mixtures were performed on a pair of 2 m Porapak N columns connected to a flame ionization detector (FID) and a nitrogen phosphor detector (NPD). Cyclopropanecarbonitrile and its isomerization products were analyzed and quantitatively determined on the NPD. The epoxy molecules and their isomerization and decomposition products were determined on the FID. Sensitivity ratios of the different compounds involved were determined on each detector using standard mixtures.

GC peak areas were integrated with a Spectra Physics Model SP4200 computing integrator and were transferred to an IBM/PC for data reduction and graphical presentation.

2.3. Reflected shock temperatures and densities

Reflected shock temperatures were calculated from the extent of conversion of the internal standard which was added to the reaction mixture, using the relation: $T = -(E/R)/\ln\{-\frac{1}{At}\ln(1-\chi)\}$ where E is the activation energy of the isomerization of cyclopropanecarbonitrile reaction, A is its preexponential factor, t is the reaction dwell time and χ is the extent of decomposition defined as: $\chi = \Sigma[\text{isomerization products}]_t/[\text{cpcn}]_0)$ (Lifshitz, Shweky, Kiefer and Sidhu 1991).

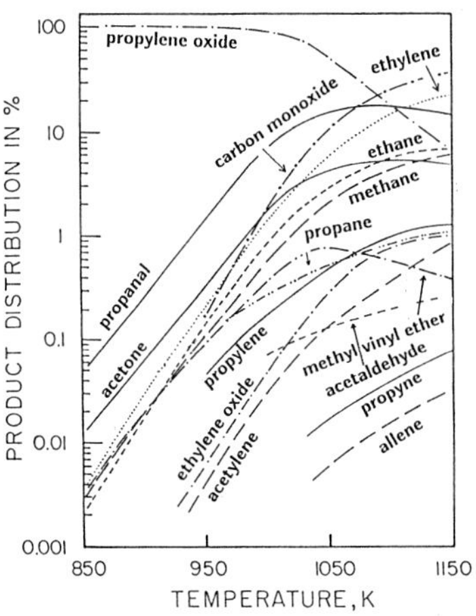

Fig. 3. Product distribution in methyl oxiran

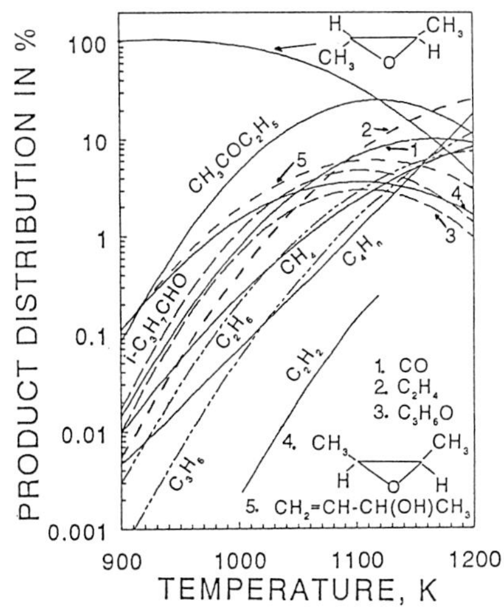

Fig. 4. Product distribution in dimethyloxiran

Reflected shock densities were calculated from the measured incident shock velocities using the three conservation equations and the ideal gas equation of state. The latter were measured with two high frequency pressure transducers placed 300 mm apart near the end plate of the driven section. A third transducer placed at the center of the end plate provided measurements of the reaction dwell times (approximately 2 ms) with an accuracy of approx. 5%. Cooling rates were approximately 5×10^5 K/s.

3. Results

3.1. Product distribution

The distributions of reaction products in the decomposition of methyloxiran and *cis*-1,2-dimethyloxiran are shown in Figs.3 and 4 respectively. The isomerization products are dominant at low temperatures. As the temperature increases the products resulting from decompositions and free radical reactions such as CO, C_2H_4, C_2H_6 and CH_4 become as important. Figs.5 and 6 show, as an example, Arrhenius plots of the isomerization first order rate constants for methyloxiran $\rightarrow$ acetone and 1,2-dimethyloxiran $\rightarrow$ methyl-ethyl ketone. These two compounds are among the major isomers in the systems. The heavy lines (at low temperatures) are Arrhenius lines of the data obtained at low temperatures. The dashed lines are extrapolations to the temperature range of the present experiment. The solid line at high temperatures in methyloxiran is an Arrhenius plot of data obtained recently by Zhiwei Huixing and Degang (1990) using the single pulse shock tube technique. The agreement of Zhiwei's et al data with both the low temperature data and the results obtained in the present experiments is not good. The reason for this discrepancy is unclear. Among the isomerizations that take place in 1,2-dimethyloxiran and that does not exist in methyloxiran is the *cis* $\rightarrow$ *trans* isomerization. The temperature dependent evolution of *trans* 1,2-dimethyloxiran in mixtures containing 1% *cis*-1,2-dimethyloxiran in argon can be seen in Fig.4 as product No 4.

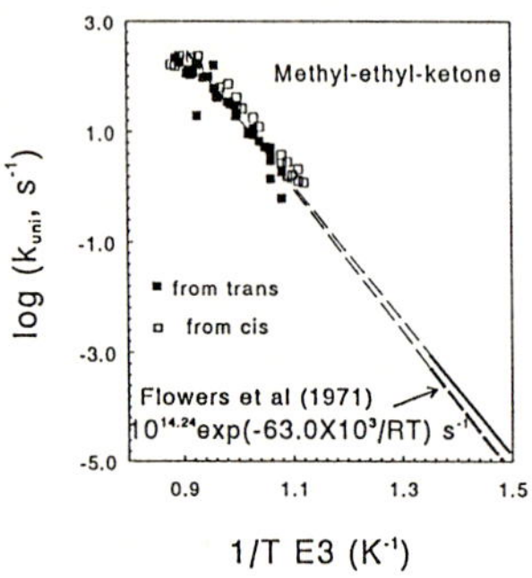

Fig. 5. Isomerization of dimethyl oxiran to M.E.K.

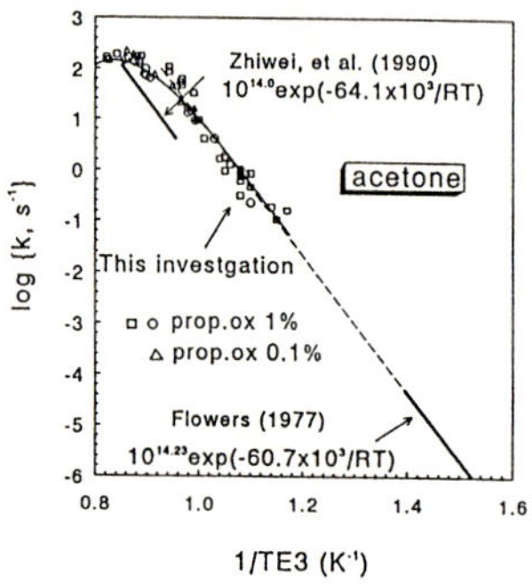

Fig. 6. Isomerization of methyl oxiran to acetone

4. Discussion

Figs.3 and 4 show a large number of isomerization and decomposition products resulting from the thermal reactions of both methyloxiran and 1,2-dimethyloxiran. Four isomers were analyzed in methyloxiran and five in 1,2-dimethyloxiran.

4.1. Isomerizations

In both methyloxiran and 1,2-dimethyloxiran all the isomerizations (except for the *cis* $\rightarrow$ *trans* isomerization in 1,2-dimethyloxiran) take place via two simultaneous processes: (1) Migration of either an H atom or a methyl group from one carbon atom to another carbon or to an oxygen atom. (2) Rupture of either a C-C or a C-O bond and formation of another C=C or C=O double bond. We have run several shocks with mixtures containing 0.5% methyloxiran and 0.5% methyloxiran-d_6 in argon in order to examine whether the isomers, particularly CH_2=CH-O-CH_3 and CH_2=$CHCH_2OH$ which were found in rather small concentrations are really obtained by unimolecular processes and thus must preserve the original skeleton of the propylene oxide. Namely, whether molecules with isotopic compositions such as CH_2=CH-O-CD_3 or CD_2=CD-O-CH_3 (m/z 61) which must be formed by recombination of CH_3 with CD_2=CDO or CD_3 with CH_2=CHO, can be found in the post-shock mixtures. Fig.7 shows, as an example, the isotope distribution of methyl-vinyl ether in the post-shock mixtures. As can be seen, no peak at m/z 61 is obtained, verifying the unimolecular nature of the production of methyl-vinyl ether. Similar

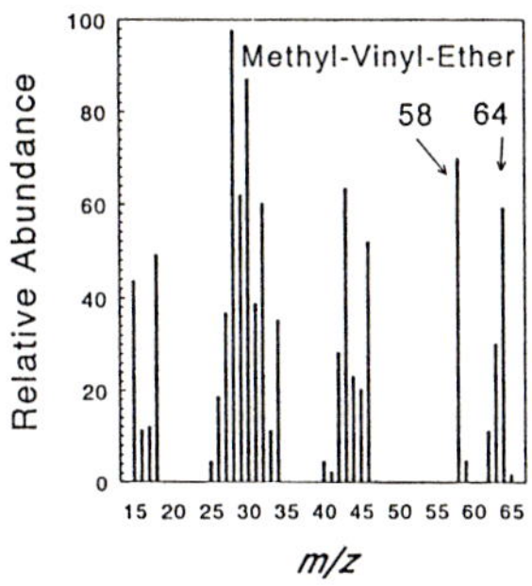

Fig. 7. Isotope distribution in methyl-vinyl ether

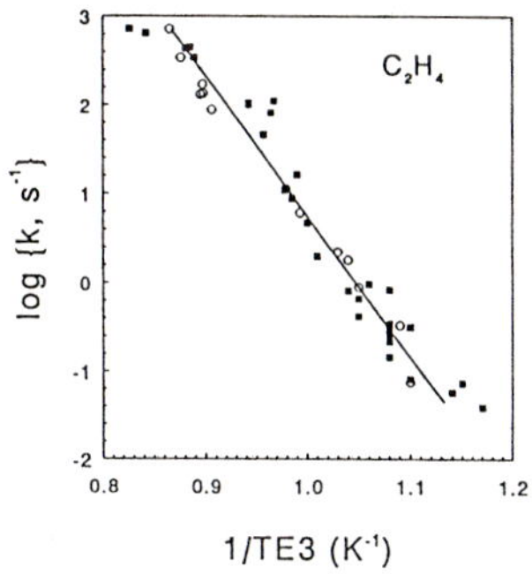

Fig. 8. Arrhenius plot for methyl oxiran → ethylene

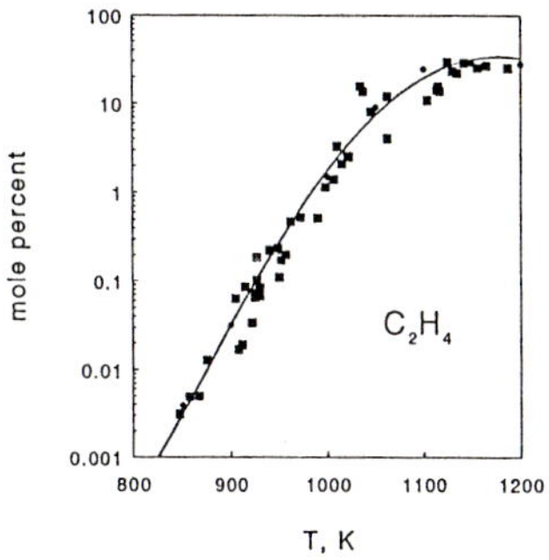

Fig. 9. Experimental and calculated mole percent of ethylene

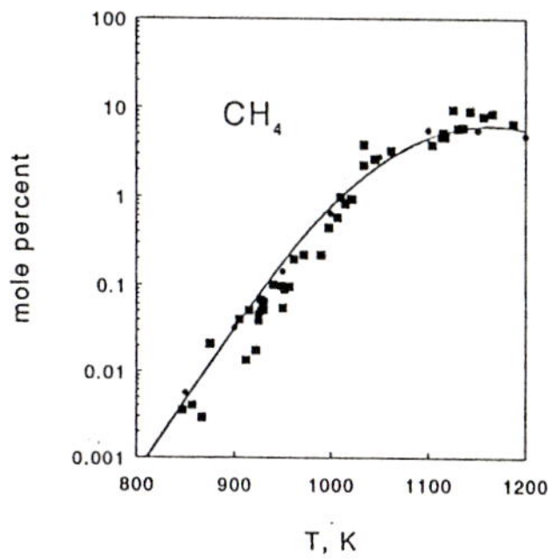

Fig. 10. Modelling of mole percent of methane

results were obtained with allyl alcohol. The transition structures leading to the production of these two isomers must be stiff. Indeed, the preexponential factors for their formation are smaller than those that characterize the production of acetone and propanal.

Fig.8 shows an Arrhenius plot of the first order rate constant of the formation of ethylene in the decomposition of propylene oxide. Its Arrhenius parameters are A = 16.5 s^{-1} and E = 72 kcal/mol.

4.2. Decompositions

As has been mentioned previously the isomers which are thermochemically more stable than the epoxides are formed with excess thermal energy just enough to allow the molecules to break down into two reactive fragments. The thermally excited propanal ($CH_3CH_2CHO\#$), for example, which is formed as the main isomerization product in methyloxiran can decompose to $C_2H_5^\bullet$ + $HCO^\bullet$ or $CH_3^\bullet$ + $CH_2CHO^\bullet$ (see Fig.1). On the other hand, thermally unexcited propanal, with an energy distribution corresponding to 1000 K, cannot yield the observed quantities of ethylene by decomposition. Ethylene can thus be produced either by the above mentioned route or by the process: $CH_3^\bullet + CH_3^\bullet \rightarrow C_2H_5^\bullet + H^\bullet$, $C_2H_5^\bullet \rightarrow C_2H_4 + H^\bullet$. If the first route is dominant then the ethylene which is formed in an experiment with 0.5% methyloxiran and 0.5% methyloxiran-d_6 in argon must be isotopically unscrambled. This is indeed what has been found. This conclusion regarding the mechanism of formation of ethylene was also supported by experiments which were carried out in the presence of large excess of toluene.

Although we did not perform experiments with isotopically labeled 1,2-dimethyloxiran it is believed that the decomposition of the thermally excited methyl-ethyl ketone is responsible for the production of $C_2H_5^\bullet$ which is the initiator of the free radical reactions in this system.

To account for the production of the plethora of isomerization and decomposition products we have constructed a reaction scheme containing 37 species and 68 elementary reactions which describes the overall decomposition of methyloxiran. Figs.9 and 10 show the experimental (squares) and calculated (circles and solid line) mole percent of ethylene and methane. As can be seen the agreement is quite good.

References

Blades AT (1968) Thermal reactions of propylene oxide. Can. J. Chem. 46:3283-3284.

Flowers, MC (1977) Kinetics of the thermal gas-phase decomposition of 1,2-epoxypropane. J. Chem. Soc. Faraday Trans. 73:1927-1935.

Flowers MC, Parker RM (1971) Kinetics of the thermal gas-phase reactions of cis- and trans-2,3-epoxypropane. J. Chem. Soc. (B) :1980-1988.

Hardwick TJ (1968) Thermal decomposition of propylene oxide. Can. J. Chem. 46:2454-2456.

Lifshitz A, Ben-Hamou H (1983) Thermal reactions of cyclic ethers at high temperatures. I. Pyrolysis of ethylene oxide behind reflected shocks. J. Phys. Chem. 87:1782.

Lifshitz A, Moran A, Bidani S (1987) Thermal reactions of acetonitrile at high temperatures. Pyrolysis behind reflected shocks. Intl. J. Chem. Kinet. 19:61.

Lifshitz A, Shweky I, Kiefer JH, Sidhu SS (1992) Rate of the retro Diels-Alder dissociation of 1,2,3,6-tetrahydropyridine over a wide temperature range. In: Takayama K, (ed) Shock Waves, Proceedings 18th Intl. Symposium on Shock Waves, Springer, Berlin, pp 825-830.

Newfeld ML, Blades AT (1963) Can. J. Chem. 41:2956.

Stull DR, Westrum Jr. EF, Sinke GC (1969) The Chemical Thermodynamics of Organic Compounds. John Wiley and Sons, N. Y.

Zhiwei Q, Huixing Y, Degang H (1990) Kinetics of the thermal gas-phase isomerization of epoxypropane in a single pulse shock tube. Acta Physico-Chimica Sinica 1:39-44.

Reactions of C_3H_3Cl with H_2, C_3H_4, C_2H_2 and C_2H_4 Behind Reflected Shock Waves

R.D. Kern, H. Chen, Z. Qin and K. Xie
Department of Chemistry, University of New Orleans, New Orleans, Louisiana 70148, USA

Abstract. Mixtures of propargyl chloride diluted in neon have been studied by dynamic sampling of the reflected shock zone by time-of-flight mass spectrometry. For the C_3H_3Cl system, benzene is detected only in trace amounts as an expected product of reactions involving C_3H_3 radicals; only C_2H_2, C_4H_2 and HCl appear as major products. However, in mixtures of C_3H_3Cl containing H_2, benzene is readily observed. Experiments on a 3% C_3H_3Cl - 5% D_2 mixture reveal a constant temporal HCl/DCl product ratio that is approximately equal to 5. It is proposed that the major channel for C_3H_3Cl dissociation yields HCl and c-C_3H_2 (singlet cyclopropenylidene). In the presence of H_2, c-C_3H_2 reacts exothermically to form the thermally excited adduct $C_3H_4^*$ which readily isomerizes to either allene or propyne as predicted by a bimolecular-QRRK calculation. The calculations reveal that formation of allene and propyne account for 89% of the product distribution in the reaction of c-C_3H_2 with H_2. Benzene production is also recorded in mixtures of C_3H_3Cl + allene, C_3H_3Cl + propyne, C_3H_3Cl + C_2H_2 and C_3H_3Cl + C_2H_4. The products observed in these mixtures over the range 1350 - 1700 K are C_2H_2, C_4H_2, HCl, benzene and trace amounts of C_6H_2. The model utilizes a 36 step mechanism and yields satisfactory agreement between the calculated and experimental results.

Key words: Chemical reaction, Benzene formation, Time-of-flight mass spectrometry

1. Introduction

One of the most important issues in the study of the pre-particle mechanism is formation of the first aromatic ring from which soot aggregates are eventually formed. Benzene formation by the reactions of propargyl radicals in allene pyrolysis (Wu and Kern 1987) was proposed and subsequently supported by work on the thermal decomposition of 1,2-butadiene (Kern et al. 1988) wherein C_3H_3 is produced initially by C-CH_3 bond fission followed by benzene formation. Recently, this proposal received additional support as an effective route to benzene formation. Propargyl radicals were recorded in rich hydrocarbon flames (Thomas et al. 1991). Benzene formation is observed in the reactions of C_3H_3Cl and C_3H_3Br with Na vapor (Alkemade and Homann 1989), where the initial reaction is C_3H_3Cl (C_3H_3Br) + Na $\rightarrow$ C_3H_3 + NaCl (NaBr). Benzene is a product in the unimolecular dissociation of 1,5-hexadiyne (Stein et al. 1990). Benzene is formed as a major product in the pyrolyses of allene and propyne using a single pulse shock tube (Hidaka et al. 1989). A detailed kinetic modeling of acetylene flames provides evidence that propargyl radicals play an important role in benzene formation (Miller and Melius 1992). Although there is considerable agreement that propargyl radical is responsible for benzene formation, the detailed mechanism leading to benzene is still not apparent. The investigation of benzene formation using propargyl chloride as a C_3H_3 radical precursor has been conducted (Alkemade and Homann 1989) over a temperature range of 623 - 673 K which is far from combustion temperatures. In this work, mixtures of propargyl chloride diluted in neon have been studied by mass spectrometric analysis of the reflected shock zone at relatively higher temperatures.

Shock Waves @ Marseille II
Editors: R. Brun, L. Z. Dumitrescu

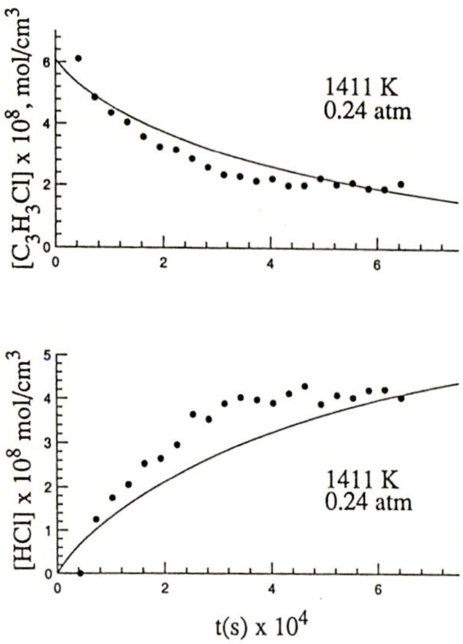

Fig. 1. 3% C_3H_3Cl mixture

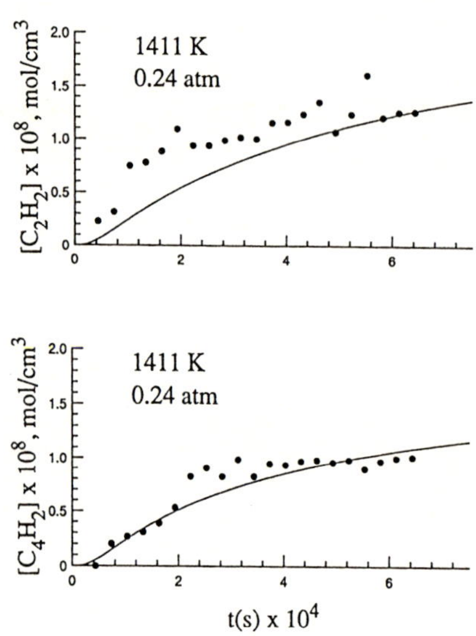

Fig. 2. 3% C_3H_3Cl mixture

2. Experimental

The reactants used are: C_3H_3Cl (Aldrich, 98%), allene (Matheson, 93%), propyne (Farchan), D_2 (Matheson, 99.5%), H_2 (Linde, 98%) and Ne (Matheson, 99.999%). C_3H_3Cl, C_3H_4A (allene) and C_3H_4P (propyne) are each purified by bulb-to-bulb distillations before diluted with neon to the desired concentrations. The mass spectrometric calibration factors for C_3H_3Cl, C_3H_4A, C_3H_4P, C_2H_2, HCl, C_4H_2, C_6H_6 and CH_4 are determined using mixtures with concentrations close to the experimental compositions at non-reacting temperatures. The calibration factor for C_6H_2 is derived from a carbon atom balance procedure (Wu et al. 1987). The apparatus and experimental procedures have been described recently (Kern and Xie 1991).

3. Results and discussion

The decomposition of a 3% C_3H_3Cl mixture takes place readily above 1400 K. Major products are C_2H_2, HCl and C_4H_2. The minor product observed is C_6H_2 along with trace amounts of C_6H_6. The amount of HCl produced is equal to that of C_3H_3Cl decomposed, indicating that Cl atoms mainly exist in HCl molecules and are balanced. However, the carbon atom balance is only 80% - 90%.

The decomposition of a 3% C_3H_3Cl - 5% H_2 mixture has the same major product distribution and carbon atom balance as those recorded in the decomposition of 3% C_3H_3Cl. The Cl atoms are also balanced. However, benzene is readily observed upon addition of 5% H_2. The results are displayed in Figs.1-4. The lines represent modeling calculations which are discussed below.

Adding 5% D_2, instead of 5% H_2, into a mixture containing 3% C_3H_3Cl yields interesting results. HCl and DCl are formed almost at the same time and the temporal ratio of HCl/DCl is constant at a value of $\sim$ 5 throughout the observation period of $\sim$ 700 μs. Peaks in the range m/e 78 - 82 are observed due to benzene and partially deuterated benzenes.

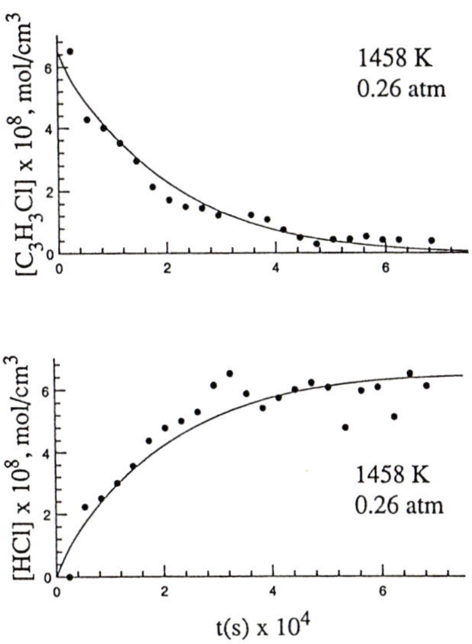

Fig. 3. 3% C_3H_3Cl - 5% H_2 mixture

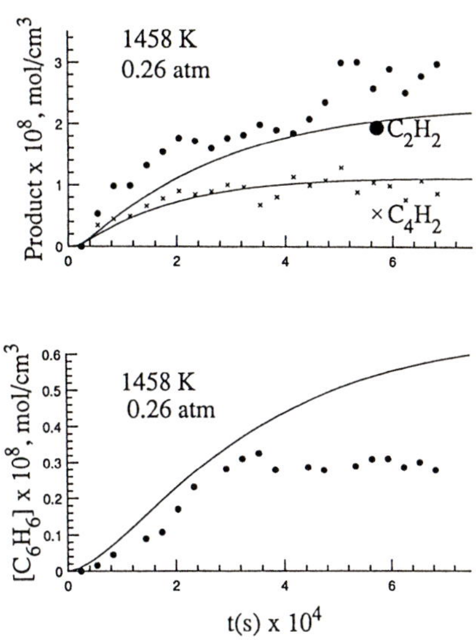

Fig. 4. 3% C_3H_3Cl - 5% H_2 mixture

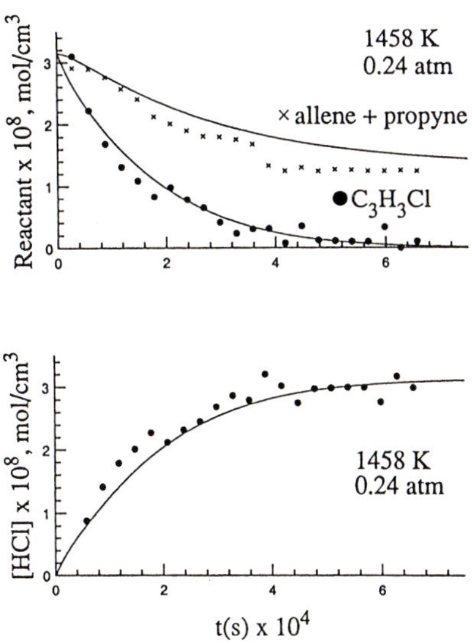

Fig. 5. 1.5% C_3H_3Cl - 1.5% allene mixture

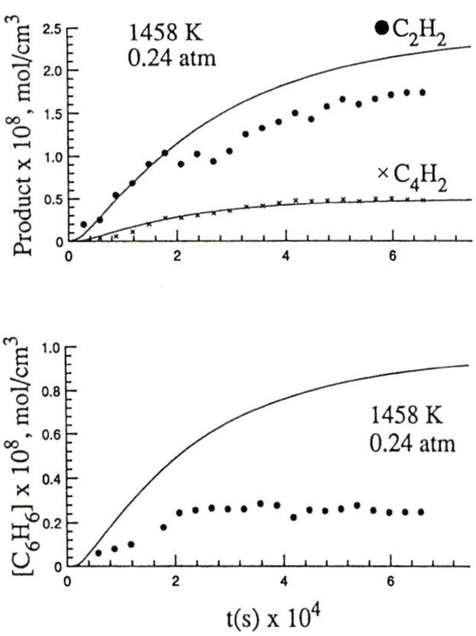

Fig. 6. 1.5% C_3H_3Cl - 1.5% allene mixture

Although benzene is recorded in trace amounts in the pyrolysis of 3% C_3H_3Cl, it is observed as a major product in the pyrolyses of 1.5% C_3H_3Cl - 1.5% C_3H_4A and 1.5% C_3H_3Cl - 1.5% C_3H_3P over the temperature range of 1350 - 1700 K. Neither allene nor propyne decomposes appreciably within this range. Thus, the pyrolyses therein must be initiated by C_3H_3Cl. The results are shown in Figs.5 and 6. Benzene is also recorded in mixtures of 3% C_3H_3Cl - 5% C_2H_2 and 3% C_3H_3Cl - 5% C_2H_4.

Table 1. Thermochemistry (at 298 K)

Species	ΔH_f^o (kcal/mol)	S^o (e.u.)	Species	ΔH_f^o (kcal/mol)	S^o (e.u.)
C_3H_4A	45.9	58.4	C_3H_4P	44.3	59.4
C_3H_3Cl	38.1	68.9	C_6H_6T	92.0	78.6
C_3H_3	81.5	60.0	c-C_3H_2	114.	56.6
C_6H_6	19.8	65.1	Cl	28.9	39.5

Table 2. Mechanisms for propargyl chloride, allene and propyne pyrolyses

	REACTIONS	A	n	E
1.	C_3H_3Cl=C_3H_3+Cl	6.6 E11	0.0	59.0
2.	C_3H_3Cl=c-C_3H_2+HCl	3.8 E11	0.0	52.0
3.	c-C_3H_2+H_2 →C_3H_4P	3.0 E10	0.0	0.0
4.	c-C_3H_2+H_2 →C_3H_4A	3.0 E10	0.0	0.0
5.	Cl+H_2=HCl+H	8.3 E13	0.0	5.0
6.	c-C_3H_2+C_3H_4P=3 C_2H_2	2.2 E13	0.0	15.0
7.	c-C_3H_2+C_3H_4A=3 C_2H_2	2.2 E13	0.0	15.0
8.	c-C_3H_2+c-C_3H_2=C_2H_2+C_4H_2	9.0 E13	0.0	13.0
9.	c-C_3H_2+C_3H_4P=2 C_3H_3	1.0 E12	0.0	4.0
10.	c-C_3H_2+C_3H_4A=2 C_3H_3	1.0 E12	0.0	4.0
11.	c-C_3H_2+c-C_3H_2=C_6H_2+H_2	4.5 E13	0.0	13.0
12.	C_3H_3+C_3H_3=C_6H_6T	4.0 E13	0.0	0.0
13.	C_6H_6T+M=C_6H_6+M	2.5 E12	0.0	0.0
14.	C_3H_3+C_3H_3=C_6H_5+H	6.3 E13	0.0	0.0

Also included are rate constants for acetylene pyrolysis (Kern et al. 1991) and for allene, propyne and ethylene (Mudipalli 1992).
Units: mol, cm^{-3}, s, kcal

The fact that the ratio of HCl/DCl is ~5 leads to the proposition of two initial parallel reactions. The main channel produces c-C_3H_2 (singlet cyclopropenylidene) and HCl.

$$C_3H_3Cl \longrightarrow c-C_3H_2 + HCl \tag{1}$$

$$C_3H_3Cl \longrightarrow C_3H_3 + Cl \tag{2}$$

Reaction 2 is required to account for the appearance of DCl. Cl atoms released from C_3H_3Cl in the presence of excess D_2 produce DCl readily at 1400 K:

$$Cl + D_2 \longrightarrow DCl + D \tag{3}$$

Without reaction 2, DCl could be still produced by the reaction of HCl + D_2 → DCl + HD. However, the exchange rate of this reaction is too slow at 1400 K (Kern and Nika 1971). Thus, reactions 2 and 3 are the main channels responsible for DCl production. Reaction 1 is proposed to be the dominant dissociation channel due to the observed ratio of HCl/DCl ~ 5. If C-Cl bond fission is the major initiating reaction, the HCl/DCl ratio would be < 1 due to the reaction of Cl + D_2 → DCl + D. This is not in accordance with the experimental value; i.e., the ratio of

HCl/DCl is ~5. Furthermore, this ratio leads to the assignment of the relative importance of reactions 1 and 2 , 83% and 17% respectively.

The proposal of two initial reactions provides an explanation for the near absence of benzene in the pyrolysis of 3% C_3H_3Cl; most of the C_3H_3Cl is converted into c-C_3H_2 and HCl and subsequent reactions of c-C_3H_2 do not produce benzene.

The experimental result that the pyrolysis of 3% C_3H_3Cl in the presence of 5% H_2 yields observable amounts of benzene is explained by the following process:

$$c - C_3H_2 + H_2 \longrightarrow C_3H_4^* \longrightarrow C_3H_4(\text{allene or propyne}) \quad (4)$$

$$c - C_3H_2 + C_3H_4(\text{allene or propyne}) \longrightarrow 2\ C_3H_3 \quad (5)$$

$$2\ C_3H_3 \longrightarrow C_6H_6 \quad (6)$$

Reactions 5 and 6 also account for benzene formation in the pyrolyses of the 1.5% C_3H_3Cl - 1.5 % C_3H_4 (allene or propyne) mixture. In the presence of H_2, c-C_3H_2 reacts exothermically to form the thermally excited adduct $C_3H_4^*$ which readily isomerizes to either allene or propyne as predicted by a bimolecular-QRRK (Dean 1985) calculation. The results from this calculation indicate that allene and propyne are the major products and account for 89% of the products from c-C_3H_2 + H_2.

The important species and their thermochemical data at 298 K are shown in Table 1.

The heat of formation and entropy at 298 K of C_3H_3Cl are obtained from group additivity. Because knowledge of the group, C-(Ct)(Cl)(H_2), is lacking, the group values of C-(C)(Cl)(H_2) are assigned to this group.The values of C_6H_6T (hexatetraene) are obtained from group additivity. c-C_3H_2 has three isomers: propadienylidene, propargylene and cyclopropenylidene. The most stable form of these three isomers is reported to be singlet cyclopropenylidene (Clauberg et al. 1992; DeFrees and McLean 1986). The heat of formation for c-C_3H_2 is 114 kcal/mol (Clauberg et al. 1992) and the entropy is calculated using its molecular properties (DeFrees and McLean 1986).

The reaction of $C_3H_3Cl \rightarrow$ c-C_3H_2+HCl is suggested to form first 3-chlorocyclopropene which then decomposes to c-C_3H_2 and HCl. This proposal stems from the study of the isomerization of allene and propyne; theoretical calculations and experimental observations provide evidence that one of the intermediates of the isomerization is cyclopropene (Yoshimine et al. 1989; Oref et al. 1988; Kakumoto et al. 1987). Analogously, C_3H_3Cl produces c-C_3H_3Cl.

The mechanism in Table 2 is composed of the 16-step mechanism of acetylene pyrolysis (Kern et al. 1991), the mechanism for allene and propyne pyrolyses (Mudipalli 1992) and rate constants derived herein for propargyl chloride decomposition. The mechanism successfully simulates the experimental profiles for C_2H_2, C_4H_2, C_3H_4 and the parent molecule. The overpredictions of benzene for mixtures of 1.5% C_3H_3Cl - 1.5% C_3H_4(A,P) represent carbon imbalances and are presumedly ascribed to soot formation due to the key role of benzene in soot formation (Kern et al. 1988). The modeling results are depicted in Figs. 1-6.

4. Conclusions

The thermal decomposition of propargyl chloride is initiated by two parallel reactions

$$C_3H_3Cl \longrightarrow C_3H_3 + Cl$$

$$C_3H_3Cl \longrightarrow c - C_3H_2 + HCl.$$

The near absence of benzene in the decomposition of propargyl chloride neat may be ascribed to the second reaction. Benzene production in the pyrolysis of mixtures of C_3H_3Cl and either allene or propyne may be explained by the successive bimolecular reactions 9, 10, 12, 13 listed in Table 2. In a mixture of C_3H_3Cl and H_2, reactions 3 and 4 lead to benzene formation.

Acknowledgments

This work is supported by the U.S. Department of Energy under contract DE-FG0585ER/13400. We appreciate helpful discussions with Professor John Kiefer.

References

Alkemade U, Homann KH (1989) Formation of C_6H_6 isomers by recombination of propargyl in the system sodium vapour/propynyl halide. Z. Phys. Chem. 161:19-34

Clauberg H, Minsek DW, Chen P (1992) Mass and photoelectron spectroscopy of C_3H_2. ΔH_f of singlet carbenes deviate from additivity by their singlet-triplet gaps. J. Am. Chem. Soc. 114:99-107

Dean AM (1985) Predictions of pressure and temperature effects upon radical addition and recombination reactions. J. Phys. Chem. 89:4600-4608

DeFrees DJ, McLean AD (1986) Ab initio molecular orbital studies of low-energy, metastable isomers of the ubiquitous cyclopropenylidene. Astrophys. J. 308:L31-L35

Hidaka Y, Nakamura T, Miyauchi A, Shiraishi T, Kawano H (1989) Thermal decomposition of propyne and allene in shock waves. Intl. J. Chem. Kinet. 21:643-666

Kakumoto T, Ushirogouchi T, Saito K, Imamura A (1987) Isomerization of allene ⇔ propyne in shock waves and ab initio calculations. J. Phys. Chem. 91:183-189

Kern RD, Nika GG (1971) A complementary shock tube technique study of the exchange of hydrogen chloide and deuterium. J. Phys. Chem. 75:171-180

Kern RD, Singh HJ, Wu CH (1988) Thermal decomposition of 1,2 butadiene. Intl. J. Chem. Kinet. 20:731-747

Kern RD, Wu CH, Young JN, Pamidimukkala M, Singh HJ (1988) Correlation of benzene production with soot yield measurements as determined from fuel pyrolyses. Energy and Fuels 2:454-457

Kern RD, Xie K (1991) Shock tube studies of gas phase reactions preceding the soot formation process. Prog. Energy Combust. Sci. 17:191-210

Kern RD, Xie K, Chen H, Kiefer JH, Sidhu SS (1991) The reaction of C_4H_2 and H_2 behind reflected shock waves. In: Takayama K (ed) Shock Waves, Proceedings. Sendai, Japan 2:729-734

Miller JA, Melius CF (1992) Kinetic and thermodynamic issues in the formation of aromatic compounds in flames of aliphatic fuels. Combust. Flame 91:21-39

Mudipalli PS (1992) Dissociation rates of propyne and allene at high temperatures and the subsequent formation of benzene. M.S. Thesis, The University of Illinois at Chicago

Oref MKI, Barzilai-Gilboa S, Lifshitz A (1988) Isomerization of cyclopropene to allene and propyne at elevated temperatures. Experimental, ab initio, and model calculations. J. Phys. Chem. 92:6924-6929

Stein SE, Walker JA, Suryan MM, Fahr A (1990) A new path to benzene in flames. 23rd Symp. Intl. on Combust. (France) 85-90

Thomas SD, Communal F, Westmoreland PR (1991) C_3H_3 reaction kinetics in fuel-rich combustion. Prepr. Pap. Am. Chem. Soc. Div. Fuel. Chem. 36:1448-1455

Wu CH, Kern RD (1987) Shock-tube study of allene pyrolysis. J. Phys. Chem. 91:6291-6296

Wu CH, Singh HJ, Kern RD (1987) Pyrolysis of acetylene behind reflected shock waves. Intl. J. Chem. Kinet. 19:975 - 996

Yoshimine M, Pacansky J, Honjou N (1989) Ab initio studies of the C_3H_4 surface. 3. Thermal isomerization. J. Am. Chem. Soc. 111:4198-4209

High-Temperature Unimolecular Decomposition of Ethyl Chloroformate: Comparison of the Secondary Competing Steps with Ethyl Formate

K. Saito, A. Oda and K. Tokinaga
Department of Chemistry, Faculty of Science, Hiroshima University, Higashi-Hiroshima 724, Japan

Abstract. The thermal decomposition of ethyl chloroformate has been investigated behind reflected shock waves. It is shown that the initiation step proceeds through a molecular elimination to produce ethylene and chloroformic acid likely as in the case of ethyl formate. It is found that in the consecutive reaction of the produced acid, decarboxylation occurs dominantly contrary to the case of formic acid in which the dehydration is more important. The decomposition mechanism is discussed on the basis of a theoretical calculation.

Key words: Unimolecular decomposition, Competing reactions, Ethyl chloroformate

1. Introduction

It is well known that ethyl formate decomposes thermally into ethylene and formic acid through a 6-center transition state (e.g., Saito et al. 1989). At high temperatures the produced formic acid decomposes consecutively through competing reactions. These are shown by the following scheme:

$$HCOOC_2H_5 \longrightarrow C_2H_4 + HCOOH \qquad (1a)$$

$$HCOOH \longrightarrow H_2O + CO \qquad (1b)$$

$$\longrightarrow H_2 + CO_2 \qquad (1c)$$

In the secondary steps, the dehydration (reaction 1b) is more important in the homogeneous reaction compared with the decarboxylation (reaction 1c), although there is a result in which those reaction rates are almost the same (Hsu et al. 1983).

In this paper we report experimental and theoretical studies on the thermal unimolecular decomposition of ethyl chloroformate. Previous kinetic investigations on this molecule were performed in a static system (Choppin and Kirby 1939) or in a stirred flow system (Lewis and Herndon 1961) at temperatures below 550 K. From the observed kinetic parameters, the decomposition is not seemed to be a simple molecular elimination, i.e., in both cases the reaction is affected by the reactor surfaces where Cl atoms and radicals are formed easily. However, if the reaction proceeds homogeneously in gas phase, the single bond fission such as Cl–C and R–O gives a high energy barrier. If the primary process is assumed to be similar to that of ethyl formate, we have chloroformic acid as a primary product and then competing secondary reactions are following as:

	ΔH_f^0/kcalmol^{-1}	
$ClCOOC_2H_5 \longrightarrow C_2H_4 + ClCOOH$	22.3	(2a)
$ClCOOH \longrightarrow HOCl + CO$	52.8	(2b)
$\longrightarrow HCl + CO_2$	– 14.9	(2c)
$\longrightarrow Cl + COOH$	80.1	(2d)

Assuming the activation energy for reaction 2a to be in the range of 40–50 kcal/mol, from the similarity to reaction 1a, the above elimination scheme seems more probable considering the Cl–C

Shock Waves @ Marseille II
Editors: R. Brun, L. Z. Dumitrescu

bond energy of of about 78 kcal/mol in the reactant. Therefore, in the present study, our aim is to know the primary reaction mechanism and its rate constant. And, if the molecular elimination is the case, next our interest is to check the relative rates of the competing reactions 2b–2d. At the present stage, since there is no information about the potential barriers for these paths, we try to evaluate the energies along with the probable reaction paths by molecular orbital calculations.

2. Experimental system

The experiments were performed in a standard 9.4–cm ID shock tube, made of stainless steel. A brief description is given here for the experimental system, since the details of the shock tube equipment and the optical system were given in a previous paper (Saito et al. 1986). The driven section (3.67–m long) was pumped down to less than 2×10^{-6} torr before each experimental run. The shock conditions for each run were calculated from the measured shock speed on the basis of the shock relations for the ideal gas. The error of the measured shock speed was less than 1% which induced maximum errors of ca. 2% in the temperature and ca. 1.3 % in the density behind the reflected shocks. A pair of MgF_2 windows (1.0 mm thickness) was mounted on the tube walls 2 cm upstream from the end plate. Time-resolved optical measurements of the reaction were performed through these windows.

The optical system used for the vacuum UV absorption was set up as follows: the resonance radiations of chlorine atom (139 nm) and nitrogen atom (174.5 nm) were produced by microwave excitation of He containing a few percent tetrachloro methane and nitrogen, respectively, in a quartz flow tube at about 10 torr. The radiation from the lamp passing through two MgF_2 windows was led to a vacuum-UV monochromator (Minuteman 302VM). Two types of photomultipliers (HTV-R431s and HTV-R1459) were used for the corresponding wavelengths to detect the absorption intensity. The time constant of the optical-electrical system was about 10 μs with an S/N ratio of about 10.

The production of carbon dioxide which is expected from a secondary reaction was monitored by IR emission from the ν_3 band. The radiation from the shock-heated gas was passed through an interference filter centered at 4.23 μm with 0.18 μm of the FWHM and detected by a HgCdTe element cooled at 77 K. All the output signals were recorded on a digital storage scope.

Mixtures used for the experiment were (0.02–0.2) mol% reactant in Ar. Reflected shock conditions were as follows: total density = $(0.12–1.07) \times 10^{-5}$ mol/cm^3 and temperature = 800–1250 K.

3. Experimental results

Fig.1 shows a typical absorption trace at 139 nm. As seen from Fig.1, the absorption intensity rises abruptly both at the incident shock front and at the reflected shock front, then it decreases gradually to a steady level. From each profile the absorption coefficient was determined as a function of the temperature. It was constant over the experimental temperature range with a value of 4.0×10^7 cm^2/mol. From the fact that the absorption coefficient is constant over a wide temperature range where the extent of the decomposition varies with the temperature, the production of chlorine atom is thought to be negligible in this experiment. The steady value of the absorption after the decay of the reactant is ascribed to the absorption by ethylene. The absorption coefficient of ethylene at this wavelength was 2.1×10^7 cm^2 /mol without temperature dependence. The first-order rate constant for the decomposition of ethyl chloroformate was then evaluated from the above absorption data. Fig.2 shows an Arrhenius plot of the rate constant at the total density of ca. $(0.5–1.0) \times 10^{-5}$ mol/cm^3. Arrhenius parameters were evaluated as: $\log(A)/\mathrm{s}^{-1}=13.0$ and $E_a=42.7$ kcal/mol including a certain amount of errors. The straight line in Fig.2 corresponds to a theoretical calculation which will be described later.

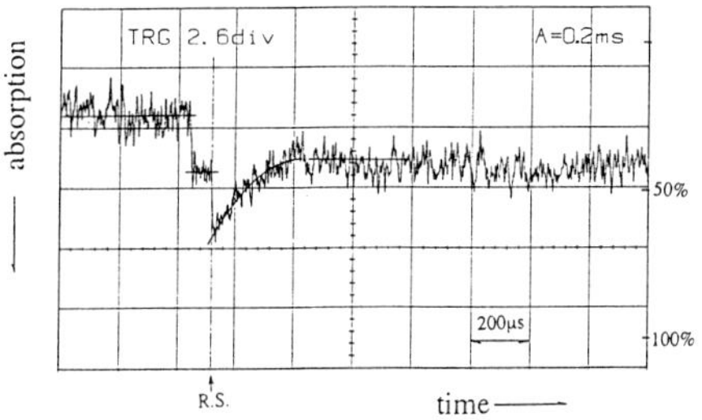

Fig. 1. Typical absorption profile at 139 nm. Conditions: T = 1028 K, total density=6.80 x 10^{-6} mol/cm^3, 0.02 mol% reactant in Ar

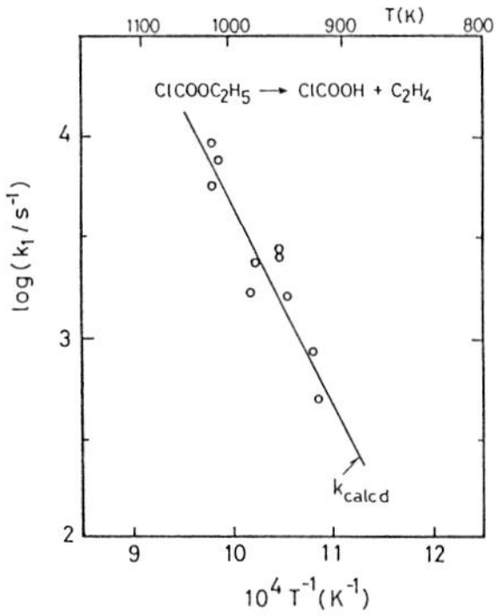

Fig. 2. Arrhenius plot of the first-order rate constant for the decay of ethyl chloroformate. The staight line corresponds to the calculated value for reaction 2a

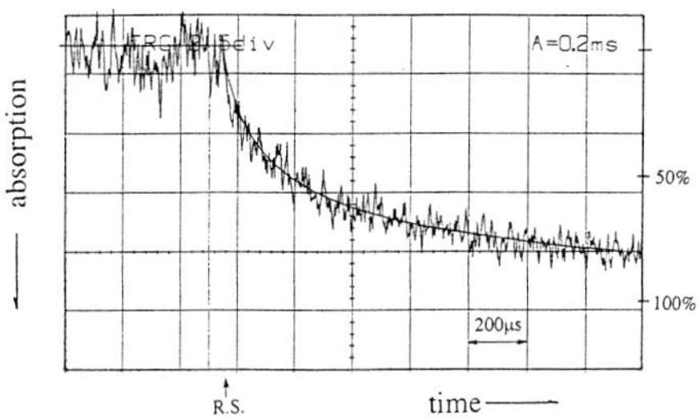

Fig. 3. Typical absorption profile at 174.5 nm. Conditions: T = 863 K, total density=8.02 x 10^{-6} mol/cm^3, 0.1 mol% reactant in Ar

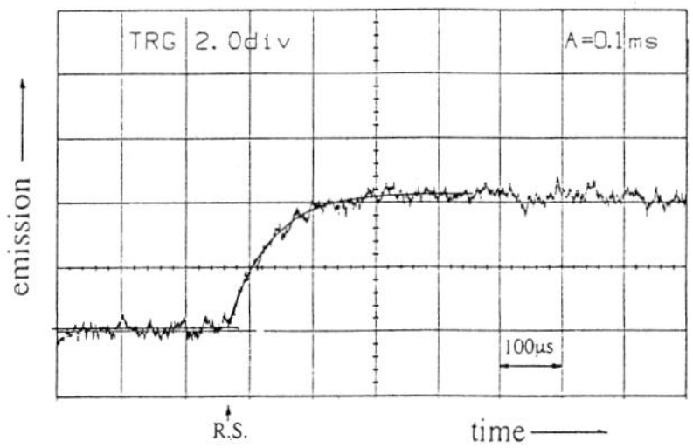

Fig. 4. Typical IR emission at 4.23 μm. Conditions: T = 1105 K, total density=9.35 x 10^{-6} mol/cm^3, 0.1 mol% reactant in Ar

The production of ethylene was monitored at 174.5 nm with the absorption coefficient of 2.4 x 10^7 cm^2/mol (Shimofuji et al. 1991). The absorption coefficient of ethyl chloroformate at this wavelength was small but not negligible and increased slightly with the temperature, e.g., it was 3.3 x 10^6 at 800 K and 4.1 x 10^6 at 1000 K. Fig.3 is a typical absorption trace at 174.5 nm. It was ascertained that the rate constant for the production of ethylene was the same as that for the decay of the reactant. This fact means that the primary step of the decomposition is molecular elimination by reaction 2a. Fig.4 shows a typical emission trace observed at 4.23 μm. From the calibration of the emission intensity by a known CO_2 mixture the amount of the produced CO_2 was mostly the same as that of the reactant within the experimental uncertainty.

4. Theoretical calculations

Ab initio molecular orbital (MO) calculations were carried out with a GAUSSIAN-90 package (Frisch et al. 1990). All equilibrium geometries and transition structures were fully optimized with the energy gradient method of the Hartree–Fock (RHF) level employing several basis sets. Vibrational frequencies were calculated by using analytical second derivatives to confirm all stationary point structures and to correct for the zero-point energy of each vibrational mode. For the evaluation of the total energies, Møller–Plesset perturbation calculations (MP2) were carried out to check the electron correlation effects. Theoretical rate constants were evaluated for possible reaction channels via conventional transition-state theory (Glasstone et al. 1941).

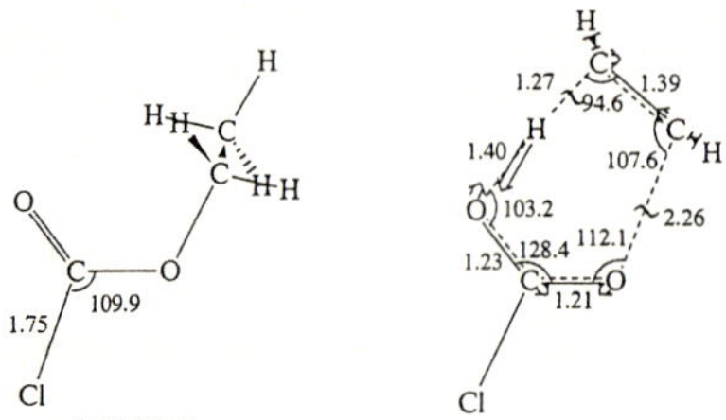

ClCOOC$_2$H$_5$ TS-2a

Fig. 5. Structures of the reactant and TS–2a optimized by HF/6–31G*. Arrows indicated on atoms are the displacement vectors

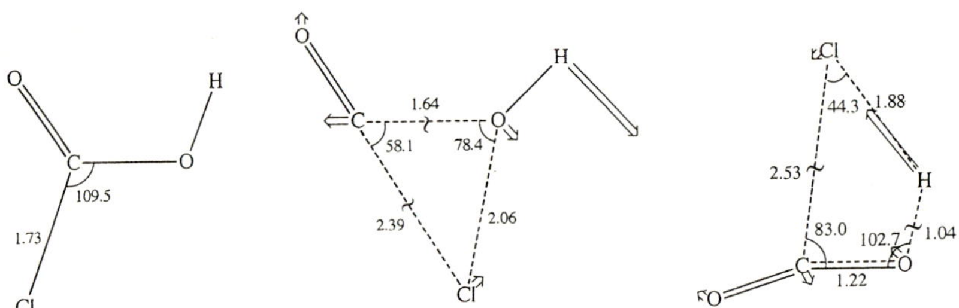

ClCOOH (trans) TS-2b TS-2c

Fig. 6. Optimized structures of chloroformic acid, TS–2b and TS–2c

Fig.5 shows structures of the reactant and the transition state (TS–2a) for reaction 2a optimized by HF/6–31G*. In TS–2a arrows indicate the displacement vector of each atom along with the reaction coordinate. It appears that the elimination occurs apparently in the direction to separate into ethylene and chloroformic acid. Fig.6 shows structures of chloroformic acid, TS–2b and TS–2c corresponding to reactions 2b and 2c, respectively. From the displacement vector indicated in Fig.6, we could image products of each reaction path. In fact, a calculation of the potential energy profile along the intrinsic reaction coordinate (IRC) shows clearly the approach to the products of each path as shown in Fig.7. It appears that reaction 2b takes a much higher potential barrier compared to reaction 2c. Comparing this result with the case of formic acid (Saito et al. 1989), the relative barrier heights of the competing steps are reverse to each other. In Fig.8, a potential diagram is shown for the reaction scheme considered in the present work.

5. Discussion

Previous kinetic investigations on ethyl chloroformate were performed in the temperature range of 380–550 K by static or stirred flow methods. Results have shown abnormally low preexponential factors such as $\log(A)/s^{-1}=10.74$ (Choppin and Kirby 1939) and 8.91 (Lewis and Herndon 1961). These results show apparently complicating features including Cl atom and radicals which are produced heterogenously at the walls. In the present experiment the reactant was highly diluted (0.02–0.2 mol%) in Ar. Also the kinetics were studied by the time-resolved measurement

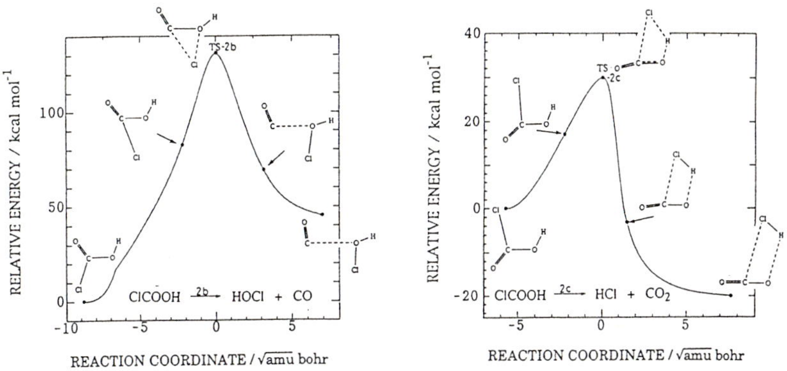

Fig. 7. Potential energy profile as a function of the reaction coordinate (IRC) for reactions 2b and 2c (HF/3–21G*)

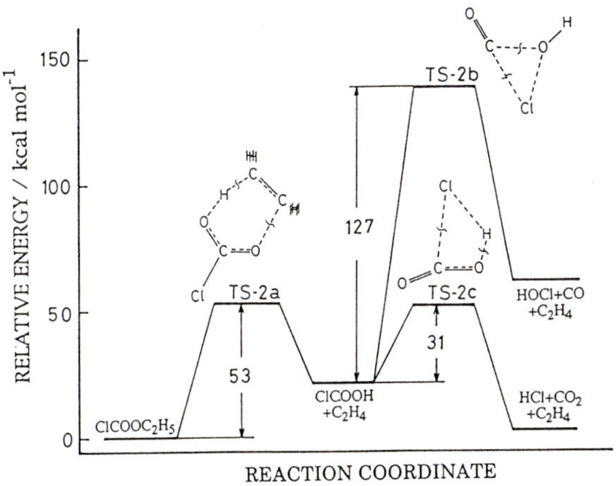

Fig. 8. Potential energy diagram of ethyl chloroformate decomposition (HF/3–21G*)

of the reactant and product species in a shock-heated mixture. Throughout the decomposition of ethyl chloroformate Cl atom was not observed by the atomic resonance absorption under the present conditions. This fact means that at temperatures around 1000 K the molecular elimination governs the initiation step because of its lower potential barrier compared to the Cl bond dissociation. Values of the theoretical rate constant for k_{2a} are in excellently agreement with the experimental data in the present temperature region, as seen in Fig.2. The Arrhenius parameters, $\log(A)/s^{-1}=13.0$ and $E_a = 42.7$ kcal/mol, seem to be reasonable for this type of reaction, e.g., for the case of ethyl formate $\log(A)/s^{-1}=12.7$ and $E_a = 48.5$ kcal/mol.

In the consecutive reaction, chloroformic acid decomposes through competing paths, 2b, 2c and 2d. According to the observation of the CO_2 production, the amount of the final concentration was comparable to the reactant. That is, reaction 2c governs the competing reactions, in agreement with the theoretical calculation as seen in Fig.8. This result is the reverse of the case of formic acid in which dehydration 1b is more important compared to decarboxylation. From the structure of TS–2c, this transition state seems to be stabilized by the strong approach of the hydrogen atom to the chlorine atom. Therefore, from the reaction dynamic aspect, the hydrogen chloride produced should be highly excited in the vibrational state. The CO_2 production rate was evaluated from the IR emission profile assuming that the intensity is expressed as a simple exponential increase. The first-order rate constant thus obtained did not increase with the total density, but rather seemed to decrease in the low-temperature region. That is, the decomposition of ClCOOH did not show a second-order feature. Arrhenius parameters are $\log(A/s^{-1}) =\sim 9$ and $E_a = 20 - 25$ kcal/mol. The preexponential factor is too small, compared with the calculated value of 13.7 for the high-pressure rate constant. Also, the activation energy is low compared to the calculated barrier height of 31 kcal/mol. One of the reasons for these abnormal results on the unimolecular reaction is thought to be that in the consecutive reactions the second reaction is affected by a chemical memory of the initiation reaction, i.e. the fresh chloroformic acid which is chemically activated goes immediately to the secondary step before it equilibrates.

References

Choppin AR, Kirby GF (1939) Homogeneous decomposition of ethyl chloroformate. J. Am. Chem. Soc. 61: 3176–3180

Frisch JMJ, Head-Gordon M et al. (1990) GAUSSIAN 90, Gaussian Inc, Pittsburgh, PA

Glasstone S, Laidler K, Eyring H (1941) The theory of rate process, McGraw–Hill, New York

Hsu DS, Shaub WM, Blackburn M, Lin MC (1983) Thermal decomposition of formic acid at high temperatures in shock waves. Nineteenth Symp (Intl.) on Combust., The Combustion Institute, Pittsburgh, pp 89–96

Lewis ES, Horndon WC (1961) Decomposition of gaseous chloroformates I. Rates of simple alkyl compounds. J. Am. Chem. Soc. 83: 1955–1958

Saito K, Shimofuji K et al. (1989) Investigations of the unimolecular behavior of relatively large molecules: shock tube experiments and molecular orbital calculations. In: Proc. 1989 Natl. Symp. on Shock Wave Phenomena, Shock Wave Research Center, Tohoku Univ. p 65

Saito K, Ito R et al. (1986) The initial process of the oxidation of the methyl radical in reflected shock waves. J. Phys. Chem. 90: 1422–1427

Shimofuji K, Saito K, Imamura A (1991) Unimolecular thermal decomposition of ethyl vinyl ether and consecutive thermal reaction of the intermediary product acetaldehyde: shock wave experiment and ab initio calculation. J. Phys. Chem. 95: 155–165

Single-Pulse Shock Tube Studies on the Reactions of Hydrogen Atoms with Acetone and Dimethylether

Wing Tsang* **and J.P. Cui**†
*Chemical Kinetics and Thermodynamics Division, National Institute of Standards and Technology, Gaithersburg, Maryland 20899
†Institute of Mechanics, Academia Sinica, Beijing, China

Abstract. Rate constants for hydrogen atom attack on acetone and dimethylether have been determined in single-pulse shock tube experiments. The following rate expressions for hydrogen abstraction have been obtained; $k(\mathrm{H+Acetone}) = 5.4 \times 10^{13} \exp(-3565/T)\mathrm{cm^3.mol^{-1}.s^{-1}}$ and $k(\mathrm{H+Dimethylether}) = 1.4 \times 10^{14} \exp(-3576/T)\mathrm{cm^3.mol^{-1}.s^{-1}}$ in the temperature range of 970-1110 K and at 2.5 atm pressure. The results are compared with lower temperature data for the same processes and evidence is presented for an upward curvature of the Arrhenius plots over the temperature range 300 to 1110 K.

Key words: Single-pulse shock tube, Chemical reactions, Hydrogen, Acetone, Dimethylether.

1. Introduction

The reactions of hydrogen atoms with organic molecules are important elementary processes during pyrolysis and combustion. There are however very few measurements at the high temperatures where the data is needed. We have developed a single-pulse shock tube technique (Robaugh and Tsang 1986) for determining rate constants and expressions in the range of 950-1200 K for such reactions and have applied it for studies on a variety of saturated and unsaturated compounds.

Table 1. Literature values for the rate constants of hydrogen atom reactions with acetone and dimethylether

	Rate constant	Temp.range, K	Reference
Acetone	$1.9\times10^{13}\exp(-3210/T)$	[298-465]	Ambidge et al. (1976)
Dimethylether	$1.3\times10^{13}\exp(-2367/T)$	[251-365]	Meagher et al. (1972)
	$1.9\times10^{13}\exp(-2593/T)$	[330-560]	Faubel et al. (1979)
	$2.6\times10^{12}\exp(-1957/T)$	[273-425]	Lee et al. (1981)

We now describe experiments dealing with hydrogen atom attack on two oxygenated species, dimethylether and acetone. Lower temperature results for the two reactions are summarized in Table 1 and plotted in Fig.1. It can be seen that in the case of dimethylether there is considerable scatter in the rate expressions. Thus, extrapolations to the high temperature region are very uncertain. Hence the need for a more direct determination. On the other hand, from Fig.1 it can be seen that the measured rate constants near room temperature are in fair agreement. This can then be regarded as a fixed point for the present considerations. Since similar methodologies were used in determining the rate expression for hydrogen atom attack on acetone, it is likely that the same considerations apply. That is, a certain degree of uncertainty in the rate expression, but fairly reliable rate constants at the mid-point of the range covered. An additional aim of this study is to obtain some experience in extrapolating kinetic data for hydrogen atom reactions over wide temperature ranges. The essence of this issue is of course the degree of curvature of the Arrhenius plots over large temperature ranges.

Shock Waves @ Marseille II
Editors: R. Brun, L. Z. Dumitrescu

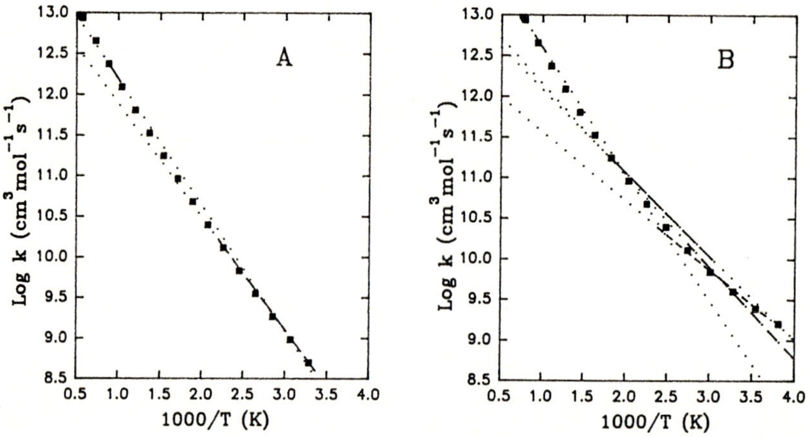

Fig. 1. Summary of past and present work on hydrogen atom attack on Acetone (A): (— —) (Ambidge et al.. 1976), and Dimethylether (B) (— —) (Meagher et al.. 1972), (- -) (Fauber et al.. 1979), (- - -) (Lee et al.. 1981). Solid lines are present results. Light dotted lines are linear extrapolations. Heavy dotted lines are best non-Arrhenius fit of all data

2. Procedure

The technique involves decomposing trace quantities of a very labile hydrocarbon, hexamethylethane in the presence of much larger amounts of a thermally stable compound such as 1,3,5 trimethylbenzene (mesitylene) and appropriate amounts of the compound of interest. The specific reactions involving hexamethylethane and mesitylene are:

Hexamethylethane	=	2 t-butyl	$= 2\ iC_4H_8 + 2H$	(1)
H+Mesitylene	=	meta-xylene	$+CH_3 \quad k(\text{disp})$	(2)
	=	3,5 dimethylbenzyl	$+H_2$	(3)

Reaction (1) releases hydrogen atoms into the system and its concentration is equal to the isobutene that is formed. The rate-determining step is the initial bond-breaking process. The subsequent decomposition of the t-butyl radical is extremely fast and for the present purposes can be regarded as instantaneous. Since the rate expression for (1) is well established as

$$k(1) = 3 \times 10^{16} \exp(-34500/T)\,\text{s}^{-1}$$

then from the yield of isobutene, one can determine the temperature for the reacting system as well as the hydrogen atom concentration released into the system. The expressions relating isobutene to rate constant for bond cleavage and the reaction temperature are

$$\log(1 - 1.03 * \text{isobutene/hexamethylethane})\ t^{-1} = k(1) = 3 \times 10^{16} \exp(-34500/T)$$

where the heating time t is of the order of 500 μs. The factor 1.03 takes into account the 3% yield of propene from t-butyl decomposition established from earlier studies (Tsang 1981). Determining the average temperature by this method removes the most serious uncertainty from single-pulse shock tube experiments.

From previous experiments (Tsang et al. 1990) we have found the rate expressions for hydrogen atom attack on mesitylene to be

$$k(3) = 3.7 \times 10^{14} \exp(-4232/T)\ \text{cm}^3.\text{mol}^{-1}.\text{s}^{-1} = k(\text{abs}), \text{m}$$

$$k(2) = 6.7 \times 10^{13} \exp(-3255/T)\ \text{cm}^3.\text{mol}^{-1}.\text{s}^{-1} = k(\text{disp})$$

The yields of meta-xylene can thus serve as a measure of the number of hydrogens that has reacted with the mesitylene. The difference in the deduced concentrations of hydrogen atom from the amount of isobutene and meta-xylene represents all the hydrogens that have abstracted hydrogens from the mesitylene and the target molecule.

An important consideration is that there are no other sources or sinks for hydrogen atoms. The primary decomposition channel from the pyrolysis of mesitylene is indeed hydrogen atoms and the 2,4 dimethylbenzyl radical. However, its rate constants for decomposition in this temperature range can be estimated to be at least 4 orders of magnitude smaller than that for the hexamethylethane. This is also the case for the decomposition of acetone and dimethylether. Hydrogen atoms can combine with radicals such as 2,4 dimethylbenzyl or methyl that is formed in the course of the experiments. Rate constants are expected to be close to collisional. However the concentrations of these radicals are much smaller than that of the stable species and the rate constants for reaction of hydrogen atoms with such compounds at these temperatures are also quite large. Therefore negligible losses of hydrogen atoms from these channels can be expected. All of these claims can be demonstrated by carrying out experiments with varying mixtures of reactants and hydrogen sources.

For the reactions of hydrogen atoms with acetone and dimethylether, the possible processes (k_{abs}) are:

$$\begin{array}{llllll} H + \mathrm{Acetone} & = & H_2 + CH_3COCH_2 & = & COCH_2 + CH_3 & k(\mathrm{abs}),a \\ H + CH_3OCH_3 & = & H_2 + CH_2O + CH_3 & = & CH_2O + CH_3 & k(\mathrm{abs}),d \\ & = & CH_3OH + CH_3 & & & \end{array}$$

where the decomposition of the organic radicals are for present purposes essentially instantaneous. In all cases, reaction leads to the formation of stable compounds and methyl radicals. There is no possibility of forming hydrogen atoms during the induced decomposition process. Methyl radicals are already present in the system and their reactions are either abstraction of a hydrogen from mesitylene or combination with the thermally stable radicals formed in the course of the reaction. Note that no methanol was detected, hence displacement cannot be important.

Single-pulse shock tubes are configured so that the stable end-products can be detected through standard analytical procedures. Every meta-xylene that is formed must be the result of the displacement process. Thus the quantity ((isobutene/m-xylene)-1) is the ratio of rates for all possible abstraction processes versus methyl displacement from mesitylene. The ratio of rate constants for abstraction from the compounds of interest and the displacement of the methyl radical from mesitylene is

$$k(\mathrm{abs}), i/k(\mathrm{disp}) = ((\mathrm{isobutene/m-xylene}) - (1{+}r))(\mathrm{Oxy})/(\mathrm{Mesitylene})$$

where Oxy is the oxygenated compound, () refers to the concentration, r is the ratio of rate constants for abstraction to displacement for H-atom attack on mesitylene and the subscript i refers to a or d for the two processes of interest. Since the rate expression for k(disp) has been given above, one can then deduce the abstraction rate constant at any temperature

3. Experimental

The experiments were carried out in a heated single-pulse shock tube. Descriptions of the apparatus and the experimental procedure can be found in earlier reports (Tsang et al. 1990). Analysis was by gas chromatography using Poropak N*) and a 30 meter wide-bore dimethylsiloxane capillary column. The former was used for the analysis of the light gases and for quantitative

*) Certain commercial materials and equipment are identified in this paper in order to specify adequately the experimental procedure. In no case does such identification imply recommendation or endorsement by the National Insitute of Standards and Technology, nor does it imply that the material or equipment is necessarily the best available for the purpose.

purposes isobutene, while the capillary column was used for the detection of meta-xylene. The dimethylether, acetone and mesitylene were obtained from Matheson, Aldrich and Wiley Scientific respectively. Gas chromatographic analysis did not reveal the presence of any impurities. Except for vigorous degassing no additional purification procedures were employed. The following mixtures were studied:

A: 3% Acetone and 0.36% Mesitylene with 100 ppm Hexamethylethane in Argon
B: 0.98% Acetone and 1.0% Mesitylene with 120 ppm Hexamethylethane in Argon
C: 2% Dimethylether and 0.5% Mesitylene with 120 ppm Hexamethylethane in Argon
D: 1.03% Dimethylether and 0.47% Mesitylene with 99 ppm Hexamethylethane in Argon

The variation in the results with the relative concentrations of the reactants serves as a critical test for the postulated mechanisms. One expects that if the mechanism postulated here is correct then the experimental $k(\mathrm{abs})_i$ determined for the two systems will in fact not be dependent of the reactant concentrations.

4. Results and discussion

All the quantitative results from the study are based on the yields of isobutene and meta-xylene. Methane and ethane are also present in large amounts, but no attempt was made to quantify the yields. Other possible products are formaldehyde and ketene. The former is insensitive to flame ionization detection, while ketene will not elute from the columns used in the present study. A very important observation is the absence of methanol. This would indicate that under the present conditions the displacement reaction $H + CH_3OCH_3 = CH_3 + CH_3OH$ does not make an appreciable contribution to the hydrogen atom induced decomposition.

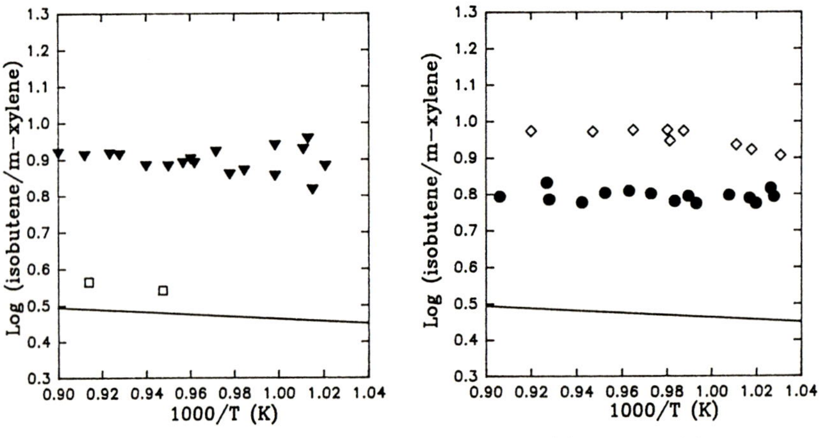

Fig. 2. Isobutene/m-xylene ratio from various experiments. Acetone:(▼) Mixture A, (□) Mixture B. Dimethylether: (◇) Mixture C, (•) Mixture D. Line is from earlier results with only m-xylene

The basic measurement from these studies is the ratio of isobutene to m-xylene. Results are summarized in Fig.2. As expected one notes the increase of this ratio as the the concentration of the target molecule is increased in comparison to that of the m-xylene. Also included are the results from earlier experiments dealing with m-xylene by itself. Since the analysis of the data are based essentially on subtracting the two quantities, one expects that the larger the difference the more accurate will be the results. As a result the two points from roughly equal amounts of dimethylether and m-xylene were not used in the analysis. However, as can be seen in the subsequent analysis, this does not really effect the results. Arrhenius plots for the abstraction rate constants can be found in Fig.3. As required, for each reaction all the data have now collapsed to a single line. The rate expressions for hydrogen atom attack to the two oxygenated compounds have been found to be

$$k(\mathrm{H}+\mathrm{Acetone}) = 5.4 \times 10^{13}\ \exp(-3565/T)\ \mathrm{cm^3.mol^{-1}.s^{-1}}$$

$$k(\mathrm{H}+\mathrm{Dimethylether}) = 1.4 \times 10^{14}\ \exp(-3576/T).\mathrm{cm^3.mol^{-1}.s^{-1}}$$

or 1.8×10^{12} and 4.6×10^{12} $\mathrm{cm^3.mol^{-1}.s^{-1}}$, respectively, at 1050 K.

In comparison with the data in Table 1, the rate parameters are all larger (see Fig.1). In addition, the extrapolated rate expressions all lead to smaller rate constants. All of these are consistent with an upward curvature of the Arrhenius plots for these processes. The exact magnitude of the curvature is unfortunately uncertain. Least squares treatment of all the data leads to the following rate expressions over the entire temperature range that has now been covered:

$$k(\mathrm{H}+\mathrm{Acetone}) \quad = 1.9 \times 10^{3}\ T^{1.78}\ \exp(-2571/T)\ \mathrm{cm^3.mol^{-1}.s^{-1}}$$

$$k(\mathrm{H}+\mathrm{Dimethylether}) = 1.3 \times 10^{13}\ T^{3.27}\ \exp(-1294/T)).\mathrm{cm^3.mol^{-1}.s^{-1}}$$

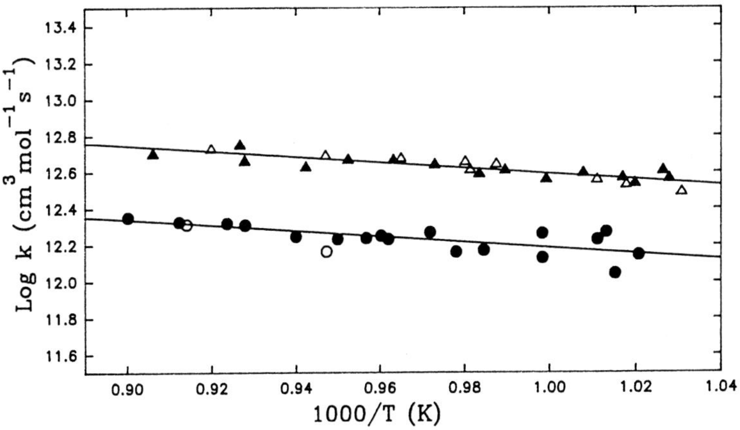

Fig. 3. Arrhenius plots for hydrogen atom attack on dimethylether(top line) and acetone(bottom line). (•) Mixture A, (o) Mixture B, (△) Mixture C and (▲) Mixture D

The quantitative significance of the $T^{1.5}$ difference in the preexponential factors is uncertain and may in fact be too large, since these can have contributions from the uncertainties in the rate expressions. The differences in absolute rate constants for the two processes are about an order of magnitude at room temperature and a factor of 2.5 at 1050 K, in favor of the abstraction of hydrogen atoms from dimethylether. These ratios should be of considerable accuracy. If one assumes an Arrhenius dependence for the ratio of rate constants then $k(\mathrm{H}+\mathrm{Dimethylether})/k(\mathrm{H}+\mathrm{Acetone})=1.6\ \exp(488/T)$. It is not unreasonable to expect that there will be considerable cancellation of the temperature dependence of preexponential factor for the two reactions. This suggests that the similarity in activation energies and the factor of 3 difference in the A-factor given earlier may be deceptive. The temperature dependence determined from the ratios of the two sets of rate constants are in accord with the approximately 8 kJ/mol difference in $C-H$ bond energies for acetone and dimethylether in favor of the latter.

References

Ambidge PF, Bradley JN, Whytock DA (1976) J. Chem. Soc. Faraday Trans. 1, 72: 1870

Faubel C, Hoyermann K, Strofer E, Wagner HGg (1979), Ber. Bunsenges. Phys. Chem. 83: 532

Lee JH, Machen RC, Nava DF, Stief LJ (1981) J. Chem. Phys., 74: 2839

Meagher JH, Timmons RB(1972) J. Chem. Phys., 57: 3175

Robaugh D, Tsang W (1986) J. Phys. Chem., 90: 4159

Tsang W (1981) Comparative rate single pulse shock tube in the thermal stability of polyatomic molecules. In: Lifshitz A (ed.) Shock Tubes in Chemistry, Marcel Dekker, p 59

Tsang W, Cui JP, Walker JA (1990) Single pulse shock tube study of the reactions of hydrogen atoms with complex aromatics. In:Kim YW (ed) Proc. 17th Intl. Symp. on Shock Tubes and Waves, AIP Conference Proceedings 208, American Institute of Physics, New York, p 63

Studies on the Reactions of Atomic Oxygen(^3P) with C_2-C_6 Alkanes at High Temperatures: Examination of the Transition State Theory

Akira Miyoshi, Kentaro Tsuchiya*, Atsumu Tezaki and Hiroyuki Matsui
Department of Reaction Chemistry, The University of Tokyo, 7-3-1 Hongo, Bunkyo-ku, Tokyo 113, Japan
*On leave from National Institute for Resources and Environment, 16-3 Onogawa, Tsukuba, Ibaraki 305, Japan

Abstract. The rate constants for the reactions of O(^3P) with a series of 7 alkanes (ethane, propane, n-butane, n-pentane, n-hexane, neo-pentane and cyclo-hexane) have been measured over the temperatures of 800-1300 K. Present data for neo-pentane and cyclo-hexane are compared with the reactive cross sections obtained by a crossed molecular beam experiment. The additivity law of the rate constant and the empirical Evans-Polanyi's relation, as well as recommended data bases for combustion modeling such as predicted by Transition State Theory (TST) calculations given by Cohen and Westberg have been examined.

Key words: Laser photolysis, ARAS, Oxygen atom, Alkane, Rate constant

1. Introduction

The reactions of O(^3P) with saturated hydrocarbons are very important in the initial stage of combustion and have been extensively studied; however, the accumulated data at high temperatures above 1000 K are not abundant except on methane. Acquisition of more reliable information for alkanes larger than methane at high temperatures will be necessary to improve the modeling of practical hydrocarbon combustion problems. By excimer laser flash photolysis behind shock waves combined with ARAS for detection of O(^3P), we have obtained rate constants for the reactions of O(^3P) with a series of linear-chain alkanes ($n = 1 - 6$), neo-pentane and cyclo-hexane at elevated temperatures ($T = 850 - 1300$ K). By summarizing these reaction rate constants, empirical rate laws or the proposed rate constants in the combustion modeling have been examined. These experimental data will be of great help in understanding fundamental features of the reactions of O(^3P) and large alkanes. A discussion on the correlation of the thermal rate constants obtained in this study and the reactive collision cross sections measured in a beam experiment (Andersen and Luntz 1980) is presented for neo-pentane and cyclo-hexane.

2. Experimental system

The experimental system is composed of a shock tube (5 cm ID and 4 m long), an atomic resonance absorption spectrometry (ARAS) and a photolysis excimer laser (ArF: 193 nm). The ArF laser was fired behind reflected shock waves with a delay time of 100 μs, where SO_2 was photolyzed to produce O(^3P). The laser beam was introduced into the shock heated sample gas through a 4×1 cm rectangular suprasil quartz window located at the end wall of the shock tube. Under a pseudo-first order condition, i.e., $[O] \ll [RH]$, the time dependence of O(^3P) concentration was monitored by ARAS with a microwave discharge lamp of 1% O_2 in He. Triplet resonance lines around 130.6 nm were separated with a VUV monochromator (f = 20 cm), and detected by a solar-blind photomultiplier tube (Hamamatsu photonics R976). The observed absorption signals were transformed into absolute concentrations by using a revised calibration curve constructed by an excimer laser photolysis in shock waves. Pseudo-first order analysis of the rate constant was provided by microcomputer, where all the digital information of the concentration of O atoms was analyzed with a nonlinear least-squares fit. The background signal for the 130.6 nm absorption was

Shock Waves @ Marseille II
Editors: R. Brun, L. Z. Dumitrescu

usually observed; this was caused by the absorption of vibrationally excited precursor molecules SO_2. As the intensity of the background absorption was kept almost steady during the time scale of the effective decay for O atoms (unless the temperature range was not too high), the signal due to absorption by O atoms was obtained simply by subtracting the background signal from the total absorption.

3. Experimental results

The rate constants for the reaction

$$CH_4 + O(^3P) \xrightarrow{k_1} CH_3 + OH \tag{1}$$

was measured over the temperature range of 1200 to 1700 K to test the reliability of our experimental data, since this reaction has been studied extensively (Ohmori et al. 1992 and references therein). It is confirmed in this study that the concentration of O atoms showed a single exponential decay and the rate constant of this study agreed very well with those of previous works. The rate constants for the reactions of O atoms with other 7 alkanes (k_2 to k_8) were determined by using the same technique over the temperature ranges of about 800 to 1300 K, i.e.,

linear-chain alkanes:

$$C_2H_6 + O(^3P) \xrightarrow{k_2} C_2H_5 + OH \tag{2}$$

$$C_3H_8 + O(^3P) \xrightarrow{k_3} C_3H_7 + OH \tag{3}$$

$$n - C_4H_{10} + O(^3P) \xrightarrow{k_4} C_4H_9 + OH \tag{4}$$

$$n - C_5H_{12} + O(^3P) \xrightarrow{k_5} C_5H_{11} + OH \tag{5}$$

$$n - C_6H_{14} + O(^3P) \xrightarrow{k_6} C_6H_{13} + OH \tag{6}$$

nonlinear alkanes:

$$neo - C_5H_{12} + O(^3P) \xrightarrow{k_7} C_5H_{11} + OH \tag{7}$$

$$cyclo - C_6H_{12} + O(^3P) \xrightarrow{k_8} C_6H_{11} + OH \tag{8}$$

Table 1. Summary of the Arrhenius parameters A and E_a in $k = A\exp(-E_a/RT)$ for the $O(^3P)$ with alkanes measured in this study

Reaction	$k = A\exp(-E_a/RT)$ A $cm^3 molecule^{-1} s^{-1}$	E_a $kJmol^{-1}$	Temp. range K
$O+CH_4$	4.70×10^{-10} (2.8)#	54.1±10.7	980-1520
$O+C_2H_6$	9.10×10^{-10} (5.7)	45.2±15.8	940-1190
$O+C_3H_8$	4.34×10^{-10} (2.3)	34.3±7.9	940-1260
$O+n-C_4H_{10}$	3.93×10^{-10} (3.3)	23.6±10.0	900-1140
$O+n-C_5H_{12}$	3.05×10^{-10} (2.1)	20.2±6.4	850-1140
$O+n-C_6H_{14}$	8.91×10^{-10} (2.1)	27.8±5.8	870-1120
$O+c-C_6H_{12}$	3.29×10^{-10} (2.5)	19.0±7.6	870-1180
$O+neo-C_5H_{12}$	1.26×10^{-9} (3.0)	37.0±8.8	880-1110

#The numbers in the parentheses indicate the uncertainty factors estimated by least-squares fit of the observed rate constants

The rate constants for the reactions (1)-(8) measured in this study are expressed by a simple Arrhenius form, $k_m = A\exp(-E_a/RT)$, and are listed in Table 1. The experimental data of our previous study for C_1-C_5 linear-chain alkanes (Miyoshi et al. 1993) are reanalyzed based on the revised calibration curve for O atoms. The present study indicates that the agreement between the experimental and the TST rate constants (Cohen and Westberg 1986) is satisfactory for CH_4 and C_2H_6. However, the agreement is not so excellent for higher alkanes, especially large differences in the TST predicted and measured rate constants for neo-pentane and cyclo-hexane are found. For the reactions of $O(^3P)+C_3H_8$, and the larger alkanes, no high temperature data above 1000 K are available. However, the present result generally agrees well with the extrapolation of that measured below 1000 K ([Tanzawa et al. 1980, Herron 1988)].

4. Discussion

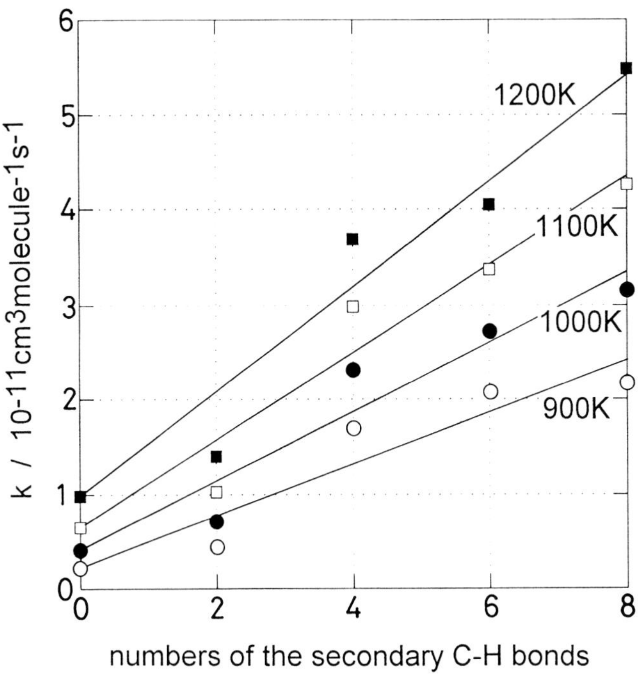

Fig. 1. The dependence of the overall rate constant on the numbers of C atoms in the secondary site of linear-chain alkanes (C_2H_6 to n−hexane) measured in this study. Present data; o: T=900 K, •: T=1000 K, □: T=1100 K, ▪ : T=1200 K

In the reactions of RH+X (X=OH, $O(^3P)$ or H), X interacts with individual C-H bonds rather than with the entire RH molecule. The competitive channels of abstraction of hydrogen atoms from the primary, secondary and tertiary sites for the alkanes have been extensively examined for the reactions with OH. The additivity rules for the rate constants of the local reaction sites are usually assumed to estimate the overall rate constants (for example: Atkinson et al. 1984). Based on the additivity rule, the rate constants for the direct abstraction reactions of a series of alkanes,

$$X + RH \longrightarrow XH + (n-R)/(s-R)/(t-R) \tag{9}$$

have been calculated by TST for X = O(^3P) (Cohen and Westberg 1986), and X = OH (Cohen 1991). The overall rate constants for these reactions are approximated by the sum of the local reaction rate constants for the attack of X to each C-H bonds, i.e., the total rate constant for the reaction (m) can be given by

$$k_m = k_{pm} + k_{sm} + k_{tm} \tag{10}$$

where, k_{pm}, k_{sm}, and k_{tm} denote the rate constants for the attack of O(^3P) on the primary, secondary, and tertiary site of the C-H bond, respectively.

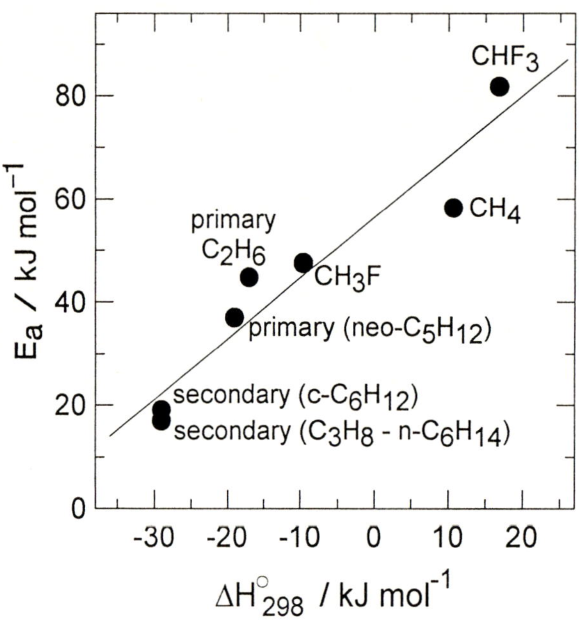

Fig. 2. The correlation of the measured activation energy and heat of reaction for O(^3P)+alkanes and fluoromethanes (Evans-Polanyi's relation)

The rate constant determined by experiments at high temperatures generally includes the contributions of these competing reaction channels, but due to the difficulty of separating the overall rate constants into those of the individual reaction channels, the validity of the relation (10) at high temperature range has not been examined well. The overall rate constants for O+linear-chain alkanes (from C_2H_6 ($m = 2$) to $n-C_6H_{14}$ ($m = 6$)) measured at 900-1200 K in this study are plotted against the numbers of the C-H bonds in the secondary site in Fig.1. Although this study has revealed that the linear correlation of the rate constants for the reaction of O(^3P) and linear-chain alkanes with the number of carbon atoms is worse than that predicted by the TST calculation, the averaged magnitudes of the rate constants kpm and ksm are estimated from the present experimental data, i.e., the slope and the intercept shown in Fig.1 are evaluated by using a least-squares analysis at fixed temperatures. The temperature dependence of the rate constants for the primary and the secondary sites in the linear-chain alkanes determined in this way revealed slightly smaller than that evaluated by TST.

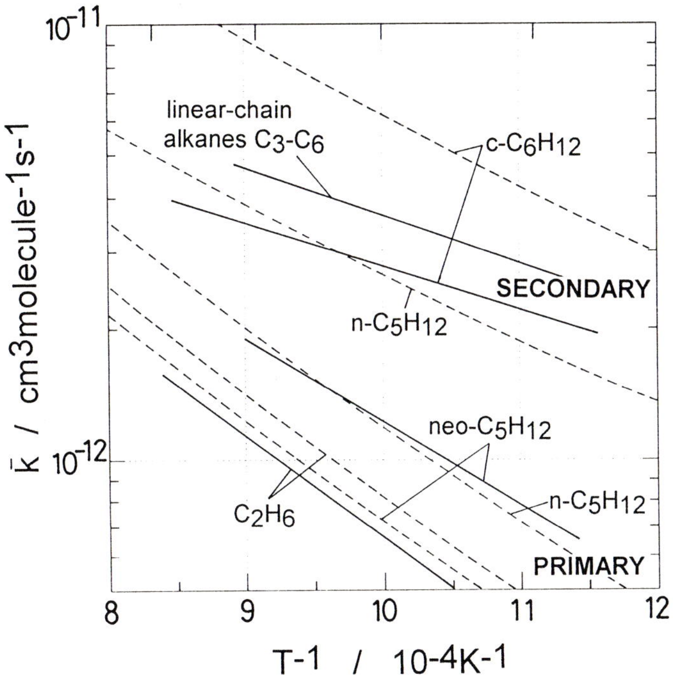

Fig. 3. A comparison of the rate constant per single C-H bond for the attack of O atom to the primary and secondary sites in alkanes. Present data: solid curve, TST calculation (Cohen and Westberg):dotted curve

An empirical relation of Evans and Polanyi for the linear relation between the effective activation energy E_a and the heat of reaction ΔH

$$E_a = \alpha\ \Delta H + \beta \tag{11}$$

has been tested for O+RH (Herron 1988) and OH+RH [Cohen 1991] at low temperature range, where, α and β are the empirically determined constants for the series of the same type of reactions. The rate constants for individual sites in the linear-chain alkanes determined in this study with an approximation of additivity law are compared with the experimental results for different alkanes and fluoromethanes. In addition to the linear-chain alkanes, *neo*-pentane and *cyclo*-hexane are used to examine the reaction rate constants for the primary and the secondary sites, respectively. The results on the estimated activation energies are summarized in Fig.2. As is indicated in this figure, the activation energies for the reactions of O+alkanes and fluoromethanes are found to have good correlation with the heat of reactions.

High temperature rate constants above 1000 K evaluated in this study are summarized in Fig.3 compared with those of TST calculation by Cohen and Westberg. Here, the rate constants are divided by the numbers of C-H bonds in the respective sites.

For C_2-C_6 linear-chain alkanes, the rate constants by Cohen and Westberg are found to be consistent with our experimental results both for primary and the secondary sites, in contrast, large difference are indicated for *neo*-pentane and *cyclo*-hexane.

The reactions of O(^3P)+*neo*-pentane and *cyclo*-hexane were also examined via a crossed molecular beam experiment, where, the cross sections for producing OH(ν, J) were measured via a LIF technique for a wide range of collision energy E_c (Andersen and Luntz 1980). These reactive cross sections represent only relative values, however, they can be used to test the relative magnitudes of the thermal rate constants for these two alkanes. The relative cross sections of the molecular beam experiment are fitted to the form

$$\alpha(E_c) = C(E_c - E_0)^n \tag{12}$$

by using nonlinear least-squares fit, where, E_0 is the threshold energy, and C and n are the constants for the respective reactions. Arrhenius parameters for the thermal rate constant (pre-exponential A factor and activation energy E_a) can be given by the energy dependence of the cross section of the form (12),i.e.,

$$A = (2/kT)^{2/3}/(\pi\mu)^{1/2} \int_{E_a}^{\infty} \{E_c\sigma(E_c)\exp[(E_a - E_c)/kT]\}\, dE_c \tag{13}$$

$$E_a = E_0 + kT\left\{(n - 0.5) + [1 + E_0/(N+1)kT]^{-1}\right\} \tag{14}$$

With these relations, the ratio of the rate constants for the reactions of O(^3P)+*cyclo*-hexane and *neo*-pentane is estimated from the beam experiment and compared with the thermal data of this study. The agreement of the ratios of the reaction rate constants is found to be satisfactory; i.e., $k_8/k_7 = 3$ (present experiment) and $k_8/k_7 = 4$ (crossed beam experiment). Meanwhile, TST calculation indicates a large difference from these experimental results, i.e., $k_8/k_7 = 10$ at 900 K. We have recently reinvestigated the rate for O+*cyclo*-hexane at room temperature via a LIF method for O atoms. The discrepancy for k_8/k_7 is suggested to be due to the rate constants employed as standard in the TST calculation.

References

Andresen P, Luntz AC (1980) The chemical dynamics of the reactions of O(^3P) with saturaed hydrocarbons. I. Experiment. J. Chem. Phys. 72: 5842

Atkinson R (1989) Kinetics and mechanisms of the gas-phase reactions of the hydroxyl radical with organic compounds. J. Phys. Chem. Ref. Data, Monograph No. 1

Cohen N (1991) Are reaction rate coefficients additive? Revised transition state theory calculations for OH+alkane reactions. Intl. J. Chem. Kinet. 23: 397

Cohen N, Westberg KR (1986) The use of transition-state theory to extrapolate rate coefficients for reactions of O atoms with alkanes. Intl. J. Chem. Kinet. 18: 99

Herron JT (1988) Evaluated chemical kinetic data for the reaction of atomic oxygen O(^3P)with saturated organic compounds in the ga phase. J. Phys. Chem. Ref. Data 17: 967

Miyoshi A, Ohmori, K, Tsuchiya K, Matsui H (1993) Reaction rates of atomic oxygen with straight chain alkanes and fluoromethanes at high temperatures. Chem. Phys. Lett. 204: 241

Ohmori K, Yoshimura M, Koshi M, Matsui H (1992) A flash photolysis study of CH_4-O_2 mixtures behind shock waves: examination of the reaction CH_3+O_2. Bull. Chem. Soc. Jpn. 65:1317

Tanzawa T, Klemm RB (1980) Paper presented at the 179th National Meeting of the American Chemical Society, Houston, p 17

Nonlinear Effects of Ultrasonic Cavitation in a Catalyst Production Process

V.Yu. Borodyllin, N.V. Malykh and V.M. Petrov
Institute of Thermophysics, Siberian Branch of Russian Academy of Sciences, Novosibirsk 630090,Russia

Abstract. A new ultrasonic synthesis method of heteropolycompounds using nonlinear effects at cavitation is discussed, that reduces process time, removes polystage and ecologically unhealthy waste.

Key words: Ultrasonic cavitation, Heteropolycompounds, Chemical synthesis

1. Introduction

The problem discussed is the development of a simple, no-waste, low power method for the synthesis of heteropolycompouds (HPCs) in chemical technology.

The interest in heteropolyacids is due to their unique properties, high molecular weight, electronic and protonic conduction, to the reduction ability with generation of coloured compounds, and to the possibility of a wide variation of the element composition.

Besides the use of HPCs in catalysis for the production of 2- propanol, tret-butanol, 2- butanol, for oxidation of acrolein in acrylic acid and for oxidation of methane, HPCs are used in analytical chemistry and clinical medicine, electronic industry.

The known methods of HPC production are the multistage processes with injection of intermediate substances in solution and following removal of impurities of foreign ions. These methods are lengthy, up to 20 days and nights, require power-intensive equipment and produce waste: salts of sodium and acids.

The purpose of the present ultrasonic method for HPC synthesis is to simplify the process, reduce the dwell time and energy expenditure and eliminate waste; in addition the method may be extended to the production of a range of substances The ultrasonic method allows also to inject additives of metals and nonmetals in solution for the production of catalyst compositions and to obtain fine HPCs: (P-Mo-V), (Si-Mo-V), (As-Mo-V), (Bi-Mo-V), (Sb-Mo-V) without injection of additives of foreign ions, as in the well-known chemical methods of HPC synthesis.

2. Nonlinear effects induced by ultrasonic cavitation

It is known from results in the field of ultrasonic cavitation (Sirotuk 1968) that the physical-chemical action of ultrasound on solutions having solid additives at ultrasound intensity $0.1 \cdots 10$ W/cm^2 leads to the grinding of particles down to fractions of one μm, mixing and heating of solutions, and also to the generation of hydrogen peroxide in water solutions, rendering easier the dissolution of some components in HPC production. Heating, crushing, mixing and activation of reagents take place due to the nonlinear effects induced by cavitation. Crushing of solid metal oxide particles is carried out by shock waves and cumulative microjets at the collapse and resonant pulsation of vapour-gas bubbles on and near the particle surfaces (Preece 1982). The dispersive action is enhanced by a higher velocity of bubble collapse. Collapse velocity is influenced by the viscosity of, the amount of dissolved gas, the temperature, and the external pressure. Mixing and heating of the solution occurs through radial oscillations of the liquid near the bubbles, acoustic and convective flows. Ultrasonic cavitation induces radiolysis of water stream in cavities. It causes chemical reactions in the solution and cavitation voids with the release of hydrogen peroxide that improves the conditions of dissolution for same reagents (Elpiner 1968).

Shock Waves @ Marseille II
Editors: R. Brun, L. Z. Dumitrescu

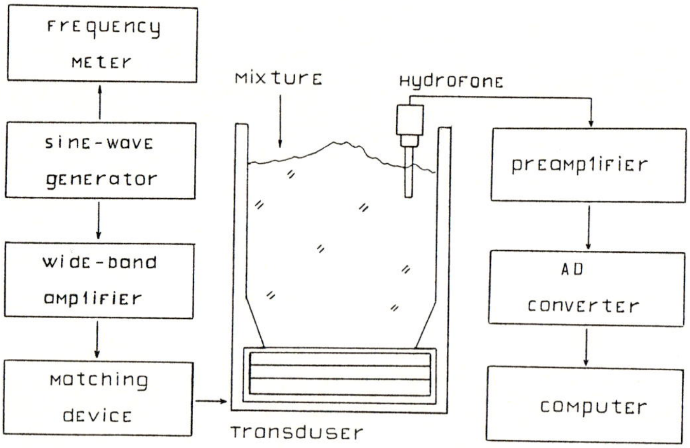

Fig. 1. Schematic of the set-up

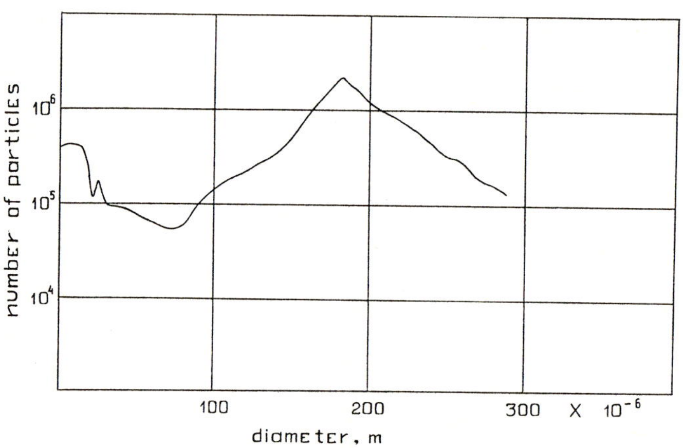

Fig. 2. Particle size distribution of silicium dioxide (450 kHz, 3 W/cm^2)

3. Experiment and results

For HPC synthesis by ultrasonic technique a wide band ultrasonic (U-3) installation was designed, consisting of a generator and piezoelectric radiator, operating in the frequency range of 50 kHz-1000 kHz with regulated power on the radiator that allows to choose optimal parameters for the best dissolution of reagents, depending on the initial dispersivity of particles and their chemical structure. A schematic of the U-3 set-up is given in Fig.1.

The electrical power of the piezoradiator was determined by the current intensity, voltage and phase angle between them. The acoustic power was calculated on the basis of the electrical one, the convertor efficiency and its transfer function. Is well known that one of the main mechanisms of chemical reaction acceleration consists in an increase of the reagent interaction area, depending on particles size and material porosity (Lorimer and Mason 1987). To investigate these matters, powders of molybdenum, vanadium and silicon oxides were studied by the methods of mercury porometry, gas absorption and granulometry before and after exposure to ultrasonic cavitation.

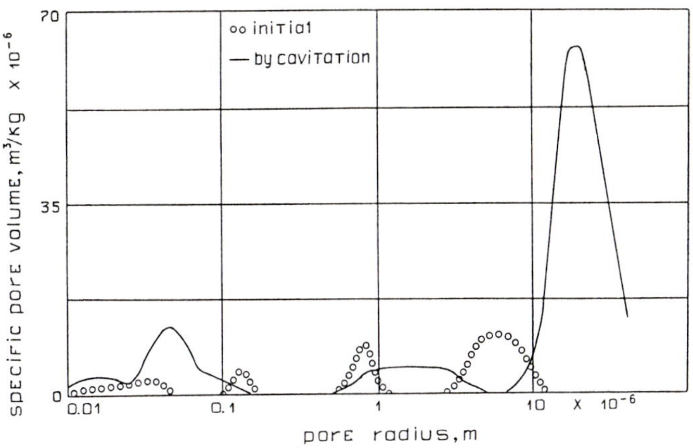

Fig. 3. Specific pore volume distribution of silicium dioxide (450 kHz, 3 W/cm^2)

Results of this study, given in Figs.2 and 3 confirm the increase of reagents interaction area caused by crushing and increase of the total pore specific volume. The technology used was developed on silicon dioxide powder, which provides one of the most rugged crystal lattices.

In Fig.2 the particle distribution is given as a function of diameters after exposure to ultrasound during an hour. The initial size of the particles was $d \geq 350\,\mu$m. After exposure the particle size decreased, all particles were in the range of $5 \div 200\,\mu$m with peaks at 180 μm and 5 μm. In Fig.3 results of mercury porometry are given confirming the increase of the total pore specific volume by 3 times and the maximal radius by 2 times. From the results obtained the maximal failure of particles is observed when the half-period of ultrasonic frequency is equal to the time of bubble collapse. Considering that the initial sizes of the bubbles are comparable with the pore radii ($500 \div 70000$ Å, Fig.4) an optimal ultrasonic frequency was chosen that could be varied during the experiment in the range of 100 kHz $\div$ 1 MHz, while the power was in the range power $1 \div 15$ W/cm^2 .

In the process of heteropolycompounds production, the temperature and solution colour were controled. The composition of heteropolyacid obtained was determined by the NMR- spectrometry method.

The new method of HPC synthesis is as follows: All components, in the form of water solution of phosphoric acid with oxides of vanadium and molybdenum, for instance, are poured simultaneously in an ultrasonic bath, where they are mixed, heated and crushed, and submitted to the activating action of hydrogene oxide. All the shown operations occur simultaneously as a result of the action of powerful ultrasound upon the solutions accompanied by cavitation in the solution that accelerates the production process of heteropolyacids (HPA) and their salts up to $1.5 \div 5$ hours and destroys waste. Thus, the following HPC compounds were obtained by ultrasonic technique: acids: HPA-2, ($H_5PMo_{10}V_2O_{40}$), molybdenophosphoric acid ($H_6P_2Mo_{18}O_{40}$), as confirmed by the NMR-analysis results given in Fig.4.

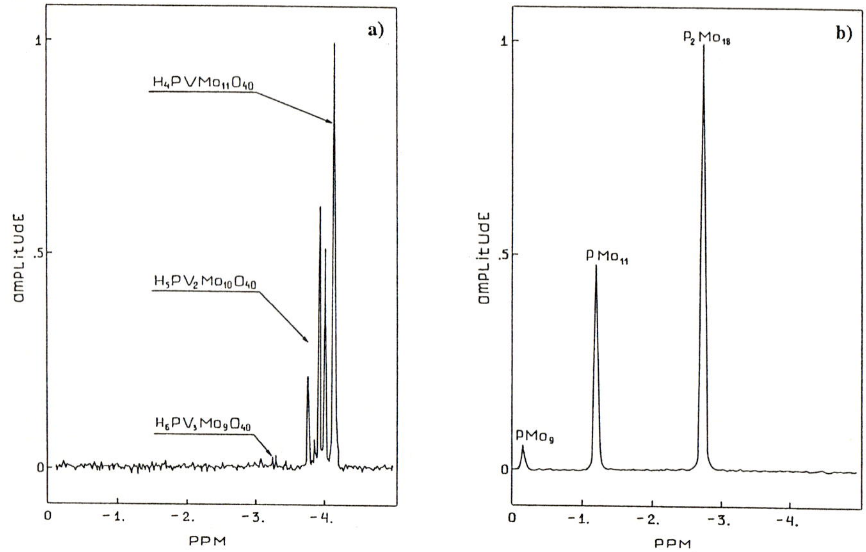

Fig. 4. Nuclear magnetic resonance spectrum of molybdenovanadiophosphoric (a) and molybdenophosphoric (b) heteropolyacids (HPA)

4. Conclusion

A new ultrasonic method for the synthesis of heteropolycompounds has been developed. Simultaneous crushing, mixing and chemical activation of reagents was obtained due to nonlinear effects induced by cavitation. This results in reduction of process time, removing ecologically unhealthy waste, broadening the range of compounds obtained.

Acknowledgment

Authors are very thankful to the Soros Fund for providing financial support to the research carried out.

References

Elpiner IE (1968) Ultrasound. Physical-chemical and biological action. Phys.-Mat. Giz, Moscow

Lorimer JP, Mason TJ (1987) Sonochemistry. The Physical Aspects. Chem. Soc. Rev. 16:239

Preece CM (1979) Erosion by cavitation. In: Preece CM (eds) Erosion. Academic Press, New-York San-Francisco London, pp 268-330

Sirotuk MG (1968) An experimental study of ultrasonic cavitation. In: Rosenberg LD (ed) Powerful ultrasonic fields. Nauka, Moscow, pp 167

Part 2: Combustion Kinetics

Combustion and Shock Waves in H_2-O_2-Steam Systems

A.Yu. Kusharin, O.E. Popov and G.L. Agafonov
Institute of Chemical Physics, Russian Academy of Science, Kosygin str.4, Moscow, 117977, Russia

Abstract. The burning velocities of stoichiometric H_2-O_2-steam mixtures with steam concentrations up to 70% (vol.) at elevated pressures and temperatures have been measured and the results compared with numerical calculations. The influence of thermal radiation and kinetic mechanism on the burning velocity was evaluated. The detonability of such mixtures at high initial pressure and temperature was checked.

Key words: Burning velocity, Steam, Thermal radiation, Detonation

1. Introduction

In recent years the influence of the steam as a diluent on the burning velocities and the flammability limits in hydrogenous mixtures is actively examined. Liu and MacFarlane (1983) measured burning velocities in H_2-air-steam mixtures at temperatures 298-523 K and steam concentration up to 15% (vol.). Koroll and Mulpuru (1986) obtained burning velocities for the same mixtures at 373 K and steam fractions up to 50% (vol.). For stoichiometric H_2-O_2 mixtures containing N_2 and steam Koroll and Mulpuru (1986) revealed that for diluent fraction less than 45% replacing N_2 with H_2O increases the burning velocity. They explained such behaviour by the high third-body efficiency of H_2O molecule in the reaction $H + O_2 + M \rightarrow HO_2 + M$. Beyond 60% the steam effect is lacking. By comparing the flammability limits in H_2-air-H_2O and H_2-O_2-H_2O mixtures Kumar (1985) disclosed that at higher temperatures steam behaves as a simple diluent. The foregoing results are refered to near-atmospheric pressures. At elevated pressures, an important region in terms of water-cooled reactors operation, experimental data involving steam is scarce. Kogarko et al. (1984) revealed that at a fixed temperature of 473 K, as initial pressure increases from 0.4 to 2.0 MPa, the limit mass concentration of the stoichiometric H_2-O_2-H_2O mixtures decreases. With a rise of pressure in steam-diluted combustible mixtures the law of ideal gas breaks down. This causes the change of thermophysical parameters in the water vapor system. Furthermore, at higher pressures the role of absorption of thermal radiation by steam molecules as well as the role of three-molecular reactions will be enhanced. Thus, one might expect some features in oxyhydrogen combustion with a hgh amount of steam.

We have carried out an experimental and theoretical investigation of the influence of steam on the burning velocity in stoichiometric H_2-O_2-H_2O flames. Initial pressures, temperatures and concentrations of saturated steam were varied through the ranges 0.15-3.0 MPa, 373-473 K, 55-70% (vol.) , respectively. We also investigated the possibility of shock waves generation in combustion of the same mixtures at initial pressure up to 2.0 MPa and temperatures 430-450 K under tube confining conditions.

2. Methods

2.1. Experimental

The burning velocities were measured by the constant-volume bomb method. Experimental arrangement and procedure are similar to those described by Kogarko et al.(1984) The reaction vessel with internal diameter 21.3 cm was placed in a 2.5 KW electric heater with a heatproof casing. In the experiments temperature differences were no more than 4 K. The combustion process was initiated by the burnout of a nichrome wire spiral in the centre of the spherical vessel.

Shock Waves @ Marseille II
Editors: R. Brun, L. Z. Dumitrescu

The bottom of the vessel was equipped with a pressure transducer with water cooling and an oscilloscope was used to record the rate of pressure growth.

Experimental data for the burning velocity S_u were obtained by pressure-time histories of combustion. The expression for $S_u = S_u(P,t)$ is identical to that of Bradley and Mitcheson (1976). Since a flame is propagated in the closed vessel, the burning velocity depends on both temperature T and pressure P, so S_u is often represented as: $S_u = S_{u_0}(T/T_0)^{\alpha}(P/P_0)^{\beta}$, where α and β are temperature and pressure exponents, respectively, and subscript 0 refers to initial conditions. Assuming that the unburnt mixture is adiabatically compressed, S_u can take the form : $S_u = S_{u_0}(P/P_0)^n, n = \beta + \alpha(\gamma_u - 1)/\gamma_u$, where γ_u is the specific heat ratio for the unburnt mixture. For getting higher accuracy of S_u measurements and widening the ranges of pressure and temperature within which S_u can be determined, a spline-interpolation procedure was used. The preceding approach was tested for stoichiometric H_2-air mixture. Initial conditions $P_0 = 0.1$ MPa and $T_0 = 298$ K were taken. As seen from Fig.1 good agreement of the present results with the experimental results of Milton and Keck (1984) is observed.

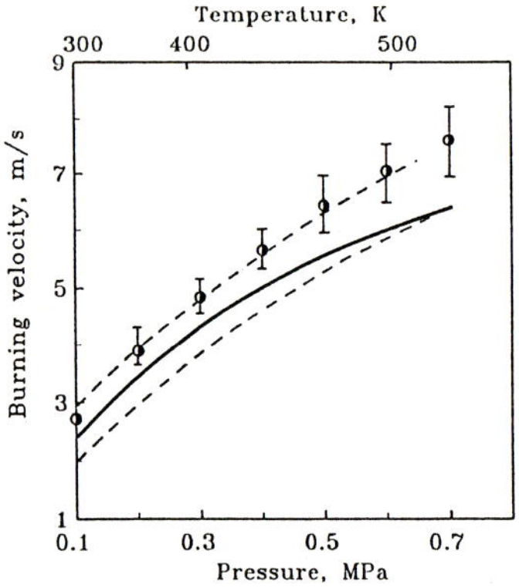

Fig. 1. Comparison between experimental and computed burning velocities of stoichiometric H_2-air mixture.
Broken lines: data of Milton and Keck.

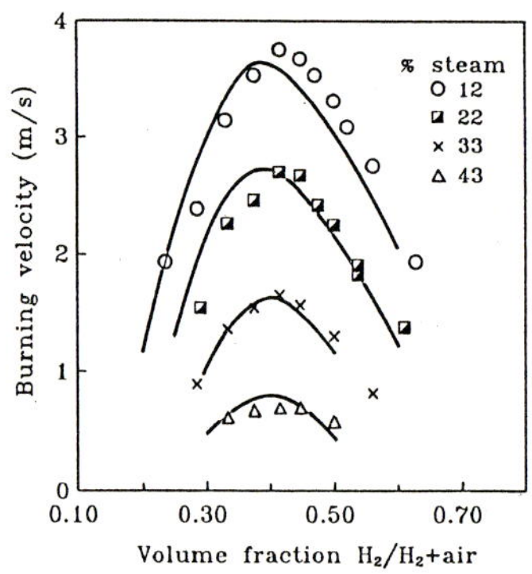

Fig. 2. Comparison between measurements (Koroll and Mulpuru) and present calculations (solid lines) of burning velocities for H_2-air-steam mixtures at 373 K and 0.1 MPa

2.2. Spherical flame propagation model

Burning velocities, radius and thickness of flame, flow velocities, temperature and species profiles were determined from the numerical solutions for laminar flame propagation in H_2-O_2-diluent mixtures. One-dimensional model calculations were made in Eulerian coordinates for freely propagating spherical flame using detailed kinetics with and without regard to thermal radiation.

2.2.1. Calculation model without regard to thermal radiation

The time-dependent conservation equations of mass, species and energy take the form:

$$\partial\rho/\partial t + \nabla_r(\rho v) = 0 \tag{1}$$

$$\rho(\partial Y_i/\partial t) = \nabla_r\left[D_{i,M}\,\rho(\nabla_r Y_i)\right] - \rho v(\nabla_r Y_i) + W_i; \quad i = 1,\ldots,N; \quad W_i = \mu_i \sum_{j=1}^{K} \nu_{ij} R_j \tag{2}$$

$$\rho C_p(\partial T/\partial t) = \nabla_r [\Lambda(\nabla_r T)] - C_p(\nabla_r T) + \sum_{j=1}^{K}(-\Delta H_j)R_j \tag{3}$$

where $\nabla_r = (1/r^2)\,\partial/\partial r\,(r^2)$; N, K are the number of components and of reactions ; ρ, T, v are, respectively, the gas density, temperature and flow velocity ; μ_i, Y_i are the molecular mass and mass fraction of the species i; C_p is the specific heat capacity; Λ and $D_{i,M}$ are the coefficients of heat conductivity and diffusion ; ΔH_j is the enthalpy of the elementary reaction j. The source term W_i depends on the reaction rate R_j. The freely propagating flame practically means that vessel walls are far from the flame front. So the boundary conditions assumes the form:

$$\begin{aligned} &(\partial T/\partial t)_{r=0} = 0 \qquad && T_{r=R_W} = T_0 \\ &(\partial Y_i/\partial r)_{r=0} = 0 \qquad && (\partial Y_i/\partial r)_{r=R_W} = 0 \end{aligned} \tag{4}$$

The H_2-O_2 reaction mechanism involving 26 elementary steps with rate parameters as well as third-body efficiencies are taken from the work of Warnatz (1984). The calculation begins with defined initial profiles of species mass fractions and temperature. By applying an implicit grid-point method to the system of unsteady equations (1-3) with boundary conditions (4) the laminar burning velocity is found when a steady state is reached as described by Spalding et al.(1971). The space derivatives are discretized on non-uniform grid. The non-linear reaction rate terms are linerized by a Taylor expansion. The time step Δt is obtained from Courant's criterion :$\Delta t < (\Delta r)_{min}/v_{max}$, where $(\Delta r)_{min}$ is chosen so that 30-35 points are in the flame front. The increase of Δr from point to point on the grid is no more than 15% . On each time step the iteration procedure is used. The correctness of calculating procedure is confirmed by Fig.2.

2.2.2. Computational model with regard to thermal radiation

In this model, a term *RAD*, denoting the radiant energy density per unit time appears in the left side of equation (3). According to Siegel and Howell (1972), the term *RAD* takes the form :

$$RAD = 4\int_0^{\infty} a_\lambda\, e_{\lambda b}\,(\lambda, T)d\lambda \; - \; \int_0^{\infty}\!\int_0^{4\pi} a_\lambda\, I_\lambda\,(\lambda,\omega)d\lambda\, d\omega \tag{5}$$

where λ is the wavelength, w the solid angle, a_λ the spectral absorption coefficient, I_λ the spectral intensity, $e_{\lambda b}$ the blackbody hemispherical spectral emissive power. Furthermore the following assumptions are made:

1) diffuse-gray gas radiation is assumed;
2) thermal radiation and absorption in gas are considered whereas scattering is ignored;
3) thermodynamical equilibrium conditions holds.

Then, for a spherical k-layer, from Eq.5:

$$(RAD)_k = 4\,a_k\,\sigma\,T^4 - 2\,a_k\,Q_R; \qquad Q_R = Q_+ + Q_-$$

where σ is the Stephan-Boltzmann constant , Q_- the incident emissive power which reaches the k-layer. Q_- and Q_+ assumes the form :

$$Q_- = Q_0\,(F_\tau)_{w-k} + (F_\alpha)_{w-k}\;; \qquad Q_+ = Q_-\;(F_\tau)_{k-k} + (F_\alpha)_{k-k}$$

$$F_\tau = F\,\exp(-\text{æ})\;; \qquad F_\alpha = \int \sigma T^4 \exp\left[-(\text{æ} - \text{æ}_*)\right] F\, d\text{æ}_*\;; \qquad \text{æ} = \int a\, ds$$

where F is the angle factor for unabsorbing medium, æ is the optical thickness, Q_0 is the radiosity:

$$Q_0 = \left\{\varepsilon\sigma T^4 + (1-\varepsilon)\,(F_\alpha)_{w-w}\right\} / \left\{1 - (1-\varepsilon)\,(F_\tau)_{w-w}\right\}$$

The absorption coefficients a_k of water vapor and the emissivity ε are dependent upon overall pressure and temperature and are calculated by the polynomials of Leckner (1972).

3. Results and discussion

In order to define the pressure and temperature dependence of the burning velocity in a H_2-O_2-steam mixture the initial temperature was varied in 20 K steps from 373 to 473 K. By application of the constant volume bomb method S_u has been measured for P up to 7.5 MPa and T up to 615 K. The variation of burning velocity with pressure at the fixed temperature of 473 K is shown in Fig.3. For all examined compositions there is a region of pressures where burning velocities are steeply decreasing. The value of the pressure exponent in this region varies between -1.5 and -1. A drastic reduction of the burning velocity with a raise in pressure in H_2-O_2-steam mixtures is qualitatively confirmed by calculations. Fig.4 presents a comparison between calculated and measured data for the oxyhydrogen mixture with 65% steam concentration and shows that the calculated burning velocity $S_u(P)$ has the same tendency to decrease as observed in the experiments. Quantitatively, predicted results are in a wholly satisfactory agreement with experimental findings only within the range of pressure from 0.15 to 0.3 MPa. At higher pressures application of the model for thermal radiation increases the burning velocity and approaches values of calculations and measurements. For the studied mixture the relative increase of burning velocity at pressures 2-5 MPa is equal to 12-18%. One of the possible causes of the disparity at elevated pressures is that a non-ideal behavior of steam and therefore a change of the kinetic parameters may occur. Therefore, a kinetic simulation was carried out of the flame propagation by varying the third-body efficiency of steam for the three-molecular reaction $H + O_2 + M \rightarrow HO_2 + M$. It turns out that the calculated burning velocities are in complete agreement with experimental values if the third-body efficiency of H_2O molecules is reduced smoothly by a factor varying from 2 at 0.5 MPa to 10 at 5 MPa. Thus, a chemical kinetic interpretation of the observed $S_u(P)$ dependence in H_2-O_2-steam mixtures is conceivable but, nevertheless, an inclusion of thermal radiation into the common model at higher pressures is necessary.

Table 1. The mean values of α, β and S_{u_0} in stoichiometric H_2-O_2-Steam and H_2-O_2-N_2 flames

Volume fraction of diluent	55.6	60	65	70
$0.2 < P < 0.8$ MPa $373 < T < 473$ K				
α_{steam}	2.23	2.65	3.01	3.28
α_{N_2}	1.26*	1.60	1.82	2.10
β_{steam}	-0.12	-0.34	-0.46	-0.54
β_{N_2}	0.26*	0.20	0.13	0.0
$T_0 = 373$ K				
P_0, MPa	0.18	0.17	0.155	0.14
$(S_{u_0})_{steam}$, m/s	3.06	2.29	1.22	0.78
$(S_{u_0})_{N_2}$, m/s	3.35*	3.03	2.50	1.76

* data of Milton and Keck (1984) are used

It is interesting to compare the pressure and temperature dependence of the burning velocity for stoichiometric H_2-O_2 mixtures containing N_2 and steam. The mean values of the temperature and pressure exponents α and β obtained in the present work at pressures $0.2 < p < 0.8$ MPa and temperatures $373 < T < 500$ K are listed in Table 1. Here also are given the burning velocities S_{u_0} at initial conditions $T_0 = 373$ K and $P_0 = 0.14$-0.18 MPa. Table 1 illustrates that H_2-O_2-steam mixtures, contrary to H_2-O_2-N_2 mixtures exhibit negative pressure exponent β. Moreover, replacing N_2 with steam in H_2-O_2-diluent mixtures increases the value of the temperature exponent α. As an example, in the same region of pressures and temperatures for a volume fraction of diluent 55.6%, calculated values of α and β are as follows: $\alpha_{N_2} = 1.26, \alpha_{steam} = 1.75; \beta_{N_2} = 0.14, \beta_{steam} = -0.16$.

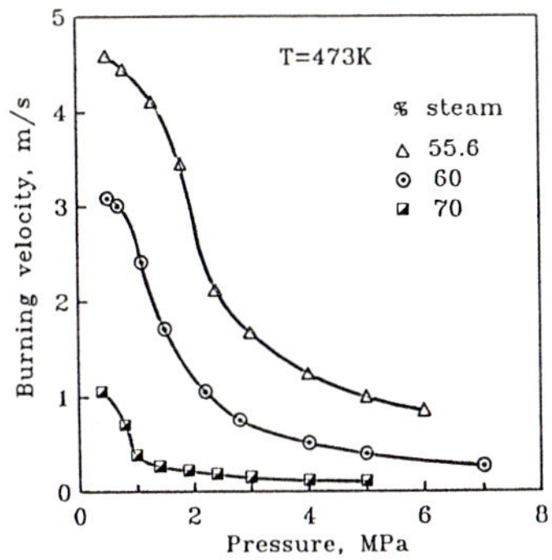

Fig. 3. Burning velocities of stoichiometric H_2-O_2-steam mixtures at elevated pressures

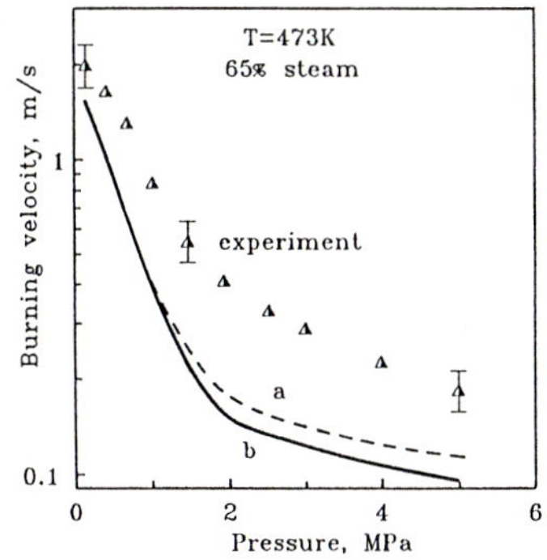

Fig. 4. Pressure dependence of burning velocity in stoichiometric H_2-O_2-steam mixture. Curves **a** and **b**: computed results with and without regard for thermal radiation, respectively

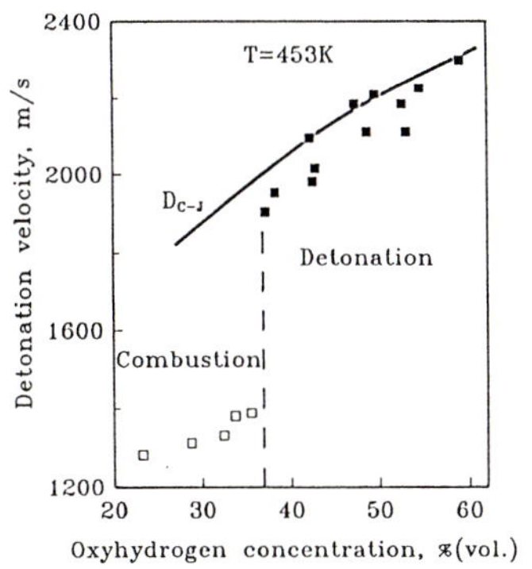

Fig. 5. Detonability limit in stoichiometric H_2-O_2-steam mixtures. Pressure of saturated steam: 1.0 MPa. A 2m-long, 2cm-ID tube is used

As shown in the experiments and calculations, the presence of a great amount of steam on the one hand, steeply decreases the burning velocity with increase of pressure, and on the other hand, more highly than in the presence of N_2, increases the burning velocity with a rise of temperature. Such $S_u(P,T)$ dependences in H_2-O_2-steam mixtures are presumably associated with the joint action of kinetic and thermal radiation mechanisms on the flame rate at elevated pressures and temperatures.

4. Detonation and shock waves in H_2-O_2-steam mixtures

The experimental results obtained in the closed spherical vessel indicated the absence of shock wave generation in the combustion of premixed H_2-O_2-steam mixtures if the steam concentration is over 50% (vol.). Moreover, in accordance with Fig.3, the increase of pressure will deteriorate the conditions for acceleration of oxyhydrogen flames in the presence of steam. Thus, shock waves in the steam-diluted combustible mixtures arise only in highly turbulent flows. Our experiments in a 2m-long, 2cm-ID steel tube revealed that at high initial pressure (up to 2.0 MPa) and temperature (430-450 K) flames can accelerate up to 1400-1500 m/s. The transition of deflagration to detonation in these mixtures is achieved at steam concentrations less than 40% (vol.). The results of the direct initiation of detonation by oxyhydrogen mixture are shown in Fig.5. The H_2-O_2-steam mixtures appeared to be detonable at steam concentrations less than 65% (vol.).

5. Conclusions

Two different modes of the combustion in the H_2-O_2-steam mixtures are considered: laminar spherical flame propagation and detonation. Both the numerical model for the flame and experiments show a drastic decrease of the burning velocity with increase of pressure. In spite of this decrease, the possibility of detonation formation in conditions of tube confinement at pressures up to 2.0 MPa and steam concentrations up to 65%(vol.) is established.

References

Bradley D, Mitcheson A (1976) The venting of gaseous explosions in spherical vessels I.Theory and II.Theory and experiments. Combust.Flame 32:201-217

Kogarko SM, Lyamin AG, Popov OE, Kusharin AYu, Dubrovin AV (1984) Determination of flame propagation limits in stoichiometric oxyhydrogen mixtures with steam. In: Proc. Special. Meeting, IAEA- TC-427.6, Vienna, pp 37-41

Koroll GW, Mulpuru SR (1986) The effect of dilution with steam on the burning velocity and structure of premixed hydrogen flames. In: Sangiovanni JJ, Kennedy LA (eds) Symp. (Intl.) Combust. 21, Pittsburgh, The Combustion Inst. pp 1811-1819

Kumar RK (1985) Flammability limits of hydrogen-oxygen-diluent mixtures. J. Fire Sci. 3:245-262

Leckner B (1972) Spectral and total emissivity of water vapor and carbon dioxide. Combust. Flame 19:33-48

Liu DDS, MacFarlane R (1983) Laminar burning velocities of hydrogen-air and hydrogen-air-steam flames. Combust. Flame 49:59-71

Milton B, Keck J (1984) Laminar burning velocities in stoichiometric hydrogen and hydrogen-hydrocarbon gas mixtures. Combust. Flame 58:13-22

Siegel R, Howell JR (1972) Thermal radiation heat transfer. McGraw-Hill Book Comp., New York

Spalding DB, Stephenson PL, Taylor RG (1971) A calculation procedure for the prediction of laminar flame speeds. Combust. Flame 17:55-64

Warnatz J (1984) Rate coefficients in the C/H/O system. In: Gardiner WC (ed) Combustion Chemistry, Springer-Verlag, NY

Comparison Between the Ignition of H_2-Air-CO_2 and H_2-Air-H_2O Mixtures by a Hot Gas Jet

N. Djebaili, R. Lisbet, C. Paillard and G. Dupré
National Center of Scientific Research and University Orléans - France

Abstract. The ignition limits of hydrogen-air mixtures diluted with steam or carbon dioxide have been studied in function of the mixture composition. Ignition was produced by a hot transient jet emerging from the downstream end of a shock tube and entering a combustion chamber. The spurted gas was heated by a reflected shock wave between 750 and 2800 K. The ejection velocity was subsonic. It was noticed that, with an initial jet at 2600 K, the ignition area of H_2-Air-CO_2 mixtures was larger than that of H_2-Air-H_2O mixtures. The ignition area was located inside the flammability range of these systems for rich mixtures. On the contrary, ignition limits and flammability limits were almost similar for lean mixtures. The chemical effect of steam on the ignition of these combustible systems are discussed.

Key words: Hydrogen combustion, Shock heating, Jet ignition, Ignition limits

1. Introduction

Hydrogen combustion is of a great interest in many applications such as ramjet burners, supersonic combustion or nuclear safety. Many data are available on the ignition mechanism of H_2-Air mixtures whereas very few are given in the literature on the ignition of these mixtures by a hot transient gas jet. Wolfhard (1958) studied the ignition of fuel-air mixtures by a jet of hot gas which was heated in a ceramic furnace. In this case, the ignition gas was continuously injected into the explosive mixture and its maximal temperature was limited to 1500 K. Enomoto et al. (1980) determined the ignitability of H_2-Air mixtures by hot air. It was found that the mixture ignited at a temperature close to self-ignition temperature. In the case of ignition by hot nitrogen, the ignition temperature is much higher and close to that occuring with ignition by a hot surface. Sakurai et al.(1985, 1986), studied the ignition of cold hydrogen spurted in hot air heated by shock waves. However, no information is available on ignition of mixtures H_2-Air-CO_2 or H_2O by a hot gas jet. The aim of the present study is to investigate the ignition of these mixtures by a transient jet of very hot gas (2000-2600 K) containing hydrogen.

2. Experimental arrangement

The experimental set-up has been built in the laboratory. It consists of a stainless steel cylindrical shock tube (4.40 m long, 52 mm ID) connected to a stainless steel cylindrical combustion chamber, 1.2 m long and 102 mm ID (Fig.1). The driven section of the shock tube is filled with a hydrogen - argon mixture and the driver section with helium. These two sections are separated by a double diaphragm. The driven section is coupled to the combustion chamber, filled with 60% hydrogen - 40% air mixture diluted by carbon dioxide or steam, by means of an injection system. This latter consists of a thin metallic plate (0.4 mm thick), pierced with a 2.5 mm diameter hole and obturated by a 12 μm thick terphane foil, which is ruptured less than 500 μs after the shock reflection at the tube end. The hydrogen - argon mixture is heated up, between 750 and 2800 K, by the reflected shock wave just before being spurted into the combustible mixture contained in the combustion chamber. The temperature behind the reflected shock is maintained approximately constant in the shock tube for several milliseconds. The pressure behind the reflected shock wave is chosen in order to obtain a subsonic velocity of the spurted gas.

Shock Waves @ Marseille II
Editors: R. Brun, L. Z. Dumitrescu

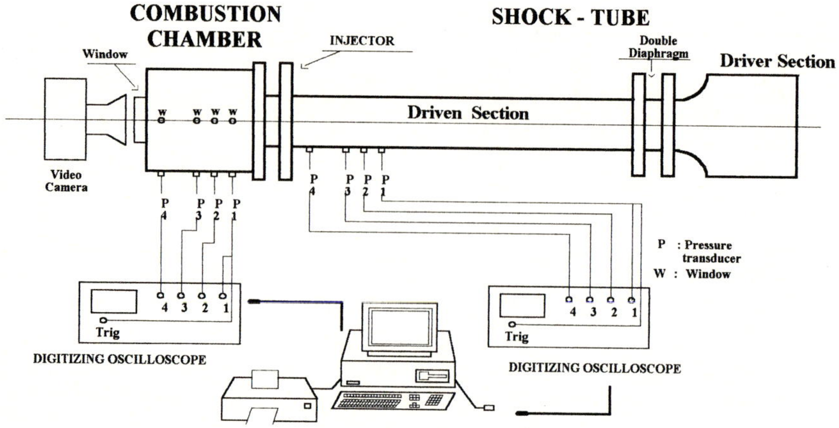

Fig. 1. Experimental apparatus

The driven section of the shock tube is equipped with four piezoelectric transducers for the measurement of the pressure behind the shock wave and of the shock wave velocity, with a rise time of 0.5 μs. The pressure rise during the combustion was measured by four piezoelectric transducers mounted flush with the inner surface of the combustion chamber. Coupled to the pressure gauges, four optical windows permitted to follow the propagation of the flame, initiated by the jet, by means of u-v sensitive photomultipliers. The gas mixtures, before admission, were prepared in two tanks by the partial pressure method and the whole experimental facility was trace-heated and insulated for reaching an initial temperature of 150°C.

3. Results and discussion

The low pressure section of the shock tube is filled up with a 60% H_2 - 40% Ar mixture which is heated up by a reflected shock wave between 700 and 3000 K. After reflection of the shock wave at the tube end, the diaphragm is ruptured with a delay that does not exceed 500 μs. The diaphragm rupture allowed the heated gas to flow through the jet orifice into a combustion chamber. The pressure generative of the jet was adjusted in order to obtain an initial subsonic velocity of the spurted gas. The combustion chamber contained H_2-Air mixtures, diluted by CO_2 or H_2O, at an initial pressure of 100$\pm$1 kPa.

3.1. Effects of the initial jet temperature on ignition

The first investigated parameter is the effect of the minimum temperature of the spurted gas on the ignition of a 18% H_2 - 82% air mixture diluted by carbon dioxide or steam. In the case of dilution with CO_2, the initial temperature is 293$\pm$2 K, and for steam, the initial temperature is 403$\pm$2 K. The lowest jet temperature able to ignite the combustible mixture without any diluent is about 750$\pm$20 K. This temperature is close to the self-ignition temperature observed by Conti and Hertzberg (1988) and by Enomoto et al.(1980). If 35% CO_2 is added to the combustible mixture, ignition occurs if the initial jet temperature is higher than 1200 K. When the amount of carbon dioxide is increased in the combustible mixture, the minimum jet temperature able to ignite the mixture increases too.

As shown in Fig.2, there is a CO_2 concentration limit (45%) above which even when the temperature of the spurted gases is raised, no ignition takes place and no flame propagates in the

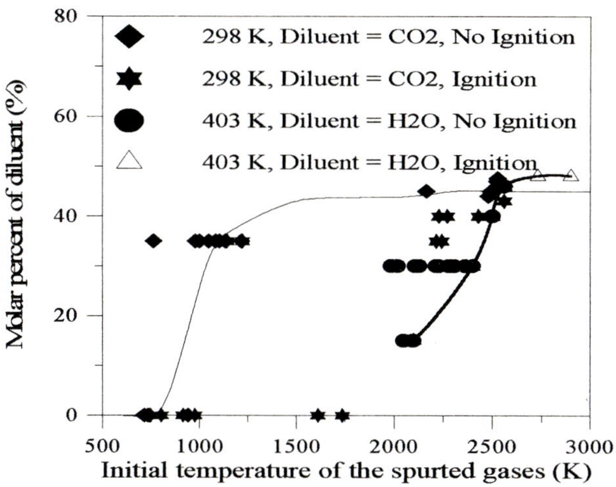

Fig. 2. Minimum ignition temperature of the spurted gas versus the concentration of diluent added to the H_2 - Air mixture initially at 100 kPa

combustible mixture. This CO_2 concentration limit defines the ignition limit by a hot transient gas jet and allows to define the critical temperature corresponding to an asymptotic temperature that is close to 2000 K.

When steam is used as a diluent, the asymptotic temperature is found to be higher, close to 2500 K. These observations show that the ignition of the given combustible mixture diluted with CO_2 is easier than when diluted with H_2O.

3.2. Ignition limit by a hot transient gas jet

The ignition limit by a hot transient gas jet has been determined for H_2-Air-CO_2 or H_2O. The combustible mixture was at an initial pressure and temperature respectively of 100±1 kPa and 403±2 K. Three equivalence ratios were chosen : $\phi = 1, \phi = 0.55$ and $\phi = 0.25$. The temperature able to produce the hot jet was about 2550±50 K, a value close to or higher than the asymptotic temperature defined earlier.

3.2.1 Ignition limits of H_2 - Air - CO_2 mixtures

Fig.3 shows the ignition limits of H_2 - Air - CO_2 mixtures by a hot gas jet determined in the present study. They are compared to the flammability limits given by Zabetakis (1965). In this case, the initial temperature of the fuel mixture is 298 K. The lower ignition limit of the mixture exempt of CO_2 is 4.5±0.5% H_2. This value is in good agreement with the flammability limit of H_2 - Air (Zabetakis 1965). For rich H_2 - Air - CO_2 mixtures, the limit was found to be equal to 66±1% H_2 and much lower than the upper flammability limit (72% H_2). For lean mixtures, ignition limits by a hot jet are the same as flammability limits.

3.2.2. Ignition limits of H_2 - Air - H_2O

Fig.4 illustrates the ignition limits of H_2 -Air - H_2O by a hot gas jet, and the flammability limits given by Langer and Stock(1988). As previously shown with lean H_2 - Air - CO_2 mixtures, the ignition limits of lean H_2 - Air - H_2O mixtures are in good agreement with the flammability limits. The same discrepancy between these two limits are observed for rich mixtures. The maximum

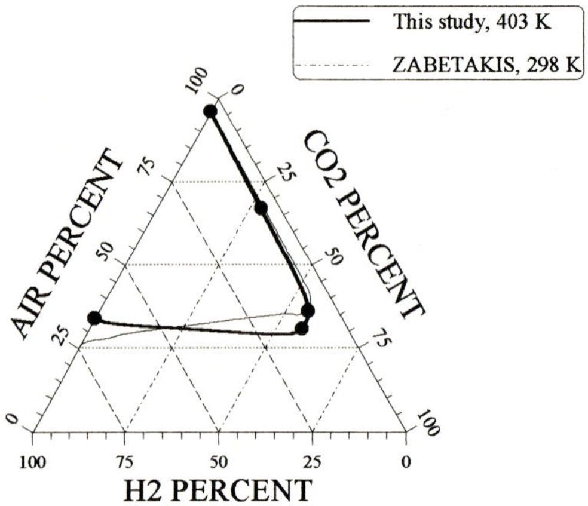

Fig. 3. Ignition limits of H_2 - Air - CO_2 mixtures by a hot gas jet

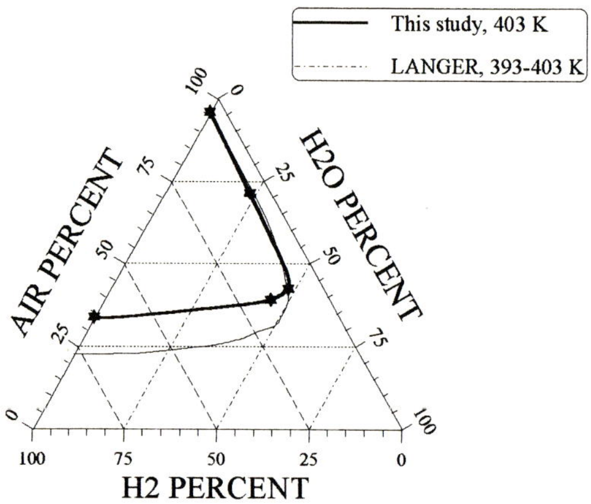

Fig. 4. Ignition limits of H_2 - Air - H_2O mixtures by a hot gas jet

concentration limit is found to be 48±1% H_2O and corresponds to a fuel equivalence ratio equal to 0.55, whereas in the case of flammability limits, the maximum steam concentration corresponds to the stoichiometric mixture.

3.2.3 Comparison of the H_2 - Air - CO_2 and H_2 - Air-H_2O diagrams

These two diagrams has been drawn for the same conditions of jet parameters. As shown in Fig.5, the ignition area in the case the of CO_2 dilution is larger than in the case of steam dilution. When carbon dioxide or steam are used as diluent, one can expect that the ignition area of H_2 - Air - CO_2 will be situated inside the ignition domain of H_2 - Air - H_2O, or at least will be the same, since the molar heat capacity of carbon dioxide is slightly higher than the molar heat capacity of steam.

This discrepancy can be explained by the reaction mechanism. The ignition and flame propagation mechanisms is dominated by chain branching steps:

$$H + O_2 \longrightarrow OH + O$$

$$O + H_2 \longrightarrow OH + H$$

These steps are competitive with the trimolecular reaction:

$$H + O_2 + M \longrightarrow HO_2 + M$$

that can be considered as a termination. The H_2O collision partner is more efficient than CO_2 (Warnatz 1984).

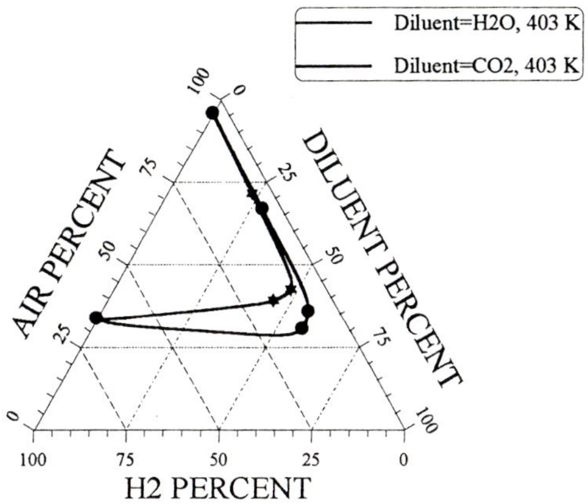

Fig. 5. Ignition limits of H_2 - Air - Diluent by a hot gas jet

This fact can explain that the ignition range is more extensive with CO_2 than with steam. The role of steam is not only due to a thermal effect but also to a chemical effect.

Conclusion

An original set-up has been built in the laboratory in order to produce a hot jet which involves hydrogen with an initial temperature, prior to ejection, varying between 750 and 3000 K. The ignition of H_2 - Air - diluent mixtures by a hot transient jet of H_2 - Ar has been studied and ignition limits domain drawn. The comparison between the ignition limits of H_2 - Air - CO_2 mixtures and H_2- Air - H_2O mixtures shows the importance of the chemical mechanism of the ignition process by a hot transient jet of hydrogen and argon.

Acknowledgements

This research was supported by "Electricité De France", Direction de l'Equipement, Service Etudes et Projets Thermiques Nucléaires (SEPTEN) under contract N°NC 9238.

References

Conti RS, Hertzberg M (1988) Auto-ignition temperature for hydrogen - air and methane - air mixtures. Journal of Fire Sciences 6: 348-355

Enomoto K, Furuhama S, Kobayachy Y (1980) Ignitability of hydrogen - air - mixture by hot surfaces and hot gases. Hydrogen Energy Progress, Proc. of the 3rd World Hydrogen Energy Conference, Tokyo, Japan, vol. 2, pp 1149-1163

Higashino F, Ishii Y, Sakurai A (1986) Measurement of ignition delay time on unsteady hydrogen jets. In: Bowen JR, Leyer J-C, Soloukhin RI (eds) Dynamics of Reactive Systems, Part II : Modeling and Heterogeneous Combustion, Progress in Astronautics and Aeronautics vol. 105, pp 407-415

Ishii Y, Higashino F, Sakurai A (1986) Ignition of spurting hydrogen jets behind a reflected shock wave. Bull. of JSME, 29, 256: 3348-3451

Langer G, Stock M (1988) Deflagration-detonation transition in H_2-steam-air mixtures under severe accident conditions. Battelle Institute Frankfurt, International Conference on Thermal Reactor Safety

Sakurai A (1985) Auto-ignition of hydrogen by a shock-compressed oxidizer. In: Bershader D, Hanson RK (eds) Proc. 15th Intl. Symp. on Shock Waves and Shock Tubes, Berkeley, California, pp 77-86

Warnatz J (1984) Critical survey of elementary reaction rate coefficients in the C/H/O systems. In: Gardiner WC Jr (ed) Combustion Chemistry, Springer Verlag, New-York, p. 187

Wolfhardt HG (1958) The ignition of combustible mixtures by hot gases. Jet Propulsion: 798-804

Zabetakis MG (1965) Flammability characteristics of combustible gases and vapors. Bulletin 627, Bureau of Mines, pp 121

Reduced Chemical Reaction Mechanism of Shock-Initiated Ignition of Methane and Ethane Mixtures with Oxygen

W.C. Gardiner, Jr.*, V.V. Lissianski* and V.M. Zamanski†
*Department of Chemistry and Biochemistry, The University of Texas, Austin, USA
†Energy and Environmental Research Corporation, Whitehouse, NJ, USA

Abstract. Reduced kinetic mechanisms were developed to describe the ignition of mixtures of small hydrocarbons in shock waves. The approach combined elementary and generalized reactions to account for the chemical processes of chain initiation, propagation, true and degenerate branching, and termination. The rate coefficients for the elementary reactions were retained at their measured values, while the rate coefficients for the generalized reactions were adjusted to match a small set of shock tube ignition delay data. Comparison with a broad range of experimental data for methane, ethane and mixtures of both indicated good agreement for the whole range of conditions studied in shock tubes so far.

Key words: Chemical kinetics, Reduced mechanism, Ignition delay

1. Introduction

One of the main applications of the shock tube to combustion chemistry has been in the measurement of ignition delays. Their dependence upon mixture composition and pre-reaction shock conditions has been used to derive insights into the chemical reaction mechanisms of ignition. Two approaches have been commonly used to interpret the data. The simpler one has been to derive algebraic correlation formulas relating ignition delay to temperature and composition, while at the other extreme, sets of coupled ordinary differential equations describing chemical reaction mechanisms consisting of up to hundreds of elementary reaction steps have been studied by large-scale numerical integration. The correlation formula approach has the disadvantage of containing only summary information; the detailed reaction mechanism, on the other hand, contains an overwhelming amount of information. In both cases, it is understood that the parameters derived from the experiments are probably not valid for extrapolations well outside of the conditions studied in the experiments.

Neither of these approaches is well suited for direct transfer of shock tube experience to practical applications in combustion technology, because either too little or far too much information is conveyed. A compromise between these extremes can be reached by using "reduced" chemical reaction mechanisms that can reproduce the main aspects of the combustion chemistry with fewer chemical steps than have to be written down to represent elementary chemical steps in a "detailed" mechanism. A number of such mechanisms have been studied over the years, most having had the object of accounting for the dependence of the laminar flame propagation speed upon upstream conditions. In the "Shell" mechanism (Halstead et al. 1975, 1977) a set of effective reactions was used to describe the high pressure ignition of complex hydrocarbon fuels as studied in a rapid compression machine. Peters (1985) suggested that one should reduce the number of reactions in a detailed mechanism by taking into account certain steady-state values of radical concentrations and evaluating reaction rates accordingly. A combination of elementary and effective reactions that we will call a "skeleton" mechanism was used by Borisov et al. (1989) to describe the ignition of methane at high temperatures. Their approach combined elementary and generalized reactions to account for the fundamental chemical processes of chain generation, propagation, true and degenerate branching, and termination, using as guidelines for the generalized reactions some well-known prototypes of real combustion reactions. Because their approach

Shock Waves @ Marseille II
Editors: R. Brun, L. Z. Dumitrescu

is based on fundamental knowledge of hydrocarbon oxidation chemistry it can also be applied for complex hydrocarbon fuels for which detailed chemistry is still not known. In this paper we report a modified form of the mechanism of Borisov et al. (1989) and explore its potential to describe the ignition of methane, ethane and their mixtures with oxygen in shock waves.

Table 1. Reactions and rate coefficients for methane and ethane ignition. All reactions are considered irreversible. Rate coefficients are given in the form $k = A\ \exp(-E/RT)$. M represents the collision partner for the unimolecular reactions

	Reaction	A/cm^{3}mol^{-1}s^{-1}	E/kJmol^{-1}	A/cm^{3}mol^{-1}s^{-1}	E/kJmol^{-1}
		Methane		Ethane	
1	$CH_4+M \longrightarrow R+U+M$	2.0×10^{17}	370		
2	$C_2H_6 \longrightarrow U+U$	6.3×10^{14}	335	7.0×10^{14}	336
3	$R+(CH_4,C_2H_6) \longrightarrow U+P$	7.9×10^{14}	63	$540\ T^{3.5}$	22
4	$U+O_2 \longrightarrow R+I$	1.3×10^{11}	56	7.9×10^{12}	56
5	$U+O_2 \longrightarrow 2R+I$	5.0×10^{13}	105	5.0×10^{13}	105
6	$U+I \longrightarrow 2R+P$	1.0×10^{14}	46	1.0×10^{14}	46
7	$R+I \longrightarrow 2U+P$	1.0×10^{15}	42	1.0×10^{15}	42
8	$U+U \longrightarrow C_2H_6$	$2.4\times10^{14}T^{-0.4}$	0	$2.4\times10^{14}T^{-0.4}$	0

2. Methane ignition

Skeleton mechanisms for CH_4 and C_2H_6 ignition are shown in Table 1.

They comprise the same steps, but some of the different rate coefficient expressions differ for the two fuels, as discussed below. The symbolic species R and U represent reactive and unreactive radicals, I represents stable intermediates, and P represents the reaction products. In reaction (3), the second reagent is either CH_4 or C_2H_6 depending on the fuel used. Steps (1) and (2) initiate chains by thermal decomposition of fuel, while the termination step (8) limits the concentration growth of chain centers by accounting for the fact that the CH_3 concentration rises to such high values under typical ignition conditions of methane and ethane that its recombination to C_2H_6 is significant. Step (5), in terms of elementary reactions well known in combustion, would represent true chain branching as $H+O_2 \longrightarrow OH+O$ and $CH_3+O_2 \longrightarrow CH_3O+O$, while steps (6) and (7) represent degenerate branching and steps (3) and (4) represent chain propagation. Sensitivity analyses and extensive comparisons with corresponding ignition profiles calculated using detailed mechanisms showed that different steps govern the rate of ignition for different compositions and temperatures; generally, computed ignition delays proved to be most sensitive to the rate coefficients of steps (4), (5) and (8). The rate coefficients of steps (1), (2), (3) and (8) for methane were left at measured values; for $CH_4+M \longrightarrow R+U+M$ we kept the $CH_4+M \longrightarrow CH_3+H+M$ expression of Bowman (1975), for $C_2H_6 \longrightarrow U+U$ we kept the expression for $C_2H_6 \longrightarrow CH_3+CH_3$ of Hidaka et al. (1985), for $R+CH_4 \longrightarrow U+P$ we kept the $H+CH_4 \longrightarrow H_2+CH_3$ expression of Roth and Just (1975), and for $U + U \longrightarrow C_2H_6$ we kept the $CH_3 + CH_3 \longrightarrow C_2H_6$ expression of Warnatz (1984). The rate coefficient expressions of steps (4), (5), (6) and (7) were adjusted by trial and error to match selected experimental ignition delay data. Comparisons with larger sets of experimental ignition delays were then made i.e., for nine mixtures with the full range of shock tube ignition delays measured by Crossley et al. (1972); and those reported by Gardiner et al. (1987), Frenklach and Bornside (1984) and Lifshitz et al. (1971). We also made comparison with non-shock-tube ignition measurements, including hot bulb (Borisov et al. 1983) ignition delays at temperatures down to 870 K. Some results of these calculations are compared to the experimental results and are shown in Figs.1–3. The agreement between computations and experiments is seen to be good over the entire range of ignition conditions, with maximum deviations of 40% at most.

Comparisons with detailed chemistry calculations (Frenklach et al. 1992) show that the skeleton mechanism qualitatively reproduces the behavior of fuel, radical, intermediate and final product concentration during the induction zones of shock-heated CH_4–O_2–Ar mixtures.

3. Ethane ignition

Rate coefficient expressions for the ethane skeleton mechanism are presented in Table 1. For reaction (3) the rate coefficient expression of $H+C_2H_6 \longrightarrow C_2H_5+H_2$ from Warnatz (1984) was used. Rate coefficient expressions for reactions (2) and (5–8) are the same as in the methane skeleton mechanism. Experimental (Burcat et al. 1972; Grillo and Slack 1976; Bowman 1970; Cook and Williams 1971, 1975; Burcat et al. 1971) ignition delay measurements were used to compare results of calculations with experiments. Analysis of the experimental data showed substantial differences in the derived dependence of ignition delays on initial conditions. For example, Burcat et al. (1971) reported an effective activation energy for ethane of 143 kJ/mol while Grillo and Slack (1976) give 57 kJ/mol; according to Burcat et al. (1971) and Cook and Williams (1971) the order of ignition delay with respect to oxygen concentration is −1.26 and −0.6, respectively. We checked the validity of the ethane skeleton mechanism on eight mixtures of different compositions. In Figs.4 and 5 some of the calculated and experimental results are compared. Computed ignition delays agree with the experimental results to within 50%.

4. Ignition of methane/ethane mixtures

The skeleton mechanism can be used to describe the ignition of mixtures of methane and ethane, and thereby can serve to describe the ignition behavior of natural gas. Usually natural gas is found to have methane fractions of 90 to 95%, the remainder being ethane and still smaller fractions of other hydrocarbons and inert species such as N_2 and CO_2. Thus the ignition mechanism of methane/ethane mixtures is essentially equivalent to the ignition mechanism of natural gas itself.

The skeleton mechanism for methane/ethane/oxygen mixtures consists of reactions (1) through (8), reaction (3) being used once with CH_4 and once with C_2H_6 as the second reagent. Rate coefficient expressions for reactions (1), (2), (3) and (5–8) were kept the same as in the methane and ethane skeleton mechanisms. The rate coefficient of reaction (4) was adjusted by comparing skeleton mechanism simulations with detailed mechanism simulations using the detailed ignition mechanism of Westbrook (1979), which had been optimized to account for the ignition of methane/ethane mixtures over the full relative composition range. The switching function

$$k_4 = 8.79 \times 10^{12}(f + 0.0145)e^{-56\,\mathrm{kJ}/RT}$$

was derived. It has the property of increasing rapidly from the smaller k_4 values of CH_4 ignition when trace amounts of C_2H_6 are added, then approaching the C_2H_6 k_4 expression gradually for large C_2H_6 fractions. In Fig.6 ignition delay calculations for mixtures with different C_2H_6 fractions are compared with experimental results of Crossley et al. (1972). The skeleton mechanism gives practically the same promoting effect as was found in the experiments. Comparisons of computed total radical concentration profiles ([U] + [R]) during the induction zone of pure methane mixtures and with ethane added were made with the sum profile of $[H]+[OH]+[HO_2]+[CH_3]$ computed with the detailed mechanism of Westbrook (1979). The agreement was good both for pure methane and for its mixtures with ethane.

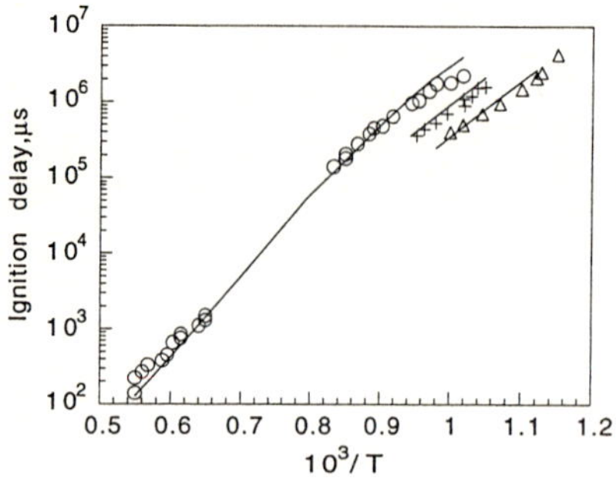

Fig. 1. Comparison of measured (Borisov et al. 1983, points) and computed (lines) ignition delays at $P = 1$ atm: (○) 6%CH_4–12%O_2–82%Ar, (+) 6%CH_4–24%O_2–70%Ar, (△) 6%CH_4–54%O_2–40%Ar

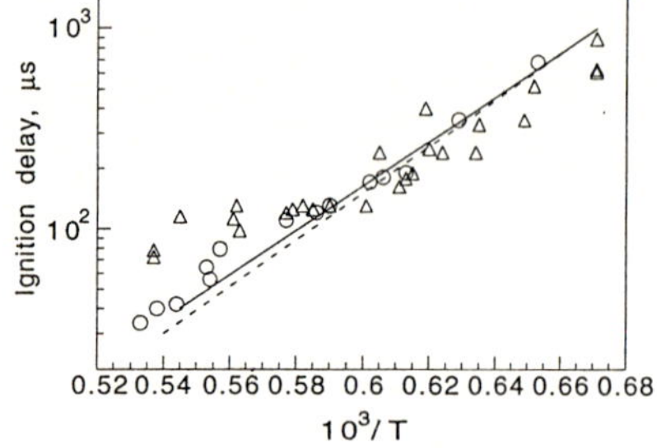

Fig. 2. Comparison of measured (points) and calculated (lines) ignition delays for 3.5%CH_4–7.0%O_2–89.5%Ar. (○ and dashed line) $\rho_5 = 7.5$ mol/m^3 (Lifshitz et al. 1971), (△ and solid line) $\rho_5 = 6.3$ mol/m^3 (Crossley et al. 1972)

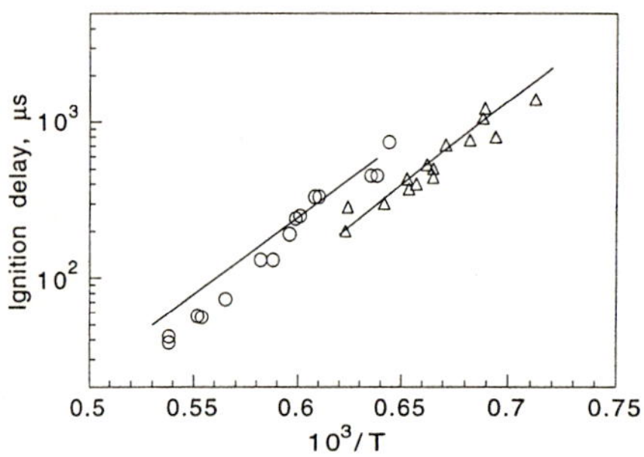

Fig. 3. Comparison of measured (points) and calculated (lines) ignition delays. (○) 1%CH_4–10%O_2–89%Ar, $\rho_5 = 2.83$mol/m^3 (Gardiner et al. 1987); (△) 9.5%CH_4–19%O_2–71.5%Ar, $\rho_5 = 2$ mol/m^3 (Frenklach and Bornside 1984)

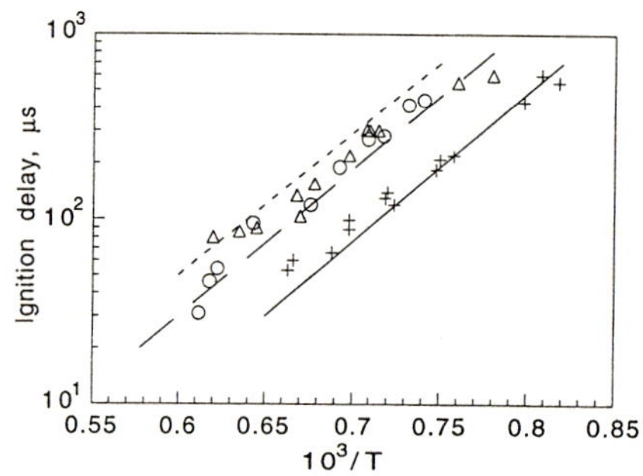

Fig. 4. Comparison of measured (Burcat et al. 1972, points) and computed (lines) ignition delays for 2%C_2H_6–7%O_2–91%Ar: (+ and solid line) $\rho_5 = 6.6$ mol/m^3, (○ and dashed line) $\rho_5 = 1.9$ mol/m^3, (△ and dotted line) 0.6%C_2H_6–2.1%O_2–97.3%Ar, $\rho_5 = 6.27$ mol/m^3

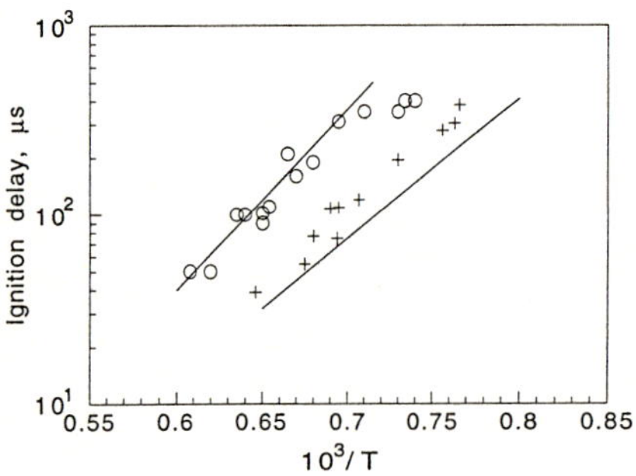

Fig. 5. Comparison of measured (Burcat et al. 1972, points) and computed (lines) ignition delays for (○) 1.06%C_2H_6–1.85%O_2–97.09%Ar, $\rho_5 = 65$ mol/m^3, (+) 3.44%C_2H_6–6.02%O_2–90.54%Ar, $\rho_5 = 84$ mol/m^3

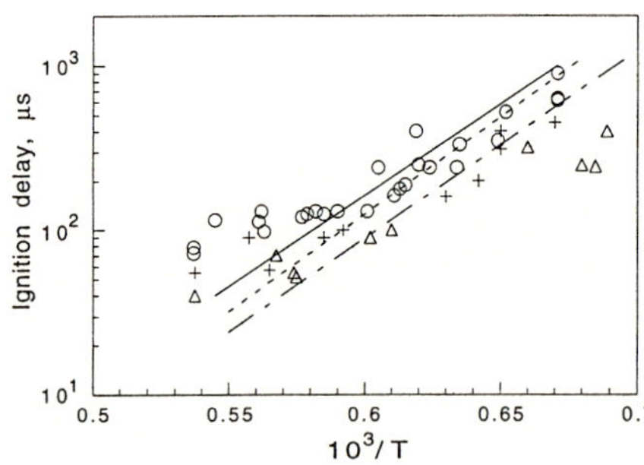

Fig. 6. Dependence of ignition delay on temperature and ethane addition. Lines – skeleton mechanism calculations, points – data of Crossley et al. (1972) at $\rho_5 = 6.3$ mol/m^3 for 3.5%CH_4–7%O_2–89.5%Ar (○). The other mixtures include additionally 0.05% (+) and 0.164%C_2H_6 (△)

5. Conclusions

Reduced "skeleton" reaction mechanisms were developed to describe ignition delays of CH_4 and C_2H_6 mixtures with oxygen in shock waves. The approach combines elementary and generalized reactions to account for the chemical processes of chain generation, propagation, true and degenerate branching and termination. Rate coefficient expressions for the generalized reactions were adjusted to describe experimental ignition delay data. Comparisons with experiment indicated that the skeleton mechanisms give the same dependences of ignition delay on mixture composition, initial temperature and pressure found experimentally and can be used for ignition delay calculations as well as for estimation of radical concentrations during ignition delay.

Acknowledgment

This research was supported by the Gas Research Institute and the Robert A. Welch Foundation.

References

Borisov AA, Dragalova EV, Lissianski VV, Skachkov GI, Zamansky VM (1983) Kinetics and mechanism of methane oxidation at high temperatures. Oxidation Communications 4:45

Borisov AA, Zamansky VM, Lissianski VV, Skachkov GI, Troshin KYa (1989) Promoted high-temperature reactions. In: 22d Symp. (Intl.) Combust. p 903

Bowman CT (1970) An experimental and analytical investigation of the high-temperature oxidation mechanisms of hydrocarbon fuels. Comb. Sci. and Techn. 2:161

Bowman CT (1975) Non-equilibrium radical concentrations in shock-initiated methane oxidation. In: 15th Symp. (Intl.) Combust. p 869

Burcat A, Crossley RW, Scheller K, Skinner GB (1972) Shock tube investigation of ignition in ethane-oxygen-argon mixtures. Comb. Flame 18:115

Burcat A, Scheller K, Lifshitz A (1971) Shock-tube investigation of comparative ignition delay times for C1–C5 alkanes. Comb. Flame 16:29

Cooke DF, Williams A (1971) Shock-tube studies of the ignition and combustion of ethane and slightly rich methane mixtures with oxygen. In: 13d Symp. (Intl.) Combust. p 757

Cooke DF, Williams A (1975) Shock tube studies of methane and ethane oxidation. Comb. Flame 24:245

Crossley RW, Dorko EA, Scheller K, Burcat A (1972) The effect of higher alkanes on the ignition of methane-oxygen-argon mixtures in shock waves. Comb. Flame 19:373

Frenklach M, Bornside D (1984) Shock initiated ignition in methane-propane mixtures. Comb. Flame 56:1

Frenklach M, Wang H, Rabinovich MJ (1992) Optimization and analysis of large chemical kinetic mechanisms using the solution mapping method – combustion of methane. Progr. Energy and Comb. Sci 18:47

Gardiner WC, Hwang SM, Rabinowitz MJ (1987) Shock tube and modeling study of methyl radical in methane oxidation. Energy and Fuels 1:545

Grillo A, Slack MW (1976) Shock tube study of ignition delay times in methane-oxygen-nitrogen-argon mixtures. Comb. Flame 27:377

Halstead MP, Kinch LJ, Prothero A, Quinn CP (1975) A mathematical model for hydrocarbon autoignition at high pressures. Proc. Roy. Soc. A 346:515

Halstead MP, Kirsch LJ, Quinn CP (1977) The autoignition of hydrocarbon fuels at high temperatures and pressures – fitting of a mathematical model. Comb. Flame 8:45

Hidaka Y, Shiba S, Takuma H, Suga M (1985) Thermal decomposition of ethane in shock waves. Intl. J. Chem. Kin. 17:441

Lifshitz A, Scheller K, Burcat A, Skinner GB (1971) Shock-tube investigation of ignition in methane-oxygen-argon mixtures. Comb. Flame 16:311

Peters N (1985) Reducing mechanisms. In: Smooke MD (ed) Numerical Simulation of Combustion Phenomena, Lecture Notes in Physics. Springer-Verlag, New York, pp 48–67

Roth P, Just Th (1975) Atom-Resonanzabsorptionsmessungen beim thermischen Zerfall von Methan hinter Stosswellen. Ber. Bunsenges. Phys. Chem. 79:682

Warnatz J (1984) Rate coefficients in C/H/O system. In: Gardiner WC (ed) Combustion Chemistry. Springer, New York pp 197-360

Westbrook CK (1979) An analytical study of the shock tube ignition of mixtures of methane and ethane. Comb. Sci. and Techn. 20:5

Comparison of Shock-Tube Ignition Characteristics of Various Fuel-Air Mixtures at High Pressures

K. Fieweger, H. Ciezki* and G. Adomeit
Institut für Allgemeine Mechanik, RWTH Aachen, Templergraben 64, 52056 Aachen, Germany

Abstract. The self-ignition behaviour of various fuel-air mixtures under engine relevant conditions has been investigated in a high-pressure shock tube without using a heat bath. Mixtures of n-heptane, benzene, iso-octane, methanol and MTBE with air were investigated. Two pressure levels for the investigation: 13 bar and nearly 40 bar have been chosen. For n-heptane, the temperature variation covers the region between 1350 K and 660 K, depending on the pressure level, while the ignition behaviour of the other fuels has been determined only above 800 K.

The ignition of n-heptane begins with a rapid pressure increase, especially at higher temperatures. Iso-octane, methanol, MTBE and especially benzene show a slow initiation of the ignition without distinct pressure peak (mild ignition) at low temperatures, which at higher temperatures changes to a rapid pressure increase after a variable time lag (strong ignition). A comparison of the ignition delay times is also presented.

Key words: Ignition delay time, Strong and mild ignition, Kinetics

1. Introduction

Information on the self-ignition of fuel-oxygen mixtures is of basic importance as well for its direct implications to practical combustion and ignition processes as for the possibility to derive basic kinetic information from these data. Hence a wealth of data has been gathered about this process. The shock tube has been prominent as a tool here because of the very short heating time by the shock wave and because the conditions achieved behind it are homogeneous. In technical systems quite generally air-fuel mixtures are used and hence direct transference of experimental results to these systems is desirable. In our case the desire to obtain data applicable to piston engine ignition and knock phenomena was the motive to develop a shock tube facility capable of handling air-fuel mixtures up to initial pressures p_5 of more than 40 bar. The gasdynamic features observed in these experiments should also be of direct relevance to corresponding engine phenomena. Hence this work fits well into the rapid compression device investigations extending their data to smaller values of ignition delay and higher temperatures. In this region the shock tube has definite advantages, concerning compression times, cleanliness, complete coverage of pressure and temperature range without the necessity to change the composition of the inert fraction of air (Park 1990).

Because commercial fuels consist of a great number of components, typical members of different fuel groups are investigated at the moment: n-heptane and iso-octane (2,2,4-trimethylpentane) for paraffins, benzene for aromatics, methanol for alcohols and MTBE (methyl-tert. butylether) for ethers. Mixtures of n-heptane and iso-octane are used as primary reference fuels to determine the octane number. N-heptane also possesses a cetane number typical for a good Diesel fuel. The others are possible additive fuel components, benzene and MTBE ameliorate knocking properties (Westbrook 1991, Leppard 1991). The data obtained for stoichometric mixtures will be presented here.

* Current address: DLR, Institut für Chemische Antriebe und Verfahrenstechnik, D-74239 Hardthausen

Shock Waves @ Marseille II
Editors: R. Brun, L. Z. Dumitrescu

2. Experimental apparatus

The apparatus used is a single–pulse high–pressure shock–tube with an inner diameter of 87 mm, which is described in various publications in detail (Ciezki 1989, 1990, 1993).In order to conduct experiments under conditions relevant to piston engine combustion the final length of 3.3 m of the driven section has been laid out for a maximum of 500 bar.

The measuring plane with four ports is located 15 mm in front of the end wall and is equipped with an acceleration compensated piezoelectronic PCB 113A pressure gauge and observation windows. A new measuring section with ten ports and four pressure gauges along the tube axis is in use since 1992.

A volume element at the shock tube axis in the measuring plane 15 mm in front of the end wall is focussed on two bifurcated fibre optics of fused silica. Band emissions are selected using interference filters. Three thin–film heat transfer gauges are mounted flush to the shock tube wall along the shock tube length in order to measure the velocity of the incident shock wave. With this information the temperature T_5 behind the reflected shock wave has been calculated by solving the gasdynamic balance equations, taking into account also the real caloric and thermic behaviour of the gases (Lee 1975, Gardiner 1984). Further information and details about the apparatus, the measuring equipment and the operating modus have been presented elsewhere (Ciezki 1990, 1993). The synthetic air delivered from Linde consists of 20 % oxygen and 80 % nitrogen and is free of hydrocarbons, CO_2 and water (less than 2 ppm). The fuels have UVASOL–grade and have been obtained from Merck and Riedel van der Haen.

3. Experimental results and discussion

The following Figs.1 – 4 show time histories of pressure and CH-band emissions at 4315 Å for n-heptane and benzene. The different traces in each figure have been obtained by varying the temperature T_5, while pressures p_5 are maintained at almost constant values. All mixtures are stoichiometric. The pressure level of p_5 is about 13 bar.

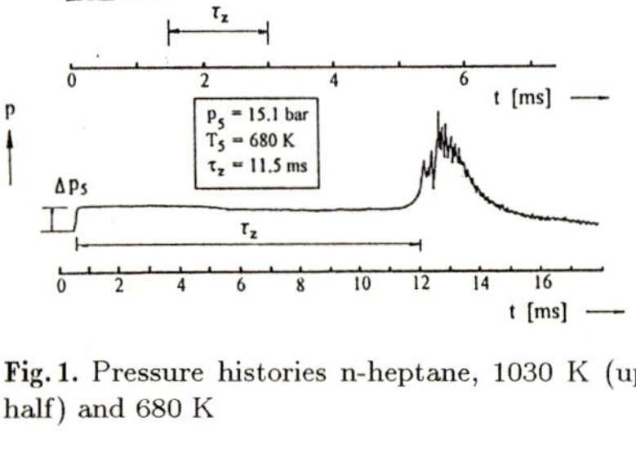

Fig. 1. Pressure histories n-heptane, 1030 K (upper half) and 680 K

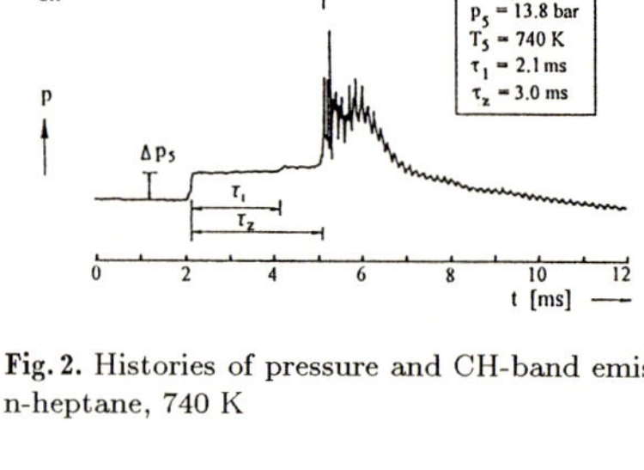

Fig. 2. Histories of pressure and CH-band emissions, n-heptane, 740 K

Figs.1 and 2 show time histories of pressure and CH-band emissions of the stoichiometric n-heptane/synthetic air-mixtures for various values of T_5. At high temperatures (see upper half of Fig.1) the first part of the main ignition is very pronounced and connected with a very rapid initial pressure rise. At the same time a maximum of the CH-band emission is observed. This is followed by a second region of slower, but nevertheless discernible pressure increase, which is particularly prominent at lower temperatures. To this average pressure record rapid pressure oscillations are

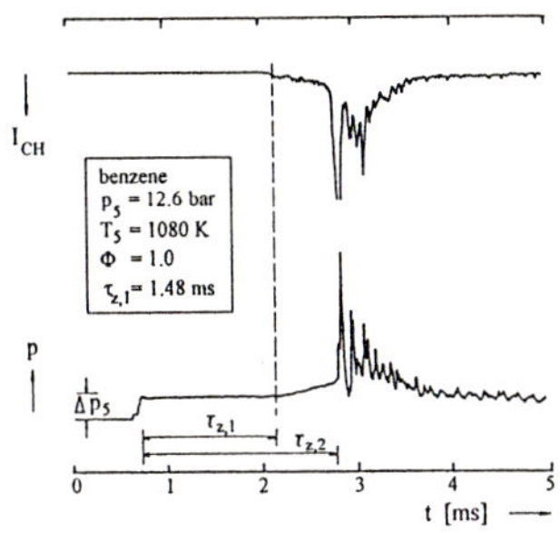

Fig. 3. Histories of pressure and CH-band emissions, benzene, 1080 K

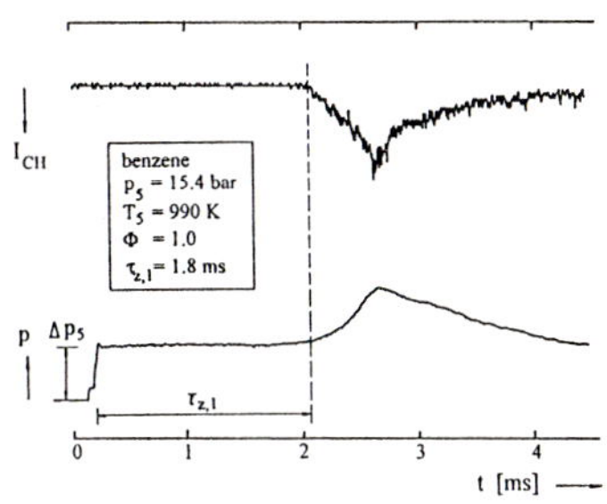

Fig. 4. Histories of pressure and CH-band emissions, benzene, 990 K

superimposed of a duration of approximately 10^{-4}s. They appear to be connected with transverse modes of pressure wave propagation (Oppenheim 1985). Lowering the temperatures T_5 (see Fig.2) the first distinct pressure peak diminishes in amplitude, while the following pressure oscillations stay prominent. For n-heptane these pressure oscillations have been observed at all conditions investigated. Fig.2 is representative for the self-ignition process in the temperature region between 700 and 1000 K and shows many important details. The incident and reflected shock wave may be seen on the first slope of pressure increase leading to region 5 with essentially constant pressure p_5. After a delay τ_1 a weak preignition is seen followed by the main ignition τ_z. This two stage ignition has been observed and investigated experimentally and theoretically in many details in recent investigations (Beckers 1953, Sitkei 1964, Park 1990, Westbrook 1991, Borisov 1967, Halstead 1977, Franck 1986, Prehn 1966, Sochet 1990, Inomata 1990, Pitz 1989, Müller 1991, Chevalier 1990). Going to lower temperatures the pressure rise caused by the preignition at τ_1 becomes larger whereas the time difference $\Delta\tau = \tau_z - \tau_1$ becomes shorter. Finally, the difference $\Delta\tau$ between the ignition delay times vanishes and the pressure trace shows only the main ignition process as can be seen in the lower half of Fig.1. The complete results of our investigations of n-heptane-air mixtures are published in detail in Ciezki (1993).

Contrary to n-heptane, the self-ignition of benzene proceeds considerably more smoothly, beginning at $\tau_{z,1}$ as can be seen in Figs.3 and 4. At lower temperatures (Fig.4) the pressure trace shows only a single distinct maximum. Pressure oscillations, which are characteristic for the n-heptane self-ignition process are not observed here. At higher temperatures (Fig.3) the ignition is divided into two parts. A moderate pressure rise occurs at the beginning of the main ignition at $\tau_{z,1}$. After a distinct time a further rapid pressure increase occurs at $\tau_{z,2}$, which finally shows a sharp peak, followed by pressure oscillations. The time history of the light emission has a similar behaviour and shows definitely that at $\tau_{z,1}$ the main ignition process begins. No preignition process has been observed. Oppenheim et al. investigated the gasdynamic processes connected with self-ignition behind the reflected shock wave. They observed two different types of ignition which they named mild and strong ignition (Oppenheim 1985, Vermeer 1972), the latter being connnected with a shock wave generated in the proximity of the back wall. The comparison of Figs.3 and 4 shows that in the temperature range between 990 K and 1080 K the transition from a mild to a strong ignition process occurs. This point of transition is indicated more precisely in Fig.5, where two pressure histories are plotted in one diagram. The temperature of these two experiments differs only by 40 K. Ignition delay time $\tau_{z,1}$ and pressure p_5 are very similar, also the first pressure slope. The development of the pressure recordings with temperature is illustrated in Fig.6 for a pressure level of 13 bar.

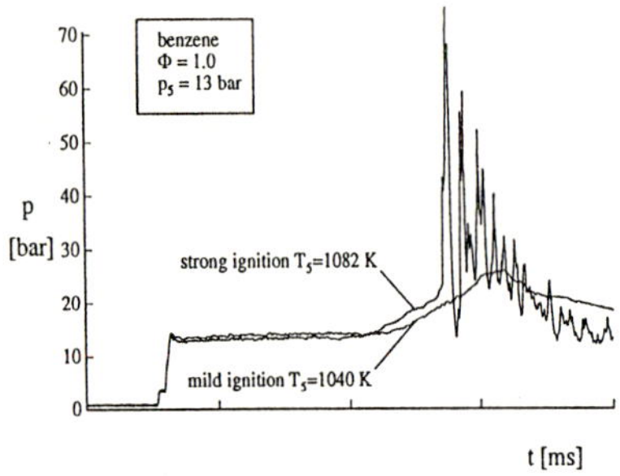

Fig. 5. Transition from mild to strong ignition for benzene

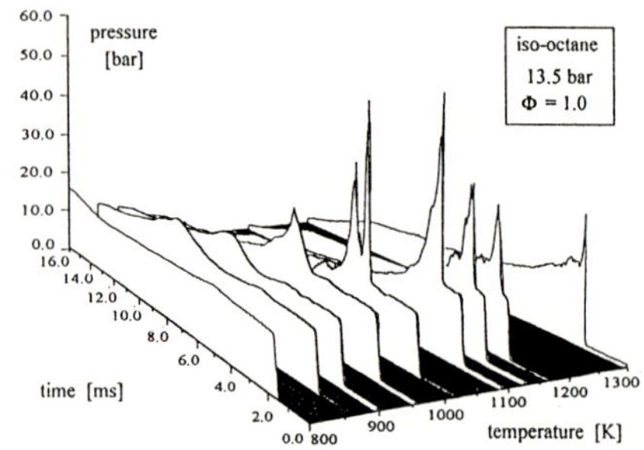

Fig. 6. Pressure histories for various temperatures, benzene, p_5 = 13 bar

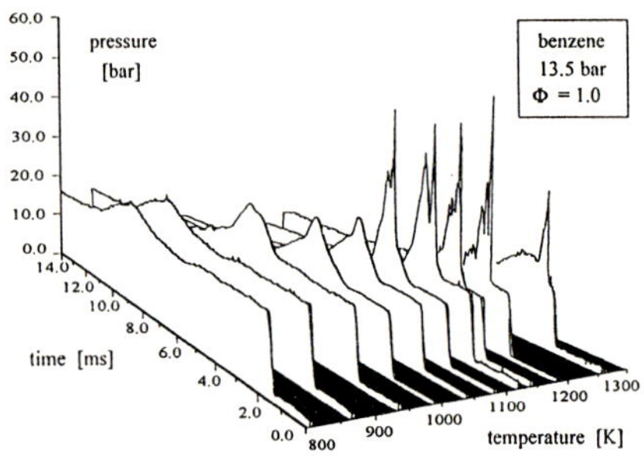

Fig. 7. Pressure histories for various temperatures, iso-octane, p_5 = 13 bar

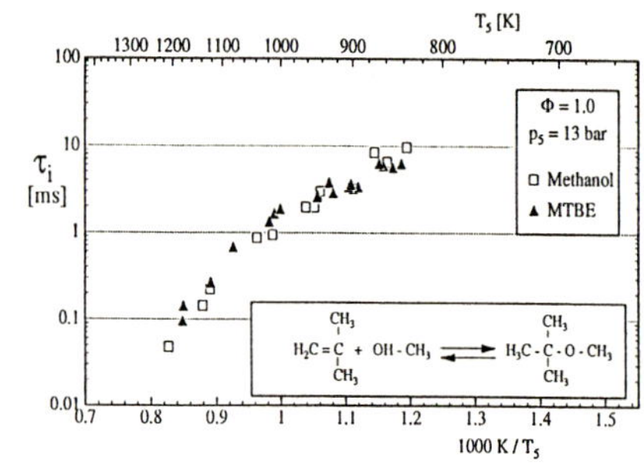

Fig. 8. Ignition delay times of methanol and MTBE; $p_5 \approx 13$ bar

The records of iso-octane show at high temperatures a rapid sudden pressure increase followed by further oscillations similar to the observations made for n-heptane. Going to low temperatures these oscillations disappear similar to benzene, which is indicative of the transition to mild ignition. The pressure histories in Fig.7 show that the ignition characteristic of iso-octane is very similar to that of benzene described above.

The pressure traces of methanol and MTBE also show features similar to those described for benzene. A slow initiation is followed by a rapid pressure increase and consecutive oscillations at high temperatures. At low temperatures the trace is smooth and posseses only one maximum.

In Figs.9 and 10 the ignition delay time τ_z ($\tau_{z,1}$ for benzene) is plotted logarithmically versus the inverse of the temperature behind the reflected shock $1/T_5$. The pressure p_5 is approximately equal to 13 bar for all the data points in Fig.9. In Fig.10 it varies from fuel to fuel but is near 38 bar. For n-heptane the ignition delay dependence upon the temperature both in the high and in the low temperature region can be expressed approximately by straight lines. In an intermediate range the dependence becomes strongly nonlinear and is represented by the well known S-shaped dependence. In this so called region of negative temperature coefficient the alkylperoxy radical reactions become dominant (see Westbrook 1991, Chevalier 1990). Increasing the pressure the S-shaped part is shifted to higher temperatures and lower ignition delay times, which is obvious when Figs.9 and 10 are compared.

The iso-octane data plotted in Figs.9 and 10 almost coincide with the n-heptane data for temperatures higher than 1000 K (13 bar) and 900 K (38 bar). Lowering the temperatures, the

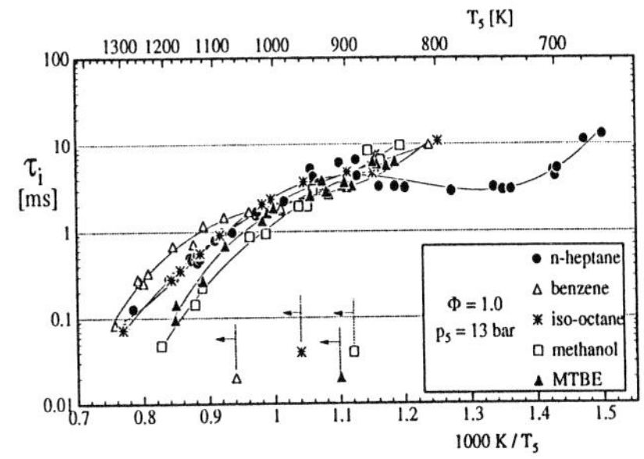

Fig. 9. Ignition delay times of various fuels; $p_5 \approx 13$ bar. The symbols with the arrows indicate transition from mild to strong ignition

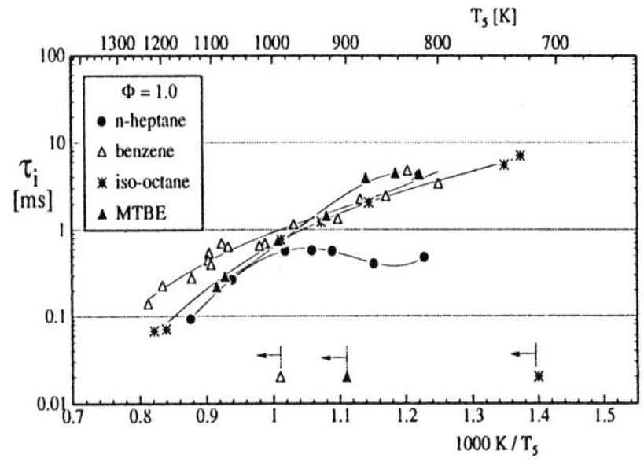

Fig. 10. Ignition delay times of various fuels; $p_5 \approx 38$ bar. The symbols with the arrows indicate transition from mild to strong ignition

ignition delay times of iso-octane become longer by a factor of about ten, which explains the large difference of the two fuels in relation to the octane number.

The ignition delay times of benzene show a monotonous dependence upon temperature. At the pressure of $p \simeq 13$ bar the delay times of benzene lie above those of n-heptane except for a small range around 1000 K. This intersection disappears for higher pressures and the differences between the ignition delay times become larger, as Fig.10 shows. This becomes particularly pronounced in the important region where n-heptane displays the S-shaped dependence. This should be of significance with respect to anti-knocking properties.

The MTBE ignition delay shows also the monotonous dependence upon temperature observed for benzene. The overall gradient, however, is steeper, which leads to ignition delay times below those of n-heptane at high temperatures and low pressures. The data overlap in a region around 900 K. At around 38 bar the high temperature values of MTBE and n-heptane are close together differing, however, considerably in the low temperature transition region. This again is indicative of the MTBE anti-knock properties. Methanol shows a very similar behaviour to MTBE comparing the low pressure data (Figs.8 and 9). This can be explained by the fact that MTBE decomposes into methanol and iso-butene (Dunphy 1989).

The point of transition from a strong ignition, where a sharp pressure peak and pressure oscillations occur, to a mild ignition, without pressure oscillations, is marked for every fuel except n-heptane in Figs.9 and 10. This point is located between 1060 K (13 bar) and 990 K (38 bar) by benzene and between 960 K (13 bar) and 715 K (38 bar) by iso-octane. MTBE shows this transition between 910 K (13 bar) and 900 K (38 bar) and methanol at 890 K (13 bar). N-heptane shows strong pressure oscillations at all investigated temperatures, as has been described above.

Acknowledgements

This work was supported by the Deutsche Forschungsgemeinschaft as a project of the Sonderforschungsbereich 224 “Motorische Verbrennung”.

References

Beckers A (1953) Mot. Techn. Z. 14: 345

Borisov AA et al. (1967) Arrests in the ignition of heptane-iso-octane-air-mixtures. Russian Journal of Physical Chemistry 41, 5: 605

Chevalier C, Louessard P, Müller UC, Warnatz J (1990) A detailed low-temperature reaction mechanism of n-heptane auto-ignition. COMODIA 90, Kyoto

Ciezki H, Adomeit G (1989) Shock-tube investigations on the ignition delay of n-heptane/air mixtures in a high pressure shock-tube under conditions relevant to Diesel engine combustion. In: Kim YW (ed) 17th Intl. Symp. on Shock Tubes and Waves, AIP Conference Proceedings Vol. 208, p 707

Ciezki H (1990) Untersuchung des Selbstzündverhaltens von n-Heptan/Luft-Gemischen unter hohem Druck. Ph. D. Thesis, RWTH Aachen, Verlag der Augustinus Buchhandlung Aachen, No. 80, ISBN 3-925038-80-9

Ciezki H, Adomeit G (1993) Shock tube investigation of self-ignition of n-heptane/air-mixtures under engine relevant conditions. Comb. Flame 93: 421-433

Dunphy MP, Simmie JM (1989) High temperature oxidation of methyl tert.-butyl ether. Combustion Science Technology 66: 157

Franck J, Griffiths JF, Nimmo W (1986) The control of spontanous ignition under rapid compression. In: 21st Symposium (Intl.) on Combustion, The Combustion Institute, Pittsburgh, p 447

Gardiner WC (ed) (1984) Combustion Chemistry. Springer Verlag, New York

Inomata T, Griffiths JF, Pappin AJ (1990) The role of additives as sensitisers for the spontaneous ignition of hydrocarbons. In: 23rd Symposium (Intl.) on Combustion, The Combustion Institute, Pittsburgh, p 1759

Lee BI, Kesler MG (1975) A generalized thermodynamic correlation based on three-parameter corresponding states. AIChE Journal 21, 3: 510

Leppard WR (1991) The autoignition chemistries of octane-enhancing ethers and cyclic ethers: a motored engine study. SAE Paper No. 912313

Martinengo A, Wagner HGg, Zunft D (1959) Untersuchungen über Selbstzündreaktionen von Kohlenwasserstoff-Luftmischungen durch adiabatische Verdichtung. Z. Phys. Chem. Neue Folge, 22: 292

Müller UC, Peters N, Linan A (1992) Global kinetics for n-heptane ignition. In: 24th Symposium (Intl.) on Combustion, The Combustion Institute, Pittsburgh, p 777

Oppenheim AK (1985) Dynamic features of combustion. Phil. Trans. R. Soc. Lond. A 31 5: 471-508

Park P, Keck JC (1990) Rapid compression machine measurements of ignition delays for primary reference fuels. SAE Paper No. 900027

Pitz WJ, Warnatz J, Westbrook CK (1988) A detailed chemical kinetic reaction mechanism for the oxidation of iso-octane and n-heptane over an extended temperature range and its application to analysis of engine knock. In: 22nd Symposium (Intl.) on Combustion, The Combustion Institute, Pittsburgh, p 893

Prehn H (1966) Untersuchung der Reaktionsvorgänge und des Selbstzündverhaltens von Kohlenwasserstoff- Luft- und Sauerstoff-Luft-Mischungen in Temperaturbereichen oberhalb 1000 K. Ph.D. Thesis, RWTH Aachen

Sitkei G (1964) Kraftstoffaufbereitung und Verbrennung bei Dieselmotoren. Springer Verlag, Berlin

Sochet et al. (1990) Autoignition of butane: a burner and a rapid compression study. In: 23rd Symposium (Intl.) on Combustion, The Combustion Institute, Pittsburgh, p 1753

Vermeer DJ, Meyer JW, Oppenheim AK (1972) Autoignition of hydrocarbons behind reflected shock waves. Combust. Flame 18: 327

Westbrook ChK, Pitz WJ, Leppard WR (1991) The autoignition chemistry of paraffinic fuels and pro-knock and anti-knock additives: a detailed chemical kinetic study. SAE Paper 912314

Investigation on Soot Formation of n-Heptane-Oxygen-Argon Mixtures in a High Pressure Shock Tube

J. Yao, K. Fieweger, M. Kleppel and G. Adomeit
Institut für Allgemeine Mechanik, RWTH Aachen, Templergraben 64, 52056 Aachen, FRG

Abstract. In this work, soot formation processes during self-ignition of n-heptane-oxygen-argon mixtures have been investigated using an optical measurement technique in a high-pressure shock tube. In order to achieve conditions relevant to Diesel engine combustion, three pressure levels (15 bar, 25 bar and 35 bar) have been chosen. The equivalence ratio was varied between 2.0 and 4.0. The soot formation process can be described by the time dependence of the mean particle diameter, the soot volume fraction and the soot production rate. These parameters are dependent on temperature, pressure and mixture composition. In order to investigate these quantities in situ, the dispersion quotient method was used. The attenuation of two laser beams (488 nm and 633 nm) caused by soot particles was recorded. From these signals the time-resolved mean diameter of the particles was determined applying the Mie-theory of scattering. Particle number density, soot volume fraction and soot production rate were also calculated.

Key words: Soot formation, Soot volume fraction, Dispersion quotient method

1. Introduction

Since a few decades the soot formation of fuel-air mixtures has been one of the main problems in combustion systems. The soot formation is usually undesired, because soot is injurious to human health as well as to the environment. For the investigation of soot formation processes a variety of measurement techniques has been developed. Generally speaking, these techniques can be divided into two categories: probe techniques and optical methods. The probe techniques are suitable for investigation of primary soot particles in the size range of 10-50 nm (Graham 1975). The optical measurement techniques are non-disturbing techniques which are the best suitable for in situ measurements during the combustion. Penner et al. (1984) give a detailed review on these techniques. The major part of the investigations on soot formation have been carried out in test engines, shock tubes and burners. The burners are used essentially to investigate gaseous fuels whereas in test engines all kinds of fuels can be investigated. In these devices however, flow, transport and mixing processes superimpose the chemical reactions. The shock tube allows to investigate only the chemical influences. So these results can be used as a basis for the kinetic modelling of the soot formation process.

2. Experimental

The single-pulse-high-pressure shock tube used in this investigation has an inner diameter of 87 mm and consists of a 6.3 m long driven section and a 5.3 m long driver section. The measuring plane is located 15 mm in front of the end wall. This device has been described in previous publications in detail (Ciezki 1989, 1990, Yao 1993). The setup of the measurement apparatus is shown in Fig.1. Two laser beams, He-Ne (633 nm) and Ar^+ (488 nm), are used to measure the extinction due to soot particles. The two detectors are silicium photodiodes. The interference filters select the beams according to the wavelength. The output signals of the detectors are stored by a transient recorder with a scan rate of 0.5 μs, and read by a PC for further evaluation.

All results were obtained with a fuel-oxygen mixture in a dilution of inert gas (96% argon, 98% argon and 98% nitrogen) in order to minimize the influence of water droplets on the extinction and to improve the quality of the signals. For further information and details about the apparatus, the measuring equipment and the operating modus see Yao (1993).

Shock Waves @ Marseille II
Editors: R. Brun, L. Z. Dumitrescu

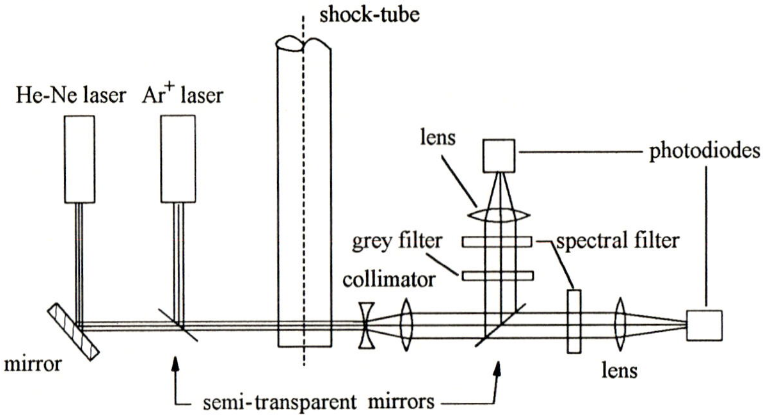

Fig. 1. Experimental setup for extinction measurements

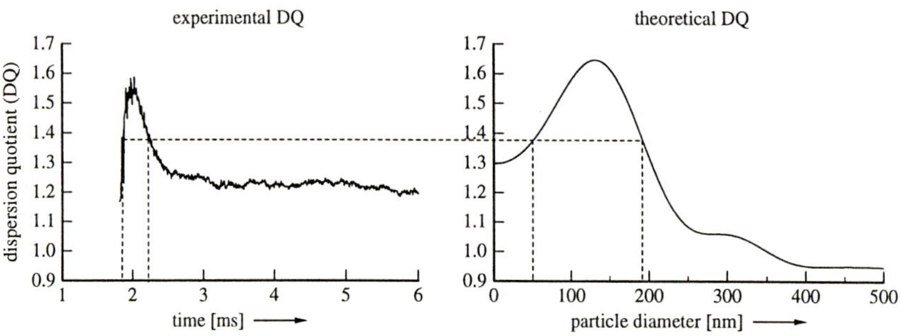

Fig. 2. Determination of the particle diameter growth by comparing experimental and theoretical DQ

3. Dispersion Quotient Method

The dispersion quotient method is applied to measure the particle size and the concentration. This method is based on the Lambert-Beer law of light extinction:

$$I = I_0 \; exp \; (-N \pi a^2 Q L) \quad , \tag{1}$$

where I_0 is the incident intensity, I the transmitted intensity, N the particle number density, a the particle radius, Q the extinction factor and L the path length of extinction. Applying Eq.1 to both wavelengths, taking the logarithm of the intensities and then the quotient of the two equations we obtain

$$DQ = \frac{Q_2}{Q_1} = \frac{\ln(I_{20}/I_2)}{\ln(I_{10}/I_1)} \quad , \tag{2}$$

where the quantities appearing on the right side are determined by experiment. The ratio of Q_2 and Q_1 is called dispersion quotient (DQ) (Kerker 1969, Lester 1975).

The value of the DQ can be computed making use of the Mie-Theory of light scattering and is depending on particle diameter, refractive index and wavelength. The dependence upon particle diameter in this manner is plotted in the right half of Fig.2. Also shown is the experimental measured DQ plotted versus the time scale. By comparing the measured DQ in Eq.2 with the computed curve as shown in Fig.2 we can determine the growth of the particle diameter with

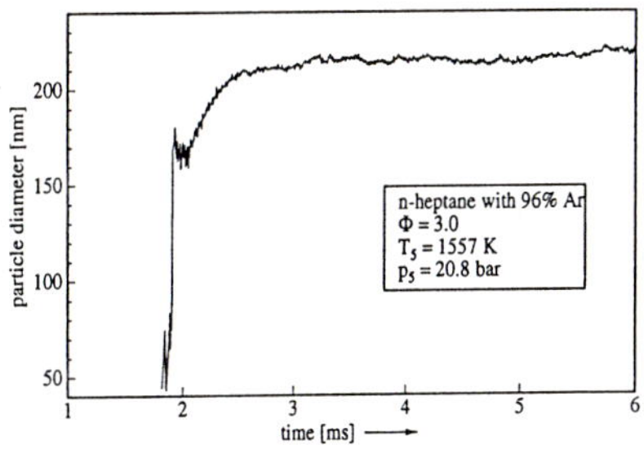

Fig. 3. Calculated particle diameter

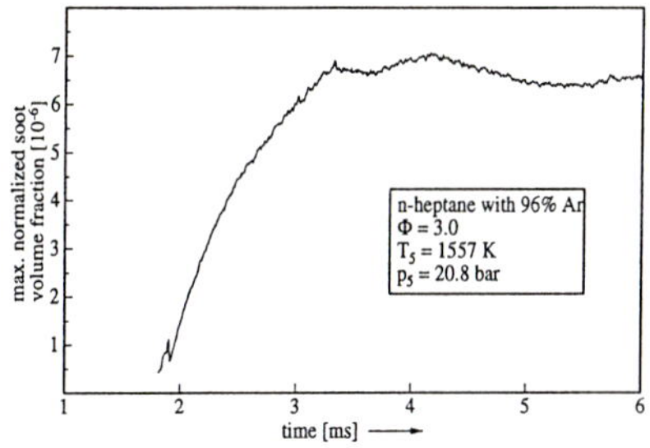

Fig. 4. Calculated soot volume fraction

time. It can be assumed that the soot particles in the shock tube grow continuously during the formation process (Wittig 1987, Müller 1991). Under this assumption the particle size can be determined uniquely. Substituting the diameter into Eq.1 the particle number density can be computed. From this the soot (particle) volume fraction f_v can then be calculated:

$$f_v = 4/3\ N\pi a^3 = \frac{4 \ln I_0/I}{3LQ/a} \quad . \tag{3}$$

Fig.4 shows the characteristic development of the calculated soot volume fraction in dependence upon time ($p_5 = 20.8$ bar, $T_5 = 1557$ K, $\Phi = 3$, 96% Argon).

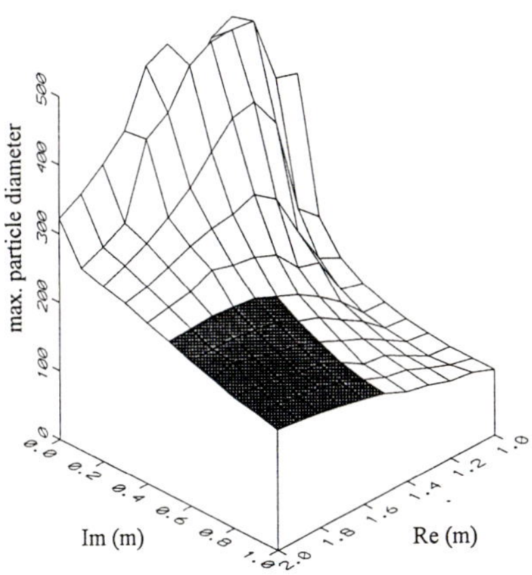

Fig. 5. Maximum particle diameter with refractive index varied. The hatched area marks values found in literature

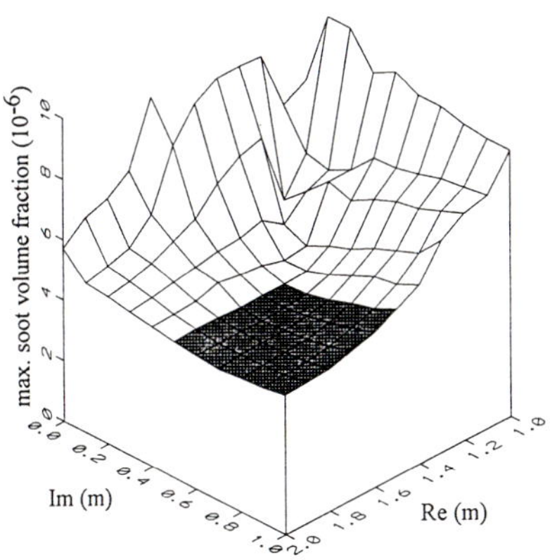

Fig. 6. Maximum soot volume fraction with refractive index varied. The hatched area marks values found in literature

For all calculations the refractive index of carbon powder ($1.96 - i0.66$), taken from Roessler (1979) and Wittig (1987) was used. The data given in the literature for the refractive index of soot vary between $1.5 \ldots 2.0$ for the real part and between $-0.5 \ldots -1.0$ for the imaginary part. In order to estimate the influence of this large variation of the refractive index on the calculated maximum particle diameter and the maximum soot volume fraction, for one experiment a set of calculations was carried out using different values for the refractive index. The real part was varied in the range of $1.0 \ldots 2.0$, and the imaginary part in range of $0.0 \ldots -1.0$. Figs.7 and 8 show the results. It can be seen that for the range found in the literature for the refractive index (hatched area) the calculated results fluctuate only in a small range.

The particle size distribution of soot may be described by a first order logarithmic distribution. The width of this distribution is described by the parameter σ. Fig.6 shows the dependence of the calculated maximum soot volume fraction on σ. For reasonable values of σ ($0.0 < \sigma < 0.5$) the assumption of monodisperse particles causes only small differences.

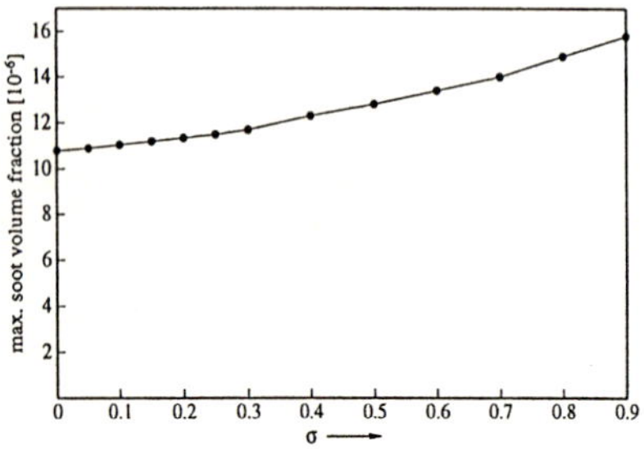

Fig. 7. Dependence of the maximum soot volume fraction on σ

4. Results

Figs.8 and 9 show the maximum soot volume fraction as a function of the temperature for different pressures with fixed equivalence ratio (Fig.8) and for different equivalence ratios with fixed pressure (Fig.9). For increasing equivalence ratios the maximum of the curve and the right flank is shifted to higher temperatures, while the left flank on the whole stays at similar temperatures. This typical "bell-shaped" dependence (Kern 1991) can be explained by production and decomposition of polycyclic-aromatic ring structures. At low temperatures there are not enough C_2-, C_3- radicals to produce aromatic ring structures (soot precursors). At high temperatures the aromatic ring structures decompose very quickly. Because of this the soot production is favoured at middle temperatures. For low temperatures, the equivalence ratio has little influence on the soot volume fraction. This behaviour can be explained by the fact, that the quantity of oxygen does not essentially influence the soot production at low temperatures (Wang 1981). For increasing pressure only the value of the maximum increases while its position is unchanged.

Besides the soot volume fraction, the soot production rate is of importance. It is defined as the soot mass produced per millisecond. We determined the initial soot production rate as well as the average production rate. The average production rate is the average value over the total interval of soot formation. In Figs.10 and 11 the average soot production rate for varied pressures with fixed equivalence ratio, and for varied equivalence ratios with fixed pressure are shown. It is seen that for increasing equivalence ratio the maximum of the curve and the right flank are shifted, and that the height of the maximum increases, while for increasing pressure only the value of the maximum increases. This corresponds to the observations made in Figs.8 and 9.

To investigate the influence of the inert gas dilution on the soot formation, especially on the soot volume fraction, the fuel-oxygen to argon ratio was varied. Furthermore a nitrogen dilution was used. Fig.12 shows the maximum soot volume fraction as a function of the temperature for different inert gases with fixed equivalence ratio. It is obvious that a decreasing percentage of the argon dilution from 98% to 96% shifts the maximum of the curve to lower temperatures, while

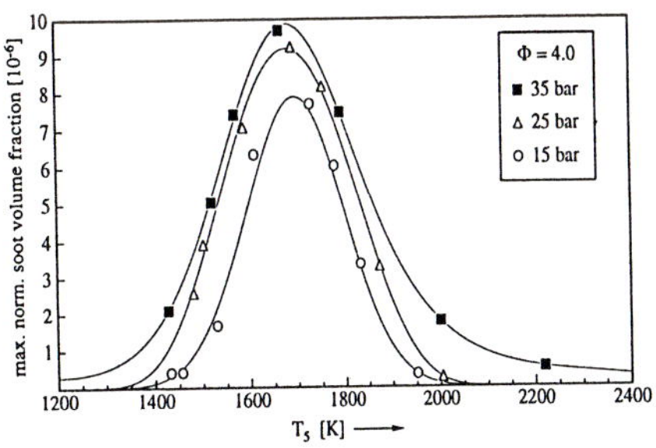

Fig. 8. Soot volume fraction versus temperature with pressure varied; 98% argon

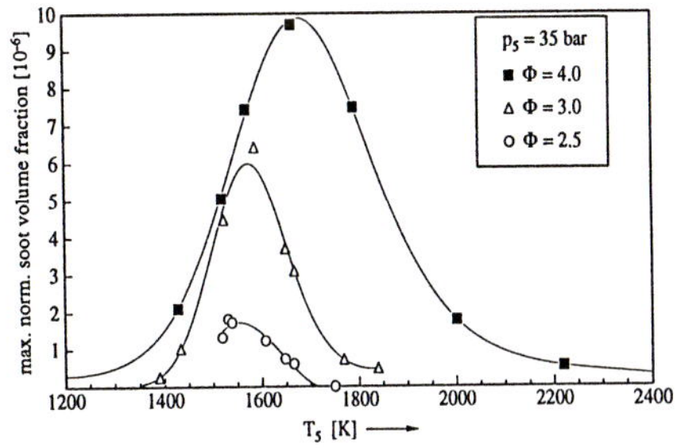

Fig. 9. Soot volume fraction versus temperature with equivalence ratio varied; 98% argon

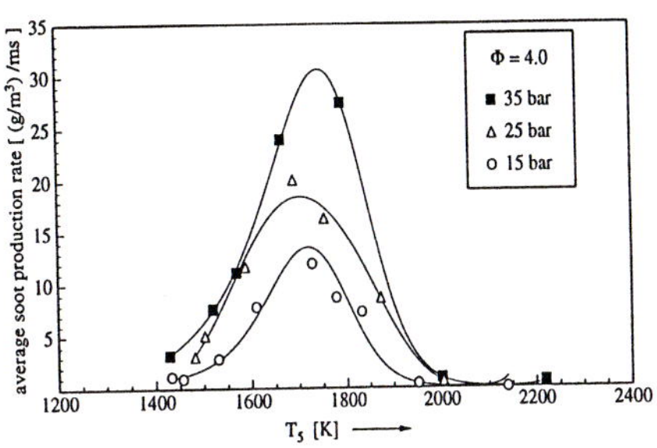

Fig. 10. Soot production rate versus temperature with pressure varied; 98% argon

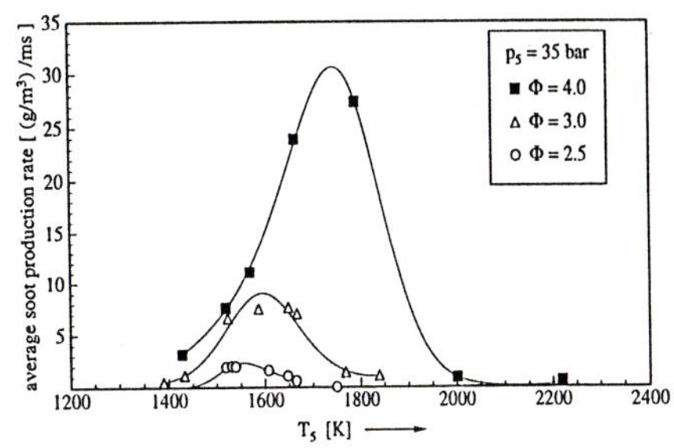

Fig. 11. Soot production rate versus temperature with equivalence ratio varied; 98% argon

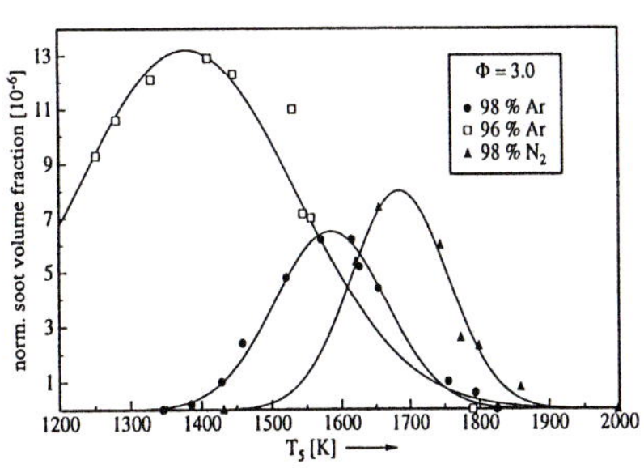

Fig. 12. Maximum soot volume fraction with the inert gas varied

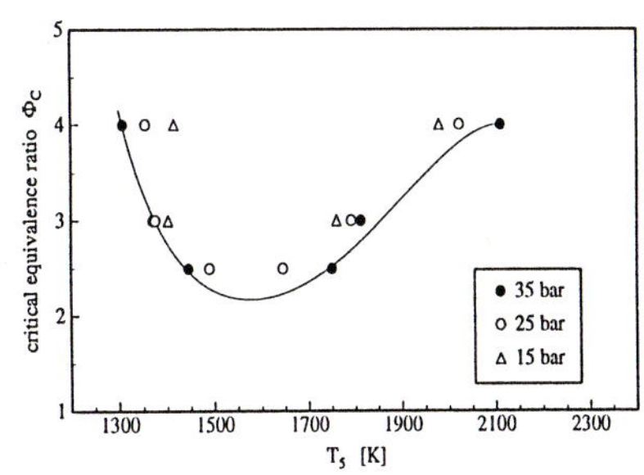

Fig. 13. Soot limits of n-heptane-oxygen-98% argon mixtures

the height of the maximum is considerably increased by a factor of about two, which corresponds to the doubling of the n-heptane fraction in the test gas.

If argon is replaced by nitrogen, the maximum of the curve is shifted to higher temperatures, and the height of the maximum also increases, but both effects are small compared to the effect that results from a smaller argon percentage. Hence the soot formation mechanism is dependent upon the inert dilution. Similar effects of the influence of the inert gas are observed for ignition delay times (Zallen 1975).

Fig.13 shows the soot limits of n-heptane-oxygen mixtures diluted with 98% argon. The soot limit is defined as the transition from sooting to non-sooting combustion. These limits can be determined by extrapolation of the curves in Figs. 8 – 11 to the temperature axis.

All these results presented above are in good qualitative agreement with data of various fuels obtained by shock tube investigations (e.g. Müller 1991). N-heptane, however, has been investigated apparently here for the first time.

Acknowledgements

This work was supported by the Deutsche Forschungsgemeinschaft as a project of the Sonderforschungsbereich 224 “Motorische Verbrennung”.

References

Ciezki H, Adomeit G (1989) Shock tube investigations of the ignition delay on n-heptane/air mixtures in a high pressure shock tube under conditions relevant to Diesel engine combustion. In: Kim YW (ed) 17th Intl. Symp. on Shock Tubes and Waves, AIP Conference Proceedings Vol. 208, p 707

Ciezki H (1990) Untersuchung des Selbstzündverhaltens von n-Heptan/Luft-Gemischen unter hohem Druck. Ph. D. Thesis, RWTH Aachen, Verlag der Augustinus Buchhandlung Aachen, No. 80, ISBN 3-925038-80-9

Evans M, Williams A (1981) Shock tube studies on the formation of soot from the combustion and pyrolysis of some hydrocarbons. Fuel 60: 1047-1056

Frenklach M et al. (1983) A conceptual model for soot formation in pyrolysis of aromatic hydrocarbons. Combust. Flame 49: 275-282

Glassman I (1988) Soot formation in combustion processes. In: 22nd Symposium (Intl.) on Combustion, The Combustion Institute, Pittsburgh, pp 295-311

Graham SC et al. (1975) The formation and coagulation of soot aerosols. In: Kamimoto G (ed) 10th Intl. Symp. on Shock Tubes and Waves, pp 621-631

Kerker M (1969) The scattering of light and other electro-magnetic radiations. Academic Press, New York

Kern RD, Xie K (1991) Shock tube studies of gas phase reactions preceding the soot formation process. Prog. Energy Combust. Sci., Vol. 17, pp 191-210

Lester TW, and Wittig S (1975) Particle growth and concentration measurements in sooting homogeneous hydrocarbon combustion systems. In:Kamimoto G (ed) 10th Intl. Symp. on Shock Tubes and Waves, pp 632-639

Müller A, Wittig S (1991) Influence of temperature and pressure on soot formation in a shock tube under high pressure conditions. In: Takayama K (ed) 18th Intl. Symp. on Shock Waves, Sendai, pp 759-764

Penner SS et al. (1984) Laser diagnostics applied to combustion systems. In: 20th Symposium (Intl.) on Combustion, The Combustion Institute, Pittsburgh, pp 1149-1176

Roessler DM, Faxvog FR (1979) Opacity of black smoke: calculated variation with particle size and refractive index. Appl. Opt. 18, 9:1399-1403

Wittig S et al. (1987) Dispersion quotient method for optical particle sizing (in German), 3. TECFLAM-Seminar, Nov. 5. 1987, Karlsruhe, Germany

Wang TS et al. (1981) Combustion kinetics of soot formation from toluene. In: 18th Symposium (Intl.) on Combustion, The Combustion Institute, Pittsburgh, pp 1149-1157

Yao J (1993) Untersuchungen zur Rußbildung von n-Heptan/Luft-Gemischen im Stoßwellenrohr. Ph. D. thesis, RWTH Aachen, VDI Verlag, Düsseldorf No. 315, ISBN 3-18-141503-0

Zallen DM, Wittig S (1975) Effects of nitrogen on the shock induced ignition of methane. In:Kamimoto G (ed) 10th Intl. Symp. on Shock Tubes and Waves, p. 640

Shock-Tube Study of High-Temperature Oxidation of Phenyl Radical

N. Fujii, H. Muratake, K. Okamura, T. Kobayashi and Y. Nosaka
Department of Chemistry, Nagaoka University of Technology, Kamitomioka, Nagaoka 940-21, JAPAN

Abstract. The kinetics of the C_6H_5-O_2 reaction have been studied using time-resolved oxygen atom resonant absorption spectroscopy and simultaneous measurements of two infrared emissions due to CO and CO_2. Nitrosobenzene was used as a source of C_6H_5 radical and the mixtures with O_2 dilute in Ar were heated over the temperature range of 1500-2000 K behind shock waves. The profiles obtained were compared with those by the computer simulations using the schemes reported previously by some researchers.

Key words: Phenyl radical, Oxidation, Kinetics, Shock wave

1. Introduction

In contrast with the studies on the oxidation of paraffin hydrocarbons, there have been only few kinetic studies on the oxidation of gaseous aromatic hydrocarbons. The oxidation of gaseous benzene, which is the most simple and typical compound among aromatic hydrocarbons, has been studied kinetically by several authors, and some mechanisms for the breakage of benzene ring by O_2 and O, and also the formation of CO, CO_2 and lower hydrocarbons were proposed (Fujii and Asaba 1973; Venkat et al. 1982; Hsu et al. 1984; Bittker 1991). Recently we studied the high temperature reaction of C_6H_6-N_2O system behind shock waves (Fujii et al. 1989; Fujii 1993) and found that the reaction kinetics could not be explained by the reaction mechanism proposed previously for the benzene oxidation (Fujii and Asaba 1973). In the present stage, the mechanism for high temperature oxidation of benzene remains largely speculative. In the present study, in order to evaluate the kinetics more precisely, the oxidation of phenyl radical was examined using a mixture of C_6H_5NO-O_2 and the results were compared with those for oxidation of C_6H_6.

2. Experimental

2.1. O atom resonance absorption measurement

Experiments were carried out by using conventional shock waves in a 46 mm ID stainless steel shock tube. The shock tube and related instrumentation have been described elsewhere (Fujii 1989). ARAS method was used to monitor the time histories of O atom concentration at 130.5 nm. The source of resonance radiation was a microwave discharge lamp in a flow of O_2 diluted with He maintained at pressure of about 5 torr. The transmitted light was measured by using a VUV monochromator and a solar-blind photomultiplier. Experiments were carried out on C_6H_5NO-O_2 mixtures highly dilute in Ar (C_6H_5NO/O_2 = 50/125-500 ppm) in the temperature range 1500-2000 K at pressures of about 1 atm.

2.2. Infrared emission measurement

Experiments were conducted using a stainless steel shock tube of 78 mm ID. A detailed description has been given elsewhere (Fujii et al. 1987). Simultaneous measurements of two infrared emissions with InSb detectors through interference filters (4.25 μm with fwhm = 0.16 μm for CO_2 and 4.87 μm with fwhm = 0.24 μm for CO) were performed on the mixtures (C_6H_5NO/O_2 = 0.1/0.3-0.5 %) over the temperature range of 1500-2400 K at pressures of 3-4 atm behind reflected shock waves.

Shock Waves @ Marseille II
Editors: R. Brun, L. Z. Dumitrescu

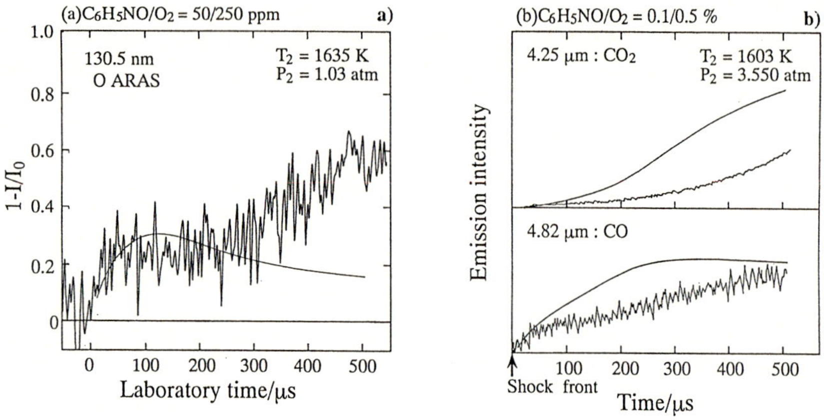

Fig. 1. Typical profiles of a) O atom absorption, and b) CO and CO_2 emissions. Smooth lines are simulated ones

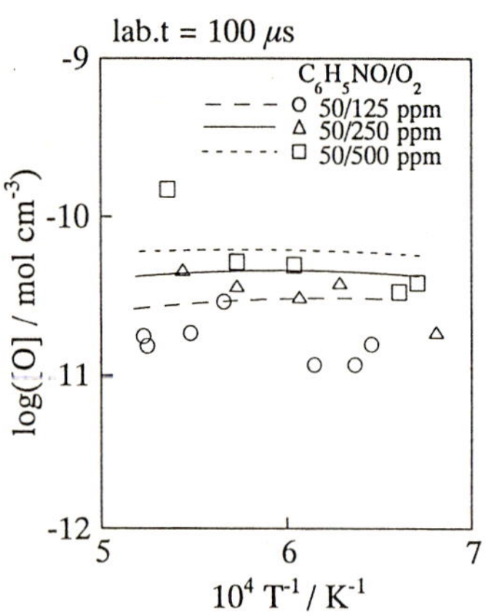

Fig. 2. O atom concentrations produced at lab. $t = 100\ \mu s$. Lines are simulated ones

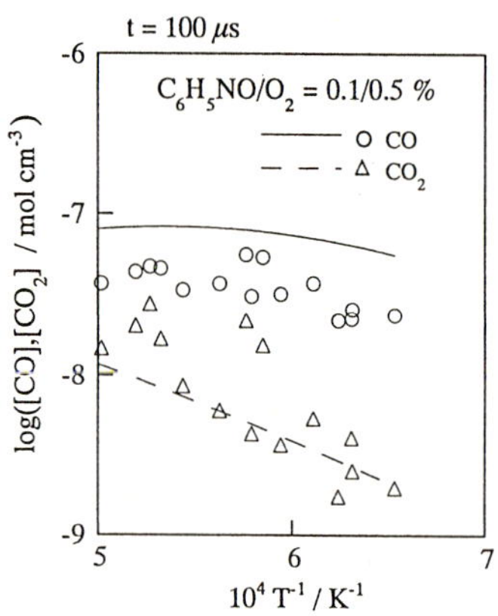

Fig. 3. Concentrations of CO and CO_2 produced at $t = 100\ \mu s$. Lines are simulated ones

3. Results and discussion

It was observed that, in contrast with the C_6H_6-O_2 system, O atom absorption increased rapidly to a peak without an induction period and then decreased gradually under the experimental conditions. A typical O atom absorption profile is shown in Fig.1a. The growth rates of CO and CO_2 measured by the 4.82 and 4.52 μm emissions, respectively were fast compared with those for the benzene oxidation. Typical emission profiles are shown in Fig.1b.

In order to characterize the initial formation rate of O atoms behind incident shock waves, the concentration of O atoms produced at certain reaction time was estimated from the absorption profile. Example of the plot of the O atom concentrations at laboratory time = 100 μs under the various conditions was shown in Fig.2. Also, to estimate the formation rates of CO and CO_2

behind reflected shock waves, the rates were similarly compared by the concentrations of CO and CO_2 produced at certain time. Example of the plot for CO and CO_2 concentrations at t = 100 μs was shown in Fig.3.

Here, experiments of 130.5 nm absorption and also of 4.82 and 4.52 μm emissions for the decomposition of C_6H_5NO were performed respectively, in order to check the interferences due to the decomposition products. The effects were found to be not so serious at the temperature below about 2000 K.

The measured profiles of the above species were compared with those by simulations. The reaction scheme used for the simulation was based upon that previously reported for the C_6H_6-O_2 system (Fujii and Asaba 1973, Bittker 1991). The schematic reaction route is as follows:

(i) $C_6H_5 + O_2 \longrightarrow 2CO$ + low hydrocarbons

(ii) low hc $\longrightarrow CO + H_2O$ (oxidation of low hc)

(iii) $H_2 - O_2$ reaction $\longrightarrow$ O, H, OH, H_2O

(iv) $CO \longrightarrow CO_2$ (oxidation)

Here, the overall oxidation process of lower hydrocarbons, reaction (ii), may have some problems, but these are not considered to be very severe under the present conditions. Accordingly, those reactions seem to be almost well established except for the ring-opening-oxidation (i). Recently, for this overall reaction (i), the following elementary reactions were proposed; their reaction rates are shown in Table 1:

(1) $C_6H_5 + O_2 \longrightarrow C_6H_5O + O$

(2) $C_6H_5O \longrightarrow CO + C_5H_5$

(3) $C_5H_5 + O \longrightarrow CO + C_4H_5$

(4) $C_4H_5 + M \longrightarrow C_2H_2 + C_2H_3 + M$

Table 1. Rate constants of elementary reactions for the C_6H_5-O_2 system

No.	Elementary reaction	log A	b	E/kJ	Ref.
(1)	$C_6H_5 + O_2 \rightarrow C_6H_5O + O$	12.32	0	31.2	Lin (1987)
(2)	$C_6H_5O \rightarrow C_5H_5 + CO$	11.40	0	183.6	Lin (1986)
(3)	$C_5H_5 + O \rightarrow C_4H_5 + CO$	14.00	0	0.0	Brezinsky (1986)
(4)	$C_4H_5 \rightarrow C_2H_2 + C_2H_3$	11.60	0	176.7	Weissman (1988)

$k = AT^b \exp(-E/RT)$ (in units of $cm^3\ mol^{-1} \cdot s^{-1}$)

In the simulation using 41 elementary reactions including the above 4 reactions, the O atom concentration profile calculated was converted to the absorption profile using a calibration curve obtained by separate shock heated N_2O-Ar experiments, also CO and CO_2 concentration were converted to 4.82 and 4.25 μm emission profiles, respectively using calibration curves. In this simulation the correction for the nonideality of shocked flow caused by the boundary layer behind the incident shock front (Fujii et al. 1979) was taken into account.

Examples of the simulated concentration profiles under both the conditions of O-ARAS experiments and IR emission experiments are shown in Figs.4a and b. In the latter experimental condition for IR emission measurements, the simulated O atom profile seems to be strange. This means that the reaction scheme used may have some problems under the high reactants concentration. The profiles simulated were compared with the experiments. An example of the comparison of the simulated O atom absorption profile with the experiments is shown by the smooth line in Fig.1a. Agreement between the two profiles is good, especially in the initial stage, in the figure.

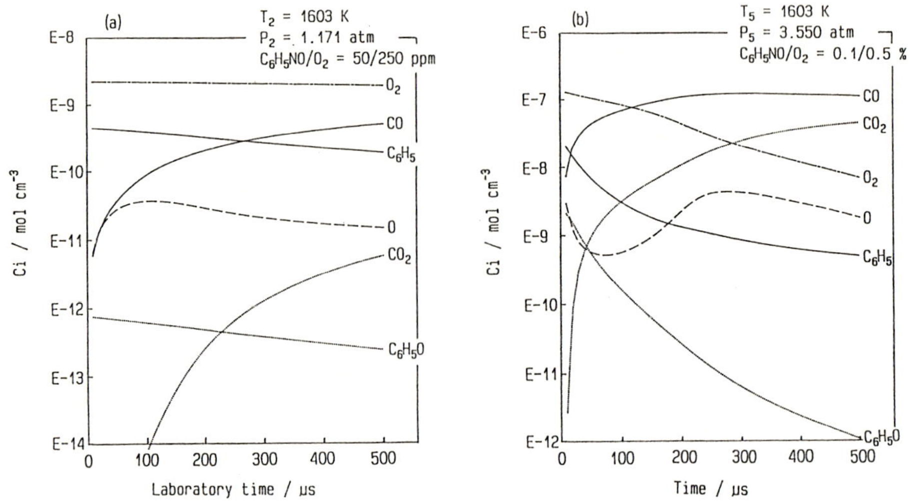

Fig. 4. Examples of simulated profiles of the main species for: a) ARAS, b) IR emission

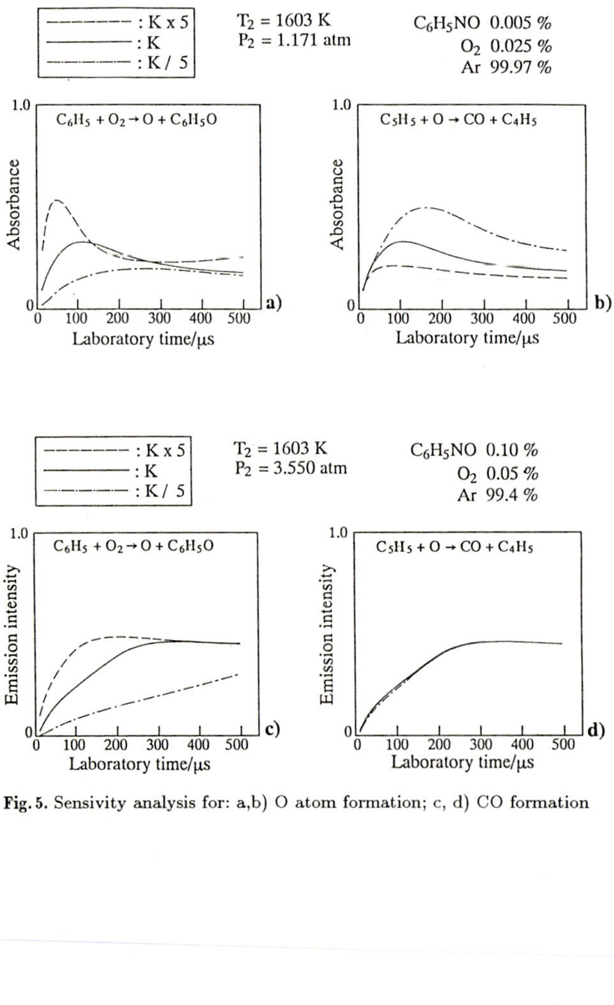

Fig. 5. Sensivity analysis for: a,b) O atom formation; c, d) CO formation

Fig.2 shows also the comparison of the simulated O atom concentration at laboratory time = 100 μs with the experiments. In the figure, agreement between the simulation and the experiments for the lean C_6H_5-O_2 case is fairly good, however the simulated O atom concentration is higher than experiments when O_2 concentration is low. This shows that some important elementary reactions including O atom, such as C_6H_5+O, are missing, especially in a rich mixture, in the reaction scheme used in the simulation.

Also, examples of the comparison of the simulated CO and CO_2 emission profiles with the experiments are shown by smooth lines in Fig.1b. The simulated emission profiles of CO and CO_2 are higher than experiments, respectively. Fig.3 shows also the comparison of the simulated CO and CO_2 concentration at time = 100 μs with the experiments. In the figure, the simulated CO concentration is higher than in the experiments. Also the figure shows that the simulated CO_2 concentration at time = 100 μs almost agreed with experiments at lower temperature but was lower than experiments at higher temperature. These results mean that there are some problems for the reaction path to CO formation in the simulation.

A conventional sensitivity test (Fujii et al. 1981) was performed as follows: The profiles of O atom and CO concentrations were computed by multiplying or dividing the standard rate constant value (Table 2) by 5, where the sensitivities of the rate constants of the above reactions (1)-(4) to O atom and CO formation are evaluated. The results of the sensitivity test showed that the calculated concentration profile of O atoms was very sensitive to the rate constants of reactions (1) and (3), and not so sensitive to those of reactions (2) and (4), as shown in Figs.5a and b. Also it was found that the calculated concentration profile of CO was very sensitive to the rate constant of reaction (1), and not so sensitive to those of reactions (2), (3) and (4), as shown in Fig.5c and d.

Accordingly, the rate constants of reactions (1) and (3) will be determined by fitting the simulation to experimental profiles of O atom and CO. However, from the present results, we found some problems in the reaction scheme. Namely, at first we should add some important elementary reactions including O atoms, such as C_6H_5+O in the simulation.

Acknowledgment

This work was partially supported by a Grant-in-Aid for Scientific Research from the Ministry of Education, Science and Culture to which we wish to express our thanks. We also express our thanks to Prof. Lin MC for his advise in the nitrosobenzene experiments.

References

Bittker DA (1991) Detailed mechanism of oxidation of benzene. Combust. Sci. and Tech. 79:49-72

Brezinsky K (1986) The high-temperature oxidation of aromatic hydrocarbons. Prog. Energy Combust. Sci. 12: 21

Fujii N, Asaba T (1973) Shock-tube study of the reaction of rich mixtures of benzene and oxygen. 14th Symp. Intl. on Combustion, The Combustion Institute pp 433-442

Fujii N, Miyama H, Koshi M, Asaba T (1981) Kinetics of ammonia oxidation in shock waves. 18th Symp. Intl. on Combustion, The Combustion Institute pp 873-883

Fujii N, Kakuda T, Takeishi N, Miyama H (1987) Kinctics of the high temperature reaction of CO with N_2O. J. Phys. Chem. 91:2144-2148

Fujii N, Sagawai S, Sato T, Nosaka Y, Miyama H (1989) Study of the thermal dissociation of N_2O and CO_2 using $O(^3P)$ atomic resonance absorption spectroscopy. J. Phys. Chem. 93:5474-5478

Fujii N, Sakatume N, Miyama H (1989) High temperature reaction of the C_6H_6-N_2O system in shock waves. Proc 1988 Natl. Symp. on Shock Waves Phenomena, Shock Wave Research Center, Tohoku University pp 77-86

Fujii N (1993) Combustion and oxidation mechanism of aromatic compounds. In: Someya T (ed) Advanced Combustion Science, Springer-Verlag, pp 144-149

Hsu DSY, Lin CY, Lin MC (1984) CO formation in early stage high temperature benzene oxidation under fuel lean condition: Kinetics of the initiation reaction, $C_6H_6 \longrightarrow C_6H_5+H$. 20th Symp. Intl. on Combustion, The Combustion Institute, pp 623-630

Lin CY, Lin MC (1986) Thermal decomposition of methyl phenyl ether in shock waves: The kinetics of phenoxy radical reactions. J. Phys. Chem. 90:425-431

Lin CY, Lin MC (1987) Kinetics of the $C_6H_5+O_2$ reaction. The Combustion Institute, Eastern Section, Fall Technical Meeting, pp 7.1-3

Venkat C, Brezinsky K, Glassman I (1982) High temperature oxidation of aromatic hydrocarbons. 19th Symp. Intl. on Combustion, The Combustion Institute, pp 143-152

Weissman MA, Benson SW (1988) Rate parameters for the reactions of C_2H_3 and C_4H_5 with H_2 and C_2H_2. J. Phys. Chem. 92: 4080-4084

Shock Tube Combustion of Liquid Hydrocarbon Sprays at High Temperatures

P. Cadman
Department of Physics, University of Wales, Aberystwyth, SY23 3 BZ, U.K.

Abstract. The liquid spray combustion of separate hydrocarbons, toluene (C_7H_8), n-tetradecane ($C_{14}H_{30}$), and ethanol (C_2H_5OH) have been investigated in reflected shock waves at pressures up to 20 bar and temperatures between 850- 2000+ K. This was done by injecting about 30 mg of the fuel as an aerosol into oxygen (1-20%)/ argon mixtures. Mixtures of these fuels with small quantities of additives such as t-butyl peroxide (TBP) 1-10%, i-propyl nitrate (IPN) 5%, and water (for ethanol only) were also investigated.

Key words: Combustion, Hydrocarbons, Liquid spray

1. Introduction

Liquid spray combustion of hydrocarbons represent very extensive and important processes in modern society. It is used in gas turbines, Diesel engines and burners. The conditions which occur in this type of combustion (1-100+ atm, 700-3000 K) are those which can be obtained using shock waves. In this study a shock tube technique has been used to investigate the liquid spray combustion of hydrocarbons under conditions similar to those pertaining in Diesel combustion in order to study (i) the fundamental processes which occur; (ii) the effect of additives on the characteristics of this process; and (iii) the production of soot and other pollutants from the combustion.

Compression ignition engines (Diesel) offer advantages over spark ignition engines in terms of economy and CO emissions into the environment but produce larger carbon emissions. The future use of lower grade fuels and the enforcement of stricter controls makes it imperative that we obtain a better understanding of the processes that occur in liquid phase combustion, the formulation of fuels from components and the influence of additives. The liquid spray combustion of individual fuels (and with additives) and in particular with regard to the combustion characteristics, ignition delays and the production of soot particulates has been studied in this work.

2. Experimental

The experimental apparatus is shown in Fig.1. The stainless steel shock tube is 6.4 cm internal diameter and has a driver section of up to 6.4 m and a driven section of 5-6 m. Plastic 'Melinex' diaphragms of about 5-20 microns thickness were used to separate the driver and driven sections. These were burst cleanly by passage of a 12 V/ 4 amp pulse through a hot wire loop between the two diaphragm layers. The same control unit was used to operate the pulse injector with a delay for synchronisation. Ten ports were used to observe the combustion, one at the end plate, five 10 cms from it, two 20 cms from the end and two at the 30 cm position. PCB pressure transducers placed on the end plate and at 10 and 70 cm away from it were used to determine the incident and reflected shock velocities and hence the incident and reflected Mach No. The reflected shock temperature and pressure were calculated using the Chemkin program in the usual way.

The main measuring station was at the 10 cm position, where two windows were used to monitor the emission from the combustion via optical fibres, narrow band pass filters and photomultipliers.

Shock Waves @ Marseille II
Editors: R. Brun, L. Z. Dumitrescu

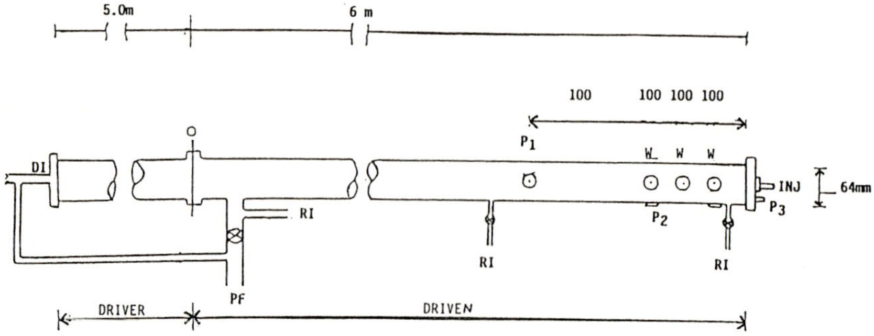

Fig. 1a. Schematic of the shock tube
W - window port; Pn - pressure transducers; INJ - fuel injector; RI - reactant gas inlet system; DI - driver gas inlet; PF - pump outlet

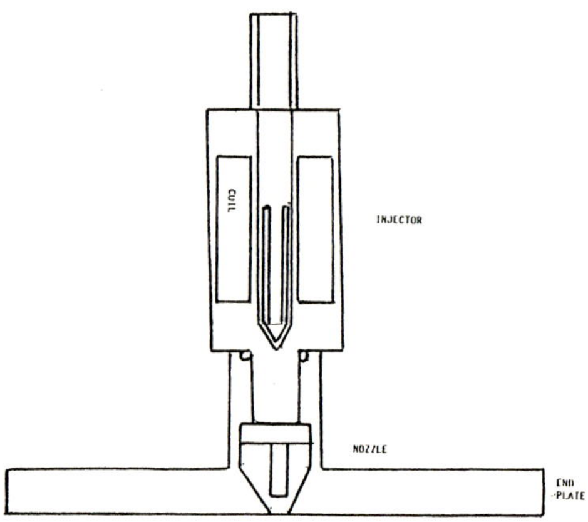

Fig. 1b. Schematic of jet fuel injection system. Backing pressure was 170 psi

One fibre incorporated a splitter, so that 454 and 546 nm radiation emission could be observed simultaneously from the same reaction volume, for use in the determination of temperature changes. A second optical fibre was used to determine the 633 nm radiation output. 488 nm laser radiation was passed through the reaction zone at the 10 cm point, to monitor both the droplet evaporation and the soot/precursor appearance. A 633 nm helium laser passed through the windows at the 20 cm position to follow the spray history and soot formation at this point. Initially another 633 nm laser passed through the tube at the 30 cm from the end wall position but no evidence was found for the spray reaching this far down the tube after compression by the shock; so, these measurements were discontinued.

The absorption intensity profile was followed by passing the laser beams through narrow band-pass filters at the appropriate wavelengths before detection by the photocells. The outputs from the three emission measurements, the two or three laser absorption measurements and the end wall and 10 cm pressure transducers were digitised and passed into a 2 kbyte 8 channel transient recorder for transfer to a microcomputer for permanent storage and analysis.

30 mg aliquots of hydrocarbon fuels of 20 msec duration were injected from the end plate up the tube in synchronisation with the shock initiation pulse. The injector is shown in Fig.1 and was controlled by an automobile petrol injector which could be opened and closed by 12 V variable length pulses in synchronisation with that to the diaphragms. This system was found to operate

satisfactorily and give aerosols of fuels with a backing pressure of argon in the range 80-120 psi. The droplet sizes obtained varied with the fuel but were typically in the range 7-70 μm.

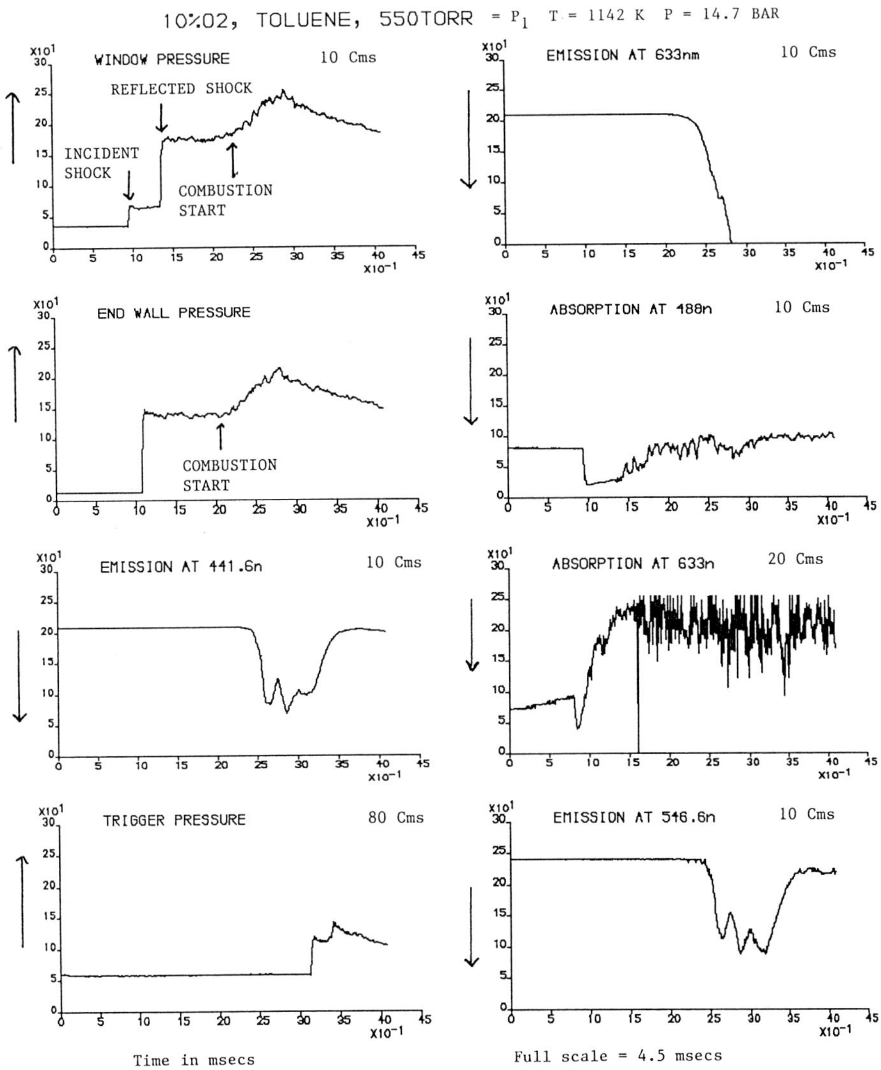

Fig. 2. More uniform combustion

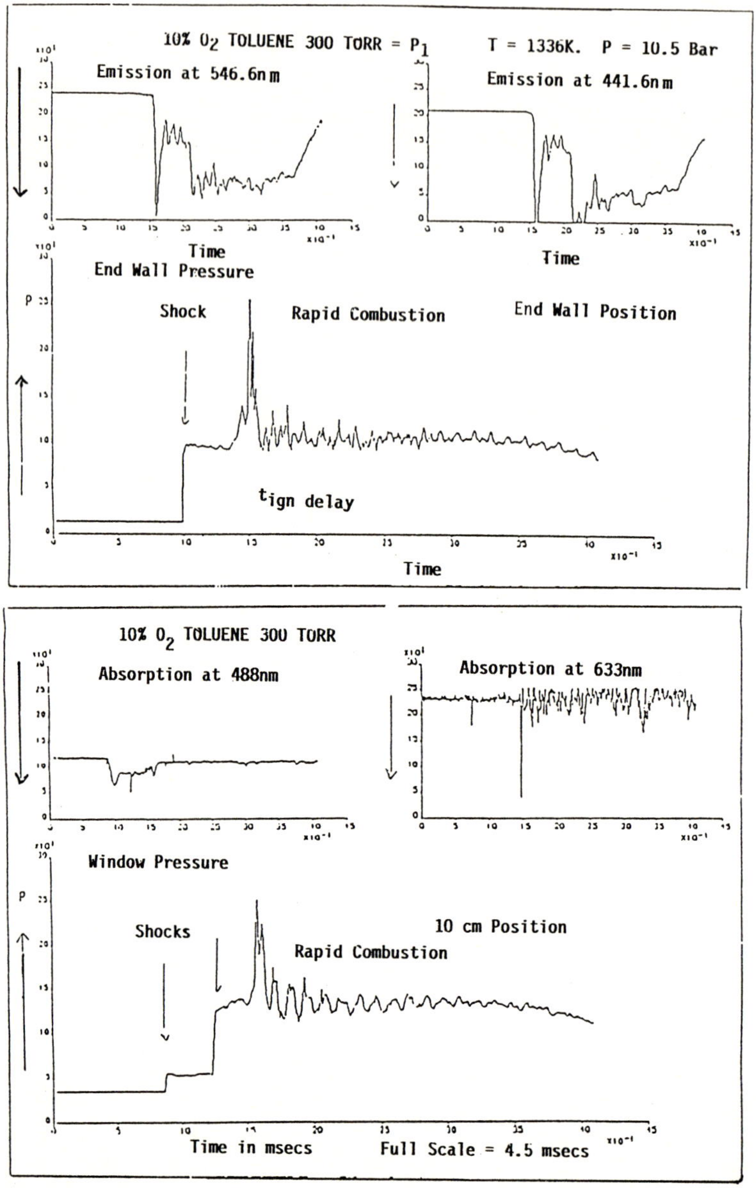

Fig. 3. Very rapid combustion

3. Results

The liquid spray combustion of toluene tetradecane, ethanol and mixtures of these fuels with various additives were studied over a pressure range of about 4-20 bar and temperatures from 950-2000 K. Typical results are shown in Figs.2 and 3. The onset of combustion is shown in both the pressure profiles and in the three different wavelength emission- time profiles. The arrival of the reflected wave is shown on the 10 cm pressure transducer and also shown on the two different laser absorption traces (488 and 633 nm). The 488 nm trace coincides with the emission and 10cm pressure profile in position.

In both Figs.2 and 3 little or no soot formation in the first few millisecs is seen for these experimental conditions. In other work this was not the case. Two general types of combustion were found depending on fuel additive, pressure and temperature. In Fig.2 the combustion pressure rose gradually in a smooth way which we can call Type I combustion. This is a more normal type of combustion. A second type of combustion (Type II) where the pressure rose very sharply as a spike followed by a whole series of smaller spikes is shown in Fig.3. The light outputs followed these pressure changes in synchronisation. The velocity of the flame through the reaction zone is high ($> 10^7$ cm/s) shown by the almost simultaneous pressure increases at the end wall and at the 10 cm position. This combustion may be due to small unsustained detonations or microexplosions occurring in various regions of the combustion mixture. This is called Type II combustion The latter was also found to be generally noiser and tended to occur at the higher temperatures.

Table 1. Production of rapid pressure rises in liquid spray combustion

Fuel	% O_2	Press. [bar]	Additive/fuel conc.	Combustion type
Toluene	10	10-12	5% IPN	Type II all conditions
Toluene	10	8-11	-	Type II>1200 K
Toluene	5	7-10	-	Type II>1430 K
Toluene	1	9-11	-	Type I only
Toluene	20	10-12	-	Type II only >1250 K
Toluene	10	10-12	Fuel Conc~1/2	Type II >1390 K
Toluene	10	10-12	Fuel Conc $sim1.2$	No Type II
Toluene	10	7- 9	1% TPB	Few Type II
Toluene	10	7-11	10% TPB	Type II>1300 K
Tetradecane	10	10-12	-	No Type II
Tetradecane	20	10-12	-	Type II >1240-1300 K not consistent
Tetradecane	20	10-12	5% IPN	Most Type II
Tetradecane	5	11-12	-	One result Type II
Tetradecane	10	15-17	-	Type II only >1300 K
Tetradecane	1	10	5% TPB	No Type II
Tetradecane	5	10-12	5% TPB	Only one Type II at 1580 K
C_2H_5OH	10	12	5 % IPN	Type II >1300 K
C_2H_5OH	10	12	10 % H_2O, 5% IPN	No Type II
C_2H_5OH	10	12	20 % H_2O	No Type II
C_2H_5OH	10	12	10 % H_2O	Only Type II >1600 K

This process may be the reason why Diesel 'knock' occurs in engines. Petrol 'knock' has been well studied but Diesel less extensively. It was found here originally in toluene combustion and first thought to be unique to this system. However, later studies also found this type of combustion in both tetradecane and ethanol systems. Table 1 shows the appropriate conditions when this Type II combustion was found. Its occurrence was not always consistent, probably due to the non-uniform conditions which exist in sprays. This type of combustion has also been seen by Adomeit and co-workers (1993) in the combustion of gas phase mixtures of various fuels such as benzene and heptane.

The ignition delay times were determined from the apparent increase in temperature (which occurs when the combustion starts) as a marker for this delay. This was found by ratioing the emissions at 454 and 546 nm. This method was found to give a better indication of this delay than using either the start of the pressure rise or the start of emission. Studies were conducted over a range of temperatures (850-2000 K), pressures (5-20 bar), oxygen/argon ratios (1-20%) for toluene (Fig.4). The ignition delay shortens as the oxygen is increased i.e. as the decrease in the fuel/oxygen ratio (i.e. leaner mixtures).

Other studies showed that increase in total pressure at the same % oxygen also had the same effect. This seemed to be mainly due to the increase in oxygen (at constant O_2/Argon) rather than just a pressure effect. The addition of 1% TBP had little effect on delay times but 10% TBP did reduce them. 5% IPN seemed to give a larger reduction than either of the two previous mixtures. Tetradecane/oxygen mixtures showed similar trends with fuel/oxygen ratio, although the ignition delays found for this fuel were less than for toluene. The temperature dependency of the ignition delays for tetradecane was also lower than for toluene, perhaps pointing to its higher cetane number rating than toluene. In fact toluene ignition delays at typical starting temperatures in Diesel engines (~700 K) would be predicted to be very long and that the engines would not start very well with this fuel.

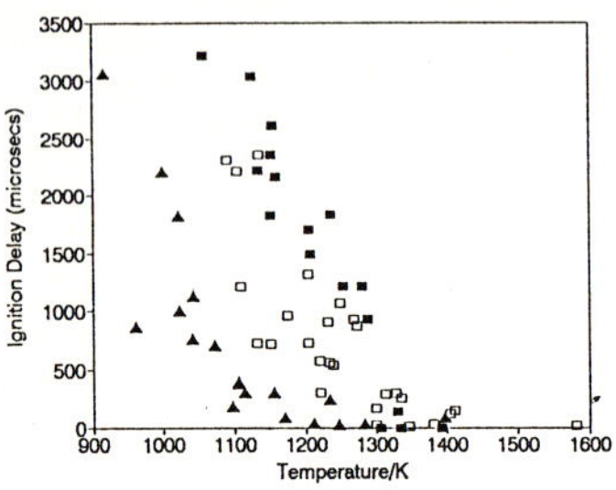

Fig. 4. Toluene, pressure=10-12 bars. Ignition delay vs. temperature.
Variation of delay at different Oxgen concentrations
□ 5% O_2; ■ 10% O_2; △ 20% O_2

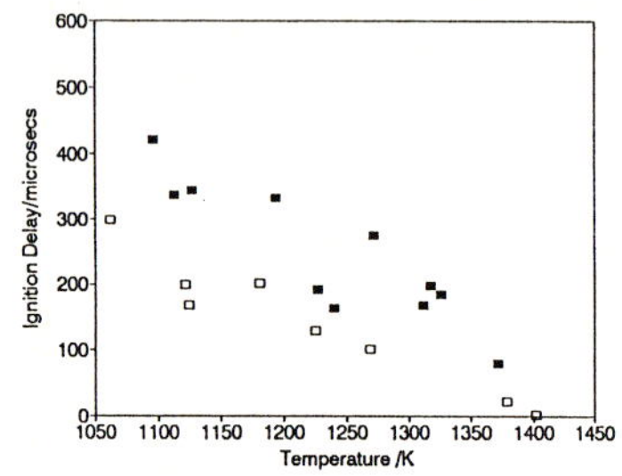

Fig. 5. Tetradecane, 20% O_2, 11 bar.
Effect of 5% IPN on ignition delay
□ 20% O_2; ■ 20% O_2/5% IPN

Fig.5 shows the effect of 5% IPN on the ignition delays of tetradecane compared to 'pure' fuel /oxygen mixtures. Ethanol delay times varied in a similar manner with pressure, temperature and oxygen as for the other fuels. It was found however, that delay times for ethanol were similar to those of tetradecane although the former has a low cetane number and is usually classifed as a bad Diesel fuel.

The addition of 10-30% water to the ethanol increased the delay times with the increase of water as expected. These mixtures, however, gave very low soot yields and also stopped the occurrence of any type II combustion except at very high temperatures, even when 5% IPN was also added to the fuel. The results found in this work suggest that this shock wave technique can be used to investigate the processes occurring in liquid spray combustion.

References

Adomeit G (1993) personal communication

Fieweger K, Ciezki H, Adomeit G (1993) Comparison of shock-tube ignition characeristics of various fuel-air mixtures at high temperatures. These Proceedings

Spray Combustion of Methanol in Reflected Shock Waves

T. Tsuboi and R. Katoh
Faculty of Engineering, Yokohama National University Tokiwadai, Hodogaya-ku, Yokohama 240, Japan

Abstract. A single-compression, "tailored" interface shock tube was used to measure the thermal energy radiated from spray combustion. Precompressed methanol was injected through a throttle nozzle into air in the reflected shock wave. Monochromatic emission and emissive powers were followed with infrared detectors at different distances from the injection nozzle. Flame radii were also determined. Results showed: (1) The thermal radiation energy increased with increasing temperature from 1000 K to 2000 K and with increasing pressure from 0.5 MPa to 2 MPa. The thermal radiation energy from soot was 1/100 lower than that of light oil. (2) The maximum soot concentration was 5×10^{-7} cm^3 soot / cm^3 soot + gas, which corresponded to 1/200 of that of light oil.

Key words: Spray combustion, Methanol combustion, Shock tube, Thermal radiation

1. Introduction

Because of diminishing petroleum resources as well as problems with exhaust gases such as NO_x, SO_x and with soot particles, methanol is suggested as one of the alternative fuels for internal combustion engines. The heat loss due to thermal radiation of soot was reported in the case of light oil to be about 1-5% of the low calorific value of fuel (Tsuboi and Nagaya 1987). Thus the necessity arises to study how much the thermal radiation of methanol decreases. Beside this we plan to use liquid fuel as the fuel of a supersonic combustion engine. We have to know the ignability of various liquid fuels. Therefore, we observed the ignition and combustion of cetane, n-pentane, n-butane and propane in liquid spray as well as of methanol.

2. Experimental

Several single-compression techniques (Pischinger et al. 1959; Affleck et al. 1968; Miyasaka et al. 1976; Zigan 1977) have been developed, in order to observe the spray combustion process. One such method is the tailored interface shock tube that we employed (Tsuboi et al. 1985). In the previous study we reported the availability of the tailored shock tube technique (Tsuboi et al. 1986, 1987, 1991). Beside this one can obtain constant pressures and temperatures for a long time by using the over-tailored condition, though one cannot obtain high temperatures (not higher than about 1000 K).

Experiments were performed in a steel shock tube (ID 97 mm) with a 7 m low-pressure section filled with dry air, a 6 m high-pressure section filled with nitrogen and helium. Temperature T_5, concentration c_5 and pressure P_5 in the reflected shock wave were calculated under real gas assumptions by using the measured incident shock velocities, based on three conservation laws. The measured pressure P_5 varied following the arrival of the reflected shock wave because of the boundary layer effect and the multireflections of the shock wave at the end wall. The set of mixture ratio and pressure ratio of the driver and driven gases was subject to minor modifications in the experiment. In this way, it was possible to obtain an almost constant pressure for about 12 ms, and in the case of over-tailored conditions for about 20 ms.

Precompressed methanol was injected through a throttle nozzle (pinhole diameter: 1 mm) into high temperature air behind the reflected shock wave. The pressures of 0.46, 1.0, 1.5, and 2.0 MPa were selected as the shock-heated air pressure $P_{5r,air}$, while the air temperature was kept within 1345 ± 30 K. The temperatures of 980, 1090, 1200, 1345, 1690, and 1975 K were selected

Shock Waves @ Marseille II
Editors: R. Brun, L. Z. Dumitrescu

as the shock-heated air temperature $T_{5r,air}$, while the air pressure was kept within 1.00 ± 0.04 MPa. The injection pressure of methanol P_{inj} was 15 MPa, the amount of fuel was 37 ± 1 mg, and the duration of fuel injection was 4.7 ± 0.1 ms.

In addition, n-pentane, n-butane and propane were injected into shock heated air at 1 MPa and 1100 K at injection pressures of 5, 10 and 15 MPa, in order to observe the ignition delay. Radii $r(z,t)$ mm, where soot particles radiate, were determined by the existence of radiation coming through the light guide. Pyroelectric detectors, with constant sensitivity to incident radiation energy at wavelengths between 1 and 15 μm, were used for measuring the emissive powers. The monochromatic emissive powers $E(\lambda, t, z)$ were measured using infrared detectors; filters used in this study were 0.63, 0.80. 1.10, 1.45, 2.17, 3.40, 3.90 and 4.20 μm. The calibration of monochromatic emissive power $E(\lambda, t, z)$ was made by using a black-body furnace. The signals of infrared detectors, pyroelectric detectors and PIN-Si photodiodes were observed simultaneously. The pressure profile of the reflected shock wave and the amount of injection nozzle opening were also measured simultaneously.

3. Results

(a) the 8 time-histories of infrared monochromatic radiation energies, (b) the 8 time-histories of pyroelectric detectors, (c) the 6 signals obtained by the binary system of PIN-Si photodiodes set for measurement of the existence of infrared radiation, and (d) the amount of injection nozzle opening and the air-pressure profile behind the reflected shock wave were measured simultaneously. The emission from evaporated methanol was measured at 3.30 and 3.82 μm. The emission from water gas and carbon dioxide at 1.50 and 2.31 μm and that from carbon dioxide at 4.27 μm were observed respectively. At the beginning and the end of combustion, and at low compressed air pressure these gas emission signals were strong. The emission from carbon dioxide among the emission at 2.31 μm was deducted by using the emission intensity from carbon dioxide at 4.27 μm. Emission intensities at 0.63, 0.80, 1.10, 2.31 μm were used for the measurement of radiation from soot particles. From these spectral intensities the particle temperatures T_p, kcl-values and the emissivities were calculated by using the Hottel-Broughton equation (Hottel et al. 1932).

The emissive powers $E_s(t,z)$ were obtained by integrating monochromatic emissive powers $E(\lambda, t, z)$. The emissive powers $E_p(t,z)$ could be obtained by differentiating the signals measured by the pyroelectric detectors (signals (b)). The flame form was obtained from the signal (c). Using the flame forms and the emissive powers one can calculate the total thermal radiation in one fuel injection by the following integration: $Q_{spect} = \int E_s(t,z) 2r(z) \mathrm{d}z/\cos(\theta)\, dt$; $Q_{pyro} = \int E_p(t,z) 2r(z) \mathrm{d}z/\cos(\theta)\, dt$. Fig.1(a,b) shows the relation between the thermal radiation energies per one injection and the compressed air temperature and also the relation between the thermal radiation and the compressed air pressure.

From these figures one finds: (1) The radiation energy Q_{spect} due to soot particles is about 1/50 smaller than that of gas radiation. The radiation of methanol spray combustion is caused by combustion gases, such as carbon dioxide and water gas. (2) The radiation energy increases with increasing air temperature from 1000 to 2000 K. (3) The radiation energy increases with increasing air pressure from 0.5 to 2.0 MPa.

The ignition of n-pentane, liquid n-butane and liquid propane was observed under the condition of $P_{5r,air}$= 1.0 MPa and $T_{5r,air}$= 1000 K behind the reflected shock wave. The n-pentane ignited relatively easily in our experimental conditions. However, the liquid n-butane could not easily ignite and the liquid propane was much more difficult. Fig.2 shows the probability of ignition for various fuels.

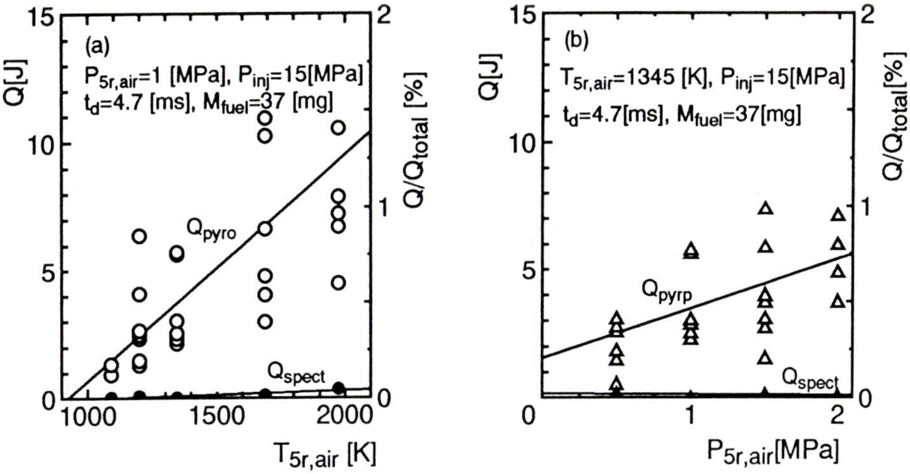

Fig. 1. (a) Relation between thermal radiation energies per injection Q and compressed air temperature $T_{5r,air}$.
(b) Relation between thermal radiation energies per injection Q and compressed air pressure $T_{5r,air}$
Q_{spect} is thermal radiation of soot particles per injection.
Q_{pyro} is thermal radiation of soot + gas per injection.
The scale of right side of figure is the ratio of thermal radiation to low calorific value of fuel

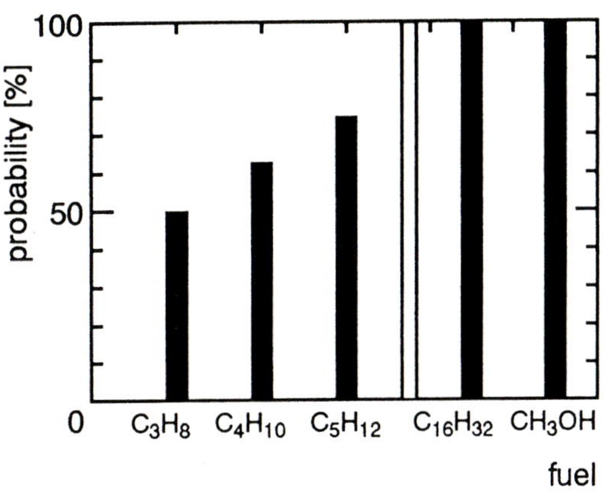

Fig. 2. Probability of ignition
C_3H_8, C_4H_{10}, C_5H_{12}:
$P_{inj} = 10\,MPa$, $M_{fuel} = 5.0\,mg$, $t_d = 5.5\,ms$,
$T_{5r,air} = 1029\,K$, $P_{5r,air} = 1.0\,MPa$
$C_{16}H_{32}$, CH_3OH:
$P_{inj} = 10\,MPa$, $M_{fuel} = 37\,mg$, $t_d = 5.5\,ms$,
$T_{5r,air} = 1095\,K$, $P_{5r,air} = 1.0\,MPa$

4. Discussion

4.1. Thermal radiation

According to the previous results for "light oil" (Tsuboi et al. 1987), the thermal radiation intensity of light oil spray combustion increased until 1500 K with increasing temperature and reached its maximum value around 1500 K and then decreased with increasing temperature. However, such a maximum value could not be observed in the case of methanol spray combustion, and the thermal radiation increased with increasing compressed air temperature monotonously from 1000 to 2000 K. The amount of thermal radiation of light oil spray combustion per injection was 7% of the low calorific value, while that of methanol was 1.5%. Besides, much of this radiation was due to gas radiation, and the thermal radiation due to soot particles was 0.05% at most. This value (0.05%) is under 1/100 of that of light oil. This agrees with the empirical fact that the soot formation in methanol combustion is small. The ratio of the gas (carbon dioxide and water gas) radiation dQ_{gas}/dt to the total radiation dQ_{all}/dt shows that, at 2-3 ms after ignition, the gas radiation was small and then it increased rapidly. At the end of combustion most of the gas radiation was due to gas radiation, not due to soot. The thermal radiation increased with increasing air pressure.

4.2. Soot concentration

The maximum concentration in methanol combustion was 5×10^{-7} cm^3 soot / cm^3 gas + soot, while that of light oil was 10^{-4} cm^3 soot / cm^3 gas + soot. The difference was 1/200. The value $m = 1.66 - 0.55\,i$ was used as the complex index of soot at 1 μm (Darzell et al. 1969).

The time history of the amount of soot V_s in flame could be calculated by integrating the soot concentration for the entire flame volume. The profile of dV_s/dt agreed with dQ_s/dt. The maximum amount of soot during combustion $V_{s,max}$ was 1×10^{-5} cm^3 and this value hardly changed with changing air temperature and air pressure. However, with increasing temperature the amount of $V_{s,max}$ increased just a little, and $V_{s,max} = 1.8 \times 10^{-5}$ cm^3 at 1975 K. This value was 0.03 - 0.05% of the fuel (methanol), assuming that the soot density is 0.9-1.8 g/cm^3 (Tsuboi et al. 1986).

4.3. Ignition delay τ_i

The ignition delay was defined as the time interval between the initiation of fuel injection and ignition. The ignition was defined as the time when the radiation was observed at any of 40 observation points. Fig.3 shows the Arrhenius plots of this ignition delay. The small circles and the dash- dot line show the result for light oil. Though it was assumed from octane-value that the ignition delay of methanol is longer than that of light oil, the experimental result showed no significant difference between methanol and light oil. It was assumed that with increasing ignition delay the mixing of air and fuel increases and the soot formation decreases, and therefore the thermal radiation decreases. Fig.4 shows the relation between the thermal radiation and the ignition delay. The thermal radiation decreased with increasing ignition delay, though the results were not so distinct compared with that of light oil because of large experimental errors.

4.4. Ignition of other fuels

It was expected before experiment that the liquid propane could easily ignite, because propane can evaporate much easier than other fuels used in our experiment. However, the results showed that it was difficult to ignite. The vapor pressure of liquid propane is high at room temperature, though the compressed air pressure is less than the boiling pressure at room temperature. The evaporating rate due to heat conduction from shock-heated air is also high. Therefore, a gaseous fuel-rich region is produced around the injection nozzle. This region is so fuel-rich that the gas mixture cannot ignite. As the result the ignition occurs only around the boundary between compressed air and fuel gas. This leads to the decrease of ignition probability. Also the liquid butane could not easily ignite. Pentane as well as methanol could relatively easily ignite, because the fuel droplets penetrate further in the compressed air, before they evaporate.

5. Conclusion

The thermal radiation of methanol combustion was measured in shock-heated air. The ignition and the combustion of n-pentane, liquid n-butane and liquid propane were also observed in reflected shock waves. The results are:

(1) The thermal radiation increased with increasing temperature from 1000 to 2000 K and with increasing pressure from 0.5 to 2 MPa. The thermal radiation due to soot particles was under 1/100 of that of light oil.

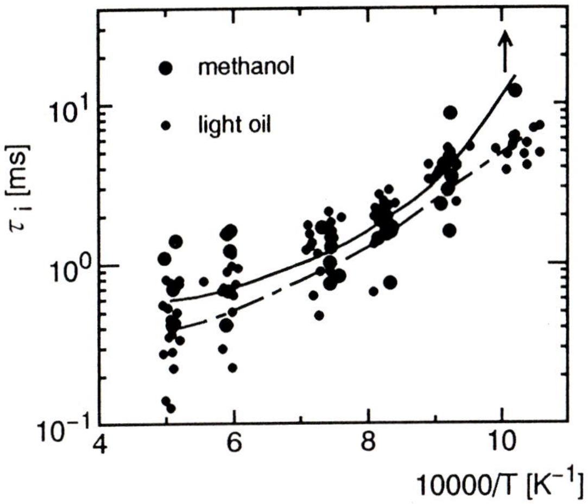

Fig. 3. Arrhenius plots of ignition delay τ_i. $P_{5r,air}$ = 1.0± 0.04 MPa, P_{inj} = 15± 0.5 MPa, M_{fuel} = 37± 3 mg, t_d = 4.7± 0.2 ms. The small circles show the ignition delay of light oil

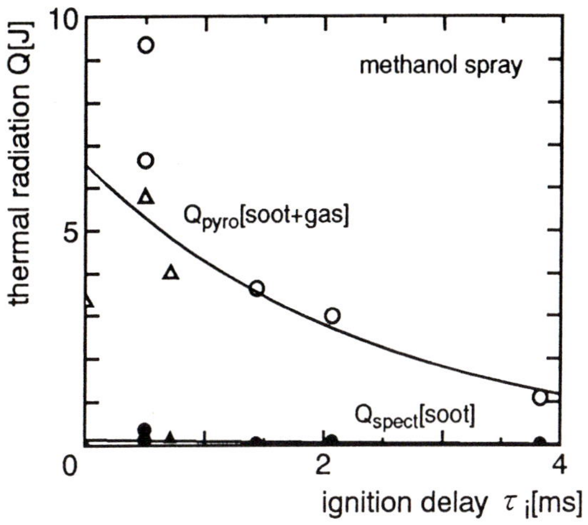

Fig. 4. Relation between Q_{pyro}, Q_{spect} and τ_i. Open and solid circles are: $P_{5r,air}$ = 1.0± 0.04 MPa, P_{inj} = 15± 0.5 MPa, M_{fuel} = 37± 3 mg, t_d = 4.7± 0.2 ms. Open and solid triangles are: $T_{5r,air}$ = 1345± 24 K, P_{inj} = 15± 0.5 MPa, M_{fuel} = 37± 3 mg, t_d = 4.7± 0.2 ms. Open circles and triangles show Q_{pyro}, while solid symbols show Q_{spect}

(2) The maximum soot concentration was 5×10^{-7} cm^3 soot / cm^3 gas + soot and it was about 1/200 of that of light oil. This corresponds to 0.03 - 0.05% of injected methanol burned after forming soot.

(3) The ignition susceptibility of the liquid petroleum spray depended on the vapor pressure of fuel, i.e. the probability of the fuel ignition decreased with increasing vapor pressure.

References

Affleck WS, Thomas A (1968) An opposed piston rapid compression machine for preflame reaction studies. Proc. Instn. Mech. Engrs. 183:365

Dalzell WH, Sarofim AF (1969) Optical constants of soot and their application to heat-flux calculations. Trans. ASME J. Heat Trans. 100:100

Hottel HC, Broughton FP (1932) Determination of true temperature and total radiation from luminous gas flame. Industrial and Engineering Chemistry, 4:166

Miyasaka K, Mizutani Y (1976) Ignition delays of spray columns behind a reflected shock. Proc. 16th Symp. (Intl.) on Combustion, pp 639-645

Pischinger A, Pischinger F (1959) Bombenversuche über die Diesel-Verbrennung unter motorischen Bedingungen. MTZ 20: 1

Tsuboi T, Fukushima M, Sato T, Oguri T (1985) Thermal radiation during spray combustion behind reflected shock waves. Proc. Intl. Symp. on Diagnostics and Modeling of Combustion in Reciprocating Engines, pp 159-165

Tsuboi T, Kamei M, Kobayashi K, Ikuta Y, Katoh R (1986) Optische Messungen der Dichte der von Azetylen Gebildeten Rußteilchen. Jpn. J. Appl. Phys. 25:1165

Tsuboi T, Nagaya K (1987) Study on dependence of thermal radiation energy of liquid spray flame on pressure and temperature. In: Bershader D, Hanson RK (eds) Proc. 16th Intl. Symp. on Shock Tubes and Waves, pp 503-509

Tsuboi T, Kurihara Y (1991) Influence of fuel injection pressure on thermal radiation energy of diesel spray combustion. In: Takayama K (ed) Proc. 18th Intl. Symp. on Shock Waves, pp 765-770

Tsuboi T, Sato T, Kaneko Y, Nakajima T, Ita S, Nagaya K (1987) Measurement of heat loss due to thermal radiation of liquid spray flame using shock tube technique. Proc. 21st Symp. (Intl.) on Combustion, pp 473-479

Zigan D (1977) Untersuchungen an einem Einhub-Dieseltriebwerk. MTZ, 20:1

Numerical Simulation of Shock Waves Through Combustible Two-Phase Media

A. Koichi Hayashi, Takayuki Fuyuto and Toshi Fujiwara
Department of Aeronautical Engineering, Nagoya University, Nagoya 464-01, Japan

Abstract. The physics of grain dust ignition and combustion in a high-temperature atmosphere behind a shock wave is numerically studied using an Eulerian description of the two-phase flow system. Corn starch is used as a dust in the present calculation. Unsteady two-dimensional conservation equations of mass, momentum, and energy are integrated in the form of Euler equations for both gas and solid phases using a TVD-upwind method. Results show that the calculation with an one-step overall corn starch/air reaction mechanism gives the two-phase C-J detonation velocity, the ignition and transition to detonation, and the two-phase detonation features; it is found that the two phases have large lags in temperature and velocity fields.

Key words: Two-phase flow, Combustion, Detonation, Numerical simulation

1. Introduction

Flame propagation and transition to detonation have been studied for many years, however the study of two-phase detonations started just recently. Recent studies of dust detonations are presented experimentally by Kauffman et al.(1979), Wolanski et al. (1983), Wolinski et al. (1991), Ohyagi et al.(1991), Li et al. (1991), Zhang et al.(1991), and numerically by Wolanski et al.(1983), Eidelman and Yang(1991). The characteristic features of two-phase detonations are revealed by their work, but the details of the phenomena and their numerical study are still open.

In the present paper the ignition and combustion of a corn starch dust-air two-phase mixture which is initiated by a shock wave are studied numerically by applying Euler two-phase equations.

2. Governing equations

In order to set up the solid-gas two-phase equations, the following assumptions are considered to simplify the system:

1. The gas mixture consists of O_2, H_2O, CO_2 as reactive species and N_2 as a diluent.
2. Solid particles are rigid spheres and collisions among particles are neglected.
3. The specific heat of particles is constant and temperature inside particles is uniform.
4. The gas specific heat is a polynomial function of temperature, but the equation of state for the ideal gas is used.
5. An artificial solid phase pressure is used to get a conservative form of Euler's equations.
6. External forces such as gravity are neglected.

With these assumptions the Euler equations are set up for both gas and solid phases considering that the heterogeneous phase can be taken as continuous. In the gas phase equations the total mass and species equations are simultaneously solved to avoid errors accumulating on a specific species. The governing equations for a two-dimensional system in the present problem are as follows:

$$\frac{\partial \boldsymbol{U}}{\partial t} + \frac{\partial \boldsymbol{F}}{\partial x} + \frac{\partial \boldsymbol{G}}{\partial y} + \boldsymbol{H} = \frac{\partial \boldsymbol{F}_v}{\partial x} + \frac{\partial \boldsymbol{G}_v}{\partial y} + \boldsymbol{P} \qquad (1)$$

where

Shock Waves @ Marseille II
Editors: R. Brun, L. Z. Dumitrescu

$$\boldsymbol{U}=\begin{pmatrix} c_g \\ c_g u_g \\ c_g v_g \\ e_g \\ c_{gi} \\ c_s \\ c_s u_s \\ c_s v_s \\ e_s \\ N_p cr \end{pmatrix},\ \boldsymbol{F}=\begin{pmatrix} c_g u_g \\ c_g u_g^2+(1-\varepsilon)P \\ c_g u_g v_g \\ \{e_g+(1-\varepsilon)P\}u_g \\ c_{gi}u_g \\ c_s u_s \\ c_s u_s^2+\varepsilon P \\ c_s u_s v_s \\ \{e_s+\varepsilon P\}u_s \\ N_p u_s \end{pmatrix},\ \boldsymbol{G}=\begin{pmatrix} c_g v_g \\ c_g u_g v_g \\ c_g v_g^2+(1-\varepsilon)P \\ \{e_g+(1-\varepsilon)P\}v_g \\ c_{gi}v_g \\ c_s v_s \\ c_s u_s v_s \\ c_s v_s^2+\varepsilon P \\ \{e_s+\varepsilon P\}v_s \\ N_p v_s \end{pmatrix},\ \boldsymbol{P}=\begin{pmatrix} \Gamma \\ \Gamma u_s \\ \Gamma v_s \\ \Gamma\frac{e_s}{c_s} \\ \omega_i \\ -\Gamma \\ -\Gamma u_s \\ -\Gamma v_s \\ -\Gamma\frac{e_s}{c_s} \\ 0 \end{pmatrix}$$

$$\boldsymbol{H}=\begin{pmatrix} 0 \\ c_s(1-\varepsilon)A_s(u_g-u_s) \\ c_s(1-\varepsilon)A_s(v_g-v_s) \\ c_s B_s \\ 0 \\ 0 \\ -c_s(1-\varepsilon)A_s(u_g-u_s) \\ -c_s(1-\varepsilon)A_s(v_g-v_s) \\ -c_s B_s \\ 0 \end{pmatrix},\ \boldsymbol{F}_v=\begin{pmatrix} 0 \\ 0 \\ 0 \\ -(1-\varepsilon)q_x \\ c_g D_i\frac{\partial Y_i}{\partial x} \\ 0 \\ 0 \\ 0 \\ 0 \\ 0 \end{pmatrix},\ \boldsymbol{G}_v=\begin{pmatrix} 0 \\ 0 \\ 0 \\ -(1-\varepsilon)q_y \\ c_g D_i\frac{\partial Y_i}{\partial y} \\ 0 \\ 0 \\ 0 \\ 0 \\ 0 \end{pmatrix}$$

$$c_g=(1-\varepsilon)\rho_g \tag{2}$$

$$c_s=\varepsilon m_s=\frac{4}{3}\pi r_s{}^3 m_s N_p \tag{3}$$

$$P=\sum_{j=1}^{N}\frac{c_{gj}}{1-\varepsilon}R_j T_g=\sum_{j=1}^{N}\frac{c_{gj}}{1-\varepsilon}\frac{\mathrm{R}}{W_j}T_g \tag{4}$$

$$h_i(T_g)=\int_{T^0}^{T_g}C_{pi}(T_g)dT+h_i^0 \tag{5}$$

$$e_g=\sum_{j=1}^{N}c_{gj}h_j-(1-\varepsilon)P+\frac{c_g}{2}(u_g^2+v_g^2) \tag{6}$$

$$e_s=c_s\left(\int_{T^0}^{T_s}C_s dT_s+h_s^0+\frac{u_s^2+v_s^2}{2}\right) \tag{7}$$

where ε, m_s, r_s, and N_p are the gas phase volume fraction, material density, radius of a solid particle, number density of solid particles, respectively. The constant-pressure specific heat for species i, C_{pi} is described by a polynomial function of temperature to evaluate the total gas phase energy e_g with the specific enthalpy $h_i=\int_{T^0}^{T}C_{pi}dT+h_i^0$ as follows:

$$C_{pi}/R_i=a_{1i}+a_{2i}T+a_{3i}T^2+a_{4i}T^3+a_{5i}T^4 \tag{8}$$

where R_i denotes the gas constant for the species i. The coefficients a_{ni} $(n=1\cdots5)$ in Eq.8 are obtained from the best fit to the data in the JANAF tables (Stull and Prophet 1971).

The molecular viscosity and thermal conductivity coefficients for species, μ_i and κ_i are calculated by the generalized kinetic theory formula with Eucken's correction. The mixture viscosity and thermal conductivity coefficients, μ and κ, are obtained from the Sutherland–Wassiljewa formula and Wilke's approximation, respectively.

The binary diffusion coefficients between species i and j, D_{ij}, are calculated by the Chapman-Cowling empirical formula and the effective diffusion coefficients D_i are evaluated by Fick's law (Hishida and Hayashi 1991). A_s and B_s in Eq.1 are the parameters related to the drag force and heat conduction between two phases.

3. Chemical reaction model

The overall reaction rate of corn starch is modeled by comparing with the experimental data of ignition delay obtained by Ohyagi et al.(1993). The overall one-step corn starch reaction is:

$$C_6H_{10}O_5 + 6(O_2 + 3.76N_2) \longrightarrow 5H_2O + 6CO_2 + 22.56N_2 \quad (9)$$

where the overall rate for this reaction (the particle radius decreasing rate) is:

$$-\dot{r}_s = u_b = A \exp\left(-\frac{E_a}{RT_g}\right) \quad [H_2O]^{s_1}[CO_2]^{s_2} \quad [m/s] \quad (10)$$

where u_b is the burning velocity, A the exponential factor of 10^{-3} [m/sec], E_a the activation energy of 7900 [J/mol], R the universal gas constant, [X] the molecular density of species X [mol/m^3], and s_1 and s_2 are constants.

The total mass of solid particles gasified per unit time and unit volume, Γ, is expressed as:

$$\Gamma = 4\pi r_s^2 u_b m_s N_p \quad [kg/m^3.sec] \quad (11)$$

where m_s is the mass of a solid particle and N_p the particle number density.

Wolanski et al.(1990) use a similar one-step reaction mechanism for corn starch, but their model does not give the ignition delay behind a shock wave nor in a high-temperature atmosphere. Hence they artificially set the ignition delay to fit experimental data. Our model predicts the experimental ignition delay data of Ohyagi et al.(1993) and Wolinski and Wolanski(1991).

4. Numerical method

The gas-solid two-phase Euler equations are solved using a second-order explicit Harten-Yee non-MUSCL modified- flux TVD- upwind scheme (Yee 1989) for the convective terms and a second-order central difference method for the diffusion terms. The grid system is 51 x 181 points with a square mesh of 1 x 1 mm. The initial conditions are that the mixture of corn starch and air (equivalence ratio 1.0) at 293 K and 1 atm flows from upstream to the shock front and air with a very small amount of solid particles is put behind the shock wave. In order to set a perturbation, a high temperature (2930 K) and high pressure (50 atm) region of 10 x 15 mm is located near the lower wall. The shock is obtained from an one-dimensional calculation of the same problem without any perturbation. As boundary conditions one assumes constant upstream values, while second order extrapolation is used downstream. Upper and lower walls conditions are slip and adiabatic. Solid particle data are: diameter 14.5 [μm]; volumetric mass, m_s 750 [kg/m^3], heat capacity, C_s 1000 [J/K.kg], heat of combustion -677 [kcal/mol] (1 atm, 293 K).

5. Results and discussion

The corn starch-air two-phase unsteady combustion flow is modeled by two-dimensional Euler equations using a TVD-upwind scheme. Since a shock-fixed coordinate system is chosen, solid particles as well as gas flow into the shock wave at a Mach number of 3.5.

Figs.1-(a) through -(j) are pressure profiles of this simulation. The figures show the formation and development of the transient combustion wave from 5.42 through 110.76 μsec. Calculation shows the initial formation of a triple shock near the wall at t=110.76 μsec, which grows further. The whole sequences clearly indicate the transverse waves moving between the upper and lower walls.

Figs.2 and 3 are temperature profiles of air and corn starch, respectively. The main features of a two-phase detonation are seen; there appears clearly a temperature lag between the two phases. The temperature rise in the gas-phase is faster than that in the solid particle-phase. Ignition of the solid particles occurs farther downstream from the shock front near the lower wall, which disturbs the shock. Then new ignitions and explosions occur to produce probable triple points.

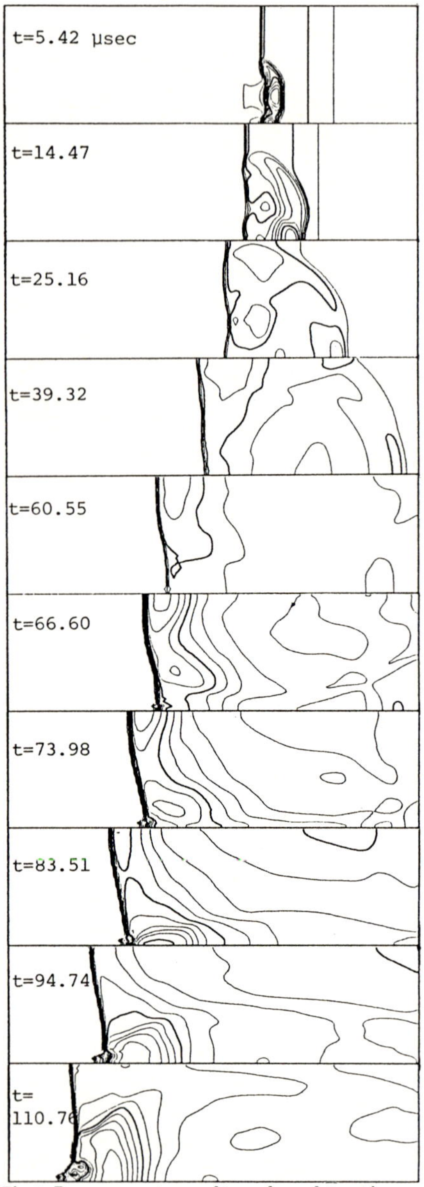

Fig. 1. Pressure contours of two-phase detonation

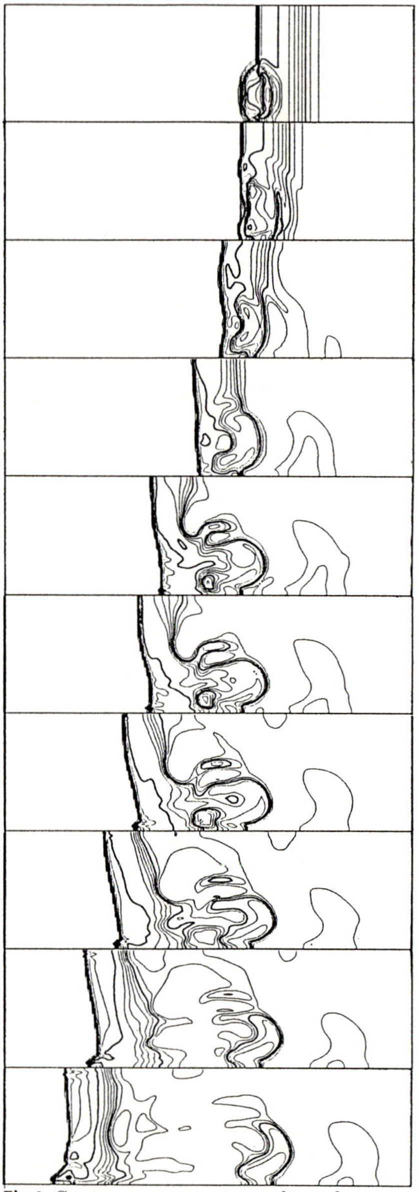
Fig. 2. Gas temperature contours of two-phase detonation

At the same time another hot spot behind the shock wave appears near the upper wall. Figs.4 and 5 show solid particle density and CO_2 mass concentration profiles. Since the solid particles are fuel and CO_2 is product, comparing the two figures provides the degree of combustion. The state upstream the detonation wave is unburnt and that downstream the wave is burnt. The two-phase detonation has a large combustion lag in comparison with a gaseous detonation.

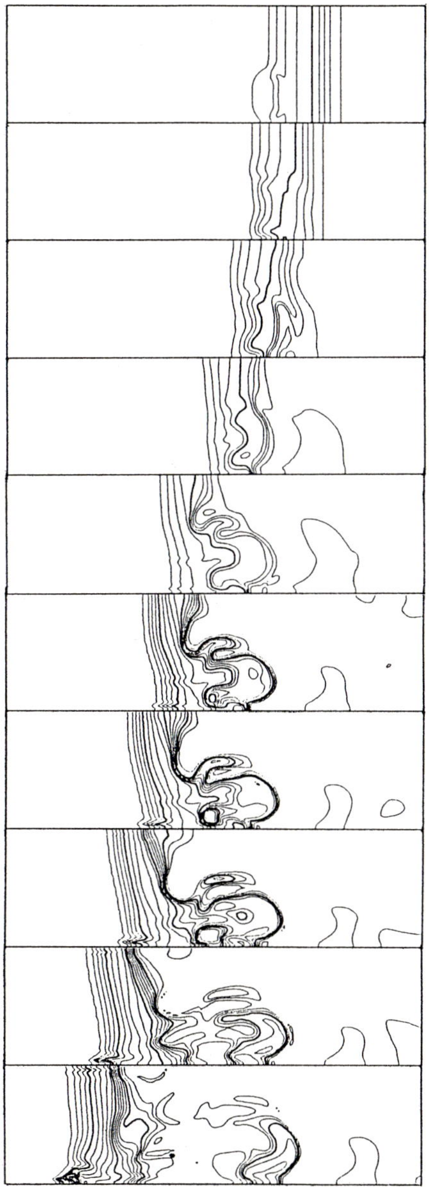

Fig. 3. Solid particle temperature contours of two-phase detonation

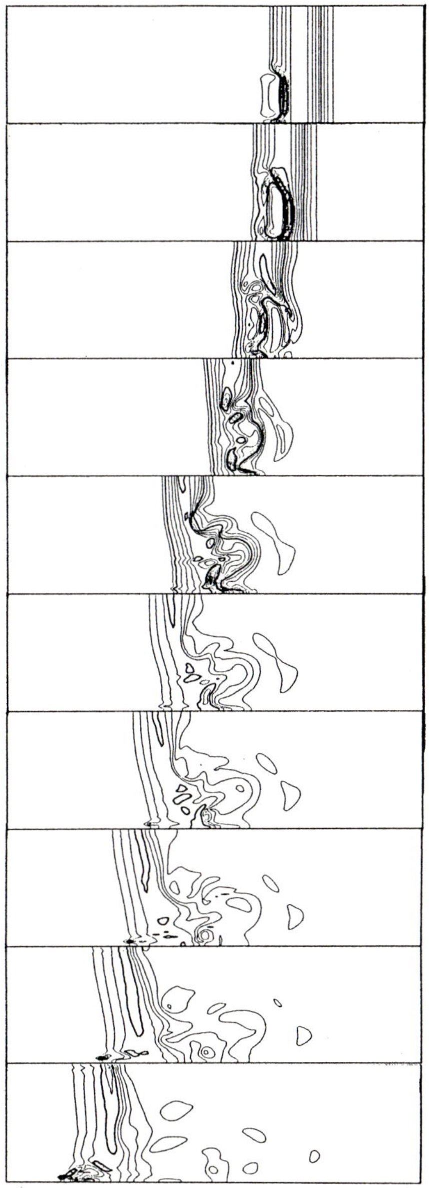

Fig. 4. Solid particle density contours of two-phase detonation

As seen in these figures, the two-phase flow exhibits lags in temperature and pressure, its combustion is inhomogeneous and has many unburnt islands. The unstable combustion occurs due to the solid-phase, which indicates the complicated phenomena of the two-phase detonation. The two-phase C-J detonation velocity in the present calculation is about 2000 [m/sec] or less. This value is close to that of Zhang et al.(1991) and may indicate that our one-step corn starch/air reaction model is good enough to predict two-phase detonation phenomena.

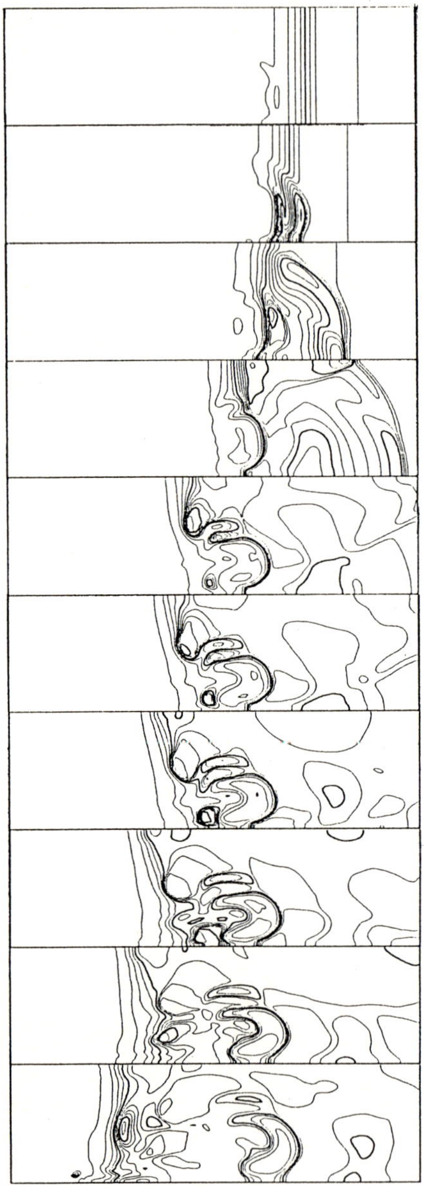

Fig.5. CO_2 mass concentration contours of two-phase detonation

5. Conclusion

Corn starch / air two-phase combustion waves are modeled by two-dimensional unsteady Euler equations using a TVD-upwind scheme. Numerical results using the modeled one-step overall corn starch/air reaction mechanism show agreement of the two-phase C-J detonation velocity with experimental data. The present simulation shows the main feature of two-phase detonation: large lags of temperature and velocity between the two phases. These generate the unstable detonation structure.

References

Eidelman S, Yang X (1991) Detonation wave propagation in combustible mixtures with variable density distributions. In: Kuhl et al. (eds) Prog. Astron. & Aeron. 154, AIAA, pp 228-251

Hishida M, Hayashi AK (1991) Numerical simulation of pulsed jet plume combustion. Ibid. pp 343-361

Kauffman CW, Wolanski P, Ural E, Nicholls JA, Van Dyke R (1975) Shock wave initiation combustion of grain dust. Proc. Intl. Symp. on Grain Dust, Manhattan Kansas, pp 164-190

Li Y-C, Alexander CG, Wolanski P, Kauffman CW, Sichel M (1991) Experimental investigations of accelerating flames and transition to detonation in layered grain dust. In: Kuhl et al. (eds) Proc. 13th ICDERS, pp 170-184

Ohyagi S, Tanaka T, Yoshihashi T, Suzuki S (1993) Proc. Japan Shock Wave Symp. pp 539-541

Stull DR, Prophet H (1971) JANAF Thermochemical Tables, 2nd ed. NSRDS-NBS37

Wolanski P,Lee D,Sichel M,Kauffman CW,Nicholls JA (1983) The structure of dust detonations. In: Bowen et al. (eds) Prog Aeron & Astron 94, AIAA, pp 241-263

Wolinski M, Kapuscinski M, Wolanski P (1991) Detonation processes in dusty mixtures of different oxygen contents. In: Kuhl et al. (eds) Proc. 13th ICDERS, pp 135-147

Yee HC (1989) NASA TM 101088

Zhang F, Greilich P, Ven A.v.d., Grönig H (1991) Two-head detonation structure in corn starch- oxygen mixtures. Ibid. pp 216-227

Ignition Delay-Times of n-Pentane in a Shock Tube

Alexander Burcat and Michael Dvinyaninov
Faculty of Aerospace Engineering, Technion - Israel Institute of Technology, Haifa, 32000 Israel

Abstract. The combustion kinetics of n-pentane was investigated in detail for the first time. Ignition delay times of n-C_5H_{12} and Oxygen mixtures diluted in Argon were measured over a wide range of compositions, pressures and temperatures. The 240 experiments were statistically correlated to an overall equation

$$\tau = 10^{-12.8}\exp[+34610/RT] \times [C_5H_{12}]^{0.29}[O_2]^{-1.10}[Ar]^{0.13}\text{sec.}$$

An effort was made to check and reduce the 325 reaction kinetic scheme.

Key words: Pentane, Oxidation, Ignition delay times, Kinetics

1. Introduction

The kinetic schemes for the combustion of hydrocarbons usually contains hundreds of elementary reactions, and there is a constant request to reduce the size of these schemes as much as possible. In the present research the combustion kinetics of n-pentane is investigated in detail and two known schemes by Warnatz (1984) and Westbrook (1988a), are considered in an attempt to keep them compatible with the experimental results of lower hydrocarbons. It was necessary to investigate in detail the kinetics of pentane, in order to establish a proper experimental and simulation basis.

The shock tube ignition delay data have been shown to be the most useful and reproducible experimental data for hydrocarbons and this method was pursued.

2. Experimental

2.1. Apparatus

Ignition delay-times were performed with n-pentane mixtures in a single pulse stainless-steel shock tube. The ignition delay-time is defined as the time interval between the pressure rise caused by the reflected shock and the onset of pressure rise due to the ignition phenomenon. The tube was described in an earlier paper (Burcat et al. 1992).

2.2. Handling of mixtures

The fuel used in this study was Riedel de-Haën spectroscopic reagent. The gases were 99% pure I.B. Miller Oxygen and Nitrogen, 99% pure Herzliya Argon, and Airco pure grade Helium, used without further purification. All gas mixtures were made manometrically in stainless steel containers where they were pressurized to $\sim$ 50 psia. To ensure thorough mixing of the gases, the mixtures were left in the containers for 48 hours before use.

2.3. Calculations

The reflected shock temperatures were calculated using standard conservation equations and the ideal gas equation of state, assuming frozen chemistry. The enthalpies used in these calculations were taken from a recent compilation (Burcat and McBride 1992).

2.4. Gas chromatography

Post-reflected shocks were quenched gasdynamically just before the ignition occurred and the gas was extracted from the end-plate region into preevacuated glass bulbs. The gas was then analysed in a Packard 427 gas chromatograph with a flame ionization detector using a Porapack N column and Helium carrier gas. The integrating facility used was Nelson 900 series.

Shock Waves @ Marseille II
Editors: R. Brun, L. Z. Dumitrescu

3. Results

240 ignition experiments were made with n-pentane, and approximately 25 experiments were quenched before ignition for gas chromatographic purposes. Table 1 lists the seven mixtures used in this research.

Table 1. Experimental conditions of pentane-oxygen-argon mixtures.

Group	Symbol	C_5H_{12} %	O_2 %	Ar%	$P \sim$ torr	No of shocks
1	△	1.01	8.05	90.94	76.	32.
2	○	0.50	8.02	91.48	76.	34.
3	+	2.01	8.01	89.98	75.	36.
4	▲	1.01	4.06	94.93	76.	31.
5	•	1.00	15.89	83.11	76.	37.
6	▽	1.99	15.90	81.11	65.	44.
7	◇	0.50	4.02	95.48	250.	26.

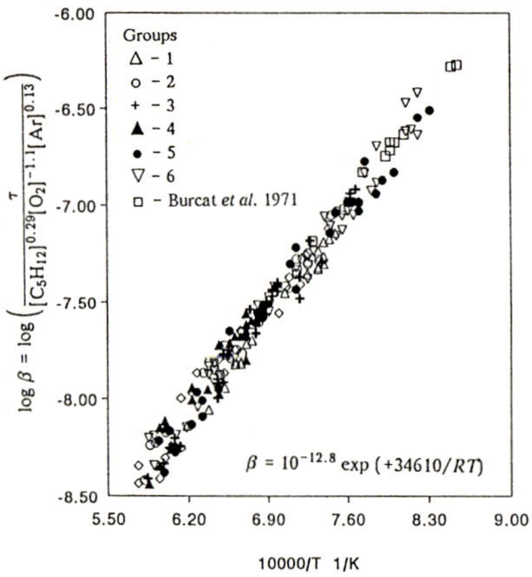

Fig. 1. Overall correlation of $\log \beta$ vs. $1/T$ for all the experiments with 2σ confidence. Experiments marked as □ are from Burcat et al. 1971.

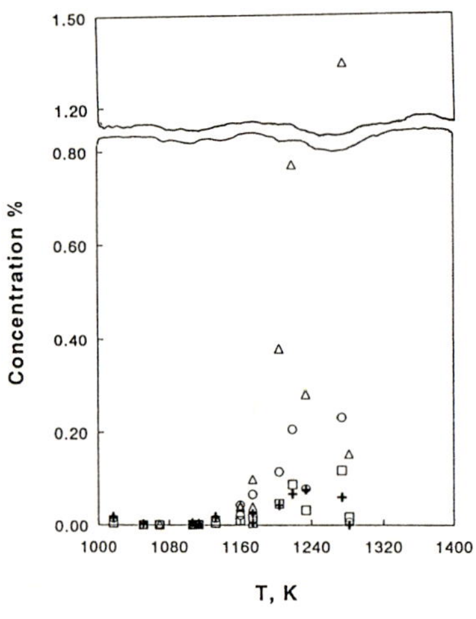

Fig. 2. Product distribution for CH_4 □, C_2H_4 △, (C_3H_8 + C_3H_6) ○, and C_4H_8 +, the main products found in the preignition region of n-pentane C_5H_{12}.

In Fig.1 all the experiments are presented after being statistically correlated to an overall line. The correlation found for the ignition delay time using Semenov's formula is

$$\tau = 10^{-12.8\pm0.4} \exp[+(34610 \pm 835)/RT] \times [C_5H_{12}]^{0.29\pm0.05}[O_2]^{-1.10\pm0.06}[Ar]^{0.13\pm0.07}\text{sec}.$$

It is worth mentioning that 10 experiments run more than 20 years ago (Burcat et al. 1971) in a different location and a different shock tube, fit very well the given correlation.

The gas chromatographic experiments run with the stoichiometric mixture # 1 only, gave us the product distribution of the stable hydrocarbon compounds available before ignition. The presence of methane, ethane, ethylene, propane, propene, allene and propyne, butane and 2-butene was recorded. Traces of acetylene, butadiene and pentene were also recorded. The species that appeared in more than trace quantities are shown in Fig.2.

4. Simulations

The kinetic scheme simulations were run with the CHEMKIN kinetic program (Kee et al. 1992). Its thermodynamic file was updated (Burcat 1992). Three schemes were used: that of Westbrook (1988a), Westbrook with equations changed according to Tsang, and Warnatz's scheme. We found that Warnatz's scheme cannot work since this is basically the propane scheme with three additional reactions. These three reactions were not enough to show the difference between pentane and propane and thus no "ignition" was observed with pentane. The reactions changed in the basic scheme of Westbrook (1988a) were:

a) In reactions 6,7, and 17, the efficiency of M was taken as in Westbrook's (1984) propane.
b) Reactions 97 and 98 that appeared in propane and omitted in pentane, were reintroduced.
c) Reaction 38 was changed from 3.2×10^{13} to 2×10^{13} according to Tsang.
d) Reaction 53 from $10^{12} \exp(-5000/RT)$ to $8.43 \times 10^{9} \exp(-3875/RT)$ (Tsang).
e) Reaction 101 from $9.55 \times 10^{13} \exp(-31000/RT)$ to $3 \times 10^{14} \exp(-32960/RT)$ (Warnatz).
f) Reaction 239 from $6.31 \times 10^{16} \exp(-81900/RT)$ to $6.9 \times 10^{11} \exp(-53390/RT)$ (Zychlinski).
g) Reaction 246 from $10^{10} \exp(-1800/RT)$ to $1.88 \times 10^{12} \exp(-1800/RT)$ (Harris and Kerr).

In addition we found that the basic pentane scheme included an additional reaction # 326:

$$C_2H_3 + O_2 \rightleftharpoons CH_2O + HCO \quad k_f = 3.98 \times 10^{12} \exp(+250./RT) \quad \text{mole/cc/sec}$$

that did not appear in the original Westbrook propane and butane schemes (Westbrook et al. 1984; Pitz and al.1984). Our procedure was the following:

1) The original Westbrook (1988a) pentane scheme was calculated and will be referred from now on as WEST with the symbol ○. This scheme was run for a variety of concentrations (stoichiometric, rich and lean) and temperatures, for pentane mixtures, as occured in representative experiments (Fig.3). The same scheme was used to compare ignition experimental conditions of methane (Fig.4), ethane, ethylene (Fig.5) and propane (Fig.6), as well as concentrations of species as found in our gaschromatographic analysis for pentane mixtures. The latter were compared for methane (Fig.8), ethylene, propane + propene and butene-2 (Fig.7) concentrations.
2) Two reactions (97 and 98) were added to the original Westbrook scheme :

$$C_3H_8 + H \rightleftharpoons H_2 + i{-}C_3H_7 \quad k_f = 8.7 \times 10^{6} T^2 \exp(-5000/RT) \ \text{mole cc}^{-1}\,\text{sec}^{-1} \qquad (97)$$

$$C_3H_8 + H \rightleftharpoons H_2 + n{-}C_3H_7 \quad k_f = 5.6 \times 10^{7} T^2 \exp(-7700/RT) \ \text{mole cc}^{-1}\,\text{sec}^{-1} \qquad (98)$$

which are important for propane but the authors assumed that they could be omitted from the pentane scheme. As it turns out they are important for pentane as well. Also the efficiency of M was changed according to (a). This Westbrook variant was marked as A and its symbol is △.
3) A third Westbrook variant scheme was made by eliminating reaction (326) described above from mechanism A. This scheme was marked as B and its symbol is •.
4) The fourth variant is the B scheme which was changed according to steps a-g described above and this variant was marked as C and its symbol is ■.
5) The fifth Westbrook variant scheme is the same as C but including reaction 326. (This is the same as the basic Westbrook pentane scheme with the changes mentioned in steps a-g above.) This variant was marked as D and its symbol is □.
6) The sixth scheme is the original Westbrook (1984) scheme for propane marked as W-PR and its symbol is ◇.

Fig.3 is the pentane kinetic simulation graph where the experimental data represented by the solid line is compared with calculations made for five different scheme variants. The dotted line shows the 2σ error limit around the experimental value. As we can see the activation energy for

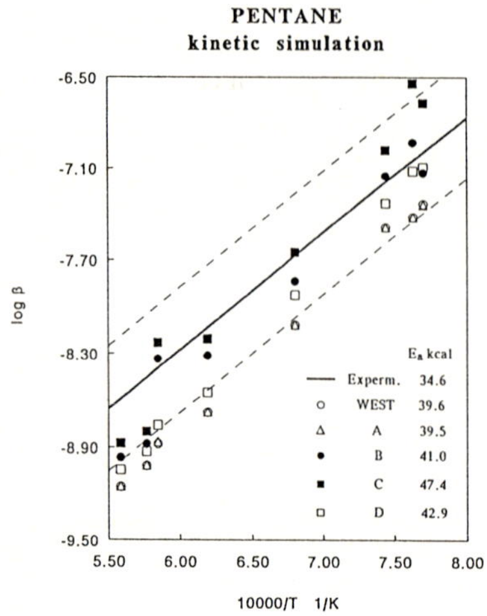

Fig. 3. log β vs. $1/T$ for five different simulation schemes for pentane. The solid line represents the experimental statistical correlation. The dotted lines are the $\pm 2\sigma$ error. WEST = ○; A = △; B = •; C = ▪; D = □.

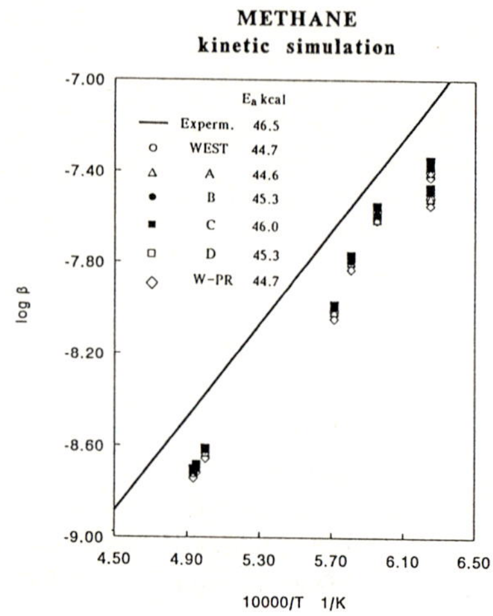

Fig. 4. *og*β vs. $1/T$ for methane using six different kinetic schemes for pentane. WEST = ○; A = △; B = •; C = ▪; D = □; W-PR = ◇. The solid line is a correlation of experiments from Lifshitz et al.1971

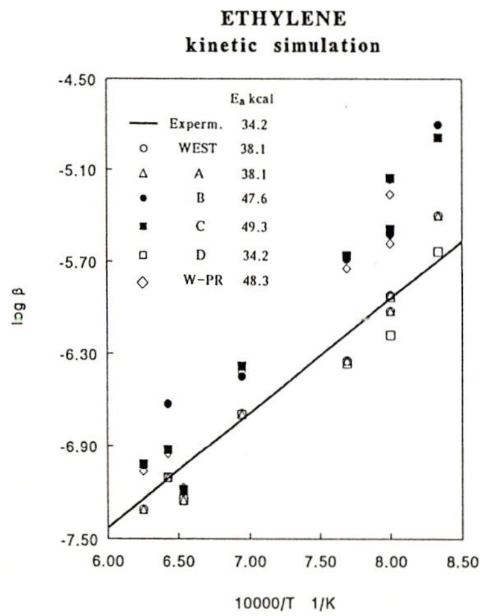

Fig. 5. log β vs. $1/T$ for ethylene using six different kinetic schemes for pentane and propane. WEST = o; A = △; B = •; C = ■; D = □; W-PR = ◇. The solid line is a correlation of experiments from Baker et al.

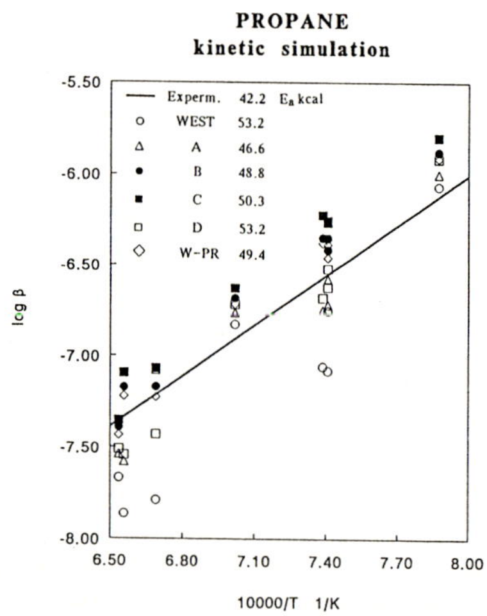

Fig. 6. log β vs. $1/T$ for propane using six different kinetic schemes for pentane and propane. WEST = ○; A = △; B = •; C = ■; D = □; W-PR = ◇. The solid line is a correlation of experiments from Burcat et al.

the different species is higher than the experimental. The best values are those of mechanism C, but the best activation energy is that of A.

The kinetic schemes for pentane were used to verify the kinetic agreement for lower hydrocarbons. Fig.4 shows that all the six schemes mentioned give the same values for methane.

Fig.5 shows the agreement of simulating ignition delay for ethylene with the different pentane schemes, and the original Westbrook propane scheme. The agreement is acceptable for schemes that included reaction 325 and not acceptable where it was lacking. A very similar picture for ethane is obtained as well.

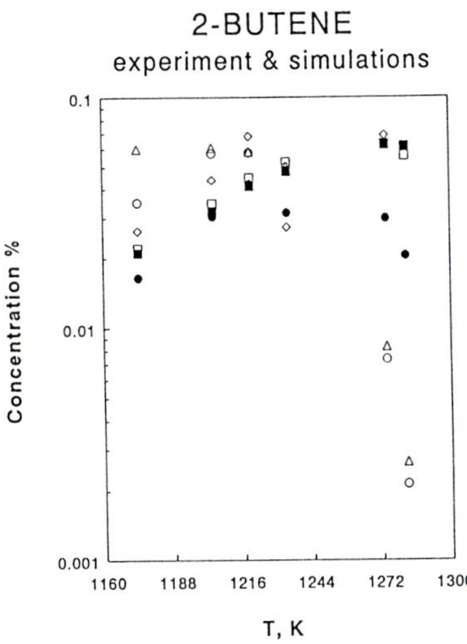

Fig. 7. Concentration % of 2-butene vs. the temperature using five different kinetic schemes for pentane and propane. Experiment = ◇ ; WEST = ○; A = △; B =•; C = ■ ; W-PR = □.

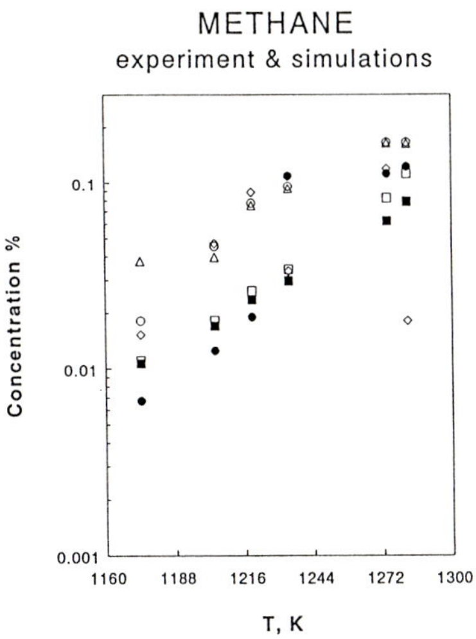

Fig. 8. Concentration % of methane vs. the temperature using five different kinetic schemes for pentane and propane. Experiment = ◇; WEST = ○; A = △; B =•; C = ■ ; W-PR = □.

Fig.6 shows the agreement of propane with the pentane scheme. As expected the agreement of the original Westbrook scheme for propane, W-PR is very good. The original Westbrook pentane scheme is very bad because of elimination of reactions 97 and 98.

When looking for concentrations in our simulations, considerable agreement is found with the gas-chromatographic data we present here for stoichiometric pentane mixtures. We have looked at methane concentrations, ethylene, propane + propene and butene-2. The best correlation is found for butene-2 in Fig.7. Here all correlations give values very close to the experimental ones.

Calculated values for methane show very good correlation with the experimental evidence. This is not surprising since the basis for Westbrook's simulation for methane were concentration values found experimentally. The methane agreement is shown in Fig.8.

The ethylene concentration correlation shows a little less agreement than methane, and propane + propylene shows no agreement at all.

The amount of allene and propyne found experimentally was at most 10^{-4} moles/cc while the calculations showed concentrations of 2-8 $\times 10^{-2}$ moles/cc. The same was found by Westbrook (1988a) as well.

6. Conclusions

It was shown that the basic Westbrook scheme for pentane needed some updates in order to fit well to a good set of experiments. His fuel consumption was too high. Westbrook (1988a) corrected it by pinpointing the 35 most important reactions in his scheme and changing their magnitude by 20% in the direction that reduces the fuel consumption. Our findings have shown that fewer reactions had to be changed, all according to literature values. Nevertheless when making any changes in the scheme of a higher hydrocarbon the agreement of the low hydrocarbons has to be checked and rechecked. In our investigation it was shown that for pentane the activation energy factor is an important parameter. This parameter is not well represented, and the closest representation is by scheme C which does not include reaction 326, while for most other parameters, scheme D (which includes 326), should be preferred. Therefore further work on the subject is necessary. It is on the basis of this investigation that a system that can simulate the kinetics of combustion of hexane isomers should be pursued.

Acknowledgement

The second author acknowledges the Technion Vice President for Research - Struck Fund, for a postdoctoral fellowship during 1992-3.

References

Baker JA, Skinner GB (1972) Shock-tube studies on the ignition of ethylene-oxygen-argon mixtures. Comb. Flame 19:347-350

Burcat A, Lifshitz A, Scheller K, Skinner GB (1970) Shock-tube investigation of ignition in propane-oxygen-argon mixtures. Combust. Symp. 13, pp 745-755

Burcat A, Scheller K, Lifshitz A (1971) Shock tube investigation of comparative ignition delay times for C_1-C_5 alkanes. Comb. Flame 16:311

Burcat A, Pitz WJ, Westbrook CK (1992) Comparative ignition of hexane and octane isomers in a shock tube. In: Takayama K (ed), Proc. 18th Intl. Symp. on Shock Waves, Springer-Verlag, pp 771-780

Burcat A, McBride B (1992) Ideal gas thermodynamic data for compounds used in combustion and air-pollution. T.A.E. Report # 675 Technion, Haifa

Harris SJ, Kerr JA (1988) Relative rate measurements of some reactions of hydroxyl radicals with alkanes studied under atmospheric conditions. Intl. J. Chem. Kinet. 20:939

Kee J, Rupley FM, Miller JA (1992) Chemkin-II: a Fortran chemical kinetics package for analysis of gas-phase chemical kinetics. Sandia Report SAND 89-8009B. U.C-706

Lifshitz A, Scheller K, Burcat A, Skinner GB (1971) Shock-tube investigation of ignition in methane-oxygen-argon mixtures. Comb. Flame 16:311-321

Pitz WJ, Westbrook CK, Proscia WM, Dryer FL (1984) A comprehensive chemical kinetic reaction mechanism for the oxidation of n-butane. 20th Comb. Symp., pp 831-843

Tsang W, Hampson RF (1986) Chemical kinetic database for combustion chemistry. Part 1: methane and related compounds. J. Phys. Chem. Ref. Data 15:1087-1279

Warnatz J (1984) Chemistry of high-temperature combustion of alkanes up to octane. 20th Combust. Symp., pp 845-857

Westbrook CK, Pitz WJ (1984) A comprehensive chemical kinetic reaction mechanism for oxidation and pyrolysis of propane and propene. Comb. Sci. Technol. 37:117-152

Westbrook CK, Pitz WJ, Thoronton MM, Malte PC (1988) A kinetic modeling study of n-pentane oxidation in a well stirred reactor. Comb. Flame 72:45-62

Westbrook CK, Warnatz J, Pitz WJ (1988) A detailed chemical kinetic reaction mechanism for the oxidation of iso-octane and n-heptane over an extended temperature range and its application to analysis of engine knock. 22nd Comb. Symp., pp 893-901

Zychlinski W, Bach G, Heinich K, Zimmerman G (1979) Chem. Tech. (Leipzig) 31:239.

A Shock Tube Study of the Reaction CH_3+O_2

M. Braun-Unkhoff, C. Naumann and P. Frank
DLR-Institut für Physikalische Chemie der Verbrennung 70503 Stuttgart, Germany

Abstract. The reaction CH_3+O_2 has been investigated at elevated temperatures in the postshock region behind reflected shocks. The thermal decomposition of low concentrations (10-50 ppm) of $(CH_3)_2N_2$ served as CH_3 source. Reaction progress was measured by applying narrow-line-width laser absorption of OH-radicals at $\lambda = 308.6091$ nm ($Q_1(5)$-line). The experiments were carried out at temperatures from 1250 to 1800 K at total pressures from 2.2 to 3 bar. For the reaction R1: $CH_3+O_2 \rightarrow CH_2O+OH$ and R2: $CH_3+O_2 \rightarrow CH_3O+O$ the following rate expressions were derived (k in $cm^3\ mol^{-1}\ sec^{-1}$):

$$k_1 = 3.1 \times 10^{10} \exp(-4402/T)$$

$$k_2 = 2.1 \times 10^{12} \exp(-12242/T).$$

Key words: Shock tubes, Chemical kinetics

1. Introduction

Methyl radicals are important intermediates formed during both the pyrolysis and oxidation of simple hydrocarbons. Their subsequent reactions are of fundamental interest for the mechanistic description of combustion processes. The reaction of methyl with O plays an important role under radical-rich conditions (i.e. flame propagation), whereas the reaction with molecular oxygen becomes dominant in radical-poor situations (i.e. during ignition and induction periods). The former reaction that produces CH_2O+H is well established (see Baulch et al. 1992), the latter involves two competing reactions:

$$\text{R1}: \quad CH_3 + O_2 \longrightarrow CH_2O + OH$$

$$\text{R2}: \quad CH_3 + O_2 \longrightarrow CH_3O + O.$$

R1 is exothermic ($\Delta H^0_{r,298} \approx 53$ kcal mol^{-1}), whereas R2 has an reaction enthalpy of about 28 kcal mol^{-1}. Although a large number of individual studies has been carried out in the high temperature regime, there is no sufficient consensus regarding both the rate constants and branching ratio for the reactions R1 and R2 (see Baulch et al.1992). Several groups have reported CH_3O+O to be the principal high-temperature channel (Bhaskaran et al. 1979, Hsu et al. 1983), whereas some authors concluded that CH_2O+OH is the dominant channel below 1900 K (Saito et al. 1986, Fraatz 1991). This suggestion is supported by a computational study predicting that reaction R1 dominates R2 at all temperatures below 2800 K (Zellner and Ewig 1988). In this work, k-values were measured by using a shock tube technique. The thermal decomposition of $(CH_3)_2N_2$ provides CH_3 radicals for temperatures $T > 1200$ K. Therefore azomethane was used as the CH_3 source.

2. Experimental section

Details of the shock tube technique have been described in more detail elsewhere (Frank and Just 1985). Only a short summary will be given. The experiments were conducted in a high purity stainless steel shock tube (ID 100 mm) consisting of a test section of about 13 m in length and a driver section of about 5 m. The measurements were performed behind the reflected shock front close to the end flange ($\sim$ 5 mm). Narrow-line-width laser absorption of OH-radicals at $\lambda = 308.6091$ nm ($Q_1(5)$-line of the $A^2\Sigma^+ - X^2\Pi$ system) was used to monitor time dependent

Shock Waves @ Marseille II
Editors: R. Brun, L. Z. Dumitrescu

OH concentrations. A SpectraPhysics 380 ring dye laser was intracavity doubled using an angle-tuned $LiIO_3$-crystal. The laser was run with Rhodamine 6G and pumped with 4-5 W (all lines) from a SpectraPhysics argon ion laser. 3 mW of single-mode UV power was obtained. The laser mode structure was observed by passing a portion of the fundamental beam into a scanning interferometer (2 GHz free spectral range). The wavelength of the visible output of the ring dye laser was monitored by a Burleigh WA 20 wavemeter. Additionally, a methane-air flame was used to ensure that the laser wavelength was tuned to the OH absorption peak. This was checked for all experiments.

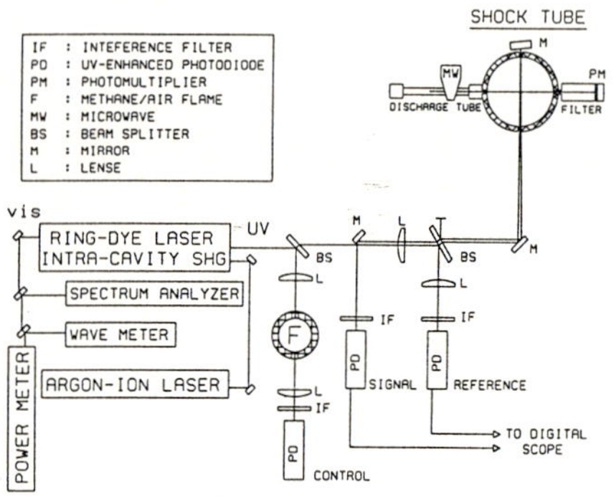

Fig. 1. Sketch of the experimental setup

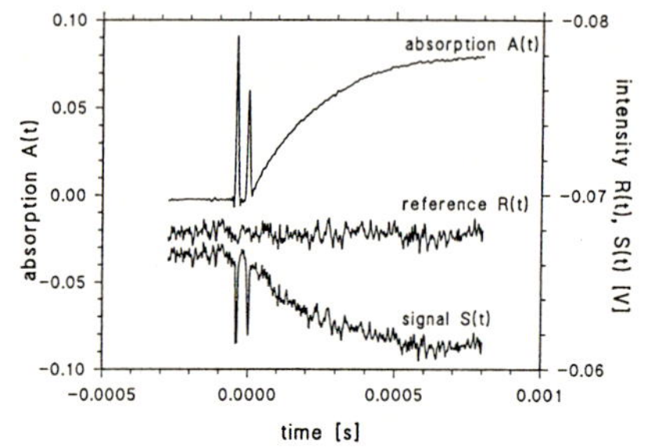

Fig. 2. Time dependence of signal and reference beam

The experimental setup is shown in Fig.1. The UV beam is coupled to the shock tube via fibre optics. The optical path length was doubled by means of a mirror. The beam is split before passing the shock tube in order to cancel laser fluctuations. The detectors were amplified UDT020UV photodiodes (Gigahertz Optik) masked with interference filters. The reference and signal are recorded on a digital storage oscilloscope (Nicolet, sampling time 1 μs). Actually, absorption $\geq 0.5\%$ can be measured precisely. A typical absorption profile is to be seen in Fig.2.

The initial concentrations were 10-50 ppm azomethane and 0.6-1.6 % O_2, both diluted with argon. Azomethane was prepared from sym-dimethyl-hydrazine by oxidation with copper-(II)-chloride. No impurities could be detected. The initial concentrations of azomethane were permanently controlled by means of a gas chromatograph equipped with a flame ionization detector (FID); no difference between measured and calculated concentrations was found. The test gases were of high purity (Ar=99.9999 %, O_2=99.9995 %).

3. Evaluation

The values for the absorption cross section that have been computed according to established procedures (Rea et al. 1987, Yuan et al. 1991) were checked in separate experiments. Mixtures of 300-2400 ppm H_2 and 140-2200 ppm O_2 were shock heated in the temperature range of 2000-2600 K at total pressures from 2 to 7.5 bar. The observed absorption profiles were reproduced by simulation using the reaction system of Masten et al. 1991). The rate coefficients of the most important reactions were considered to be well known. From the comparison of observed and calculated absorption profiles the temperature and pressure dependence of the absorption cross section σ can be derived.

A typical absorption profile is shown in Fig.3. After an induction period of about 200 μs the OH concentration growths rapidly because of chain-branching reactions. Some further 200-300 μs

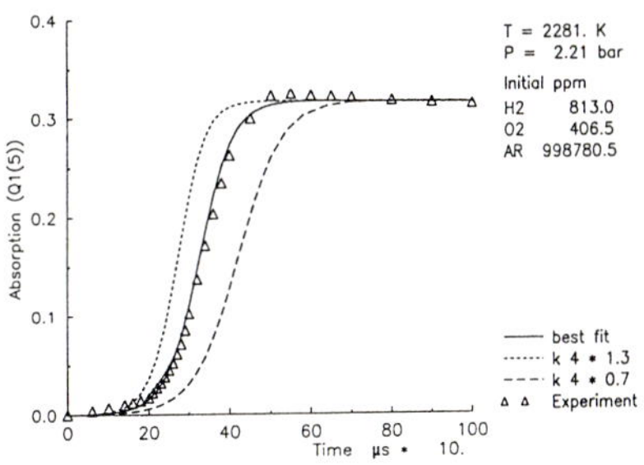

Fig. 3. H_2/O_2 system : a typical OH absorption profile

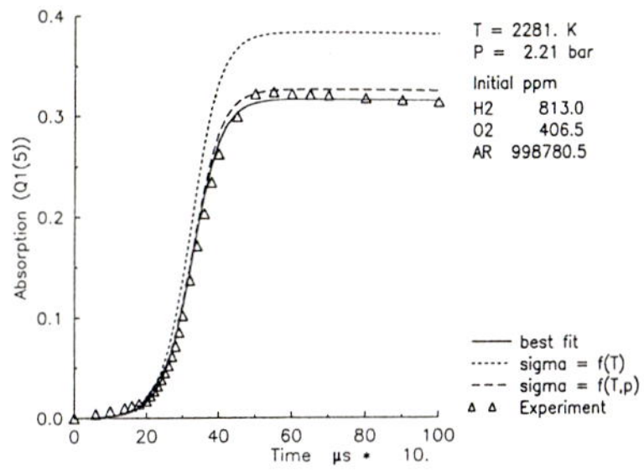

Fig. 4. H_2/O_2 system: OH absorption profile showing the temperature and pressure dependence of the absorption cross section σ

later a quasi-stationary equilibrium is achieved. Variations of the rate coefficient of the reaction:

$$R4: \quad H + O_2 \longrightarrow OH + O$$

by $\pm 30\%$ do not agree with the observed profiles. However, with respect to the absorption cross section σ, no influence on the absorption profile at later reaction times can be noticed. The absorption cross section σ depends on temperature and pressure as may be derived from Fig.4.

4. Results and discussion

OH radicals are produced directly via reaction R1 and on the other hand by the reaction of H atoms with highly excessive molecular oxygen via the reaction sequence R2/R3/R4:

$$\begin{aligned} R1: &\quad CH_3 + O_2 \longrightarrow CH_2O + OH \\ R2: &\quad CH_3 + O_2 \longrightarrow CH_3O + O \\ R3: &\quad CH_3O \longrightarrow CH_2O + H \\ R4: &\quad H + O_2 \longrightarrow OH + O. \end{aligned}$$

In former investigations, there is considerable disagreement regarding both the rate constants and branching ratios for the reactions R1 and R2 (see for example discussion in Baulch et al. 1992). OH radicals are generated furthermore -because of the excess of O_2- in subsequent reaction steps from H atoms according to reaction R4. Secondary reactions of methyl with OH:

$$R5: \quad CH_3 + OH \longrightarrow \text{products}$$

have to be considered also. There is still a good deal of uncertainty about the possible product channels (see for example discussion in Baulch et al. 1992, Bott and Cohen 1991, Oser et al. 1991). Up to now there are no reliable kinetic data available for combustion temperatures.

The experiments covered the temperature range 1250-1800 K at total pressures from 2.2 to 3 bar. A large reaction scheme consisting of 31 reactions with 24 species was adopted for calculating the absorption profiles. Only the important reactions will be discussed.

In all experiments, the absorption rises immediately without any induction period; this is true also for measurements conducted at the lowest temperatures of the investigated temperature range. This indicates that at the beginning of the observation time interval the OH species must be produced directly by an initiation reaction. Fig.5 shows results from an experiment at the lower limit of the investigated temperature range.

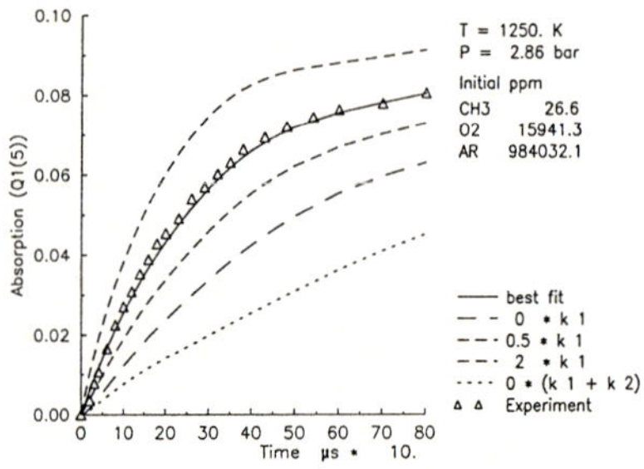

Fig. 5. CH_3-O_2 system: k_1 and k_2 sensitivity of the measured OH absorption profile

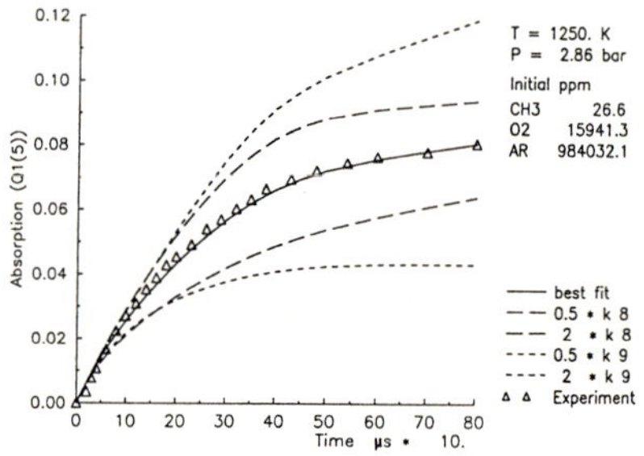

Fig. 6. CH_3-O_2 system: k_8 and k_9 sensitivity of the measured OH absorption profile

There is good agreement between calculated and measured absorption profiles. Under these conditions, OH radicals are generated dominantly by the reaction R1 (long-dashed curve: $k_1 * 0$); the short-dashed curves show the sensitivity of the absorption profile with respect to variations of the k_1 values by a factor of 2 or 0.5, respectively. If the second initial reaction R2/R3 (and also R1) is excluded, the calculated OH absorption profiles are much too low (dotted line, Fig.5). Under this condition, the OH species are generated from the reaction of H atoms with O_2 according to reaction R4, whereas H radicals stem from the pyrolysis of methyl:

$$\text{R6}: \quad CH_3 + CH_3 \longrightarrow C_2H_5 + H$$

$$\text{R7}: \quad C_2H_5 \longrightarrow C_2H_4 + H.$$

A reliable estimation of the additionally produced amount of OH can be made because the rate expressions for R6 and R7 are considered to be well known (Frank and Braun-Unkhoff 1987). It should be remarked that the OH profiles are sensitive to variations of the k-value of the reaction R4. Because the rate coefficient for R4 is considered to be known precisely no changes were made (Masten et al. 1991). At very low temperatures, both the recombination reactions R8 and R9:

$$\text{R8}: \quad CH_3 + CH3(+M) \longrightarrow C_2H_6(+M)$$

$$\text{R9}: \quad H + O_2 + M \longrightarrow HO_2 + M$$

do influence the measured OH absorption (see Fig.6). A variation of the k-values of either reaction R8 (long-dashed curves) or R9 (dotted curves) by a factor of 2 or 0.5, respectively, results in OH absorption profiles that are considerably much too low or much too high regarding the experimental ones. The measured OH profiles could be reproduced better if -for low temperatures- the falloff expression for reaction R8 was lowered by about 20-30% with respect to the values given by Baulch et al. (1992). Up to now, no direct experiments were conducted for this reaction in the temperature and pressure regime of the present work. For reaction R9 the Arrhenius expression given by Baulch et al. (1992) was used.

OH profiles were calculated for different product channels of the reaction CH_3OH:

$$\text{R5a}: \quad CH_3 + OH + M \longrightarrow CH_3OH + M$$
$$\text{R5b}: \quad CH_3 + OH \longrightarrow CH_3O + H$$
$$\text{R5c}: \quad CH_3 + OH \longrightarrow CH_2^* + H_2O.$$

As an example, Fig.7 shows the influence of these reactions on the OH profiles measured in the (CH_3+O_2) system. For the reaction R5a the Arrhenius expression that has been derived by Dombrowski et al. (1991) from the thermal decay of methanol was used. The k-values of the reaction R5b were taken from the study of Bhaskaran et al. (1979) who investigated the reaction of ethane with O_2 behind reflected shock waves by direct measurement of H and O atoms via the ARAS technique. Hack et al. (1988) investigated the reverse reaction R5c; they concluded that the dominant products are CH_3 and OH.

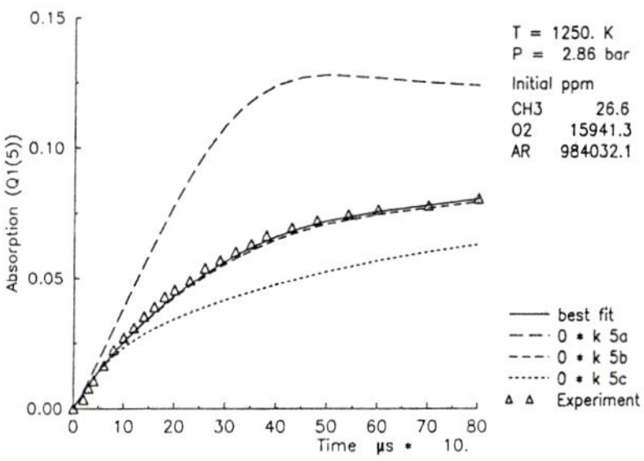

Fig. 7. CH_3-O_2 system: influence of the reactions R5a - R5c on the measured OH absorption profile

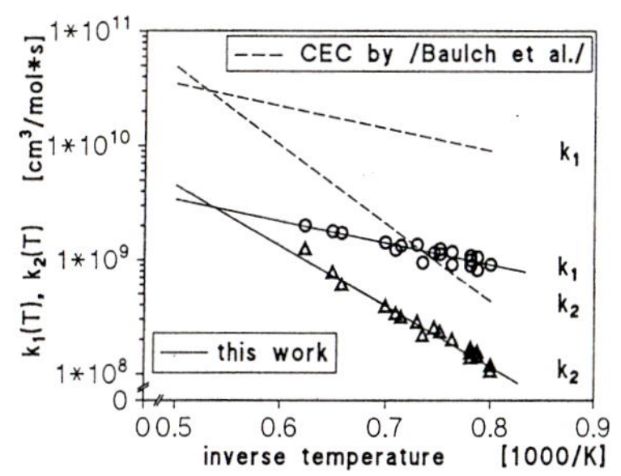

Fig. 8. Arrhenius plot of the reactions R1 and R2

At low temperatures, reaction R5a is a very effective sink for OH because - under these conditions - the formed methanol is relatively stable and does not react any further (Fig. 7, long-dashed curve). Reactions R5b and R5c do not consume OH, on the contrary they are OH sources: the formed H atoms react immediately with excessive O_2 giving OH. Sensitivity analysis revealed that some more reactions, especially R10-R12:

$$\text{R10}: \quad H_2O + O \longrightarrow 2OH$$
$$\text{R11}: \quad CH_2O + O \longrightarrow CHO + OH$$
$$\text{R12}: \quad CH_2O + OH \longrightarrow CHO + H_2O$$

are necessary to reproduce the experimental absorption profiles for temperatures $T > 1400$ K at the end of the observation time. The k-values of the reactions R11, R12 were lowered by about 30-50% with respect to those given in the data evaluation by Baulch et al. (1992). Again, up to now, there are no direct measurements for R11, R12 in the temperature and pressure regime of the present work. The Arrhenius expression given by Masten et al. (1991) was used for reaction R10.

According to measured and calculated OH absorption profiles, the following rate expressions were derived for the reactions R1 and R2 (see Fig. 8):

R1 : $CH_3 + O_2 \longrightarrow CH_2O + OH$:

$k_1\,[cm^3mol^{-1}s^{-1}] = 3.1 \times 10^{10} \times \exp(-4402/T)$

R2 : $CH_3 + O2 \longrightarrow CH_3O + O$:

$k_2\,[cm^3mol^{-1}s^{-1}] = 2.1 \times 10^{12} \times \exp(-12242/T)$

The rate expressions that have been obtained in the present work are in excellent agreement with those given by Fraatz (1991). The recommended k-values of the CEC data evaluation given by Baulch et al. (1992) are higher than those derived in the present work: about one order of magnitude for reaction R1 and about a factor of 5-6 for reaction R2 (see Fig. 8).

Acknowledgements

The authors wish to thank N.Ackermann and M.Kallfass for technical assistance. We are also very grateful to Dr.C.Wahl and M.Kapernaum for preparing and analyzing azomethane.

References

Baulch DL, Cobos CJ, Cox RA, Esser C, Frank P, Just T, Kerr JA, Pilling MJ, Troe J, Walker RW, Warnatz J (1992) Evaluated kinetic data for combustion modeling. J. Phys. Chem. Ref. Data 21 No.3:566

Bhaskaran KA, Frank P, Just Th (1979) High-temperature methyl radical reactions with atomic and molecular oxygen. In: Lifshitz A, Rom J (eds) 12th Intl. Symp. on Shock Tubes and Waves p 503

Bott JF, Cohen N (1991) A shock tube study of the reaction of methyl radicals with hydroxyl radicals. Intl. J. Chem. Kinet. 23: 1017

Dombrowsky Ch, Hoffmann A, Klatt M, Wagner HGg (1991) Ber. Bunsenges. Phys. Chem. 95: 1685

Fraatz W (1991) Ph.D. Thesis, University of Göttingen

Frank P, Braun-Unkhoff M (1987) A shock tube study on the reaction $CH_3+CH_3=C_2H_5+H$. In: Grönig H (ed) Proc. 16th Intl. Symp. on Shock Tubes and Waves p 379

Frank P,Just T(1985)High-temperature reaction rate for $H+O_2=OH+O$ and $OH+H_2=H_2O+H$. Ber. Bunsenges. Phys. Chem. 89:181

Hack W, Wagner HGg, Wilms A (1988) A study of the reaction of CH_2 (a^1A_1) with H_2O in the gas phase. Ber. Bunsenges. Phys. Chem. 92: 620

Hsu DSY, Shaub M, Craemer T, Gutman D, Lin MC (1983) Kinetic modeling of CO production from the reaction of CH_3+O_2 in shock waves. Ber. Bunsenges. Phys. Chem. 87: 909

Masten DA, Hanson RK, Bowman CT (1991) Shock tube study of the reaction $H+O_2=OH+O$ using laser absorption. J.Phys.Chem. 94: 7119

Oser H, Stothard N, Humpfer R, Grotheer H (1991) Direct measurement of the reaction CH_3+OH at ambient temperature in the pressure range 0.3-6.2 mbar. J. Phys. Chem. 96: 5359

Rea EC Jr, Chang AY, Hanson RK (1987) Shock tube study of pressure broadening of the $A^2\Sigma^+ - X^2\Pi(0,0)$ band of OH by Ar and N_2. JQSRT 37(2): 117

Saito K, Ito R, Kakumoto T, Imamura A (1986) The initial process of the oxidation of the methyl radical in reflected shock waves. J. Phys. Chem. 90:1422

Yuan T, Wang C, Yu CL, Frenklach M (1991) Determination of the rate coefficient for the reaction H+O−2=OH+O by a shock tube/laser absorption/detailed modeling study. J. Phys. Chem. 95:1258

Zellner R, Ewig F (1988) Computational study of the $CH_3=O_2$ chain branching reaction. J. Phys. Chem. 92:2971

Shock-Controlled Chemical Processing

A.T. Mattick, D.A. Russell, A. Hertzberg and C. Knowlen
Department of Aeronautics and Astronautics, University of Washington, Seattle, Washington, USA

Abstract. A continuous-flow chemical reactor is described which uses shock waves to effect pyrolysis of hydrocarbons for the commercial manufacture of olefins. In this reactor, heat is added to an inert carrier gas, which is cooled to sub-pyrolysis temperatures by expansion to supersonic speed, and mixed with a supersonic flow of feedstock. Deceleration of the mixture by a standing shock wave initiates pyrolysis. Short reaction durations and high pyrolysis temperatures result in higher olefin yields than are attainable with conventional reactors. A simulation of ethane pyrolysis using the shock wave reactor predicts a 20% increase in ethylene yield and a 15% decrease in energy consumption compared to conventional reactors.

Key words: Pyrolysis, Standing shock wave, Reacting flow, Ethylene

1. Introduction

The manufacture of many commercially important chemicals entails pyrolysis of hydrocarbon feedstocks, an energy-intensive process which is now carried out by passing the feedstocks (components of oil or natural gas) through tubes in a furnace. The yields of valuable products depend on the temperature history of the feedstock, and for olefin manufacture, yields are maximized by the application of high temperatures for short durations (Orriss and Yamaguchi 1987). Despite the continuing evolution of conventional pyrolysis technology toward higher temperatures and lower residence times, the high heat fluxes required to achieve optimal yields are beyond the capabilities of existing heat transfer materials. This paper discusses a new approach for pyrolysis, the shock wave reactor (SWR), in which the temperature history of reactants is controlled by gasdynamic means rather than by wall heat transfer (Mattick et al. 1991, Hertzberg et al. 1993). The reactor uses a standing shock wave in a supersonic flow to achieve a precisely-controllable temperature pulse that can be tailored to maximize the yield of desired products. An initial examination of this approach by Mattick et al. (1991) established the basic operating criteria and analyzed a point design for ethylene production by ethane pyrolysis. That exercise indicated that ethylene yields may be as much as 20% higher than with conventional pyrolysis, although the energy consumption in the point design was 50% higher. The present paper addresses gasdynamics issues of reactor operation, and presents a more detailed assessment of its capabilities for olefin production. This assessment demonstrates that the high yields of the initial study may indeed be reached, with an energy consumption lower (by 15% or more) than with conventional pyrolysis reactors.

2. The shock wave reactor

The primary aim of the SWR is to make use of the rapid temperature recovery as a supersonic flow of reactant encounters a shock. This approach was motivated by the routine use of chemical shock tubes to create precisely-controllable reaction conditions in the study of kinetic rates. The present approach departs from shock-tube concepts in that it is a continuous-flow process, using a standing shock wave. In contrast with the shock tube, where the pressure of a driver gas is used to energize reactions, the SWR uses the kinetic energy in a supersonic reactant flow itself to provide the reaction heat.

The supersonic flow is a mixture of a high-enthalpy carrier gas (steam) and feedstock. As shown in Fig.1, the carrier gas is heated to a stagnation temperature T_{c0} in excess of the minimum

Shock Waves @ Marseille II
Editors: R. Brun, L. Z. Dumitrescu

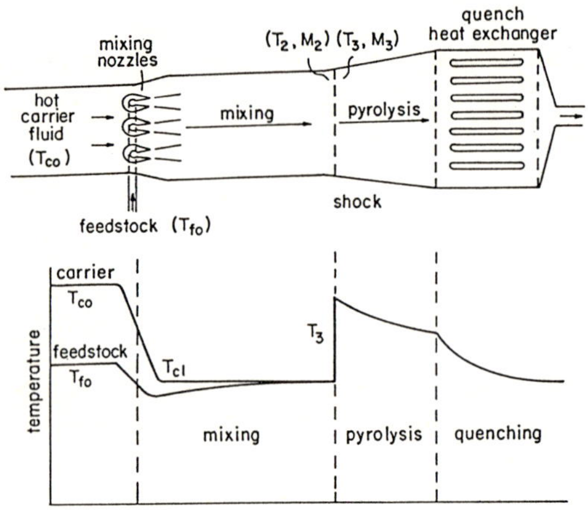

Fig. 1. Shock wave reactor and gas temperature profile

pyrolysis temperature T_p, and expanded in an array of nozzles to supersonic Mach number M_{c1}, and temperature $T_{c1} < T_p$. The feedstock is heated to $T_{f0} < T_p$ and expanded to supersonic speed and temperature T_{f1} using nozzles which are interleaved with the carrier nozzles. Both gases issue into a chamber where they mix supersonically, coming to a temperature $T_2 < T_p$ and Mach number M_2. The supersonic mixture then encounters a standing shock wave, established by suitable adjustment of feed pressure, back pressure and channel geometry, and is decelerated to a Mach number $M_3 < 1$, and temperature $T_3 > T_p$, initiating rapid pyrolysis. Because the endothermic reactions proceed adiabatically, the temperature decreases downstream of the shock. Divergence of the walls of the pyrolysis channel acts to slow the reacting flow and helps maintain the temperature above T_p. When the reaction has proceeded to the desired degree, the mixture is rapidly quenched to freeze the product distribution. This may be accomplished with conventional "transfer-line" heat exchangers (Mol 1983), or by re-accelerating the flow to near sonic speed prior to heat transfer.

This approach separates the processes of heat addition and pyrolysis. Little or no pyrolysis of the feedstock occurs prior to the shock, and the post-shock temperature history can be controlled to optimize yields of desired products by means of the carrier stagnation properties, channel geometry, and mass ratio of carrier fluid to feedstock. A particular advantage of the SWR is that the walls of the mixing and pyrolysis zones can be kept at modest temperatures. A difficult problem in commercial pyrolysis is the formation of coke on the pyrolysis tubes, which is exacerbated at high wall temperatures (Ranzi et al. 1985). The short pyrolysis durations (50 msec and below) of the SWR may also mitigate formation of wall deposits, since there is a finite induction time (10's of msec) for coke precursors and the deposition rate is observed to rise to a maximum at times of 200-500 msec (Albright and Tsai 1983).

A key feature of the SWR is the use of mixing nozzle arrays to produce parallel supersonic flows of feedstock and carrier gases. The technology of such multi-nozzle flows and the theoretical understanding of the supersonic mixing process has been advanced by the development of high-power gasdynamic and chemical flow lasers (Russell 1974). As discussed by Mattick et al. (1991), the characteristic mixing length will be on the order of 100 times the nozzle spacing, so that a mixing section 1-2 meters in length would be adequate for nozzle dimensions of about 1 cm. The transit time in this section would be of order 1 msec under expected operating conditions.

Although the SWR was conceived to be a "wall-less" reactor, in the sense that wall heat transfer does not govern the chemical reaction history, boundary layers nevertheless affect reactor operation. In particular, it is well known that a stationary, normal shock reverts to a spread-out compression region as the boundary-layer thickness upstream of the shock increases. Measurements by Merkli (1976) show that the re-compression length increases roughly linearly with boundary layer thickness, reaching a maximum of about 10 channel heights when the boundary layers merge. For expected SWR mixing channel conditions, the full boundary layer height is estimated to be 2 cm at the end of a 1 m long mixing channel (just upstream of the shock). With a channel height of 20 cm, the re-compression length is expected to be about 50 cm. Although the risetime for the temperature would still be very short, different fluid elements would experience different temperature histories, reducing the control over chemical yields. This difficulty can be mitigated by using boundary layer bleed upstream of the shock. In a commercial-scale facility with a channel height of > 0.5 m, the bleed would be a very small fraction of the mass flow and should have little effect on efficiency.

A second concern is the variation of flow properties (temperature and density) in the growing, turbulent boundary layer in the pyrolysis zone. In fact, the property variation in a turbulent boundary layer is only significant over a small fraction of the full layer thickness near the wall. Assuming that boundary layer bleed is used upstream of the shock, it is estimated that the properties will be uniform over 90-95% of the flow area of a commercial scale SWR.

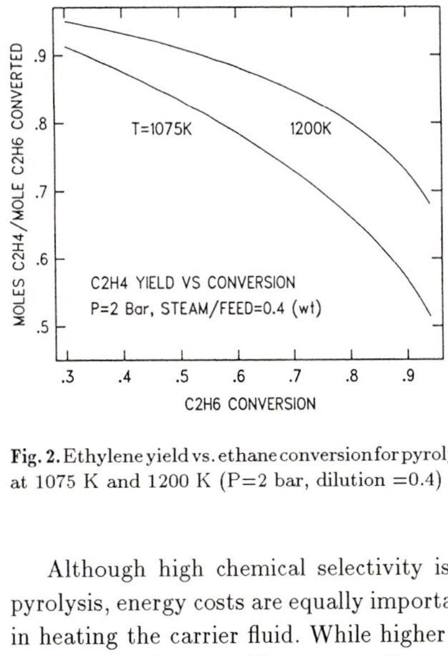

Fig. 2. Ethylene yield vs. ethane conversion for pyrolysis at 1075 K and 1200 K (P=2 bar, dilution =0.4)

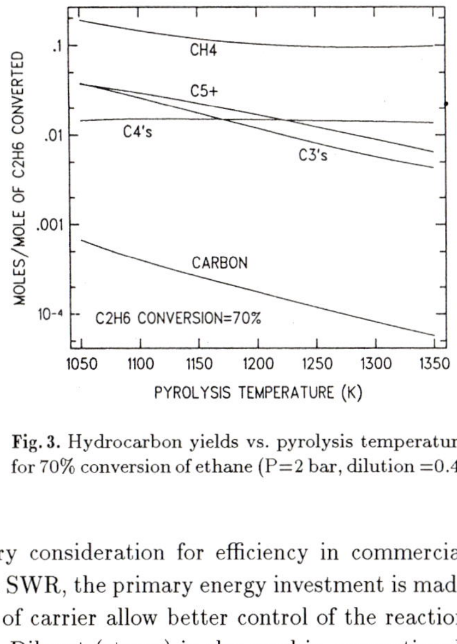

Fig. 3. Hydrocarbon yields vs. pyrolysis temperature for 70% conversion of ethane (P=2 bar, dilution =0.4)

Although high chemical selectivity is a primary consideration for efficiency in commercial pyrolysis, energy costs are equally important. In the SWR, the primary energy investment is made in heating the carrier fluid. While higher fractions of carrier allow better control of the reaction temperature history, this increases the energy cost. Diluent (steam) is also used in conventional, tube-in-furnace pyrolysis, for the purpose of lowering the hydrocarbon partial pressure (for better selectivity) while keeping the total pressure above atmosphere (typically 2 atm), and also to reduce fouling of pyrolysis tube walls. In ethylene manufacture, for example, steam dilution (steam mass/feedstock mass) ranges from 0.4:1 for ethane feedstock up to 1:1 for liquid feedstocks.

For the SWR the primary purpose of the carrier fluid is to provide the enthalpy for the reaction, and the dilution must be sufficient to maintain temperature levels for high chemical selectivity while the pyrolysis proceeds to the desired conversion of feedstock. In design studies of the use of the SWR for ethane pyrolysis, it has been found that the dilution must be above 4:1 in order to achieve olefin yields appreciably above those obtained with conventional pyrolysis. Thus, the specific energy consumption (kJ/kg of product) depends critically on the ability to recover energy remaining in the carrier after the reaction is completed. Today's olefin manufacturing plants achieve an energy recovery of 94-95% (Tsai et al. 1985). At this recovery level, it appears that even at 5:1 dilutions, the specific energy consumption for SWR pyrolysis will be comparable to or less than that of conventional plants, owing to the higher yield of valuable products.

An approach for reducing energy consumption in the SWR is to use a smaller amount of carrier at a higher stagnation temperature, to provide the required pyrolysis energy. In order to prevent excessive post-shock temperatures which will over-crack the feedstock, a system of multiple oblique shocks can be used to decelerate the mixture, resulting in a reduced temperature rise in comparison with a single normal shock. The subsonic diffuser plays a relatively more important role in recovering temperature to maintain an optimal temperature history. Pressure losses are also reduced, facilitating energy recovery. This approach is used in the example of the next section. Alternatively, a supersonic diffuser (converging section) can be placed upstream of the shock to effect a more gradual temperature rise. Pyrolysis will proceed in both the supersonic and subsonic sections, and the channel geometry can be chosen to optimize the reaction history.

Preliminary studies of this approach indicate that, with the use of carrier temperatures in the range 1800-2000 K, the dilution can be reduced to below 2:1. A third approach is to add oxygen to the carrier, so that partial oxidation of the feedstock proceeds in parallel with pyrolysis to provide part of the heat of reaction. An initial examination indicates that using an oxygen-to-feedstock ratio of only 0.2:1, the steam dilution could be reduced to about 2.5:1, and the specific energy consumption would be reduced by half as compared to the conventional technology. Partial oxidation has been of increasing interest for olefin manufacture, particularly in the use of methane as a feedstock (Burch and Tsang 1990).

3. Application to ethylene manufacture

A series of design studies have been carried out to assess the potential advantages of using the shock wave reactor for commercial pyrolysis. These studies have focused on the production of ethylene using ethane as a feedstock. Although the SWR is capable of utilizing a variety of feeds, ethane was chosen for the present example, due to the relative simplicity of chemical modelling. The specific energy consumption in conventional ethane pyrolysis is about 16 MJ (furnace heat) per kilogram of ethylene produced.

The improvement in selectivity that results from the use of higher pyrolysis temperatures is illustrated in Figs.2 and 3. The results were obtained using a mechanistic kinetic model developed by Sundaram and Froment (1978) for simulation of ethane and propane pyrolysis. The model includes 49 reactions and 20 species. Plotted in Fig.2 are the computed carbon fractions of ethylene in the product vs. ethane conversion for constant temperature pyrolysis of ethane at 1075 K (effective average temperature in conventional pyrolysis) and at 1200 K. This simulation used a steam dilution of 0.4 and a total pressure of 2 atm. Because of the falloff in ethylene selectivity at higher conversions, conventional plants typically convert about 60% of the feed per pass, the remaining ethane being recycled to the pyrolysis tubes after chemical separation. It is evident that pyrolysis at 1200 K gives significantly better selectivity. At 70% conversion, the ethylene fraction is 15% higher at 1200 K than at 1075 K. Fig.3 plots mole fractions of other hydrocarbons in the product vs. temperature for an ethane conversion of 70%. The most significant result is that the fraction of carbon and C_5's, which are the primary sources of fouling, decreases substantially with pyrolysis temperature at fixed conversion. The residence time for 70% conversion drops from about 500 msec at 1075 K to 12 msec at 1200 K.

Fig.4 shows a point design illustrating the operating parameters of the SWR for the pyrolysis of ethane. This example uses a multiple-shock system to initiate pyrolysis, and includes processes for heating feedstock and carrier as well as for quenching and energy recovery. A steam dilution

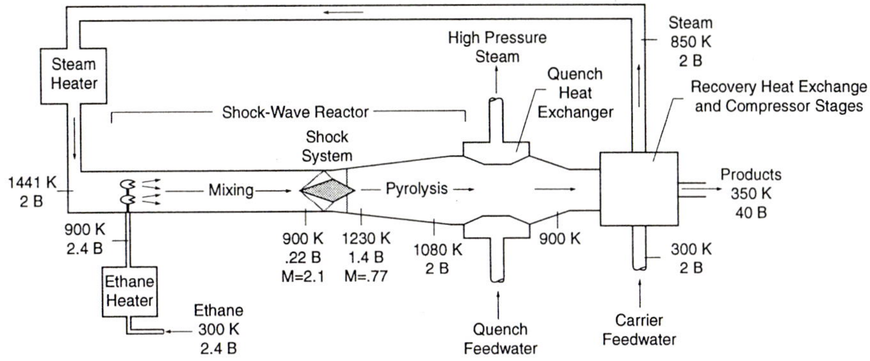

Fig. 4. Point design of a shock wave reactor for ethylene manufacture using ethane feedstock

of 4.5:1 (4.5 kg-steam/kg-ethane) was chosen to keep the peak steam temperature within limits attainable with available high-temperature tubing. The steam is heated by conventional (tube-in-furnace) means to 1441 K at 2.0 bar, and then expanded to Mach 2.09, which lowers the temperature to 944 K. Ethane, heated to 900 K and then expanded via interleaved nozzles to 750 K, Mach 2.2, mixes with the steam to produce a mixed stream at Mach 2.14 and 900 K. Wedge-shaped bodies placed in the flow decelerate the mixture to Mach 0.77 by a combination of oblique and normal shocks, and the temperature rises to 1230 K, initiating pyrolysis. Using the model of Sundaram and Froment (1978), 70% of the ethane is consumed over a 52 msec pyrolysis duration, and the ethylene yield is 63%. (0.63 moles of ethylene per mole of ethane). By coincidence the recovery pressure is very nearly equal to the steam feed pressure (2 bar), due to a combination of heat absorption and increase in mole number during pyrolysis.

The 1080 K product is quenched to 900 K by heat transfer with boiling feedwater (at 60 bar) in the quench heat exchanger. Energy remaining in the quench exhaust is recovered in a secondary, counterflow heat exchanger, and via 5 compression/heat-exchange stages, depicted simply as the "energy recovery system" in Fig.4. These stages remove water from the product and pressurize it to 40 bar, comparable to the pressures used in olefin plants for cryogenic separation of product species. The recovered energy is transferred to the incoming carrier fluid. Steam generated in the quench heat exchanger is superheated and expanded in a turbine to provide power for the compressors. The product stream, at 40 bar, leaves the energy-recovery unit at 350K with a water vapor content of 0.01 kg water/kg of hydrocarbon. The heat recovery/compression scheme is similar to that used in current plants.

Using this approach, 71% of the energy required to produce the 1441 K carrier steam (starting from liquid water at 298 K) is provided by heat recovery from the product stream, and the remainder, 10.7 MJ/kg of ethylene product, is provided by furnace heating. The ethane feedstock requires 3.0 MJ of preheat and 1.6 MJ of compressor work is required, both per kg of ethylene. The latter, however, can be provided by utilizing the high-pressure steam generated by the quench heat exchanger. Thus the energy consumption for this example is estimated to be 13.7 MJ per kg of ethylene product, 15% lower than for conventional cracking plants.

This point design illustrates the potential of the shock wave reactor to achieve higher single pass yields in the pyrolysis unit of an olefins plant, while consuming an energy less than or comparable to conventional tube-in-furnace units. The higher yield may also reduce the energy costs and the size of the equipment for chemical separation operations downstream of the reactor. Another advantage is that the low wall temperatures in the reactor, short residence times, and small surface to volume ratio of the SWR will lead to lessened coking of reactor walls. Further reductions in the specific energy consumption have been found in SWR designs which utilize oxidative pyrolysis and a supersonic diffuser to effect part of the pyrolysis. In the latter case energy consumption was found to be as low as 4.3 MJ/kg-ethylene, or 30% of that with conventional pyrolysis. Details of these designs and of the energy recovery system of the above example will be provided in a future publication.

4. Summary

The shock wave reactor represents a radical departure from conventional pyrolysis technology in that the reaction history is uncoupled from the process of heat addition. The use of gasdynamics allows precise control of reaction temperature and duration, which can be tailored to optimize yields of valuable products. A design example using the SWR for ethylene production with ethane feedstock demonstrates single pass yields that are 20% higher than in conventional plants, with a lower energy consumption. Variations of the basic SWR process that utilize oxidative pyrolysis or very-high-temperature steam may reduce energy consumption by 50%. The latter approach may even provide a new avenue to olefin production from methane, which would have great cost advantages compared with currently-used feedstocks.

Acknowledgment

This study was supported by the U.S. Department of Energy, Grant # DE-FG06-93ER12136.

References

Albright LF, Tsai TC (1983) Importance of surface reactions in pyrolysis units. Chapter 10 in: Albright LF et al. (eds) Pyrolysis: Theory and Industrial Practice, Academic Press NY, pp 233-254

Burch R, Tsang SC (1990) Investigation of the partial oxidation of hydrocarbons on methane coupling catalysts. Appl. Catal. 65: 259-280

Hertzberg A, Mattick AT, Russell DA (1993) Apparatus for initiating pyrolysis using a shock wave. U.S. Patent 5,219,530

Mattick AT, Hertzberg A, Russell DA (1991) Shock controlled reactors. In: Takayama K (ed) Shock Waves, Proc. 18th Intl. Symp. on Shock Waves, pp 1289-1294

Merkli PE (1976) Pressure recovery in rectangular constant area supersonic diffusers. AIAA J 14(2): 168-172

Mol A (1983) Transfer line exchangers. Chapter 18 in: Albright LF et al. (eds) Pyrolysis: Theory and Industrial Practice, Academic Press NY, pp 451-471

Orriss R, Yamaguchi H (1987) Idemitsu's Chiba ethylene plant proves modern technology. Oil and Gas J Mar 9: 27-30

Ranzi E, Dente M, Pierucci S, Barendregt S, Cronin P (1985) Coking simulation aids on-stream time. Oil and Gas J Sept 2: 49-52

Russell DA (1974) Fluid mechanics of high-power grid-nozzle lasers. AIAA Paper 74-223

Sundaram KM, Froment GF (1978) Modeling of thermal cracking kinetics 3: radical mechanisms for the pyrolysis of simple paraffins, olefins and their mixtures. Ind. Eng. Chem. Fundam. 17: 174-182

Tsai FW, Che SC, Minet RG (1985) Why retrofit furnaces? Hydrocarbon Proc. Aug: 41-47

Part 3: Non-Equilibrium Flow

Hypervelocity Carbon Dioxide Flow Simulation

R.G. Morgan, A.J. Neely and M.L. Scott
Department of Mechanical Engineering The University of Queensland, Australia 4072

Abstract. This paper reports on a series of experiments performed in the superorbital expansion tunnel at the University of Queensland. Carbon dioxide was used as the test gas, and flow velocities up to 13 km/sec were recorded. Static pressure and flat plate heat transfer measurements in the test section indicated that steady test flows of the order to 10 microseconds were obtained. Flow condition was estimated using an equilibrium analysis, accounting for shock reflection from the tertiary diaphragm, and for secondary shock attenuation during its passage through the test gas. The flow properties thus obtained were used to estimate the theoretical flat plate heat transfer using an equilibrium reference enthalpy correlation, and measurements were made on a small flat plate model instrumented with thin film heat transfer gauges. The shape of the profiles against distance was consistent with steady laminar boundary layer flow for the duration of the test time, but the absolute values of measured heat transfer were significantly lower than the correlation's. The reason for this is not clear at present, but is consistent with the findings of other researchers using carbon dioxide at lower enthalpies. The experiments confirm that conditions suitable for further aerodynamic testing have been created.

Key words: Expansion tube, Carbon dioxide, Hypervelocity aerodynamics

Notation

H total flow enthalpy of expanded test gas
M test gas Mach number
p static pressure
ρ density
T static temperature
U_e equiv. flight velocity for same total enthalpy
U_{s1} primary shock speed
U_{s2} secondary shock speed
U_{s3} tertiary shock speed

1. Introduction

From a practical point of view, the behavior of carbon dioxide in hypervelocity flow is of interest because it is the major constituent of the Martian atmosphere. Mars is a prime target for space exploration, and it is expected that, from the later years of this century, visits to the surface of that planet will take place with increasing frequency. This will require an understanding of the aerodynamics of flight through the Martian atmosphere at speeds which may go as high as 16 km/sec.

There is currently very little data available for carbon dioxide in high speed aerodynamic flows (Gupta et al. 1991). Studies in a reflected shock tunnel (Ebrahim and Hornung 1975) at speeds up to 5 km/sec have confirmed that important aerodynamic parameters, such as shock shape and density ratio are strongly influenced by flow enthalpy. Expansion tube studies up to 7 km/sec (Miller 1975, 1978), have investigated the influence of enthalpy on shock standoff and stagnation point heat transfer. Tauber et al. (1993) investigated the validity of flat plate empirical correlations for surface heat transfer in high speed carbon dioxide flows. These studies are very significant for the design of craft for flight in the Martian atmosphere, and emphasise the pressing need for more data in the higher velocity flow regime.

Fundamental studies of carbon dioxide plasma behind shocks with speeds of up to 17 km/sec have been performed in a non-reflected double-diaphragm free-piston driven shock tunnel, (Ebrahim 1975), but the flows were not suitable for aerodynamic testing. The expansion tube

Shock Waves @ Marseille II
Editors: R. Brun, L. Z. Dumitrescu

concept has the capability to produce flows of the same enthalpy, but with a much more realistic gas composition (Tamagno 1990).

The expansion tube facility at the University of Queensland (Morgan and Stalker 1991), extends the usable operating envelope of ground testing to include superorbital flight velocities, and when operated with carbon dioxide test gas it can give useful information relating to Martian flight. It is envisaged that the facility will be of interest for the purposes of CFD validation, for the investigation of the validity of empirical calculation techniques, and for the study of generic shaped scale models.

Carbon dioxide is also of interest for fundamental research, because it dissociates through two basic reactions, namely

$$CO_2 \longrightarrow CO + O$$
$$CO \longrightarrow C + O$$

which have very different dissociation energies, 3.3 and 11.1 ev/molecule respectively. Therefore the effects of the first reaction on, say, shock density ratios, should be largely complete before the second commences. Such an effect is of major importance, and should readily be observed through macroscopic effects on simple experimental configurations. Future experiments are planned to investigate this effect. The carbon monoxide dissociation reaction is of further additional interest, because it has the same molecular weight, and a very similar dissociation potential to that of nitrogen. Dissociating nitrogen has been effectively studied in shock tunnels before (e.g. Hornung 1972), and it will be revealing to investigate the influence of a similar reaction in a more complicated chemical situation.

2. The facility

A schematic of the facility is shown in Fig.1, and its operation is described in the paper by Neely and Morgan (1993), included in these Proceedings. The strength of the primary shock through the test gas is enhanced by means of a compound driver, consisting of a free-piston compressor coupled to a shock heated helium section. Enthalpy and total pressure addition across the unsteady expansion is optimised by using helium accelerator gas. For efficient use of the facility, the shock speeds in the three driven sections must be set at appropriate values. The primary shock speed must be overtailored so that the shock-heated helium in region 2, Fig.1, has a higher speed of sound than the expanded driver gas in region 3, otherwise the shock heated driver tube is counter productive. Also, a significant increase in flow velocity must occur across the unsteady expansion between regions 2 and 7, otherwise region 6 would be better driven directly from the expanded driver gas in region 3, again rendering the shock heated secondary driver tube redundant. Finally, the unsteady expansion which processes the test gas between regions 8 and 12 must add sufficient velocity and total enthalpy to the flow so that hypervelocity flow conditions are achieved in the test section. If these guidelines are followed, then the facility is able to produce test flows at superorbital velocities with conditions suitable for aerodynamic testing, and with higher total enthalpies than can be produced by any other combination of the shock tube components. Morgan and Stalker (1991), outline a procedure for optimising the tunnels performance. A problem with expansion tubes in the past has been the difficulty in obtaining a range of useful test conditions. The work of Paull (1993) has identified noise from the driver as being a major cause of corruption of the test flow, and has identified a mechanism for isolating the test gas from the disturbances. This requires maintaining an increase in sound speed across the interface from the expanded driver gas to the shock heated gas in the first driven tube. Fortunately, this condition is fundamental to the correct operation of the compound driver used in the superorbital facility. However, the facility has multiple gas interfaces and diaphragms, and

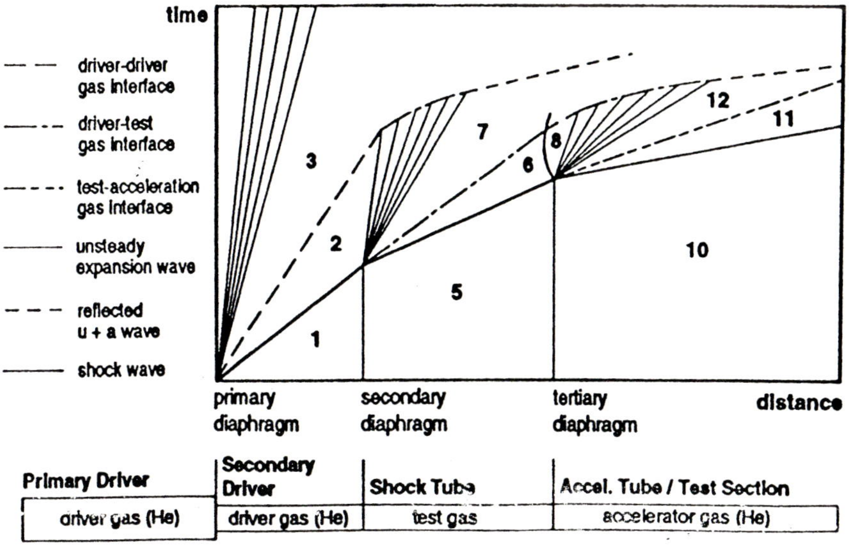

Fig. 1. Wave diagram illustrating Superorbital Expansion Tube operation

has many potential sources and amplifiers of noise. It is still not yet clear just what range of test conditions will be achievable with satisfactory flow quality.

The original test conditions were obtained using air as the test gas, this paper presents the first results of running the facility with carbon dioxide.

Table 1. Experimental and calculated conditions

Run No.	U_{s1} km/s	U_{s2} km/s	U_{s3} km/s	p kPa	T K	M	ρ kg/m^3	H MJ/kg	U_e km/s
239	6.0	6.5	11.4	1.49	4329	7.9	9.5e-4	79	12.5
240	5.7	6.1	11.1	1.19	4246	8.2	8.4e-4	72	12.0
246	5.8	7.9	12.6	2.86	4564	7.6	1.5e-3	99	14.0

3. The experiments

The expansion tube was run for a range of operating conditions, and the results of selected runs are given in the table. Test conditions were estimated using the technique of Neely and Morgan (1993), which uses the experimentally measured shock speeds, and accounts for viscous attenuation of the secondary shock, shock reflection from the tertiary diaphragm, and assumes equilibrium gas composition throughout. The final velocity of the expanded test gas is taken to be equal to the measured tertiary shock speed, as a viscous analysis in this region indicated that the accelerator gas/test gas interface had reached maximum separation from the shock.

In Fig.2 sample pressure/time traces are shown for a condition which has a tertiary shock velocity of 11.4 km/sec, and which has a total enthalpy corresponding to a flight velocity of 12.5 km/sec. It is noted that good agreement is given between the predicted and experimental pressures. The duration of steady pressure in the test section is seen to be of the order of 15 microseconds. The two pressure traces shown are both in the acceleration tube, one being at the exit station, and the other located 345 mm upstream. The arrival of the unsteady expansion terminates the test flow in this instance.

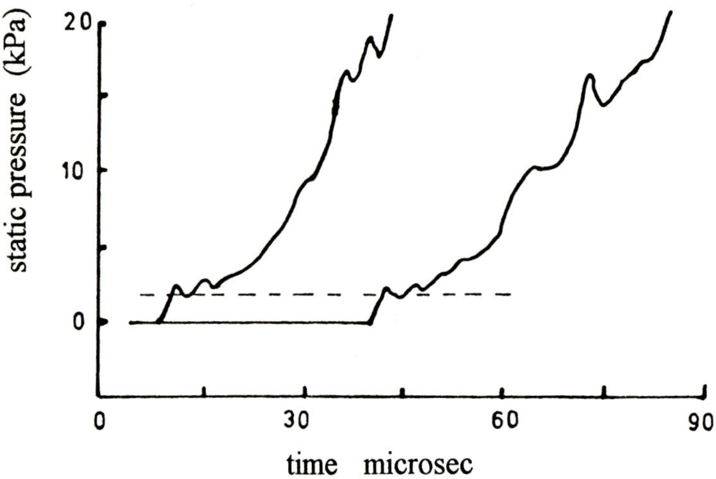

Fig. 2. Static pressure

In Fig.3 profiles of measured surface heat transfer are plotted for two runs with calculated total enthalpies of 99 and 72 MJ/kg. Also shown are the heat transfer levels predicted by the empirical reference enthalpy correlations of Eckert (1955). The formulation of Gupta et al. (1991) was used to estimate the viscosity of the gas at the reference condition. The agreement with the predictions is remarkable, considering the correlation was developed for a much lower velocity flow regime. Some influence of real gas effects is included in the formulation, as the flow properties calculated at the reference conditions include equilibrium dissociation composition. The results are consistent with the work of Tauber et al. (1993), which showed the correlations over predicting by about 25% at velocities up to 7 km/sec. The heat transfer gauges were of thin film construction, coated with a layer of glass of unknown surface catalycity. However, this is not thought to have reduced the measured heat transfer significantly, as the diffusion time for species across the boundary layer for these conditions would preclude significant heat transfer through surface recombination. Because kinetic energy makes up a minimum of 80% of the total enthalpies for the test conditions considered, it may be considered that catalytic effects could account for no more than 20% of the surface heat flux.

The good agreement between measurement and prediction adds credibility to the assumption of equilibrium chemistry across the unsteady expansion fan. It is also significant to note that the role of chemical kinetics takes on a more fundamental role in the expansion tube than in a reflected shock tunnel, where recombination during the steady nozzle expansion serves to determine the condition of the expanded test gas, but not its total enthalpy. The 'enthalpy multiplication' feature of unsteady expansions works only on the static component of enthalpy, through its influence on sound speed, so that if recombination does not occur the high potential final enthalpy will not be reached. Seeing that the total enthalpy is contained primarily in the kinetic energy behind the tertiary shock, it follows that the very high shock speeds measured would not have been possible without significant recombination across the expansion. This was confirmed by

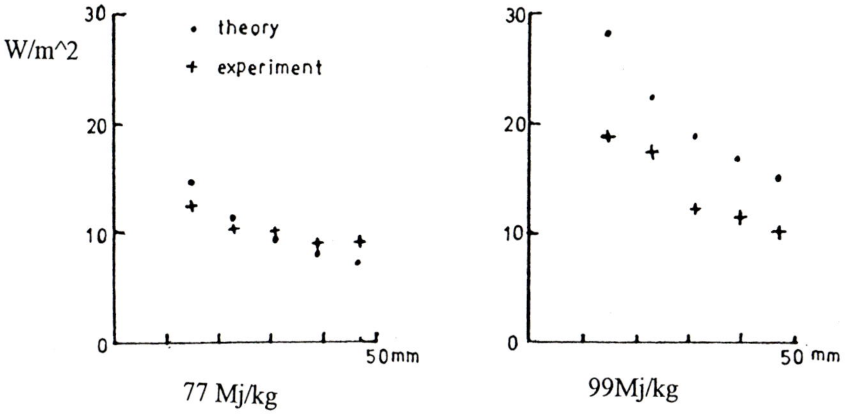

Fig. 3. Flat plate heat transfer

repeating the analysis across the unsteady expansion using frozen conditions. Without the energy release through recombination of the products dissociated by the secondary shock, and the shock reflecting from the tertiary diaphragm, the static pressure and temperature were seen to drop to zero in the expanding gas.

4. Conclusions

The superorbital expansion tube has been shown to produce a flow of usable test gas in carbon dioxide with flow durations of the order of 15 microseconds. With flow velocities of 13 km/sec this gives a slug length of the order of 200 mm, with which small models of about 60 mm length may be started. Static pressure and zero incidence flat plate heat transfer measurements agree well with predictions, based on an equilibrium flow model, and empirical heat transfer correlations. The enthalpy range at which such correlations can be experimentally tested has been increased, and the trends observed are consistent with the findings of other workers at lower enthalpies. A major role for the facility type is apparent for studying the aerothermodynamics of Martian entry.

Acknownedgements

This work was supported by the Australian Research Council.

References

Ebrahim NA, Hornung HG (1975) High enthalpy non equilibrium carbon dioxide nozzle and wedge flows: experiment and calculations. AIAA Journal 13, 7

Ebrahim NA (1975) Non equilibrium in high enthalpy carbon dioxide flows. PhD Thesis, Australian National University

Eckert ERG (1955) Engineering relations for friction and heat transfer to surfaces in high velocity flow. J. of the Aero. Sci., Aug.: 585-587

Gupta RN, Lee KP, Moss JN, Sutton K (1991) A viscous shock layer analysis of the Martian aerothermal environment. AIAA Paper 91-1345, AIAA 26th Thermophysics Conference, Hawaii

Hornung HG (1972) Nonequilibrium dissociating nitrogen flows over spheres and circular cylinders. J.Fluid Mech. 53:149-176

Miller CG, Moore JA (1975) Flow establishment times for blunt bodies in an expansion tube. AIAA Journal 13, 12

Miller CG (1978) Operational experience in the Langley expansion tube with various test gases. AIAA Journal 16, 3

Morgan RG, Stalker RJ (1991) Double-diaphragm driven free-piston expansion tube. In: Takayama K (ed) shock waves, Proc. 19th ISSW, Sendai, Japan, Springer, pp 1031-1038

Neely AJ, Morgan RG (1993) Hypervelocity aerodynamics in a superorbital expansion tube. These Proceedings

Paull A (1993) Noise reduction in argon driven expansion tubes. These Proceedings

Tamagno J (1990) Hypervelocity real gas capabilities of GASL's expansion tube (HYPULSE) facility. AIAA Paper 90-1390

Tauber ME, Yang L, Paterson J (1993) Flat surface heat transfer correlations for Martian entry. J.Spacecraft and Rockets 30, 2: 164-169

Experiments on Hypervelocity Dissociating Flow over Spheres

Chihyung Wen and Hans Hornung
Graduate Aeronautical Laboratories, California Institute of Technology

Abstract. This paper presents the experimental results of measurements of the stagnation point heat transfer rates experienced by spherical models of 1, 2, 3, 4 and 6 inch diameter in nitrogen, air and carbon dioxide at stagnation enthalpies ranging from 4 to 22 MJ/kg and stagnation pressures from 25 to 80 MPa. The experimental results are compared with existing results obtained in the ballistic range facility at NASA Ames and also with numerical calculations. The experimental results obtained by optical differential interferometry were compared with the images constructed from flowfields computed using the method of Candler (1988). Good agreement of fringe pattern and shock shape was observed. A novel flow visualization technique using sodium seeding to increase the sensitivity of conventional shadowgraphic techniques by resonant enhancement of the refractivity of the medium was also used. The resonantly-enhanced shadowgraph suggests a possible way of visualizing the vortical structure induced by the strong entropy layer.

Key words: Hypervelocity, Blunt body flow, Experiment

1. Introduction

Most of the experimental results on heat transfer rates through highly cooled boundary layers on spheres have been obtained from measurements made in shock tubes. These experiments correctly simulate the stagnation enthalpy which is the parameter of major interest in heat transfer measurements. However, the Reynolds number and Mach number are always low. The free piston shock tunnel T5 can simulate the Reynolds number to about one million and stagnation enthalpy to about 25 MJ/kg. The actual size of the model is also much bigger than that used in the shock tube, so that nonequilibrium effects can be studied without the constraint of the size of the model. Numerical methods for the computation of blunt body flow fields have also become very much quicker, so that even interferograms of three-dimensional chemically reacting flows can be computed with reasonable speeds. For laminar flow, the viscous case is also accessible, so that heating rates may also be compared with experimental measurements and theoretical models, see e.g. Candler (1988).

In the work presented here the aim was to gather new data on blunt body flows and to compare them with numerical computations. In addition, a goal was to look for one of the interesting features of hypervelocity blunt-body flows, namely the vorticity concentration associated with the edge of the high-entropy layer generated in the nose region by the curved shock with resonantly-enhanced shadowgraphy.

2. Experiment

2.1. Facility

The GALCIT free piston shock tunnel T5 was used for the experiments. The facility uses the principle of free-piston adiabatic compression of the driver gas of a shock tunnel to achieve the high shock speeds and densities required to generate high enthalpy and reaction scaling. The shock tunnel has the additional essential feature that the test duration is sufficiently short to avoid destruction of the machine by melting, yet long enough to provide good measurements during the steady-flow period. A more detailed description of T5 and its performance envelope, flow quality and repeatability may be found in Hornung (1992).

Shock Waves @ Marseille II
Editors: R. Brun, L. Z. Dumitrescu

2.2. Freestream conditions

For the experiments, the flow was generated through a contoured nozzle of 300 mm exit diameter and 30 mm diameter throat. The reservoir specific enthalpy is determined from the measured shock speed and the measured reservoir pressure, using an equilibrium calculation. The nozzle flow is then computed from this and the measured reservoir pressure by using an axisymmetric inviscid non-equilibrium flow code developed by Rein (1989). For the present experiments, the reservoir enthalpy varied between 4.5 MJ/kg to 22.0 MJ/kg, giving free stream velocities in the range from 2.5 km/s to 5.6 km/s. The Reynolds number based on the model diameter ranged from 7.26×10^4 to 1.60×10^6. Air, nitrogen and CO_2 were used as test gases. The latter gas is of interest in Martian atmospheric entry, where the atmosphere is composed primarily of CO_2. It also exhibits more dramatic real-gas effects than air or nitrogen, because of the large fraction of energy that goes into vibration and dissociation.

2.3. Instrumentation, data reduction and accuracy

Four Medtherm coaxial thermocouples of type E (chromel-constantan) were mounted flush with the surface of the sphere to measure the heat transfer distribution. They are 1.6 mm in diameter and are mounted at the stagnation point and 20°, 40° and 60° from the stagnation point. Only stagnation point heat transfer rate measurements will be presented here. The junction was formed by gently sanding the surface to be exposed to the flow. The cold junction was the feed-through plate for signal cables between the test section and the laboratory.

The one-dimensional semi-infinite slab theory was used to reduce the time history of the surface temperature to heat flux, given that the test time is short compared to the heat penetration time into the gauge (see Schultz and Jones, 1973):

$$\dot{q}(t) = \sqrt{\frac{\rho c k}{\pi}} \int_0^t \frac{dT(\tau)}{d\tau} \frac{d\tau}{\sqrt{t-\tau}},$$

where the thermal product $\sqrt{\rho c k}$ for the gauge material was determined to be 8919 W/(mK). The indirect method was used to process these data numerically, in which integration was performed first and then the differentiation. The accuracy was assessed to be of the order of 15% and the response time of 0.1 microsecond was advertised by the manufacturer, Medtherm Corp., more than adequate for the test time of T5, which is about 1.5 ms.

2.4. Flow visualization

Two flow visualization techniques were used for the experiment. Differential interferometry using a Wollaston prism with a divergence angle of 34" was used throughout most of the experiments. In the last five shots, resonantly-enhanced shadowgraphy with sodium seeding was used to see the boundary layer structure along the wall of the model in the first place. For more information about the resonantly-enhanced technique see Germain, Cummings and Hornung (1993). The flow was seeded by painting a very thin layer of a saturated saline solution in a small spot at the nose of the sphere. This was then left to dry, so that a thin film of salt was left to be ablated during the test. The light source used was a tunable dye laser pumped by a neodymium-YAG laser. The wavelength is about 589 nm with a bandwidth of 3 GHz which is tuned to be just off the center of one of the sodium D-lines.

3. Computational method

Candler's code was used to construct the inviscid reacting flow over a sphere. This program uses a finite-volume method. The flow field is described by coupled partial differential equations for the conservation of species, mass, mass-averaged momentum, vibrational energy of each diatomic species and total energy. The steady-state solution to these fully coupled equations is obtained for different gases characterized by a two-temperature model using an implicit Gauss-Seidel line relaxation technique. The code is also set up for viscous flow. The inviscid code was used to obtain the conditions at the edge of the boundary layer for the theoretical heat flux predictions and to construct the computational differential interferograms for comparison with the experimental pictures. For more details, see Hornung, Wen and Candler (1993). The viscous code was used to compute the heat transfer rate along the wall of the model.

4. Results and discussion

Figs.1(a) and 1(b) show the comparison of experimental and computational interferograms for the cases of air and CO_2. The comparison shows excellent agreement.

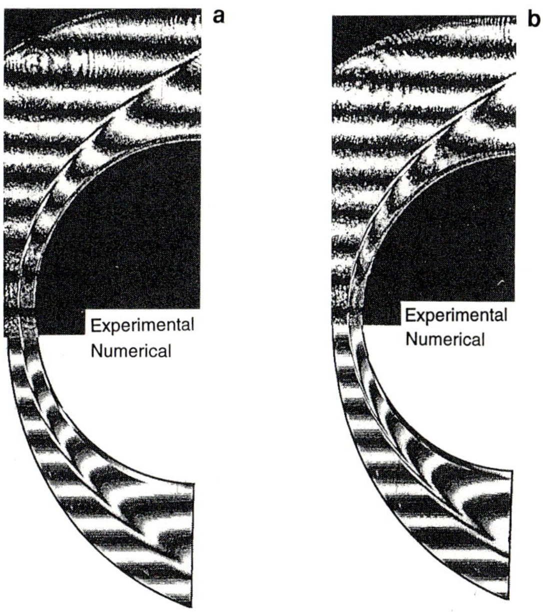

Fig. 1. Comparison of experimental and computed differential interferograms of air and CO_2 flow over a sphere of 4 in diameter. (a): air, freestream: 4.86 km/s, 0.0199 kg/m^3, 16.0 MJ/kg. (b): CO_2, 3.49 km/s, 0.0326 kg/m^3, 11.35 MJ/kg

One of the main purposes of the present experiments was to measure stagnation point heat transfer rates in air, N_2 and CO_2. For laminar boundary layer flow, Lees (1956) and Fay and Riddell (1958) have predicted the following dependence:

$$St_s \sim \frac{1}{\sqrt{Re_\infty}},$$

where St_s is the Stanton number at the stagnation point and Re_∞ is the freestream Reynolds number based on the model diameter. Stanton number and Reynolds number are defined as:

$$St_s = \frac{\dot{q}_s}{\rho_\infty u_\infty (h_0 - h_w)},$$

$$Re_\infty = \frac{\rho_\infty u_\infty D}{\mu_\infty}.$$

A viscosity model for reacting gases developed by Blottner et al. is used to determine species viscosity. The viscosity of the gas mixture is then calculated using Wilke's semi-empirical mixing rule.

The measured stagnation point heat transfer rate was normalized to form the Stanton number, and a plot of this against the Reynolds number is shown in Fig.2. Also shown in Fig.2 are the theoretical predictions of Fay and Riddell's correlation of the equilibrium boundary layer for air, N_2 and CO_2. The agreement of the present measurements with theoretical predictions of Fay and Riddell is good. In Fig.3(a) and Fig.3(b) the experimental data for air and CO_2 are compared with the numerical results and the existing data obtained by the ballistic range facility in NASA Ames (Yee et al. 1961). In Fig. 3(b), Sutton and Graves' (1971) correlation for CO_2 is also plotted. All the data follow the theoretical predictions though data show more scatter for the CO_2 case. It seems that the experimental data for CO_2 agree better with Fay and Riddell's equilibrium correlation than with Sutton and Graves' correlation.

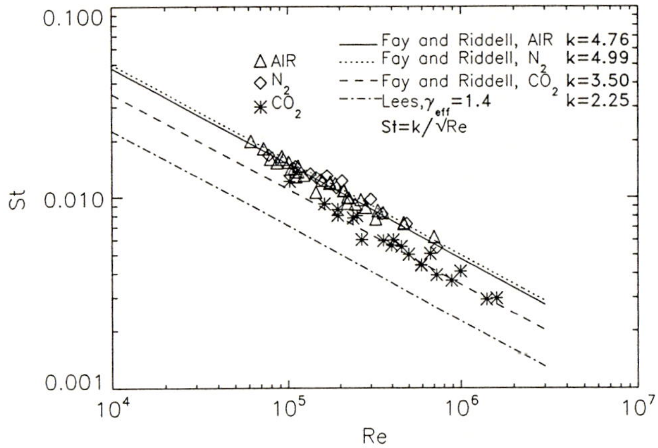

Fig. 2. Stagnation-point heat flux in dimensionless form, and comparison with Lees' theory, and Fay and Riddell's correlation

It is known that there are strong entropy gradients generated in the nose region of a blunt body in hypervelocity flows. The high-entropy layer essentially wets the body as it flows downstream. According to Crocco's theorem, the edge of the entropy layer is also a region of strong vorticity. Fig.4 shows computational vorticity contours of a typical CO_2 run. Strong vorticity gradients are observed. Fig.5(a) shows a resonantly-enhanced shadowgraph of the CO_2 flow with the same conditions of Fig.4. As may be seen, streamlines are marked by the sodium seeding and they are starting to roll up at the side of the sphere. Fig.5(b) shows these vortical structures on a larger scale. These vortical structures may be the by-product of the entropy layer. Further experiments are needed before definite conclusions may be reached.

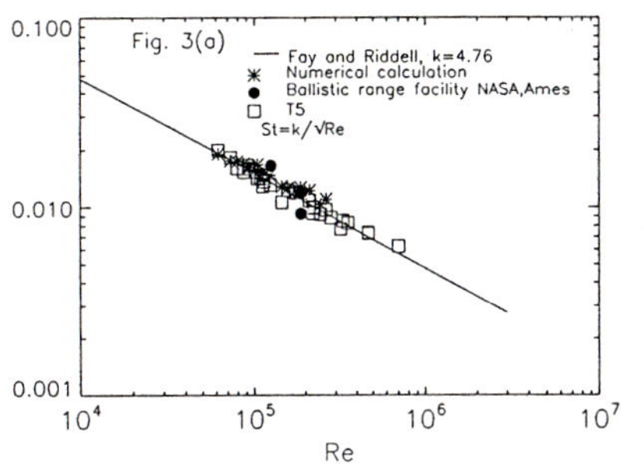

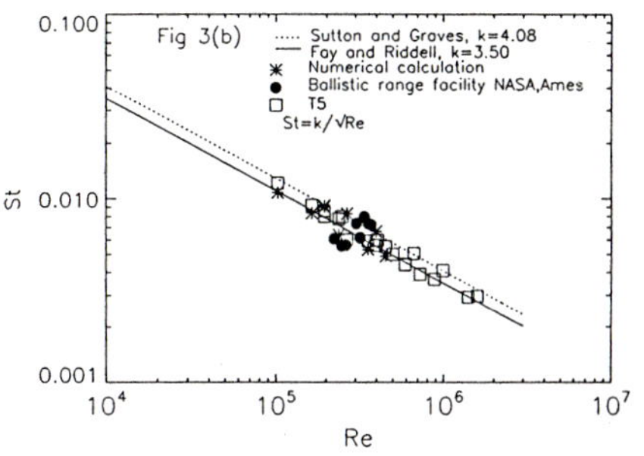

Fig. 3. Comparison of experimental stagnation-point heat flux of T5 and NASA Ames' ballistic range facility, numerical results and theoretical predictions for (a): air, and (b): CO_2

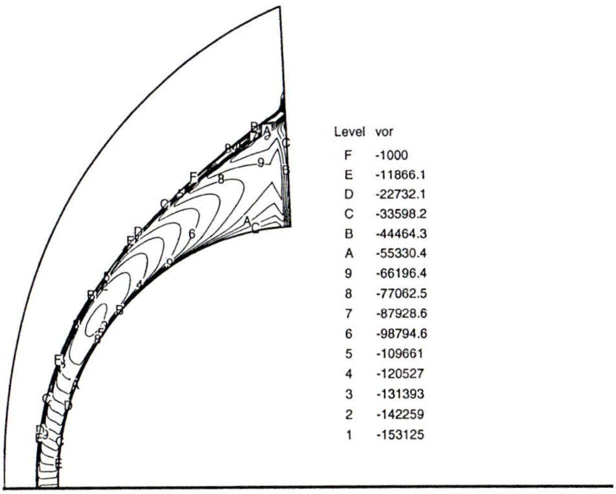

Fig. 4. Result of a computation of the vorticity field over a sphere of 4 in diameter. Freestream condition: CO_2, 3.55 km/s, 0.081 kg/m^3, 11.27 MJ/kg

5. Conclusions

(1) Very good agreement was observed between the experimental differential interferographic fringe patterns and the numerical ones.

(2) Good agreement was observed among the measured stagnation point heat transfer rates, numerical computation results and theoretical predictions. This provides another calibration of T5.

(3) The resonantly-enhanced shadowgraphic technique may provide a promising tool for studying the vorticity-interaction problem in hypervelocity blunt-body flows.

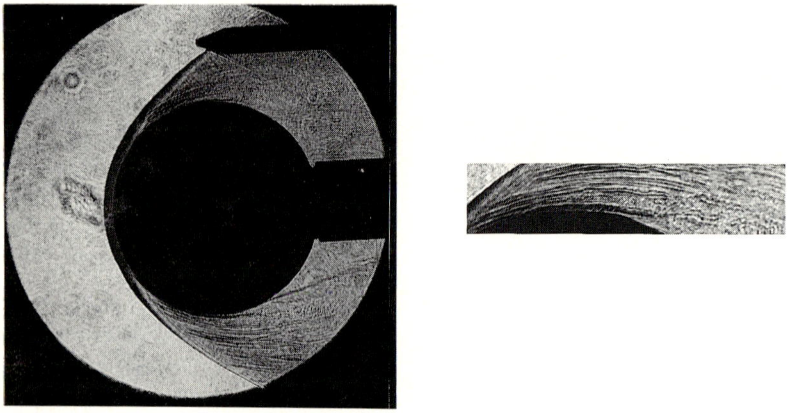

Fig. 5. Resonantly-enhanced shadowgraph of CO_2 flow over a sphere, at the same conditions as those of Fig. 4

Acknowledgement

This work was supported by AFOSR Grant F49610-92-J-0110 (Dr. L. Sakell) and AFOSR grant F49620-93-1-0338 (Dr. J. Tishkoff).

References

Candler GV (1988) The computation of weakly ionized hypersonic flows in thermo-chemical nonequilibrium. Ph.D. Thesis, Stanford University

Fay JA, Riddell FR (1958) Theory of stagnation point heat transfer in dissociated air. Jour. Aero. Sci. 25, 2, Feb

Germain P, Cummings E, Hornung HG (1993) Transition on a sharp cone at high enthalpy; new measurements in the shock tunnel T5 at GALCIT. AIAA Paper 93-0343, Reno

Hornung HG (1992) Performance data of the new free-piston shock tunnel at GALCIT. AIAA Paper 92-3943, Nashville

Hornung HG, Wen C, Candler, GV (1993) Hypervelocity flow over spheres. Acta Mechanica, to appear

Lees L (1956) Laminar heat transfer over blunt bodies at hypersonic flight speeds. Jet Propulsion 26, 4, April

Rein M (1989) SURF: A program for calculating inviscid supersonic reacting flows in nozzles. GALCIT Report FM 89-1

Schultz DL, Jones, TV (1973) Heat transfer measurements in short duration facilities. AGARD Report 165

Sutton K, Graves, AR (1971) A general stagnation-point convective-heating equation for arbitrary gas mixtures. NASA TR R-376 November

Yee L, Bailey HE, Woodward HT (1961) Ballistic range measurements of stagnation-point heat transfer in air and in carbon dioxide at velocities up to 18,000 feet per second. NASA TN D-777 March

Raman Scattering Measurements of Vibrational Relaxation in Expanding Nitrogen

W.D. Gillespie* **and S.P. Sharma**[†]
*Aero/Astro Engineering Dept., Stanford University, Stanford, CA 94305, USA
[†]Aerothermodynamics Branch, NASA-Ames Research Center, Moffett Field, CA 94035, USA

Abstract. Spontaneous Raman scattering spectroscopy is used to study vibrational relaxation behavior in a nonequilibrium expansion. A sample of vibrationally excited, undissociated N_2 is produced in a reflected shock tunnel and rapidly expanded through a two-dimensional nozzle. The flow characteristics of the short-lived steady expansion are confirmed by comparing holographic interferometry measurements directly to 2-D flowfield calculations. The vibrationally relaxing gas is probed on the axis of the nozzle by a single 250 mJ pulse from a 248 nm KrF excimer laser. The Raman scattered light is dispersed and detected by an intensified CCD array. Analysis of the Raman spectra shows that the relative populations of the first few vibrational levels do not deviate significantly from a Boltzmann distribution. Pointwise measurements of the vibrational energy are reported for corrected area expansion ratios of up to 5.5 for stagnation conditions of 102 atm, 2800 K and 102 atm, 5600 K. Good agreement is found between the measured vibrational state distributions and detailed numerical simulations based on the Landau-Teller model and on an anharmonic oscillator model. Relaxation times are deduced by comparison to the Landau-Teller solutions and found to be within the uncertainty of Millikan and White's correlation for vibrational relaxation under shock heating conditions.

Key words: Raman spectroscopy, Shock tunnel, Vibrational relaxation.

1. Introduction

The vibrations of gas molecules can play a significant role in the chemistry, radiation, and thermal behavior of high temperature flows. Since a finite time is required for molecular vibrations to become excited, rates for equilibration of vibrational energy must be considered in many rapidly changing flow environments.

Although vibrational *excitation* rates have been established in N_2 over a wide range of temperatures (Millikan and White, 1963) using the well-known Landau-Teller (1936) model for vibrational relaxation, vibrational *de-excitation* rates have been subject to uncertainty for many years (Hurle 1971). Previously available measurements in expanding flows have indicated that de-excitation occurs much faster than excitation (observed behind shock waves). An empirical factor, $\phi = \tau_{shock}/\tau_{expansion}$, was defined to describe the observed increase based on the Landau-Teller formulation. Reported values for ϕ range from 1 to 150 in N_2.

In an effort to explain the observed relaxation in expansions, investigators (eg. Bray 1968; Center and Caledonia 1972) have challenged the harmonic oscillator approximation made in the Landau-Teller model. The use of a more realistic anharmonic oscillator model results in relaxation which can be faster when excited vibrational levels are populated. The enhanced relaxation is a result of more efficient vibration-translation (V-T) energy transfer from the more closely spaced higher vibrational levels and the participation of exothermic vibration-vibration (V-V) exchanges. Non-resonant V-V exchanges result in non-Boltzmann vibrational distributions when the ratio of vibrational "temperature" to kinetic "temperature" becomes significantly greater than 1. These conditions are indeed found in many nonequilibrium expansions.

Although quantitative agreement with measurements in CO expansions has been achieved using the anharmonic oscillator model (Bender 1978; Ruffin 1993), the calculations fail to predict the degree of accelerated relaxation observed in N_2 expansions. Of course other factors may affect

Shock Waves @ Marseille II
Editors: R. Brun, L. Z. Dumitrescu

the relaxation, especially the presence of impurities which act as very efficient collision partners for vibrational energy transfer, but no conclusive explanation has been found for the discrepancies.

In the present experiment, vibrational population distributions in the supersonically expanding N_2 flow are measured with Raman spectroscopy. The measurements are compared to calculations based on both the Landau-Teller (harmonic oscillator) and anharmonic oscillator models for vibrational relaxation.

2. Experimental apparatus

The shock tunnel used for this experiment is described in detail by Sharma and Park (1988). Briefly, it consists of a 10 cm diameter electric arc-driven shock tube at NASA Ames Research Center with a 84 cm long driver and a 460 cm long aluminum driven section. An aluminum insert near the end of the shock tube containing a 2-D nozzle and large ports for optical access (Fig.1) effectively converts the tube into a reflected shock tunnel. Note, however, that no secondary diaphragm is used. Arc-heated helium is used to drive a shock wave in prepurified N_2 (total estimated impurities including leakup and outgassing <10 ppm). The shock is reflected at the nozzle insert, providing reservoir conditions as shown in Table 1. Two test conditions are considered, denoted by Case A and Case B, for which steady test flows lasting up to 300 μsec are achieved.

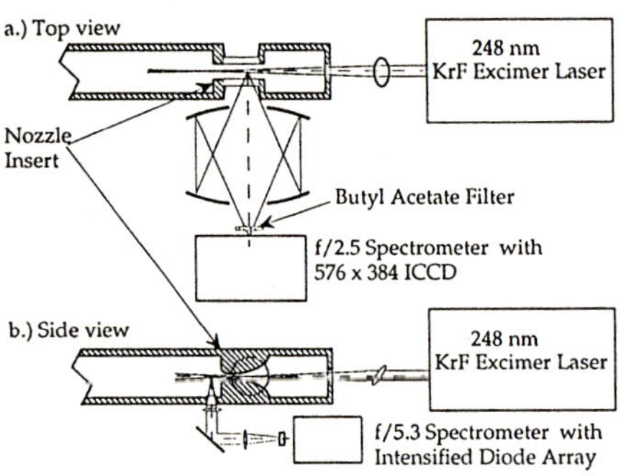

Fig. 1. Experimental arrangement for spontaneous Raman scattering measurements.

Table 1. Shock tunnel test conditions.

Case	Speed (km/sec)	Load (Torr)	P_0 (atm)	T_0 (K)
A	1.6	400	102	2800
B	2.6	100	102	5600

Interferograms of the nozzle flow are made using a Lumonics HLS2 Ruby laser (693 nm) using standard dual-pulse holographic interferometry. Further details are given by Gillespie (1993).

Raman scattering measurements are made using a narrow-band Lambda Physik EMG 150 ET KrF excimer laser aligned with the axis of the shock tube (Fig.1). The laser is triggered once during the test time, and the linearly polarized, 250 mJ laser pulse passes through a tilted lens which has been adjusted to create horizontal and vertical laser sheets in the nozzle and reservoir respectively. The Raman scattered light from the nozzle is collected at 90° by an f/2.5 imaging system and focused into a specially designed f/2.5 imaging spectrometer equipped with a gated, microchannel plate intensified CCD camera. The Raman scattered light from the reservoir is focused by a pair of lenses onto a f/5.3 McPherson 0.3m spectrometer and detected by an intensified diode array.

3. Procedure

In nonequilibrium nozzle flow, the conditions at any point in the expansion are dependent on the flow history up to that point. Thus, to deduce relaxation times from pointwise measurements in the steady expansion, the thermodynamic conditions in the entire flow must be known. Since it is impractical to measure all of the flow properties at every point, numerical simulations are required for data analysis. In the present study, the necessary computations are verified against pressure and density measurements made in the reservoir and the nozzle flow.

Shock speed, pressure, and radiation are monitored for every run (sample traces are given by Gillespie, 1993). These measurements confirm the wave diagram for the shock tunnel and establish the length of test flow and reservoir conditions for the nozzle. Raman scattering measurements in the reservoir region complement the pressure and shock speed measurements with direct temperature determination behind the reflected shock. These measurements establish the stagnation conditions to be input into flowfield calculations.

Interferograms quantify the density field in the nozzle test flow from the throat to the exit, verifying the steadiness and uniformity of the flow. Since the bulk properties are insensitive to the relaxation behavior in the expansion, it is not possible to draw conclusions about the vibrational energy from the interferograms, but they do serve to confirm nozzle flowfield calculations.

A two-dimensional (2-D) Navier-Stokes calculation is performed to include boundary layer effects and effects due to vertical and streamwise velocity components in the rapidly expanding nozzle. The 2-D code utilizes the Landau-Teller relaxation equation to predict energy transfer from the vibrational to translational modes and is described by Ruffin (1993). Interferograms are simulated from the 2-D computational results using a code developed by Yates (1992) and are compared directly to the experimental interferograms.

A detailed computation of the vibrational relaxation mechanism for an anharmonic oscillator requires the solution of the vibrational master equations. Solving the vibrational master equations for the full 2-D flow is not feasible, however, due to the large number of equations and grid points required for such a calculation. A quasi 1-D description is desired for detailed calculations, but a corrected nozzle geometry must be used to compensate for 2-D inviscid effects and boundary layer effects. The required corrected geometry is generated by selecting a streamline near the axis of the nozzle in the inviscid core flow from the experimentally verified 2-D calculation.

The rates used in the 1-D computation for vibrational energy transfer collisions are from a modified version of SSH theory (see Ruffin 1993). The empirical constant which determines the ground state V-V rate is chosen by matching the computed rates to the experimental data of Valyansky et al. (1990), and the V-T rates are scaled to match Millikan and White's correlation. The quasi 1-D fluid dynamic equations and vibrational master equations are solved in a coupled manner using a space marching, implicit method.

The vibrational population distributions are measured on the axis of the supersonic nozzle with spontaneous Raman scattering spectroscopy. Raman scattering is an inelastic scattering process in which light incident on a molecule is shifted in frequency as the molecule undergoes a transition to a new vibrational and/or rotational state. Anharmonicity and vibrational-rotational coupling in the N_2 molecule causes its Raman spectrum to contain many lines, each corresponding to a ro-vibrational level. The vibrational levels are spectrally resolved in the present measurement to reveal vibrational population distributions, but the finer rotational structure is not resolved.

The measured population distributions are compared to the distributions computed using the quasi 1-D code. Vibrational temperatures are defined to infer Landau-Teller relaxation times by varying the parameter ϕ in the calculation until agreement with the measurements is achieved.

4. Results and discussion

An interferogram of the test flow for Case A is shown in Fig. 2 along with a simulated interferogram based on the 2-D computation. The fringes in the interferograms correspond to lines of constant density. Agreement between the measurement and calculation is very good, particularly along the centerline where Raman measurements are made. The quasi 1-D calculations also reproduce the bulk properties along the centerline when the corrected nozzle geometries deduced from the 2-D code are used. The actual and corrected geometries can be found in Ruffin (1993) or Gillespie (1993).

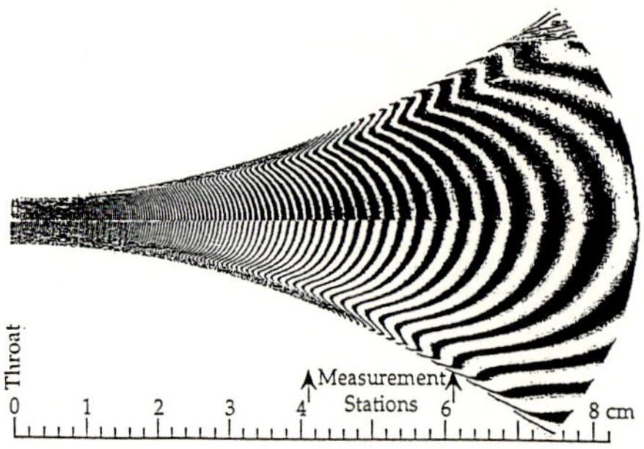

Fig. 2. Comparison of experimental (top) and synthesized (bottom) interferograms for Case A.

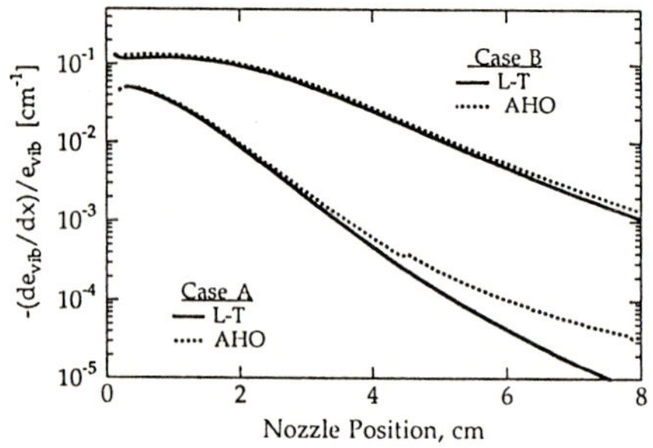

Fig. 3. Quasi 1-D computation for transfer of vibrational energy.

The rate of energy transfer out of the vibrational mode as computed in the quasi 1-D formulation is plotted in Fig.3 for Cases A and B. The Landau-Teller rate and the anharmonic oscillator rate are both shown in the figure for comparison. For Case B, the energy transfer rates for both models are nearly identical, indicating that the contributions to relaxation due to V-V transfers and overpopulation of the upper states are essentially insignificant for this case. For Case A, however, the V-V relaxation becomes the dominant process near the exit of the nozzle, resulting in greater relaxation using the anharmonic oscillator model. Note, however, that the rates diverge only where the magnitudes of the energy transfer rates are very small. Thus the observable overall relaxation is not predicted to be significantly faster than Landau-Teller for either of the conditions studied.

A sample Raman spectrum for Case A from a point 6.1 cm from the throat of the nozzle is shown in Fig.4. The signal at each of the vibrational peaks is primarily from the labeled transition, with a smaller contribution from the neighboring vibrational transitions and a very small contribution from the Raman O branch. The vibrational population distribution deduced from the spectrum shown in Fig.4 is plotted in Fig.5 along with the distribution obtained for the anharmonic oscillator model and a Boltzmann best fit to the experimental data. A slight overpopulation of high vibrational levels ($v>10$) is predicted by the quasi 1-D computations. The populations in these levels for the temperature ranges considered are so small, however, that detection of the predicted overpopulations is well beyond the capability of single-shot spontaneous Raman scattering measurements. The anharmonic oscillator solution and the Boltzmann best fit fall right on top of one another, and the experimental data agrees very well for the first few vibrational levels, and within the error bounds for the higher levels where the signal is low. Although the 4th and 5th vibrational levels do appear to be slightly overpopulated in this plot, the signal levels are so low ($S/N<1$) that deviations from Boltzmann are not meaningful.

A bandfit is performed on the Raman spectrum of Fig.4, yielding a vibrational temperature. The best fit values are given in Fig.4. Bandfitting results for a series of measurements are shown as symbols in Fig.6. The curves in Fig.6 are from the quasi 1-D calculations showing translational temperature and vibrational temperatures computed using both the Landau-Teller formulation, and the anharmonic oscillator model.

The relaxation rates inferred from the measurements are compared to Millikan and White's curve fit and to previous experimental values in Fig.7 (taken essentially from von Rosenberg et

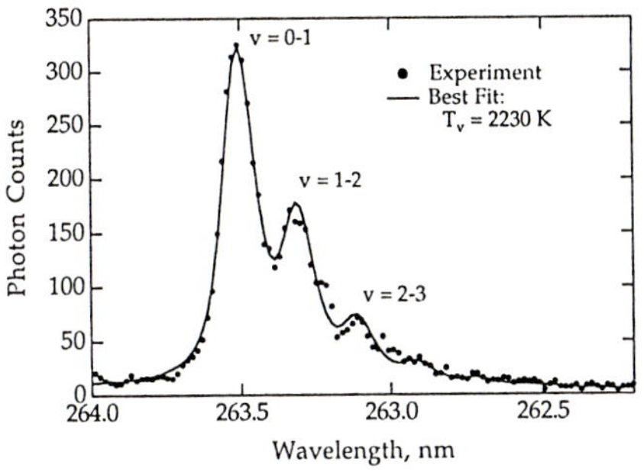

Fig. 4. Experimental Raman Spectrum for Case A from 6.1cm downstream of the nozzle throat.

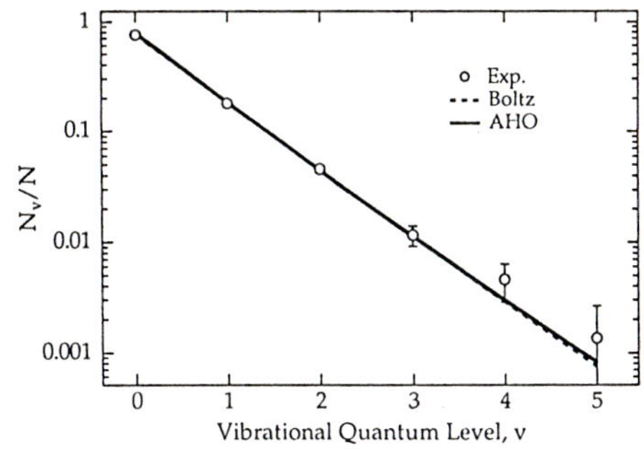

Fig. 5. Vibrational population distribution inferred from spectrum of Fig. 4.

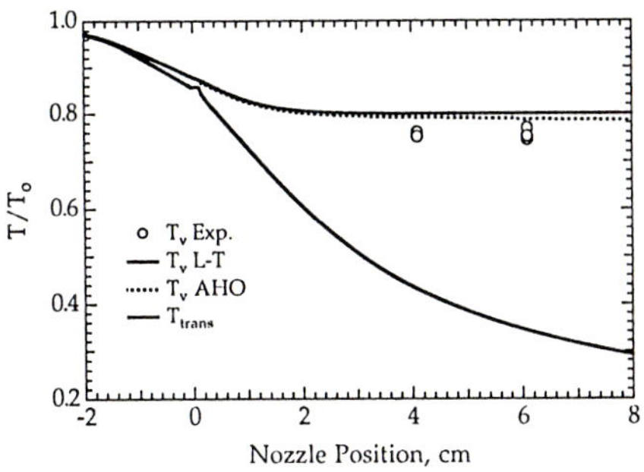

Fig. 6. Measured and computed vibrational temperatures in the nozzle expansion for Case A. L-T: Landau-Teller; AHO: Anharmonic Oscillator.

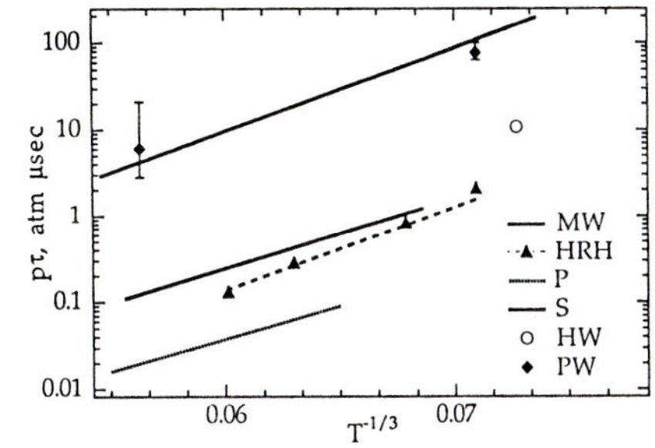

Fig. 7. Inferred relaxation rates compared to previous measurements. MW: Millikan and White (1963); HRH: Hurle et al. (1964); P: Petrie (1965); S: Sebacher (1966); HW: Holbeche et al. (1967); vRTT: lower limit for von Rosenberg et al. (1971); PW: Present Work.

al. 1971). Neglecting the small difference in relaxation for the anharmonic oscillator, the present measurements are in good agreement with relaxation from shock heating measurements. Millikan and White estimate a 50% error for their curve fit, which covers the present measurements.

The source of discrepancy between the Raman scattering measurements where $\phi \approx 1$ and earlier works where $\phi \gg 1$ most likely lies in one or more of the factors listed below:

1. All but the recent Raman measurements were limited to indirect determination of vibrational temperatures through coupling with a thermometric seed species (compromising the gas purity) or reliance on uncertain electronic transition moments. Raman scattering, by contrast, is a direct method giving easily interpretable results for vibrational population distributions.
2. Raman measurements are made pointwise on the axis of the flow. Previous measurements were line-of-sight, integrating through a cold, albeit thin, boundary layer.
3. Even a small level of impurity (>10 ppm) can alter observed relaxation times. Reliability and convenience for gas purity and facility cleanliness are improved through use of modern high quality N_2 sources and cleaner high-vacuum pumping systems.
4. The present technique measures both reservoir and nozzle static temperatures, not relying solely on calculations of reservoir conditions based on insensitive shock speed measurements. Reservoir conditions and nozzle temperatures are measured with the same technique, minimizing the effect of systematic errors in measurement.

5. Holographic interferometry coupled with detailed 2-D computations are used to account for 2-D and boundary layer effects in the expansion.

5. Conclusions

The dramatic increase in vibrational relaxation rates for a cooling flowfield reported in most previous measurements is not reproduced by these experiments. Values for the empirical factor, ϕ, are measured to be less than 1.5, indicating that the Landau-Teller model describes the observed relaxation to within the accuracy of Millikan and White's curvefit for shock heating cases. The technique used for the present experiments represents a major improvement over indirect measurements.

Acknowledgement

The authors gratefully acknowledge the collaboration of Dr. Steve Ruffin in this study. Thanks also to Prof. Edward H. Piepmeier for design of the f/2.5 spectrometer, Scott Meyer for many helpful discussions, and Dr. Leslie Yates for generation of the simulated interferogram.

References

Bender DJ (1978) Measurement of vibrational population distributions in a supersonic expansion of carbon monoxide. Physics of Fluids 21: 1073-1085

Bray KNC (1968) Vibrational relaxation of anharmonic oscillator molecules: relaxation under isothermal conditions. J. Phys. B 1: 705-717

Center RE, Caledonia GE (1972) Anharmonic effects on the rate of relaxation of vibrational energy in rapidly expanding flows. J. Chem. Phys. 57: 3763-3770

Gillespie WD (1993) Raman scattering measurements of vibrational relaxation in expanding nitrogen. Stanford University PhD Thesis

Holbeche TA, Woodley JG (1966) Spectrum-line-reversal measurements through unsteady rarefaction waves in vibrationally relaxing gases. Royal Aircraft Est. Tech. Memo. No Aero 937

Hurle IR, Russo AL, Hall JG (1964) Spectroscopic studies of vibrational nonequilibrium in supersonic nozzle flows. J. Chem. Phys. 40: 2076-2089

Hurle IR (1971) Nonequilibrium flows with special reference to the nozzle-flow problem. In: Stollery JL, Gaydon AG, Owen PR (eds) Proc. 8th Intl. Shock Tube Symp., London

Landau L, Teller E (1936) Zur Theorie der Schalldispersion. Physikalische Zeitschrift der Sowjetunion 10:34-43

Millikan RC, White DR (1963) Systematics of vibrational relaxation. J.Chem.Phys. 39:3209-3213

Petrie SL (1965) 1965 Heat Transfer and Fluid Mechanics Institute. Stanford Univ Press, Stanford, CA, pp 282-300

Ruffin SM (1993) Vibrational energy transfer of diatomic gases in hypersonic expanding flows. Stanford University PhD Thesis

Sebacher DI (1966) 1966 Heat Transfer and Fluid Mechanics Institute. Stanford Univ Press, Stanford, CA, pp 315-334

Sharma SP, Park C (1988) Operating characteristics of a 60- and 10 cm electric arc-driven shock tube, Parts I and II. J. Therm. and Heat Transf., 4:259-272

Valyansky SI, Vereschagin KA, Volkov AYu, Smirnov VV, Fabelinsky VI (1990) CARS measurement of intra-mode V-V exchange rate constants. Chemical Physics, 140:59-64.

von Rosenberg, Jr CW, Taylor RS, Teare JD (1971) Vibrational relaxation of CO in nonequilibrium nozzle flow, and the effect of hydrogen atoms on CO relaxation. J. Chem. Phys. 54: 1974-1987

Yates LA (1992) Images constructed from computed flowfields. AIAA Paper 92-4030, AIAA 17th Aerospace Ground Testing Conference, Nashville, TN July 6-8

Comparisons of CFD with CARS Measurements in Hypervelocity Nitrogen Flows

R.R. Boyce*, **D.R.N. Pulford***, **A.F.P. Houwing***, **Ch. Mundt**† **and R.J. Sandeman***
*Department of Physics & Theoretical Physics, Faculty of Science, Australian National University, Canberra A.C.T. 0200, Australia
†Deutsche Aerospace, Munich, Germany

Abstract. Rotational and vibrational temperatures determined from broadband single pulse Coherent Anti-Stokes Raman Scattering (CARS) experiments 7.2~ 7.5 MJ/kg) pulsed hypervelocity blunt body flow are presented. Comparing these temperatures with nonequilibrium chemistry nozzle and blunt body CFD codes shows good agreement, despite indicating certain inadequacies in the methods used.

Key words: CFD code validation, CARS temperature measurement, Hypervelocity nitrogen

1. Introduction

The work presented in this paper is part of a collaborative CFD code validation project, in which the Australian National University T3 free-piston hypervelocity shock tunnel (Stalker 1972) is being used to generate data from real gas blunt body shock layers for validation purposes. The CFD code in question is the coupled Euler/boundary layer code of Mundt (1992). To extend previous work, comparing shock shapes and surface heat transfer rates with the code (Boyce and Mundt 1991, 1991), to include the measurement of flow-field parameters, Coherent Anti-Stokes Raman Scattering (CARS) was applied to the incident freestream and inside the shock layer for real gas nitrogen blunt body flows (Boyce et al. 1992). The CARS technique has been applied extensively in the past but had not been applied to shock tunnel blunt body flows until the work of Boyce et al. (1992). That work represented the first high spatial resolution, high enthalpy shock layer CARS measurements. This paper concerns the comparison of CFD predictions with rotational and vibrational temperatures calculated from such measurements.

2. Theoretical considerations

2.1. Calculation of the freestream conditions

For the shock tunnel experiments here, the shock-heated gas has a high total enthalpy and so ideal gas assumptions are not valid. The conditions in this nozzle reservoir are calculated from the measured shock speed and initial gas pressure using the one-dimensional isentropic equilibrium shock tube program ESTC (McIntosh 1968). The calculation of the subsequent two-dimensional nonequilibrium chemistry/vibration expansion flow through the nozzle has been done using the computer code SURF (Rein 1989), in which vibrational freezing at a specified temperature is used to account for vibrational relaxation effects. In the current study, a vibrational freezing temperature T_v was determined with the one-dimensional nonequilibrium code STUBE (Vardavas 1984). A vibrational relaxation length L_v was calculated at each point along the nozzle centreline using the Landau-Teller model, and compared with the distance L_f the flow moves for the vibrational energy to fall to $1/e$ of its local value. Sudden freezing occurs when the condition $L_v \ll L_f$ changes rapidly to $L_v \gg L_f$, with the $L_v = L_f$ point defining T_v.

Previous work (Hurle 1971) has shown that the Landau-Teller model underpredicts vibrational deexcitation rates, resulting in the overprediction of vibrational freezing temperatures. Hence the present work uses the mean of the Landau-Teller T_v and the freestream temperature at thermal equilibrium, for the vibrational freezing point. In this way, it is expected that the uncertainty due

Shock Waves @ Marseille II
Editors: R. Brun, L. Z. Dumitrescu

to this assumption in the predicted freestream will be no more than ±10% for the pressure and temperature, and ±0.5% for the density and velocity.

2.2. Computational modelling of the shock layer flow

The CFD code (herein called the DASA code) has been described in detail by Mundt (1992), and is briefly summarised below. It is a coupled Euler/2nd-order boundary layer code in which the boundary layer on the blunt body is calculated by solving the non-equilibrium chemistry second order boundary layer equations, and the inviscid part of the shock layer is calculated by solving the non-equilibrium chemistry Euler equations. Two-species finite-rate chemical reactions are modelled, and vibration and rotation are assumed to be in equilibrium with translation. For the coupling, an equivalent mass source distribution is calculated from the boundary layer solution and impressed on the Euler calculation as a boundary condition at the body, accounting for the displacement effect of the boundary layer. The resulting equivalent inviscid flow is again coupled to the boundary layer method.

2.3. Coherent Anti-Stokes Raman Scattering

The theoretical description of CARS is well established and has been described in several recent reviews (Eckbreth 1988, Clark and Hester 1988). The pump and Stokes beams, at frequencies ω_1 and ω_2 respectively, interact through the third order non-linear susceptibility $\chi^{(3)}(\omega_3)$ of the medium to generate a polarization field which acts as a source for anti-Stokes radiation at frequency $\omega_3 = 2\omega_1 - \omega_2$. $\chi^{(3)}$ consists of resonant and non-resonant components. The former is related to the product $N\Delta j$ of the species number density and the population difference between the levels involved in the ro-vibrational transition j. The latter is contributed by electrons and remote resonances. If the frequency difference $\omega_1 - \omega_2$ lies close to a Raman active resonance of the medium (in this case, transitions between identical rotational lines in adjacent vibrational levels in molecules), then a resonance enhanced CARS signal results, and if ω_2 is sufficiently broadband to span the ro-vibrational spectrum of the molecule of interest, then the CARS spectrum will reflect that spectrum. The CARS intensity (for plane waves and a phase mismatch $\Delta k = k_2 + k_3 - 2k_1$ close to zero, where k_i are the wave vectors of the beams), is given by

$$I_3 = \left(\frac{4\pi^2\omega_3}{c^2 n_3}\right)^2 I_1^2(\omega_1)\, I_2(\omega_2)\, |3\chi^{(3)}(\omega_3)|^2\, z^2$$

where c is the speed of light, I_i are the incident laser intensities, z is the distance over which the phase matching occurs, and n_3 is the refractive index at ω_3. An analysis of the CARS spectrum obtained yields the rotational and vibrational temperatures and the species concentration (if compared to a reference signal) at the point of beam overlap.

Assuming Boltzmann distributions based on different rotational (T_r) and vibrational (T_v) temperatures, it is possible to relate the population difference Δj to those temperatures. This allows I_3 to be calculated to within the factor N, provided the nonresonant component is small (or can be accounted for), and the value of z is known. For broadband CARS, $I_3(\omega_3)$ is a convolution of the pump laser, Stokes laser and Raman line profiles and the instrument function of the detector system. If these convolutions are appropriately accounted for, T_r and T_v can be determined by least squares fitting (treating z, T_r and T_v as free parameters) the normalised (area = 1) experimental spectrum to similarly normalised theoretical spectra. Normalised spectra may be compared since the present resonant N_2 CARS signal is much greater than the nonresonant background.

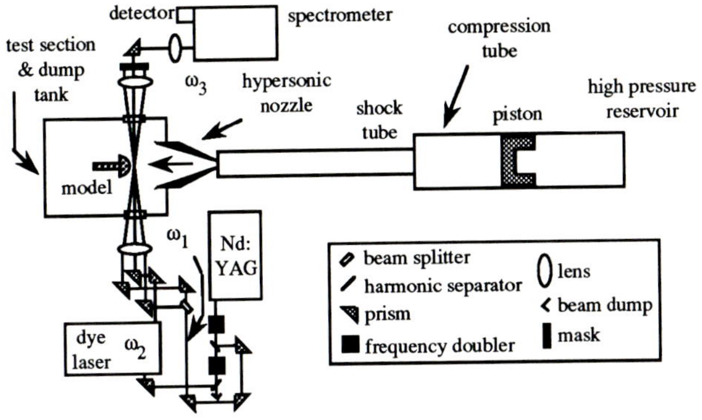

Fig. 1. Free-piston shock tunnel & CARS optical arrangement

3. The experimental arrangement

3.1. The shock tunnel facility

The experiments were performed on the Australian National University's T3 free piston shock tunnel (Stalker 1972) (see Fig.1) operated in reflected mode. The reservoir of high pressure nitrogen at 7.2 MJ/kg enthalpy, expands through a conical nozzle to provide a hypervelocity flow in the test section where the model is mounted. A Pitot survey of the freestream flow indicated that steady flow (constant Pitot to reservoir pressure ratio) commenced approximately 650 μs after shock reflection at the nozzle entrance. A comparison of the measured pressure ratio to that calculated with SURF, showed that the effective nozzle area ratio at the model location was 130 ± 5. The test conditions, calculated as described in 2.1, are summarised in Table 1. Experimental uncertainties quoted there account for both area ratio and reservoir uncertainties, and the T_v uncertainty. The model used (Boyce and Mundt 1991), is an axisymmetric hyperboloid with nose radius 15.18 mm, length 75 mm, and base diameter 100 mm. It was mounted at 0° incidence along the nozzle centerline.

3.2. Optical setup

The optical arrangement (Fig.1) was a folded box phase-matching configuration giving an interaction volume 4 mm along and 0.2 mm perpendicular to the optical axis, resulting in very high spatial resolution. The CARS system used a frequency doubled narrowband Nd:YAG laser (Quanta Ray GCR-4, 532 nm after doubling, 0.02 nm FWHM, 10 Hz pulse rate, 8.5 ns pulse), which provided the two pump beams and the pump source for a broadband dye laser Stokes beam (607 nm, 6.4 nm FWHM). The CARS signal was dispersed by a 1 m spectrometer and imaged onto a gated dual microchannel plate image intensifier, fibre optically coupled to a CID camera. The Nd:YAG laser triggering was controlled to place the laser pulse within the period of steady flow on the model. The camera, with a 90 ns detection window, was triggered simultaneously with the laser. The interaction region location was determined by fitting an aluminium template to the model in the plane perpendicular to the optical axis, and then burning it with the pump beams.

Table 1. Shock tube, nozzle reservoir & freestream conditions

Shock tube/Nozzle reservoir		Freestream conditions	
Primary shock speed (m/s)	2910±50	Velocity (m/s)	3550±50
Shock tube fill (kPa)	92.4±0.2	Static pressure (kPa)	5.024±0.710
Reservoir pressure (MPa)	20.3±1.5	Temperature (K)	751.2±80
Reservoir temperature (K)	5610±130	Density (kg/m^3)	0.0225±0.0016
Res. stagnation enthalpy (MJ/kg)	7.2±0.2	Dissociation fraction	0.0001

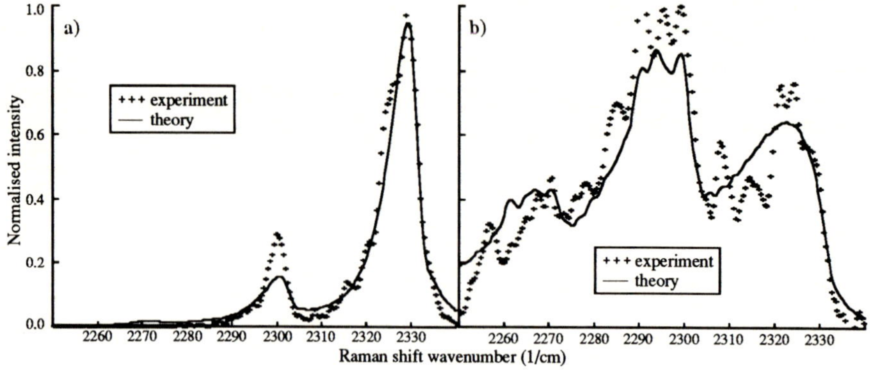

Fig. 2. (a) CARS spectra in freestream; (b) CARS spectra in the shock layer

4. Results

Several N_2 CARS spectra in the freestream and shock layer were produced for the conditions of Table 1. Fig.2a presents the broadband freestream CARS signal. The two features in the experimental spectra represent rotationally averaged vibrational states of the N_2 ground electronic state $X^1\Sigma_g$. The large and small peaks are produced by scattering from the $(v', v'') = (1,0)$ and (2,1) Raman transitions, respectively. Fig.2b presents the average of three broadband shock layer CARS spectra. The three partly overlapping peaks in this figure correspond to the (1,0), (2,1) and (3,2) transitions. Fig.3 displays the bow shock, body contour and inviscid shock layer temperature contours as calculated by the DASA code. The small boxes shown represent the spatial uncertainty in the CARS interaction region locations. Strong temperature gradients exist normal to the body, with a 1400 K temperature variation near the CARS interaction point, over a 3.2 mm shock layer thickness.

The raw data recovered from the 2-D detector consists of CARS signals which have profiles that depend upon the Raman line, the pump beam, and the broadband dye laser profiles; shot to shot dye laser mode structure fluctuations; the detection system instrument function; and background noise. To determine T_r and T_v using the method described in 2.3, these factors are accounted for as follows: The least squares fitting of theoretical spectra to the experimental spectra is done after dividing the raw data by a Gaussian fit to the dye laser profile, and after convoluting the theoretical spectra with the pump line width and an instrument function (accounting for the spectrometer dispersion, distortion introduced by the image intensifier and the pixel dimensions on the CID chip) for the detector system. The theoretical spectra account for the Raman line profiles by using the isolated line model (Hall 1979), which assumes pressure dominated linewidth broadening. The pressures used were taken from the SURF and DASA codes. No attempt was made to account for background noise, or for the mode noise of the dye laser other than to average three shock layer shots, since this structure was well below the resolution limit of the spectrometer and could not be measured. The best fitting theoretical spectra obtained in this way are represented by the solid lines in Figs.2a and 2b. The rotational and vibrational temperatures obtained from the CARS spectra are presented in Table 2 along with values predicted by the SURF and DASA codes. The theoretical freestream vibrational temperature shown is the Landau-Teller T_v. Uncertainties quoted for the CFD values are based on the location uncertainty of the CARS point, not the freestream uncertainties. Due to the single shot nature of the experiment, shot to shot dye laser mode structure variability is believed to contribute the largest errors to temperature determination, with other random sources of noise originating from intensity fluctuations in the dye and pump laser beams, and nonlinear background noise from self-heating in the CID chip. The CARS measurement errors quoted in Table 2 represent the commonly accepted ±10% uncertainty for single shot CARS (Clark and Hester 1988). No error analysis has yet been made for the current application of CARS.

Table 2. Temperature comparison: CARS vs theoretical predictions

Freestream temperatures		Shock layer temperatures	
Rot. temperature (K, CARS)	850±100	Rot. temperature (K, CARS)	3730±400
Rot. temperature (K, Theory)	751.2±80	Vib. temperature (K, CARS)	4000±400
Vib. temperature (K, CARS)	1985±200	Noneq. temperature (K, CFD)	3420±130
Vib. temperature (K, Theory)	2040		

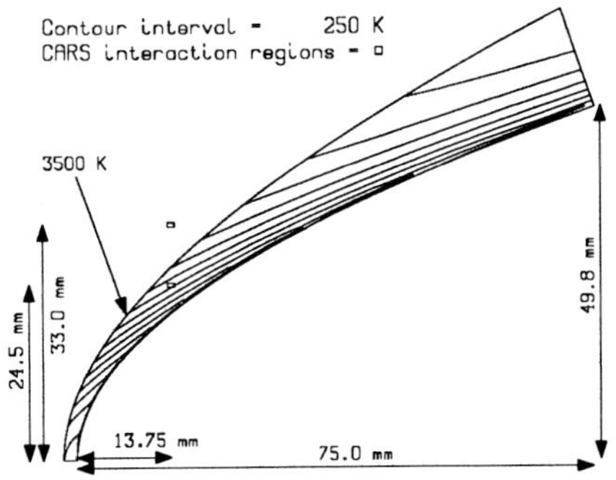

Fig. 3. CFD temperature contours, and CARS interaction region locations

Comparison of the measured freestream temperatures with those predicted by the codes shows a higher measured rotational temperature and a slighltly lower measured vibrational temperature than the Landau-Teller T_v. This T_v comparison is better than Hurle (1971) would predict, and if applied to the nozzle calculation instead of the value of 1450 K used, would decrease the freestream translational temperature, worsening that comparison. The effect of the freestream translational temperatures on the subsequent shock layer calculations would be an underprediction of shock layer translational temperatures, and in comparison to the present CARS measurements, this is the case. Further reasons for the current shock layer comparisons are: the assumption of equilibrium vibration by the DASA code, which would give a lower translational temperature than the expected nonequilibrium vibration behaviour (shown in Boyce and Mundt 1991 for similar flow conditions); the neglect by the DASA code of the frozen vibrational energy component that exceeds the thermal equilibrium value; and the uncertainty in the location of the CARS interaction region. The reason for this uncertainty is the slight thermal expansion of T3 and the CARS optical tables during each day. Table 2 shows a higher measured shock layer vibrational temperature than rotational temperature. This is not correct physically, as a nonequilibrium vibration shock layer should give a vibrational temperature between the freestream value and the shock layer rotational temperature. Thus these measurements indicate that the ±10% uncertainties assumed are of the correct order of magnitude.

No attempt has yet been made to quantitatively investigate the effect on the computations of assuming the measured vibrational freezing temperature to be correct. This represents future work, and is important because the accurate knowledge of freestream conditions is essential in a code validation exercise. In addition, finite rate vibrational relaxation must be included in the DASA code before it can be validated at the current flow conditions. However, given the preceding discussion and the experimental uncertainties involved, the agreement between the code and the experiment is good.

5. Conclusions

Using data fitting techniques and accepted CARS theory, the first measurements of local rotational and vibrational temperatures in the freestream and shock layer of a pulsed hypervelocity blunt body nitrogen flow have been made. Despite reasonably large experimental uncertainties, comparison of computational fluid dynamics predictions with the measurements indicate certain inadequacies in the CFD models used - the use of the Landau-Teller model in the nozzle expansion, and the assumption of vibrational equilibrium in the shock layer. The comparison highlights the shortcomings of relying on numerical predictions of freestream conditions for code validation purposes. Apart from those inadequacies, the agreement is good, and indicates that CARS can be used successfully for code validation.

Acknowledgements

We express our thanks to Mr Paul Walsh for his expert operation of T3, and to Dr Martin Rein for permission to use his code SURF. This work has been financially supported by the Australian Research Council, the Sir Ross and Sir Keith Smith Fund and NASA contract NAGW 1467.

References

Boyce RR, Mundt Ch (1991) Aeronautical CFD studies of reentry flows. Proc. Intl. Aerospace Congress 1991, v.3, p 23-31

Boyce RR, Mundt Ch (1991) Computational fluid dynamics code validation using a free piston hypervelocity shock tunnel. In: Takayama K (ed) Proc. 18th Intl. Symp Shock Waves, Sendai, pp 1127-1132

Boyce RR, Pulford DR, Newman DS, Houwing AFP, Sandeman RJ (1992) Preliminary CARS measurements in hypervelocity nitrogen blunt body flows for CFD validation purposes. Proc. 7th Natl. Space Eng. Symp., Canberra

Clark RJH, Hester RE (1988) In: Advances in Non-Linear Spectroscopy (Advances in Spectroscopy Vol. 15). John Wiley and Sons Ltd, pp 193-237

Eckbreth AC (1988) In : Laser Diagnostics for Combustion Temperature and Species. Abacus Press (now Gordon and Breach), Kent, England, UK, pp 3-24

Hall RJ (1979) CARS spectra of combustion gases. Combustion and Flame 35: 47

Hurle IR (1971) Nonequilibrium flows with special reference to the nozzle-flow problem. In: Stollery JL, Gaydon AG, Owen PR (eds) Proc 8th Intl. Shock Tube Symp., London

Lordi JA, Mates RE, Moselle, JR (1966) Computer program for the numerical solution of nonequilibrium expansions of reacting gas mixtures. CAL Report AD-1689-A-6

McIntosh MK (1968) Computer program for the numerical calculation of frozen and equilibrium conditions in shock tunnels. ANU Internal Report

Mundt Ch (1992) Calculation of hypersonic, viscous, non-equilibrium flows around reentry bodies using a coupled boundary layer/Euler method. AIAA Paper 92-2856

Rein M (1989) SURF : A Program for calculating inviscid supersonic reacting flows in nozzles. GALCIT FM 89-1

Stalker RJ (1972)Development of a hypervelocity wind tunnel. Aero. J. Roy. Aero. Soc. 76:374

Vardavas I (1984) Modelling reactive gas flows within shock tunnels. Aust. J. Phys. 37:157

Research on Nonequilibrium Phenomena at NASA Ames's E.A.S.T. Facility

Surendra P. Sharma
NASA Ames Research Center, Moffett Field, CA 94035, U.S.A.

Abstract. The paper reviews the recent experimental and theoretical research activities on nonequilibrium phenomena in hypersonic flows conducted at NASA Ames's E.A.S.T. facility. The activities include: 1) assessment of our present capability of computing nonequilibrium radiative emission in the light of recent measurements in air and nitrogen in the facility, 2) research on expanding flows: measurements of nonequilibrium populations of 9 lower vibrational levels of nitrogen in an expanding flow, measurement of Landau-Teller correction factor for expanding flows and development of a 1-D nozzle code accounting for detailed state-by-state vibrational nonequilibrium, 3) study of flow phenomena in shock tubes using a time-accurate axisymmetric fluid code, and 4) analytical investigation of coupled rotational-vibrational-dissociation phenomena.

Key words: Shock tube, Hypersonic, Nonequilibrium flows

1. Introduction

Thermo-chemical nonequilibrium processes in hypersonic flows during the relaxation period preceded by sudden compression or expansion affect the fluid flow in three major ways (Howe 1985; Sharma 1990; Park 1985): (i) The individual share of rotational, vibrational, electronic and translational modes contributing to the internal energy pool of the fluid no longer remains uniquely defined,and becomes a transient quantity, approaching its final value. The approach to the final value of energy content in these modes from their initial values is, as a rule, not linear, and a characteristic overshoot or undershoot is always present. The distribution within each of these energy modes also, in most cases, remains in a state of nonequilibrium accompanied by associated dissociation and ionization processes. (ii) Since these kinetic and chemical phenomena absorb energy, the bulk properties of the fluid also change. These changes in bulk properties can be seen in modifications of certain flowfield parameters such as shock standoff distance and pressure distribution around a vehicle. Such a variation in pressure fields and the resulting forward movement of center of pressure on the vehicle (positive pitching moment), was observed on the Space Shuttle (Compton 1982; Reid 1985). (iii) Interaction of the flowfield with the surroundings such as radiative heat transfer, flow separation, and shock and boundary layer interaction are also effected by this nonequilibrium. In fact laboratory data indicate that nonequilibrium radiation emitted by air may be 2 to 15 times larger than the equilibrium value (Allen 1962; Tier 1961; Shirai 1980; Sharma 1990).

To determine the effects of chemical nonequilibrium mentioned above over a realistic hypersonic vehicle, it would be desirable to conduct an experiment in which all aspects of fluid flow are simulated. Such an experiment is extremely difficult to set up. The only practical alternative is to develop a theoretical model of the phenomena and to simulate the desired flow including the thermo-chemical nonequilibrium, and compare the results with the experiments conducted in the facilities under conditions where only a portion of the flow phenomena is simulated. Once such a physical model is validated against available experimental data, it can be used to simulate and study any other flowfield for which experimental data are not available. Three types of experimental data are needed to assist the aerospace community in this model development process: 1) data which will enhance our phenomenological understanding of the relaxation processes, 2) data on reaction rates for the relevant reactions, and 3) data on bulk properties, such as spectral

Shock Waves @ Marseille II
Editors: R. Brun, L. Z. Dumitrescu

radiation emitted by the gas, for a given set of aerodynamic conditions. The scientific group at NASA Ames's Electric Arc-driven Shock Tube (E.A.S.T.) Facility is engaged in conducting theoretical and experimental research with a goal of enhancing our phenomenological understanding of the nonequilibrium relaxation processes. This paper summarizes the present ongoing research activities at the E.A.S.T. facility. The activities discussed are:

A. Radiation emission measurements behind a normal shock and radiation code development.
B. Experimental investigation of vibrational and rotational nonequilibrium in expanding flows.
C. Experimental and theoretical investigation of shock tube flow.
D. Theoretical modelling of coupled rotational vibrational dissociation phenomena in diatomic molecules.

2. E.A.S.T. facility

2.1. The shock tube

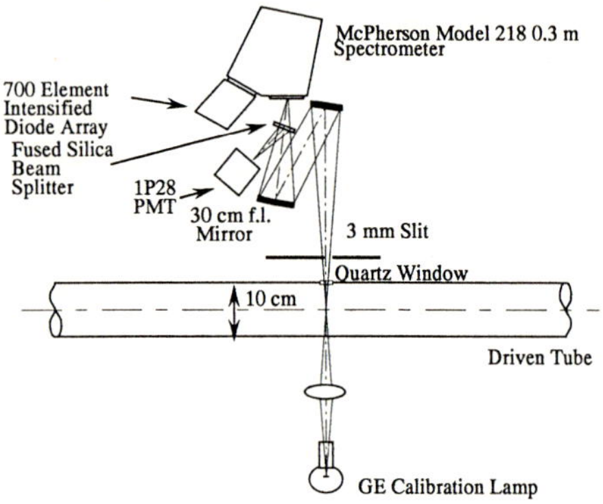

Fig. 1. Schematics of emission measurement collection optics

NASA Ames's electric arc driven shock tube facility has been in existence since the 1960's at NASA Ames Research Center, Moffett Field, CA, U.S.A. The operating characteristics of the facility are described by Sharma (1988). The facility at NASA Ames Research Center consists of one driver system and two parallel driven tubes. One is a 10 cm i.d. tube 12 m in length, and the other is a 60 cm i.d. tube 21 m in length. The driver can be operated in two configurations: 1) a 17.7 cm conical drive configuration with a 10.16 cm exit (driver volume = 632 cm^3), and 2) a variable length (34 - 137 cm) 10-cm i.d. cylindrical configuration (driver volume = 2,669 to 10,752 cm^3). The length of the cylindrical drivers can be varied by using a Lexan filler plug.

Energy to the driver is supplied by a 1.24 megajoule 40 kV capacitor energy storage system. The 6-tier capacitor bank has 220 capacitors. By using different combinations of series-parallel connections the capacitance of the bank can be varied from 149 μF to its maximum value of 6,126 μF (1,530 μF for 40 kV operation). Nominal total system inductance exclusive of the load (arc) is 0.26 μH and the resistance is 1.6 mΩ.

Using the two different driven tubes, by varying the driver/driven gas combination, driver charge pressure and preset capacitor bank voltages, shock velocities in the range of 3.0 - 50.0 km/sec have been obtained. In order to minimize the level of impurities, contact of the test gas with steel and any material containing carbon or hydrocarbons has been minimized. Past experience shows that the spectra of the test gas, which was in contact with steel wall and carbon (burned mylar diaphragm), were overwhelmed by spurious emission spectral lines of iron and CN-violet (Sharma 1988). For this reason, at present aluminum diaphragms are used.

2.2. Diagnostic systems and theoretical tools:
The diagnostic system available at the facility consists of: 1) emission measuring devices, 2) flow-visualization techniques, and 3) laser based non-intrusive diagnostics.

2.2.1. Emission measurements
The emission diagnostic system comprises of: 1) a linear intensified 700 active element diode array and a 2-D intensified CCD array with 576×384 active elements, both gateable within a time range of 30 ns - 2.5 μs and both with a 2000-8000 Å spectral response. The diode arrays are mounted at the image plane of a f#/5.3 McPherson Model 218 0.3 m spectrometer. A typical optical set-up for this measurement is shown in Fig.1. A special f#/1.5 spectrometer has also been designed for this purpose. For the measurement of the total radiation (2500-10000 Å) a special custom-designed spectrometer is available. Photomultiplier tubes (PMTs) are used to record the total radiation from the test as well as from the driver gas as they pass through the test section. The signal from the PMTs is used to estimate the test time and to trigger the diode array system at a given moment during the test history.

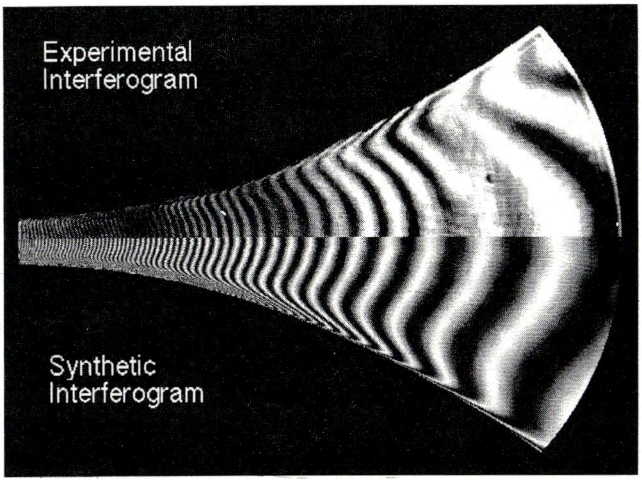

Fig. 2. Experimental and synthetic holograms of an expanding flow

2.2.2. Flow visualization
A Nd:YAG laser doubled to 532 nm, with a 25 ns laser pulse, is used to record holograms of a desired flow field. Both beams, the object and reference, fall onto an Agfa 8E56 film plate with their path lengths matched within a few centimeters. An etalon is used in the laser path just ahead of the beam splitter to ensure that the coherent length of laser light is larger than the path differential between the two beams. The film is exposed twice: once during the test and the second reference exposure is performed soon before or after the test. Interferograms are reconstructed from the single-plate, double-exposure holograms on a secondary reconstruction optical set-up using an Argon ion laser. Such an interferogram of an expanding flow recorded at the facility is shown Fig.2. Also shown in the figure is a synthetic interferogram generated from the flowfield computations using a 2-D code available at the facility (Sharma 1992)

2.2.3. Non-intrusive measurements of thermodynamic parameters
Spontaneous Raman scattering diagnostics:

Spontaneous Raman scattering, in which the molecules in each of vibrational and rotational energy levels modulate the incident laser frequency, can provide the needed data on the nonequilibrium populations at any given instant. The vibrational population distribution at various nozzle stations is measured with spontaneous Raman scattering. A KrF excimer laser (Lambda Physik EMG 150) delivering 250 mJ in a 25 nsec pulse at 248 nm probes the test gas (Fig.3). The linewidth is 0.003 nm (manufacturer's information), and direct measurements of Rayleigh scattered light show that the linewidth is indeed smaller than the resolution of the spectrometer. The excimer beam is focused into a 6 x 0.1 mm sheet in the nozzle by a 2000 mm lens tilted by nearly 45° to the beam. An f/5.3 spherical concave mirror collects the Raman scattered light, and a cylindrical lens corrects for astigmatism. Suprasil II windows and lenses were chosen to minimize fluorescence from the optics. This collection scheme images the focal volume of the beam onto the slit of a 0.3 m McPherson 218 spectrometer. The spectrometer is equipped with a 3600 ln/mm grating and an intensified CCD camera which provide spectral resolution of 0.1 nm. Background emission is reduced by gating the intensifier during the laser pulse. A butyl acetate filter of 1 cm pathlength placed near the entrance slit essentially eliminates the Rayleigh scattered light entering the spectrometer.

Atomic oxygen concentration measurements:

Under nonequilibrium conditions the concentration of atomic oxygen can be measured by probing the ground state with single or multi-photon techniques. The lowest energy transition out of the ground state of oxygen is $^3P \rightarrow {}^3S$. The ground state is split into three levels at 0, 158 and 227 cm^{-1}, so the $^3P \rightarrow {}^3S$ transition consists of lines at 130.217, 130.486, and 130.603 nm. This transition is strongly allowed and the lines are relatively broad for concentrations greater than 10^{16} cm^{-3}, therefore the absorption measurements are made far in the wings.

The output beam of an ArF excimer laser is shifted in a hydrogen Raman shifter to provide a tunable light source in the 130 nm region. The beam of an ArF excimer laser is focused into a hydrogen Raman cell. The 6th anti-Stokes line generated by a 193 nm pump beam is at 130 nm. Cooling the cell with liquid nitrogen reduces the Q-branch width and concentrates population into the J=0 and 1 levels, thus improving conversion efficiency. An aperture selects the 130 nm beam and a narrow band filter reduces stray light. A MgF_2 flat and solar-blind photo multiplier monitor the reference intensity, and the transmitted beam is directed through the test section. A monochromator with a solar-blind photomultiplier measures the transmitted intensity.

Fluid and radiation codes

The NonEQuilibrium AIr Radiation (NEQAIR) program written by Park (1988) at NASA Ames is extensively used to compute radiative emission from flowfields. Two computer codes have been written to couple the one-dimensional fluid equations with NEQAIR: 1) STRAP (Shock Tube RAdiation Program) for the shock tube flow behind a normal shock, and 2) SPRAP (Stagnation Point RAdiation Program) for the stagnation streamline in the shock layer over a blunt body. Recently these programs have been improved by Whiting (1990) and combined into a single code. The improvements consist of: 1) generalization of chemical reaction schemes, 2) improvement in the accuracy of partition functions and radiation calculation schemes, and 3) inclusion of the species C, CO, CN, and C_2 in the excitation calculations. All of these codes assume $T_{elec} = T_e = T_v$ and $T_R = T$ in the fluid computations with the average temperature, $T_a = \sqrt{TT_v}$ in the rate equations.

3. Research activities

3.1. Radiative emission measurements and related code developments

In order to understand the basic physics of the nonequilibrium radiation phenomena emission measurements are being conducted. The data so acquired are being used to modify the existing physical model of nonequilibrium radiation. Both, nonequilibrium and equilibrium radiation emitted by shock compressed test gas at 6.2 km/sec, p_1 = 1.0 torr nitrogen and 10.2 km/sec,

$p_1 = 0.1$ torr air in the 3200-7000 Å spectral range were recently measured (1990-1992). The nonequilibrium spectrum corresponds to the point of peak total radiation. Measured vibrational and rotational temperatures were lower than the values predicted by the present models (Sharma 1991), with the rotational temperature lagging significantly (8800 K (exp) vs. 15,000 K (theory)). In an attempt to understand the nature of the nonequilibrium, NONEQ2 code (Whiting 1990) has been extensively used to simulate the experimental spectra. The results of the comparisons between theory and experiments can be summarized as below.

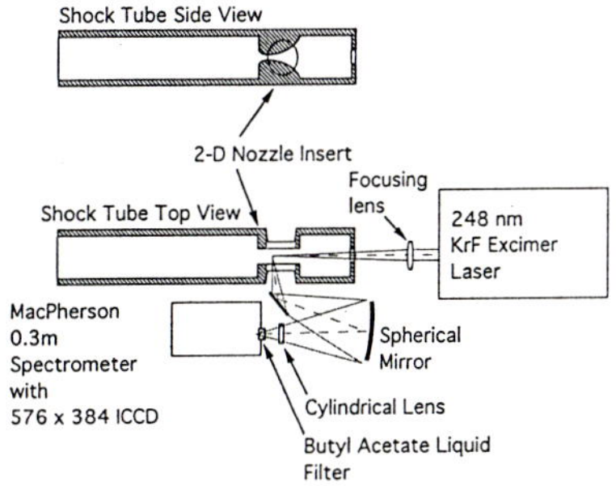

Fig. 3. Layout for Raman scattering diagnostics

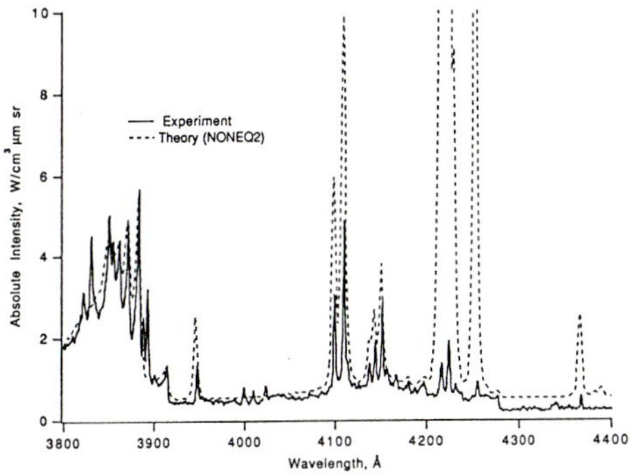

Fig. 4. Emission spectra at 10.2 km/sec in 0.1 torr air

3.1.1. Equilibrium radiation

At mid-range velocities ($\sim$ 6.2 km/sec, 1.0 torr) the present physical model is able to simulate the emission spectra with reasonable accuracy provided the impurities like CN-violet are accounted for (Moreau and Bourquin 1993, Sharma 1993). On the other hand, at velocities higher than 10.0 km/sec and low densities, the code predicts much higher radiative emission from most of the atomic systems, and lower emissions from the neutral species (such as, $N_2(2^+)$) than measured experimentally (Fig.4). The discrepancy between the computations and the experiments does not seem to be uniformly distributed over any particular atomic system, not even in the same multiplet system. For example, the computed emission from the $4p^4S^o \rightarrow 3s^4P$ nitrogen triplet system at 4151.46, 4143.42 and 4137.63 Å falls, on average, within a factor of 5 of the experimental data. However, for the $4p^4P^o \rightarrow 3s^4P$ multiplet system the computed values average about 50 times higher than the experimental values. Two possible explanations are proposed: 1) The QSS formulation in the code may not be adequate, and/or 2) The published f-numbers of the individual states used for the multiplet systems in the code may not be accurate.

3.1.2. Nonequilibrium radiation

The simulation capability of the present physical for the $N_2^+(1^-)$ band system is excellent (Sharma 1993). At mid-range velocities ($\sim$ 6.2 km/sec, 1.0 torr) the model is able to simulate with reasonable accuracy the spectra of other band systems and N and O atomic systems provided the slow relaxation of the vibrational and rotational temperature is accounted for (Moreau and Bourquin 1993). However, at higher velocities ($\sim$ 10.2 km/sec, 0.1 torr) the discrepancy between the theory and experimental values grows much larger; the model predicting larger atomic and smaller molecular (neutral species, such as $N_2(2^+)$) radiation. Possible explanation for these discrepancies are similar to those mentioned for the equilibrium case.

3.2. Relaxation phenomena in expanding flows

Using spontaneous Raman scattering as the diagnostic tool, the vibrational and rotational relaxation in expanding flows are investigated. The test gas used is prepurified nitrogen (99.9995%).

The investigations are conducted in a simple 2-D nozzle geometry. For this purpose a 2-D quadratic nozzle profile plug is installed in the 10 cm driven tube and the shock tube is operated in a reflected shock tunnel mode. The experiments have already been conducted for two sets of test conditions: 1) Low enthalpy: U_s= 1.6 km/sec, T_0=2800 K, P_0=102 atm, 2) High enthalpy: U_s=2.6 km/sec, T_0=5600 K, P_0=102 atm. For both tests, spontaneous Raman data are recorded at various locations along the axis between −0.55 cm to 6.10 cm as measured from the throat. Raman spectra for three such stations are shown in Fig.5. The columns of pixels on the 2-dimensional ICCD array are binned to form 'superpixels', each of which is designated by a symbol in the spectrum. The abscissa is scaled to approximate the number of photons detected for each 'superpixel' so that, for example, at x=-0.6 cm we see that a total of about 30,000 Raman scattered photons are detected.

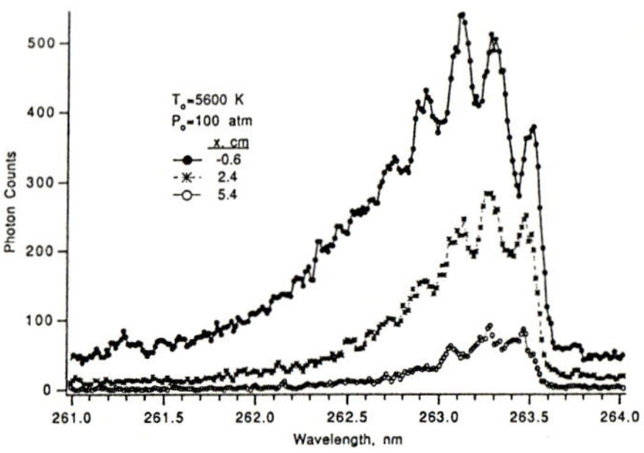

Fig. 5. Raman spectra taken at three stations along the nozzle centerline. Distances are from throat

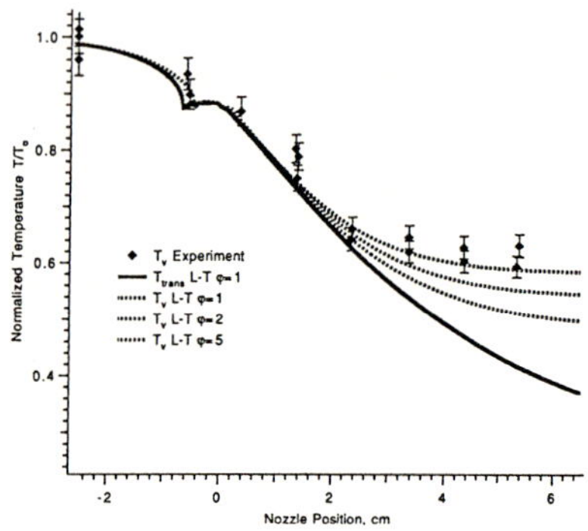

Fig. 6. Normalized temperatures along centerline

The Raman spectra are analyzed first by deducing vibrational population distributions. The overlap between the Q-branches of adjacent vibrational levels is accounted for by computing the contributions from each level in a given wavelength interval. Since these contributions are dependent on the rotational temperature, an estimate for the rotational temperature based on the 2-D flowfield computations is made (Sharma et al. 1992). A separate set of vibrational level overlap contributions is constructed for each measurement station. For each run, an entire matrix of overlap factors is made for the wavelength intervals corresponding to each peak of the spectrum. This matrix is inverted to yield the intensities of all vibrational levels. Instead of deducing a single vibrational temperature from the vibrational distributions, better accuracy is obtained using a band shape-fitting technique and assuming a Boltzmann distribution for $v = 0$ to 8.

The results for the vibrational temperature measurements are shown in Fig.6 for all seven stations in the nozzle. The experimental data do not fall along a single line, yet the overwhelming evidence shows a relaxation rate that is reasonably well described by the simple Landau-Teller relaxation and certainly is not much faster than Landau-Teller. Based on these results, an upper limit of 1.5 can be placed on ϕ. The value of ϕ measured in the present experiment is considerably lower than previous estimates which lie in the 5 - 70 range. This study differs from previous work with nitrogen in two important ways. First, spontaneous Raman scattering provides a direct measurement of the vibrational population. In the past experiments, the flow was either seeded with thermometric species such as CO or sodium, or an electron beam was used to excite the nitrogen into an electronically excited state where it would radiate. Second, it is well known that the presence of even a small amount of impurity can significantly increase the relaxation rate and the purity of the test gas is very important. In this experiment, seeding is not required to monitor the flowfield. The present experiment therefore represents the first direct measurement of vibrational energy distribution in a spectroscopically clean flow.

3.3. Shock tube flow

Experimental investigation and physical modelling of various processes, such as boundary layer and reflected shock interaction, occurring in shock tube flows are being conducted in the facility. In experimental investigations, axial shock velocity profiles, static and stagnation pressures, arc driver parameters, and heat transfer data are recorded. In order to understand the physics of various phenomena a time-dependent axisymmetric simulation of the shock tube has been developed (Wilson 1993). In this simulation the fluid equations are solved by using an explicit finite-volume form of the Harten-Yee upwind TVD scheme which gives second-order spatial and temporal accuracy. At present the gas model considers the three major species (N_2, N, and He) and accounts for finite-rate chemical processes. A separate equation for vibrational energy is included so that vibrational nonequilibrium effects can be assessed. The full Navier-Stokes viscous terms are included in the axisymmetric simulations.

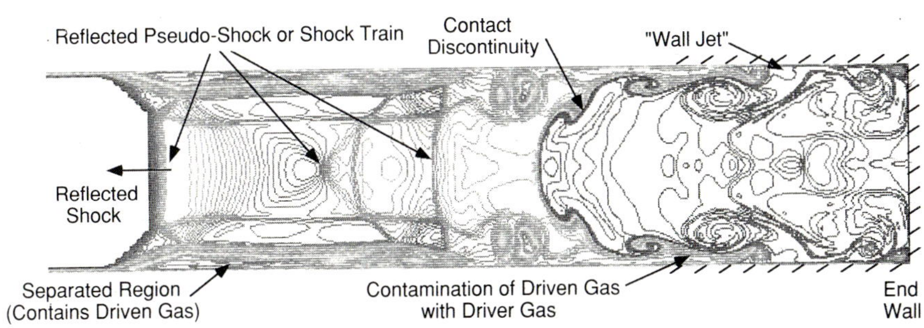

Fig. 7. Computed temperature contours in the reflected shock region of a shock tube

The code is being used to simulate the flowfield of the E.A.S.T. facility for various operating conditions. Computed temperature contours of such a simulation is shown in Fig.7. The temperature contours correspond to the time when the reflected shock approaches the contact discontinuity and interacts with it. As can be seen from the figure the flowfield is very complex. The penetration of the driver gas along the walls is clearly seen; the bubble of separated flow under the bifurcated shock has carried hot driven gas with it far from the end wall. Such details as pseudo shocks can also be seen. This powerful axisymmetric code captures various interesting features of the complex shock tube flow and can be an excellent tool, when used in conjunction with experimental investigations, in our ongoing research activity on shock tube flows.

3.4. Coupled rotation-vibration-dissociation phenomena

Using a coupled rotation-vibration-dissociation model (Sharma 1992) the rotational and vibrational relaxation times for molecular hydrogen as a function of final temperatures (500-5000 K), in a hypothetical scenario of sudden compression, are computed. At time $t = 0$, the gas is suddenly heated by an unspecified means so that the translational temperature acquires a new value T. The box is henceforth maintained at an isothermal condition. Due to the suddenly changed T, collisional processes occur to vibrationally excite or de-excite to relax to a condition appropriate to T. Some of the collisions lead to a free-bound (dissociation or recombination) transition

$v, J \rightleftharpoons A + B$. There are a total of 15 vibrational states starting from $v = 0$ to $v = 14$, and a total of 348 rotational states with 39 rotational states at $v = 0$ and 5 rotational states at $v = 14$. It is further assumed that the ionization level is small and the bound-bound collisional transitions occur only when both colliding partners are molecules. In this manner we can focus our study to the effects of the resonance transitions in V-V energy transfer, a dominant process in flowfields around hypersonic vehicles.

The state-to-state rate coefficients for this study have been determined using a quasi-classical trajectory method. This requires as input a potential energy function which describes the interaction of the two hydrogen molecules. The potential used in the present work is a multi-parameter analytic representation of the results of extensive *ab initio* electronic structure calculations, and is expected to faithfully reflect the properties of the accurate potential. The final states of the target were assigned using the histogram method while a summation over the final states of the projectile were performed. This yields rates for all processes of the form:

$$H_2(v, J) + M \rightarrow H_2(v', J') + M \quad \text{and} \quad H_2(v, J) + M \rightarrow H + H + M \tag{1}$$

This determines all state-to-state rate coefficients except the ones for free to bound transitions. These are generated from the bound to free transition rate coefficients by using microscopic reversibility. Finally, to ensure that the proper equilibrium is reached, all bound-bound state-to-state rate coefficients for endoergic transitions are replaced with the values calculated using microscopic reversibility from the reverse transitions.

The relaxation times can be computed by monitoring the time evolution of $\rho_{v,J}$ and ρ_v, for a given vibrational and rotational quanta. However, in order to visualize the bulk rotational relaxation behavior, the quantity usually measured in an experiment, we follow the time evolution of the average rotational/vibrational energy of the molecules. Computed values of the relaxation parameter $\tau \times P_0$ for a temperature range of 500-5000 K are plotted in Fig.8 along with some available experimental data. In the 200-1200 K temperature range Yoder's and Lench and Grönig's data show an increase in the relaxation parameter with increase in temperature, as do the computed values.

The computed data from this present study, marked with thick solid line with square symbols in the figure, are in reasonable agreement with Lensch and Grönig's data. Although, the computed values at T=500 K (dotted line with square symbols) are slightly lower than their data. It should be remembered that the cross sections for this temperature were obtained by extrapolation, instead of actual trajectory computations; extrapolations could be responsible for this discrepancy. Gallagher and Fenn's data are slightly higher, however, their general trend is the same as the computed data and Lensch and Grönig's data. No reliable data are available for temperatures above 700 K.

For temperatures above 700 K and below 2000 K, the computations show that the rotational relaxation parameter, $\tau \times P_0$ continues to increase with increase in temperature. At around 2000 K, with 2% of the molecules dissociated, $\tau \times P_0$ achieves a maximum of 2.5×10^{-8} atm.sec. At this point the dissociation starts to play an important role in the relaxation process. The energy lost in each dissociation is 36110.8 cm^{-1} and it occurs at the expense of the energy in the rotational mode. As a result the equilibration process of the rotational energy slows down. From Fig.8 we can see that the relaxation parameter $\tau \times P_0$ decreases rapidly, as the temperature increases from 2000 to 5000 K. At around 5000 K about 99% of the molecules are dissociated and the value of $\tau \times P_0$ decreases to 7×10^{-11} atm.sec.

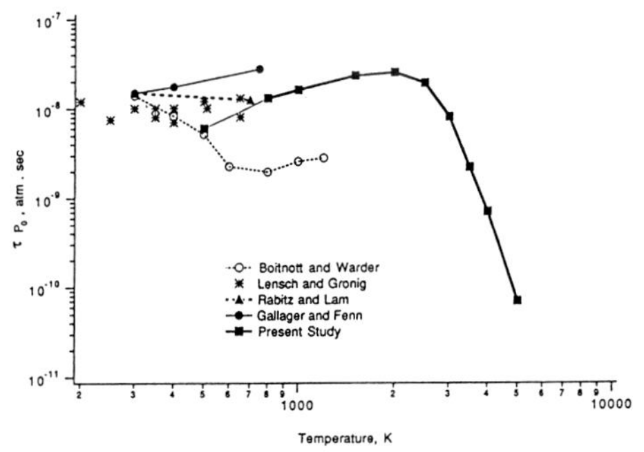

Fig. 8. Rotational relaxation parameter for hydrogen as as function of temperature

4. Conclusions

The research group at NASA Ames's Electric Arc-driven Shock Tube (E.A.S.T.) Facility is engaged in conducting theoretical and experimental research with a goal of enhancing our understanding of the phenomenological nonequilibrium relaxation processes. Recently emission spectra at shock velocities of 6.2 km/sec in 1.0 torr nitrogen and at 10.2 km/sec in air were recorded. The emission data are now being used to improve the capabilities of present radiation models. Using spontaneous Raman scattering as a diagnostic tool, vibrational and rotational relaxation processes in expanding flows are being investigated. Recent experiments show that the Landau-Teller correction parameter ϕ for vibrational relaxation in nitrogen is on the order of 1.0-1.5, which is much smaller than currently reported values of 5-70. A powerful axisymmetric code to simulate the time-dependent behavior of shock tube flow has been developed. Using a coupled rotation-vibration-dissociation model the rotational and vibrational relaxation times for molecular hydrogen as a function of final temperatures (500-5000 K), in a hypothetical scenario of sudden compression, are computed. The state-to-state rate coefficients for this study have been determined using a quasi-classical trajectory method.

References

Allen RA, Rose PH, Camm JC (1962) Nonequilibrium and equilibrium radiation at super-satellite re-entry velocities. AVCO Everett Res. Lab. Rep. 156

Arnold JO, Cooper DM, Park C, Prakash SG (1980) Line-by-line transport calculations for Jupiter entry probes. In: Olstad WB (ed) Progress in Aeronautics and Astronautics, Vol. 69, Entry Heating and Thermal Protection, AIAA, New York, pp. 52-82

Boitnott CA, Warder RC Jr (1971) Shock tube measurements of rotational relaxation in hydrogen. The Physics of Fluids 14, 11: 2312-2316

Camm JC, Kivel RL, Taylor RL, Teare JD (1959) Absolute intensity of nonequilibrium radiation in air and stagnation heating at high altitudes, AVCO Everett Res. Lab. Rep. 93

Compton HR, Scallion WJ, Suit WT, Schiess JR (1982) Shuttle entry performance and stability and control derivatives extracted from flight measurement data. AIAA Paper 82-1317

Gallagher RJ, Fenn JB (1974) Rotational relaxation of molecular hydrogen. The Journal of Chemical Physics 60, 9:3492-3499

Howe JT (1985) Introductory aerothermodynamics of advanced space transportation systems. Journal of Spacecraft and Rockets 22, 1: 19-26.

Lensch G, Grönig H (1977) Experimental determination of rotational relaxation in molecular hydrogen and deuterium. In: Ahlborn B, Hertzberg A, Russell D (eds) Proc. 11th Intl. Symp. Shock Tubes & Waves, Univ. Wash. Press, pp 132-139

Park C (1985) Problems of rate chemistry in the flight regimes of aeroassisted orbital transfer vehicles. In: Nelson HF (ed) Progress in Astronautics and Aeronautics: Thermal Design of Aeroassisted Orbital Transfer Vehicles Vol. 96, AIAA, NY pp. 511-537

Park C (1988) Assessment of a two-temperature kinetic model for dissociating and weakly ionizing nitrogen. Journal of Thermophysics and Heat Transfer 2, 1: 8-16

Rabitz H, Lam S-H (1975) Journal of Chemical Physics 63:3532-3542

Reid R C (1985) A flight test challenge: aeroassist for reusable, space-based transportation. Paper presented at 9th Aerospace Testing Seminar, Los Angles, CA, October

Sharma S (1993) Assessment of nonequilibrium radiation computation methods for hypersonic flows (Invited article), International Journal of Modern Physics C - Physics and Computers (in press). Also NASA TM-103994, January 1993

Sharma SP, Ruffin S M, Gillespie WD, Meyer S A (1992) Nonequilibrium vibrational population measurements in an expanding flow using spontaneous Raman scattering. AIAA Paper 92-2855, July

Sharma SP, Gillespie WD (1991) Nonequilibrium and equilibrium shock front radiation measurements. J. Thermophys. & Heat Transfer 5, 3: 257-265. Also AIAA Paper 90-0139

Sharma SP, Park C (1990) A survey of simulation and diagnostic techniques for hypersonic nonequilibrium flows. J. Thermophys. & Heat Transfer 4, 2: 129-142. Also AIAA Paper 87-0406

Sharma SP (1990) Operating characteristics of a 60- and 10-cm electric arc-driven hock tube-Part I: the drive. J. of Thermophys. & Heat Transfer 4, 3: 259-265. Also AIAA Paper 88-0142

Sharma S (1992) Rotational relaxation of molecular hydrogen at moderate temperatures. AIAA Paper 92-2854, presented at the 27th Thermophysics Conference, Nashville, TN

Shirai H, Park C (1980) Experimental studies of radiative base heating of a Jovian entry model. In: Olstad WB (ed) Entry Heating and Thermal Protection: Progress in Astronautics and Aeronautics Vol. 69, AIAA, NY, pp. 148-171

Teare JD, Georgiev S, Allen RA (1961) Radiation from the nonequilibrium shock front. AVCO Everett Res. Lab. Rep. 112

Yoder MJ (1971) Ph.D. Thesis, University of Michigan

Whiting E E, Park C (1990) Radiative heating at the stagnation point of the AFE vehicle. NASA TM 102829, November

Wilson GJ, Sharma SP, Gillespie WD (1993) Time-dependent simulations of reflected-shock / boundary layer interaction. AIAA Paper 93-0480

Numerical Simulation of Reacting Flows Related to the HEG

K. Hannemann, S. Brück and G. Brenner
DLR - Institute for Fluid Mechanics, Bunsenstraße 10, D-37073 Göttingen, Germany

Abstract. In order to obtain a good quality i.e. parallel and uniform flow in the test section of the High Enthalpy Shock Tunnel Göttingen (HEG) the facility was equiped with an axisymmetric contoured nozzle. The nozzle flow was investigated numerically at an off-design condition using nitrogen as test gas. Numerical simulations assuming the flow to be either in chemical/thermal equilibrium or in chemical/thermal nonequilibrium were carried out. For the case of chemical/thermal equilibrium a 'turbulent' simulation was additionally performed. The development of the turbulent boundary layer along the nozzle wall was compared with the predictions of the Edenfield correlation. The Pitot pressure at the nozzle exit resulting from the 'turbulent' simulation was compared with Pitot-rake measurements taken in HEG. Additional to the nozzle simulations numerical simulations of shock-on-shock interactions, created when an oblique shock impinges on a bow shock formed in front of a circular cylinder were performed to accompany experiments in HEG. To study the basic influence of dissociation, nitrogen has also been used as test gas for these investigations.

Key words: Shock/shock interaction, Nozzle flow, High-enthalpy flow

1. Numerical method

The numerical simulations were performed using the implicit DLR Euler/Thin-Layer Navier-Stokes code NSHYP (Brenner and Prinz 1992, Brück, Brenner and Kortz 1993). The three-dimensional, time-dependent thin-layer Navier-Stokes equations in conservation form for a multicomponent mixture of perfect gases are considered. For the simulations discussed in this paper the fluid is assumed to be nitrogen. The nondimensionalized equations governing the motion of the fluid written in cartesian coordinates are:

$$\frac{\partial \hat{q}}{\partial t} + \frac{\partial \hat{E}}{\partial \xi} + \frac{\partial \hat{F}}{\partial \eta} + \frac{\partial \hat{G}}{\partial \zeta} = \frac{1}{Re_{\infty,L}} \frac{\partial \hat{S}}{\partial \zeta} + Q, \tag{1}$$

where

$$\hat{q} = J^{-1} \left[\rho_{\mathrm{N}_2},\ \rho_{\mathrm{N}},\ \rho u,\ \rho v,\ \rho w,\ e,\ e^{\nu}_{\mathrm{N}_2}\right]^T . \tag{2}$$

The spatial order of the evaluation of the inviscid fluxes $\hat{E}$, $\hat{F}$, $\hat{G}$ is formally of second order and based on the TVD upwind discretisation of Yee (1989) and on a non-MUSCL approach. The source term Q describes the increase and decrease of species concentration and vibrational energy. The temperature dependence of the forward reaction velocities is described by the modified Arrhenius formula $k^f = aT^b exp(-E_a/k_B T)$. Here k_B is the Boltzmann constant. The parameters a and b as well as the activation energy E_a for the relevant nitrogen reactions were taken from Park and Menees (1978). The determination of the vibrational energy of molecular nitrogen is based on the assumption of a harmonic oscillator. The transition of energy between the vibrational modes and the translational modes is approximated using the Landau-Teller formulation. The vibrational relaxation times were obtained from correlations of Millikan and White (1963) and Park (1989).

The calculation of the transport properties is based on the hypotheses of Stokes, Fourier and Fick. The viscosity of the single species is obtained from the curve fits of Blottner, Johnson and Ellis (1971), the mixture viscosity from Wilke's mixture rule. The thermal conductivity is obtained

Shock Waves @ Marseille II
Editors: R. Brun, L. Z. Dumitrescu

based on the Eucken correction. A constant Lewis number of $Le = 1.4$ is used to calculate the diffusion coefficient. Because of the short run time of HEG, the model and the nozzle wall hardly change their temperature during a test. Thus, it was assumed that the fully catalytic walls have a constant temperature of $\overline{T}_{Wall} = 300$ K. At the subsonic inflow of the nozzle the velocity component in streamwise direction u is first extrapolated. Due to the low velocities it is assumed that the flow remains in chemical and thermal equilibrium. To obtain the complete state of the gas it is further assumed that the fluid expands isentropically from the stagnation region to the position of the inflow boundary.

The simulation of the turbulent nozzle flow has been performed using the algebraic turbulence model of Baldwin and Lomax (1978).

Table 1. Freestream conditions

Simulation	# 46	—	# 49	—
v_∞ [km/s]	5.608	5.823	5.823	5.770
ρ_∞ [kg/m^3]	0.001586	0.00164	0.00164	0.00209
T_∞ [K]	739.3	770.0	770.0	946.0
p_∞ [Pa]	370	401	401	652
α_∞ [ ]	0.113	0.136	0.136	0.110
Ma_∞ [ ]	9.877	9.990	9.990	8.83

2. Numerical simulation of the flow in the HEG nozzle

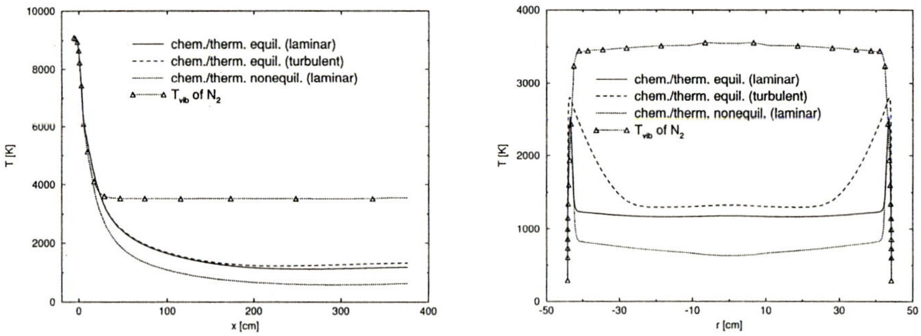

Fig. 1. Temperature distribution along the centerline (left) and at the nozzle exit (right)

The length of the axisymmetric contoured nozzle of the HEG (see Eitelberg, McIntyre and Beck 1992) is $L = 3.75$ m (Hannemann 1990). It has a throat radius of $r = 1.1$ cm and an area ratio of $A_{exit}/A_{throat} = 1600$. The reservoir condition considered here is: $\rho_{res} = 30.6$ kg/m^3, $h_{res} = 16.8$ MJ/kg, $(H_i/RT_0 = 213)$, $p_{res} = 940$ bar, $T_{res} = 9100$ K, $\alpha_{N_2} = 0.864$, $\alpha_N = 0.136$. In Fig.1 the translational/rotational temperature and the vibrational temperature of N_2 are plotted along the centerline of the nozzle and at the nozzle exit. The difference in temperature resulting from the different simulations is based on the fact that, if finite-rate chemistry and vibrational relaxation are taken into account, the chemical reactions as well as the vibrational relaxation freeze during the expansion. At the nozzle exit the translational/rotational temperature resulting from the nonequilibrium simulation amounts to 53 % of the temperature resulting from the equilibrium

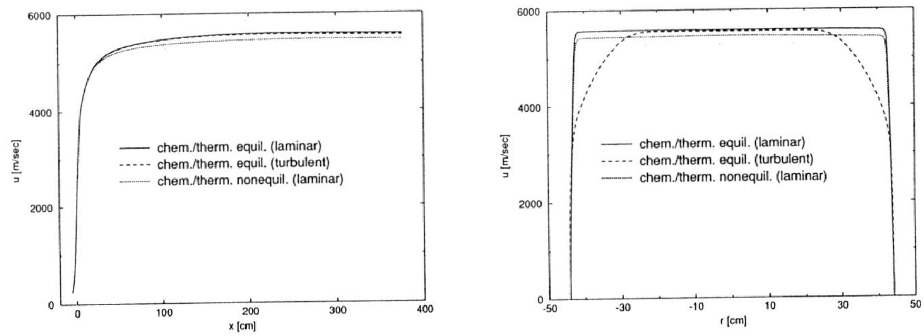

Fig. 2. Velocity distribution along the centerline (left) and at the nozzle exit (right)

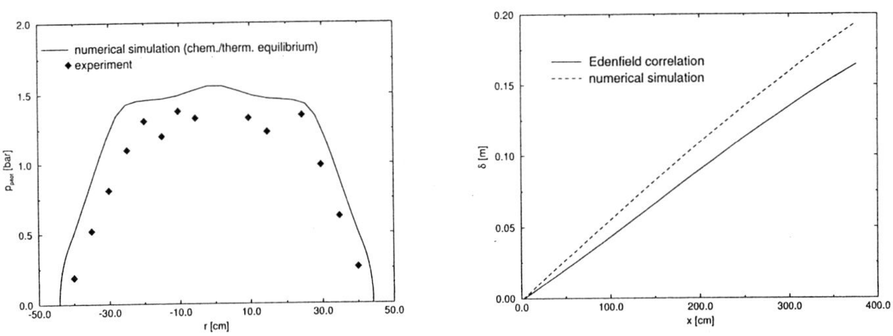

Fig. 3. Comparison of measured and computed Pitot pressure at the nozzle exit (left) and comparison of the boundary layer thickness predicted by the Edenfield correlation and the boundary layer thickness obtained numerically

simulation. For the simulation considering a turbulent boundary layer it was assumed that the transition point lies directly downstream of the throat. At the nozzle exit the boundary layer thickness amounts to approximately 20 cm leaving a uniform core region of 50 cm in diameter.

The u-velocity distributions shown in Fig.2 show that the exit velocity is only reduced by 2 % (115 m/sec) when switching from an equilibrium flow to a nonequilibrium flow. For the present reservoir conditions Pitot rake measurements have been performed in HEG (see Eitelberg 1993). A comparison of these values with the Pitot pressure at the nozzle exit resulting from the numerical simulation assuming the flow to be in chemical/thermal equilibrium is plotted in Fig.3. In the core region the Pitot pressure obtained experimentally is approximately 10 % lower than the numerical data. On the centerline the numerical simulation shows an increase of Pitot pressure caused by a compression wave which focuses at this point. This compression wave is present due to the fact that an off-design condition of the nozzle is considered here. On the right-hand side of the plot the boundary layer thickness is shown and agreement is very good between numerical simulation and experiment. Due to the fact that the experimental Pitot pressure profile is slightly asymmetric the boundary layer thickness is 15 % larger on the left hand side of the plot. In the design procedure of the HEG nozzle (see Hannemann 1990) the final nozzle geometry was obtained by using the empirical boundary layer correlation of Edenfield (1972). For that reason the development of the boundary layer thickness predicted by Edenfield (1972) using the correlation $\delta/x = 1.1(0.195M^{0.375}/Re_{x_\infty}^{0.166})$ was compared with the result of the turbulent nozzle simulation. Keeping in mind that the use of a turbulence model causes uncertainties it can be seen in Fig.3

that the correlation provides a relatively good approximation of the boundary layer thickness obtained numerically. At the nozzle exit the difference amounts to 14 %.

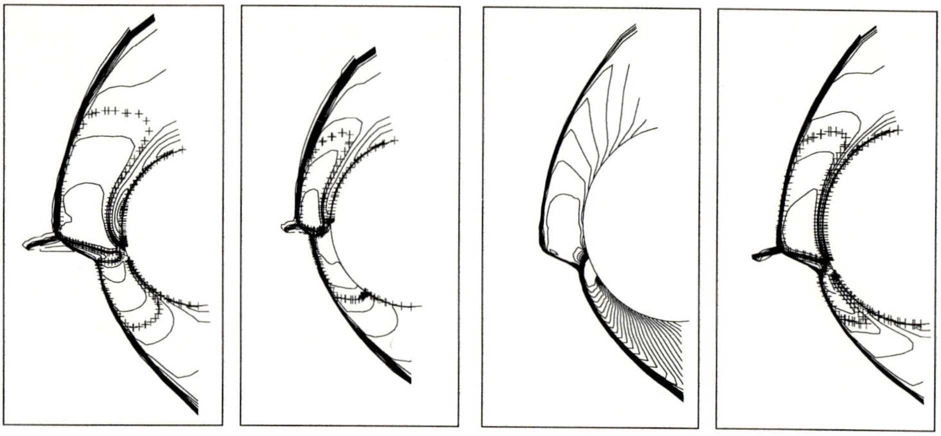

Fig. 4. Case 1: Mach number distribution for a frozen (left) and a nonequilibrium flow (second from left); Case 2: pressure and Mach number distribution (second from right and right)

3. Numerical simulation of shock-on-shock interactions

It is well known that shock-on-shock interactions in front of blunt bodies caused i.e. by the interaction of a ramp shock with an inlet strut or a bow shock with a winglet can lead to severe heat and pressure loads (see e.g. Edney 1968). One effect which has not been taken into account in these investigations is that at hypersonic flight velocities the flowfield around a blunt body is highly influenced by the dissociation effects occurring behind the bow shock. In Fig.4 (case 1) lines of constant Mach number are shown resulting from the numerical simulation of a shock-shock interaction which was classified by Edney (1968) as type IV interaction. The freestream conditions for all simulations discussed here are listed in Table 1.

The left plot reveals the flow field obtained by assuming the chemical reactions to be frozen. It can clearly be seen in Fig.4 that if finite rate chemistry is taken into account the dissociation behind the bow shock tends to decrease the shock stand-off distance. Due to the fact that the position of the impinging shock was held constant the interaction region moved upward. This example shows that the type of interaction can be changed due to dissociation effects. To gain more insight in how much the finite rate chemistry of the dissociation processes influences the various types of shock-on-shock interactions in front of a circular cylinder, experiments have been performed in HEG using nitrogen as test gas (Kortz, McIntyre and Eitelberg 1993). For this purpose a circular cylinder was traversely mounted in the test section together with a 10° wedge type shock generator.

In the experimental procedure heat-flux and wall pressure were measured in the middle part of the cylinder. Additionally, holographic interferometry was used for flow visualization. For comparison, interferograms have also been calculated using the numerical data (see Brück, Brenner and Kortz 1993). Numerical simulations have been performed for three different types of interactions. For two of them refered to as HEG test #46 and #49 experimental data is available. The flowfield resulting from test #46 shows that the impinging shock intersects the upper part of the strong bow shock (see Fig.5). According to Edney's calssification a type IV shock-shock

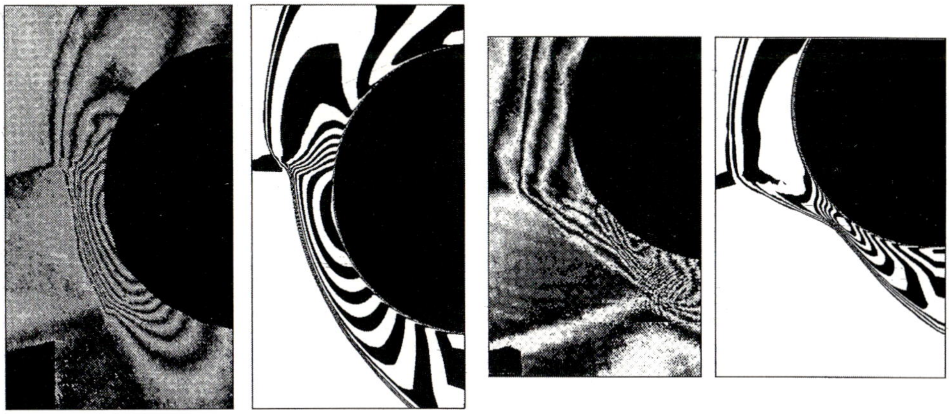

Fig. 5. Comparison of interferograms obtained experimentally and numerically for HEG test #46 (left) and test #49 (right)

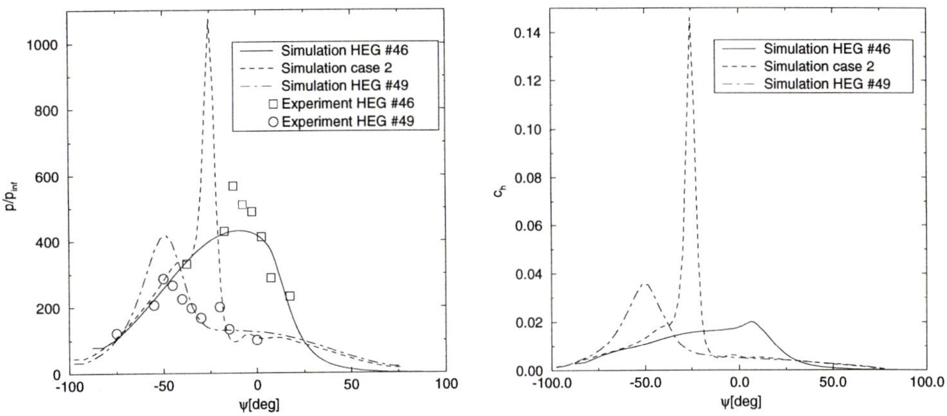

Fig. 6. Pressure distribution (left) and heat flux distribution (right) on the cylinder surface

interaction would be achieved. The strong bow shock is divided in two parts which are linked by a weak shock with a small supersonic region dowstream. The surface pressure distribution shows a maximum which corresponds to the stagnation point. Here the experimental values show a peak pressure which was not detected by the numerical simulation (see Fig.6). The comparison of the measured and calculated interferogram shows that the shock stand-off distances agree very well. Additionally the fringe distribution in the interaction area turns out to be similar.

Compared to the numerical result the experiment reveals a rather long fringe distance behind the bow shock. One possible reason for this discrepancy can be found in the modelling of the finite rate reactions, which may predict too short relaxation times. Note that the uncertainty involved in the reaction rate data is a key problem when simulating reactive hypersonic flows. Test #49 reveals a type III shock-shock interaction. Only in the upper part of the flow field a subsonic region is present. The shear layer between this area and the supersonic flow behind the weak shock is curved around the lower cylinder surface. It impinges on the surface in the area of the stagnation point. The interferograms in Fig.5 reveal a qualitative agreement. The experimental results show in the lower part a bow shock which is not curved like the bow shock in the numerical simulation. However, behind the bow shock similar structures are present. As for test #46 the influence of the expansion resulting from the end of the wedge can again be seen in the experimental interferogram. Due to the fact that this interaction is embedded in supersonic flow its influence is only local. Again, the surface pressure distribution (Fig.6) shows qualitatively the same behaviour as the experimental values. Due to the fact that the differences of the flow structures between test #46 and test #49 are relatively large, a third complementary numerical simulation (case 2) has been performed using an intermediate incident shock position.

As can be seen in Fig. 4, a typical type IV shock-shock interaction occurs for the conditions of case 2. Due to the small shock stand-off distance a well-developed supersonic jet does not exist. However, the wall pressure distribution shows a peak at the stagnation point which is now much higher than the peaks revealed by the interactions discussed previously. The heat flux coefficient plotted in Fig.6 shows for all investigated flow fields the same behaviour as the pressure distribution.

Acknowledgements

The authors wish to thank Dr. G. Eitelberg and S. Kortz for the stimulation to investigate the shock-on-shock interactions and for many discussions. Further we have to thank them for making the experimental data available to us.

References

Baldwin BS, Lomax H (1978) Thin-layer approximation and algebraic model for separated turbulent flows. AIAA Paper 78-257

Brenner G, Prinz U (1992) Numerical simulation of chemical and thermal nonequilibrium flows after compression shocks. AIAA Paper 92-2879

Brück S, Brenner G, Kortz S (1993) Numerical simulation of shock-shock interactions in nonequilibrium flow. DLR-IB 221-93 A 09

Edenfield EE (1972) Design of a high Reynolds number Mach number 8 countoured nozzle for the hypervelocity wind tunnel (F). AEDC-TR-72-48

Edney B (1968) Anomalous heat transfer and pressure distributions on blunt bodies at hypersonic speed in the presence of an impinging shock. FFA Report 115, The Aeronautical Research Institute of Sweden

Eitelberg G, McIntyre TJ, Beck WH (1992) The High Enthalpy Shock Tunnel in Göttingen. AIAA Paper 92-3942

Eitelberg G (1993) Experimental verification of real gas effects in high enthalpy flows. These Proceedings

Hannemann K (1990) Design of an axisymmetric contoured nozzle for the HEG. DLR-FB 90-04

Kortz S, McIntyre TJ, Eitelberg G (1993) Experimental investigation of shock-on-shock interactions in the High Enthalpy Shock Tunnel Göttingen (HEG). These Proceedings

Millikan RC, White DR (1963) Systematics of vibrational relaxation. J. Chem. Physics 39

Park C, Menees GP (1978) Odd nitrogen production by meteroides. J. Geophysical Research 83, 68

Park C (1989) Nonequilibrium Hypersonic Aerothermodynamics. John Wiley, New-York

Yee HC (1989) A class of high-resolution explicit and implicit shock-capturing methods. VKI Lecture Series 1989-04

Measurement of Heterogeneous Recombination Probabilities in Shock Tube Simulation of Space Shuttle-Type Reentry

N.N. Koudriavtsev
Moscow Institute of Physics and Technology, Moscow Region, Dolgoprydny, 141700, Russia

Abstract. A critical review of shock tube techniques for measuring the heterogeneous recombination probability of atoms on space vehicle surfaces while modeling natural re-entering conditions is presented. Results of the determination of the heterogeneous recombination probability for various types of thermal protection systems of Space Shuttle-type vehicles are given and analyzed.

Key words: Heterogeneous recombination probability, Thermal protection system, Space Shuttle

1. Introduction

Shock tube facilities are widely used for gasdynamics and high temperature kinetics experiments. It is convenient to use the shock tube technique for modelling the heat transfer to the thermal protection systems of Space Shuttle-type vehicles during reentry into the rarefied Earth atmosphere in the altitude range H=50-80 km, where both gasdynamic and chemical kinetic effects are responsible for thermal loading.

2. Theory

The heat flux to the surface of a solid body in high enthalpy dissociated gas flow consists of two terms (neglecting radiative heat transfer): convective - Q_{conv} and chemical - Q_{chem}, arising due to the surface heat release by the heterogeneous recombination of atoms. The total heat flux to the surface depends on the heterogeneous recombination probability (HRP $-\gamma_w$) if the Damköhler number for heterogeneous recombination Da_w is in the range:

$$10^{-3} < \mathrm{Da}_w < 5.0, \tag{2.1}$$

$\mathrm{Da}_w = \tau_d/\tau_w$, where τ_d is the time of particle diffusion across the boundary layer, and τ_w the characteristic time of surface recombination.

If the Damköhler number Da_w is less than 10^{-3} the heat flux arising due to heterogeneous recombination is negligible in comparison with the convective one. In the opposite limiting case of high Damköhler number: $\mathrm{Da}_w > 5.0$ all atoms near the surface are able to recombine and generate a significant recombination heat flux, but thereby no dependence on heterogeneous recombination probability is revealed.

Within the range of Damköhler numbers of Eq.2.1, that approximately corresponds to heterogeneous recombination probability of oxygen and nitrogen atoms: $10^{-4} < \gamma_w < 0.3$ ($\mathrm{Da} \sim \gamma_w$) the value of γ_w may by reliably obtained from the heat flux measurements. This range of γ_w is important for the Space Shuttle applications, because of the recombination catalytic activity of its thermal protection systems - High Temperature Reusable Surface Insulation with borosilicate glass coating and reinforced carbon-carbon at real surface temperatures during vehicle reentry in the range $T_w = 1000 - 1800$ K.

Shock Waves @ Marseille II
Editors: R. Brun, L. Z. Dumitrescu

3. Generation of high-enthalpy dissociated gas flows

In shock-tube methods a hypersonic, strongly dissociated gas flow is created behind the incident shock wave (SW) propagating in the shock tube channel (ISW-Incident Shock Wave method) (Nasser and East 1980, Berkut et al. 1985), or after expansion through a supersonic nozzle of the twice heated and compressed gas in the stagnation region behind the SW reflected from the end-plate (SE-supersonic expansion method) (East et al. 1980, Doroshenko et al. 1991).

For maximum sensitivity in the measurement of heterogeneous recombination probabilities the diatomic gas has to be fully dissociated. A typical duration of the quasi-stationary gas flow is 100-200 μs for the ISW method and 200-1000 μs for the SE method. To remove the influence of gas-phase chemical kinetics on heterogeneous recombination probability the boundary layer near the solid body has to be chemically frozen. The evaluation of the value of the Damköhler number for gas-phase chemical reactions guaranteeing their freezing is (Nasser and East 1980):

$$\mathrm{Da}_g < 3 \times 10^{-4} \qquad (3.1)$$

$\mathrm{Da}_g = \tau_d/\tau_g$, where τ_g is the characteristic time of gas-phase chemical reactions. Evaluations according to Eq.3.1 give the upper limit of the pressure near the surface at about 100-200 Torr for the recombination of nitrogen, oxygen and hydrogen atoms.

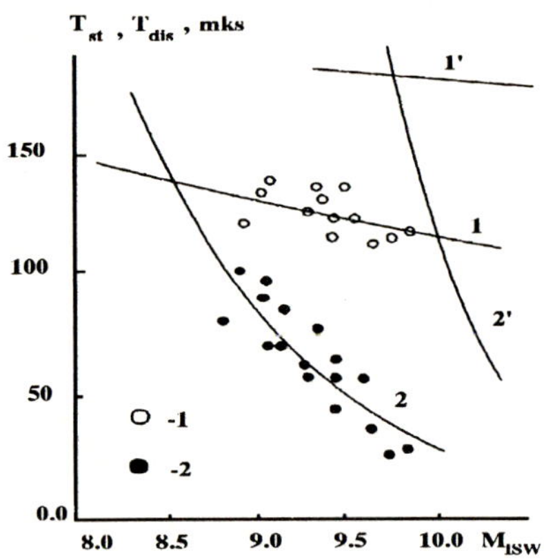

Fig. 1. Theoretical calculation and experimental determination of quasi-steady flow time (1- for oxygen, and 1'- for nitrogen- containing mixtures) and the characteristic dissociation time (2- for oxygen, and 2'- for nitrogen- containing mixtures) *vs.* incident shock Mach number M_{isw}. Initial compositions of the mixtures: 20% O_2 + 80% Ar, and 10% N_2 + 90% Ar, initial pressure in the shock tube 1 torr

3.1. Incident shock wave method

In this method the test model is flowing supersonically behind the incident shock wave propagating in the shock-tube channel. The pressure limit near the test surface providing frozen boundary layer restricts the initial pressure in the shock tube to a value of about 1.0 Torr. At such low initial pressures the elementary one-dimensional theory of shock tube operation fails because of the interaction of the inviscid core flow with the boundary layer behind the SW. As a result the SW decelerates and the contact surface accelerates to diminish the length of the heated gas volume so the time of quasi-stationary flowing of the model will decrease.

The results of theoretical calculations and experimental determination of the quasi-stationary flow time in conditions required for nitrogen and oxygen atoms recombination measurements are represented in Fig.1 vs. incident SW Mach number M_{isw}. In this figure the characteristic dissociation time τ_{dis} behind the incident shock wave for nitrogen and oxygen molecules is also shown. The quasi-stationary flow time decreases slightly with M_{isw}, whereas the dissociation time τ_{dis} decreases with M_{isw} very sharply due to the strong accceleration of the dissociation processes with temperature and pressure at higher Mach numbers.

From Fig.1 it is evident that at SW Mach numbers $M_{isw} > 8.5$ for oxygen and $M_{isw} > 10$ for nitrogen the time of quasi-stationary flow τ_{st} is greater than the characteristic dissociation time τ_{dis} and so during a time interval $\delta\tau = \tau_{st} - \tau_{dis}$ we have a flow over the model of a quasi-stationary fully dissociated gas. The typical duration of quasistationary flow $\delta\tau$ in the incident shock tube method is 100-300 μs.

3.2. Supersonic expansion method

In this method the gas is twice compressed and heated behind the incident and reflected shock waves and then expanded in a supersonic nozzle. The duration of quasi-stationary hypersonic flow is controlled by the stagnation region being destroyed by disturbances generated in the interaction of the reflected SW and contact surface and propagating into the stagnation region, as well as by the expansion wave coming from the nozzle exit and propagating in the opposite direction. Typically, the duration of quasi-stationary gas flow $\delta\tau$ is about 200-500 μs.

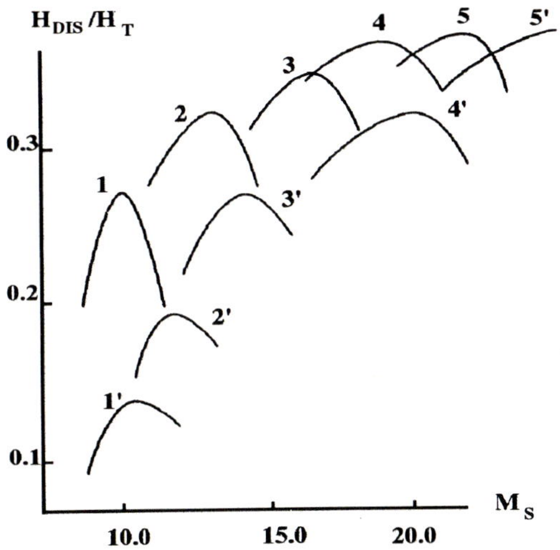

Fig. 2. The ratio of dissociation enthalpy H_{dis} to total gas enthalpy H_t *vs.* incident SW Mach number M_{isw} for mixtures of oxygen + argon (Curves 1-5), and nitrogen + argon (Curves 1'-5'). For the oxygen mixtures Curve 1 corresponds to 20% O_2 contamination, 2- 40%, 3- 60%, 4- 80%, 5- pure O_2. For nitrogen mixtures: 1- 5%, 2- 10%, 3- 20%, 4- 40%, 5- pure N_2. Pressure in the tube is 1 torr

Due to twice heating the gas, the required incident SW velocity for providing complete dissociation may be lower in comparison with the ISW method. A specially countoured supersonic nozzle is used to produce the hypersonic flow of dissociated gases with constant cross-section parameters. The recombination of atoms in the expanding gas flow is to be frozen. This is achieved by restricting the nozzle throat height and the pressure in the stagnation region.

In both methods dilution of diatomic molecules by high-weight noble gases such as Ar is used. This allows to reduce the required shock wave velocity and also to improve shock tube operation: increasing the duration of quasi-stationary flow generation and diminishing the spatial fluctuations in the dissociated gas. Typical concentration of diatomics in the mixture is 5.0-20%. Sensitivity decrease because of the atom contamination restriction is not serious. In Fig.2 the ratio of dissociation enthalpy H_{dis} to total gas enthalpy H_t is represented vs. incident SW Mach number

M_{isw} for oxygen + argon and nitrogen + argon mixtures with diatomic contamination varying from 5 to 100%. This ratio defined the recombination surface heat flux and so the sensitivity of the determination of heterogeneous recombination. Results of Fig.2 confirm the possibility of diminishing the incident SW velocity in diluted mixtures without seriously decreasing the HRP determination sensitivity. For each mixture an optimal ISW velocity or Mach number- M_{opt} exists. At $M < M_{opt}$ the gas is not fully dissociated; conversely, at $M > M_{opt}$ the total enthalpy of the gas is increased and so is the convective part of the surface heat flux.

Table 1. Incident shock-wave method

Parameters	Initial	Behind ISW	Q_{wt}, W/cm^2	Q_{wr}, W/cm^2
Pressure, torr	0.5–2.0	50–250	80–400	4.0–100
Temperature, K	300	4000–9000		
Mach No of ISW or flow	8.0–20	2.0–8.0		

Table 2. Supersonic expansion method

Parameters	Initial	Behind RSW	After SE	At BL outer limit	Q_{wt} W/cm^2	Q_{wr} W/cm^2
Pressure, torr	0.1–50	$10^2 - -10^4$	3.0–10.0	10.0–50.0	50–200	1.0–50
Temperature, K	300	5000–10000	500–800	2500–5000		
Mach No. ISW flow	6.0–15.0	0	6.0*	1.5*		

ISW - incident SW; RSW - reflected SW; SE - supersonic expansion; BL - boundary-layer; *optimal conditions; Q_{wt}, Q_{wt}, total and recombinative heat fluxes

4. Tested samples and heat flux measurements

Test models in the form of flat plates with sharp leading edges are usually used in shock tube experiments. The dimensions of these models are several centimeters in the ISW method (limited by shock tube channel cross section) and approximately one order of magnitude greater in the SE method, providing testing of thermal protection system elements of real vehicles, which have typical dimensions of 15 x 15 cm. For preheating the samples to temperatures of 1000-2000 K in order to simulate re-entry conditions, a powerful continuous laser (Berkut et al. 1985) or an electrical heater (Doroshenko et al. 1991) are used. The sample is fixed in an aerodynamic holder situated in the shock tube channel. The surface temperature increase during interaction with the high-enthalpy hypersonic gas flow does not exceed several Kelvin's in shock-tube pulsed experiments, which is less than 0.1% of the initial temperature. So, high sensitivity, time-resolved techniques are to be used to measure surface temperatures for retrieving the surface heat flux.

The surface temperature is measured by two techniques: thin film thermocouple probes (Nasser and East 1980, East et al. 1980), or optically, by recording the increase of radiation from a thin surface layer due to its heating during pulsed experiments (Berkut et al. 1985, Doroshenko et al. 1991). To achieve high temporal resolution on the microsecond time-scale thermocouple gauges must have a thin film metal strip of about 0.1 μm thickness deposited on an insulated plate as the sensitive element. Then a coating of electrically insulating material such as quartz with approximately the same thickness is deposited above it. Finally a thin film of the investigated material, also about 0.1 μm, is deposited on this insulating layer. The resolution time of these gauges is about several microseconds (Nasser and East 1980, East et al. 1980).

These thermocouple gauges may be utilized when the surface temperature does not exceed 500-600 K. At higher temperatures the adhesive properties of these thin layers are not sufficient to mechanically sustain this complex system. In addition, the electrical contact problems of the sensitive thin metal layer and/or between thin metal layers and connecting wires are very serious due to different thermal expansion of metal and isolating films, and their are destroyed during experiments.

Table 3.

Material	P_{bl}, torr	T_{bl}, K	U_{bl}, km/s	Q_w, W/cm^2	γ_w
Pt	79–90	3.0–4.0	2.6–2.8	150–200	$(1.3\pm0.4)10^{-2}$
Al_2O_3	82–86	3.1–3.3	2.8–2.9	134–139	$(2.5\pm0.7)10^{-2}$
SiO	75–96	3.2–5.5	2.4–2.8	111–162	$(2.0\pm0.5)10^{-2}$
Quartz	85–87	4.7–4.8	2.6–2.7	125–135	$1.0\ 10^{-3}$

P_{bl}, T_{bl}, U_{bl} are parameters at the outer limit of the boundary layer

It should be also mentioned, that the recombination catalytic activity of the surfaces of the bulk materials and deposited thin films of the same materials may be significantly different. Hence, serious limitations exist in utilizing thin-film thermocouple probes in shock tube experiments on heterogeneous recombination measurements at high surface temperatures. In (Berkut et al. (1985), Doroshenko et al. (1991) an optical method of nonstationary surface temperature measurements has been developed and is utilized in routine experiments for heterogeneous recombination measurements (Berkut et al. 1989a, b, 1986).

It was theoretically shown, that, neglecting radiation scattering in solid samples, the nonstationary radiation intensity from the surface in a quasi-stationary high-enthalpy gas flow is proportional to the time of flow exposure $I \sim Q_w \tau_{ex}$ if $K_\lambda \delta \ll 1$, or $I \sim Q_w \tau_{ex}^{1/2}$ in the opposite limit case $K_\lambda \delta \gg 1$, where Q_w is the surface heat flux, τ_{ex} the duration of surface exposure to gas flow, K_λ the spectral absorption coefficient, δ the thickness of the heated layer in pulsed experiments. From these relations, the value of the surface heat flux Q_w may be retrieved on the basis of measurements of nonstationary increase of the radiation intensity from the surfaces.

The calibration procedure of these heat transfer gauges may be performed in two ways. In the first method two of those gauges are situated in practically the same place on the solid sample with one of them being reference coated with a material providing high $\mathrm{Da}_w > 5.0$, or low $\mathrm{Da}_w < 10^{-3}$. For these reference gauges the heat flux may be easily calculated and no information about surface catalytic activity is required. This calibration procedure is often used for thermocouple gauges (Nasser and East 1980, East et al. 1980). In the second method a special calibration procedure completing each test is to be done. In this calibration procedure the surface, maintained at the same parameters as in the test is blown with a noble gas, when the surface heat flux is easily determined theoretically. This calibration procedure is used for the optical method (Berkut et al. 1989a, b, 1986).

Typical shock-tube operating parameters for the measurement of nitrogen, oxygen and hydrogen atom heterogeneous recombination probabilities are presented in Tables 1 and 2.

5. Experimental results

Results of oxygen-atom HRP measurements at surface temperature $T_w = 300$ K are presented in Table 3.

Results of catalytic activity testing of the "Buran" thermal protection system coating (borosilicate glass with ingredients of SiB_4) are represented in Fig.3 as well as Space Shuttle coating, measured in plasma jet facilities (Scott 1980, Steward and Rakich 1982), and flight experiments (Rakich et al. 1982). Results for pure borosilicate glass are also represented (Wood and Wise 1961) for comparison.

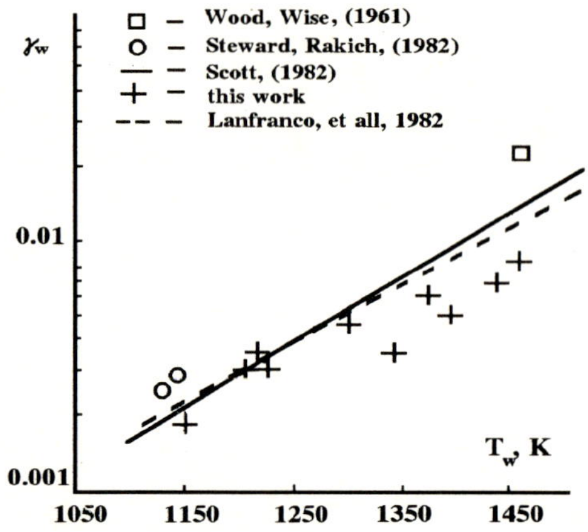

Fig. 3. Results of high-temperature catalytic activity measurements on the coating of space vehicles

References

Berkut VD, Kovtun VV, Koudriavtsev NN, Novikov SS, Sharovatov AI (1985) Determination of the time-resolved probabilities of heterogeneous recombination of atoms in shock-tube experiments. Intl. J. Heat and Mass Transfer 12, 5: 567-575

Berkut VD, Koudriavtsev NN, Novikov SS (1986) Surface thermophysical properties due to the accommodation of chemical energy of supersonic dissociated gas flow. Report of Inst. for High Temperatures, Moscow (in Russian)

Berkut VD, Doroshenko VM, Kovtun VV, Koudriavtsev NN, Novikov SS, Smirnov NV, Sharovatov AI (1989a) Measurements of the atom heterogeneous recombination probabilities on heated surfaces under supersonic stream conditions. Sov. J. Chem. Phys.(in Russian) 8, 10:1345-1353

Berkut VD, Doroshenko VM, Zhdanok SA, Kovtun VV, Koudriavtsev NN, Novikov SS, Sharovatov AI (1989b) Determination of probability of atom heterogeneous recombination on heated to high temperatures (T_w =1000 K) surfaces of solid bodies flowed by shock wave dissociated gas. Report of Institute of Heat and Mass Transfer, Belorussian Academy of Science, Minsk (in Russian) 8, 30

Nasser EAM, East RA (1980) A shock tube investigation of heat transfer from dissociated hydrogen to catalytic surfaces. Intl. J. Heat and Mass Transfer, 23, 4: 515-521

Doroshenko VM, Koudriavtsev NN, Smirnov NV (1991) Heat release on a solid surface in hypersonic flow and recombination coefficient measurements. Proc. 13-th International Colloquium on Dynamics of Explosions and Reactive Systems, Nagoya, Japan, p.224

East RA, Stalker RJ, Baird JP (1980) Measurements of heat transfer to flat plate in a dissociated high enthalpy laminar air flow. J. Fluid Mech. 97, 4: 979-999

Rakich JV, Steward DA, Lanfranco MJ (1982) Results of a flight experiment on the catalytic efficiency of the Space Shuttle heat shield. AIAA Paper 82-944

Scott CD (1980) Catalytic recombination of nitrogen and oxygen on high temperature reusable surface insulation. AIAA Paper 80-1477

Steward DA, Rakich JV (1982) Catalytic surface experiments on the Space Shuttle. AIAA Paper 82-186

Wood BJ, Wise H (1961) The interaction of atoms with solid surfaces. In: Proc. Raref. Gas. Dynam. Symposium N.-Y., Academic Press p.51-59

Front Structures of Strong Shock Waves in Air

J. Koreeda*, H. Yanagisawa*, K. Maeno*, H. Honma*, S.A. Bystrov†, V.I. Ivanov‡ and F.V. Shugaev‡

* Dept. of Mechanical Engineering, Chiba University, Yayoi-cho 1-33, Inage-ku, Chiba 263, Japan
† Dept. of Microelectronics, Nuclear Research Institute, Moscow State University, Moscow 119899, Russia
‡ Dept. of Physics, Moscow State University, Moscow 119899, Russia

Abstract. Front structures of strong shock waves in air were experimentally investigated by means of the laser Schlieren technique. Shocks in air at 13 Pa were generated with Mach numbers ranging from 2 to 34. The detector response to a single pulse of a ruby laser was obtained. This enabled the response correction of the Schlieren signals and the correction is greater at high Mach numbers ($M > 16$). Density profiles were obtained by integrating the corrected Schlieren signals. We present data of the reciprocal shock thickness for the investigated Mach numbers.

Key words: Shock structure, Laser Schlieren, Strong shocks

1. Introduction

Shock structures have been studied theoretically, for example, by solving the Boltzmann equation. Experimentally, one can measure the intensity of a laser beam deflected or diffracted from the shock front, for example. In the latter case, the intensity and position of the incident laser beam on the detector depends on the form of the index of refraction within the shock wave. From the mathematical point of view there are difficulties associated with the so-called ill-posed problem. Tikhonov and Arsenin (1986) showed that in some cases the ill-posed problem can be solved by a regularization method. In the present paper, we obtained Schlieren signals of strong shock waves in air. The ill-posed problem was not solved, but a more simplified analysis was made. The density profile in the shock front was obtained by a direct integration of the Schlieren signal with response correction. As it will be shown, the inclusion of the detector response in the experimental Schlieren signal is of fundamental importance when analyzing strong shock waves, or other fast flow disturbances in general.

At high Mach numbers ($M > 16$) there is a lack of experimental data concerning the shock structure. We conducted experiments to investigate aspects of the shock structure, like the density profile and the shock thickness at high Mach numbers.

2. The experiment

A free-piston, double diaphragm shock tube generates the shock waves. Shock Mach numbers range from 2 to 34 in low density air (13 Pa). The test section has a square cross section of 4×4 cm. To investigate the shock structure we considered the laser Schlieren technique as illustrated in Fig.1. A He-Ne laser beam with Gaussian profile (max. power= 10 mW, $\lambda = 632.8$ nm, beam dia.= 0.8 mm) intersected the test section and was detected by a detector with a double split photodiode. Then, the Schlieren signal was recorded in a digital oscilloscope and the data fed into a personal computer for data analysis. Kiefer and Hajduk (1980) and Dove and Teitelbaum (1978) presented discussions about laser Schlieren detectors and data analysis including diffraction and interference effects in the laser beam.

One important aspect in the analysis of strong shock waves by the laser Schlieren technique is the detector response. For shocks with "low" Mach numbers ($M \approx 2$) the detector showed to be fast enough to obtain the correct signal. But in the case of strong shock waves ($M > 16$) the signal must be obtained in a few μs and therefore the response time becomes a relevant factor.

Shock Waves @ Marseille II
Editors: R. Brun, L. Z. Dumitrescu

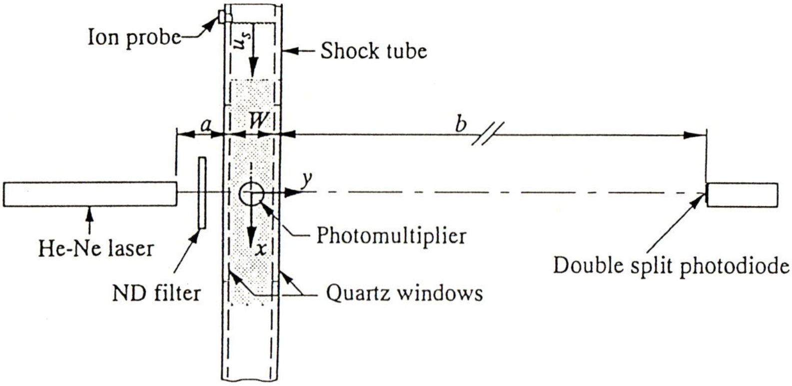

Fig. 1. Laser Schlieren layout

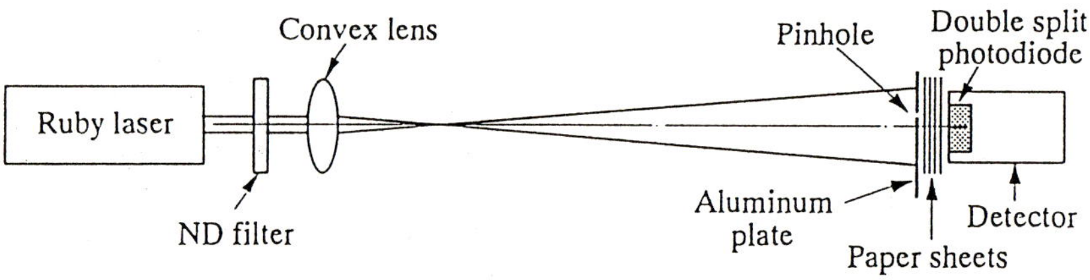

Fig. 2. Detector response experiment

Hence, a detailed analysis of the detector response function becomes necessary. To accomplish this, we examined the detector response due to a single pulse of a ruby laser (max. power= 1 MW, $\lambda = 694.3$ nm, pulse duration (half peak width)= 80 ns). The layout of this experiment is shown in Fig.2. An aluminum plate with a pinhole was placed in front of the detector. In this way, the laser pulse hit only one of the halves of the split photodiode. As the output signal of the detector is the amplified difference between the two halves of the split photodiode, shooting the laser pulse on one of these halves simulated a fast Schlieren signal. To avoid damaging the detector, the pulse diameter was greatly enlarged and sheets of paper were used to attenuate the incident light. The laser power was set to a minimum and air breakdowns were minimized.

Before the shock experiments, a calibration curve was obtained by sliding the detector parallel to the shock tube, i.e., in the x direction and then recording the micrometer reading and the detector signal. This curve is shown in Fig.3. All the Schlieren signals were within the linear range of this curve.

3. Data analysis

3.1. Detector response

In Fig.4, we have the output signal of the detector due to a single pulse of the ruby laser. The axial power distribution of the incident laser pulse was assumed to have a Gaussian shape. Let us observe the shapes of the incident laser and the detector output. We see that the detector response for the "climbing" part is very good. But the detector response in the decreasing portion of the

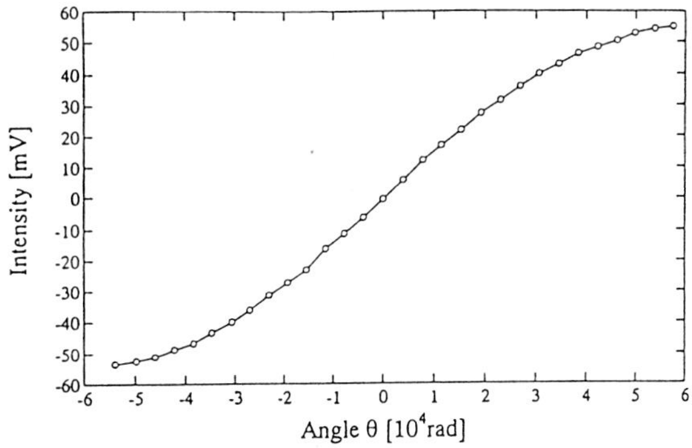

Fig. 3. Calibration curve of the detector. a = 70 mm, b = 1280 mm, intensity reduction of ND filter = 1/8

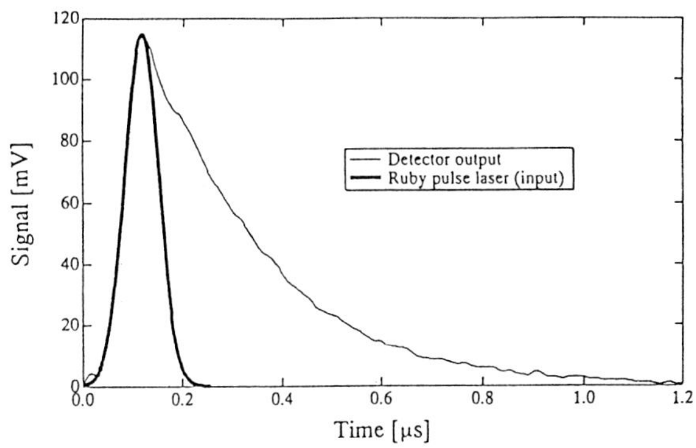

Fig. 4. Detector response to a single laser pulse

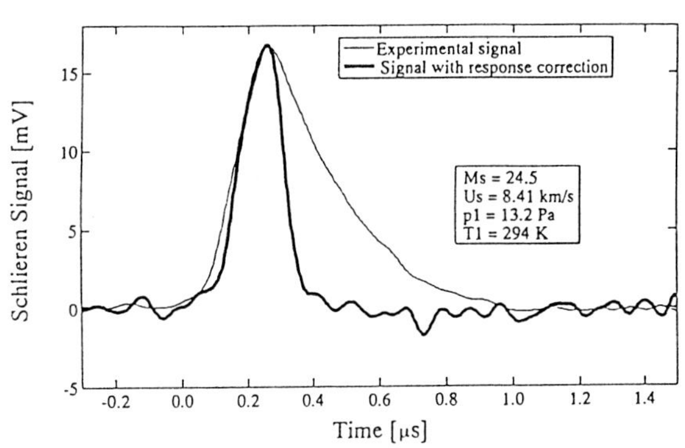

Fig. 5. Response correction of the Schlieren signal

curve is not so good. Nevertheless, the correct signal can be obtained if we know the relation between the incident and the output signal. Let us consider the following differential equation:

$$dY/dt + \alpha Y = F(t) \tag{1}$$

where $Y(t)$ is the detector output signal, $F(t)$ is the response-corrected Schlieren signal and α is the reciprocal time constant of the detector. Considering the laser pulse signal as a "fast" Schlieren signal, then we can find α such that the detector output signal is curve-fitted by the calculated $Y(t)$. In this way we found $\alpha = 4.2\ \mu s^{-1}$, that is, the detector response time is $1/\alpha = 0.238\ \mu s$.

The Schlieren signals $Y(t)$ of the shock waves (detector outputs) were data smoothed and fed into Eq.1 and then we obtained the response-corrected Schlieren signals $F(t)$. Fig.5 shows an example for shock Mach number $M_s = 24.5$. Shocks around $M_s = 2$ gave rise to Schlieren signals practically invariant under Eq.1, but at higher Mach numbers the experimental signals were "squeezed" after the response correction.

3.2. Density profile

As the distance b in Fig.1 varied between 900 mm and 1280 mm, the diffraction of the laser beam was not taken into account. To calculate the density profile we assumed that the laser beam is deflected as a single light ray. In this case, and also assuming a plane shock, the density can be obtained from the integration of the density time derivative

$$d\rho/dt = \theta u_s/KW \tag{2}$$

where ρ is the density, ρ_1 is the density ahead of the shock, θ is the deflection angle of the laser beam, K is the Gladstone-Dale constant, W is the distance traversed by the laser beam inside the test section and u_s is the shock velocity. The angle θ is related to the intensity of the Schlieren signal by using the calibration curve shown in Fig.3.

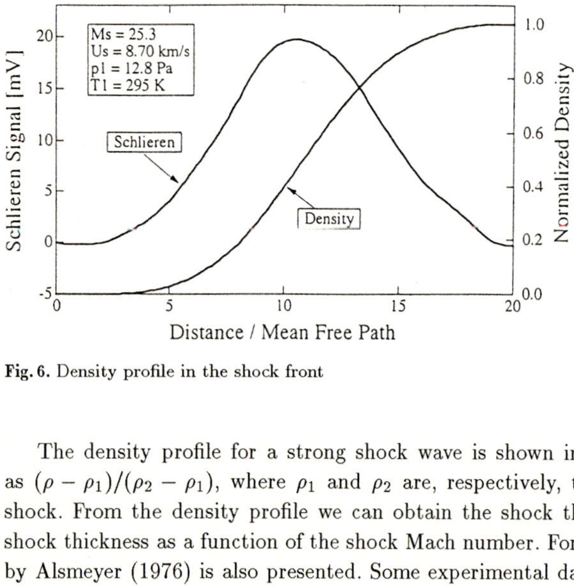

Fig. 6. Density profile in the shock front

The density profile for a strong shock wave is shown in Fig.6. The ordinate is normalized as $(\rho - \rho_1)/(\rho_2 - \rho_1)$, where ρ_1 and ρ_2 are, respectively, the densities ahead and behind the shock. From the density profile we can obtain the shock thickness. Fig.7 shows the reciprocal shock thickness as a function of the shock Mach number. For comparisons, the curve fit obtained by Alsmeyer (1976) is also presented. Some experimental data at "low" Mach numbers are also plotted to give support to the results pertaining to strong shocks.

For pressures ahead of the shock higher than 26 Pa, apparently the shocks were much thicker than the results for 13 Pa. This is because the finite laser beam diameter becomes of the same order of magnitude as the shock thickness. Hence, it was important to generate the shock waves in 13 Pa.

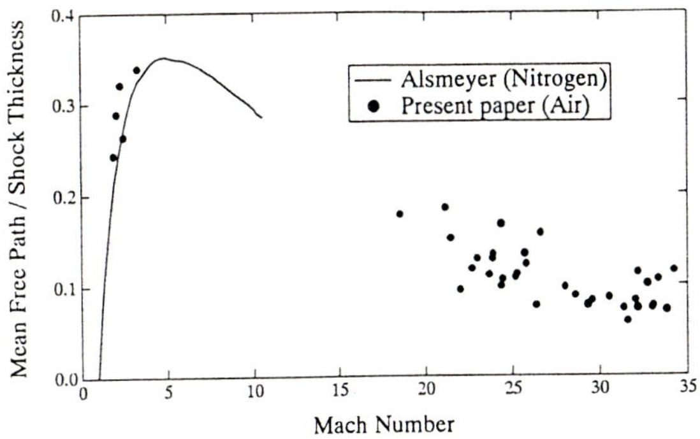

Fig. 7. Reciprocal shock thickness versus shock Mach number

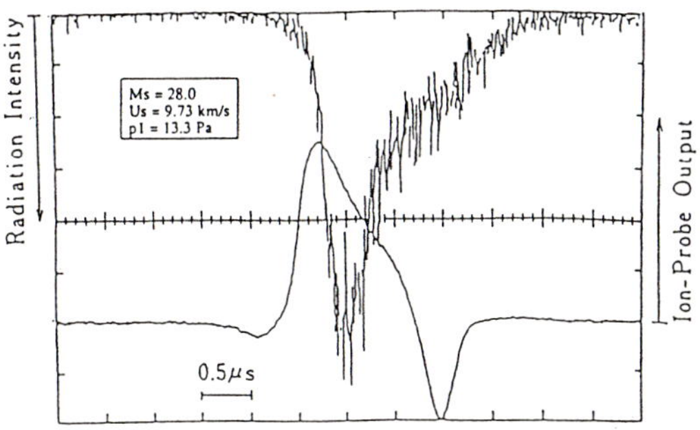

Fig. 8. Profiles of radiation intensity and ion-probe output

4. Final remarks and perspectives

In this paper we obtained the thickness of shocks with Mach numbers between 16 and 34. The analysis of the Schlieren signal was simple but took into account the detector response. Furthermore, if the detector response is not considered, it is not possible to obtain correct density profiles at high Mach numbers. As the Schlieren signal is proportional to the density gradient, information can be reliably obtained when analyzing front structures of shock waves.

In the case of weak shock waves, curved shock fronts can appear. As the shock number increases, the shock front becomes flat. The results presented here were obtained supposing plane shocks. This is the case of strong shocks and it can be confirmed by the shape of the Schlieren signal, that is, there is no distinguishable "negative" spike due to laser energy removal due to the passage of the curved shock, as predicted by Dove and Teitelbaum (1978).

Here, the density profile of strong shocks was experimentally obtained. There were no difficulties posed by the natural radiation emanating from the hot air behind the shock. Using the laser Schlieren technique it was possible to define clearly the shock front. This enables us to investigate the position of the radiation profile in the shock front. In previous studies, Honma and Iizuka (1992, 1991) and Iizuka and Honma (1991) observed radiation emission behind strong shock waves in air by using an image converter camera and a pair of photomultipliers. We also compared the results to numerical analysis. In these studies there was a difficulty in determining the position of the shock front in relation to the radiation profile. Also, for our experimental conditions, some theoretical analyses suggest that the shock thickness and the pulse width of the radiation just behind the shock front have the same order of magnitude. Furthermore, the translational nonequilibrium zone merges with the vibrational and the electronic nonequilibrium ones, where the later is related to radiative characteristics.

Fig.8 exhibits the signals from an ion probe and a photomultiplier. If we consider the steep rise of the ion probe as corresponding to the shock front, we can easily define the front. However, this rise does not correspond to the front in an exact sense. We need to catch the profile of electrically neutral species in the shock front. Simultaneous observations of density and radiation profiles are in progress.

References

Alsmeyer H (1976) Density profiles in argon and nitrogen shock waves measured by the absorption of an electron beam. J. Fluid Mech. 74:497-513

Dove JE, Teitelbaum H (1978) Interaction of a laser beam with a curved shock front; the location of the time origin in laser Schlieren experiments. In: Ahlborn B, Hertzberg A, Russell D (eds) Proc. 11th Intl. Symp. on Shock Tubes and Waves, Seattle, pp 474-481

Honma H, Iizuka H (1991) Experimental and numerical studies of radiation emission from high-temperature air behind 10 km/s shock waves. SAE Technical Paper Series 912025, reprinted from Intl. Pacific & Space Tech. Conf. and 29th Aircraft Symp. Proc., Gifu, Japan

Honma H, Iizuka H (1992) A shock tube study on radiation emission from high-temperature air in hypersonic flows. Aerothermochemistry of Spacecraft, IUTAM, Marseille

Iizuka H, Honma H (1991) A numerical analysis for radiation profile behind strong waves about 10 km/s in low-density air, In: Takayama K (ed) Proc. 18th Intl. Symp on Shock Waves, Sendai, pp 711-716

Kiefer JH, Hajduk (1980) Rate measurement in shock waves with the laser-Schlieren technique. In: Lifshitz A, Rom J (eds) Proc. 12th Intl Symp. on Shock Tubes and Waves, Jerusalem, pp 97-110

Tikhonov AN, Arsenin VY (1986) The methods of solution of ill-posed problems. Nauka, Moscow (In Russian)

Non-Equilibrium Flow Behind Strong Shock Waves in Air

A. Viviani, R. Savino and S. Cozzolino

Dipartimento di Ingegneria, Seconda Università di Napoli, via Roma 29, I-81031 Aversa (CE), Italy

Abstract. The non-equilibrium aerothermochemical model developed by Viviani (1993), on the base of the Coupled-Vibration-Dissociation-Flow (CVDF) model for a simple diatomic vibrating and dissociating gas introduced by Viviani (1992a) and proved to be consistent by linearization and irreversible thermodynamic arguments (Viviani 1992b), is herein applied to the study of the non-equilibrium flow evolution behind strong shock waves. Tests and comparisons with other models are given for various free-stream conditions corresponding to different areas of the flight corridor of re-entry vehicles.

Key words: Aerothermochemistry, Non-equilibrium, Reacting mixture, Shocks

1. Introduction

Modelling of high speed aerodynamics requires to account for a large number of non-equilibrium internal exchanges of masses and energy. Although current Computational Fluid Dynamics studies on hypersonic flows incorporate in numerical codes an increasing number of physico-chemical non-equilibrium processes, the complexity of the phenomenologies involved calls for parallel basic researches for the development of theoretical self-consistent models, maybe simpler but more general than very sophisticated models based on ad-hoc formulations. Papers dealing with mathematical models of the non-equilibrium internal processes have been reviewed by Tirsky (1993) who reports the present status of the art. To formulate adequate mathematical models of the vibration and dissociation processes, suitable combinations of phenomenological and kinetic theories have been recently used (Viviani, 1992a, 1992b; Viviani and Savino, 1992). Viviani and Savino (1992) proposed an Extended Irreversible Thermodynamics approach for deriving, at different orders, the field equations of a multireacting gas mixture with an arbitrary number of nonequilibrium internal processes; Viviani (1992a,b) considered the influence of the flow velocity on the rates of non-equilibrium internal processes, by introducing the model of coupled vibration-dissociation-flow (CVDF) for a diatomic vibrating and dissociating gas, and indicated a method to proof correctness and consistency of non-linear non-equilibrium thermochemical models; with this method some inconsistency with classical irreversible thermodynamics has been detected in the model developed by Thivet et al. (1991). The CVDF model has been then extended by Viviani (1993) to the more general case of the air, on accounting also for the works of the Marseille group (Brun et al. 1990; Druguet et al. 1992). The air is modelled as a multi-component, multi-temperature, chemically reacting gas mixture, with three dissociation and three exchange reactions; Oxygen and Nitrogen are both vibrating and dissociating in non-equilibrium, and exchange reactions lead to the formation of Nitric Oxide which, in turn, vibrates and dissociates in nonequilibrium, contrary to what is generally assumed; cross-coupling effects between the different rate processes and the influence of the air flow on those processes are all taken into account. Herein we shall present results obtained with application of the above model to the air flow behind strong shock waves for various free-stream conditions, giving comparison with other models.

2. Results

To investigate the nonequilibrium coupling effects, included in the thermochemical model, at various conditions achieved during re-entry trajectories, we considered different upstream conditions: ($M_\infty = 25,\ H = 75$ km; $M_\infty = 15,\ H = 60$ km; $M_\infty = 8,\ H = 45$ km).

Shock Waves @ Marseille II
Editors: R. Brun, L. Z. Dumitrescu

Table 1. Upstream and initial conditions for the three Altitude-Mach cases investigated

Re-entry conditions	$M = 25$ ($H = 75$ km)		$M = 15$ ($H = 60$ km)		$M = 8$ ($H = 45$ km)	
Freestream properties	∞	initial	∞	initial	∞	initial
T (K)	208	25523	247	11040	264	3576
p (N/m^2)	2.4	1741	22	5760	149	11108
u (m/s)	7235	1215	4726	805	2606	468

Table 2. Characteristic temperatures for the molecular species

Molecular species / Charact. temperature (K)	O_2	N_2	NO
θ_r	4.17	5.78	4.93
θ_v	2256	3372	2720
θ_D	59500	113000	75500

Table 3. Computed distance behind the shock where the various partial equilibrium conditions are being to be reached (thermodynamic affinity A <10^{-2}), for the different kinetic models, in the case M=25, H=75 km

Kinetic model	Partial equilibrium conditions					
	$A_{V_{NO}} \to 0$	$A_{V_{O_2}} \to 0$	$A_{V_{N_2}} \to 0$	$A_{NO} \to 0$	$A_O \to 0$	$A_N \to 0$
UVD ($T_{V_{NO}} = T$)	-	2.15 cm	7.5 cm	18 cm	48 cm	53 cm
CVD ($T_{V_{NO}} = T$)	-	2.7 cm	7.1 cm	7.8 cm	52.4 cm	60 cm
CVDF ($T_{V_{NO}} = T$)	-	3.06 cm	7.58 cm	8 cm	54.8 cm	62 cm
CVDF ($U_{NO} = \infty$)	4.7 cm	3.1 cm	7.6 cm	8 cm	55 cm	62 cm

The values of the free-stream temperature, pressure and velocity, and the initial values just behind the shock, deduced from the Rankine-Hugoniot relations, are given in Table 1. For each case, in order to assess the various mechanisms of mass and energy exchanges and the effects of the gas velocity on the internal processes, we calculated the relaxation regions behind the shock with three different thermo-kinetic models of increasing complexity:

1) chemical reactions and vibrational excitation processes are independent, i.e. the coupling factors are all unitary and the coupling terms in the equations for the vibrational rates are equal to zero (Uncoupled Vibration-Dissociation model, UVD);
2) coupling between vibration and dissociation are taken into account independently for Oxygen and Nitrogen, while the vibrational degree of freedom of Nitric Oxide is considered in equilibrium (CoupledVibration-Dissociation model, CVD);
3) the influence of the flow velocity is taken into account (Coupled Vibration-Dissociation-Flow model, CVDF), for both cases of equilibrium and nonequilibrium of the Nitric Oxide vibrational process; in the non equilibrium case, all coupling effects between energy and mass exchanges are considered.

In case of coupling between vibration and dissociation (second and third case) we assumed the nonpreferential models for the dissociations of Oxygen and Nitrogen ($U_{O_2} = U_{N_2} = \infty$, see Marrone and Treanor, 1963), while both preferential and nonpreferential models were investigated for the nonequilibrium process of Nitric Oxide; the used values of the characteristic rotational, vibrational and dissociation temperatures of the molecular species are given in Table 2; for the backward constants of the chemical reactions we considered the temperature dependences suggested by Park (1990); the average relaxation times were determined using the Millikan-White (1963) semiempirical formula, corrected by the additional term proposed by Park (1990).

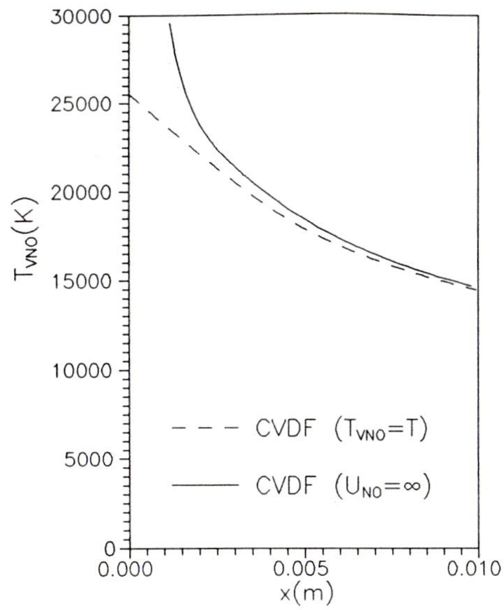

Fig. 1. Profiles of the vibrational temperature T_{vNO} for the CVDF model in the cases of equilibrium ($T_{vNO} = T$) and nonequilibrium ($U_{NO} = \infty$) of the vibrational degree of NO, at M_∞=25, H = 75 Km

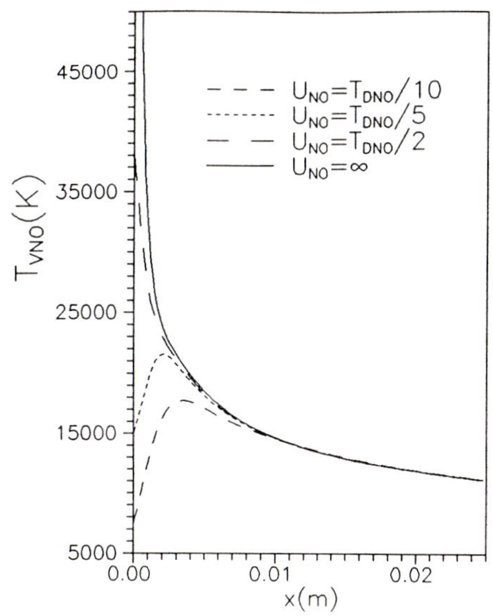

Fig. 2. Profiles of the vibrational temperature T_{vNO} for the CVDF model and different values of (U_{NO}), at M_∞=25, H = 75 km

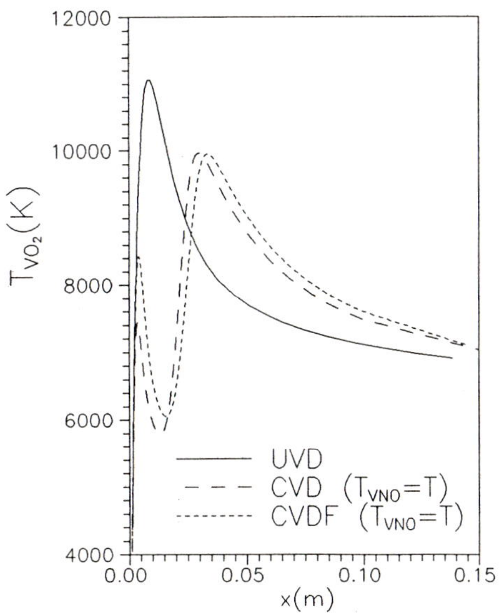

Fig. 3. Comparison between the profiles of the vibrational temperature T_{vO_2} for the three models (UVD, CVD, CVDF), at M_∞=25, H = 75 km

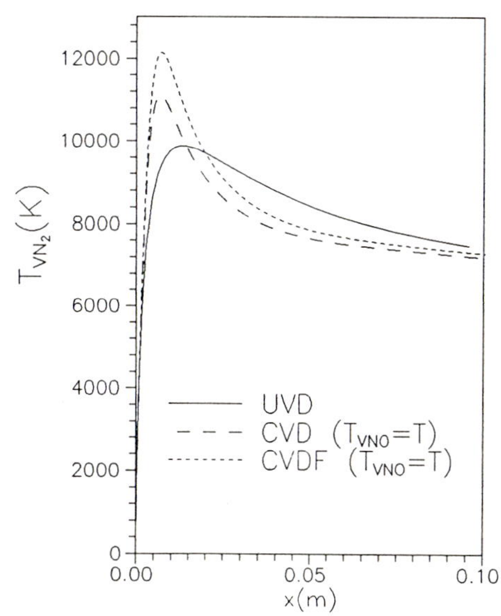

Fig. 4. Comparison between the profiles of the vibrational temperature T_{vN_2} for the three models (UVD, CVD, CVDF), at M_∞=25, H = 75 km

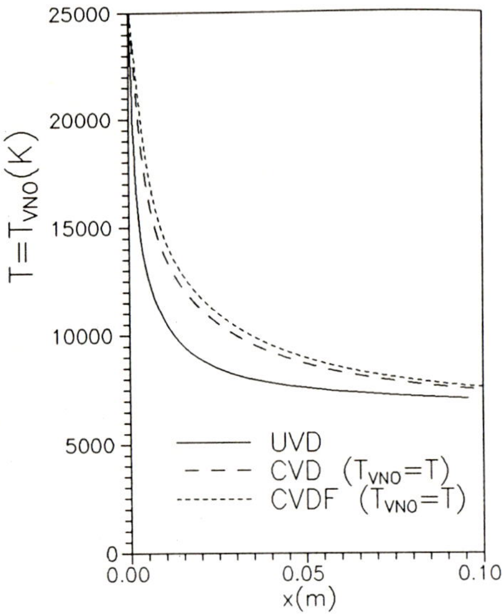

Fig. 5. Comparison between the profiles of the vibrational temperature $T_{v\mathrm{NO}}$ for the three models (UVD, CVD, CVDF), at M_∞=25, H = 75 km

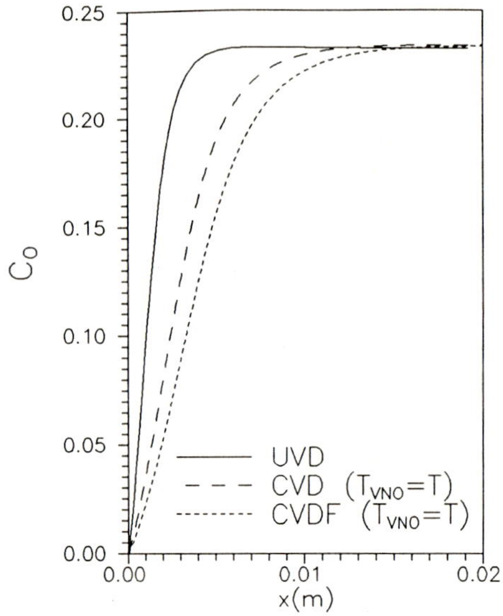

Fig. 6. Comparison between the profiles of mass concentration C_O for the three models (UVD, CVD, CVDF), at M_∞=25, H = 75 km

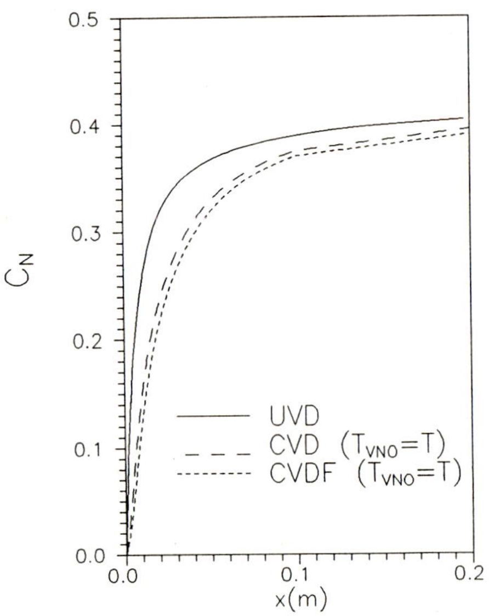

Fig. 7. Comparison between the profiles of mass concentration C_N for the three models (UVD, CVD, CVDF), at M_∞=25, H = 75 km

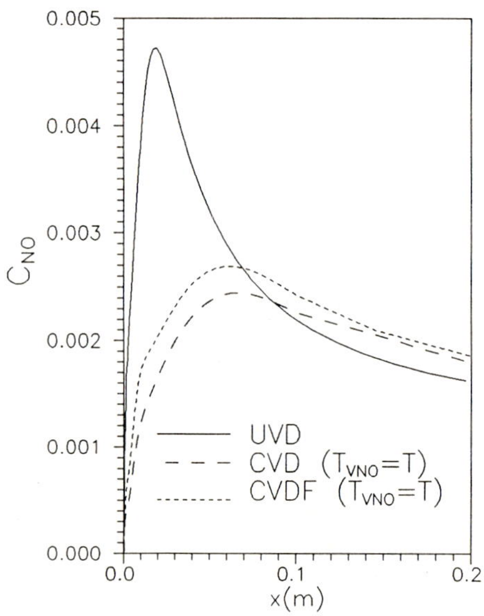

Fig. 8. Comparison between the profiles of mass concentration C_NO for the three models (UVD, CVD, CVDF), at M_∞=25, H = 75 km

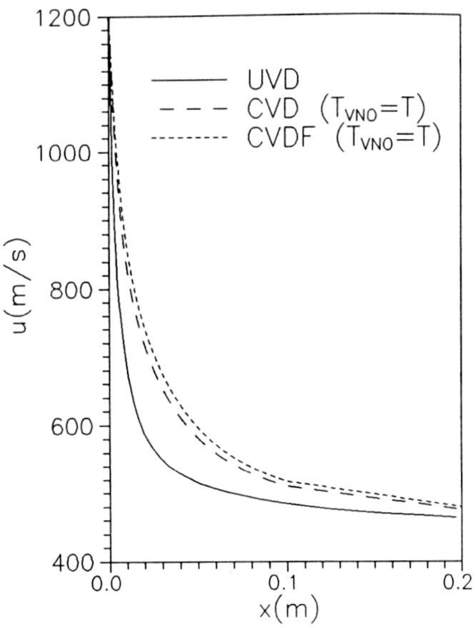

Fig. 9. Comparison between the velocity profiles for the three models (UVD, CVD, CVDF), at M_∞=25, H = 75 km

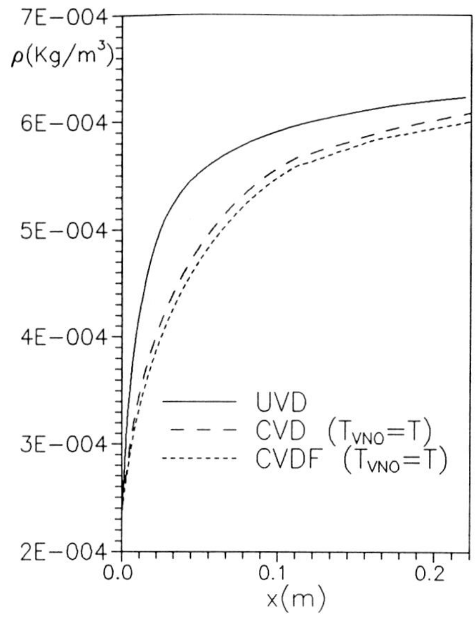

Fig. 10. Comparison between the density profiles for the three models (UVD, CVD, CVDF), at M_∞=25, H = 75 km

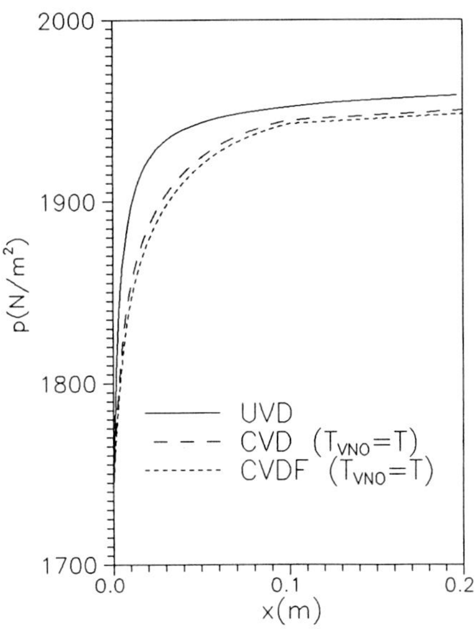

Fig. 11. Comparison between the pressure profiles for the three models (UVD, CVD, CVDF), at M_∞=25, H = 75 km

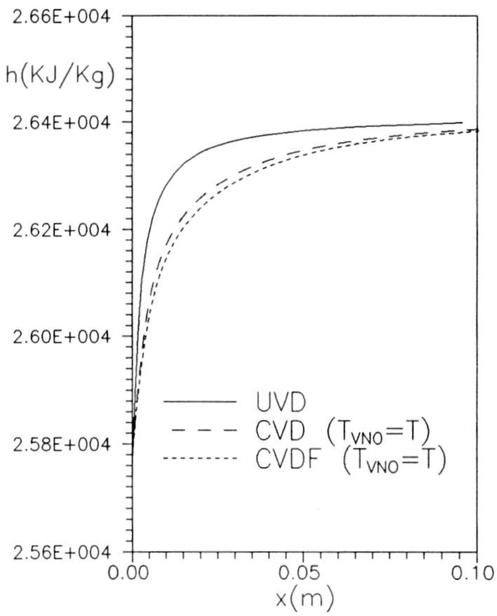

Fig. 12. Comparison between the enthalpy profiles for the three models (UVD, CVD, CVDF), at M_∞=25, H = 75 km

Table 4. Values of the Coupling Factors (V) for the different chemical reactions, computed at the conditions of Table 3. In the case of the CVDF model with $T_{V_{NO}} = T$ the values are almost the same as those for $U_{NO} = \infty$

	Kinetic model	Partial equilibrium conditions					
		$A_{V_{NO}} \to 0$	$A_{V_{O_2}} \to 0$	$A_{V_{N_2}} \to 0$	$A_{NO} \to 0$	$A_{O} \to 0$	$A_{N} \to 0$
$V^{(1)}$	CVD ($T_{V_{NO}} = T$)	2.47	1.1	0.98	0.99	1	1
	CVDF ($U_{NO} = \infty$)	1.87	1.09	0.99	1	1	1
$V^{(2)}$	CVD ($T_{V_{NO}} = T$)	4.19	2.16	1.54	1.45	1.01	1
	CVDF ($U_{NO} = \infty$)	2.36	2.06	1.52	1.48	1.01	1
$V^{(3)}$	CVD ($T_{V_{NO}} = T$)	1	1	1	1	1	1
	CVDF ($U_{NO} = \infty$)	0.96	0.98	1	1	1	1
$V^{(4)}$	CVD ($T_{V_{NO}} = T$)	4.19	2.16	1.54	1.45	1.01	1
	CVDF ($U_{NO} = \infty$)	2.44	2.09	1.52	1.47	1.01	1
$V^{(5)}$	CVD ($T_{V_{NO}} = T$)	2.47	1.1	0.98	0.99	1	1
	CVDF ($U_{NO} = \infty$)	1.94	1.1	0.99	0.99	1	1
$V^{(6)}$	CVD ($T_{V_{NO}} = T$)	10.4	2.45	1.53	1.44	1.01	1
	CVDF ($U_{NO} = \infty$)	4.73	2.31	1.51	1.47	1.01	1

Results for ($M_\infty = 25$, $H = 75$ km) are given in Table 3 and 4, and in Figs. 1-12. In Table 3 are reported the distances at which each of the six internal processes is reaching its partial equilibrium state, and in Table 4 are shown the values of the coupling factors evaluated at conditions of Table 3.

References

Brun R, Colas P, Gubernatis P, Zeitoun D (1990) Practical physico-chemical models for high-speed air flow-field computations. Submitted to Shock waves

Druguet MC, Zeitoun D, Imbert M, Brun R (1992) Chemical and vibrational non-equilibrium inviscid hypersonic nozzle flows. In: Abgral R et al.(eds) Hypersonic Flows for Reentry Problems III, p 1035 (Springer Verlag)

Marrone PV, Treanor CE (1963) Chemical relaxation with preferential dissociation from excited vibrational levels. Phys. Fluids 6: 1215

Millikan RC, White D (1963) Systematics of vibrational relaxation. J. Chem. Phys. 39: 3209

Park C (1990) Nonequilibrium Hypersonic Aerothermodynamics, John Wiley & Sons, N-Y

Thivet F, Perrin M.Y, Candel S (1991) A unified nonequilibrium model for hypersonic flows. Phys. Fluids A 3: 2799

Tirsky AG (1993) Up-to-date gasdynamic models of hypersonic aerodynamics and heat transfer with real gas properties. Ann. Rev. Fluid Mech. 25: 151

Viviani A (1992a) Coupling between flow velocity and non-equilibrium internal processes in hypersonic flow aerothermochemistry. IUTAM Symposium on Aerothermochemistry of Spacecrafts and Related Hypersonic Flows, Marseille (France), 1-4 September 1992

Viviani A (1992b). The Coupled Vibration Dissociation Flow model for non-equilibrium vibrating and dissociating diatomic gas. Development and proof of consistency. Internal Report (Nov. 1992), Seconda Università di Napoli - Facoltà di Ingegneria, Aversa (Italy)

Viviani A (1993). A unified and consistent model of thermal and chemical non-linear non-equilibrium for hypersonic high-temperature air flow. Internal Report (May 1993), Seconda Università di Napoli - Dipartimento di Ingegneria, Aversa (Italy) (in Italian)

Viviani A, Savino R (1992) Extended irreversible thermodynamic modelling of high temperature hypersonic flows. IUTAM Symposium on Aerothermochemistry of Spacecrafts and Related Hypersonic Flows, Marseille (France), 1-4 September

Level Kinetics of the Dissociation and Vibrational Relaxation Behind Shock Waves

E. Nagnibeda
St.Petersburg University

Abstract. In this paper the structure of the relaxation zone behind shock waves in nonequilibrium gases is discussed. The dissociation and relaxation kinetics are numerically studied in the framework of the level approach on the basis of the kinetic equations for the level populations. The results allow establishing the limits of applicability for quasi-stationary distributions and tracing their formation behind shock waves. Different initial distributions are considered and anharmonic molecular oscillations are taken into account. It is shown that in the case of nonequilibrium initial flow with strong vibrational excitation the anharmonic effects may be of importance behind a shock wave. They result in the disturbance of the distribution canonical invariancy and in a non-monotonous dissociation rate dependence on the temperature and on the distance from the shock wave front.

Key words: Gas kinetics, Dissociation, Vibrational relaxation.

1. Introduction

In the early stages of relaxation theory the chemical kinetics was studied in the framework of equilibrium thermodynamics. In the sixties, the first models of the coupling between vibrational relaxation and dissociation were proposed (Treanor and Marrone 1962). Most of the models of this coupling which appeared in the later years, use a two-temperature approximation with rotational - translational and vibrational temperatures or a multi-temperature approach with different temperatures of various kinds of vibrations. Multi-temperature distributions are known to be valid in conditions when the resonant energy transfers occur more often than nonresonant ones. But these kinds of energy exchange rates become similar to each other and to the dissociation rate at high temperatures. In molecular nitrogen such a temperature is about 8000 K. Moreover multi-temperature distributions take place only in the case when the internal energy of each kind of vibrations can be described by the model of a harmonic oscillator. But this model appears not to be valid on the high vibrational levels near the dissociation threshold and it is necessary to take into account anharmonic effects in this case. So, only a level approach may ascertain the limits of validity of multi-temperature distributions.

The aim of this paper is to study the structure of the relaxation zone behind a shock wave in the case of both equilibrium and strong nonequilibrium initial flow.

2. Level vibrational kinetics

The relaxation zone structure is studied on the basis of the equations for vibrational level populations $N_i(x)$ and conservation equations for density $\rho(x)$, velocity $u(x)$ and temperature $T(x)$:

$$u\,\frac{dN_i}{dx} = F_i^{rel} \tag{1}$$

$$\begin{aligned} \rho_0 u_0 &= \rho u \\ \rho_0 u_0^2 + p_0 &= \rho u^2 + p \\ h_0 + \frac{u_0^2}{2} &= h + \frac{u^2}{2} \end{aligned} \tag{2}$$

Shock Waves @ Marseille II
Editors: R. Brun, L. Z. Dumitrescu

$$F_i^{rel} = N_{i+1}\left(\sum_j K_{i+1,i}^{j,j+1} N_j + K_{i+1,i}\right) n - N_i\left(\sum_j K_{i,i+1}^{j+1,j} N_{j+1} + \sum_j K_{,i-1}^{j,j+1} N_j + \right.$$

$$\left. + K_{i,+1} + K_{i,i-1}\right) n + N_{i-1}\left(\sum_j K_{i-1,i}^{j+1,j} N_{j+1} + K_{i-1,i}\right) n \qquad (3)$$

Here ρ_0, u_0, T_0 - are the parameters in the undisturbed flow,$h = \frac{7}{2}RT + e_v$, $e_v = \frac{1}{m}\sum_i N_i\epsilon_i$, N_i are referred to the number density n, $K_{i,i+s}^{j,j+f}$ - are the numbers of energy transfers $(i,j \rightarrow i+s, j+f)$ in unit time.

The molecular vibrational spectrum ϵ_i is simulated as a Morse oscillator:

$$\epsilon_i = E_1 i - \Delta E i(i-1),\ E_1 = \omega_e - 2x_e\omega_e,\ \Delta E = \omega_e x_e,$$

x_e, ω_e are spectroscopical constants. Translational and rotational degrees of freedom are assumed to be in equilibrium with the temperature $T(x)$.

As vibrational distribution in the initial undisturbed flow $N_i(0)$ we consider both equilibrium and nonequilibrium distributions:

$$N_i(0) = \frac{1}{Z(T_v)}\exp\left(-\frac{\epsilon_i}{kT_v}\right)$$

$Z(T_v)$ is the vibrational statistical sum, T_v is the vibrational temperature (at equilibrium $T_v = T$ and in the nonequilibrium case $T_v \neq T$).

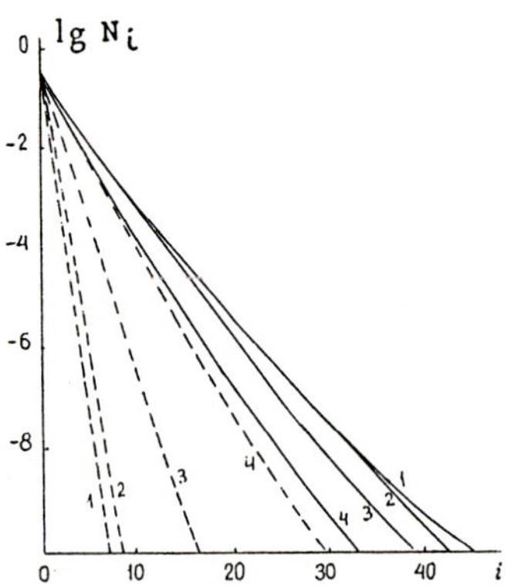

Fig. 1. Level populations in N_2 behind a shock wave with M = 7.6, $T_v = T_0$ = 1000 K (full lines) and M = 4.2, T_0 = 300 K, T_v = 5000 K (dashed lines)

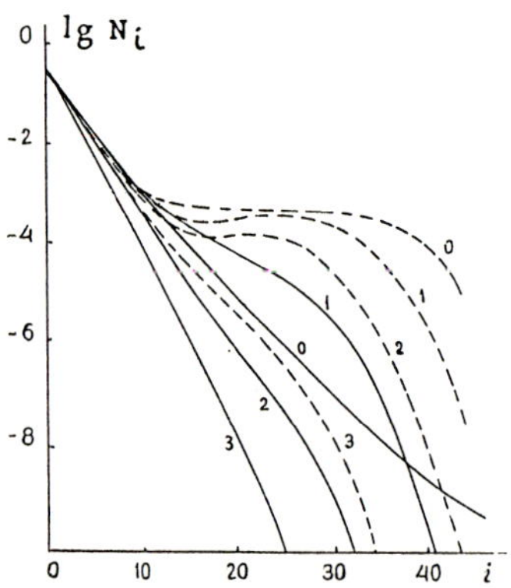

Fig. 2. Level populations in N_2 behind a shock wave with M = 3.3, T_0 = 500 K, T_v = 5500 K

The system (1), (2) for 50 levels in molecular nitrogen was solved by Gear's method (Gear 1971) which proved itself to be a good method for rigid systems. Transitions probabilities $K_{i,i+s}^{j,j+f}$ are taken from the SSH-theory modified for anharmonic oscillators as well as from Billing's trajectory calculations data (Billing and Fisher 1979).

The level populations N_i versus i are to be found in Figs.1,2. Different numbers: 0,1,2,3,4 on the curves correspond to the initial distributions (0) and to different distances s_n/ℓ $(n = 1,2,3,4)$ from the shock wave front (ℓ - is the free path length in the initial flow). In Fig.1 s_n/ℓ are equal to 2, 30, 650, 5300 at $n = 1,2,3,4$ respectively and in Fig.2 s_n/ℓ are equal to 3500, 15 10^3, 75 10^3 (full lines) and 1100, 2800, 25 10^3 (dashed lines) at $n = 1,2,3$.

Fig.1 depicts the relaxation zone structure in the case of equilibrium and weak nonequilibrium initial flow ahead of the shock wave. It is seen that in such conditions the conservation of the initial Boltzmann form of distribution takes place. This property is well known for the system of harmonic oscillators: the so-called canonical invariancy of the distribution.

Fig.2 depicts the level populations in the relaxation zone in the case of strong nonequilibrium initial flow. In such a situation the disturbances of an initial form of distribution are shown. In fact in condition $T_{v_1} > T_1$ (T_1, T_{v_1} are the temperatures behind the shock front at the beginning of the relaxation zone) in a system of anharmonic oscillators the stream of vibrational quanta to upper levels takes place and leads to distributions differing from two-temperatures ones.

Finally it might be well to point out that the relaxation zone structure depends on two parameters: the first of them, $\beta = T_{v_0}/T_0 = T_{v_1}/T_0$ is connected with the nonequilibrium state in the initial flow, and the second one, $\gamma(M) = T_1/T_0$ is connected with the shock wave intensity. (One can see that $\beta = 1$ corresponds to equilibrium undisturbed flow). It is shown that if $\beta \gg \gamma$ anharmonic effects are of substantial influence and level populations differ from multi-temperature distributions. The numerical solution obtained shows three stages of the relaxation zone. The first stage is the forming of a quasi-stationary distribution because of intensive VV-vibrational energy exchanges. The second one is a quasi-stationary stage when the distribution may be represented as a combination of a Treanor distribution on lower levels, a plateau-distribution on middle levels and a Boltzmann distribution on upper levels. And the third stage is the evolution of the quasi-stationary distribution towards equilibrium by VT vibrational - translational energy transfers.

Fig.3 shows the temperatures and density changes in the relaxation zone.

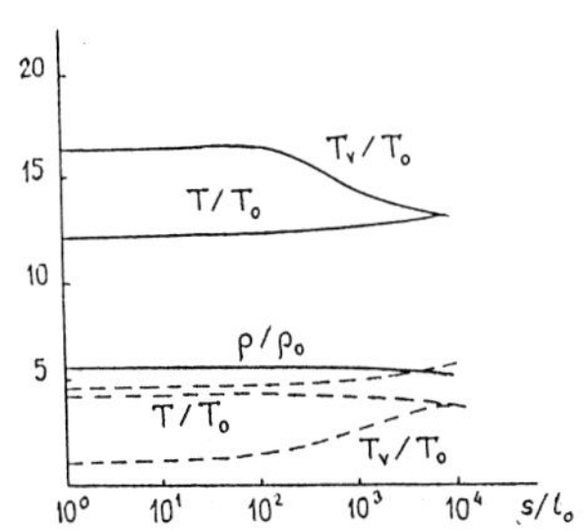

Fig. 3. Temperatures T/T_0, T_v/T_0 and density ρ/ρ_0 behind a shock wave in the case of nonequilibrium (full lines) and equilibrium flow before shock wave (dashed lines)

3. Dissociation kinetics in a nonequilibrium gas

In this case of dissociation and vibrational relaxation behind a shock wave Eqs.1 take the following form:

$$u\frac{dN_i}{dx} = F_i^{rel} + F_i^{diss} \tag{4}$$

$$F_i^{diss} = (n + n_a)(n_a^2 K_i^{rec} - N_i K_i^{diss}) \tag{5}$$

Here n_a is the atom density, K_i^{diss}, K_i^{rec} are dissociation and recombination rate constants for the i-th level.

If we consider the initial stage of dissociation when the recombination may be neglected and dissociation takes place only on the upper level, F_i^{diss} may be written as follows:

$$F_i^{diss} = \begin{cases} 0, & \text{if } i < r \\ K_{r-1,r}N\ N_{r-1} + \sum\limits_{m=0}^{r-2} K_{r-1,r}^{m+1,m} N_{r-1}\ N_{m+1}, & \text{if } i = r. \end{cases}$$

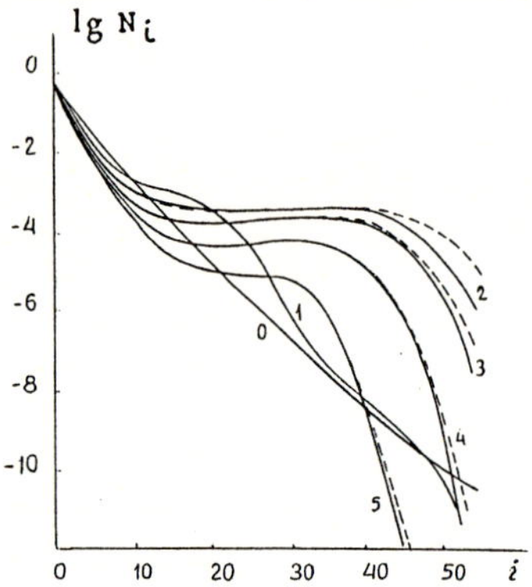

Fig. 4. Level populations in N_2 at different moments of time t_m ($m = 0, 1, \ldots 5$) in the case of dissociation and vibrational relaxation (full lines) and only vibrational relaxation (dashed lines); $T_v = 5500$ K, $T_0 = 500$ K

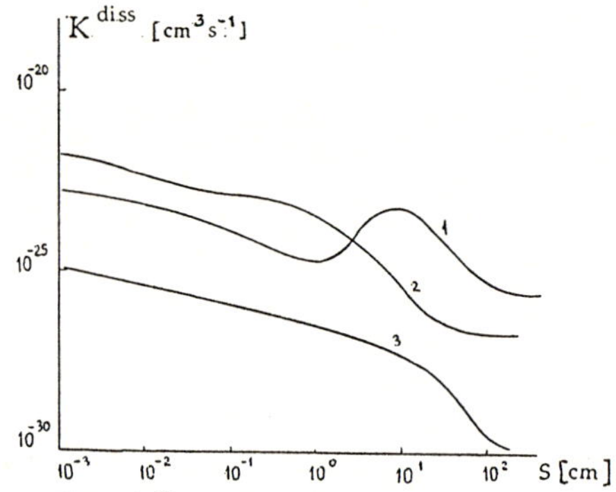

Fig. 5. K^{diss} in nonequilibrium N_2:
1: $T_{v0} = 5500$ K, $T_0 = 500$ K, $M = 3.3$
2: $T_{v0} = 5500$ K, $T_0 = 500$ K, $M = 5.0$
3: $T_{v0} = 4000$ K, $T_0 = 1000$ K, $M = 3.3$

Solution of the system (1), (4), (5), (2) (Kovalev and Nagnibeda 1991) shows the impact of dissociation on the upper levels. This problem was solved also for a homogenous gas (N_2 and CO). In Fig.4 the level populations in N_2 at different moments of time are shown: t_n are equal to 120, 1000, 6000, 2 10^4, 1.7 10^5 (μs) at $n = 1, 2, \ldots 5$. Full lines correspond to dissociation in the relaxing gas, dashed lines correspond only to vibrational relaxation. The most considerable impact of dissociation takes place during the quasi-stationary stage of process (2nd line).

The non-monotonous character should be emphasized of the dissociation rate constant K^{diss} changing as a function of s/l_0 and of the temperature T in the case of strong vibrational excitation in undisturbed flow ($T_{v_0} \gg T_0$). Actually the intensive vibrational quanta stream to upper levels because of VV-energy exchanges at the beginning of the relaxation zone results in a considerable increase of K^{diss} and, after that, VT-exchanges lead to a decrease in the number of molecules at upper levels and hence to the decrease of K^{diss} (Fig.5). This effect diminishes with increasing M because it leads to an increase of T_1 respectively and the condition $\beta \gg \gamma$ appears not to hold.

References

Billing GD, Fisher ER (1979) VV and VT rate coefficients in N_2 by a quantum-classical model. Chem. Phys. 43: 395

Gear CW (1971) DIFSUB for solution of ordinary differential equations. Comm. Assoc. Comput. Math. 14: 185

Kovalev II, Nagnibeda EA (1991) Numerical investigation of the dissociation rate in vibrational nonequilibrium gas. In: Proc. 10th Conf. on Rarefied Gas Dynamics. Moscow. Energ. Inst. 1: 132 (in Russian)

Treanor CE, Marrone PV (1962) Effect of dissociation on the rate of vibrational relaxation. Phys. Fluids 5: 1022

Vibration-Dissociation Relaxation Phenomena in Shock-Heated Nitrogen

J.G. Méolans, M. Mouti, F. Lordet and A. Chauvin
Laboratoire IUSTI/MHEQ, Université de Provence, Centre Saint-Jérôme, 13 397 Marseille, France

Abstract. An analysis is presented of nonequilibrium phenomena behind a plane shock wave with emphasis on the vibrational relaxation and dissociation of pure diatomic nitrogen. In the first part, the temperature range is 600 K - 2500 K and dissociation processes are neglected. The population of each vibrational level is computed by solving relaxation and conservation equations. The populations obtained are compared to those calculated using a Treanor or a non-equilibrium Boltzmann distribution. The vibrational cross-section includes orientation and anharmonicity effects. In the second part, the temperature range is 2500 K - 5500 K and dissociation is taken into account. The vibrational transition probabilities associated with the atom-molecule interaction are deduced from the cross-section used in the first part. Three approaches are presented. It is assumed that dissociation must occur preferentially from the higher vibrational states: the Marrone and Treanor probability model is extended and employed to anharmonic oscillators.

Key words: Plane shock wave, Vibrational relaxation, Dissociation, Vibrational populations

1. Introduction

In the present work, on each side of the shock, the gas is assumed to be at low density and relaxation processes take place essentially by binary interaction between the molecules or between molecules and atoms. The prevalence of short range forces in collisions is admitted. The rotational relaxation of molecules is very fast compared to the vibrational relaxation time and to the time necessary for the establishment of a dissociation equilibrium; so a local equilibrium is associated for the rotation (Stupochenko et al. 1967). The non-equilibrium relaxing flow must verify the system of gas dynamic equations of the Hugoniot-Rankine type which express conservation of mass, mementum and energy. This system must be supplemented by relaxation equations which describe the kinetics for establishing vibration-dissociation equilibrium.

2. Vibrational relaxation: $600 < T < 2500$ K

In the first part, the vibrational processes are of the following form:

$$\mathrm{N}_2(v_1) + \mathrm{N}_2(v_2) \longleftrightarrow \mathrm{N}_2(v_1') + \mathrm{N}_2(v_2') \tag{1}$$

Neglecting herein the dissociation process, the specific enthalpy h of the gas and the pressure P used in the conservation equations are expressed as usual for a pure diatomic gas.

The oscillator model used in the basic calculation is a Morse anharmonic oscillator.

2.1. Relaxation equations

2.1.a) In the first approach, the relaxation equations are the balance equations of the molecular populations for each vibrational level n_{v_1} including all the vibrational processes, that is to say:

$$\frac{\partial n_{v_1}}{\partial t} + \frac{\partial n_{v_1} u}{\partial x} = \frac{Z_{(\mathrm{N}_2-\mathrm{N}_2)}}{n_{\mathrm{N}_2}} \sum_{v_2, v_1', v_2'} \left(-n_{v_1} n_{v_2} P_{v_1 \to v_1'}^{v_2 \to v_2'} + n_{v_1'} n_{v_2'} P_{v_1' \to v_1}^{v_2' \to v_2} \right) \tag{2}$$

where u is the macroscopic velocity of shocked gas in the shock fixed system and $Z_{(\mathrm{N}_2-\mathrm{N}_2)}$ is the number of collision made by one molecule per second. n_{N_2} is the molecule number density.

$P_{v_1 \to v_1'}^{v_2 \to v_2'}$ is the transition probability of a molecule from vibrational level v_1 to v_1' colliding with another molecule from the vibrational level v_2 to v_2'. These quantities are related to the average inelastic cross section $\bar{\sigma}_{v_1 \to v_1'}^{v_2 \to v_2'}$:

Shock Waves @ Marseille II
Editors: R. Brun, L. Z. Dumitrescu

$$Z_{(N_2-N_2)} P_{v_1 \to v_1'}^{v_2 \to v_2'} = 2\, n_{N_2} \left(\frac{2kT}{\pi \tilde{M}_{(N_2-N_2)}} \right)^{1/2} \bar{\sigma}_{v_1 \to v_1'}^{v_2 \to v_2'} \tag{3}$$

where T is the temperature and $\tilde{M}$ is the reduced mass.

2.1.b) The second approach is a simplified one based on the same considerations as in the first approach, but neglecting the V-V processes to highlight their influence.

2.1.c) In the third approach, assuming a regime strongly dominated by V-V processes, one obtains the vibrational population distributions of Treanor (Brun and Zappoli 1977). The well-known mean relaxation equation governs the average quantum number $\bar{V}$ and allows to determine the macroscopic quantity K which depends on the anharmonicity and on the non-equilibrium and which appears in the vibrational distribution populations. So it is possible to calculate all the mean parameters and to deduce the vibrational populations.

2.1.d) In the fourth approach, by using an harmonic oscillator model, the T-V de-excitation probability is proportional to the upper vibrational quantum number of the molecule. The resulting relaxation equation governing the vibrational energy per mass unit is the Landau-Teller equation (Landau and Teller 1936). The associated vibrational population distribution is the non-equilibrium Boltzmann distribution expressed at the vibrational temperature T_v. So one may again determine the mean parameters and calculate the vibrational populations.

2.2. Cross section calculation: method and results

A generalised first-order distorted wave method developed previously (Méolans and Chauvin 1991) is used; and one obtains the cross-section averaged over a Boltzmann distribution for the rotational states and over a Maxwellian distribution for the velocities :

$$\bar{\sigma}_{v_1 \to v_1'}^{v_2 \to v_2'} = \bar{\sigma}_{v\, v'} = \chi_{v_1, v_1'} \cdot \chi_{v_2, v_2'}\, W\left(Q^{*2}\right) \bar{\sigma}_p \tag{4}$$

The factor $\chi_{v_1,v_1'} \cdot \chi_{v_2,v_2'}$ describes the orientation effects of the interaction potential. It is independent of temperature and decreases slowly with the level (from 1.4 to 0.8 for T-V processes). $W\left(Q^{*2}\right)$ is a factor close to 1 depending on the most efficient dimensionless wave number Q^*. $\bar{\sigma}_p$ is the expression obtained with a spherical intermolecular potential by Schwartz and Herzfeld (1954).

2.3. Numerical method

2.3.a) In the first approach, the numerical computation requires the transformation of Eq.2 . From Eq.2 and the mass conservation equation, one deduces:

$$\frac{d}{dt}\left(\frac{n_{v_1}}{n_{N_2}}\right) = Z_{(N_2-N_2)} \sum_{v_2, v_1', v_2'} \left[-\left(\frac{n_{v_1}}{n_{N_2}}\right)\left(\frac{n_{v_2}}{n_{N_2}}\right) P_{v_1 \to v_1'}^{v_2 \to v_2'} + \left(\frac{n_{v_1'}}{n_{N_2}}\right)\left(\frac{n_{v_2'}}{n_{N_2}}\right) P_{v_1' \to v_1}^{v_2' \to v_2} \right] \tag{5}$$

The populations are calculated by solving a first-order differential system using the fourth-order Runge-Kutta method.

2.3.b) In the second approach, the numerical method is the same as in the first one.

2.3.c) In the third approach, $\bar{e}_v$ and $\bar{V}$ are expressed as a functions of T and K. The conservation equations are differentiated and after substitutions, a system of differential equations is obtained:

$$\frac{dM_n}{dt} = f_n\left(u, T, \bar{e}_v, P, n_{N_2}, K\right) \tag{6}$$

where M_n $(n = 1, 2, \ldots 6)$ is successively equal to u, T, $\bar{e}_v$, P, n_{N_2}, K. Thus, the Runge-Kutta method can be applied by integrating several macroscopic parameters.

2.3.d) In the fourth approach, the calculation is very easy since $\bar{e}_v$ is obtained directly from the integration of the Landau-Teller equation.

2.4. Results concerning the first part

In front of the shock: $P_1 = 140$ Pa, $T_1 = 278$ K, Mach= 6

* For the first five levels, the populations calculated with the different models are close to each other.

* The Boltzmann distribution gives significant differences compared to results obtained with the anharmonic approach for the higest levels (Figs.1, 2, 3).

* In the front of the shock, only n_0 is excited. During the relaxation, the first five levels store a significant amount of vibrational energy.

* The thermal energy is transformed and stored into vibrational energy through T-V processes. The V-V contribution is opposite to the T-V one and its influence increases for higher levels (Figs.1, 2, 3).

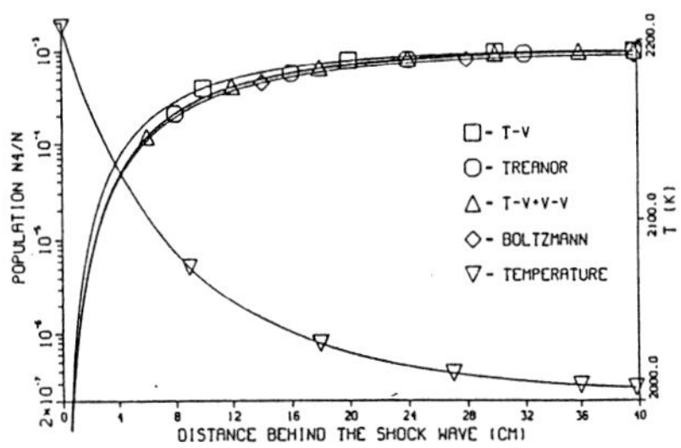

Fig. 1. Vibrational population (n_4/n) in N_2 as a function of the shock distance. Behind the shock front: $P = 16315$ Pa, $T = 2207$ K, $u = 1908$ m.s^{-1} and Mach = 6

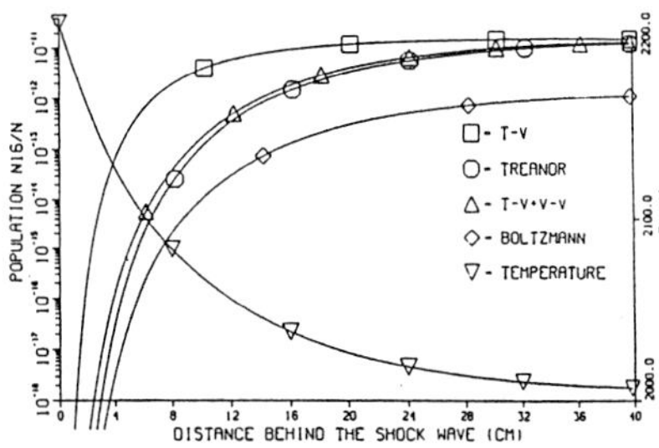

Fig. 2. Vibrational population (n_{16}/n) in N_2 as a function of the shock distance. Behind the shock front the conditions are the same as in Fig.1

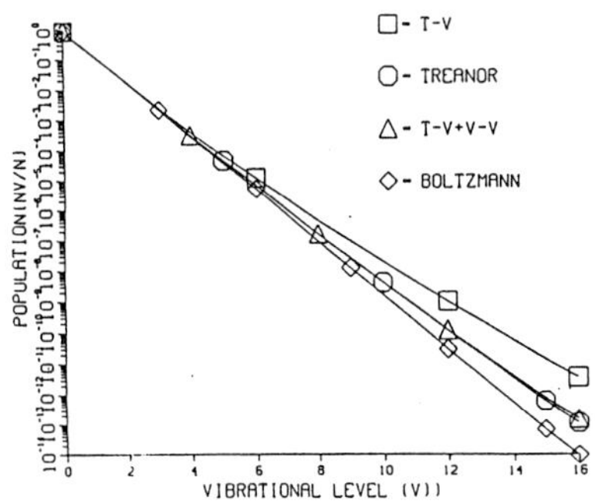

Fig. 3. Vibrational population (n_v/n) in N_2 behind the shock. $T = 2068$ K, $u = 365.27$ m.s^{-1} and the distance from the shock wave is 9.7 cm

3. Vibration-dissociation relaxation: $2500 < T < 5500$ K

Taking into account the dissociation process in the present temperature range, Eq.1 must be supplemented by the dissociation reactions and by N_2-N TV processes:

$$N_2(v_1) + N_2(v_2) \longleftrightarrow 2N + N_2(v_2') \qquad N_2(v_1) + N \longleftrightarrow 2N + N \tag{7}$$

$$N_2(v_1) + N \longleftrightarrow N_2(v_1') + N \tag{8}$$

The enthalpy and pressure of the gas are now respectively:

$$h = Y_{N_2}\left[C_{p(N_2)_{trans-rot}}\,T + \bar{e}_v\right] + Y_N\left[C_{p(N)_{trans}}\,T + h_N^0\right] \tag{9}$$

$$P = \rho \hat{R} T\left[Y_{N_2}/\hat{M}_{N_2} + Y_N/\hat{M}_N\right] \tag{10}$$

where C_p is the specific enthalpy at constant pressure; h_N^0 is the standard enthalpy of formation of N; $\hat{R}$ is the molar gas constant; $\hat{M}$ is the molar mass; Y_{N_2} and Y_N are respectively the mass fractions of N_2 and N: $Y_{N_2} = 2\,n_{N_2}/(2\,n_{N_2} + n_N)$ and $Y_N = n_N/(2\,n_{N_2} + n_N)$

3.1. The kinetic equations

Due to reactions (7) and (8), the kinetic equations are more complex than Eq.2, i.e:

$$\begin{aligned}\frac{\partial n_{v_1}}{\partial t} + \frac{\partial n_{v_1} u}{\partial x} = \frac{Z_{(N_2-N_2)}}{n_{N_2}} &\left[\sum_{v_2,v_1',v_2'}\left(-n_{v_1} n_{v_2} P_{v_1\to v_1'}^{v_2\to v_2'} + n_{v_1}' n_{v_2}' P_{v_1'\to v_1}^{v_2'\to v_2}\right)\right.\\ &\left.+\sum_{v_2,v_2'}\left(-n_{v_1} n_{v_2} P_{v_1\to d}^{v_2\to v_2'} + n_N^2\, n_{v_2'} P_{d\to v_1}^{v_2'\to v_2}\right)\right]\\ &+ Z_{(N_2-N)}\left[\sum_{v_1'}\left(-n_{v_1} P_{v_1\to v_1'} + n_{v_1'} P_{v_1'\to v_1}\right) + \left(-n_{v_1} P_{v_1\to d} + n_N^2 P_{d\to v_1}\right)\right]\end{aligned} \tag{11}$$

$$\begin{aligned}\frac{\partial n_N}{\partial t} + \frac{\partial n_N u}{\partial x} = 2\,Z_{(N_2-N_2)} &\sum_{v_1,v_2,v_2'}\left[\frac{1}{n_{N_2}}\left(-n_N^2 n_{v_2'} P_{d\to v_1}^{v_2'\to v_2} + n_{v_1} n_{v_2} P_{v_1\to d}^{v_2\to v_2'}\right)\right]\\ &+ 2\,Z_{(N_2-N)}\left[\sum_{v_1}\left(-n_N^2 P_{d\to v_1} + n_{v_1} P_{v_1\to d}\right)\right]\end{aligned} \tag{12}$$

$Z_{(N_2-N_2)}\,P_{v_1\to v_1'}^{v_2\to v_2'}$ is defined in the previous Section. The inelastic frequency of vibrational collisions $Z_{(N_2-N)}\,P_{v_1\to v_1'}$ characterizing the atom-molecule interaction is easily deduced from Eqs.3 and 4, adapted to the interaction atom-molecule. $Z_{(N_2-N)}\,P_{v_1\to d}^{v_2\to v_2'}$ and $Z_{(N_2-N)}\,P_{v_1\to d}$ are expressed in Section 3.2.

3.2. Probability of dissociation

The probability $P_{v_1\to d}^{v_2\to v_2'}$ is associated with the transition of a molecule from the v_1 vibrational level to the dissociated state d colliding with another molecule whose vibrational level changes from v_2 to v_2'. Now, it is assumed that during a dissociative interaction, a change of the internal state of N_2 in level v_2 cannot occur. Then, v_2 is equal to v_2' and Eq.7 becomes:

$$N_2(v_1) + M \longleftrightarrow 2N + M \tag{13}$$

where M represents successively N_2 and N. The dissociation processes are governed by the following equation:

$$\left\{\frac{\partial [N_2]_{v_1}}{\partial t} + \frac{\partial [N_2]_{v_1} u}{\partial x}\right\}_D = \sum_{M=N_2,N} -k_{f_{v_1}}^{(N_2-M)}[N_2]_{v_1}[M] + \text{backward reaction term} \tag{14}$$

where $[N_2]$ and $[M]$ are the molar concentrations. Comparing Eq.14 to the dissociative terms of Eq.11 one obtains the relationships:

$$Z_{(N_2-M)}\,P_{v_1\to d}^{N_2-M} = n_M k_{f_{v_1}}^{(N_2-M)}/\mathcal{N} \tag{15}$$

$P_{v_1\to d}^{N_2-M}$ represents either $P_{v_1\to d}^{v_2\to v_2'}$ or $P_{v_1\to d}$ depending on whether M is N_2 or N; $\mathcal{N}$ is the Avogadro number. Marrone and Treanor introduced a nondimensional coefficient, the physical meaning of which is:

$$\mathcal{P}_{v_1\to d}^{\mathrm{N_2-M}} = \left(\frac{dn_{v_1}}{dn_{\mathrm{N_2}}}\right)_{\mathrm{DM}} = \left(\frac{d[\mathrm{N_2}]_{v_1}}{d[\mathrm{N_2}]}\right)_{\mathrm{DM}} \tag{16}$$

where the subscript "DM" characterises the evolution resulting of the dissociative process only with a partner M. On the other hand, as it is well known:

$$\left(\frac{d[\mathrm{N_2}]_{v_1}}{dt}\right)_{\mathrm{D}} = \sum_{\mathrm{M=N_2,N}} \left(\frac{d[\mathrm{N_2}]_{v_1}}{dt}\right)_{\mathrm{DM}} = -\sum_{\mathrm{M=N_2,N}} k_{f_{v_1}}^{(\mathrm{N_2-M})}[\mathrm{N_2}]_{v_1}[\mathrm{M}] \tag{17}$$

The global dissociative kinetics is described by a reaction similar to Eq.13, replacing $\mathrm{N_2}(v_1)$ by $\mathrm{N_2}$, so:

$$\left(\frac{d[\mathrm{N_2}]}{dt}\right)_{\mathrm{D}} = \frac{d[\mathrm{N_2}]_{\mathrm{D}}}{dt} = \sum_{\mathrm{M=N_2,N}} \left(\frac{d[\mathrm{N_2}]}{dt}\right)_{\mathrm{DM}} = -\sum_{\mathrm{M=N_2,N}} k_f^{(\mathrm{N_2-M})}[\mathrm{N_2}][\mathrm{M}] \tag{18}$$

Using Eqs.16,17,18 and reminding that $k_{f_{v_1}}$ does not depend on the non-equilibrium one obtains:

$$k_{f_{v_1}}^{(\mathrm{N_2-M})} = k_{f_{\mathrm{eq}}}^{(\mathrm{N_2-M})} \left(\mathcal{P}_{v_1\to d}^{\mathrm{N_2-M}}\right)_{\mathrm{eq}} \{[\mathrm{N_2}]/[\mathrm{N_2}]_{v_1}\}_{\mathrm{eq}} \tag{19}$$

The coefficient $\mathcal{P}_{v_1\to d}^{\mathrm{N_2-M}}$ is derived from the phenomenological description of Marrone and Treanor (1963). So $\left(\mathcal{P}_{v_1\to d}^{\mathrm{N_2-M}}\right)_{\mathrm{eq}}$ can be easily deduced; using Eq.19, the rate for each vibrational level is:

$$k_{f_{v_1}}^{(\mathrm{N_2-M})} = k_{f_{\mathrm{eq}}}^{(\mathrm{N_2-M})} \left(Q(T)/Q(-U)\right) \exp\left\{(\varepsilon_{v_1}/k)\left(1/U + 1/T\right)\right\} \tag{20}$$

$Q(A)$ is the vibrational partition function at "temperature A". Then Eq.15 gives the frequencies of collision. This result does not require in any way the harmonic oscillator assumption. Finally, the backward probability $P_{d\to v_1}^{\mathrm{N_2-M}}$ is deduced by applying the principle of detailed balancing.

We have chosen to describe the dissociation process assuming an equal coefficient for any level v_1. Thus U is taken equal to infinity (non-preferential assumption), then $Q(-U)$ is equal to the total number of vibrational levels of the molecules.

3.3. Numerical method

Y_{v_1} is the mass fraction of molecules on the v_1th vibrational level, expressed as:

$$Y_{v_1} = 2n_{v_1}\Big/\left(2\sum_{v_1} n_{v_1} + n_{\mathrm{N}}\right) = 2n_{v_1}/\left(2n_{\mathrm{N_2}} + n_{\mathrm{N}}\right); \qquad Y_{\mathrm{N_2}} = \sum_{v_1} Y_{v_1} \tag{21}$$

Using Eqs.11 and 12 and the equation of mass conservation, one obtains:

$$\begin{aligned} \frac{dY_{v_1}}{dt} = \frac{Z_{(\mathrm{N_2-N_2})}}{Y_{\mathrm{N_2}}} & \left[\sum_{v_2,v_1',v_2'} \left(-Y_{v_1}Y_{v_2}P_{v_1\to v_1'}^{v_2\to v_2'} + Y_{v_1'}Y_{v_2'}P_{v_1'\to v_1}^{v_2'\to v_2}\right)\right. \\ & \left. + \sum_{v_2,v_2'} \left(-Y_{v_1}Y_{v_2}P_{v_1\to d}^{v_2\to v_2'} + 2Y_{\mathrm{N}}^2 Y_{v_2'}\left(2n_{\mathrm{N_2}} + n_{\mathrm{N}}\right) P_{d\to v_1}^{v_2'\to v_2}\right)\right] \\ & + Z_{(\mathrm{N_2-N})}\left[\sum_{v_1'} \left(-Y_{v_1}P_{v_1\to v_1'} + Y_{v_1'}P_{v_1'\to v_1}\right) + \left(-Y_{v_1}P_{v_1\to d} + 2Y_{\mathrm{N}}^2\left(2n_{\mathrm{N_2}} + n_{\mathrm{N}}\right)P_{d\to v_1}\right)\right] \end{aligned} \tag{22}$$

$$\begin{aligned} \frac{dY_{\mathrm{N}}}{dt} = \frac{Z_{(\mathrm{N_2-N_2})}}{Y_{\mathrm{N_2}}} & \left[\sum_{v_1,v_2,v_2'} \left(-2Y_{\mathrm{N}}^2 Y_{v_2'}\left(2n_{\mathrm{N_2}} + n_{\mathrm{N}}\right) P_{d\to v_1}^{v_2'\to v_2} + Y_{v_1}Y_{v_2}P_{v_1\to d}^{v_2\to v_2'}\right)\right] \\ & + Z_{(\mathrm{N_2-N})}\left[\sum_{v_1} \left(-2Y_{\mathrm{N}}^2\left(2n_{\mathrm{N_2}} + n_{\mathrm{N}}\right)P_{d\to v_1} + Y_{v_1}P_{v_1\to d}\right)\right] \end{aligned} \tag{23}$$

Thus Y_{v_1} and Y_{N} are calculated by solving a first-order differential system as in the first part. For each step the mean physical parameters and the reduced populations may then be deduced.

3.4. Results and conclusion concerning the second part

Herein we assume in front of the shock: $P_1 = 390$ Pa, $T_1 = 271$ K, Mach = 10

* The reduced populations increase strongly immediately behind the shock wave. Then the vibrational excitation processes are counterbalanced by dissociation: thus the population curves become very smoothly decreasing (Figs.4,5).

* Excluding the strong excitation zone, the populations obtained with the harmonic approach are close to those obtained with the anharmonic one when V-V processes are neglected (Figs.4,5).

* The influence of V-V process increases with the level and involves the anharmonicity effects (Figs.4,5,6).

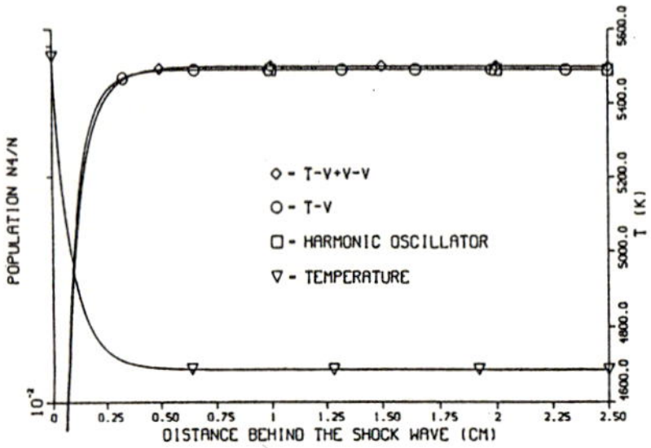

Fig. 4. Vibrational population (n_4/n) in N_2 as a function of the shock distance. Behind the shock front: $P = 45434.9$ Pa, $T = 5525$ K and $u = 549$ m.s^{-1}

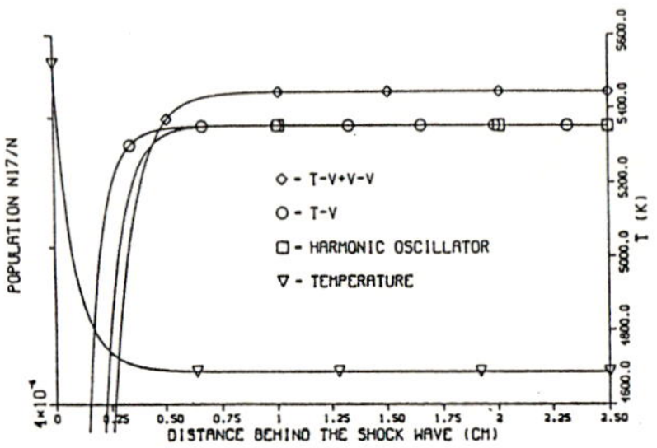

Fig. 5. Vibrational population (n_{17}/n) in N_2 as a function of the shock distance. Behind the shock front the conditions are the same as in Fig.4

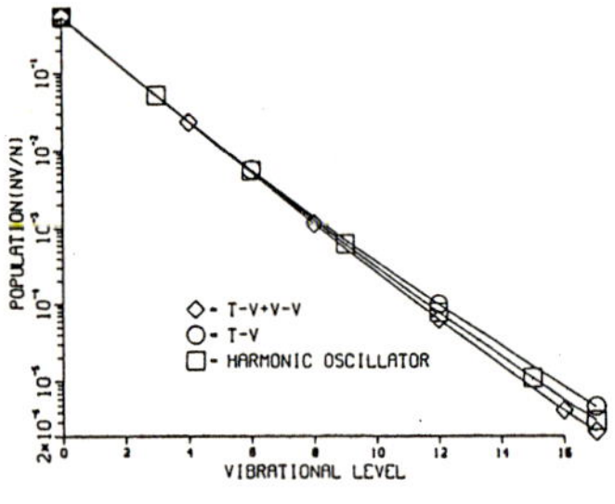

Fig. 6. Vibrational population (n_v/n) in N_2 behind the shock: $T = 4686.4$ K, $u = 479.7$ m.s^{-1} and the distance from the shock wave is 1.99 cm

References

Brun R, Zappoli B (1977) Model equations for a vibrationally relaxing gas. Phys. Fluids 20: 1441-1448

Landau L, Teller E (1936) Theory of sound dispersion. Phys. Zeit. Sowj. 10: 34-43

Marrone PV, Treanor CE (1963) Chemical relaxation with preferential dissociation from excited vibrational level. Phys. Fluids 6:1215-1221

Méolans JG, Chauvin AH (1991) Scattering cross sections for homonuclear diatomic molecule collisions: influence of the angular potential dependence. AIAA Paper-91-1340, 26th Thermophysics Conf., Honolulu, Hawaii

Schwartz RN, Herzfeld KF(1954) Vibrational relaxation times in gases. J. Chem. Phys. 20:1591-1599

Stupochenko YV, Losev SA, Osipov AI (1967) Relaxation in shock waves. Springer-Verlag, NY

Two-Temperature Chemical Kinetics. Dissociation of Diatomic Molecules Behind a Strong Shock Wave

Stalii A. Losev and Oleg P. Shatalov
Institute of Mechanics, Moscow State University, Moscow, Russia

Abstract. Experimental data about the vibrational temperature during the process of dissociation of diatomic molecules are discussed. The experiments allowed to substantiate the two-temperature model of the dissociation rate constant $K_d(T, T_v)$. Constancy of the vibrational temperature T_v, revealed in experiments during the quasistationary stage of the dissociation permits to present the rate constant as a function of T (translational temperature) only, with T_v=const, and to recommend an expression for K_d to the range of high and extremely high temperatures, where the experimental data are absent.

Key words: Shock wave, Vibrational non-equilibrium, Dissociation rate, Chemical kinetics

Attempts to describe the dissociation processes of molecules behind the front of a shock wave at high and extremely high temperatures, when the limiting stage is the vibrational excitation of these molecules, count more than thirty years. The hypothesis put forward by Montroll and Schuler (1957) about continuous evolution of Boltzmann distributions sequences from initial to equilibrium states during a vibrational relaxation process was extended to the dissociation process (Hammerling, Teare, Kivel 1959). Thereafter, the number of papers and approaches on this problem has grown. The dissociation model, described by Losev, Generalov (1961) and called β-model, is based on the assumption that the molecule dissociates from the vibrational levels situated near (less than βkT) the dissociation limit. According to this model the population distribution of the lower vibrational levels is Boltzmanian, with the temperature T_v ($\beta \sim 1$, with T the translational temperature, T_v the vibrational temperature, k the Boltzmann constant). In the papers of Treanor and Marrone (1962) and Marrone and Treanor (1963) two models were considered: the equal probability of the dissociation from any vibrational level and the preferential dissociation from the upper levels (both models assume the existence of the vibrational temperatures). In the later works (see, for example Macheret, Rich (1993), Smekhov, Zhluktov (1993)) a number of real properties such as anharmonicity, nonadiabatic character of collisions at high temperatures, rotation of the molecule, disturbance of Boltzmann distribution near the dissociation limit, etc. has been included in the dissociation models.

Theoretical models permit to describe the dissociation process at high temperatures, where there are no experimental data. For this aim, theoretical models are linked with available experimental data. As usual the theoretical model contains such parameters as effective interaction section, steric factor, interaction potential parameters, etc., which allow to link the model and the experimental data. But this procedure reserves for theory the function of qualitative description of the process only, and deprive the theory of quantitative value. Therefore it is not necessary to use the complex theoretical models in order to calculate the dissociation rate constant at high temperatures. In this paper one of the simplest models - the β model mentioned above (Losev, Generalov 1961) is used for the description of the dissociation processes in air (N_2, O_2 and NO molecules).

The data available at present permit to distinguish along the temperature abscissa three regions with a particular character of the dissociation of diatomic molecules in each of them.

At relatively low temperatures, which do not exceed 2θ (θ the characteristic vibrational temperature, corresponding to the energy of the first vibrational level of the molecule), the process of the vibrational relaxation is mainly accomplished before the beginning of the dissociation. Here the dissociation is one-temperature process, i.e. $T_v = T$. It is known that except some specific

Shock Waves @ Marseille II
Editors: R. Brun, L. Z. Dumitrescu

molecules a few collisions are sufficient to achieve rotational-translational equilibrium. Therefore we may consider that $T_r = T$ during the process of dissociation (T_r is the rotational temperature). Numerous articles and reviews (see for example Baulch et al. (1976), Krivonosova et al. (1987)) contain data about the dissociation rate constants in this temperature region.

At high temperatures the characteristic dissociation time becomes of the order of the vibrational relaxation time ($\tau_v \sim \tau_d$), the dissociation proceeds while vibrational relaxation is not complete, here $T_v < T$. In a number of experimental works, carried out in the sixties, a relatively slow rise of dissociation rate constant vs. temperature T was observed at these temperatures. Namely, the dissociation rate constant, presented in generalized Arrhenius form ($k_d = AT^n \exp(-D/kT)$), reveals a negative temperature dependence of the preexponent factor, the index n grows up with the temperature increase. At these temperatures the process of dissociation requires a two-temperature description.

The experimental data are absent in the last temperature region, where the temperature is still higher than in the previous one. However, it is possible to assume that the dissociation rate constant will increase owing to rotational decreasing of dissociation energy (Jaffe 1985, Kuznetsov 1982), molecular decomposition from lower vibrational levels, and increase of the interaction energy of the particles. The characteristic time of dissociation in this region (of the order of 100000 K for air molecules) is a few unit τ_0 (τ_0 the mean time between two collisions).

In this paper the macroscopic (not multilevel) approach is used. In such an approach the variables are the concentrations of gas components $[A_j]$ and the vibrational temperature as a measure of the vibrational energy. Chemical kinetics equations will be as follows:

$$d[A_j]/dt = k_d(T, T_v)[Aj][M], \tag{1}$$

where $k_d(T, T_v)$ is the dissociation rate constant, which in general depends on both temperatures: T and T_v, [M] is the concentration of the collider. There are some works devoted to the measurement of the vibrational temperature during the dissociation of diatomic molecules. Emission or absorption spectroscopic methods were used to determine the vibrational temperature as a function of the relative population of two vibrational levels. In particular, such measurements were carried out in nitrogen, dissociating behind the front of a shock wave at $T \leq 17000$ K (Yalovik, Losev 1972). Experiments were done in a shock tube, at shock wave velocities of 3-7 km/s; the absorption of vacuum ultraviolet ($\lambda = 120$ and 127 nm) was measured. Analogous experiments were carried out in oxygen. Vibrational temperature was determined from two-channel quartz ultraviolet absorption ($\lambda = 190$ and 230 nm) at $T \leq 10000$ K (Shatalov 1973). Recording of IR-emission of NO ($\lambda = 4.87$ and 5.22 μm) and measurement of the relative intensities on these two channels allows to determine the values of the vibrational temperature of dissociating molecules NO at different instants behind the front of the shock wave at $T = 4000 - 7500$ K (Koshi et al. 1979). Experiments in N_2 and O_2 gave information about the population of vibrational levels up to $v = 6$, experiments in NO - up to $v = 8$. Processing of experimental oscillograms together with conservation laws, state equation, Dalton's law and kinetic equations similar to Eq.1 makes it possible to determine the distribution of all other gas parameters behind the front of the shock wave: translational temperature T, density ρ, concentration of gas components $[A_j]$ and the dissociation rate constants $k_d(T, T_v)$. There is qualitative agreement of the results of these three investigations. When the translational temperature behind the front of the shock wave exceeds some characteristic value, which depends upon the sort of dissociating molecules, the vibrational temperature goes up to this value and then remains constant during the process of dissociation. These characteristic values for N_2, O_2, NO are shown in Table 1. Such a behavior of the vibrational temperature corresponds to the quasistationary stage of dissociation process. At this stage, according to the known CVD model, increase of vibrational energy of molecule during the vibrational relaxation is equilibrated by decrease of this energy owing to dissociation, so $d\varepsilon/dt = 0$ (ε vibrational energy). The two-temperature dissociation rate constant, according to the above-mentioned model, may be written as:

$$k_d(T,T_v) = k_d^0(T) \cdot Z(T,T_v)$$
$$Z(T,T_v) = \frac{1-\exp(-\theta/T_v)}{1-\exp(-\theta/T)} \cdot \exp\left[-\frac{D-\beta kT}{k} \cdot \left(\frac{1}{T_v} - \frac{1}{T}\right)\right] \tag{2}$$

Parameters θ and D are known and given in Table 2 together with the parameter β, derived from cited experimental works, and the temperature ranges, where these parameters were determined. $k_d^0(T)$ values are also known, they are the dissociation rate constants at low temperatures, where $T_v = T$ and $Z = 1$. In the present paper, the ladder model (Nikitin 1974) was used to approximate the experimental data for $k_d^0(T)$:

$$k_d^0(T) = A\exp(-D/kT)\,[1-\exp(-\theta/T)] \tag{3}$$

However modern aero-cosmic problems need information about the reaction rate constants at high and extremely high temperatures, where there are no experimental data. Here the extrapolation to these temperatures was done according to the β-model with the following assumptions: 1) the constancy of the vibrational temperature T_v, revealed experimentally up to $T \sim 10^4$ K, remains the same in the whole temperature range; 2) the parameter β is a constant up to temperature $T \sim D/k\beta$; 3) at temperatures $T > D/k\beta$ the nonequilibrium factor $Z(T,T_v)$ is equal to 1. The latter assumption corresponds to extremely high temperatures, where dissociation occurs from the lowest level and $\tau_d \sim \tau_0$. Because of the approximate character of the model the application limits of the last correspond to temperatures lower than values satisfying the condition $\beta kT = D$.

Table 1. Vibrational temperature of diatomic air molecules during quasistationary dissociation

M	T_v, K	T, K
O_2	6000±500	>6500
N_2	7500±500	>8000
NO	5000±500	>5500

The approach presented here allows to use a one-temperature description of the dissociation process instead of a two-temperature one. In order to determine the one-temperature rate constant, values of T_v and β from Table 2 may be substituted into the expression for $k_d(T,T_v)$ in Eq.2. k_d values determined in this fashion are represented in Table 3. Thus the considered model permits to analyse the process of the dissociation of the O_2, N_2 and NO molecules, taking into account vibrational nonequilibrium at temperatures as high as a few tens of thousand degrees, where experimental data are absent.

Table 2. O_2, N_2 and NO dissociation energy D/k, characteristic vibrational temperature θ, parameter β and temperature range T over which β was determined

	O_2	N_2	NO
D/k, K	59380	113200	75500
θ, K	2240	3354	2700
β	1.5	3	3
T, K	6500-10000	8000-17000	5500-7500

Table 3. Equilibrium ($T_v = T$) and nonequilibrium ($T_v < T$) dissociation rate constants of O_2, N_2 and NO

T, K	Reaction, k_d, cm^3/mol.s	Note
	$O_2+O_2 \longrightarrow O_2+O+O$	
$T \leq 6000$	$k_d = k_d^0 = 2.2 \cdot 10^{16} \exp(-59380/T)[1-\exp(-2240/T)]$	$T_v = T$
$6000 < T < 30000$	$k_d = 7.64 \cdot 10^{10} \exp(T/4000)$	$T_v = 6000$ K
$T > 30000$	$k_d = k_d^0$	$T_v \leq 6000$ K
	$N_2+N_2 \longrightarrow N_2+N+N$	
$T \leq 7500$	$k_d = k_d^0 = 3.2 \cdot 10^{16} \exp(-113200/T)[1-\exp(-3354/T)]$	$T_v = T$
$7500 < T < 30000$	$k_d = 1.6 \cdot 10^8 \exp(T/2500)$	$T_v = 7500$ K
$T > 30000$	$k_d = k_d^0$	$T_v \leq 7500$ K
	$NO+NO \longrightarrow NO+N+O$	
$T \leq 5000$	$k_d = k_d^0 = 6.1 \cdot 10^{15} \exp(-755000\ /T)[1-\exp(-2700/T)]$	$T_v = T$
$5000 < T < 20000$	$k_d = 3.5 \cdot 10^7 \exp(T/1667)$	$T_v = 5000$ K
$T > 20000$	$k_d = k_d^0$	$T_v \leq 5000$ K

References

Baulch DL et al. (1976) Evaluated kinetic data for high temperature reactions. Vol.3. Butterworths, London

Hammerling P, Teare JD, Kivel B (1959) Theory of reaction from luminous waves in nitrogen. Phys. Fluids 2:422

Jaffe RL (1985) Rate constants for chemical reactions in high-temperature non-equilibrium air. AIAA Paper 1038

Koshi M, Bando S, Saito M, Asabe T (1979) Dissociation of nitric oxide in shock waves. In: 17-th Symp. (International) on Combustion, (Leeds 1978) Pittsburg, p.553

Krivonosova OE et al. (1987) Recommended data on chemical reaction rate constants of molecules, consisting of N-O atoms. In: Smirnov BM (ed) Khimiya Plazmy. Energoatomizdat, Moscow, 14, pp 3-23

Kuznetsov NM (1982) Kinetics of monomolecular reactions. Nauka, Moscow (in Russian)

Losev SA, Generalov NA (1961) To the investigation of excitation of vibration and dissociation of oxygen molecules at high temperature. Docl. Acad. Nauk SSSR (in Russian) 141:1072

Macheret SO, Rich JW (1993) Theory of nonequilibrium dissociation rates behind strong shock waves: Classical model. AIAA Paper 93-2860

Marrone PV, Treanor CE (1963) Chemical relaxation with preferential dissociation from excited vibrational level. Phys. Fluids 6:1215

Montroll EM, Schuler KE (1957) Studies in nonequilibrium rate processes. I. The relaxation of a system of harmonic oscillators. J. Chem. Phys. 26:454

Nikitin EE (1974) Theory of elementary atomic and molecular processes in gases. Clarendon Press, Oxford

Shatalov OP (1973) On the dissociation of oxygen molecules in absence of vibrational equilibrium. Fizika Gorenija i Vzriva (in Russian) 5:699

Smekhov GD, Zhluktov SV (1993) Dissociation of two-atomic molecules at extremely high temperatures. Khimicheskaya Physica (in Russian) 12:337

Treanor CE, Marrone PV (1962) Effect of dissociation on the rate of vibrational relaxation. Phys. Fluids 5:1022

Yalovik MS, Losev SA (1972) Kinetics of vibrational excitation and dissociation of nitrogen at high temperature. In: Losev S.A. (ed) The problems of physico-chemical kinetics of high-speed gas flows (in Russian). Moscow Univ. Press, Moscow, pp 4-34

Experimental Investigation of Non-Equilibrium Temperatures in a Plane Shock Wave

Yury V. Romanenko, Oleg P. Shatalov and Igor E. Zabelinsky
Institute of Mechanics, Moscow State University, Moscow, Russia

Abstract. The radiation peak of O_2, N_2 and Ar behind the front of a shock wave was experimentally investigated in a shock tube; the possible connection of this radiation with a "longitudinal" translation temperature in accordance to the phenomenological model of Mott-Smith (1951) is discussed.

Key words: Shock wave, Radiation, "Maxwellization", Longitudinal temperature

We report on an experiment that studied the radiation characteristics of a heated gas behind the front of a plane shock wave. The investigations were made in N_2, O_2 and Ar with the initial pressure $P_1 = 1 \cdots 20$ Torr and Mach number $M_s = 5.6 \cdots 10$. The experiment was carried out in a cylindrical shock tube of 50 mm ID. The radiation intensities were measured with the help of a photomultiplier, optical filters and a monochromator BM-50. The space resolution of the optical system was 0.5 mm. The electronic measurement section had a cut-off frequency of 10 MHz. Measurements of the speed of the shock wave were done by piezoelectric gauges. Signals from PM and gauges were written on a transient recorder DL922, and were entered into an IBM PC/AT computer. The shock speed and initial pressure were measured to an accuracy of 2%. The purity of investigated gases was 99.999% for N_2 and 99.99% for Ar. Commercially pure oxygen was used. Oxygen was purified at the temperature of liquid nitrogen.

The measurements of radiation intensity in argon, nitrogen and oxygen revealed an explicit peak immediately behind the front of the shock wave in the range of Mach numbers (Fig.1a). Earlier a peak of radiation has been observed in different gases. Traditionally the existence of this peak was connected with the presence of easily excitable impurities in such gases (Allen 1962). Present measurements revealed the strongest intensity for argon and the weakest one for oxygen (Fig.1a). When the Mach number was increased the radiation behind the shock front became stronger and the peak becomes undistinguishable. This phenomenon is most clearly observed in argon. The peak radiates within the visible range in N_2 and Ar (Fig.1c) and within the ultra-violet range in O_2 (Fig.1b). That leads to an assumption that the peak radiation is least linked with impurities in gases and mostly related to the main components of the gas. Radiation in the ultra-violet and visible ranges of the spectrum corresponds to transitions between excited electronic states of particles. In the limit of the resolution of the experimental methods the radiation peak is observed immediately behind the front of the shock wave. It is difficult to connect this radiation with some relaxation processes at gas parametrs behind the front of shock wave (P_2 and T_2).

For the excitation of O_2, N_2, Ar electronic states, a temperature much greater than T_2 is needed. The existence of such a high, but nonequilibrium "longitudinal" temperature directly within the front of the shock wave is infered from the pioneering work by Mott-Smith (1951) and is connected with the process of "maxwellization" (interaction of low pressure gas moving with velocity V_s and high pressure gas). This particular mechanism explaining the origin of the radiation peak immediately behind the front of the shock wave is considered and discussed in the present work. "Maxwellization" process is the worst in argon, because argon is better described by the model of hard spheres. As a result argon has a higher "longitudinal" temperature than nitrogen and oxygen and more intensive radiation behind the front. Therefore, further investigations were carried out in argon. A two-step ionisation model of the noble gas is often assumed:

Shock Waves @ Marseille II
Editors: R. Brun, L. Z. Dumitrescu

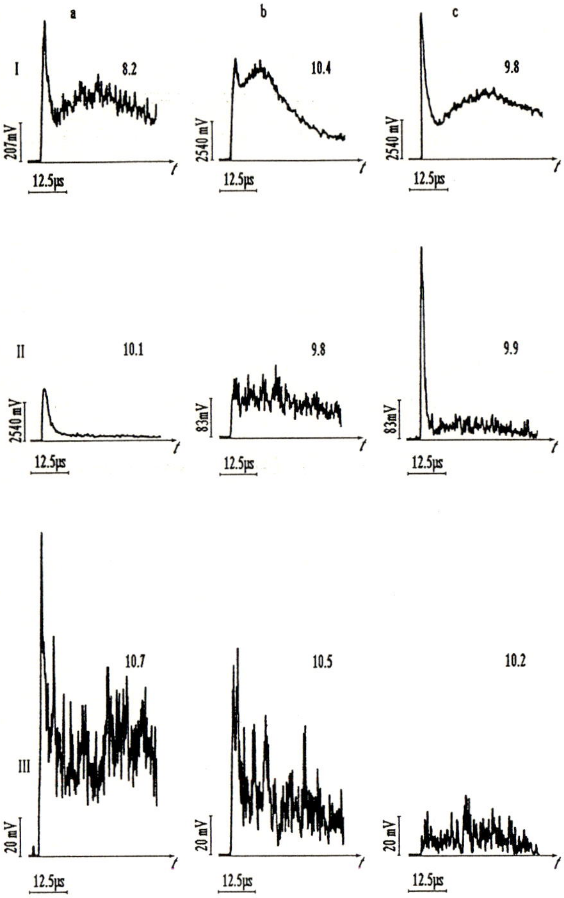

Fig. 1. Oscillograms of radiation behind the front of a shock wave:
(a): $\lambda = 200 \cdots 800$ nm;
(b): $\lambda = 200 \cdots 400$ nm;
(c): $\lambda = 400 \cdots 800$ nm;
I - Ar; II - N_2; III - O_2; $P_1 = 1$ Torr. Numbers above oscillograms are M_s values

$$\mathrm{Ar} + \mathrm{Ar} \longrightarrow \mathrm{Ar}^* + \mathrm{Ar} \qquad - 11.6\mathrm{eV} \tag{1}$$

$$\mathrm{Ar} + \mathrm{Ar}^* \longrightarrow \mathrm{Ar} + \mathrm{Ar}^+ + \mathrm{e}^- \qquad - 4.3\mathrm{eV} \tag{2}$$

We proposed that the process (1) also occurs in the front of a shock wave, where according to the conservation laws in nonelastic atomic collisions, in particular, atoms in a metastable level $4s'[1/2]_0$ with energy $E = 11.71$ eV are produced in the following process:

$$\mathrm{Ar}\{^1S_0\} + \mathrm{Ar}\{^1S_0\} \longrightarrow \mathrm{Ar}\{4s'[1/2]_0\} + \mathrm{Ar}\{^1S_0\} \tag{3}$$

Immediately behind the front there are processes of further excitation of the argon metastable level:

$$\mathrm{Ar}\{4s\prime[1/2]_0\} + \mathrm{Ar}\{^1S_0\} \longrightarrow \mathrm{Ar}\{nl[J]_0\} + \mathrm{Ar}\{^1S_0\} \tag{4}$$

where n, l, J are quantum numbers of the atomic excited level. Levels $\mathrm{Ar}\{nl[J]_0\}$ decay rapidly with emission which forms a bright peak of radiation immediately behind the shock front.

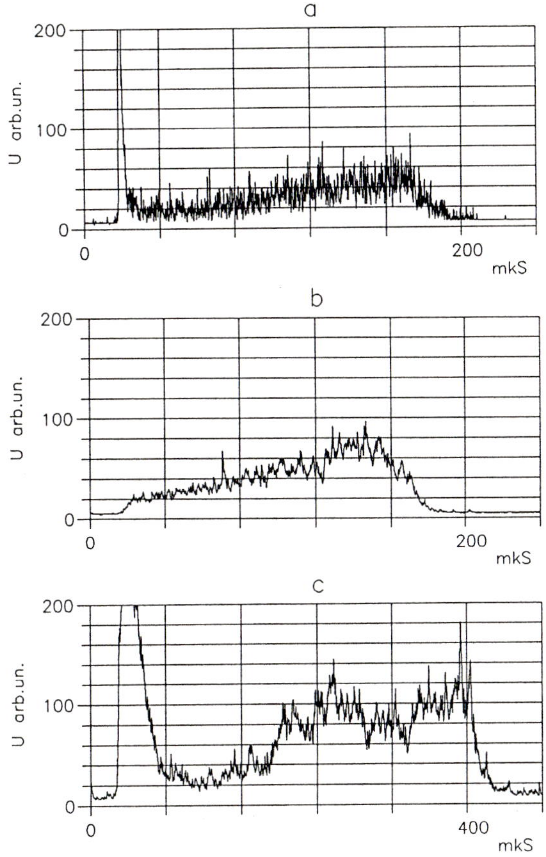

Fig. 2. Oscillograms of radiation behind the front of a shock wave:
(a):Ar, $P_1 = 5$ Torr, $M_s = 8.3$, $\lambda = 451.073$ nm, $\Delta\lambda = 3.3$ nm;
(b):Ar, $P_1 = 5$ Torr, $M_s = 8.6$, $\lambda = 738.398$ nm, $\Delta\lambda = 3.3$ nm;
(c):Ar, $P_1 = 20$ Torr, $M_s = 5.6$, $\Delta\lambda = 200 \cdots 800$ nm

In Fig.2a the radiation peak is seen at $\lambda = 451.073$ nm, here $M_s = 8.3$, $P_1 = 5$ torr and $\Delta\lambda = 3.3$ nm. We proposed that this radiation is due to the following process:

$$\mathrm{Ar}\{5p[1/2]_0\} \longrightarrow \mathrm{Ar}\{4s'[1/2]_1\} + \hbar\nu \qquad (\lambda = 451.073\,\mathrm{nm}) \tag{5}$$

Levels with another moment must show a different behaviour. For example, the level $\mathrm{Ar}\{4p'[3/2]_2\}$ is unlikely to form in the front in the process (4) because of the impossibility of the moment change in this reaction. Such a level may occur in further processes and interactions. The measurement of radiation in transition (6)

$$\mathrm{Ar}\{4p'[3/2]_2\} \longrightarrow \mathrm{Ar}\{4s[3/2]_1\} + \hbar\nu \qquad (\lambda = 738.398\,\mathrm{nm}) \tag{6}$$

points to the absence of increase of this radiation immediately behind the shock front (Fig.2b). Thus a spectrum analysis of the radiation peak in Ar at $M_s = 8.3$, and $P_1 = 5$ Torr demonstrates the presence of high-lying levels with energies 12-15 eV behind the front of the shock wave. Preliminary quantitative estimates of the radiation intensity show, that the last one in the transition $5p[1/2]_0 \longrightarrow 4s'[1/2]_1$ in the conditions of Fig. 2a is approximately 10^3 times greater than it would follow from routine shock wave equations. So this allows to suppose the existence of a high-temperature region in the front of the shock wave, where the additional metastable levels are produced. The radiation peak in our experiments was observed behind the front of the shock wave even at low Mach numbers ($M_s = 5.6$, temperature behind the shock front $T_2 = 3.1 \cdot 10^3$ K). If we assume that spectral properties are similar to those at $M_s = 8.3$, then the appearance of metastable atoms with the energy $E = 11.71$ eV is possible only if there is a region inside the front of shock wave with high "longitudinal" temperature. According to the Mott-Smith phenomenological model the maximum "longitudinal" temperature would be $T_x = 25 \cdot 10^3$ K, as given by the following equation (Losev et al. 1989) in the center of the shock front at $x = 0$:

$$T_x = (m/2k) \cdot \left[(2kT_1/m)^{1/2} + V_s + (2kT_2/m)^{1/2}\right]^{1/2} \tag{7}$$

This temperature is high enough to produce argon in the $4s'[1/2]_0$ level inside the shock front.

Experiments of this kind allow to distinguish in future between Mott-Smith and other models of the shock wave front structure (Bashlikov and Velikodny 1989; Cercignani 1975; Genich et al. 1990).

References

Allen RA, Camm JC, Keck JC (1962) Radiation of hot nitrogen. J.Q.S.R.T. 1:269

Bashlikov AM, Velikodny VY (1989) Nonequilibrium processes in the front of shock wave. Letters in Journal of Technical Physics (in Russian) 15, No 5:24

Cercignani C (1975) Theory and application of the Boltzmann equation. Scottish Academic Press. Edinburgh and London

Genich AP, Kulikov SI, Manelis GV, Chereshnev SA (1990) Molecular velocity distribution function in a shock wave front in gas mixtures. Mechanica Jidkosti i Gasa (in Russian) 2:144

Losev SA, Romanenko YV, Shatalov OP, Zabelinsky IE (1989) Investigation of air radiation parameters behind the front of incident shock wave. Preprint No 3836 (in Russian). Institute of Mechanics, Moscow State University, Moscow

Mott-Smith HM (1951) The solution of the Boltzmann equation for a shock wave. Phys. Rev. 82:885

Nonequilibrium Flow Around an Injection Jet into a Hypersonic Flow

Yumiko Inoue* **and A. Koichi Hayashi**†

* Research Engineering Department, Nagoya Aerospace Systems, Mitsubishi Heavy Industries, ltd. † Department of Aeronautical Engineering, Nagoya University, Nagoya 464-01, Japan

Abstract. Flowfields of jets with injected angles of 45° and 90° into a main hypersonic flow of $M_\infty = 20$ at 70 km altitude are simulated numerically by solving the two-dimensional Reynolds-averaged full Navier-Stokes equations with a N_2/O_2 chemical kinetic model and two-temperature model. The Harten-Yee type TVD scheme and the central difference scheme are used to calculate the convection and the viscous terms of the governing equations, respectively. The results capture the bow and separation shocks near the injector. The pressure of the 45° jet is higher than that of the 90° jet, but the temperature of the 45° jet is lower than that of the 90° jet. The moment of the 45° jet is not different from that of the 90° jet. The jet leaning toward the main flow by 45° is bent suddenly downstream due to the main flow and has a weak impingement on the wall downstream the injector.

Key words: Hypersonic flow, Injection jet, Two-temperature model

1. Introduction

Transversely injected jets into supersonic and hypersonic flows have been studied by many scientists, since the flowfield around the jet is applicable to many applications including the fuel injection system of SCRAM jet engine combustors and Reaction Control System (RCS) jets; one of the attitude control systems for a hypersonic flight vehicle. It is due to the jet interaction with the main flow that strong shock waves, turbulent boundary layers, boundary layer separation regions, and injected secondary flows appear near the injector of a side jet. When a jet is injected into the hypersonic flow, $20 \le M_\infty \le 30$, from the blunt body, the total temperature of the jet is lower than 1000 K and the maximum temperature in the outer flowfield is about 20000 - 30000 K. In this case, the dissociation, ionization, and recombination reactions start among molecules, atoms, ions, and electrons in the flowfield where the temperature is a few hundreds up to several tens of thousands Kelvin.

In order to study such flowfields, the thermally and chemically nonequilibrium flow will be analyzed using multi-temperature models.

In the present study, the flowfields of a N_2 jet injected with 45° and 90° angles into a main flow of Mach 20 - 30 are simulated numerically to evaluate the moment acting upon the body and the heat flux on the body surface. The governing equations consist of the conservation equations of the species, two-dimensional Reynolds-averaged full Navier-Stokes equations, total energy conservation equation, vibrational-electronic energy conservation equations with a two-temperature model, and a N_2/O_2 chemical reaction mechanism. These equations with real gas effects are integrated numerically using the Harten-Yee type explicit TVD scheme for the convection terms and a central difference scheme for the viscous terms.

2 Modeling

2.1. Governing equations

The governing equations consist of the two-dimensional conservation equations of the species, momentum, total energy and vibrational-electronic energy in conservative form:

$$\frac{\partial \boldsymbol{U}}{\partial t} + \frac{\partial \boldsymbol{F}}{\partial x} + \frac{\partial \boldsymbol{G}}{\partial y} = \frac{\partial \boldsymbol{F}_v}{\partial x} + \frac{\partial \boldsymbol{G}_v}{\partial y} + \boldsymbol{S}_c + \boldsymbol{S}_v \tag{1}$$

Shock Waves @ Marseille II
Editors: R. Brun, L. Z. Dumitrescu

where the conservative vector $\boldsymbol{U}$, the convection and viscous terms in the x-direction $\boldsymbol{F}$ and $\boldsymbol{F}_v$, and the chemical and the relaxational source terms $\boldsymbol{S}_c$ and $\boldsymbol{S}_v$ are:

$$\boldsymbol{U}=\begin{pmatrix}\rho_i\\ \rho u\\ \rho v\\ e\\ e_v\end{pmatrix},\quad \boldsymbol{F}=\begin{pmatrix}\rho_i u\\ \rho u^2+p\\ \rho uv\\ (e+p)u\\ e_v u\end{pmatrix},\quad \boldsymbol{F}_v=\begin{pmatrix}\rho D_i\frac{\partial Y_i}{\partial x}\\ \tau_{xx}\\ \tau_{xy}\\ u\tau_{xx}+q_x\\ q_{vx}\end{pmatrix},$$
$$\boldsymbol{S}_c=\begin{pmatrix}\dot\omega_i\\ 0\\ 0\\ 0\\ 0\end{pmatrix},\quad \boldsymbol{S}_v=\begin{pmatrix}0\\ 0\\ 0\\ 0\\ Q_T-P_e\left(\frac{\partial u}{\partial x}+\frac{\partial v}{\partial y}\right)\end{pmatrix};\qquad (i=1,\cdots N_s) \tag{2}$$

The convection and viscous terms in the y-direction are similar to the terms in the x-direction. The variables in Eq.2 are

$$e=\sum_{j=1}^{N_s}\rho_j h_j+\frac{\rho}{2}(u^2+v^2) \tag{3}$$

$$p=\sum_{j=1}^{N_s-1}\rho_j R_j T+\rho_e R_e T_v, \tag{4}$$

$$\tau_{xx}=\lambda\frac{\partial u}{\partial x}+2\mu\frac{\partial u}{\partial x},\qquad \lambda=-\frac{2}{3}\mu \tag{5}$$

$$q_x=\kappa\frac{\partial T}{\partial x}+\kappa_v\frac{\partial T_v}{\partial x}+\rho\sum_{j=1}^{Ns}D_j h_j\frac{\partial Y_j}{\partial x} \tag{6}$$

$$q_{vx}=\kappa_v\frac{\partial T_v}{\partial x}+\rho\sum_{j=1}^{Ns}D_j h_{vj}\frac{\partial Y_j}{\partial x} \tag{7}$$

$$Q_T=\sum_s^{\text{mol}}\rho_s\frac{\epsilon^*_{v,s}-\epsilon_{v,s}}{<\tau_s>}+2\rho_e\frac{3}{2}R_0(T-T_v)\sum_s^{N_s-1}\frac{\nu_{e,s}}{M_s}-\sum_s^{\text{ion}}\dot{n_{e,s}}\hat{I}_s+\sum_s^{\text{mol}}\dot\omega_s\hat{D}_s \tag{8}$$

$$C_p=\sum_{s=1}^{Ns}C_p^s \tag{9}$$

$$C^s_{p\,tr}=\begin{cases}\frac{7}{2}R_s & (s=\text{molecules})\\ \frac{5}{2}R_s & (s=\text{atoms})\\ 0 & (s=\text{electrons})\end{cases} \tag{10}$$

$$C^s_{pv}=C^s_p-C^s_{ptr} \tag{11}$$

where Y_j is the mass fraction of j-species and N_s is the number of chemical species considered in this study. R_s is the gas constant of s-species. The sum from 1 to N_s-1 is expressed as the sum of the heavy particles. When real gas effects are considered, the specific heat of each species C_p^s is approximately expressed by a fourth-order polynomial in the translational temperature T.

2.2. Transport properties

In the two-temperature model, the transport properties are evaluated using Yos's formula cited by Lee (1984). The molecular viscosity coefficient μ is calculated by the following formula:

$$\mu = \sum_{s=1}^{N_s-1} \frac{m_s X_s}{\sum_{r=1}^{N_s-1} X_r \Delta_{sr}^{(2)}(T) + X_e \Delta_{se}^{(2)}(T_e)} + \frac{m_e X_e}{\sum_{r=1}^{N_s-1} X_r \Delta_{er}^{(2)}(T_V)} \tag{12}$$

The thermal conductivity coefficients κ and κ_v are evaluated from each energy mode of the molecules. The translational conductivity coefficients κ_{tr} of the heavy particles are:

$$\kappa_{tr} = \frac{15}{4} k \sum_{s=1}^{N_s-1} \frac{X_s}{\sum_{r=1}^{N_s-1} a_{sr} X_r \Delta_{sr}^{(2)}(T) + 3.54 X_e \Delta_{se}^{(2)}(T_e)} \tag{13}$$

where a_{sr} is

$$a_{sr} = 1 + \frac{(1 - M_s/M_r)(0.45 - 2.54 M_s/M_r)}{(1 + M_s/M_r)^2} \tag{14}$$

The rotational energy mode conductivity κ_r is

$$\kappa_r = k \sum_{s=\mathrm{mol}} \frac{X_s}{\sum_{r=\mathrm{mol}} X_r \Delta_{sr}^{(1)}(T) + X_e \Delta_{se}^{(1)}(T_e)} \tag{15}$$

The translational-rotational thermal conductivity coefficient κ in Eq.6 is obtained by the sum of κ_{tr} and κ_r as

$$\kappa = \kappa_{tr} + \kappa_r \tag{16}$$

The vibrational energy mode conductivity is assumed to be the same as the rotational energy mode conductivity. The electronic energy mode conductivity is calculated by the following equation:

$$\kappa_e = \frac{15}{4} k \frac{X_e}{\sum_{r=1}^{N_s} 1.45 X_r \Delta_{er}^{(2)}(T_V)} \tag{17}$$

The vibrational-electronic thermal conductivity in Eqs.6 and 7 is:

$$\kappa_v = \kappa_{vib} + \kappa_e \tag{18}$$

Modified collision integrals in Eqs. 12, 13, 15, and 17 are evaluated by the following equations:

$$\Delta_{sr}^{(1)}(T) = \frac{8}{3} \left[\frac{2 M_s M_r}{\pi R_0 T (M_s + M_r)} \right]^{1/2} \pi \overline{\Omega_{sr}^{(1,1)}} \tag{19}$$

$$\Delta_{sr}^{(2)}(T) = \frac{16}{5} \left[\frac{2 M_s M_r}{\pi R_0 T (M_s + M_r)} \right]^{1/2} \pi \overline{\Omega_{sr}^{(2,2)}} \tag{20}$$

where the collision integrals $\overline{\Omega_{sr}^{(1,1)}}$, $\frac{\Omega_{sr}}{\sigma_{sr}^2}$, and $\frac{\Omega_{se}}{\sigma_{sr}^2}$ are:

$$\overline{\Omega_{sr}^{(1,1)}} = \overline{\Omega_{sr}^{(2,2)}} = \Omega_{sr} \tag{21}$$

$$\frac{\Omega_{sr}}{\sigma_{sr}^2} = 1.147 \left(\frac{T}{T_{\epsilon,sr}} \right)^{-0.145} + \left(\frac{T}{T_{\epsilon,sr}} + 0.5 \right)^{-2.0} \tag{22}$$

$$\frac{\Omega_{se}}{\sigma_s^2} = 1.147 \left(\frac{T_V}{T_{\epsilon,s}} \right)^{-0.145} + \left(\frac{T_V}{T_{\epsilon,s}} + 0.5 \right)^{-2.0} \tag{23}$$

where $\sigma_{sr} = (\sigma_s \sigma_r)/2$ and $T_{\epsilon,sr} = \sqrt{T_{\epsilon,s} T_{\epsilon,r}}$. The effective temperature $T_{\epsilon,s}$ and collision diameter σ_s are obtained from Gnoffo (1989).

The effective molecular diffusion coefficient of the s-th species, D_s is evaluated from the binary diffusion coefficient, D_{sr} and D_{se} as follows:

$$D_s = \frac{1 - Y_s}{\sum_{r=1,r\neq s}^{N_s} X_r / D_{sr}} \tag{24}$$

$$D_{sr} = \frac{kT}{p\Delta_{sr}^{(1)}(T)} \tag{25}$$

$$D_{er} = \frac{kT_V}{p\Delta_{er}^{(1)}(T_V)} \tag{26}$$

The ion diffusion is modeled by an ambipolar diffusion:

$$D^a_{\text{ion}} = 2D_{\text{ion}} \tag{27}$$

The electron diffusion is evaluated by the assumption of the zero conduction current, as:

$$D_e = m_e \frac{\sum_{s=\text{ion}} D^a_s X_s}{\sum_{s=\text{ion}} m_s X_s} \tag{28}$$

2.3. Vibrational-electronic relaxation

The source terms of vibrational-electronic energy equations are expressed as the vibrational and electronic relaxation. The first term in Eq.8 evaluates the energy exchange between the vibrational and translational energies by the molecular collisions. The vibrational relaxation characteristic time, $< \tau_s >$ is calculated by the Millikan and White model and the correction term by Park cited from Gnoffo (1989):

$$< \tau_s > = \tau_s^{MW} + \tau_s^P \tag{29}$$

$$p\tau_s^{MW} = \frac{\sum_{j=1}^{N_s-1} n_j \exp\left[A_s(T^{-1/3} - 0.015\mu_{s,j}^{1/4}) - 18.42\right]}{\sum_{j=1}^{N_s-1} n_j} \tag{30}$$

$$\tau_s^P = (\sigma_s \bar{c}_s n_s)^{-1} \tag{31}$$

The second term in Eq.8, which is used by Appleton and Bray (1964), shows the energy exchange between heavy particles and electrons by elastic collisions. The effective electronic collision frequency of the s-th species, ν_{es} is obtained as follows;

$$\nu_{e,s} = \begin{cases} n_s \sigma_{e,s} \left(\frac{8kT_e}{\pi m_e}\right) & (s = \text{ion}) \\ \frac{8}{3}\left(\frac{\pi}{m_e}\right)^{1/2} n_s e^4 \frac{1}{2kT_e^{3/2}} \ln\left(\frac{k^3 T_e^3}{\pi n_e e^6}\right) & (s = \text{molecule}) \end{cases} \tag{32}$$

The third and fourth terms in Eq.8 express the source terms of the vibrational and electronic energy by the ionization and dissociation, respectively. $\hat{D}_s$ is the vibrational energy loss per unit mass of molecules which dissociate with the rate $\dot{\omega}_s$. $\hat{I}_s$ is the first ionization energy of the s-th species per mole and $\dot{n}_e$ the mole ionization rate of the s-th species.

2.4. Chemical reaction model

In the present analysis, a N_2/O_2 chemical kinetic model is applied using 11 species (N_2, O_2, N, O, NO, N_2^+, O_2^+, N^+, O^+, NO^+, e) and 21 elementary reactions. This reaction model is proposed by Park (1991). The forward rate constants of each reaction are given in the following modified Arrhenius form:

$$k_{fj} = C_{fj} T_q^{n_j} \exp(-E_{fj}/kT_q). \tag{33}$$

T_q is the dummy temperature as $T_q = \sqrt{T^q T v^{q-1}}$ and Arrhenius constants are obtained from Park (1991). Table 1 shows 21 elementary reactions and q for each reaction. The backward rate constants are calculated from the concentration equilibrium constant K_{cj} as $k_{bj} = k_{fj}/K_{cj}$.

Table 1. N_2/O_2 chemical kinetic model (in cm^3–g·mole–sec–cal units)

	Reaction	q	C_f	n_f	E_f/κ
1	$O_2+M_1 = 2O+M_1$	0.5 E0	2.00 E0	-1.50	5.950 E04
	M_1=[N_2]+[O_2]+5[N]+5[O]+[NO]+[n_2^+]+[O_2^+]+5[N^+]+5[O^+]+[NO^+]				
2	$N_2+M_2=2N+M_2$	0.5 E0	7.00 E21	-1.60	11.32 E04
	M_2=[N_2]+[O_2]+4.29[N]+4.29[O]+[NO]+ [N_2^+]+[O_2^+]+4.29[N^+]+4.29[O^+]+[NO^+]+1714[e]				
3	$NO+M_3=N+O+M_3$	0.5 E0	5.00 E15	0.00	7.550 E04
	M_3=[N_2]+[O_2]+22[N]+22[O]+22[NO]+[N_2^+]+[O_2^+]+22[N+]+22[O^+]+[NO^+]				
4	$O+NO=N+O_2$	1.0 E0	8.40 E12	0.00	1.945 E04
5	$O+N_2=N+NO$	1.0 E0	6.40 E17	-1.00	3.840 E04
6	$O_2^++O=O_2+O^+$	1.0 E0	4.00 E12	-0.09	1.800 E04
7	$NO^++O=NO+O^+$	1.0 E0	2.75 E13	0.01	5.100 E04
8	$N_2+O^+=N_2^++O$	1.0 E0	9.10 E11	0.36	2.280 E04
9	$NO^++N=O+N_2^+$	1.0 E0	7.20 E13	0.00	3.500 E04
10	$NO^++O_2=NO+O_2$	1.0 E0	2.40 E13	.41	3.260 E04
11	$NO^++O=N^++O_2$	1.0 E0	1.00 E12	0.50	7.720 E04
12	$O_2^++N=N^++O_2$	1.0 E0	8.70 E13	0.14	2.860 E04
13	$O^++NO=N^++O_2$	1.0 E0	1.40 E05	1.90	2.660 E04
14	$O_2^++N_2=N_2^++O_2$	1.0 E0	9.90 E12	0.00	4.070 E04
15	$NO^++N=O^++N_2$	1.0 E0	3.40 E13	-1.08	1.280 E04
16	$NO^++O=O_2^++N$	1.0 E0	7.20 E12	0.29	4.860 E04
17	$N+O=NO^++e+e$	1.0 E0	8.80 E08	1.00	3.190 E04
18	$O+O=O_2^++e+e$	1.0 E0	7.10 E02	2.70	8.060 E04
19	$N+N=N_2^++e+e$	1.0 E0	4.40 E07	1.50	6.750 E04
20	$O+e=O^++e+e$	0.0 E0	3.90 E33	-3.78	15.85 E04
21	$N+e=N^++e+e$	0.0 E0	2.50 E34	-3.82	16.86 E04

2.5. Numerical method

The governing equations in the physical domain (x,y) are transformed into the computational domain (ξ,η) to be solved on a uniform grid system. The Harten-Yee type TVD (Yee 1989) and central difference schemes are applied to the convection and viscous term, respectively.

Initial and Boundary Conditions

In this analysis, the main flow conditions are set up at the altitude of 70 km; the pressure p_∞ of 5.221×10^{-6} MPa and temperature T_∞ of 219.59 K. The Mach number of the main flow M_∞ is 20. The pressure of the N_2 jet p_j is 6.201×10^{-3} MPa, the jet temperature T_j 300.0 K, and the jet Mach number M_j unity. The pressure ratio between the jet and the main flow p_j/p_∞ is 1.188×10^3. The injector width is 1.8 m. The blunt nose radius is 20 m. Two cases of the injection angles are calculated: the angle of the injection δ in the first case is 90° from the wall while in the second case it is 45° from the wall with the jet inclined upstream.

The grid points of 193 × 51 in x- and y-direction, respectively are used in the physical region shown in Fig.1. The grid points are clustered around the injector to resolve the flowfield accurately. The wall boundary condition is non-slip and adiabatic.

3. Results and discussion

In order to calculate the case with chemical reactions and vibrational-electronic relaxations, the flowfields without chemical reactions and relaxations are simulated until the flow becomes steady, then they are set for the initial conditions of the case with chemical reactions and relaxations. The steady flow is judged from the condition that the order of magnitude of the L_2-norm of the residual is -11 ~ -12 and the maximum temperature of the whole flowfield is not changed.

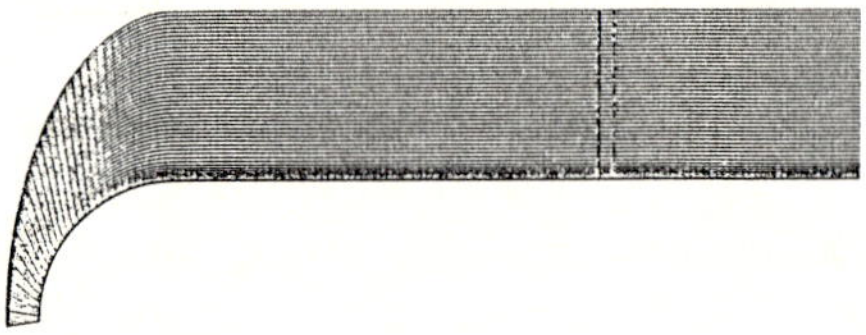

Fig. 1. Computational domain

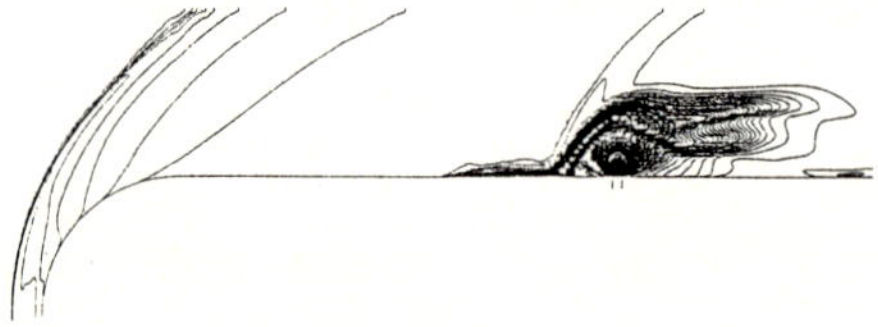

Fig. 7. Density contours (Case 1, $\delta = 90°$)

Fig. 2. Pressure contours (Case 1, $\delta = 90°$)

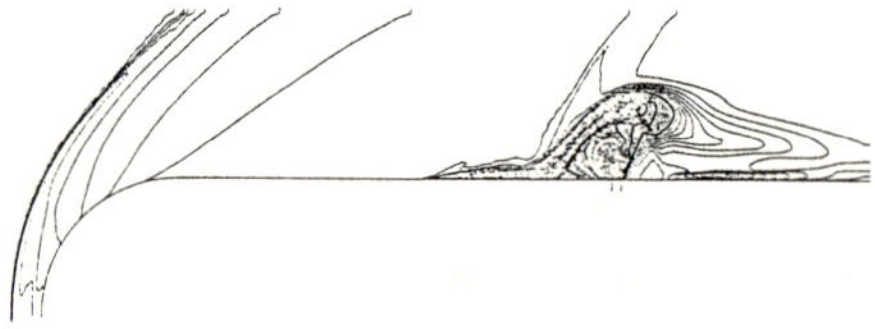

Fig. 8. Density contours (Case 2, $\delta = 45°$)

Fig. 3. Pressure contours (Case 2, $\delta = 45°$)

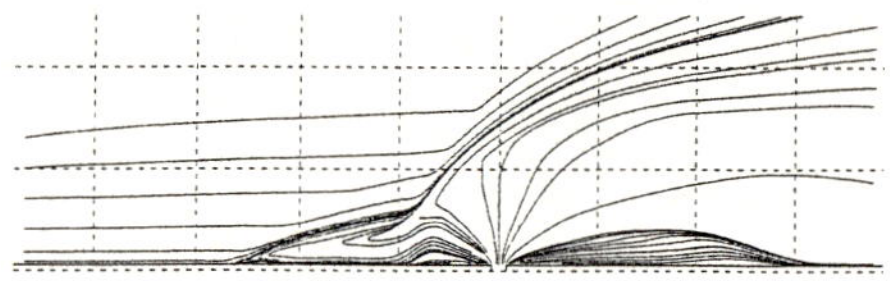

Fig. 9. Streamline diagrams (Close up near the injector, Case 1)

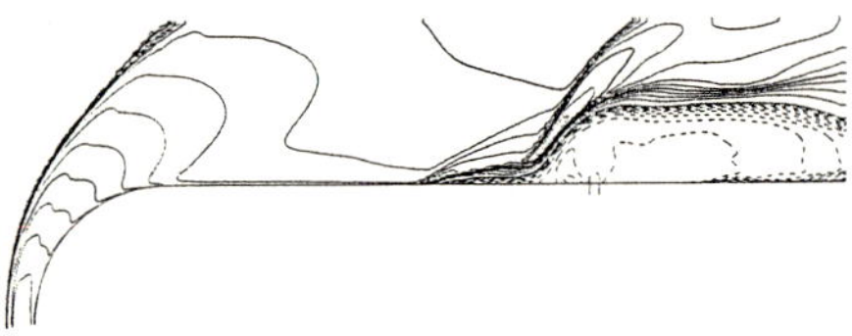

Fig. 4. Temperature contours (Case 1, $\delta = 90°$)

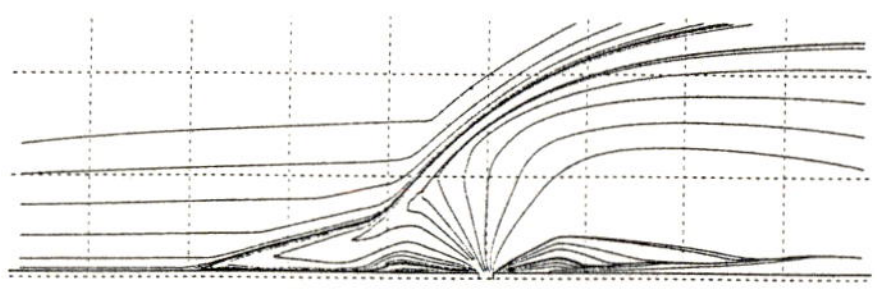

Fig. 10. Streamline diagrams (Close up near the injector, Case 2)

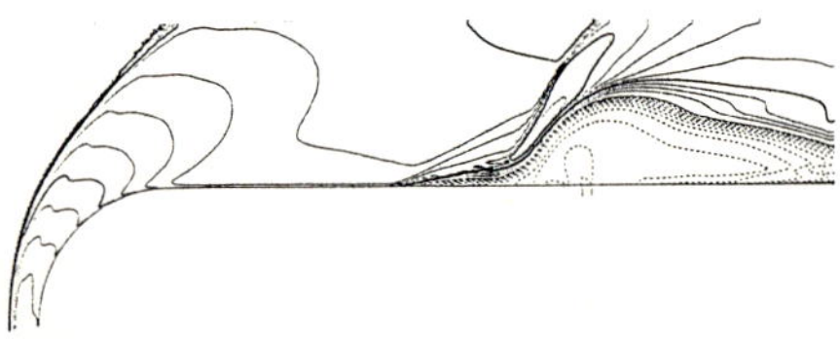

Fig. 5. Temperature contours (Case 2, $\delta = 45°$)

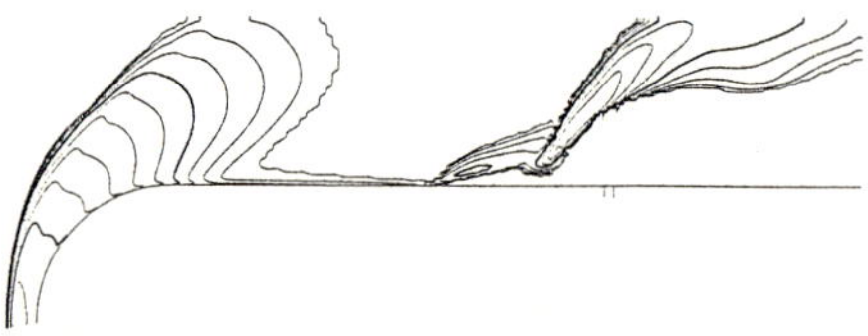

Fig. 11. Electron mole fraction contours (Case 1, $\delta = 90°$)

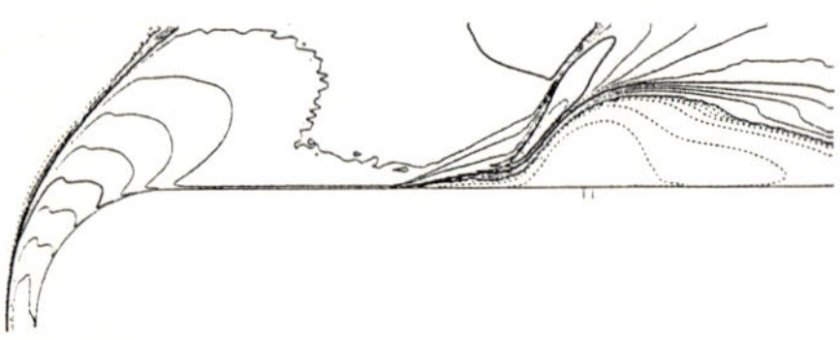

Fig. 6. Vibrational temperature contours (Case 2, $\delta = 45°$)

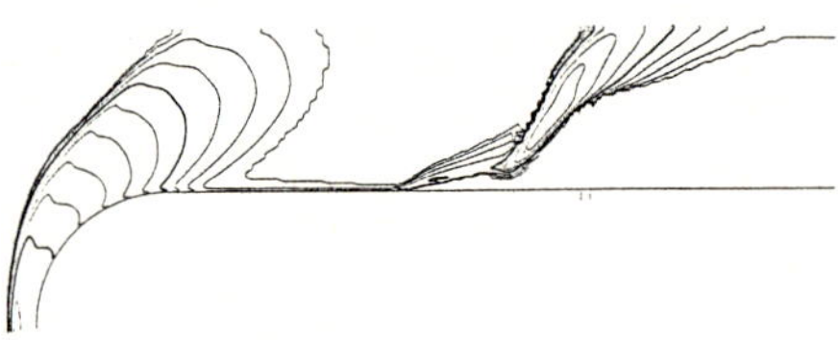

Fig. 12. Electron mole fraction contours (Case 2, $\delta = 45°$)

Properties of jets

The results of the numerical analysis are illustrated in Figs.2 - 13. Figs.2 and 3 show the pressure contours in Case 1 (injection angle $\delta = 90°$), and Case 2 ($\delta = 45°$), respectively. Temperature contours in Cases 1 and 2 are shown in Figs.4 and 5, respectively. Vibrational temperature contours in Case 2 are shown in Fig.6. Density contours in both cases are shown in Figs.7 and 8, respectively. Figs.9 and 10 show the streamline diagrams in Cases 1 and 2. These results clearly show that the injected jet expands rapidly, collides with the main flow and is bent parallel to the wall by the main flow. The three strong shock waves: the bow shock by the blunt nose, the bow shock by the injected jet and the separation shock are also seen in the figures. The separation point of the boundary layer in Case 2 is located about 10 m upstream from the point in Case 1. The reason is that the barrier due to the inclined jet is located further upstream from the injector than that due to the transverse jet. Comparing Fig.2 with Fig.3, the pressure in the separation and bow shock region in Case 2 is stronger than that in Case 1. From Figs.4 and 5, however, the temperature in the bow shock in Case 2 is about 1000 K lower than that in Case 1. It is explained as follows: the jet in Case 2 collides harder with the main flow than that in Case 1. The bow shock in Case 2 is stronger than that in Case 1 and, therefore, the pressure and the density in the bow shock are larger than in Case 1. But the rate of increase of the density is larger than that of the pressure because the jet comes into "a head-on collision" with the main flow and the jet gas is resisting to be swept away. Therefore the temperature in Case 2 is lower than that in Case 1 due to satisfying the equation of state. As far as the thermal nonequilibrium is concerned, Fig.6 shows that the vibrational-electronic energy equilibrates with the translational energy in the shocks and does not equilibrate with that in the expanded jet. From Figs.7 and 8, a slightly higher density region is seen downstream the injector in Case 2. The reason is that the 90° jet flows away along the wall downstream the injector, but the 45° jet collides with the wall and has a weak impingement with the wall. Comparing Figs.9 and 10, there are weakly recirculating regions in both cases, and the jet in Case 2 is immediately bent towards the wall. It is considered that the reattachment of the 45° jet is stronger than that of the 90° jet.

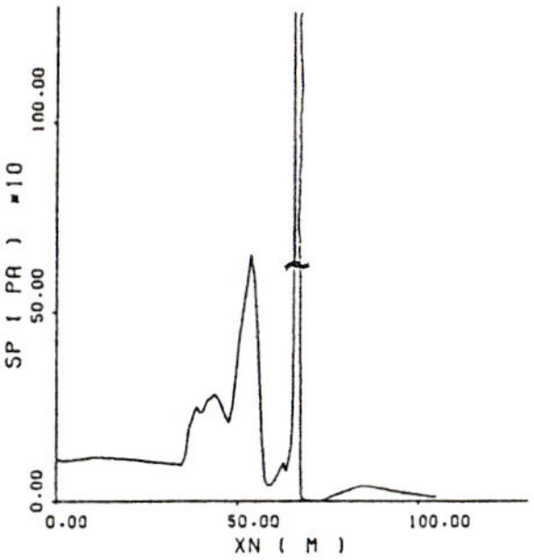

Fig. 13. Wall pressure profile (Case 2, $\delta = 45°$)

Chemical reactions

The mole fraction of oxygen atoms is of the order $10^{-2} \sim 10^{-3}$ upstream the injector in both cases which are not shown in the figures. Figs.11 and 12 show the mole fraction contours of the electrons e in Cases 1 and 2, respectively. The mole fraction of the electron is of the order 10^{-9} in the bow shock due to the jet in both cases. However the mole fraction of electrons in the separation shock due to the inclined jet is somewhat larger than that due to the transverse jet. These results support the results of the temperature, indicating the necessity to conside the wall material for the proof of corrosion and the location of electronic devices which may be jammed by the ions and electrons.

Moments

Fig.13 shows the wall pressure profiles in Case 2 which have a high pressure region upstream near the injector. This figure shows the appearance of reattachment of the 45° jet to the wall. The moment is calculated from the pressure profile $\mathcal{M}_1$ and is shown in Table 2. When the position of the center of gravity of the body is calculated from the data of the body shape, the center of gravity is located 21.8 m upstream the injector. $\mathcal{M}_2$ is the moment of the jet obtained by the calculation without the pressure profile.

Table 2. Moments around the center of gravity (in N·m/m units)

δ	90°	45°
$\mathcal{M}_1$	3.945×10^4	3.791×10^4
$\mathcal{M}_2$	-18.66×10^4	-18.62×10^4

The positive moment has nose down direction. $\mathcal{M}_1$ of the 45° jet is not different from that of the 90° jet. The high wall pressure region in the bow and separation shocks upstream the injector is located near the center of gravity. Hence the high wall pressure in this region does not influence the moment around the center of gravity.

4. Conclusions

Flowfields of jets with injected angles of 45° and 90° in a main hypersonic flow of $M_\infty = 20$ at 70 km altitude are simulated numerically by solving the two-dimensional Reynolds averaged full Navier-Stokes equations with a N_2/O_2 chemical kinetic model and two-temperature model. The Harten-Yee type TVD scheme and central difference schemes are used to calculate the convection and the viscous terms of the governing equations, respectively. The results capture the bow and separation shocks near the injector. The pressure of the 45° jet is higher than that of the 90° jet, but the temperature of the 45° jet is lower than that of the 90° jet. The moment of the 45° jet is not different from that of the 90° jet. The jet inclined towards the main flow at 45° is bent suddenly downstream due to the main flow and has weak impingement with the wall downstream the injector.

References

Appleton JP, Bray KNC (1964) The conservation equation for a nonequilibrium plasma. J.Fluid Mech. 20: 659-672

Gnoffo PA, Gupta RP, Shinn JL (1989) Conservation equations and physical models for hypersonic air flows in thermal and chemical nonequilibrium. NASA TP-2867

Inoue Y, Hayashi AK (1992) Numerical analysis of chemical reactive flows around injection jet into a hypersonic flow. In: Proceedings of the 18th ISTS

Lee JH (1984) Basic governing equations for the flight regimes of aeroassisted orbital transfer vehicles. AIAA Paper 84-1729

Park C (1991) Chemical-kinetic problems of future NASA missions. 1. Earth entries: a review. AIAA Paper 91-0464

Yee HC (1989) A class of high-resolution explicit and implicit shock-capturing schemes for inviscid and viscous hypersonic flows. NASA TM-101088

Propagation Behaviour of Strong Shock Waves with Chemical Reactions in a Shock Tunnel Nozzle

M. Nishida and M.G. Lee
Department of Aeronautics and Astronautics, Kyushu University, Fukuoka 812, Japan

Abstract. The unsteady starting process in the hypersonic nozzle of a free-piston shock tunnel is simulated using an inviscid flow model with equilibrium air chemistry. High-temperature effects are introduced using the equivalent gamma method and the thermodynamic properties of equilibrium air are given by curve fits. Results show the propagation behavior of strong shock waves in the starting process and the unsteady starting characteristics of an axisymmetric nozzle.

Key words: Shock wave, Chemical equilibrium flow, Nozzle starting process, TVD scheme

1. Introduction

Reflected-type shock tunnels have been one of the most successful types of a hypersonic impulse test facility that can economically produce hypersonic flows. In recent years there has been a renewal of interest in the impulse test facilities as a means of providing the flow conditions that hypersonic vehicles encounter. Among the various types of shock tunnels, a free-piston shock tunnel that employs the compression process of a driver gas by a heavy piston can easily produce high-enthalpy and high-pressure gas in the reservoir section upstream of a nozzle throat. Thus, this type of shock tunnel has been used successfully in various aerodynamic experiments related to reentry aerodynamics and SCRAM jet engines, etc (Morrison et al. 1990). However, the extremely short running time may severely disturb the versatility of this facility. Hence, the starting process and shock propagation behavior in this facility have been one of the interesting problems. Concerning the starting process of a conventional type shock tunnel nozzle, Prodromou and Hillier (1991) conducted numerical simulations using a Godunov-type scheme and compared their numerical solutions with experimental results (Amann and Reichenbach 1973). The present authors also calculated the starting process including the bursting of a secondary diaphragm by using the non-MUSCL type TVD scheme (Lee and Nishida 1992).

In the present work, the unsteady starting process in the hypersonic nozzle of a free-piston shock tunnel is numerically simulated using an inviscid flow model with equilibrium air chemistry. The high-temperature effects are introduced into the numerical calculations using the equivalent gamma method which makes the numerical calculation easy.

2. Flow model

The wave diagram in a nozzle of reflected-type shock tunnel that represents the starting process is shown in Fig.1. When an incident shock wave reaches the end-wall of the shock tube section, the shock reflects and produces a constant property region. The region behind the reflected shock is called Region 5. In Region 5 flow is nearly at rest, even though the reflected shock moves upstream. By expanding the gas of Region 5 through a nozzle, a hypersonic flow can be generated. A secondary diaphragm initially separates the shock tube section gas from the nozzle gas. After the bursting of the secondary diaphragm by the high pressure behind the reflected shock, the test gas following the primary shock wave expands through the nozzle throat into the nozzle.

A sketch of the density and pressure variations in the nozzle is given in Fig.2 where P.S. means the primary shock, C.S. the contact surface, S.S. the secondary shock. A secondary shock wave is generated since there is a discontinuity between the pressures in Region 7 (R7) and in Region 6

Shock Waves @ Marseille II
Editors: R. Brun, L. Z. Dumitrescu

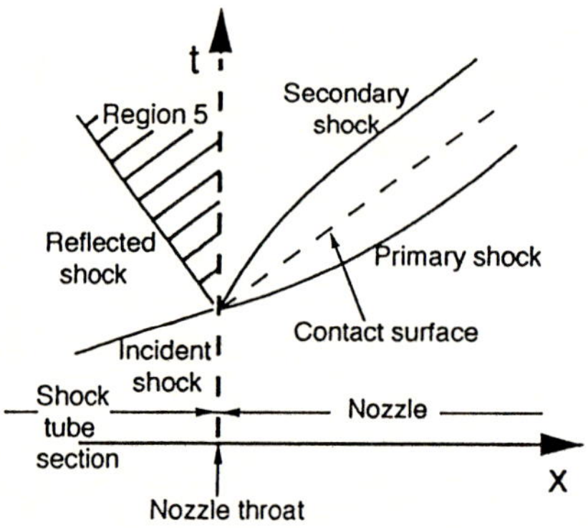

Fig. 1. Wave diagram for the nozzle starting process in a shock tunnel

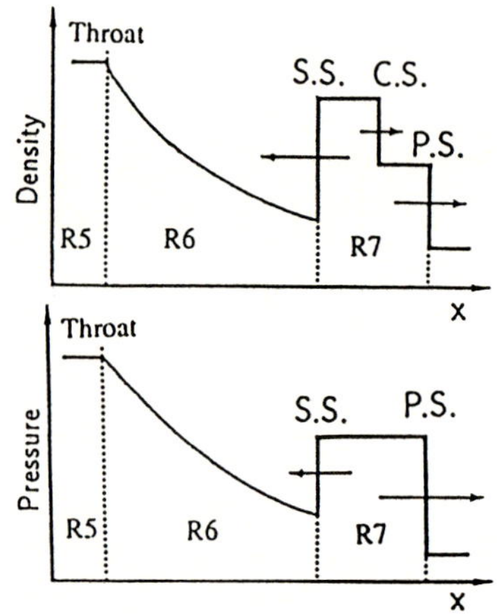

Fig. 2. Sketch of the density and pressure variations in the nozzle

(R6). The moving direction of this secondary shock wave is upstream relative to the nozzle main flow, but it is drifted downstream by the supersonic main flow. When this secondary shock or the tail of unsteady expansion waves passes through the nozzle exit, the shock tunnel is started and the steady test flow for aerodynamic experiments is available. The steady flow is terminated by the arrival of the driver gas at the nozzle exit. It is of great importance to shorten the starting time since it reduces the actual measurement time.

3. Air chemistry

The chemical equilibrium air considered here is assumed to be composed of nine chemical species (N_2,O_2,NO,N,O,NO^+,N^+,O^+,e^-). The chemical reactions of high-temperature air including nine chemical species are given as follows:

(1)	$O_2 \rightleftharpoons O+O$	(2)	$N_2+O_2 \rightleftharpoons 2NO$
(3)	$N_2 \rightleftharpoons N+N$	(4)	$NO^+ + e^- \rightleftharpoons NO$
(5)	$N^+ + e^- \rightleftharpoons N$	(6)	$O^+ + e^- \rightleftharpoons O$

The concentrations of these species are determined by solving a set of nonlinear equations, which are derived from the change and elemental conservation and the law of mass action, using Prabhu and Erickson's code (1988).

4. Numerical algorithm

The gas temperature behind a strong shock is highly elevated and consequently not only vibrational excitation but also chemical reactions may take place. In the present study, it is assumed that thermochemical equilibrium is attained immediately behind the shock. The ratio of specific heats for such a high temperature gas is not constant any more and it is given as a function of two thermodynamic properties, which leads to difficulty in numerical calculations, i.e., the difficulty in estimating the flux Jacobian. In order to avoid such difficulties, the concept of equivalent γ was introduced to make the numerical calculation convenient (Grossman and Walters 1989). The principal feature of their method lies in the particular treatment of the general equation of state.

4.1. Equation of state for an equilibrium gas

In general, thermodynamic properties are interrelated by two thermodynamic variables in a thermochemical equilibrium flow, that is, the general equation of state is written as

$$p = p(\rho, e) \tag{1}$$

where p is the pressure, ρ the density and e the internal energy. Though the above equation can not be expressed in the form of a simple explicit function, the following relation may be obtained especially for a calorically perfect gas:

$$p = (\gamma - 1)\rho e \tag{2}$$

It is assumed that an expression similar to the above relation holds for a high temperature gas, that is

$$p = (\tilde{\gamma} - 1)\rho e \tag{3}$$

where $\tilde{\gamma}$ is the equivalent gamma that is not a constant and is given as a function of two thermodynamic variables, that is, $\tilde{\gamma} = \tilde{\gamma}(\rho, e)$. The equivalent gamma is obtained by curve fits constructed from statistical thermodynamic calculations. The pressure increment can be obtained from Eq.1 to construct the numerical flux function. Grossmann and Walters (1989) presented the expression for the pressure increment, that involves Δe, $\Delta \rho$ and $\Delta \tilde{\gamma}$. Although the thermodynamic properties other than $\tilde{\gamma}$ may vary over many orders of magnitude, $\tilde{\gamma}$ is a slowly varying fuction of the thermodynamic variables for a high temperature gas. It actually stays in the range of 1 to 5/3 (Grossman and Walters 1989). Hence, assuming that the derivative of $\tilde{\gamma}$ is sufficiently small, they derived $\Delta \tilde{\gamma}$ by neglecting terms of order $(\Delta \tilde{\gamma})^2$. Thus, the pressure increment can be evaluated, that is employed in the Harten-Yee scheme using the Roe's approximate Riemann solver for equilibrium high-temperature gases.

4.2. Governing equations and numerical procedure

The flow considered here is axisymmetric and inviscid since the essential features of the flowfield during the starting process can be sufficiently understood with an inviscid flow model. The numerical method for spatial discretization employs a second order non-MUSCL TVD scheme of Harten-Yee type (Yee 1989). The time integration is done by the explicit operator splitting technique (fractional step method) of Strang type.

For a chemical equilibrium flow analysis, in general, the governing equations are coupled with the law of mass action. To solve the law of mass action for possible equilibrium chemical reactions requires an iterative procedure at each computational grid point, which is a considerably time-consuming task, so that the iteration method is actually impossible. Srinivasan et al. (1987) presented curve fits for the thermodynamic properties of equilibrium air. Their code can determine the thermodynamic properties in a non-iterative way. In the present calculation, Srinivasan et al.'s curve fits have been used in the computation of the thermodynamic properties of equilibrium air flows. The computation of concentrations employing Prabhu and Erickson's code has been conducted using the flow properties determined from the flowfield calculation.

4.3. Initial and boundary conditions

Alhough the ratios of pressure and density across a shock wave in a calorically perfect gas are expressed by simple functions such as the Rankine-Hugoniot relation, there is no such a simple relation for chemically reacting gases. Therefore we must solve iteratively the equations for an equilibrium flow across a normal shock wave to acquire the initial conditions for the shock jump. Thus, the initial conditions have been determined.

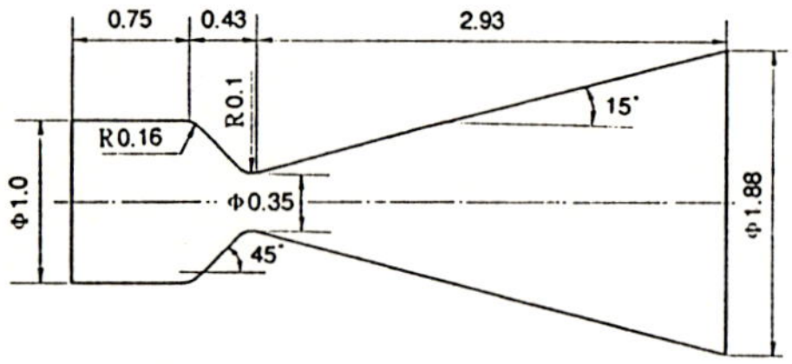

Fig. 3. Nozzle geometry

Reflection conditions on the nozzle wall, symmetry conditions on the nozzle axis, and inflow-outflow conditions on the left and right sides of the computational domain are imposed.

In Fig.3 is shown the geometry of the axisymmetric nozzle considered here that is based on the JPL nozzle, which is shown in a scale nondimensionalized by the shock tube diameter. The nozzle throat diameter is 0.35 and hence the norminal Mach number of the nozzle is 5.

The computational grid system uses 100×40 points. Elliptic partial differential equations (Poisson equation) are used to generate body-oriented grids that permit a one-to-one mapping.

5. Results and discussion

The present calculations were performed for a nonstationary shock wave of $M_s = 20$ travelling initially in the low pressure section of the shock tube where $p_1 = 101.4$ kPa and $T_1 = 300$ K. The constant log-density contours which show the time-developing flowfields are given in Fig.4. The time given in the figure is nondimensional, namely $t^* = ta_1/R$, where R is the radius of the shock tube and a_1 is the speed of sound in front of the incident shock. In the figure, P.S., S.S. and R.S. mean the primary shock, secondary shock and reflected shock, respectively. It can be seen that there are several vortex structures behind the reflected shock which moved towards upstream. The vortex behind a reflected shock has been also observed in the numerical results obtained by Jacobs (1991). However, his numerical calculations adopted a constant ratio of specific heats, that is, $\gamma = 1.3$ in order to approximate some of the high temperature effects. In Fig.4, the reflected shock also shows a Mach reflection at the wall and central axis of the nozzle. The upstream-facing secondary shock wave is generated near the nozzle axis as shown in Fig.4(b) and gradually develops towards the nozzle wall with increase in time. A weak shock wave generated at the nozzle throat is seen. The shock wave alternatively reflected from the centerline runs towards the wall of the nozzle and reaches the nozzle exit as shown in Fig.4(f). The shape of the primary shock is considerably distorted as shown in Fig.4(d) whereas the secondary shock shows no such a shape distortion. At present, it is uncertain whether the shape distortion of the primary shock is due to numerical instability or to a physical phenomenon.

The distributions of temperature, density and pressure along the nozzle axis are presented in Fig.5, where the solid and dashed lines represent the distributions at $t^* = 1.173$ and 3.204, respectively. In the figure of the temperature distribution (Fig.5(a)), the primary shock (P.S.), secondary shock (S.S.) and reflected shock (R.S.) are shown. At $t^* = 1.173$ the reflected shock looks considerably sharp but the primary shock is smeared. The secondary shock can be clearly identified at this instant. In the region from the nozzle throat to the secondary shock, the temperature, density and pressure decrease towards downstream. This is due to the flow expansion from the nozzle throat. An oscillation behind the reflected shock can be observed in all distributions. At present we have no satisfactory explanation of these oscillations. There exists an increase in temperature around x/L=0.4 and this position corresponds to the place where the shock wave generated from the nozzle throat reflects from the nozzle axis. Such an increase is also identified in the density and pressure distributions.

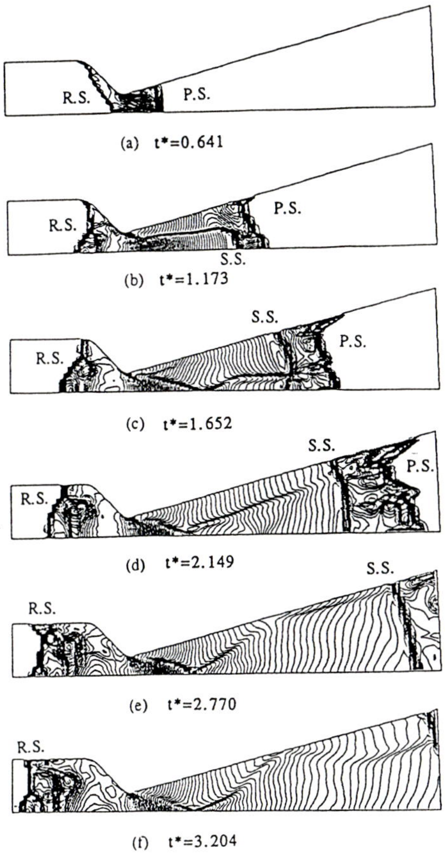

Fig. 4. Time development of flowfield; log-density contours, $M_s=20$.

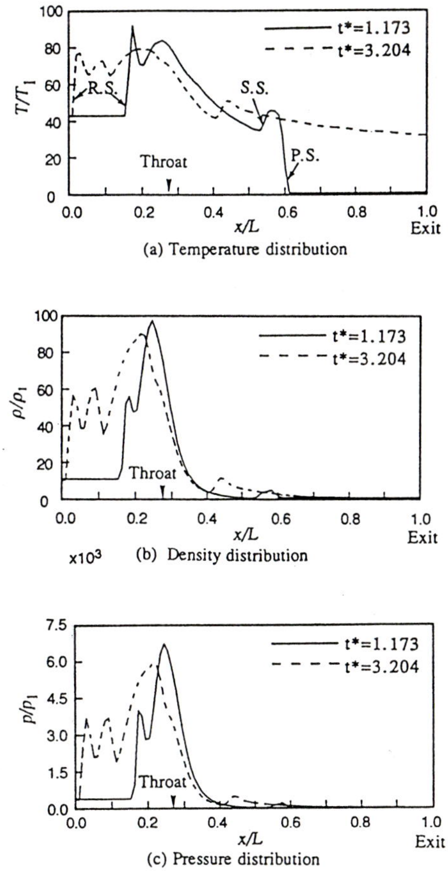

Fig. 5. Distribution of temperature, density and pressure along nozzle axis, $M_s=20$.

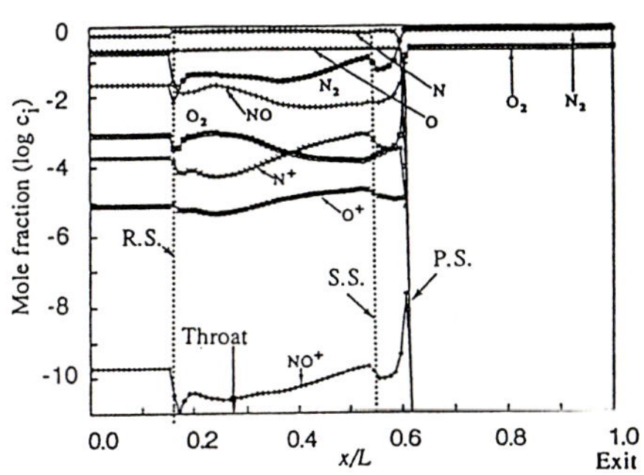

Fig. 6. Concentration distribution along nozzle axis at $t^*=1.173$, $M_s=20$.

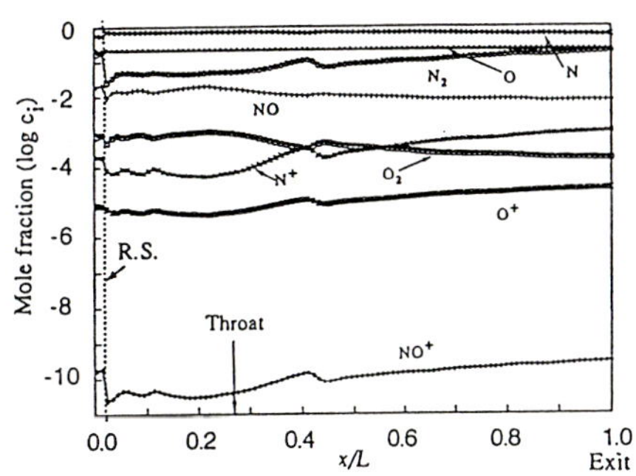

Fig. 7. Concentration distribution along nozzle axis at $t^*=3.204$, $M_s=20$.

Figs.6 and 7 show the mole fraction along the nozzle axis at the same stages as in Fig.5. The primary shock, secondary shock and reflected shock also are identifyed by P.S., S.S. and R.S., respectively. As shown in Fig.6, oxygen and nitrogen molecules dissociate immediately behind the primary shock wave and produce O and N. Ion species also are produced. It can be seen that although there are no obvious variations in the concentrations of N, O and NO across the the secondary shock wave, the changes of concentrations in the other species are obviously observed. Behind the reflected shock, dissociation of N_2 and O_2 occurs. Correspondingly the concentration of N is increased and the concentration of O is appreciably elevated.

Fig.7 indicates the mole fraction at $t^* = 3.204$ when the quasi-steady flow was established along the nozzle axis. The slight variations of species except O, N and NO species are shown at the position where the weak shock generated from the nozzle throat reflects from the nozzle axis.

6. Concluding remarks

Numerical simulations have been conducted using an inviscid flow model with equilibrium air chemistry to investigate the transient nozzle flow in a shock tunnel. Results show the propagation behavior of the primary shock, the secondary shock and unsteady expansion waves. The present computational results of the unsteady nozzle starting process lead to the following conclusions:

(1) Several vortices are observed behind the reflected shock which moved upstream undergoing Mach reflection.

(2) A weak shock is generated from the nozzle throat. The distributions of the temperature, density and pressure are considerably changed across the position where the shock reflects from the nozzle axis.

(3) The changes of mole fraction distribution are obviously shown across the reflected, primary and secondary shocks.

References

Amann HO, Reichenbach H (1973) Unsteady flow phenomena in shock tunnel nozzles. In: Bershader D, Griffith W (eds) 9th Intl. Symp. Shock Tubes and Waves, Stanford, USA, pp 96-112

Grossman B, Walters RW (1989) Analysis of flux-split algorithm for Euler's equations with real gases. AIAA J. 27:524-531

Jacobs PA (1991) Simulation of transient nozzle flow in a shock tunnel and a high Mach number nozzle. In: Dwyer HA (ed) Proc. 4th Intl. Symp. on Comput. Fluid Dyn., Davis, USA, pp 533-538

Lee MG, Nishida M (1992) Computation of an unsteady starting process in a hypersonic nozzle. In: Chang KS, Choi DH (eds) Proc. 5th Asian Congress of Fluid Mech., Taejon, Korea, pp 948-951

Morrison WRB, Stalker RJ, Duffin J (1990) New generation of free piston shock tunnels. In: Kim YW (ed) Proc. 17th Intl. Symp. on Shock Tubes and Waves, Lehigh, USA, pp 582-587

Prabhu RK, Erickson WD (1988) A rapid method for the generation of equilibrium chemical composition of air to 15000K. NASA TP-2792

Prodromou P, Hillier R (1991) Computation of unsteady nozzle flows. In: Takayama K. (ed) Proc. 18th Intl. Symp. on Shock Waves, Sendai, Japan, pp 1113-1118

Srinivasan S, Tannehill JC, Weilmuenster KJ (1987) Simplified curve fits for the thermodynamic properties of equilibrium air. NASA RP-1181

Yee HC (1989) A class of high-resolution explicit and implicit shock capturing methods. NASA TM-101088

Studies of Nonequilibrium Effects in Nozzle Flow Under High Reservoir Conditions

Le Jialing and Dong Weizhong
China Aerodynamics Research and Development Center, Mianyang, Sichuan, China

Abstract. A set of equations is presented, governing quasi-one-dimensional chemical, vibrational and ionization nonequilibrium nozzle flow in high reservoir conditions, with v-d coupling and electron excitation effects. In order to remove the problem of the small space marching step size under high reservoir pressure, the method of chemically frozen, thermal equilibrium flow with an equilibrium species correction is introduced. Numerical solutions were obtained for an arc-jet wind-tunnel and a shock wind-tunnel. The influence of different reservoir pressures and of various chemical models, v-d coupling effects and electron excitation effects is studied.

Key words: Nozzle flow, Nonequilibrium flow, High-enthalpy flow

1. Introduction

With the development of flight vehicles, the stagnation enthalpy encountered in hypersonic flight results in flowfield temperatures becoming high enough to dissociate chemical species in air. In China, new hypersonic testing facilities are being designed and built in order to simulate nonequilibrium phenomena such as dissociation, ionization, vibrational excitation and electron excitation of the air around flight vehicles, and to validate the methods of calculation for nonequilibrium flow. In addition to high stagnation enthalpy, extreme reservoir pressures are needed to simulate hypersonic flow. Also, in the nozzle expansion, the test gas remains in nonequilibrium. Presently, the composition of the air in the test section is still uncertain because of the nonequilibrium effects in the nozzle flow, so it is quite difficult to explain the data and the results of aerodynamic force and heat-transfer measurements obtained in the tests, and to know the properties of test gas. Therefore, it is very important to study nonequilibrium effects for nozzle flows under high reservoir conditions.

Table 1. Nozzle data

Nozzle	L_1 (cm)	L_2 (cm)	A^* (cm^2)	θ_1 (°)	θ_2 (°)	D_e (cm)	T_0 (K)	P_0 (MPa)
#1	3.5	21.5	1.0	42	11.3	8.6	4800	5
#2	3.5	296.5	1.0	42	11.3	120	8000	100, 200

2. Numerical method

2.1. Governing equations

For a steady quasi-one-dimensional flow, the nondimensional governing equations for chemical, vibrational and ionization nonequilibrium flow with v-d coupling and electron excitation are as follows:

$$\frac{d\nu_j}{dx} = (\rho v)^{-1} S_{cj}, \qquad (j = 1, \dots s, s+1) \tag{1}$$

$$\frac{de_{vj}}{dx} = v^{-1} S_{vj} \tag{2}$$

Shock Waves @ Marseille II
Editors: R. Brun, L. Z. Dumitrescu

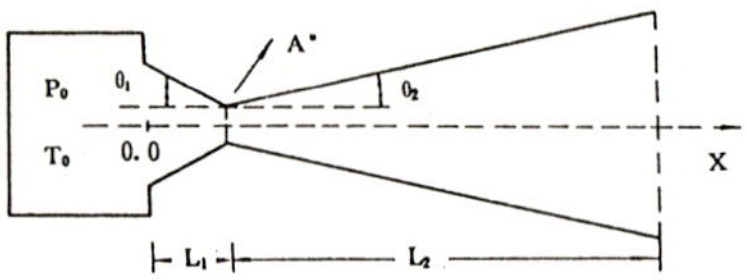

Fig. 1. Nozzle configuration

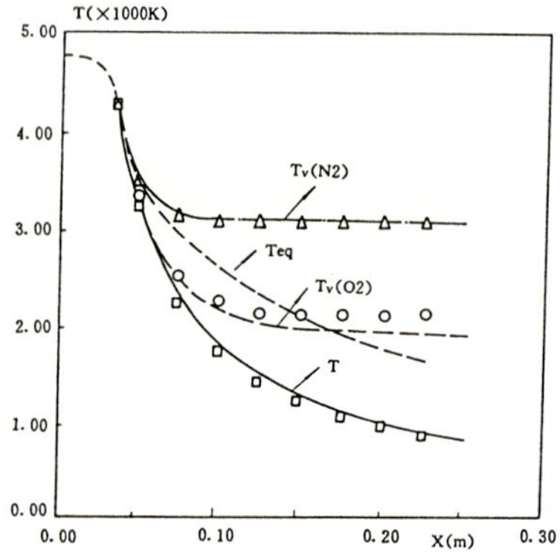

Fig. 2. Temperatures; P_0 = 5 MPa, T_0 = 4800 K. Moss model; points are data of Sagnier (1991)

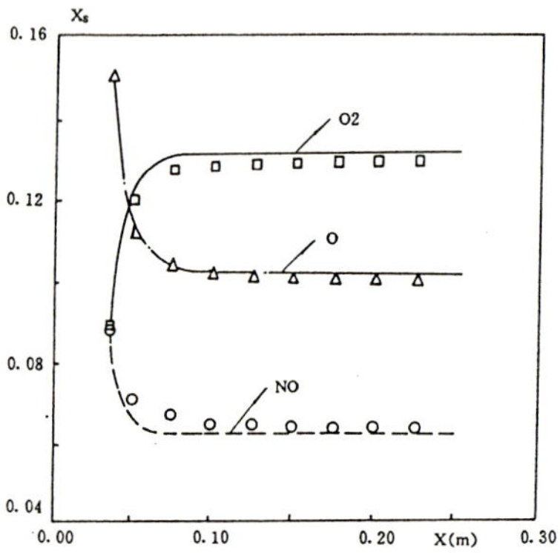

Fig. 3. Species molar fraction; P_0 = 200 MPa, T_0 = 8000 K. Moss model; points are data of Sagnier (1991)

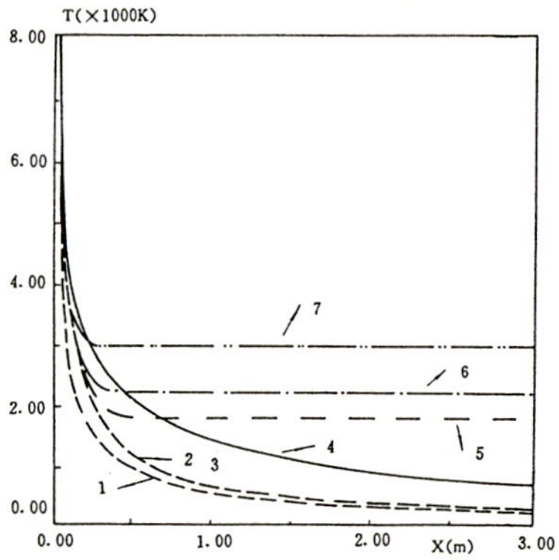

Fig. 4. Temperatures; P_0 = 200 MPa, T_0 = 8000 K. Dunn-Kang model; lines 1 to 6 plot respectively the values of $T_{f.eq}, T, T_e, T_{eq}, T_v(O_2), T_v(NO), T_v(N_2)$

$$\frac{dT_e}{dx} = -\frac{2T_e}{3v}\frac{dv}{dx} + \frac{2}{3\rho v \nu_e}(Q_{el} + Q_{in}) \tag{3}$$

$$\frac{dv}{dx} = \frac{1}{1 - M_f^2}\frac{D_v}{D} \tag{4}$$

$$\frac{dp}{dx} = -\rho v \frac{dv}{dx} \tag{5}$$

$$\frac{dT}{dx} = \left[\left(\frac{p}{v} - \rho v\right)\frac{dv}{dx} + \frac{p}{A}\frac{dA}{dx} - \rho \nu_e \frac{dT_e}{dx} - \frac{1}{v}\left(T\sum_{j=1}^{s} S_{cj} + T_e S_{ce}\right)\right] \Big/ \left(\rho \sum_{j=1}^{s} \nu_j\right) \tag{6}$$

where

$$M_f = V/a_f \tag{7}$$

$$a_f^2 = \frac{\sum_{j=1}^{s}\left[\frac{5+2(n_j-1)}{2} + C_{ej}\right]\nu_j}{\sum_{j=1}^{s}\left[\frac{3+2(n_j-1)}{2} + C_{ej}\right]\nu_j} T \sum_{j=1}^{s}\nu_j + \frac{5}{3}T_e\nu_e \tag{8}$$

$$D_v = \frac{1}{\rho}\left\{\sum_{j=1}^{s}\left[\frac{5+2(n_j-1)}{2}+C_{ej}\right]\nu_j\right\}\left(T\sum_{j=1}^{s}S_{cj}+T_eS_{ce}-\frac{pv}{A}\frac{dA}{dx}\right)$$

$$-\frac{1}{\rho}\left(\sum_{j=1}^{s}\nu_j\right)\left\{T\sum_{j=1}^{s}\left[\frac{5+2(n_j-1)}{2}S_{cj}\right]+\frac{5}{2}T_eS_{ce}+\sum_{j=1}^{s}[(n_j-1)(e_{vj}S_{cj}+\rho\nu_jS_{vj})\right.$$

$$\left.+S_{cj}\left(\Delta h_j^0+e_j\right)]\right\}+\frac{2}{3\rho}\left[\sum_{j=1}^{s}(n_j-1+C_{ej})\nu_j\right](Q_{el}+Q_{in}) \tag{9}$$

$$D = a_f^2\left\{\sum_{j=1}^{s}\left[\frac{3+2(n_j-1)}{2}+C_{ej}\right]\nu_j\right\} \quad \text{and} \quad p=\rho\left(T\sum_{j=1}^{s}\nu_j+T_e\nu_e\right) \tag{10}$$

where ν_j is the molar number of species j per unit mass, e_j the excitation energy of the electron of species j and $C_{ej} = \partial e_j/\partial T$ (Clarke et al. 1964), S_{cj} the chemical source term, S_{vj} the vibrational source term, Q_{el} and Q_{in} the production terms of free electron energy due to elastic and inelastic collisions of free electrons with heavy particles (Matsuzaki 1988). The solution of Eqs.1-6 gives $P, T, T_e, V, T_{vj}(e_{vj})$and ν_j. The results are obtained by using the standard Runge-Kutta-Gill method.

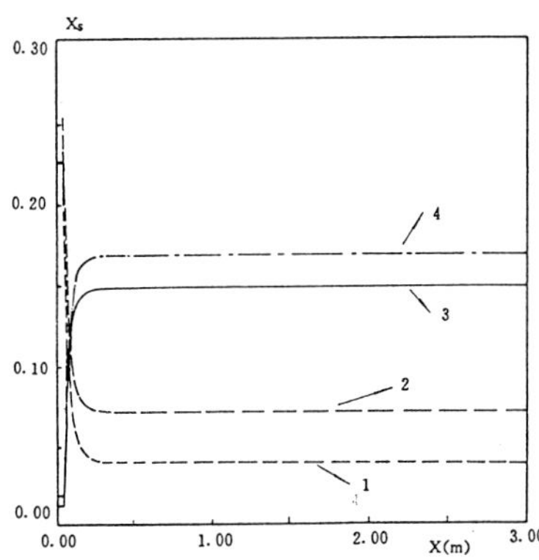

Fig. 5. Species molar fraction, Dunn-Kang model. Lines 2 and 3: x_O and x_{O_2} at P_0 = 100 MPa; lines 1 and 4: at 200 MPa

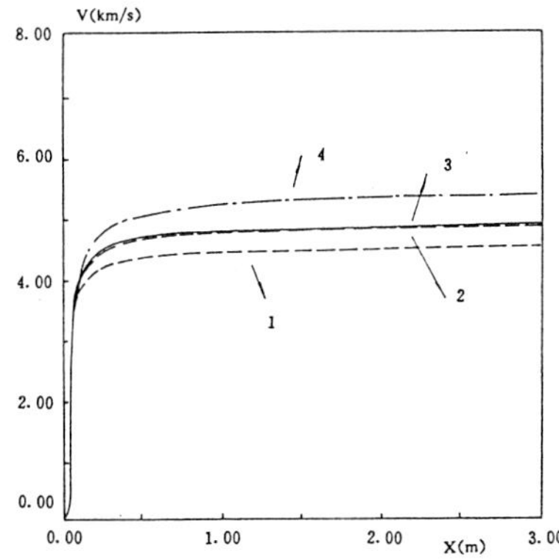

Fig. 6. Velocity; lines 1, 2 and 4: $V_{f.eq}$, V and V_{eq} at P_0 = 200 MPa; line 3: V at 100 MPa

2.2. Chemical and $v-d$ coupling model

In this study, the high-temperature air is assumed to consist of the following 11 species: O_2, N_2, NO, O, N, O_2^+, N_2^+, NO^+, O^+, N^+ and e. The rate processes considered for the above species are taken from the Moss, Park, Gardiner and Dunn-Kang model (Sagnier et al.1991). The real reaction constants (with the v-d coupling nonpreferential model) are given by:

$$k_{fi} = k_{fi}^{\infty}\prod_{j=1}^{s}V_j^{A_{ij}}, \qquad k_{bi} = k_{fi}^{\infty}/K_i \qquad \text{and} \qquad k_{fi}^{\infty} = a_iT^{bi}\exp(-c_i/T) \tag{11}$$

where V_j is the v-d coupling factor, A_{ij}=0 or 1 according to whether or not the i-th reaction is affected by the coupling process. K_i is the equilibrium constant of the i-th reaction, which is calculated by the partition function of species (Vincenti 1965).

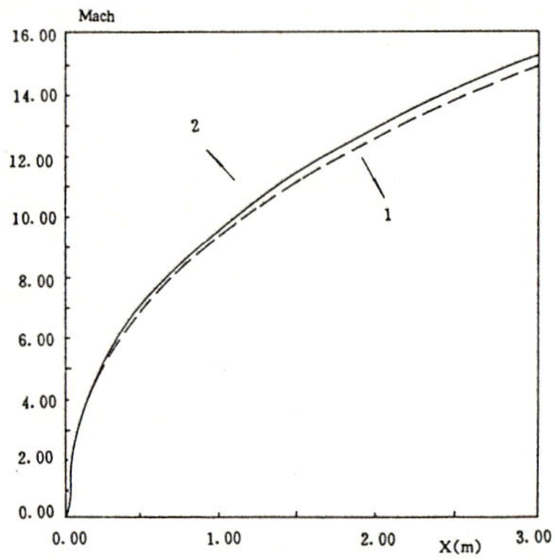

Fig. 7. Mach number; line 1, P_0 = 200 MPa; line 2,P_0 = 100 MPa

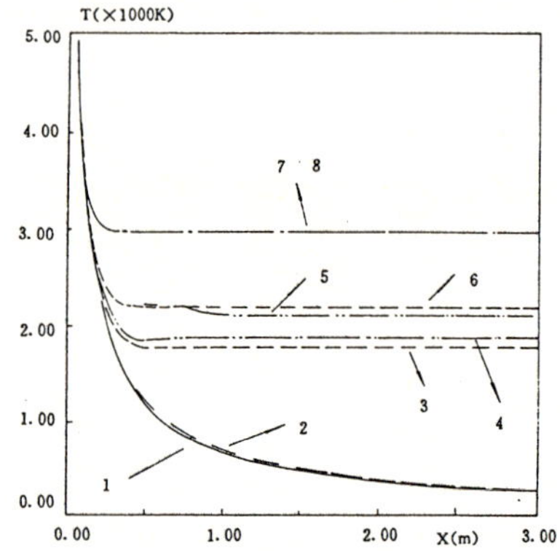

Fig. 8. Model effect on temperatures; P_0 = 200 MPa. Lines 1, 4, 5, 7: $T, T_v(O_2), T_v(NO), T_v(N_2)$, Park; Lines 2,3,6,8: $T, T_v(O_2), T_v(NO), T_v(N_2)$, Dunn-Kang

The nonpreferential model is employed for the v-d coupling effect (Sagnier 1991):

$$S_{vj} = \frac{e_{vj}^{\infty} - e_{vj}}{\tau_j} - \left(\frac{d\nu_j}{dt}\right)_f \frac{E_j\left(T, T_{vj}\right) - e_{vj}}{\nu_j} + \left(\frac{d\nu_j}{dt}\right)_b \frac{G_j(T) - e_{vj}}{\nu_j} \tag{12}$$

where $\left(\frac{d\nu_j}{dt}\right)_f$ and $\left(\frac{d\nu_j}{dt}\right)_b$ are production and depletion terms, $E_j(T, T_{vj})$ and $G_j(T)$ are the average vibrational energies of, respectively, a dissociating molecule j and of a molecule j immediately after recombination.

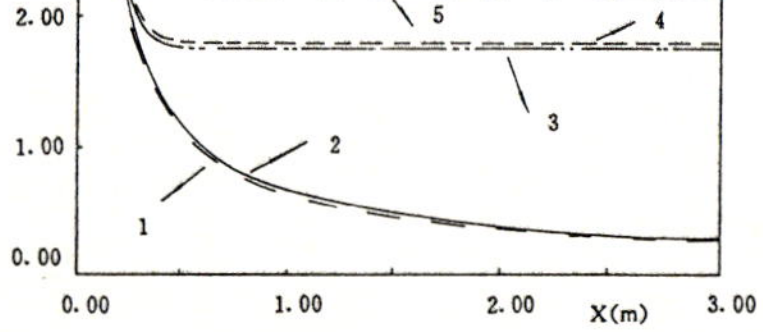

Fig. 9. Model effect on temperatures; P_0 = 200 MPa. Lines 1, 3, 5, 7: $T, T_v(O_2), T_v(NO), T_v(N_2)$, Gardiner Lines 2,46,8: $T, T_v(O_2), T_v(NO), T_v(N_2)$, Dunn-Kang

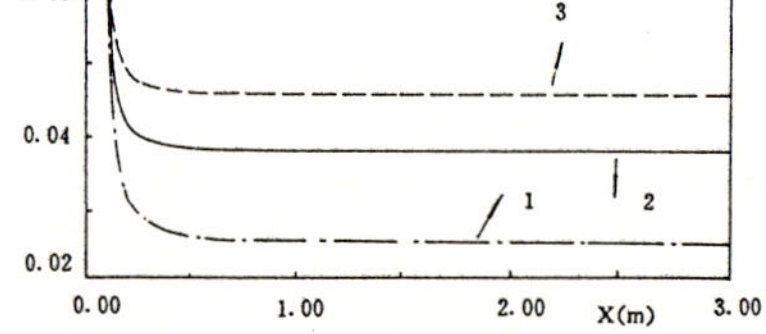

Fig. 10. Chemical model effect on molar fraction of oxygen atom; P_0 = 200 MPa. Lines 1, 2, 3: Park, Dunn-Kang, Gardiner models

2.3. The method of chemically-frozen - thermal equilibrium flow with an equilibrium species correction

Under high reservoir pressure, from the reservoir up to the throat, the expansion is weak enough for the flow to stay in a near-equilibrium state (Sagnier et al. 1991), and a small space marching

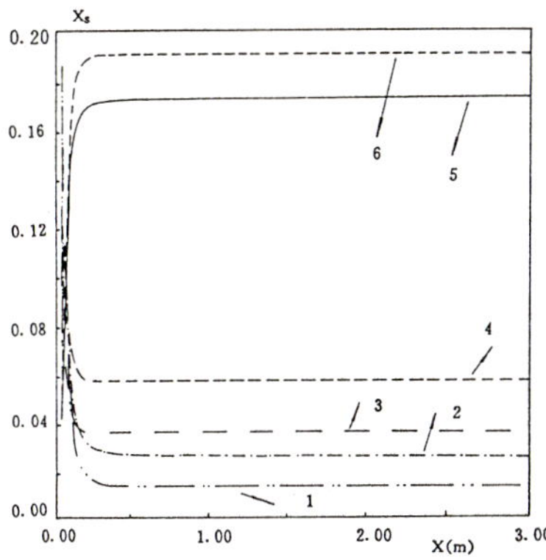

Fig. 11. Electronic excitation effect on molar fractions; P_0 = 200MPa, Park model. Lines 1,3,6: x_{NO}, x_O, x_{O_2}, without electron excitation; lines 2, 4, 5: with electron excitation

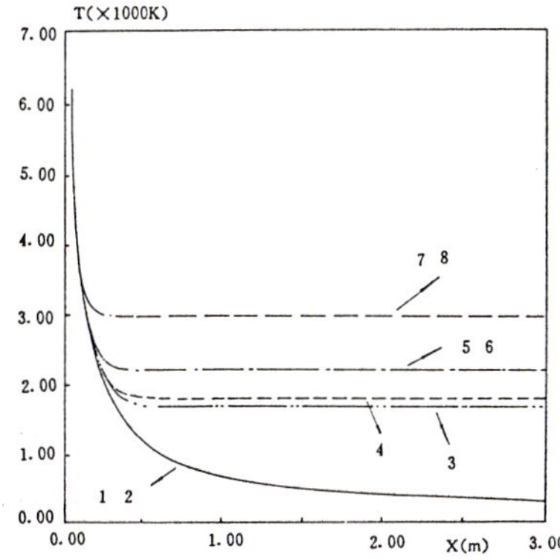

Fig. 12. $v - d$ coupling effect on temperatures; P_0 = 200 MPa, Dunn-Kang model. Lines 1, 3, 5, 7: $T, T_v(O_2), T_v(NO), T_v(N_2)$, without coupling; lines 2, 4, 6, 8: with coupling

step size must be taken. There are two methods of calculation for equilibrium flow, the first is an empirical formula method, however possibly inducing large errors for high reservoir conditions. The second is to solve the equations directly according to the properties of thermochemistry, however, practically it is not possible to find analytic expressions for the derivatives of the unknown ν_j (Anderson 1989). In the present study, first we solved the governing equations in the assumption of chemicallly frozen-thermal equilibrium flow, then chemical species were solved alone according to the local values of $p(x)$ and $T(x)$ by the equilibrium method (Vincenti et al. 1965).

3. Numerical results

3.1. The arc-jet wind-tunnel case

In order to verify the numerical methods, first the nozzle flow in the arc-jet wind-tunnel was computed (nozzle #1, see Table 1 and Fig.1). Figs.2 and 3 show temperatures and molar fractions of species. Results compare pretty well with those of Sagnier et al. (1991). The nonequilibrium temperatures (T, T_{vj}) deviate quickly from the equilibrium temperature T_{eq}. At the nozzle exit, there are great differences between T and T_{eq}, or T and T_{vj} (shown in Fig.2). The freezing of O_2, NO and O is observed at about 7 cm (Fig.3).

3.2. The shock wind-tunnel case

After the code was carefully checked, we have made an investigation for a high-enthalpy, high reservoir pressure nozzle flow (nozzle #2, see Table 1 and Fig.1). The diameter of the exit is about 120 cm. Equilibrium values were calculated by using the data of Srinivasan et al. (1987). Hereafter some of the results will be given. Nonequilibrium computation was started with the initialization of our method 1 mm downstream of throat.

3.2.1. Influence of different reservoir pressures

Fig.4 shows the temperatures along the nozzle (P_0 = 200 MPa, T_0 = 8000 K). $T_{f.eq}$ is the value for chemicallly-frozen - thermal equilibrium flow with an equilibrium species correction. Downstream of the throat, nonequilibrium temperatures (T, T_e, T_{vj}) are, for a certain distance, close to $T_{f.eq}$, which means that the present method of calculation for the region upstream of the throat is reliable and successful. With an increase in P_0 from 100 MPa to 200 MPa, T and T_e increase, but the vibrational temperature T_{vj} decreases. The molar fractions of O_2 and O change from 0.1462 and 0.0685 to 0.1659 and 0.0374 respectively at the exit (as shown in Figs.4 and 5). It means that the high reservoir presure can slightly suppress the nonequilibrium of the nozzle flow. Figs. 6 and 7 show the velocity and frozen Mach number along the nozzle. The velocity keeps at first equilibrium values, then detaches from them, and finally lies between the equilibrium value and $V_{f.eq}$. At the exit, the frozen Mach number is about 15.

3.2.2. Influence of different chemical models

Figs.8-10 show the temperature and molar fractions of the species. The model that is closest to equilibrium is Park's, the farthest one is Gardiner's. The molar fraction of atomic oxygen can vary from 0.025 to 0.045, depending on the chemical models, at the exit. The effect on $T_v(O_2)$ and $T_v(NO)$ is more important than on $T_v(N_2)$ because the O_2 and NO chemistry is more active. The situation is reversed for the vibrational temperatures because the chemical energy storage is high and so there is less energy available for vibration, and the lower translational temperature for the Landau-Teller law imply lower frozen vibrational temperatures.

3.2.3. Influence of electron excitation

It appears that the influence is weak on temperatures, but important on the molar fractions of O, O_2 and NO (shown in Fig.11). The results show that the recombination rate decreases because of electron excitation.

3.2.4. Influence of the $v - d$ coupling effect

$v-d$ coupling effect has a certain influence on $T_v(O_2)$ because O_2 chemistry is more active (shown in Fig.12).

4. Conclusions

The present numerical method of calculation for nozzle flows in high reservoir conditions is successful. In order to determine accurately the gas compositions at the nozzle exit, consideration of nonequilibrium effects and selection of the thermochemistry model are important. The high reservoir pressure can slightly reduce the nonequilibrium of the nozzle flow. If the reservoir pressure is lower, or the reservoir temperature is higher, $v - d$ coupling effect would have a stronger influence on flow properties.

References

Anderson JD (1989) Hypersonic and high temperature gas dynamics. McGraw-Hill

Clarke JF, McChesney M (1964) The dynamics of real gases. Butterworths, London

Mastsuzaki R (1988) Quasi-one-dimensional aerodynamics with chemical, vibrational and thermal nonequilibrium. Trans. Japan Soc. Aero. Space Sci. 243-258

Sagnier P, Marraffa L (1991) Parametric study of thermal and chemical nonequilibrium nozzle flow. AIAA J. 334-343

Srinivasan S, Tannehill JC, Weilmuenster KJ (1987) Simplified curve fits for the thermodynamic properties of equilibrium air. NASA-RP-1181

Vincenti WG, Kruger CH (1965) Introduction to physical gas dynamics. John Wiley

Chemically Reacting Multi-Nozzle Plume Flows

T.P. Roberts
Fluid Gravity Engineering Ltd, Chiltlee Manor, Haslemere Road, Liphook, Hampshire. GU30 7AZ, UK

Abstract. The computation of a 3D chemically reacting multi-nozzle plume from a 4-nozzle cluster has been undertaken using the Navier-Stokes solver TINA. The chemical composition of the plume consisted of CO, CO_2, O, O_2, OH, H, H_2, H_2O, N_2, Cl and HCl, with a Mach 3.8 outer freestream air flow. The technique of space-marching was used to speed up the computation in regions of fully supersonic flow. Results indicate significant levels of thermodynamic non-equilibrium in the regions of the mixing layer between the edge of the plume and the freestream air flow. In addition, marked 3D features are observed in shock shape and in general flow properties due to the interaction of the individual plumes from each of the four nozzles.

Key words: Plume flow, Nozzle flow, Chemically-reacting gas

1. Introduction

In recent years, the development of faster computer processors has enabled more ambitious CFD computations. This has naturally led into the area of modelling chemically reacting flowfields, although these are still generally 2D planar or axisymmetric. A common system but rarely modelled is that of the plume generated by a multi-nozzle configuration. The aim of the work described in this paper has been to model the 3D plume flowfield exhausting from a typical four-nozzle rocket motor cluster, for a vehicle travelling through the atmosphere at Mach 3.8. To the authors knowledge there is little other published work in this area (Cain et al. 1992, Holcomb 1989).

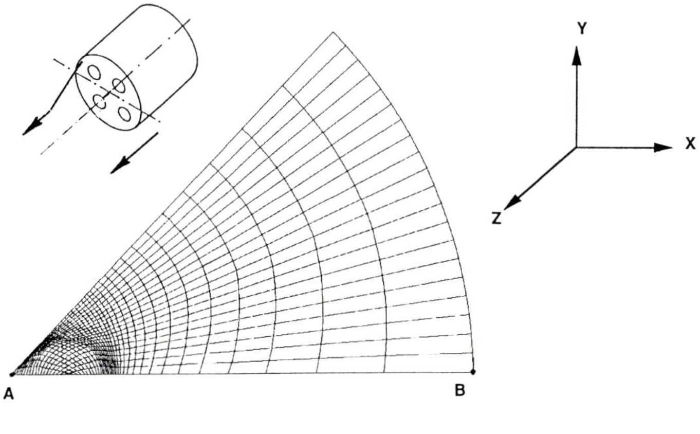

Fig. 1. Diagram of multi-nozzle and computational grid

A diagram of the set-up is illustrated in Fig.1. The interaction of the air flowing past the base of the motor and the plume gases has been considered, although the processing of the airstream by the bow shock from the front of the vehicle has been neglected. The specific aims of the work discussed has been to address both the physical aspects of the flowfield, and an approach to reduce the large computation times required for such a problem.

Shock Waves @ Marseille II
Editors: R. Brun, L. Z. Dumitrescu

2. Computations

The equilibrium conditions in the combustion chamber based on the minimisation of Gibbs Free Energy, were computed using the Gordon and MacBride code for a specified temperature, pressure and elemental composition. The extensive database returns a large number of species in the mixture, some of which are relatively low in terms of mass fraction. Immediate removal of low mass fraction level species must be avoided since it is not obvious as to whether they may play an important part in the overall reaction scheme at positions further downstream in the nozzle or plume.

To investigate this further a sensitivity analysis was carried out. From the Gordon and MacBride calculation, estimates were obtained of the chemical composition at the throat, and the nozzle exit temperature and pressure. With these starting conditions a computation was then performed with one of the simple single cell CHEMKIN chemical kinetic packages to compute the change in composition with respect to time. Species and reactions were then gradually removed and the computation repeated. The minimum reaction and species set at which there was little difference in composition change with respect to time, as compared to the more comprehensive set, was then chosen as the set to use.

Further modifications to this deck were made at a later stage when more extensive nozzle flow calculations were carried out. The final set-up consisted of 11 species : CO, CO_2, O, O_2, OH, H, H_2, H_2O, N_2, Cl and HCl, and 15 associated reactions. The set-up used is not being offered as the definitive one but it is reasonably comprehensive. If computation time was not of importance extra species that might be considered are Cl_2 and HCN, and various associated reactions.

The computation of the 3D plume flow first requires the computation of the flow through the nozzle. This is an axisymmetric problem and was computed using the FGE Navier Stokes solver TINA (Netterfield 1992). Brief details of this code are as follows:
The scheme employs a point implicit time-marching algorithm, which includes linearisation of all fluxes and source terms. The inviscid fluxes are computed using the Riemann Solver approach of the Roe form. Briefly, this entails modelling the inviscid fluxes at the cell interfaces using an upwinding technique based on Eigenvalues analogous to the wave velocities in the simple shock tube problem i.e. $u, u+c$ and $u-c$.

Extension of the scheme to second order accuracy requires the implementation of flux limiters to maintain its TVD (Time Variation Diminishing) property. The viscous terms in the solver are based on the thin layer equations. The code has a two-temperature, (translational and vibrational) thermodynamic model and a full non-equilibrium chemical kinetics model. The computed vibrational temperature is an average value over all molecular species present.

The relaxation model which determines the level of vibrational non-equilibrium is of the Landau-Teller form incorporating experimentally determined relaxation constants (Milikan and White) (see Candler 1990) for each species present. The chemistry model uses forward rate expressions of the standard Arrhenius form. Due to the lack of data reverse rates have been specified in terms of the forward rate and an equilibrium constant. The equilibrium constant is computed from the pressure, temperature and Gibbs Free Energy change for the reaction.

The computation was performed using a grid with 60 axial cells and 20 radial cells. The subsonic inflow at the exit of the combustion chamber was posed by extrapolating upstream for static pressure. From the specified total pressure in the combustion chamber and isentropic flow relations this can be used to yield the Mach number. Specified total temperature and inflow angle can be used to yield all the other flowfield parameters at the inflow.

The nozzle flow computation was performed with the viscous terms switched off, since the boundary layer at the nozzle exit is relatively thin, and therefore will have little effect on the freestream flow in this region. To include the boundary layer growth could present difficulties in the plume computation since it is impractical to use a highly stretched grid in the radial direction for the plume computation that would be directly compatible with that used for the nozzle. Therefore, the values at the plume startline must be interpolated from that at the outflow of the nozzle calculation. A simple linear interpolation is adequate unless there is a rapid change in flowfield conditions across the nozzle, such as would be the case if the boundary layer was computed.

The grid used for the plume computation is as indicated in Fig.1, and extends in the axial Z direction. The large number of cells required for such a problem makes it expensive in terms of computation time to stretch the grid extensively in any particular direction or use a large number of cells over a small volume in an effort to completely resolve regions of significant viscous interaction. However, as far as possible the grid has been set up to try and accurately model these regions. The grid represents a 45 degree sector of the plume, and by cyclic symmetry can be used to re-create the whole of the plume flow.

Two inflow conditions are specified at the upstream boundary of the grid. Firstly, the outflow of the 2D axisymmetric nozzle flow forms the inflow for the semicircular region. Secondly the freestream air forms the inflow to the outer radial regions. For the regions immediately surrounding the semicircular nozzle exit, i.e. the base region, a reflection condition has been imposed at the upstream boundary. The grid consisted of 700 cells in the X-Y plane by 73 cells in the axial direction corresponding to a total of 51100 cells. The computation performed has considered the viscous interaction in all 3 dimensions and assumed that the flow is laminar.

Some of the vibrational relaxation data and viscosity data were lacking and this has been estimated from data of other species. With respect to viscosity this is probably not important in that the variation between species is relatively small and the net effect on the flowfield is minimal. For the vibrational relaxation the importance of this will depend very much on the extent of thermodynamic nonequilibrium observed in the results.

In the absence of the base region the plume would be wholly supersonic. Relative to the vehicle this includes the mixing layer between the external freestream and the exhaust gases of the plume, since the vehicle is moving at a Mach number of 3.8. This would allow a space-marching computational approach to be used to compute the flowfield, which is significantly quicker than a fully time-marching approach. However, the base region generates areas of recirculation with embedded subsonic flow, which makes such an approach more difficult.

The method that was adopted was therefore as follows: Firstly the flowfield was computed over n axial cells using a space-marching version of TINA. The choice of n was such that at the n-th cell the flow across the whole plane of the plume would be supersonic in the marching direction. The method of space-marching was that a plane of cells was time-marched to convergence, and then the solver space-marched onto the next axial plane of cells. This provides an initial estimate of the flowfield, accepting that the method in some regions possessed an element of inaccuracy due to the presence of subsonic flow. The next stage in the computation is then to re-compute the flowfield up to the n-th axial position using the fully time-marching version of TINA. From this point the solver is then space-marched to the end of the grid in the same fashion previously described.

3. Discussion of results

A number of plots of contour data have been made and their position with respect to the plume is as follows: Using the diagram in Fig.1 axial contour plots in the $X-Z$ plane refer to the surface described by the extension of the line $A-B$ in the Z direction. For such plots the figures will illustrate axial sections through the complete plume. Contour plots in the $X-Y$ plane refer to the surface described by an $X-Y$ plane through the plume at constant Z, with reference to the axes system in Fig.1. For these plots the diagram will illustrate sections through 90 degrees of the plume. For all plots spatial dimensions are in metres. For clarity of the shaded contour levels the maximum value on the scale for the parameter being plotted is often lower than the maximum levels in the plume.

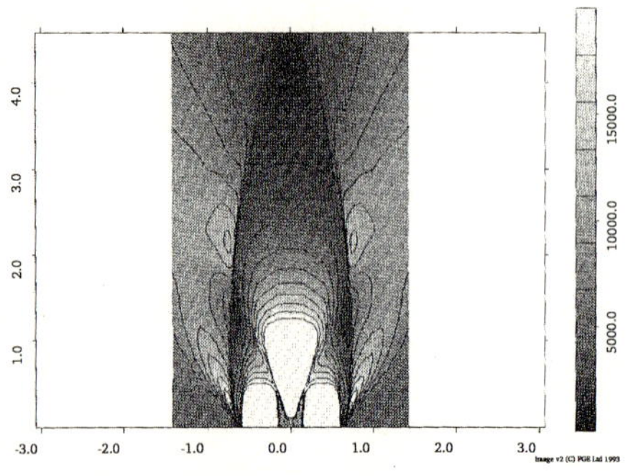

Fig. 2. Axial contour plot of pressure in $X - Z$ plane

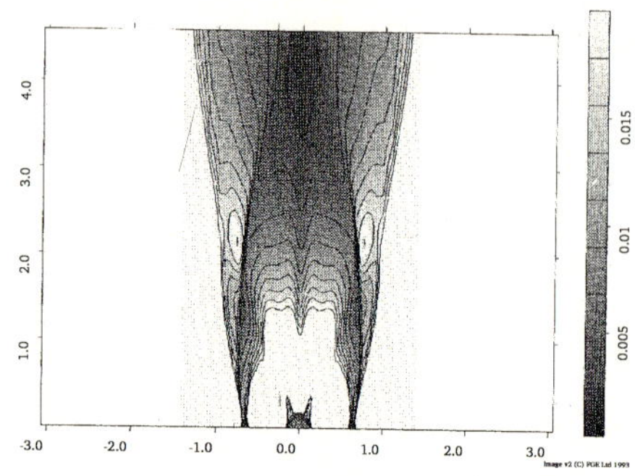

Fig. 5. Axial contour plot of density in $X - Z$ plane

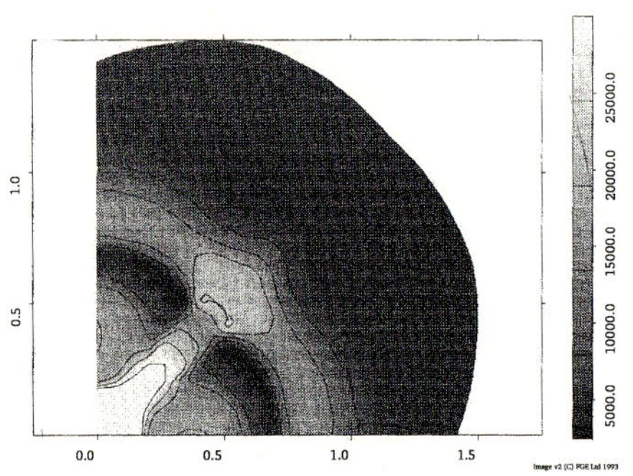

Fig. 3. Contour plot of pressure in $X - Y$ plane at Z =0.523 m

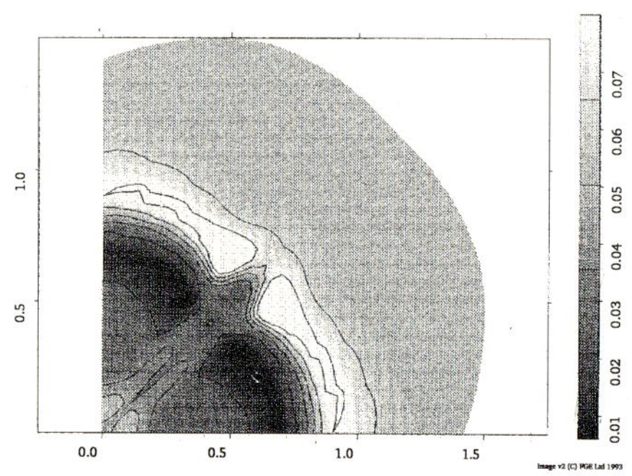

Fig. 6. Contour plot of density in $X - Y$ plane at Z =0.523 m

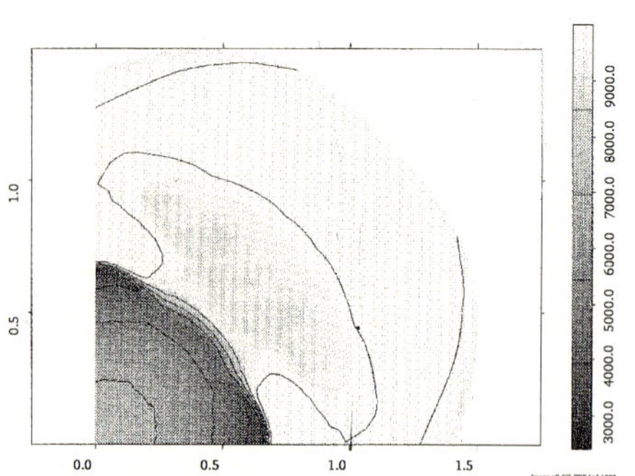

Fig. 4. Contour plot of pressure in $X - Y$ plane at Z =1.976 m

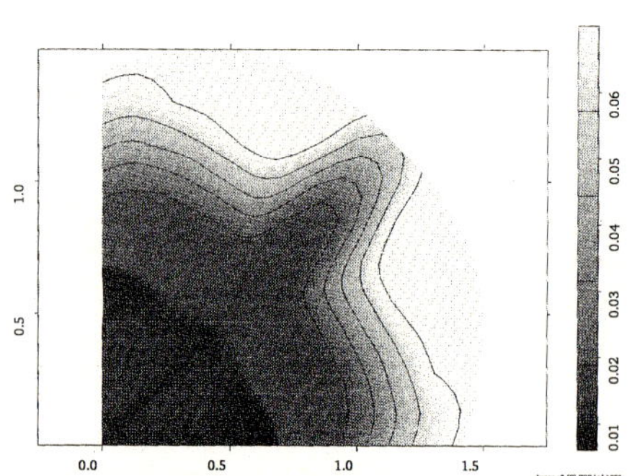

Fig. 7. Contour plot of density in $X - Y$ plane at Z =1.976 m

Fig.2 illustrates an axial plot of static pressure in the $X - Z$ plane. The figure shows the relatively complex shock structure. Firstly, there is a barrel shock for the flow expanding from each individual nozzle. In the figure this is not well defined but the beginning of the barrel shock can be seen at the edges of the distinct high-pressure core of the individual plumes.

At a larger radial position a second barrel shock can be clearly observed which corresponds to the plume structure as a whole. A third shock can be observed originating shortly downstream from the origin and propagating downstream in a radial and axial direction. This is visible in the figure at the edge of the high pressure region centred on the axis of the overall plume. This shock is generated as the individual plumes impinge on each other and align themselves with the axis of the overall plume.

Further details of the shock structure can be understood from examining Fig.3 which is a pressure contour plot in the $X - Y$ plane at an axial position of 0.523 metres. Firstly, as can be seen, the barrel shock for the overall plume is not spherical but has flattened sides, the centres of which lie at a position where adjacent individual impinging plumes have forced the gas outwards. Additionally at these points there is a peak in pressure. The second point of interest in Fig.3 is the shock originating from the origin. As can be seen, the shock for the full 360 degrees of the plume would be cross-like in shape.

In Fig.4 a similar pressure contour plot has been made at an axial position of 1.976 metres, where the individual plumes have mixed more thoroughly. The plot indicates that the overall barrel shock has become more square, and that the central cross-like shock has disappeared. Similar plots for density are given in Figs.5 to 7.

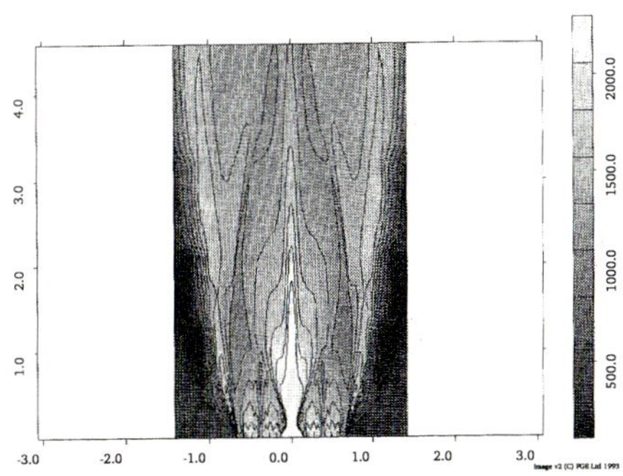

Fig. 8. Axial contour plot of vibrational temperature in $X - Z$ plane

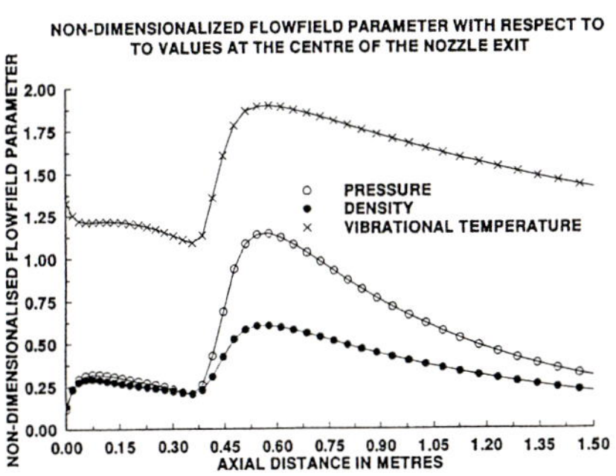

Fig. 9. Centreline flowfield values non-dimensionalized with respect to values at the centre of the nozzle exit

In Fig.8 an axial contour plot in the $X - Z$ plane is given for the vibrational temperature T_v. A point of major significance is that in the region of interaction between the individual plumes there is a significant rise in the vibrational temperature. A similar result was observed for the density and temperature.

These aspects are illustrated more clearly in Fig.9 where the individual quantity is plotted as a function of the corresponding value at the centre of the nozzle exit. This result is not surprising since considerable compression of the flowfield is occurring in this central region.

This is important in terms of the significant differences in flowfield parameters and emission characteristics to that which would be observed for the plume of a single equivalent area ratio nozzle. In Figs.10, 11 and 12 contour plots of vibrational temperature minus translational tem-

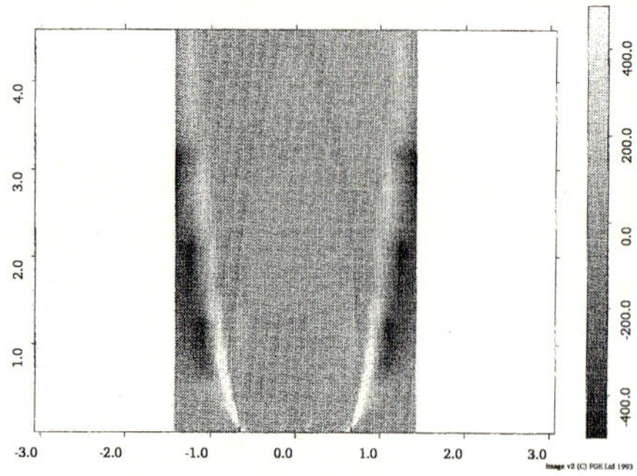

Fig. 10. Axial contour plot of $(T_v - T)$ in $X - Z$ plane

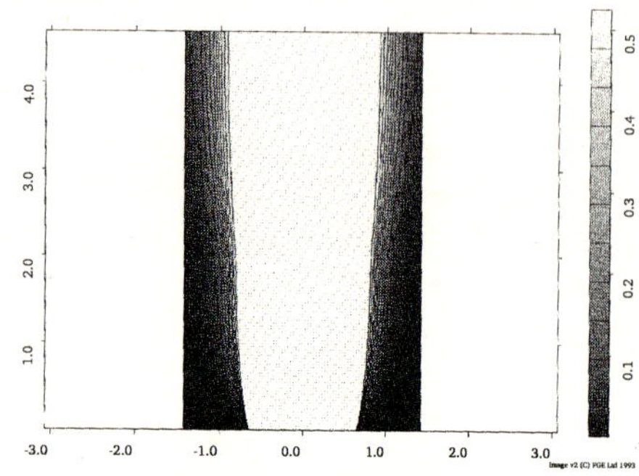

Fig. 13. Axial contour plot of CO mass fraction in $X - Z$ plane

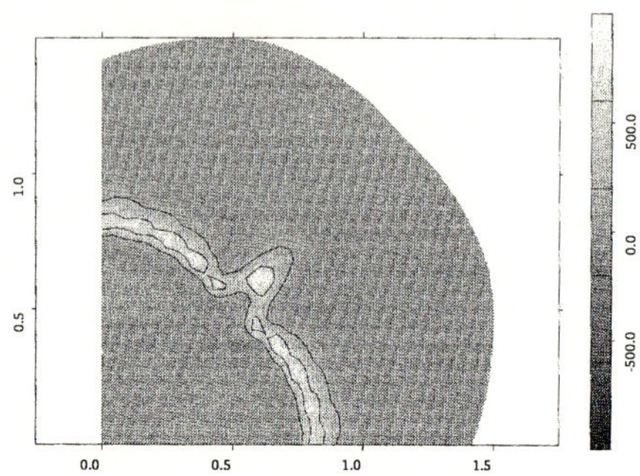

Fig. 11. Contour plot of $(T_v - T)$ in $X - Y$ plane at Z=0.523 m

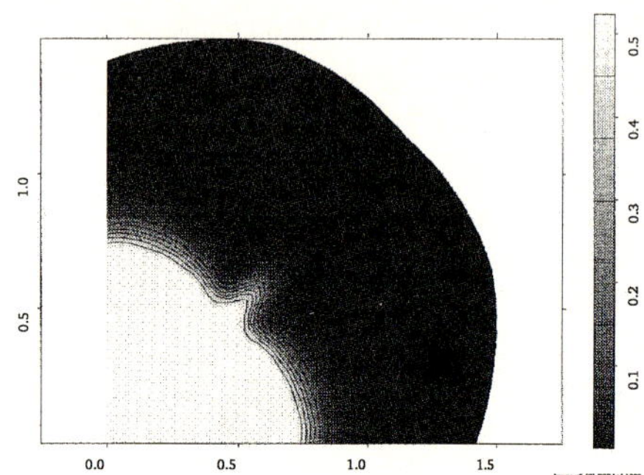

Fig. 14. Contour plot of CO mass fraction in $X - Y$ plane at Z=0.523 m

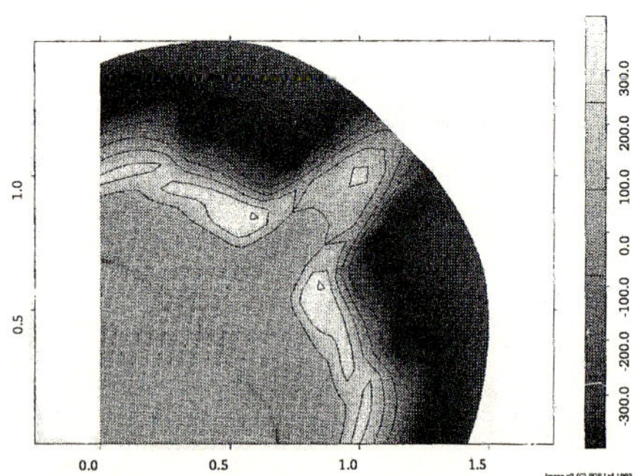

Fig. 12. Contour plot of $(T_v - T)$ in $X - Y$ plane at Z=1.975 m

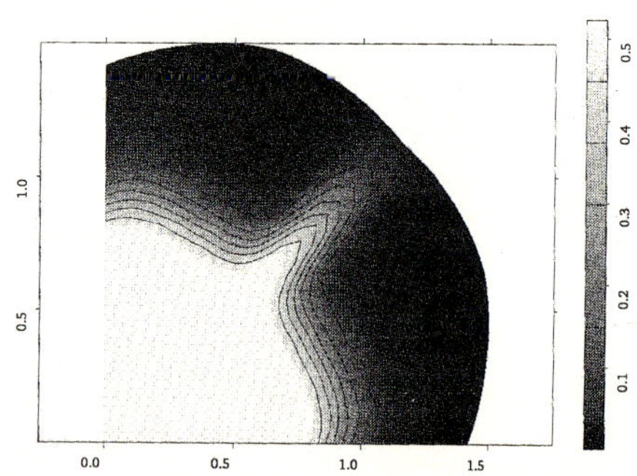

Fig. 15. Contour plot of CO mass fraction in $X - Y$ plane at Z=1.976 m

perature $(T_v - T)$ have been plotted for the $Y - Z$ plane, and $X - Y$ plane at two axial stations. Significant departure from equilibrium occurs in the location of shocks and the mixing layer. The relatively slow vibrational relaxation rates are not able to follow the rapid changes in translational temperature that occur in these regions.

Particularly interesting is the sudden reversal in the position of non-equilibrium that occurs in the outer region of the mixing layer, in that the translational temperature exceeds the vibrational temperature. This is due to a rapid rise in temperature of the freestream air gases as they undergo a compression at the edge of the plume. The slow relaxation rates of the vibrational energy do not allow the vibrational temperature to follow at the same rate.

Figs.13 to 15 indicate mass fraction levels of CO in the plume in the $X-Z$ and $X-Y$ planes. As expected the results indicate that there is little change in mass fraction levels due to chemical reactions. For such a flow where the freestream velocity is relatively high the typical Damkohler number, (equal to the ratio of the convection time to the reaction time), will be high. There is evidence though of significant changes in mass fraction levels at the edge of the plume due to the mixing of the exhaust gases and the air in the freestream.

There may be some level of numerical diffusion contributing to the extent of the mixing but the use of an Eigenvalue limiter of 0.01 should have ensured that this effect is small. Notice the larger extent of mixing that occurs in the vicinity of the $Y = X$ symmetry line due to interaction of the gases of the individual plumes.

It is interesting that the extent of this is greater at the further downstream position, as can be seen by comparing Figs.14 and 15. A similar trend can be seen in the $X - Y$ plots of density in Figs.6 and 7. The computation has assumed laminar flow, this choice being made at this stage more for simplicity than for any other reason, although in practice it is likely that the plume will be turbulent.

4. Conclusions

It has been shown that a complex three dimensional chemically reacting flowfield has been effectively modelled using the FGE Navier Stokes solver TINA. With respect to the numerics the application of space-marching in regions of the flowfield which were wholly supersonic in the marching direction, significantly reduced the computation time.

Limitations in the scheme are attributable to the lack of vibrational relaxation data for some of the species. Additionally the complexity of the problem makes it impractical in terms of computation time for parametric studies to obtain a high degree of resolution in the mixing layer. The use of some form of adaptive gridding or more cells would assist in this, but both will impose additional computation time. However, despite the modelling limitations some interesting conclusions can be drawn for regions in the core of the flow as well as in the mixing layer.

Firstly, there are significant differences between the vibrational and translational temperature in the region of the outer barrel shock and mixing layer indicating that there is a significant departure from thermodynamic equilibrium.

Secondly the mass fraction contour figures indicate that for laminar flow there is a significant level of mixing between the streams of the plume gases and the freestream air although the central core remains quite uniform. Regions of highest density, pressure and vibrational temperature are found not only near the exits of the nozzles, but also in a region around the centreline of the overall plume shortly downstream from the base of the motor.

Although some 3D aspects of the plume gradually disappear along the plume, such as the central cross-shaped shock, others are enhanced. For example the barrel shock becomes squarer in shape and the effect of the individual plume interaction on density and mass fraction levels becomes more marked, as can be seen in the vicinity of the $Y = X$ line on the $X - Y$ plane contour plots.

Acknowledgements

This work has been carried out with the support of the Procurement Executive, Ministry of Defence.

References

Cain T, Morris NM, Roberts TP, Jones TV (1992) A comparison of numerical predictions and experiments on under-expanded Jets. AIAA Paper 92-4027

Candler GV (1990) Computation of thermo-chemical nonequilibrium Martian atmospheric entry flows. AIAA Paper 90-1695

Holcomb JE (1989) Three-dimensional Navier Stokes rocket plume calculations. AIAA Paper 89-1986-CP

Netterfield MP (1992) Validation of a Navier Stokes solver for thermochemical nonequilibrium flows. AIAA Paper 92-2878

Nonequilibrium Hypersonic Flow Around a Blunt Body

E. Schall and D. Zeitoun
Lab. I.U.S.T.I./S.E.T.T. (U.R.A CNRS 1168), Dept. M.H.E.Q Université de Provence - St Jérôme 13397 Marseille Cedex 20, France

Abstract. The laminar, viscous, nonequilibrium flow around an axisymmetric blunt body has been computed. A fully implicit finite difference scheme of second order was used. The air was considered as a reactive mixture of the five species N_2, O_2, NO, N, O, in which the diatomic species, N_2, O_2 are in vibrational nonequilibrium. A range of Mach numbers from 14 to 18 has been investigated. The numerical results have been compared with those obtained by other workers and are in agreement with ballistic range data concerning the standoff shock distance at $M = 15.35$.

Key words: Hypersonic flows, Chemical and vibrational nonequilibrium, Blunt body

1. Introduction

A high velocity flow around a body leads to the formation of a strong bow shock wave. Conversion of the high kinetic energy modes and consequently significant chemical and thermal nonequilibrium occurs around the body. To treat such conditions, the air is usually considered as composed of the five following species: N_2, O_2, NO, N and O. This model assumes seventeen chemical reactions, whose reaction rates and equilibrium constants are given by Park (1989). The diatomic species N_2 and O_2 are in thermal non-equilibrium and NO is chosen to be in a thermal equilibrium state (Stupochenko et al. 1967). The subject of the present work is to put into evidence the influence of the non-equilibrium processes on the flowfield parameters. The harmonic oscillator assumption is used and the vibrational exchanges take into account only the translation-vibration transfers through the Laudau-Teller theory. A global relaxation time models the exchanges of the molecules with the other species of the mixture (Candler and MacCormack 1988). The ionization of atoms has been neglected in the chosen Mach number range. No vibration-dissociation coupling is considered either. This non-equilibrium flow is described by a system of ten equations. In order to solve this system, an implicit finite difference method, following Mac Cormack (1985), is implemented. This method is based on a predictor-corrector scheme, with second order accuracy. In the implicit operator, an Eulerian non-dissipative flux-splitting technique (Josyula and Chang 1992) is used. After discretization, the block-pentadiagonal matrix system is solved with a Gauss-Seidel line relaxation method (Zeitoun 1991). The first case presented is at Mach 15.35 which corresponds to Lobbs' experiments (1964). This case has already been computed by several authors (Candler and MacCormack 1988, Flament 1990, Josyula and Chang 1992). The following cases are at Mach numbers 14 to 18 and an altitude of 37 Km. A code-to-code comparison is also made with a recent paper by Josyula and Shang (1992) in which Roe's flux-difference split scheme is used. These authors take also into account a thermal nonequilibrium assumption for the NO molecule.

2. Governing equations

In a cylindrical coordinate system, the axisymmetric conservative equations for an air mixture in chemical and thermal nonequilibrium are the Navier-Stokes equations with, in addition, the mass evolution equations for each species and those for the vibrational energy of each diatomic species. These equations can be expressed in vector form as:

Shock Waves @ Marseille II
Editors: R. Brun, L. Z. Dumitrescu

$$\frac{\partial U}{\partial t}+\frac{\partial\left(F_e+F_v\right)}{\partial x}+\frac{\partial\left(G_e+G_v\right)}{\partial y}+\frac{1}{y}H_a=\Omega \tag{1}$$

The vector of conservative quantities is given by:

$$U=\left(\rho_s,\, \rho u,\, \rho v,\, \rho E,\, \rho_i e_{vi}\right)^T \tag{2}$$

where ρ_s is the density of each species, the subcript $s = 1, 2, \cdots 5$ represents the five species N_2, O_2, NO, N, O of the air mixture and $i = 1, 2$ refers to the diatomic species N_2, O_2, (NO is supposed to be in thermal equilibrium), ρ is the density of the mixture, u and v the components of the velocity V, and ρE is the total energy per unit volume, containing translational, rotational and vibrational energies, the latent chemical energy of the species and the kinetic energy of directed motion; ρE is given by :

$$\rho E=\sum_{s=1}^{5}\rho_s c_{v_s} T+\sum_{i=1}^{2}\rho_i e v_i+\rho_{NO} e_{vNO}+\sum_{s=1}^{5}\rho_s h_s^{for}+0.5\rho V^2 \tag{3}$$

In Eq.1 the vectors F_e, G_e, F_v and G_v represent, respectively, the Eulerian convective and viscous fluxes, and can be written as follows:

$$F_e+F_v=\begin{bmatrix}\rho_s u-\rho D_{sm}\frac{\partial Y_s}{\partial x}\\ \rho u^2+P-\tau_{xx}\\ \rho uv-\tau_{xy}\\ \left(\rho E+P-\tau_{xx}\right)u-v\tau_{xy}+q_x\\ \rho_i u e_{vi}-\rho e_{vi} D_{im}\frac{\partial Y_i}{\partial x}+q_{vix}\end{bmatrix},\quad G_e+G_v=\begin{bmatrix}\rho_s v-\rho D_{sm}\frac{\partial Y_s}{\partial y}\\ \rho uv-\tau_{xy}\\ \rho v^2+P-\tau_{yy}\\ \left(\rho E+P-\tau_{yy}\right)v-u\tau_{xy}+q_y\\ \rho_i v e_{vi}-\rho e_{vi} D_{im}\frac{\partial Y_i}{\partial y}+q_{viy}\end{bmatrix}$$

H_a is the axisymmetric vector given by

$$H_a=\begin{bmatrix}\rho_s v-\rho D_{sm}\frac{\partial Y_s}{\partial y}\\ \rho uv-\tau_{xy}\\ \rho v^2-\tau_\theta\\ \left(\rho E+P-\tau_{yy}\right)v-u\tau_{xy}+q_y\\ \rho_i v e_{vi}-\rho e_{vi} D_{im}\frac{\partial Y_i}{\partial y}+q_{viy}\end{bmatrix}$$

$\tau_{xx}, \tau_{yy}, \tau_{xy}$ are the components of the stress tensor. According to Fick's first law of diffusion the heat flux vector can be written as follows:

$$\vec{q}=-k_m\overrightarrow{\nabla T}-\sum_{s=1}^{5}h_s\rho D_{sm}\overrightarrow{\nabla Y_k}-\sum_{i=1}^{2}(\rho D_{im}Y_i)\overrightarrow{\nabla e_{vi}}$$

In this expression, k_m is the thermal conductivity of the mixture and is computed by Wilke's mixing rule, while the binary diffusion coefficients D_{sm} are given by a Chapman and Cowling empirical formula (White 1974). As concerns the vector $\omega_s = (\omega_s, 0, 0, 0, \omega_i)^T$, it contains the chemical source terms ω_s, due to the net rate production of each species, and the vibrational source terms ω_i of the species N_2 and O_2. These source terms are the result of the chemical and vibrational nonequilibrium modeling and are described hereafter.

3. Chemical and vibrational nonequilibrium modelling

In the five species air mixture, each species is considered as a perfect gas. The chemical source terms are derived from the reactions that occur between these species, being given by the following 17-reaction scheme:

$$
\begin{aligned}
N_2 + M &= 2N + M \\
O_2 + M &= 2O + M \\
NO + M &= N + O + M \\
N_2 + O &= NO + O \\
NO + O &= N + O_2
\end{aligned}
\tag{4}
$$

with M=N, O, NO, N_2, O_2. The first three reactions are dissociation-recombination type and the fourth and fifth are exchange reactions. The magnitude of the mass production or removal is obtained from:

$$
\omega_s = M_s \sum_r (\nu''_{sr} - \nu'_{sr}) \left\{ K_{f_r} \prod_s \left(\frac{\rho_s}{M_s}\right)^{\nu'_{sr}} - K_{b_r} \prod_s \left(\frac{\rho_s}{M_s}\right)^{\nu''_{sr}} \right\}
\tag{5}
$$

ν'_{sr}, ν''_{sr} refer to the stœchiometric coefficients, while k_f and k_b are the forward and backward reaction rates, as given by Park (1989) and are calculated with the translational temperature. In the vibrational modelling, only vibration-translation energy exchanges are taken into account. The source terms in the vibrational equations are classically described by the Laudau-Teller formulation and are given by:

$$
\omega_{i(T-V)} = \frac{\rho_i e^0_{v_i} - \rho_i e_{v_i}}{\tau_i^{VT}} + e_{v_i}\omega_s
\tag{6}
$$

with $e^0_{v_i}$ the vibrational energy at equilibrium; since it takes into account all the vibrational-translational exchanges, the global relaxation time is expressed as:

$$
\frac{1}{\tau_i^{VT}} = \sum_{s=1}^{5} \frac{\xi_s}{\tau_{is}^{VT}}
\tag{7}
$$

where ξ_s is the molar fraction of each species s; τ_i^{VT}, the relaxation time between the diatomic molecules i and any species s can be found in Druguet (1992).

4. Computation procedure

To correctly describe the dissipative effects in the boundary layer which develops along the body wall and the strong flow gradients around the symmetry axis, the mesh distribution must be quite stretched and consequently the time step for integration be very small. In order to overcome this difficulty and others, due to the stiffness of the chemical and vibrational equations, the system of ten governing equations (1) is solved by a strongly implicit finite-difference scheme like that used by MacCormack (1985). The physical domain (x, y) is transformed into a rectangular one (ξ, η) and for each node (i, j) of this domain the system of equations is written at each time step $(n+1)\Delta t$ as:

$$
\frac{U^{n+1} - U^n}{\Delta t} + \frac{D}{D\xi}(F_e + F_v)^{n+1} + \frac{D}{D\eta}(G_e + G_v)^{n+1} + \frac{1}{y}H_a^{n+1} + H_s^{n+1} = 0
\tag{8}
$$

F representing one of the vectors of this equation, it is linearized to yield the expansion:

$$
\Phi^{n+1} = \Phi^n + \left(\frac{\partial \Phi}{\partial U}\right)^n \delta U^n
\tag{9}
$$

where $\left(\frac{\partial \Phi}{\partial U}\right)$ is the jacobian matrix and $\delta U^n = U^{n+1} - U^n$. Under these conditions, and after using the non-dissipative flux-splitting technique proposed by Candler and MacCormack (1988), the system (8) is equivalent to :

$$\left\{I+\Delta t\left\{\frac{D_{+}A_{-}}{\delta\xi}+\frac{D_{-}A_{+}}{\delta\xi}+\frac{D^{2}S_{V}}{\delta\xi^{2}}+\frac{D_{+}B_{-}}{\delta\eta}+\frac{D_{-}B_{+}}{\delta\eta}+\frac{D^{2}R_{V}}{\delta\eta^{2}}+\frac{1}{y}C_{a}+C_{s}\right\}\right\}\delta U^{n}=\Delta U^{n} \quad (10)$$

In this equation, D, D_+, D_- are respectively central, forward and backward difference operators, while A_-, A_+, S_V, B_-, B_+, R_V, C_a, C_s are the jacobian matrices of F_e^-, F_e^+, F_v, G_e^-, G_e^+, G_V, H_a and H_s. C_a contains only the jacobian of the non-dissipative part of the vector H_a. After discretization,the system (10) may be written as a pentadiagonal matrix linear system:

$$\widehat{B}_{i,j}\delta U_{i,j+1}^{n}+\widehat{A}_{i,j}\delta U_{i,j}^{n}+\widehat{C}_{i,j}\delta U_{i,j-1}^{n}+\widehat{D}_{i,j}\delta U_{i+1,j}^{n}+\widehat{E}_{i,j}\delta U_{i-1,j}^{n}=\Delta U_{i,j}^{n} \quad (11)$$

At each time step, a predictor-corrector scheme is used, and the system (11) is solved by a Gauss-Seidel line or point relaxation method, with alternating sweeps in the backward and forward ξ directions. The conservation vector U is updated by

$$U_{i,j}^{n+1}=U_{i,j}^{n}+\delta U_{i,j}^{n} \quad (12)$$

Knowledge of the vector U allows to obtain the following variables: ρ_s, $(s=1,\cdots 5)$, ρ, u, v, $(i=1,2)$. After that, the temperature is deduced from Eq.3 using a Newton-Raphson iteration, while the pressure is found through the state equation of the gas mixture. Associated to this algorithm, the boundary conditions of the computational domain are given as undisturbed freestream values on the upstream boundary. A zero-extrapolation is imposed for the supersonic downstream boundary. A non-slip condition for the velocity components, zero normal pressure gradient, isothermal and non-catalytic conditions are enforced on the body surface.

5. Conditions of numerical simulation

The upstream flow condition corresponding to $M_\infty = 15.35$ are: $T_\infty = 293$ K, $P_\infty = 664$ Pa, $Rs = 6.35$ mm, $T_W = 1000$ K.

The upstream flow condition corresponding to $M_\infty = 14$ to 18 are: $T_\infty = 252$ K, $P_\infty = 432$ Pa, $Rs = 6.60$ mm, $T_W = 555.5$ K.

The number of the mesh points is of (50 x 50) with a refinement near the wall for a better boundary layer capturing, near the symmetry axis, and around the shock position. In order to prevent oscillations near the shock, a second and fourth order adaptative dissipation technique (Jameson and Schmidt 1981) is used ($\epsilon_2 = 0.65, \epsilon_4 = 0.04$). The minimum grid spacing in the radial direction and along the body surface are respectively equal to $\delta\eta = 0.16 \times 10^{-4}$ and $\delta\xi = 1. \times 10^{-3}$. The CPU time is about to 1.3×10^{-3} sec/pt/iter on a CRAY YMP-2E computer for the complete set of equations.

6. Discussion of the results

• M_∞=15.35. This case has often been used in the literature (Candler and MacCormack 1988, Flament 1990, Josyula and Shang 1992) for code validation of chemical and thermal nonequilibrium flows. Fig.1 shows the evolution of the translational and vibrational temperatures along the stagnation streamline. The vibrational nonequilibrium region of each diatomic species is clearly visible, and also the boundary layer in a thermal equilibrium state. In the core flow, the relaxation zone of O_2 is longer than that of N_2. In Fig.1 the translational temperature profile is also compared to those given by others authors. A shorter shock standoff distance is noticed. This last remark seems to be in agreement with experimental results, as shown in Fig.2, where Candler's results are also represented.

• M_∞=14 to 18. In this section, numerical results are discussed for three values of the freestream Mach number at an altitude of 37 Km in order to compare our numerical results with other

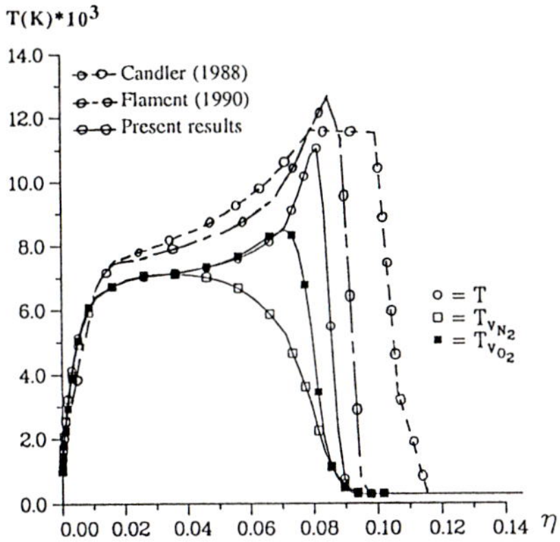

Fig. 1. Comparison of stagnation streamline temperature profiles

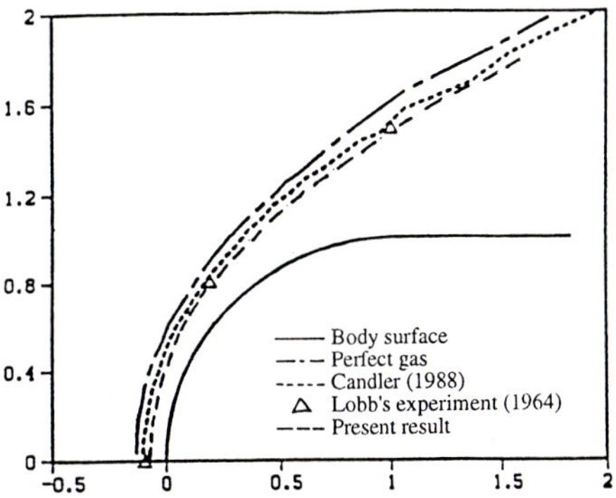

Fig. 2. Comparison of shock standoff

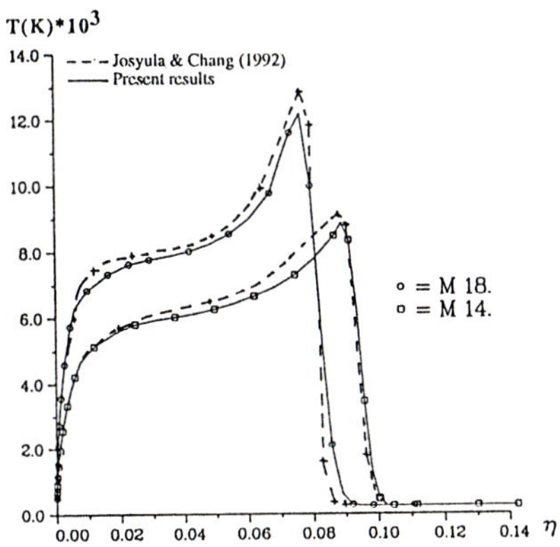

Fig. 3. Temperature distribution along symmetry axis

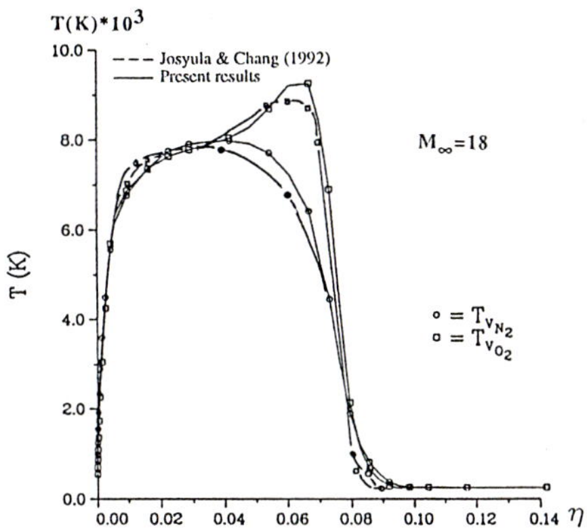

Fig. 4. Vibrational temperature along stagnation streamline

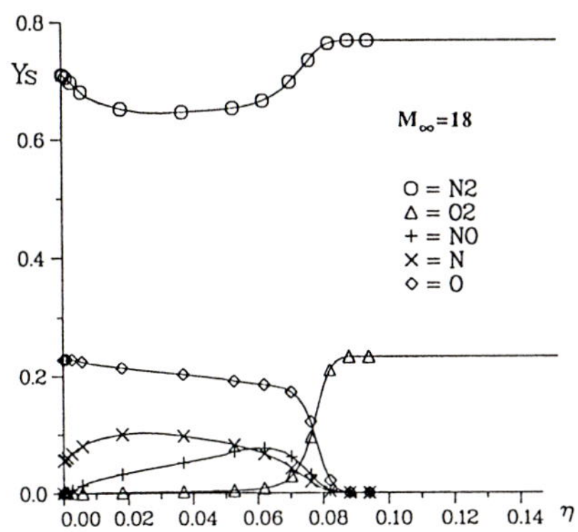

Fig. 5. Mass fractions along stagnation streamline

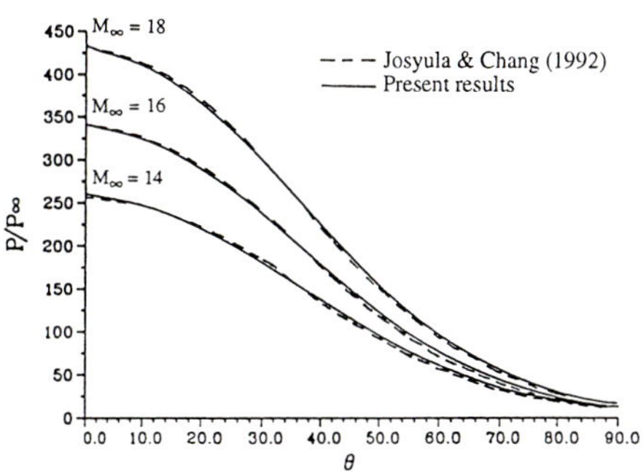

Fig. 6. Wall pressures at different Mach numbers

authors. This code-to-code comparison goes on in Fig.3, where the translational temperature distributions along the stagnation streamline for two values of the Mach number ($M_\infty = 18$ and 14) are drawn. The shock standoff distance is about the same. The slight differences existing between the two computations can be explained by the choice of thermal equilibrium for the NO molecule in our case. The agreement is better at the lower Mach number concerning the thermal boundary layer. Some slight discrepancies can also be noted in the vibrational temperature evolutions of N_2 and O_2 along the stagnation line, as seen in Fig.4. On Fig.5 the dissociation of diatomic species N_2 and O_2 behind the bow shock wave leads to the apparition of species O, N and NO. On Fig.6 excellent agreement can be noticed despite the different choice of the numerical method and of the vibrational model.

6. Conclusions

Hypersonic air flow in chemical and vibrational nonequilibrium around a hemisphere-cylinder model has been simulated numerically by a finite-difference implicit method. The present results were compared with other computations in the same upstream conditions corresponding to available experimental data, and more particularly, with the recent numerical work of Josyula and Shang (1992). From these comparisons, some conclusions can be drawn: A good agreement is found for the wall pressure, as well as for the translational and vibrational temperature evolutions along the stagnation streamline. The thermal nonequilibrium of the NO molecule has very slight effect on these evolution.

Acknowledgements

The authors would like to thank the "Centre de Calcul de l'I.M.T." for allowing them access to the Cray YMP-2E computer.

References

Candler G, MacCormack RW (1988) The computation of hypersonic ionized flows in thermal and chemical nonequilibrium. AIAA Paper 88-0511

Druguet MC (1992)Contribution à l'étude des écoulements Euleriens hypersoniques en deséquilibre thermochimique. Thèse de Doctorat, Université de Provence, Marseille

Flament C (1990) Ecoulements de fluides visqueux en déséquilibre chimique et vibrationnel, Thèse de Doctorat, Paris VI

Jameson A, Schimdt W (1981) Numerical solution of the Euler equations by finite volume methods using Runge-Kutta time stepping scheme. AIAA Paper 81-1259

Josyula E, Shang JS (1992) Computation of hypersonic flowfield in thermal and chemical nonequilibrium. AIAA Paper 92-2874

Lobb R (1964) Experimental measurement of shock detachment distance on sphere in air at hypervelocity. In: Wilson WC (ed) High temperature aspects of hypersonic flow, Pergamon Press, NY, pp.519-527

MacCormack RW (1985) Current status of numerical solution of the Navier-Stokes equation. AIAA Paper 85-0032

Park C (1989) Assessment of a two-temperature kinetic model for ionizing air. AIAA Paper 87-1574

Stupochenko YV, Losev SA, Osipov AI (1967) Relaxation in Shock Waves. Springer-Verlag, NY, pp. 259-322

White FA (1974) Viscous Fluid Flow. McGraw-Hill, NY, pp 28-36

Zeitoun D (1991) Chemical and vibrational nonequilibrium flowfields. Journal of Comp. Methods in Applied Mechanics and Eng. 90:687-692

A New Model for Rotation-Vibration-Dissociation Coupling in a Multi-Component Viscous Shock Layer

Sergey V. Zhluktov* **and Gennadiy D. Smekhov†**
*Institute of Computer Aided Design, Russian Academy of Sciences, 2-d Brestskaya 19/18, 123865 Moscow, Russia
†Institute of Mechanics at Moscow State University, Michurinskiy pr. 1, 119899 Moscow, Russia

Abstract. A new dissociation model accounting for equilibrium excitation of rotational modes of the molecules and nonequilibrium excitation of vibrational modes is presented. The model has no unknown empirical parameters, and the uncertainty of the model is determined only by uncertainty of the physical quantities characterizing collisions of the particles. The model can be approximated by a simple expression, convenient in gas-dynamic applications. Numerical calculations within the scope of the multi-component nonequilibrium full viscous shock layer model with the use of suggested formulas for dissociation rate constants show that the data of this complex model fit well with experimental data for the shock detachment distance. A comparison of various models accounting for the influence of vibrational nonequilibrium on dissociation rates is carried out. The comparison has shown that the "T_v, T" model of Park gives underestimated values of the vibrationally nonequilibrium dissociation rate constants.

Key words: Vibrational nonequilibrium, Dissociation rate constant, Adiabatic principle

1. Introduction

The development of the new maneuvering transatmospheric vehicles requires creating a dissociation model for extremely high temperatures $T = 30000 \cdots 40000K$ occurring behind strong shock waves. The "T_v, T" dissociation model of Park (1986) is fully empirical and has no theoretical basis. The model of Marrone and Treanor (1963) used by Zhluktov and Tirskiy (1990) arouses difficulties, when employed in computations, due to an indefiniteness in the parameter U. The adiabatic dissociation model of Smekhov and Zhluktov (1992, 1993) accounts for the probability of dissociation of a molecule from each vibrational level by means of the adiabatic principle. Due to the use of the adiabatic principle the disintegration of diatomic molecules can be described at temperatures of the order of magnitude about half the dissociation energy, or higher as well. In the present paper the adiabatic model is combined with the CVDV approach of Treanor and Marrone (1962), accounting for the reverse influence of reactions on vibrational relaxation, and applied to the flows past hypersonic blunt bodies.

2. Adiabatic dissociation model

For gases consisting of weakly interacting particles we can assume that each energy level can be split into independent modes, describing the translational motion of the mass center of the molecule, rotational motion of the molecule, and vibrational motion of the molecular atoms respectively. Besides, we will assume that the molecule under consideration is in the ground electronic state. The adiabatic dissociation model yields the following expression for the macroscopic dissociation constant of a diatomic molecule:

$$K_D = \sqrt{\frac{8R_AT}{\pi\mu}}\sigma\frac{N_A}{Z_v(T_v)}\frac{1}{(kT)^2}\exp\left[-\frac{D}{kT}\right]\Phi, \tag{1}$$

Shock Waves @ Marseille II
Editors: R. Brun, L. Z. Dumitrescu

$$\Phi = \frac{1}{2}\sum_{v=0}^{v_{\max}} \Phi_v, \quad \Phi_v = (D - E_v)(D - E_v + 2kT)\times$$
$$\times \exp\left[-(T_v^{-1} - T^{-1})E_v/k - \beta\sqrt{D - E_v}\right], \quad \beta = \sqrt{\mu/2N_A}/\bar{h}\alpha \tag{2}$$

where T and T_v are temperatures of the active (translational, rotational) and vibrational modes respectively; R_A is the universal gas constant, N_A is the Avogadro number, k and $\bar{h}$ are the Boltzmann and Planck constants; μ is the reduced mass of colliding particles; σ is an effective cross-section of molecular disintegration from the states near dissociation limit, which is close to the gas-kinetic cross-section (Smekhov and Zhluktov 1992); $Z_v(T_v)$ is the vibrational partition function (with quite a good accuracy it can be taken as that of harmonic oscillator cut-off at the energy of dissociation); D is the dissociation energy of the molecule; $v_{\max}$ is the maximum vibrational level corresponding to the border of a continuous spectrum; $E_v = kv\Theta(1 - \gamma v)$ is the vibrational internal energy, v is the vibrational quantum number, Θ is the characteristic vibrational temperature; $\gamma = k\Theta/4D$ is the nonharmonicity constant ($\gamma = 0$ when the molecule is considered as a rigid rotator/harmonic oscillator); α (Å^{-1}) is the parameter of long-acting interaction potential $V = V_0\exp(-\alpha r)$, r (Å) is the distance between interacting particles. Thus the physical constants Θ, D, μ, the parameters of interaction with the collision partner σ, α, and the translational and vibrational temperatures define the constant of dissociation unambiguously. The main peculiarity of the model defined by Eqs.1,2 is the finiteness of the value of the constant at low temperatures. In nonequilibrium conditions, when $T_v < T$, the dissociation rate constant is less than that in the equilibrium case. However the lowest value of K_D (when $T_v \to 0$) remains finite, its value corresponding to the dissociation from the ground vibrational state (the first term Φ_0 in the sum (2)).

Eq.2 is too cumbersome for gas-dynamic computations due to the necessity to sum Φ_v over all the levels of the anharmonic oscillator for each T and T_v. Yet the model (Eqs.1,2) can be simplified by using the harmonic oscillator model and distinguishing a certain vibrational level E_v^* whose dissociation effectiveness is equal to the average effectiveness of all the vibration-rotation spectrum:

$$\Phi = \frac{1}{2}(D - E_v^*)(D - E_v^* + 2kT)\exp\left[-\beta\sqrt{D - E_v^*}\right] Z_v(T_m) \tag{3}$$

$$T_m^{-1} = T_v^{-1} - T^{-1} \tag{4}$$

Assuming that $E_v = 0$, we obtain

$$\Phi = \frac{1}{2}D(D + 2kT)\exp(-\beta\sqrt{D})Z_v(T_m) \tag{5}$$

this being the simplest approximation of Eq.2. It should be remarked that the function $\Phi_v(E_v)$ has a maximum at a certain $\tilde{E}_v : \Phi_v(\tilde{E}_v) > \Phi_v(0)$, and $\Phi_v \to 0$ as $\tilde{E}_v \to 0$. Thus, approximation (5) is reasonable enough and corresponds to the assumption of equal dissociation probability from all the vibrational levels. At the same time the expression of Eq.5 for Φ can be improved. In the present paper it is proposed to obtain Φ from the formula

$$\Phi = \frac{1}{2}Dx(Dx + 2kT)\exp(-\beta\sqrt{Dx})Z_v(T_m) \tag{6}$$

where

$$x = 1 - 0.8\varepsilon(T_m)/D, \quad \varepsilon = k\left[\frac{\Theta}{(\exp(\Theta/T_m) - 1)} - \frac{D/k}{(\exp(D/kT_m) - 1)}\right] \tag{7}$$

In other words, we assume that $E_v^* = 0.8\varepsilon(T_m)$ where ε is the energy of a cut-off harmonic oscillator. Both models (Eqs.1,2) and (Eqs.1,6,7) contain two parameters of intermolecular interaction: σ and α. These parameters have a clear physical sense: σ is approximately equal to the gas-kinetic cross-section, α defines the long-acting part of the interaction potential. Because of this it is proposed to use published values of σ_{ij} and α_{ij}, $i,j = O_2$, N_2, NO, O, N (Glushko 1973, Gordeev et al. 1985) in both the models. The two- temperature dissociation rate constant thus defined contains no empirical parameters, and the uncertainty of the constant is determined by uncertainty of experimental data for σ_{ij} and α_{ij}.

3. Analysis of thermal nonequilibrium dissociation constants

It is interesting to discuss differences between various models of dissociation accounting for thermal nonequilibrium. At present, the model (Eqs.1,2) is the most "physical" because it allows for centrifugal dissociation due to taking into account the excitation of rotational modes. Since our interest concerns vibrational nonequilibrium, let us consider the ratio V of the vibrationally nonequilibrium dissociation constant to the thermally equilibrium one, i.e.,

$$K_D(T_v,T) = K_D(T,T) \times V(T_v,T) \tag{8}$$

In the present work numerical calculations of the quantity $W = V_{mod}/V_{ad}$ are submitted, where V_{mod} is a value of V computed with the use of a certain simplified model, and V_{ad} is the V value obtained within the scope of the adiabatic model (Eqs.1,2). The following simplified models have been considered: 1) the approximate model (Eqs.1,6,7); 2) the approximate model (Eqs.1,5), (this model expressed in terms of V coincides with the model of Hammerling and Teare (1959)); 3) the model of Treanor and Marrone (1963) with $U_i = T_{D_i}/2$ $(i = O_2, N_2, NO)$ 4) the same model with $U_i = T_{D_i}/6$; 5) the model of Park (1986) in which the dissociation constants are functions of the quantity $T_a = \sqrt{T_v T}$.

In Table 1 the maximum and minimum values of W for all the mentioned models are represented. The temperature range 1000 K$< T_v, T <$ 50000 K is considered (T_v and T were varied independently). For every molecular species all five colliding partners were considered. From the table one can see that the discrepancy of model (Eqs.1,6,7) with model (Eqs.1,2) lies in the range defined by the multiple $2.5^{\pm 1}$. The largest disagreement of the model (Eqs.1,5) is defined by the multiple $5^{\pm 1}$. The model of Treanor and Marrone is less accurate, and the smaller is the value of U, the larger is the error of the model. The largest discrepancy (more than 1000 times lower than model (Eqs.1,2)) can arise by the model of Park. So, the parameter T_a does not reflect the real physical processes of vibrational and rotational excitation contrarily to what is stated by Park (1986).

Table 1. $W = V_{mod}/V_{ad}$

Model	max or min	O_2	N_2	NO
Eqs.1,6,7	max	1.91	2.23	1.53
Eqs.1,6,7	min	0.638	0.401	0.619
Eqs.1,5	max	5.34	4.91	3.03
Eqs.1,5	min	0.719	0.337	0.773
Marrone, $U = T_D/2$	max	11.8	8.92	6.60
Marrone, $U = T_D/2$	min	0.240	0.281	0.312
Marrone, $U = T_D/6$	max	28.0	21.3	15.3
Marrone, $U = T_D/6$	min	$1.26 \cdot 10^{-2}$	$1.47 \cdot 10^{-2}$	$1.70 \cdot 10^{-2}$
Park	max	5.39	2.93	1.65
Park	min	$5.73 \cdot 10^{-6}$	$1.16 \cdot 10^{-11}$	$2.32 \cdot 10^{-7}$

4. Multi-component nonequilibrium full viscous shock layer equation set

In order to calculate hypersonic flows past an axisymmetrical body, the multi-component, nonequilibrium, full viscous shock layer model (MNFVSL) (Zhluktov and Tirskiy 1990) is used. Five-component air (O_2, N_2, NO, O, N) is considered. It is assumed that dissociation rate constants are described by Eqs.1,6,7. For the recombination rate constants the following relationships are applied

$$K_{R_{ij}} = K_{D_{ij}}(T,T)/K_{C_i}(T,T) \tag{9}$$

where $K_{D_{ij}}$ are computed using Eqs.1,6,7 under the assumption of thermal equilibrium ($T_v = T$); K_{C_i} are the equilibrium constants. For the equilibrium constants and constants of exchange reactions the data of Kang and Dunn (1973) are used. It is assumed that recombination and exchange reactions do not depend on the vibrational temperature.

Following Treanor and Marrone (1962), the "dissociation-vibration coupling" is accounted for as follows:

$$\frac{d}{dt}(x_i\varepsilon_i(T_v)) = x_i(\varepsilon_i(T) - \varepsilon_i(T_v))/\tau_i - E_{D_i}\frac{d}{dt}(x_i)_D + E_{R_i}\frac{d}{dt}(x_i)_R \tag{10}$$

This is a one-dimensional equation of vibrational relaxation in non-viscous gas and is given only to demonstrate the way the influence of reactions on vibrational relaxation is taken into account. Here $x_i = n_i/n$ is the molar fraction of the i-th species (n_i is the number of the i-th molecules in a unit volume, n is the total number of particles in a unit volume); ε_i is the energy of the cut-off harmonic oscillator (Eq.7); τ_i is the relaxation time, for which the relationship of Park (see Candler and Park 1988) is used:

$$\tau_i = \tau_{L_i} + (n < V_i > \sigma_v)^{-1}, \quad \sigma_v = 10^{-17}(50000/T)^2, sm \tag{11}$$

where τ_{L_i} is the relaxation time obtained from the Landau-Teller model; $< V_i >$ is an average heat velocity of the i-th molecules; σ_v is the effective cross-section of vibrational excitation, common for all molecules; $\frac{d}{dt}(x_i)_D$ and $\frac{d}{dt}(x_i)_R$ are the number of dissociations and that of recombinations of molecules in unit time; E_{D_i} and E_{R_i} are the mean quantities of energy lost and gained by the molecules in one dissociation event and one recombination event respectively.

The Rankine-Hugoniot shock-slip relationships are used on the shock wave. On the body surface no-slip and no-flow conditions are used for momentum equation. Radiation equilibrium of the surface is assumed for the energy conservation equation. For the equations of species mass conservation the extreme cases of a non-catalytic surface and a fully-catalytic one are used. For the equation of vibrational relaxation the following wall boundary conditions are considered:

$$T_v|_w = T_w, \quad \partial T_v/\partial y|_w = 0 \tag{12}$$

where y is the coordinate normal to the body surface. The latter condition corresponds to the assumption of zero flux of vibrational energy to the non-catalytic surface.

There is some uncertainty in the values of E_{D_i} and E_{R_i}. A comparison of approximate values $E_{D_i}(T_v,T) = \varepsilon_i(T_m)$ (Eq.7) with rigorous ones obtained using the adiabatic approach has shown that for O_2 colliding with O_2, N_2, NO, O, N the ratio $\varepsilon(T_m)/E_D(T_v,T)$ lies in the interval $(0.596 \cdots 1.31)$; for N_2 in $(0.505 \cdots 1.21)$; for NO in $(0.631 \cdots 1.17)$ when T and T_v are independently varied from 1000 K to 50000 K. Thus for all molecules of a nonionized air $\varepsilon(T_m) \times 0.75 < E_D < \varepsilon(T_m) \times 2$. Investigations have shown that a variation of the quantities $E_{D_i}(T_v,T)$ has little effect on the macroscopic flow characteristics. Two cases: $E_{D_i}(T_v,T) = \varepsilon_i(T_m) \times 2$ and $E_{D_i}(T_v,T) = \varepsilon_i(T_m) \times 0.75$ were considered for the flow past a sphere of radius 1m, at free-stream velocity 8000 m/s and altitudes 70 km and 80 km within the scope of the MNFVSL model. It was assumed that the average energy gained in a single recombination is $E_{R_i} = D_i/2$, this corresponding to $\varepsilon_i(T_m)$ as $T_m \to \infty$ or, what is the same, $T_v = T$. The discrepancy between the heat flux values obtained in this two cases is not greater than 6%, the maximum discrepancy between shock detachment distances is about 5%, and only vibrational temperature values may differ from each other by 17%. Thus, it is recommended to assume $E_{D_i} = \varepsilon_i(T_m)$ and $E_{R_i} = D_i/2$. The latter value corresponds to the reasonable suggestion that recombining molecule gains, on average, half the dissociation energy. Thermally equilibrium and nonequilibrium calculations of the flow past sphere at the above-mentioned conditions are carried out. In the case of thermal nonequilibrium the recommended dissociation-relaxation kinetics is used. The calculations have demonstrated that vibrational nonequilibrium has a relatively small effect on macroscopic flow characteristics: within 18% for the heat flux to the body surface and within 8% for the shock detachment distance.

In Table 2 the results of shock detachment distance calculations with use of Eqs.1,6,7 are compared with experimental data of Lobb (1964), who has measured the shock detachment distance y_s on spheres of radius $R = 0.635$ sm fired in air at various pressures and free-stream velocities. The free-stream temperature was constant: $T = 293$ K. For the comparison four experimental values of the shock detachment distance at the stagnation point are chosen. It can be easily seen from the table that calculated values, when thermal nonequilibrium is supposed, agree well with experimental ones (within 5%-accuracy), and thermally equilibrium values of y_s differ from nonequilibrium ones by no more than 8%.

Table 2. y_s/R

Pressure (torr)	Velocity (m/s)	Lobb (1964)	$T_v \neq T$	$T_v = T$
10	3978	0.102±0.002	0.101	0.0929
10	5534	0.082±0.003	0.0811	0.0776
5	5751	0.089±0.002	0.0848	0.0797
20	5272	0.081±0.002	0.0795	0.0771

5. Conclusions

Thus, a new kinetic model, describing an effect of the nonequilibrium vibrational excitation and equilibrium rotational excitation on the dissociation rate constants, is combined with the CVDV-model, accounting for the reverse influence of chemical reactions on vibrational relaxation, and with the MNFVSL gas-dynamic model, which enables one to obtain fast aero- and thermodynamic characteristics of the flow in the domain between the surface of a hypersonic blunt body and the sought-for shock wave. The main peculiarities of the kinetic model used are the finiteness of the dissociation constants at low vibrational temperatures, corresponding to dissociation from the vibrational ground state, and that it accounts for dissociation due to violent rotation. The distinctive property of the complex model is its definiteness, i.e., there are no unknown empirical parameters which must be determined by fitting to experiment. Some uncertainties in the average energy lost in the single dissociation have only a small effect on the flow characteristics. Results of gas-dynamic calculations by means of the model are in good agreement with experimental data.

The numerical investigations carried out enable one to infer a more general conclusion that vibrational nonequilibrium is not so strong an effect in hypersonic flows as it was assumed earlier (Marrone and Treanor 1963, Park 1986, Zhluktov and Tirskiy 1990). It is obviously due to the fact that the vibrational nonequilibrium occurs on the rotation-equilibrium background. And if the lagging of the vibrational temperature behind the translational one near a shock (where $T_v < T$ as usual) results in a sufficient decrease of the dissociation rate, then a fast relaxation of the rotational modes with the translational ones partly compensates the previous process. So the resulting difference between the thermally equilibrium and nonequilibrium dissociation constants is not so large as in the case when the models of Treanor and Marrone (see Zhluktov and Tirskiy 1990) and Park (1986) are used.

References

Candler G, Park Ch (1988) The computation of radiation from nonequilibrium hypersonic flows. AIAA Paper 88-2678

Glushko VP (ed.) (1973) Thermodynamic and thermophysical properties of combustion products. Academy of Sciences USSR, VINITI, Moscow

Gordeev OA et al. (1985) Reviews of species thermophysical properties. USSR Academy of Sciences, IVT, Moscow, No. 5(55)

Hammerling P et al. (1959) Theory of radiation from luminous shock waves in nitrogen. Phys. of Fluids 2, 4: 422-426

Kang SW, Dunn MG (1973) Theoretical and experimental studies of reentry plasmas. NASA CR - 2232

Lobb RK (1964) Experimental measurement of shock detachment distance on spheres fired in air at hypervelocities. In: Nelson WC (ed.) The High Temperature Aspects of Hypersonic Flow. Pergamon Press, MacMillan Co., New York, pp 519-527

Marrone PV, Treanor CE (1963) Chemical relaxation with preferential dissociation from exited vibrational levels. Phys. of Fluids 6, 10: 1215-1221

Park Ch. (1986) Assessment of two-temperature kinetic model for dissociating and weakly ionizing nitrogen. AIAA Paper 86-1347

Smekhov GD, Zhluktov SV (1992) The rate constants for adiabatic dissociation of diatomic molecules. Khimicheskaya Fizika 11, 9: 1171-1179

Smekhov GD, Zhluktov SV (1993) Dissociation of diatomic molecules at extremely high temperatures. Khimicheskaya Fizika 12, 3: 337-339

Treanor CE, Marrone PV (1962) Effect of dissociation on the rate of vibrational relaxation. Phys. of Fluids 5, 9: 1022-1026.

Zhluktov SV, Tirskiy GA (1990) The effect of vibration-dissociation coupling on the heat transfer and aerodynamic resistance in hypersonic flows past blunt bodies. Izvestija Akademii Nauk USSR, Mekhanika Zhidkosti i Gaza 3: 141-151

Viscous Shock Layer Analysis Using a Three-Temperature Model

Y. Sakamura and M. Nishida

Department of Aeronautics and Astronautics, Kyushu University, Fukuoka 812, Japan

Abstract. A viscous shock layer analysis of thermochemical nonequilibrium flows over a blunt body is carried out using a three-temperature model, in which translational-rotational, vibrational, and electron temperatures are considered. The result is compared with the result for a two-temperature (translational-rotational and vibrational-electron temperatures) model. Two limiting cases of the wall boundary condition for vibrational temperature are treated in this analysis to investigate the effects of the wall condition on other temperatures and wall heat flux.

Key words: Nonequilibrium viscous shock layer, Three-temperature model

1. Introduction

The atmospheric entry of space transportation systems at high altitudes generates a very strong shock wave. Such a shock wave causes vibrational excitation, dissociation and ionization of the shock layer gas, and the flow will be in thermal and chemical nonequilibrium due to very low density. Although the rotational temperature is quickly equilibrated with the translational temperature of the heavy particles, the vibrational and electron temperatures deviate from the translational temperature. It is well known that the vibrational and electron temperatures play important roles in a high temperature gas because they control the rates of chemical reactions. In the present viscous shock layer (VSL) analysis of the thermochemical nonequilibrium flows over a blunt body, three temperatures (translational-rotational, vibrational, and electron temperatures) are considered to investigate the effect of thermal nonequilibrium on the flow properties. The result is compared with that for a two-temperature model in which thermal equilibrium between the vibrational and electron energy modes is assumed. Moreover, two limiting cases of the wall boundary condition for the vibrational temperature are treated in this analysis to investigate the effects of the wall condition on other temperatures and wall heat flux.

2. Numerical analysis

2.1. Governing equations

The governing equations are the axisymmetric VSL equations formulated for a multicomponent gas flow in thermochemical nonequilibrium. They are introduced under the following assumptions: 1) The rotational temperature of molecules is fully equilibrated with the translational temperature of the heavy particles, and all the translational-rotational temperatures of species are equal; 2) All the vibrational temperatures of the molecules are equal because of the high efficiency of vibration-vibration energy transfer; 3) The gas in the shock layer does not emit nor absorbs radiation. The present VSL equations are composed of continuity, momentum, translational-rotational energy, vibrational energy, electron energy and species mass conservation equations (Nishida and Sakamura 1993). The harmonic oscillator model is employed for the vibrational energy and the average vibrational energy removed by dissociation is assumed to be 30% of the dissociation energy (Sharma et al. 1992). In the formulation of the electron energy equation, the inertial and viscous stress terms are neglected. The transport properties are evaluated with Yos' formula (Gupta et al. 1990).

Shock Waves @ Marseille II
Editors: R. Brun, L. Z. Dumitrescu

Table 1. Air chemistry

Chemical reaction			T_f	T_b	Chemical reaction			T_f	T_b
$N_2+M_1^a$	$\rightleftharpoons$	$N+N+M_1$	T_1^c	T	NO^++O	$\rightleftharpoons$	N^++O_2	T	T
$N_2+M_2^b$	$\rightleftharpoons$	$N+N+M_2$	T_1	T	O_2^++N	$\rightleftharpoons$	N^++O_2	T	T
N_2+e^-	$\rightleftharpoons$	$N+N+e^-$	T_2^d	T_3^e	$NO+O^+$	$\rightleftharpoons$	N^++O_2	T	T
O_2+M_1	$\rightleftharpoons$	$O+O+M_1$	T_1	T	$O_2^++N_2$	$\rightleftharpoons$	$N_2^++O_2$	T	T
O_2+M_2	$\rightleftharpoons$	$O+O+M_2$	T_1	T	O_2^++O	$\rightleftharpoons$	O^++O_2	T	T
$NO+M_1$	$\rightleftharpoons$	$N+O+M_1$	T_1	T	NO^++N	$\rightleftharpoons$	O^++N_2	T	T
$NO+M_2$	$\rightleftharpoons$	$N+O+M_2$	T_1	T	NO^++O_2	$\rightleftharpoons$	O_2^++NO	T	T
N_2+O	$\rightleftharpoons$	$NO+N$	T	T	NO^++O	$\rightleftharpoons$	O_2^++N	T	T
$NO+O$	$\rightleftharpoons$	$N+O_2$	T	T	O^++N_2	$\rightleftharpoons$	N_2^++O	T	T
$N+O$	$\rightleftharpoons$	NO^++e^-	T	T	NO^++N	$\rightleftharpoons$	N_2^++O	T	T
$N+N$	$\rightleftharpoons$	$N_2^++e^-$	T	T	$N+e^-$	$\rightleftharpoons$	$N^++e^-+e^-$	T_e	T_e
$O+O$	$\rightleftharpoons$	$O_2^++e^-$	T	T	$O+e^-$	$\rightleftharpoons$	$O^++e^-+e^-$	T_e	T_e

[a] $M_1 = N, O, N^+, O^+$ [b] $M_2 = N_2, O_2, NO, N_2^+, O_2^+, NO^+$

[c] $T_1 = \sqrt{T \cdot T_{vib}}$ [d] $T_2 = \sqrt{T_e \cdot T_{vib}}$ [e] $T_3 = \sqrt{T \cdot T_e}$

2.2. Boundary conditions

In the VSL analysis, boundary conditions both on the wall and behind the shock are needed. On the wall, no-slip conditions are used. The wall temperature T_w is determined by the radiative equilibrium condition (Nishida and Sakamura 1993). For the electron temperature T_e, adiabatic wall condition is assumed (Nishida 1972). The species wall boundary condition is the noncatalytic wall (NCW) condition. For the vibrational temperature T_{vib}, the two limiting cases, i.e., $(T_{vib})_w = T_w$ and $(\partial T_{vib}/\partial y)_w=0$ are considered to investigate the effect of vibrational energy accommodation at the wall, where y is the coordinate normal to the wall. Species concentrations, vibrational and electron temperatures behind the shock are set to be equal to their freestream values, and other properties are calculated using the shock jump relations.

2.3. Air chemistry

It is assumed that the freestream gas is air (76.5 % N_2+23.5 % O_2) and the shock layer gas is composed of eleven species (N_2, O_2, NO, N, O, N_2^+, O_2^+, NO^+, N^+, O^+ and e^-). In this analysis, 24 chemical reactions (Park 1991) are considered, as represented in Table 1. To take account of the effects of the vibrational and electron temperatures on the rates of the reactions, Park's two-temperature model (Park 1990) is extended to the three-temperature model. The forward and backward reaction rates (indicated by k_f and k_b, respectively) are expressed as a function of the "effective"temperatures T_f and T_b, respectively, being given by the following expressions:

$$k_f(T_f) = CT_f^s \exp(-\theta_d/T_f), \qquad k_b(T_b) = k_f(T_b)/K_{eq}(T_b)$$

The values of C, s, and θ_d are obtained from Park (1991). The effective temperatures are shown in Table 1.

2.4. Energy exchange

The energy transfer rate per unit volume between translational and vibrational energy modes Q_{T-V} is written in the following form:

$$Q_{T-V} = \sum_{i=M} \rho_i \frac{e_{vib,i}(T) - e_{vib,i}(T_{vib})}{\langle\tau_i^{MW}\rangle + \langle\tau_i^P\rangle}$$

where ρ_i is the density, $e_{vib,i}$ the vibrational energy per unit mass, $\langle\tau_i^{MW}\rangle$ the averaged relaxation time evaluated using the expression of Millikan & White (1963), and $\langle\tau_i^P\rangle$ the collision-limited value at high temperatures proposed by Park (1985). In calculating the collision-limited value, the effective cross section for the vibrational excitation is set to be 10^{-20} m^2.

The rate of the translational energy transfer between heavy particles and electrons Q_{T-e} is given by

$$Q_{T-e} = \sum_{i \neq e} n_e \frac{2m_e}{m_i} \nu_{e,i} \frac{3}{2} k(T - T_e)$$

where k is the Boltzmann constant, m_i the mass, n_i the number density, $\nu_{e,i}$ the collision frequency between electrons and other particles (Gupta et al. 1990).

Using an energy loss factor $g_{rot,i}$, the energy transfer rate between the rotational and electron energy modes is written as

$$Q_{R-e} = \sum_{i=M} n_e \frac{2m_e}{m_i} g_{rot,i} \nu_{e,i} \frac{3}{2} k(T - T_e)$$

The factor $g_{rot,i}$ is the ratio of $Q_{\mathrm{R-e}}$ to $Q_{\mathrm{T-e}}$ for molecular species i. In the present work, we assume $g_{rot,i} = 10$ for homonuclear molecules and $g_{rot,i} = 100$ for heteronuclear molecules (Ladzinis and Petrie 1974, Massey et al. 1969).

Only the vibrational energy of nitrogen is considered for the coupling with the electron energy because the nitrogen-electron coupling is much stronger than the coupling between the electron and other molecules (Lee 1980). Using the relaxation time τ_{N_2-e} taken from the work of Lee (1980), the $e - V$ energy transfer rate is given by the following conventional form:

$$Q_{e-V} = \rho_{N_2} \frac{e_{vib,N_2}(T_e) - e_{vib,N_2}(T_{vib})}{\tau_{N_2-e}}$$

3. Results and discussion

In the present analysis, calculations are carried out for an axisymmetric hyperboloid with the nose radius of 1 m, which is considered as approximating the windward plane of symmetry of a reentry body. Fifty-one grid points are put between the body and shock in the direction normal to the body. In the direction tangential to the wall, 51 grid points are also set uniformly between $s = 0$ and $s = 5$ m, where s is the coordinate measured along the body surface. The freestream conditions are given as follows: $M_\infty = 25$, $\rho_\infty = 3.17 \times 10^{-6}$ kg/m^3, $T_\infty = 181$ K at the altitude of 90 km; $M_\infty = 25$, $\rho_\infty = 2.00 \times 10^{-5}$ kg/m^3, $T_\infty = 181$ K at the altitude of 80 km. Figs.1a and 1b show the temperature profiles along the stagnation streamline under the conditions at the altitude of 80 km, that are obtained using the three-temperature model and two-temperature model, respectively. These results have been obtained for $(T_{vib})_w = T_w$. Fig.1c is the three-temperature result using the adiabatic wall condition for T_{vib}, i.e., $(\partial T_{vib}/\partial y)_\mathrm{w} = 0$. In these figures, $y/y_{sh} = 0$ and 1 indicate the wall and shock positions, respectively. Here y_{sh} is the shock layer thickness. It is seen in these three figures that all the translational-rotational temperature profiles are almost the same throughout the shock layer, namely, both the wall conditions for T_{vib} and temperature models considered in the present analysis do not affect the translational-rotational temperature. In the case of Fig. 1a, the equilibration of T_e and T_{vib} is very rapid due to the strong coupling between the vibrational and electron energy modes. The deviation in the region close to the wall is owing to the conflict of the wall boundary condition, i.e., $(T_{vib})_w =$

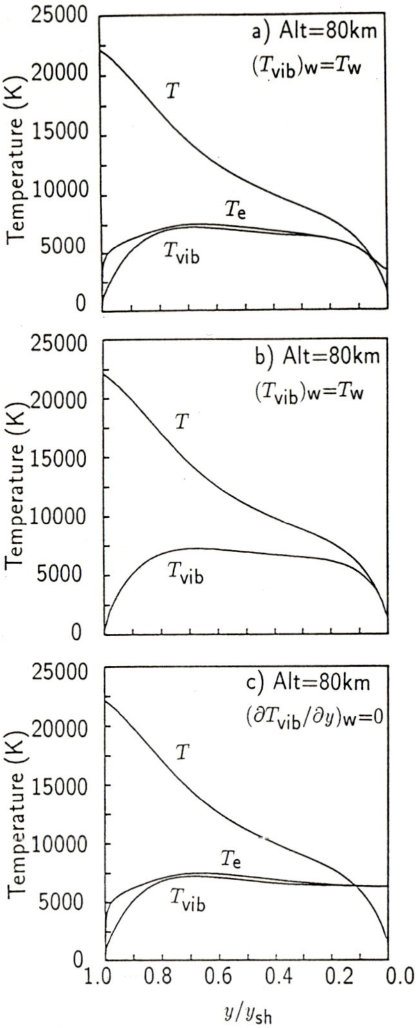

Fig. 1. Temperature distributions along the stagnation streamline, Alt = 80 km, M_∞ = 25, R_n = 1 m:
a) Three-temperature model, $(T_{vib})_w = T_w$,
b) Two-temperature model, $(T_{vib})_w = T_w$,
c) Three-temperature model, $(\partial T_{vib}/\partial y)_w = 0$.

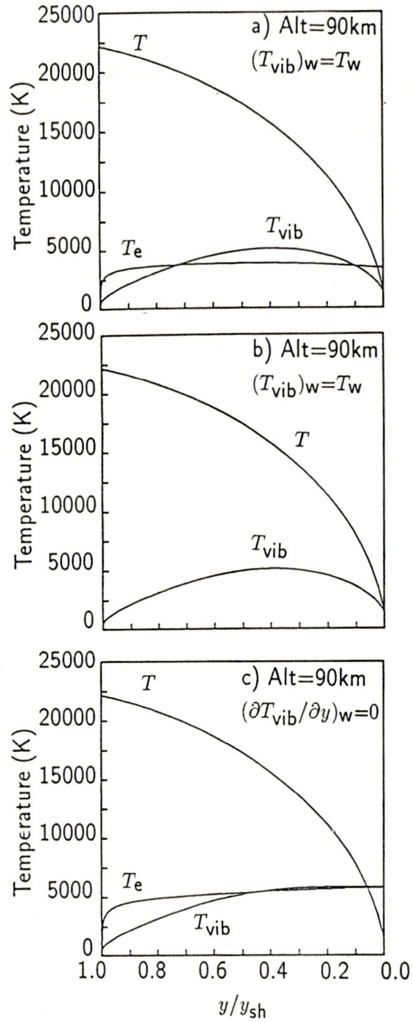

Fig. 2. Temperature distributions along the stagnation streamline, Alt = 90 km, M_∞ = 25, R_n = 1 m:
a) Three-temperature model, $(T_{vib})_w = T_w$,
b) Two-temperature model, $(T_{vib})_w = T_w$,
c) Three-temperature model, $(\partial T_{vib}/\partial y)_w = 0$.

T_w and $(\partial T_e/\partial y)_w = 0$. The difference between $(T_e)_w$ and $(T_{vib})_w$ in Fig.1a is about 1300 K. The deviation close to the wall does not appear in Fig.1c, in which the wall condition for T_{vib} is consistent with that for T_e.

The results at the altitude of 90 km are shown in Figs.2a, 2b, and 2c. The thermal equilibration among T, T_{vib}, and T_e is not attained, but the profiles of the translational-rotational temperature in Figs.2a and 2b are almost the same. Though the effect of the wall condition for T_{vib} on T_e is limited to the region close to the wall at the altitude of 80 km, the wall condition for T_{vib}

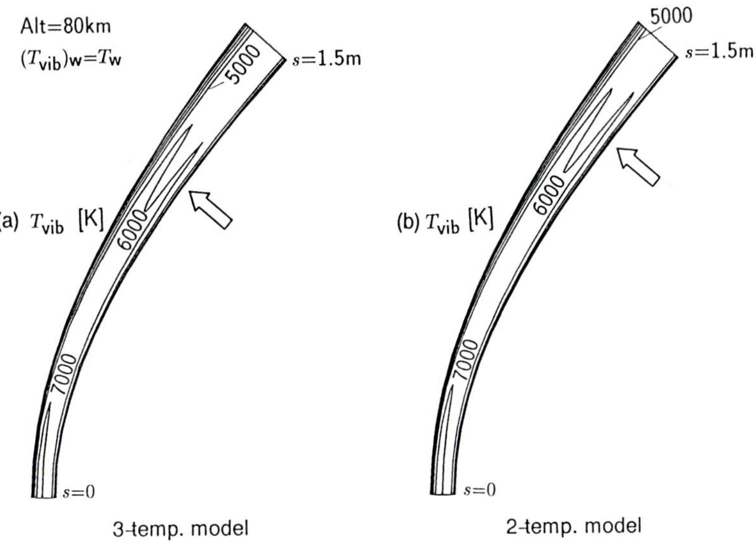

Fig. 3. Vibrational temperature contours, Alt = 80 km, $M_\infty = 25$, $R_n = 1$ m: a) Three-temperature model, $(T_{vib})_w = T_w$, b) Two-temperature model, $(T_{vib})_w = T_w$.

considerably affects the distribution of T_e throughout the shock layer at the altitude of 90 km. The coupling between vibrational and electron energy modes is not so strong due to the very low density that the wall conditions are more dominant for the distribution of temperatures than the energy couplings in the shock layer.

The vibrational temperature contours for both temperature models are shown in Figs.3a and 3b. The conditions at the altitude of 80 km are used as the freestream conditions and the wall condition for T_{vib} is $(T_{vib})_w = T_w$. These figures show that T_{vib} for the two-temperature case is slightly higher than that for the three-temperature case throughout the shock layer. This is because, for the two-temperature model, the translational-rotational energy is transferred to the vibrational-electron energy mode not only by the translation-vibration energy exchange but also by the translation-electron and rotation-electron energy exchanges. Moreover, an increase in T_{vib} near the wall seen in the regions marked with arrows is due to the formation of vibrationally excited molecules in recombination of atoms, which occur in downstream cooler region adjacent to the wall.

Table 2. Effects of temperature models and wall conditions for T_{vib} on stagnation wall heat flux

Altitude, km	$q_w(s=0)$, MW/m^2		
	3-temp.[a]	2-temp.[a]	3-temp.[b]
90	0.132	0.132	0.121
80	0.138	0.137	0.128

[a] $(T_{vib})_w = T_w$, [b] $(\partial T_{vib}/\partial y)_w = 0$

Table 2 shows the calculated wall heat flux at the stagnation point. From this table, we can see the effects of the temperature model and the wall boundary condition for T_{vib} on aerodynamic heating. The wall heat fluxes for the three-temperature cases coincide with the results for the two-temperature cases within 1%. The difference of q_w between the two wall boundary conditions for T_{vib} is about 8%.

4. Conclusions

The main conclusions of the present study are summarized as follows:

(1) The translational-rotational temperature distribution in the shock layer is not affected by both the temperature-model and the wall boundary condition for the vibrational temperature.

(2) Under the conditions at the altitude of 80 km, the equilibration of the vibrational and electron temperatures is very rapid. The influence of the wall boundary condition for the vibrational temperature on the temperature profiles appears merely close to the wall.

(3) Under the conditions at the altitude of 90 km, the influence of the wall boundary condition is more dominant in the shock layer than that of the energy transfers between different energy modes due to the highly low density.

References

Gupta RN, Yos JM, Thompson RA, Lee K-P (1990) A review of reaction rates and thermodynamic and transport properties for an 11-species air model for chemical and thermal nonequilibrium calculations to 30000 K. NASA RP-1232

Lazdinis ŠS, Petrie SL (1974) Free electron and vibrational temperature nonequilibrium in high temperature nitrogen. Phys. Fluids 17:1539-1546

Lee J-H (1980) Electron-impact vibrational excitation rates in the flowfield of aeroassisted orbital transfer vehicles. In: Moss JN and Scott CD (eds) Progress in Astronautics and Aeronautics Vol. 102: Thermophysical Aspects of Reentry Flows, AIAA, New York, pp.197-224

Massey HSW, Burhop EHS, Gilbody HB (1969) Electronic and Ionic Impact Phenomena. Vol.2 2nd ed., Oxford Univ. Press, pp 350-357

Millikan RC, White DR (1963) Systematics of vibrational relaxation. J. Chem. Phys. 39:3209-3213

Nishida M (1972) Nonequilibrium viscous shock layer in a partially ionized gas. Phys. Fluids 15:596-602

Nishida M, Sakamura Y (1993) Nonequilibrium viscous shock layer analysis using a three temperature model. In: Shizgal B, Weaver, DP (eds) Rarefied Gas Dynamics, Proceedings of the 18th International Symposium on Rarefied Gas Dynamics, AIAA, New York (to be published)

Park C (1985) Problems of rate chemistry in the flight regimes of aeroassisted orbital transfer vehicles. In: Nelson HF (ed) Progress in Astronautics and Aeronautics Vol. 98: Thermophysical Aspects of Reentry Flows. AIAA, New York, pp 511-537

Park C (1990) Nonequilibrium Hypersonic Aerothermodynamics. Wiley, New York, pp 35-38, and p 326

Park C (1991) Chemical-kinetic problems of super-escape velocity Earth entries: A review and extension. (This paper is a part of AIAA Paper 91-0464)

Sharma SP, Huo WM, Park C (1992) Rate parameters for coupled vibration-dissociation in a generalized SSH approximation. J. Therm. Heat Transfer 6:9-21

Huygens Probe Radiative Environment

Marc Baillion*, **Gilles Taquin*** **and Jacques Soler**†
*Aerospatiale Division Espace et Défense Département Aerodynamique BP 2 78133 Les Mureaux Cédex France
†SEGIME 142 rue Pierre Curie 78130 Les Mureaux France

Abstract. The Cassini mission is a NASA/ESA program planned to explore the Saturnian system, and consits of a Mariner Mark II Orbiter sending the on-board Huygens probe to Titan during its fly-by of Saturn. The Huygens probe uses aerobraking and decelerates through Titan atmosphere from 6190 m/s at 1250 km altitude down to 400 m/s near 170 km. This extremely high velocity results in dissociation and ionization in the forebody shock layer. Moreover, the relaxation times of chemical and thermal processes involved are of the same magnitude than the characteristic time of the flowfield, the shock layer being so in full nonequilibrium. As the Titan atmosphere major constituents are N_2 (more than 77% in molar fraction), CH_4 (from 0 up to 3.5%) and Argon which may exist up to 20%, non-equilibrium reactions result in CN creation which is known to be a strong radiator. Dedicated preliminary calculations show that, for certain flight conditions, the radiative heat flux may exceed the convective one. In the thermal protection design process, accurate radiative heat flux calculations are needed in order to insure a conservative design and to avoid mass penalties of the thermal protection. It is also desirable to validate our radiative calculations by comparing our results with experiments. Previous shock tube tests have been performed by C.S. Park (1991) in a Titan representative atmosphere gas medium.

This paper presents the methodology used to build a database of radiative heat flux for the Huygens probe heat shield design. First, a parametric study of the aerothermochemical modelisations in the shock tube conditions is performed and used to assess the radiative characteristics of the flowfield. Comparisons with measurements are then undertaken to check the sensitivity of CN transitions to relaxation process modelisation (Millikan and White 1963, Wood and Springfield 1963), chemical reaction rates, catalysis assumptions and type of catalysis or dissociation coupling phenomenum. The best suited thermophysic modelisation is then used to calculate radiative heat flux on the Huygens probe for typical entry conditions. The Millikan and White relaxation law, associated to a set of ionization reactions driven by translational temperature is found to give the results which are the closest to the experimental measurements.

Key words: Titan, CN radiation, Nonequilibrium

1. Presentation of the work - Goals and contents

1.1. Scope of the work

Chung S. Park (1991) has performed shock tube experiments with a gas mixture representative of a Titan atmosphere in order to measure radiative emissivities. When the upstream quiet gas comes across the shock wave, its kinetic energy and the different modes of internal energy (rotational, vibrational, electronic) are excited. This enthalpy increase results in the starting up of a complex chemical process which transforms, via very quick reaction rates, the upstream gas mixture in a plasma which evolves towards an equilibrium state where the new chemical species are stable. In a similar trend, the storage of internal energy is shared in such a way that it tends towards an equilibrium partition, at a rythm imposed by the exchanges between the different energy modes. These relaxation processes result in the creation of strong radiator molecules such as CN, at very high electronic temperature in the first part of the downstream flowfield which is in thermal and chemical non-equilibrium, and then, in an equilibrium zone where radiation vanishes.

The condition of C.S. Park's tests are: $p = 266$ Pa; $V = 5750$ m/s; $T = 300$ K and the upstream chemical composition is 95.7% N_2, 3% CH_4 , 1% Ar, 0.3% H_2 in molar fractions. The only

Shock Waves @ Marseille II
Editors: R. Brun, L. Z. Dumitrescu

experimental output in C.S. Park's thesis consists of the measurements of several monochromatic emissivities behind the shock, corresponding to the main transitions of the strongest radiators in the plasma. We shall focus only on the main transitions of CN.

The scope of our work is to numerically simulate the plasma, and then to calculate the quantities equivalent to the experimental data. The comparison with the experimental values will give us an appreciation on the deviation between the total calculation process used to get emissivities, and the actual behaviour of the plasma in shock tube conditions.

1.2. Methodology

We first have to solve the basic equations of motion and the conservation equations of mass and energy for the different species. The species number densities, used with pressure and temperatures are then used to compute the emissivities. It is then clear that comparison experiment/calculation is available to test the validity of the whole numerical process, but the difficulty is that a discrepancy from any one of the numerical steps may be compensated by another one, and thus not be detected.

2. Theoretical approach

A nonequilibrium chemistry is assumed in the flowfield. A 3-temperature model has been used and different relaxation schemes have been considered.

Chemical kinetics

We assume at first that the flowfield is in vibrational equilibrium with the translation. We consider S chemical species reacting together according to r reactions such as:

$$\sum_{j=1}^{S} \gamma'_{ij} M_j \underset{k_{bi}^{eq.vib.}}{\overset{k_{fi}^{eq.vib.}}{\rightleftharpoons}} \sum_{j=1}^{S} \gamma''_{ij} M_j$$

where γ''_{ij} and γ'_{ij} are the stœchiometric coefficients for the species j during the reaction i. Furthermore, $k_{fi}^{eq.vib.}$ and $k_{bi}^{eq.vib.}$ represent the chemical rates for the forward and backward reactions, respectively. The chemical rate of the species j, taking into account the reaction i, is given by:

$$\left[\frac{dc_j}{dt}\right]_i = \beta_{ij} \left[k_{fi}^{eq.vib.} \prod_{j=1}^{S} c_j \gamma'_{ij} - k_{bi}^{eq.vib.} \prod_{j=1}^{S} c_j \gamma''_{ij} \right]$$

where $\beta_{ij} = \gamma''_{ij} - \gamma'_{ij}$ and c_i is the molar concentration of the species j $(= \rho_j / M_j)$.

The total chemical rate of species j is obtained by summation over the r reactions. These chemical rates are linked together by the following formula: $K_i = k_{fi}^{eq.vib.} / k_{bi}^{eq.vib.}$ where K_i is the equilibrium constant for the reaction i. Typically, one of the two reaction rates is given by an Arrhenius law such as $k_f = aT^b \exp(-c/T)$ and the equilibrium constant can be calculated with the partition functions of the different chemical species implied in the reaction (Vincenti and Kruger 1965).

Vibration-dissociation coupling modelisation

If we now assume vibrational nonequilibrium, we can modify the previous formulation in order to take into account vibration-dissociation coupling. Indeed, it has been experimentally checked that, when two atoms recombine for producing a diatomic molecule, this one is in a more excited vibrational state than the other molecules inside the flowfield. Reversally, the more a diatomic molecule is vibrationaly excited, the more probably it can dissociate. For a diatomic species dissociation reaction i, we write:

$$k_{fi} = k_{fi}^{eq.vib.} \prod_{j=1}^{S} V_j^{A_{ij}} \quad \text{and} \quad k_{bi} = k_{bi}^{eq.vib.}$$

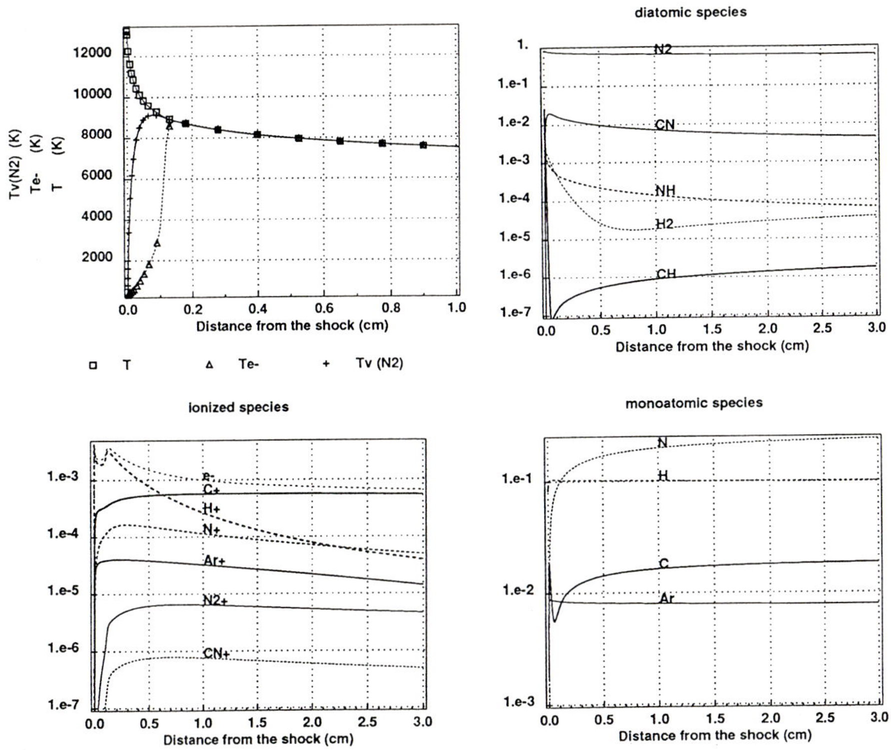

Fig. 1.

The exponent A_{ij} is a binary coefficient equal to 1 if the species j is present during the reaction i and if we consider its vibrational nonequilibrium; if not, A_{ij} is equal to 0. The term V_j corresponds to the vibration dissociation coupling factor for the species j. If we assume the equi-probability of dissociation whatever the vibrational energy level considered (Treanor and Marrone 1962, 1963), we obtain:

$$V_j = \frac{Q_V(T) \cdot Q_V(T_{mj})}{N_j \, Q_V(T_{vj})}$$

where Q_V is the partition function associated with the vibration. T_{mj} is a mean temperature given by $1/T_{mj} = 1/T_{vj} - 1/T$. N_j represents the number of vibrational levels for the species j.

Translation-vibration energy exchanges

The rate of change in the population of the vibrational states at low temperature is well described by the Landau-Teller formulation where it is assumed that the vibrational level of a molecule can change only one quantum at a time (Vincenti and Kruger 1965). The resulting energy exchange rate is:

$$Q_{T-V,j} = \rho_j \frac{e_{vj}^{eq}(T) - e_{vj}}{\tau_j}$$

where $e_{vj}^{eq}(T)$ is the vibrational energy per unit mass in local equilibrium with the translational temperature and τ_j is the vibrational relaxation time for the species j.

$$\left[e_{vj}^{eq}(T) = \frac{\hat{R}}{\hat{M}_j} \cdot \frac{\theta_{vj}}{\exp(\theta_{vj}/T) - 1}\right]$$

When we take into account the effect of the dissociation on the vibrational energy history, this equation is modified according to:

$$Q_{T-V,j} = \rho_j \frac{e_{vj}^{eq}(T) - e_{vj}}{\tau_j} - \rho_j \frac{E_v(T, T_{vj}) - e_{vj}}{C_j} \left[\frac{dC_j}{dt}\right] f + \rho_j \frac{G_v(T) - e_{vj}}{C_j} \left[\frac{dC_j}{dt}\right] b$$

where C_j is the molar concentration of the species j. The second term represents the loss of vibrational energy due to the dissociation whereas the third term represents the gain of vibrational energy due to the recombination. E_v and G_v are developed in Treanor and Marrone (1962, 1963). An expression developed by Millikan and White (1963) yields the vibrational relaxation time, τ_j. Two other laws have been checked: Wood and Springfield (1963) and Park's modification: at high vibrational temperature, the vibrational ladder climbing process due to heavy particles collisions is diffusive in nature, Park (1986, 187) proposes an empirical bridging function between the Landau-Teller and diffusive rates. A second modification is made to take into account the limiting cross section at high temperatures. Park proposes a new relaxation time $\tau_{c,j}$ to correct this unadequacy. Hence, we get $\tau_j = < \tau_j > M.W. + \tau_{c,j}$.

Translation-electron energy exchanges

The energy transfer rate between the heavy particle and the electron translational modes is given by Lee (1984):

$$Q_{T-e} = 3\hat{R}\rho_e(T - T_e)\sqrt{\frac{8\hat{R}T_e}{\pi \hat{M}_e}} \sum_{j \neq e} \frac{\rho_j \hat{N}_0}{\hat{M}_j^2} \sigma_{ej}$$

where σ_{ej} is the collision cross sections for electron-species interactions.

Electron-vibration energy exchanges

The rate of energy transfer between electron translational modes and nitrogen vibrational modes is given by:

$$Q_{e-V,j} = \rho_e \frac{\hat{M}_j}{\hat{M}_e} \frac{e_{vj}^{eq}(T_e) - e_{vj}}{\tau_{e,j}} \qquad \text{for} \quad j = N_2$$

where $e_{vj}^{eq}(T_e)$ is the vibrational energy per unit mass in local equilibrium with the electronic temperature. The relaxation time of electron-vibration energy transfer $\tau_{e,j}$ is given by Lee (1984).

Conservation equations set

The resolution of the different conservation equations (mass, momentum, species mass fraction,vibrational and electronic energy) describes the evolution of the characteristic thermochemical parameters along any flowfield streamline. The vibrational energy conservation equation may be written under the form

$$\frac{de_{vj}}{dx} = \frac{e_{vj}^{eq}(T) - e_{vj}}{u\tau_j} + \frac{E_v(T, T_{vj}) - e_{vj}}{uC_j} \left[\frac{\partial C_j}{\partial t}\right]_f - \frac{G_v(T) - e_{vj}}{uC_j} \left[\frac{\partial C_j}{\partial t}\right]_b + \frac{\rho_e}{\rho_j} \frac{\hat{M}_j}{\hat{M}_e} \frac{e_{vj}^{eq}(T_e) - e_{vj}}{u\tau_{e,j}}$$

The three first terms of the equation represent $Q_{T-v,j}/(\rho_j u)$, the last term corresponds to $Q_{e-v,j}/(\rho_j u)$. The electronic energy conservation equation may be written under the form

$$\frac{3}{2} \frac{\hat{R}}{\hat{M}_e} \frac{dT_e}{dx} = -\frac{p_e}{\rho_e u} \frac{du}{dx} + \frac{Q_{T-e}}{\rho_e u} - \frac{\sum_s \left[\frac{\partial C_s}{\partial t}\right]_f \hat{I}_S}{\rho_e u}$$

3. Comparisons calculations-experiment

3.1. Theoretical hypothesis set

The above theoretical hypotheses have been checked in order to simulate the plasma behaviour behind the shock. Each one of the theoretical hypotheses described above is considered as an option in our numerical approach, which is switched on/off in the simulations in order to check the sensitivity of the radiative properties of the plasma with respect to the physical process of interest. The final goal is to decrease $(\varepsilon - \varepsilon_{\text{exp}})$ where ε is a monochromatic emissivity integrated over the spectral area corresponding to the slit area, and ε_{exp} is the experimentally-measured value.

3.2. Comparisons calculation/experiment

Fig.1 shows the three temperatures and the molar fractions of the different species behind the shock for a particular set of hypotheses. The thermal and chemical non-equilibrium zone extends 2 mm downwards the shock. The flowfield is not too far from equilibrium at 2 cm downstream. The peak emissivity occurs roughly between 1.2 and 1.5 mm from the shock, in the nonequilibrium region and is predominant for the $\Delta V = 0$ transitions of the CN violet system.

3.2.1. Polyatomic species dissociation rate. The dissociation rate of the polyatomic species CH_4 was found to have a certain influence on the emissivity level in the equilibrium region. The reaction rates of dissociation for CH_3, CH_2, CH, expressed in an Arrhenius form, are very fast (Nelson et al. 1991, Park 1982). In our tentative to simulate the shock tube plasma, we either computed the complete sequence of CH_4 dissociation with associated finite rate, or we supposed a full dissociation of CH_4 into (CH_3+H) immediately behind the shock. These two methods are referred as respectively "actual dissociation" and "CH_4 dissociated behind the shock". Fig.2 shows that the polyatomic dissociation rate (CH_4 actual dissociation in the plasma or the full dissociation of CH_4 at the shock) has no influence on the peak emissivity level, but that it changes a bit the overshoot location. The main influence occurs after peak emissivity, where the CH_4 immediate dissociation at the shock hypothesis tends to reach quicker the equilibrium, and is, as a consequence, closer to the experimental values.

The numerical reaction rates of the CH_4 dissociation are very high ($\sim 10^{27}$ cm^3/mole.s). The above results could mean that they are still unsufficiently high, though they are already numerically difficult to handle.

3.2.2. Relaxation law. Fig.3 shows ε for the three different relaxation schemes. The Wood and Springfield relaxation law (1963) is obviously not suitable for our case, as it gives a too low level of emissivities in the non-equilibrium region. Park's modification tends to give intermediate relaxation time between Millikan and White and Wood and Springfield and the resulting ε has a similar behaviour. Peak emissivity occurs when the three temperatures coincide. As the capability of vibrational and electronic temperatures to rise up to the translational level is inversely proportional to the corresponding relaxation times in the energy exchanges processes, if τ_{vt} or τ_{et} is large, the distance from the shock to equilibrium is large. Moreover, the translational temperature decreases as we go far away from the shock. So if τ is large, the location where temperatures equilibrate shifts farther from the shock, to a lower temperature level, and results in a lower emissivity overshoot. This phenomenon is still amplified by the continuous decrease of the CN molar fraction from 1 mm downward the shock. The Millikan and White modelisation is found to give emissivities which are the closest to the experimental values.

3.2.3. Vibration dissociation coupling. Fig.4 shows the influence of the vibration dissociation coupling modelisation. As the coupling has a significant influence on the emissivities level, it will be accounted for in our modelisation.

3.2.4. Specific heat ratio behind the shock. We consider here the ratio of specific heats behind the shock γ_2. The first solution was to impose a certain value for γ_2 or to iterate on γ_2 in order to fit with the specific heat ratio of the gas mixture behind the shock. Fig.5 shows the influence of $\gamma_2 = 1.4$ or $\gamma_2 = 1.3$ behind the shock. The overshoot distance from the shock remains unchanged, but the level slightly increases for $\gamma_2 = 1.3$ because this latter value gives a higher pressure behind the shock, and then a greater number density of CN ($n = p/kT$).

3.2.5. Vibrational energy sharing. The vibrational energy sharing has been checked: N_2 species alone was first considered to relax, the vibrational energy of all the poly- or di-atomic species being then in equilibrium with the translational mode. Then some cases were performed with all poly- or di-atomic species in vibrational nonequilibrium. As the results are slightly the same, we will consider that all the poly- and di-atomic molecules relax like N_2.

3.2.6. Ionization process. As the translational temperature of the plasma is very high, ionization occurs. The driving parameters of the ionization reactions have been assessed: catalysts of ionization reactions are either the free electrons alone (except for $Ar \leftrightarrow Ar^+ + e^-$ catalysed by Ar and e^-) or free electrons plus heavy particles. Furthermore, these reactions are either driven by the translational temperature or by the electronic temperature.

By taking into account the heavy neutral species as catalysts for electron impact ionization reactions, we obviously increase the electron number density behind the shock, because heavy particles are always available to create electrons. At the opposite, when electrons alone are catalysts, a certain time elapses from shock crossing time to create electrons, which finally catalyse their own creation. The process is then amplified and known as the avalanche process. Electrons increase the high energy levels population via non elastic collisions, this results in emissivity increase.

It finally comes out, that within the cross-checked theoretical models, two of them seem to be attractive:

1. Millikan and White plus Park's modification relaxation law, neutral species + electrons as catalysts for ionization reactions (see Fig.6)
2. Millikan and White relaxation law + electrons as catalysts for ionization reactions.

For the two models: $\gamma_2 = 1.3$, vibration-dissociation coupling is modelized, CH_4 is dissociated at the shock and ionization reactions are driven by translational temperature (the cases with ionization reactions driven by electronic temperature are very far from experimental measurements). However Model No.2 gives better results, especially in the overshoot region.

4. Best estimate theoretical hypothesis set

A final adjustment of model No.2 was found to still increase our numerical accuracy in the overshoot region, and consists in making several iterations in order to fit γ_2 to the specific heat ratio of the gas mixture behind the shock. The reference model for the Huygens probe radiative heat flux calculations is then:

Millikan and White relaxation law;
electrons as catalysts for ionization reactions;
ionization reactions driven by translational temperature;
γ_2=1.35;
CH_4 dissociated at the shock.

Our numerical accuracy on the overshoot region lies within ±3% for the main CN violet transitions, which is a fairly good result (see fig.7) γ=1.3 vibration / dissociation coupling

Te and Tv relaxation *(Millikan & White + Park)*

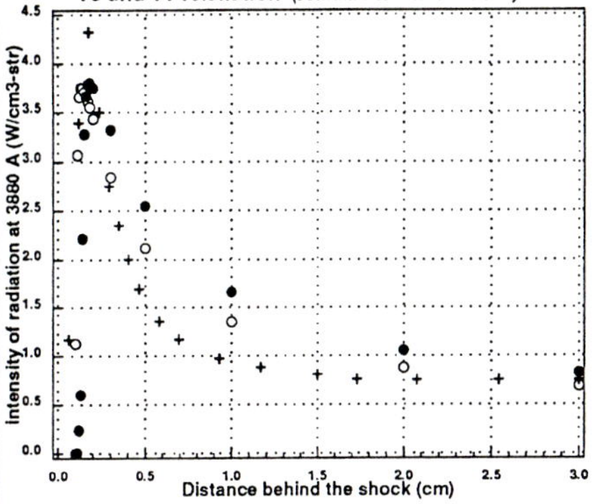

○ CH4 dissociated behind the shock
● CH4 actual dissociation
+ experiment

Figure n°2

γ=1.4 no vibration/dissociation coupling

CH4 dissociated behind the shock

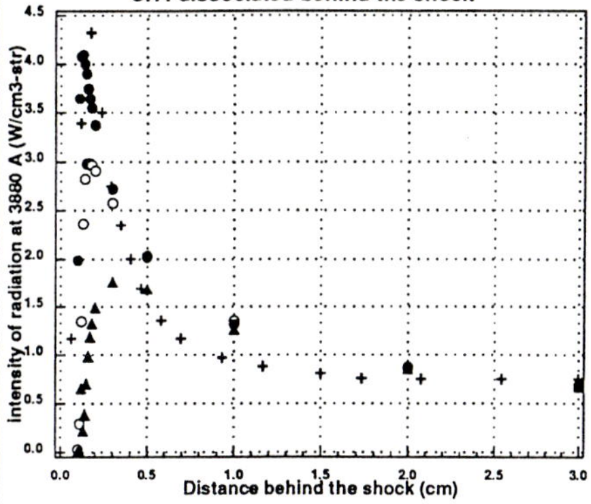

○ *(Millikan & White + Park)*
● *(Millikan & White)*
▲ *(Wood & Springfield)*

Figure n°3

γ=1.3 CH4 actual dissociation

Te and Tv relaxation *(Millikan & White)*

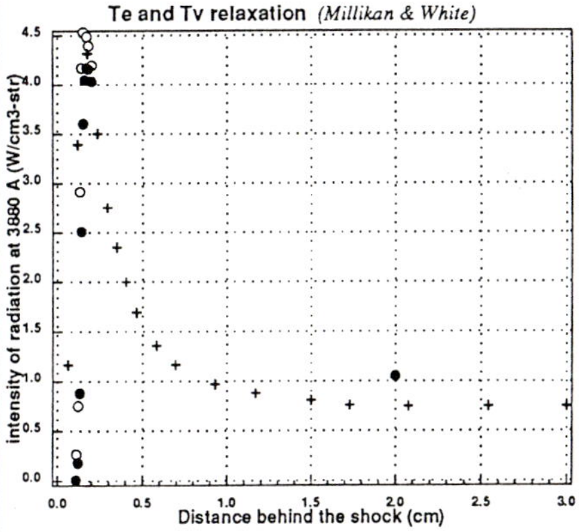

○ vibration/dissociation coupling
● no coupling

Figure n°4

vibration/dissociation coupling - actual CH4 dissociation

Te and Tv relaxation *(Millikan & White + Park)*

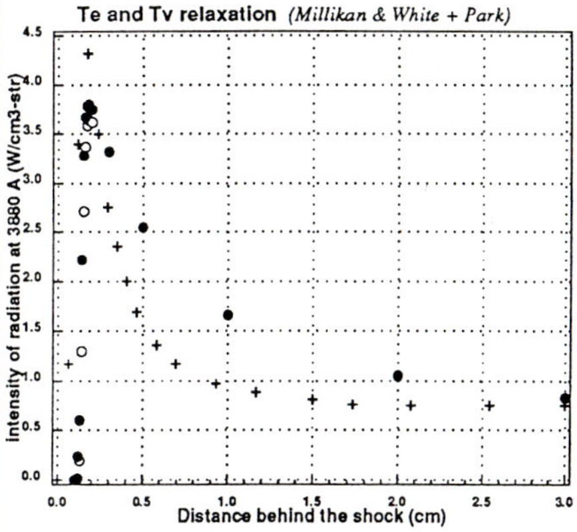

○ γ=1.4
● γ=1.3

Figure n°5

γ=1.3 vibration / dissociation coupling

Te and Tv relaxation *(Millikan & White + Park)*

CH4 dissociated behind the shock

e- and neutral species catalysis for ionization reactions

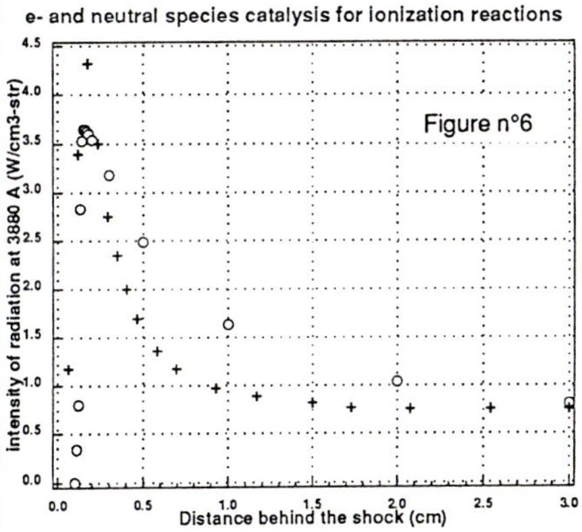

Figure n°6

γ=1.35 vibration / dissociation coupling

Te and Tv relaxation *(Millikan & White)*

CH4 dissociated behind the shock

e- catalysis for ionization reactions

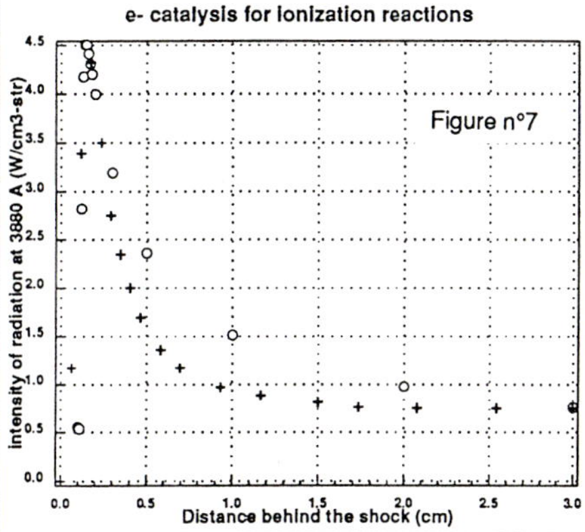

Figure n°7

5. Conclusion

Analysis of C.S. Park's shock tube experiments on a Titan representative atmosphere have enabled us to find a best estimate set of kinetic hypotheses which is in good agreement with the measurements. This model will be used as a reference for the Huygens probe radiative heat fluxes calculations. The numerical calculations show that Argon presence has a dramatic effect on the radiative heat flux level: because Argon does not react with other species, it contributes to increase the shock layer temperature, and moreover, Argon is a strong catalyst for electron creation. At the present time, scientists predict a maximum of 20% Argon in molar fraction in Titan's atmosphere. This results in a 60% increase of radiative heat flux. Unfortunately, the initial gas mixture in C.S. Park's thesis is very poor in Argon. An experimental evidence of this effect, including different initial gas mixtures, is then needed, in order to characterize the Titan atmospheric entry radiative heat flux calculations. A dedicated shock tube campaign is planned to take place at the Université de Provence facility TCM2, in mid-1993.

References

Candler G, Park C (1988) The computation of radiation from non-equilibrium hypersonic flow. AIAA Paper 88-2678

Lee JH (1984) Basic governing equations for the flight regimes of Aeroassisted Orbital Transfer Vehicles. AIAA Paper 84-1729

Marrone PV, Treanor CE (1963) Chemical relaxation with preferential dissociation from excited vibrational levels. Physics of Fluids 6, 9:1215-1221

Millikan RC, White DR (1963) Systematics of vibrational relaxation. Journal of Chemical Physics, 39, 12: 3209-3213

Nelson HF, Park C, Whiting EE (1991) Titan atmospheric composition by hypervelocity shock-layer analysis. Journal of Thermophysics and Heat Transfer 5, 2: 157-165

Park CS (1991) Studies of radiative emission from the simulated shock layer of the Huygens probe. PhD Thesis, Stanford

Park C (1982) Radiation enhancement by non-equilibrium during flight through the Titan atmosphere. AIAA Paper 82-0878

Park C (1986) Assessment of two-temperature kinetic model for dissociating and weakly-ionizing nitrogen. AIAA Paper 86-1347

Park C (1987) Assessment of two-temperature kinetic model for ionizing air. AIAA Paper 87-1574, Honolulu

Treanor CE, Marrone PV (1962) Effect of dissociation on the rate of vibrational relaxation. Physics of Fluids 5, 9: 1022-1026

Vincenti WG, Kruger CH (1965) Introduction to Physical Gas Dynamics. John Wiley and Sons

Wood AD, Springfield JF, Pallone AJ (1963) Determination of the effects of chemical and vibrational relaxation on an inviscid hypersonic flowfield. AIAA Paper 63-441

DSMC Computation of Radiative Heat Flux During Huygens Entry into the Titan Atmosphere

Torkil S. Mogstad
McDonnell Douglas Aerospace, Huntington Beach, CA USA

Abstract. Designers of thermal protection systems (TPS) for planetary hypersonic entry vehicles must rely on computational methods to predict the total heat flux generated within the shock layer. Methods for predicting the convective heat flux are readily available, however, predicting the radiative heat flux, in particular the nonequilibrium part, is non-trivial. We demonstrate the suitability of the direct simulation Monte Carlo (DSMC) method for calculating radiative heat flux. We use the Cassini/Huygens Probe entry into the Titan atmosphere as a demonstration case. The shock layer radiation is driven predominantly by chemically generated CN. Unfortunately, measured cross-sections for the excitation of CN are not available; however, by using shock tube measurements of electronic transition moments coupled with synthetic emission spectrum calculations, we are able to indirectly estimate this cross-section. Our results, using this cross-section, indicate that at the trajectory point of maximum heating the radiative heat flux is approximately 20% of the convective one.

Key words: Titan entry, Radiative heat flux, Monte Carlo method

1. Introduction

The radiative heat transfer from a high temperature shock layer to the surface of a large, blunt hypervelocity planetary entry vehicle is an important engineering consideration in the design of a thermal protection system. At super-orbital entry velocities, the radiative heat transfer to the stagnation region can exceed the convective heating rates.

When heating rates are calculated, it is important that the potential nonequilibrium nature of the flow is taken into account. The shock layer is said to be in thermal nonequilibrium when the internal energies do not form a Boltzmann distribution at the heavy particle translational temperature. Chemical equilibrium exists if the species concentrations are a function of local pressure and temperature alone; if "history" effects of convection and diffusion affect local species concentrations, the point is in local chemical nonequilibrium. Chemical equilibrium occurs when the chemical reaction rates are significantly faster than the time scales of the local flow so the species conservation equations reduce to a balance between production and destruction of the species due to chemical reactions (Mitcheltree and Gnoffo 1990).

Computational methods to estimate convective and equilibrium radiative heating are readily available; however, calculation of the nonequilibrium radiative heating is non-trivial. Nonequilibrium effects are especially important in the rarefied and transitional flow regimes where the direct simulation Monte Carlo (DSMC) method is particularly well suited; however, the method is also applicable well into the continuum regime, and it has been verified against both shock tube data (AVCO) and flight data (FIRE II) (Bird 1986, 1987). In this paper we demonstrate the usefulness of the DSMC method by calculating the total heat flux to the stagnation point of a hypervelocity probe during entry into the atmosphere of the Saturnian moon Titan.

Shock Waves @ Marseille II
Editors: R. Brun, L. Z. Dumitrescu

2. The DSMC method

The DSMC method, developed by Graeme A. Bird, involves the simultaneous computation of thousands of simulated molecules in physical space. The method is easily applied to multi-dimensional hypersonic flow to determine the flow distribution, but to calculate the thermal radiation generated in the stagnation region, a quasi one-dimensional approach is employed (Bird 1986, 1987). The stagnation streamline is modeled as a constant cross-sectional area flow with undisturbed freestream molecules entering at one end and a solid surface at the other end. Initially the flow is undisturbed freestream and an unsteady shock wave propagates from the wall. At some stage, molecules are removed from the sides of the computational tube region along the stagnation streamline, so that the exit number flux equals the free stream entry flux. The flow then settles down to a steady state and as long as the section over which the molecules are removed corresponds with the region between the shock and the surface, and momentum and energy are conserved, the steady flow corresponds with the flow along the stagnation streamline of the blunt body flow within the shock layer.

The shock standoff distance must be provided. The computational region of interest is then divided into a computational grid with grid cell dimensions on the order of the local mean free path or less. The DSMC method moves a number of simulated molecules through the computational grid, where each simulated molecule represents a large number of real molecules. After each time step, which is of the order of the time between collisions or smaller, the particles are collided according to cross-sections and relative velocities. A fraction of the collisions are regarded as inelastic for which new values of the translational and internal energies are calculated according to a Larsen-Borgnakke scheme. The fraction of inelastic collisions can be chosen to match the real gas relaxation rate as a function of temperature, but given the lack of precise data at extremely high temperatures, constant relaxation collision numbers of 5 and 50 are used for rotational and vibrational modes, respectively. The time parameter in the simulation may be identified with real time and the flow is always calculated as an unsteady flow. However, the flow will approach a steady state as the numerical sample increases with time.

The DSMC method has provisions for a series of interactions that affect the stagnation area flow, such as chemical reactions (including ionizing reactions), surface interactions (catalytic effects and variable thermal accommodation), and electronic excitation followed by radiation emissions and possibly self-absorption. The DSMC output consists of the flow properties themselves, impingement forces, and convective and radiative heat transfer to the surface. The chemical reaction rate coefficients correspond to the Arrhenius format. Thermal radiation is the result of several processes, but for a neutral gas the emissions are characterized predominantly by bound-bound radiation due to radiative transitions between two quantized energy levels of atoms and molecules.

The DSMC method employs a simplified approach for calculating the radiation; for a fraction of the collisions, the electronic states are chosen from the equilibrium distribution appropriate to the effective temperature based on the relative translational energy in the collision. The fraction is set equal to the ratio of the average cross-section for electronic state excitation to the elastic cross-section. Separate fractions are specified for collisions with neutrals, ions, and electrons. Each molecule is assigned the equilibrium distribution of states. This is physically reasonable because each simulated molecule represents an extremely large number of real molecules. It is computationally feasible because of the comparatively small number of electronic states that are available to each electronic species, but there is a significant penalty in terms of computer memory requirements. The reason for accepting this penalty is that very sparsely populated states may contribute significantly to the radiation and there would be sample size difficulties if the usual procedures were followed.

3. Electronic transitions

Electronic transitions of diatomic molecules are governed by the Einstein transition probability of spontaneous emission, A_{nm}, which in the case of dipole radiation, is expressed as follows (Herzberg 1965):

$$A_{nm} = \frac{64\pi^4\nu_{nm}^3}{3\hbar} |\boldsymbol{R}^{nm}|^2 \tag{1}$$

We see that the transition probability is proportional to the square of the electronic transition moment, $\mathbf{R}^{nm}$. If initially there are N_n^0 atoms (or molecules) in the state n and if no new atoms in this state are produced, it follows immediately from the definition of the transition probability that there will be an exponential decrease of the number of atoms in state n; thus

$$N_n = N_n^0 e^{-A_{nm}t} \tag{2}$$

After a time $\tau = 1/A_{nm}$ the number of atoms left in the state n is $1/e$ of the initial number. This time is the mean life of the radiative state n.

The integrated intensity due to spontaneous emission from a single rotational line of a diatomic molecule is given by Arnold et al. (1969):

$$E = \frac{16\pi^3 c N_u \nu^4}{3(2J'+1)} \left[\left|R_e^{\nu'\nu''}\right|^2\right] \left[S_{J''A''}^{J'A'}\right] \times 10^7 \qquad [\mathrm{W/cm^3 \cdot sr}] \tag{3}$$

where

$\left|R_e^{\nu'\nu''}\right|^2$ = square of the electronic transition moment

$S_{J''A''}^{J'A'}$ = line intensity factor

N_u = number of molecules in the upper state.

The number of particles in the upper state, N_u, is a function of total particle concentration, N, the partition function, Q, the various transition probabilities, and the internal temperatures. Hence, if the species concentration, N, is known, N_u can be calculated. Consequently, if the density and composition of a radiating gas are known, as well as the electronic transition moment, it is possible to calculate a synthetic spectrum for the electronic transitions. Such a calculation has been made by Arnold et al. (1969), for a 50% CO_2 - 50% N_2 gas mixture heated to 6000 K. Among the band systems considered were CN (violet) and CN (red) transitions. Conversely it is possible to experimentally obtain the transition moment for a gas by measuring the emitted spectrum (Arnold and Nicholls 1973).

4. The Titan atmosphere

Based on data gathered by Voyager, an atmospheric model for Titan has been developed. The most complete model is the Titan Atmosphere Engineering Model (Lellouch and Hunten 1987) which provides composition, densities, and temperatures for altitudes below about 1200 km. The main constituent is nitrogen, with 0-20% argon, and 3-5% methane. The presence of argon has been debated, but it remains a likely constituent. Several analyses of radiative heat flux generated during Titan atmospheric entry have been presented in the literature (Park 1982, Nelson et al. 1991). The predominant radiator is CN which is formed from C and N_2, where C is created after breakdown of the CH_4. The CN diatomic radical can exhibit several electronic transitions, but the two predominant ones are CN (violet), or $B^2\Sigma^+ - X^2\Sigma^+$, and CN (red), or $A^2\Pi - X^2\Sigma^+$.

5. DSMC simulation of Titan atmospheric entry

For a probe mass of 270 kg, a probe diameter of 2.5 m, entry angle of $-75°$, and an initial entry velocity of 6.19 km/s, the point with the maximum heat flux is at an altitude of about 250 km. At this point, the probe has slowed down to approximately 4.5 km/s, which corresponds to a Mach number of 19.8. The local atmospheric temperature and density are assumed to be 145 K and 6.70×10^{21} molecules/m^3, respectively. The chemistry is described by 30 reactions; no ionization is included because of the relatively low entry velocity. Furthermore, CH_4 is set to dissociate directly to CH (leaving out the intermediary steps $CH_4 \rightarrow CH_3 \rightarrow CH_2 \rightarrow CH$) which is supported by the extremely high dissociation rate of methane. The reactions included are as follows (Table 1):

Table 1. Chemical reaction set. Arrhenius reaction rate $= CT^n \exp(-E/RT)$

Reaction:	C [m^3/molecule·s]	n	E [J/molecule]
$CH_4+M \longrightarrow CH+H_3+M$	3.75×10^{-6}	-1.87	7.3×10^{-19}
$CH+M \longrightarrow C+H+M$	1.88×10^{-11}(†)	-1.00	5.69×10^{-19}
$CN+M \longrightarrow C+N+M$	1.67×10^{-7}(†)	-2.00	1.24×10^{-18}
$N_2+M \longrightarrow 2N+M$	6.17×10^{-9}(†)	-1.60	1.56×10^{-18}
$H_2+M \longrightarrow 2H+M$	2.45×10^{-11}(†)	-1.23	7.17×10^{-19}
$N_2+C \longrightarrow CN+N$	1.85×10^{-16}(†)	-0.11	3.17×10^{-19}
$CN+N \longrightarrow N_2+C$	5.55×10^{-17}(*)	-0.11	0.
$H_2+C \longrightarrow CH+H$	3.0×10^{-16}(†)	0.00	1.59×10^{-19}
$CH+H \longrightarrow H_2+C$	6.17×10^{-17}(*)	0.00	0

† (Nelson et al. 1991), * Calculated from partition function

The DSMC radiation analysis requires data input for each electronic transition, including the electronic excitation collision (EEC) number, the energy of the electronic state, the degeneracy of the state, and the mean radiative lifetime of the state. The EEC number is a measure of how effectively particles (neutrals, ions, electrons) collisionally excite the emitting species, in this case CN. Normally, electrons and ions are much more efficient than neutrals (by about one order of magnitude) in exciting the radiating species. However, in the DSMC run reported on here, there are no charged particles, and the EEC number is for the neutrals in the flow. The EEC number represents the ratio of the elastic collisional cross-section to the excitation cross-section, hence an EEC number of 10 means that one in every 10 collisions involving CN leads to excitation of the CN if the total collisional energy exceeds the energy required to reach the excited state. Unfortunately, the available data on excitation cross-sections is not only incomplete, but often based on experiments at energies different from those appropriate to the entry flow.

5.1. Initial heat flux calculations

Since the EEC collision number is not explicitly available from the literature, we initially wanted to establish the radiative flux sensitivity to variations in EEC number. For flux calculations 1 through 3 we thus used EEC numbers of 10, 100 and 1000. An EEC number of 10 is considered to be extremely low. In a previous analysis of excitation behind an AOTV bow shock (Bird 1987), EEC numbers of 100 were used for collisions between electrons/ions and neutrals; the EEC number for neutral-neutral collisions (like here) is expected to be even higher. The initial gas mixture mole fractions were 4% methane, 20% argon and 76% nitrogen. The shock standoff distance was set between 10 and 11 cm, and the vehicle wall temperature was fixed at 1500 K. The radiative lifetimes for the electronically excited CN states were taken from AVCO Air Radiation

Tables (Allen 1966); 5.12×10^{-6} s for CN (red) and 8.5×10^{-8} s for CN (violet). The results are shown in Table 2. Note that the flux numbers were derived from the total volumetric radiation within the shock layer by using the slab approximation which assumes that half of the radiation reaches the vehicle wall, and half is radiated to free space. Comparing calculations 1 to 3, it is evident that CN (violet) emissions are quite sensitive to the EEC number, varying linearly with the excitation efficiency. The CN (red) emissions are not nearly as sensitive, although they do decrease with increasing EEC (i. e., decreasing excitation efficiency).

We then performed a fourth flux calculation, essentially identical to Calculation 3, except that the radiative lifetimes were replaced by a more recent set, and the CH_4 fraction was decreased to 3%. The lifetimes are for the (0,0) transition, which according to the Franck-Condon Principle, is the predominant one in most cases. The new radiative lifetimes are 1.65×10^{-5} s for CN (red) (Arnold and Nicholls 1972) and 7.1×10^{-8} s for CN (violet) (Danylewych and Nicholls 1978). Compared to Calculation 3, we see that for a tripling of the CN (red) radiative lifetime, the corresponding emissions decrease by about 50% from 80 to 38 kW/m^2, which is according to the expected trend (one would expect the total emissions to decrease for increased radiative lifetimes, since the excited molecules travel longer before the emissions start). The CN (violet) is essentially unchanged at around 10 kW/m^2.

Table 2. DSMC calculated CN emitted radiative flux to Huygens stagnation point

	CN (red) [kW/m^2]	CN (violet) [kW/m^2]	CN (total) [kW/m^2]
Calculation 1: EEC=10	350	950	1300
Calculation 2: EEC=100	265	100	365
Calculation 3: EEC=1000	80	10	90
Calculation 4: EEC=1000 New radiative lifetimes	38	9.5	47.5

5.2. Calibrating the CN electronic excitation cross-section

So far we have completed DSMC simulations of entry into the Titan atmosphere for three different EEC numbers between 10 and 1000. In order to determine a reasonable EEC number, we need to calibrate the DSMC radiative yield against experimental data if possible. Arnold et al. (1969) report on a line by line calculation of spectra from diatomic molecules, based on the density of the radiating gas and electronic transition moments from the literature. The calculations, which were performed for a 50-50 mixture of CO_2 and N2 heated to 6000 K, considered, among others, the band systems CN (violet) and CN (red). The square of the dimensionless transition moment was assumed constant for each band system, and the CN (violet) and CN (red) values used were 0.56 and 0.52, respectively. The gas species concentrations were taken from Woodward (1967).

Using their line-by-line calculation routine, Arnold et al. (1969) obtained optically thin integrated emission intensities for a 1 cm path length for several vibrational (ν', ν'') transitions. The (0,0) emissions, which are the predominant ones, are reported as 0.307 and 0.124 W/cm^2·sr for CN (violet) and CN (red), respectively. The transition moments used by Arnold et al. (1969) have, however, since been revised. Because the emission intensities are directly proportional to the square of the transition moments (see Eq.3), it is important that correct transition moments are utilized. Shock tube determination of electronic transition moments have yielded (squared) values of 0.90 and 0.32 for CN (violet) and CN (red), respectively (Arnold and Nicholls 1972, 1973). Hence the CN electronic emission intensities should be adjusted accordingly to:

$$(0.90/0.56)\ 0.307 = 0.493 \text{ W/cm}^2\cdot\text{sr for CN (violet)}$$
$$(0.32/0.52)\ 0.124 = 0.0763 \text{ W/cm}^2\cdot\text{sr for CN (red)}$$

For a 1 cm path length, and integrated over 4π steradians, this corresponds to 6.2×10^6 W/m^3 (violet) and 9.6×10^5 W/m^3 (red). In order to calibrate the EEC number against the experimentally determined transition moments, we need to simulate the 50-50 CO_2-N_2 gas mixture using the DSMC method. No shock is included this time; we only simulate the gas mixture (N_2, CO, O_2, C_2, CN, NO) in equilibrium at 6000 K, subject to CN (violet) and CN (red) electronic transitions. By comparing the DSMC calculated emission intensities for several EEC numbers against the adjusted intensities reported by Arnold et al. (1969), we can adjust the EEC number until the intensities match.

The CN radiative lifetimes were the same as those used in Calculation 4 of Table 2. We already know that the CN (violet) emission intensity is very sensitive to the EEC number, being roughly linearly proportional. By fine tuning the EEC number, it is possible to exactly match either CN (red) or CN (violet) at the expense of the other. Similarly, it is possible to exactly match the total emission intensity. Our objective was to match the total radiative heat flux as closely as possible. In order to fine tune the EEC number for total CN (0,0) emissions, we ran several calibration cases, finally settling on an EEC number of 450. The results from the final calibration are shown in Table 3.

Table 3. Calibrated emission intensities for CO_2-N_2 Mixture [W/m^3]

	CN (red)	CN (violet)	CN (total)
Arnold et al.	0.96×10^6	6.2×10^6	7.16×10^6
DSMC (EEC=450)	2.09×10^6 (+118%)	4.96×10^6 (-20%)	7.05×10^6 (-1.5%)

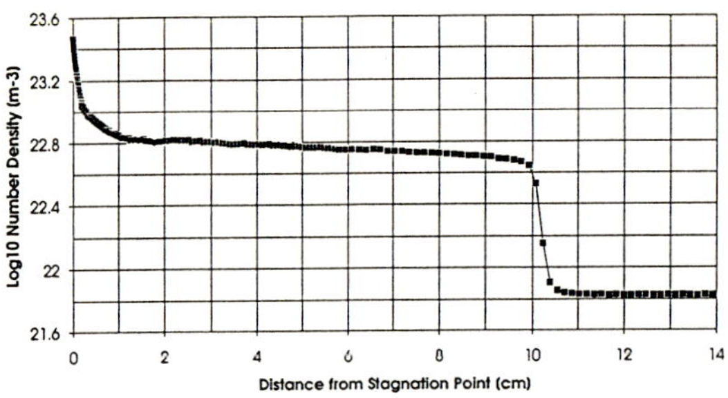

Fig. 1. Total number density across Huygens shock layer

5.3. Final heat flux calculation

Confident that we now have calibrated the EEC number for total CN (0,0) emissions in neutral-neutral collisions against experimental data, we execute a final DSMC run. The CN radiative lifetimes are the same as for Calculation 4, and the total density and temperatures within the shock layer are shown in Figs.1 and 2; notice how well the shock gradients are captured. The shock layer flow regimes vary from transitional at the shock (Kn$\sim$ 0.03) to continuum at the stagnation point (Kn$\sim$ 0.002). The translational temperature exhibits an overshoot and peaks at about 10,400 K before gradually leveling off and reaching equilibrium with the rotational temperature. The vibrational temperature stays significantly lower across the entire shock layer, reaching equilibrium with the other temperatures only immediately before the stagnation point. Notice that no temperature jump is evident at the stagnation point, i. e., the flow density at the wall is collisional enough that the temperatures of the flow and the wall converge. The species mole fractions are presented in Fig.3, showing that N_2 is about 2% dissociated at the most. CN-formation from the reaction between N_2 and C shoots up as C becomes available from the dissociation of CH_4. Close to the stagnation point, however, the CN concentration drops dramatically because (1) the CN formation reaction essentially ceases as the flow temperature

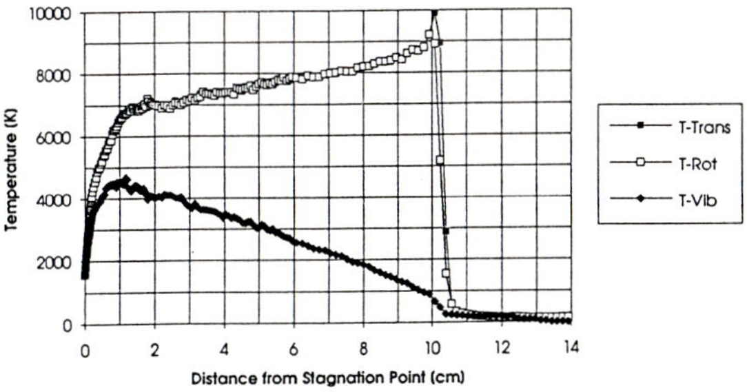

Fig. 2. N_2 temperatures across Huygens shock layer

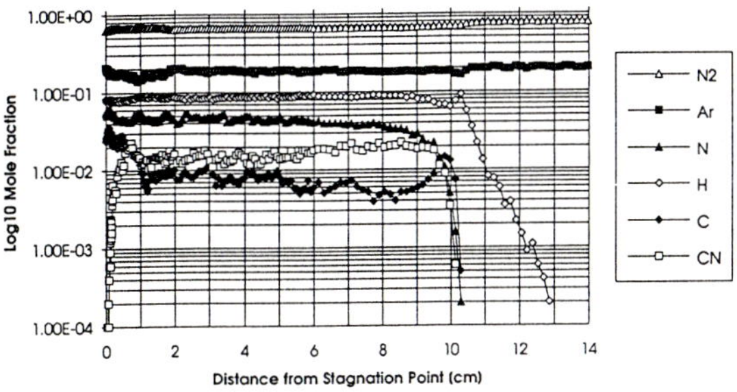

Fig. 3. Species Mole fractions across Huygens shock layer

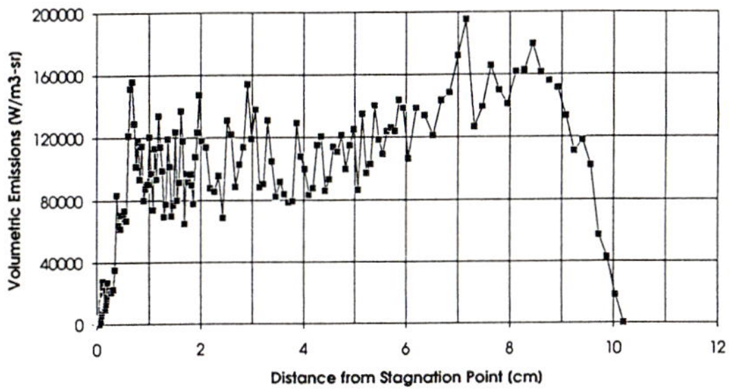

Fig. 4. CN volumetric emissions across Huygens shock layer

plummets, and (2) the CN is consumed in a reaction with N. The volumetric radiation emissions within the shock layer, shown in Fig.4, are driven both by flow temperature and CN concentration. The peak emissions lag the shock temperature peak by about 2-3 cm. Furthermore, by comparing emissions with the CN mole fraction in Fig.3 it is evident that the radiation is essentially directly proportional with the CN concentration throughout the layer. This demonstrates how important the flow chemistry is in radiation calculations. The success of any computational method depends to a very great degree on correct chemical reaction rates. The stagnation point radiative heat fluxes are listed in Table 4.

Table 4. DSMC Calculated CN radiative flux to Huygens stagnation point

	CN (red) [kW/m^2]	CN (violet) [kW/m^2]	CN (total) [kW/m^2]
Final Case: EEC=450	53	18	71

Note that no radiative self-absorption has been assumed for any of the cases, making the radiative heat flux conservative from a TPS design standpoint. It should be noted that this DSMC calculation considers only the (0,0) electronic transitions. Although this is by far the strongest transition, we would in the future like to include transitions for additional vibrational states. The convective heat flux to the stagnation point is calculated to approximately 350 kW/m^2 for a thermally accommodating, non-catalytic surface at a wall temperature of 1500 K.

6. Conclusion

In conclusion, we have demonstrated that DSMC can be a very useful numerical tool for predicting entry flow heat fluxes in the thermally and chemically nonequilibrium regime. Our calculations indicate that the radiative heating accounts for roughly 17% of the total heating at the chosen trajectory point of the Huygens Probe entry into the Titan atmosphere. This compares well with results reported by Nelson et al. (1991). The DSMC one-dimensional streamline approach is computationally relatively non-intensive (all runs were made on an AST 486/33 PC) and provides an important alternative to more rigorous codes.

References

Allen RA (1966) Air radiation tables: spectral distribution functions for molecular band systems. NASA CR-557

Arnold JO, Whiting EE, Lyle GC (1969) Line by line calculation of spectra from diatomic molecules and atoms assuming a Voigt line profile. J. Quant. Spectrosc. Radiat. Transfer 9:775

Arnold JO, Nicholls RW (1972) A shock tube determination of the electronic transition moment of the CN red band system. J. Quant. Spectrosc. Radiat. Transfer 12:1435

Arnold JO, Nicholls RW (1973) A shock tube determination of the CN ground state dissociation energy and the CN violet electronic transition moment. J. Quant. Spectrosc. Radiat. Transfer 13:115

Bird GA (1986) Direct simulation of typical AOTV entry flows. AIAA Paper 86-1310

Bird GA (1987) Nonequilibrium radiation during re-entry at 10 km/s. AIAA Paper 87-1543

Danylewych LL, Nicholls RW (1978) Intensity measurements and transition probabilities of the CN violet ($B^2\Sigma - X^2\Sigma$) band system. Proc. Roy. Soc. Lond. A. 360:557

Herzberg G (1965) Molecular Spectra and Molecular Structure, 1. Spectra of Diatomic Molecules. D. Van Nostrand Company, Inc., Princeton Toronto London New York

Lellouch E, Hunten DM (1987) Titan atmosphere engineering model. ESLAB 87/199, Space Science Department, ESA/ESTEC

Mitcheltree R, Gnoffo P (1990) Thermochemical nonequilibrium issues for Earth reentry of Mars mission vehicles. AIAA Paper 90-1698

Nelson HF, Park C, Whiting EE (1991) Titan atmospheric composition by hypervelocity shock-layer analysis. J. Thermophysics and Heat Transfer 5:157

Park C (1982) Radiation enhancement by non-equilibrium during flight through the Titan atmosphere. AIAA Paper 82-0878

Schadee A (1967) The relation between the electronic oscillator strength and the wavelength for diatomic molecules. J. Quant. Spectrosc. Radiat. Transfer 7:169

Woodward HT (1967) Predictions of shock-layer radiation from molecular band systems in proposed planetary atmospheres. NASA TN D-3850

Radiation from Air Shock Layers with Chemical and Ionizing Reactions

M.A. Gallis and J.K. Harvey
Department of Aeronautics, Imperial College of Science Technology and Medicine, London, England

Abstract. This paper deals with the evaluation of thermal radiation from hypersonic shock waves in real air, using the Direct Simulation Monte Carlo Method. The method uses the Maximum Entropy method to calculate the chemical and ionising reactions and measured and calculated cross sections to calculate the excitation probability without assuming equilibrium distribution for the electronic states. The new method is validated against experimental results from a one dimensional test case in real air at Mach 29. The radiation from the shock layers formed in front of a *Mars-net* type re-entry vehicle flying at a number of altitudes and speeds is also examined.

Key words: Radiation, Chemical reactions, Ionization, Direct Simulation Monte Carlo

1. Introduction

It is generally agreed that during hypersonic re-entry, radiation from shock layers results from the de-excitation of the electronic states of atoms and molecules. Photons originating from the electronic states of air species have energy that allocates them mostly in the UV band of the spectrum, but radiation in the IR and visible band also occurs.

Attempts to simulate thermal radiation for the rarefied hypersonic regime (e.g. Bird 1987) have so far used the assumption of an equilibrium distribution of the electronic states. However, this assumption is inappropriate in this regime where large areas of non-equilibrium are found.

Predicting the correct radiation does not only depend on the choice of an appropriate electronic excitation model. Since electronic excitation is due to electron impact it is essential to be able to predict accurately the fraction of electrons in the flowfield (for chemically reacting flows) and their temperature. This in turn requires accurate prediction of the energy exchange and ionising processes in the flowfield.

2. Prediction of the flowfield

The conditions of chemical and thermal nonequilibrium which prevail in the area of rarefied hypersonic shock layers must be taken into consideration for a physically realistic simulation of the flowfield. In this study the Maximum Entropy method (ME) (Gallis 1993) was used for the simulation of the chemical, ionising reactions and energy exchange under conditions of non- equilibrium. Unlike the conventional Borgnakke and Larsen type approaches the Maximum Entropy (ME) approach does not assume equilibrium conditions for the distribution of energy but it calculates the most probable post-collision distribution of the energy states for each type of interaction. Avoiding the assumption of an equilibrium distribution of internal energy states provides better foundation for the calculation of nonequilibrium electronic excitation.

The derivation of the most probable distribution is based on the entropy associated with each distribution. The distribution of energy states likely to cause a reaction is given as a function of the vibrational energy only. For monatomic particles where no vibrational energy is present the distribution is associated with the translational energy of the particles. The derived distributions include three disposable parameters which need to be determined independently for each reaction by comparison with experimental data.

Shock Waves @ Marseille II
Editors: R. Brun, L. Z. Dumitrescu

Once charged species appear within the flow the generation of Coulomb forces is unavoidable. If the degree of ionisation is high enough these forces can be strong enough to control the diffusion of the charged particles. Under the conditions met in our flowfields the charged particles diffuse under conditions of ambipolar diffusion. Thus it can be assumed that on average the flow is neutrally charged and that the electron current equals the ion current. This coupling of the electron and ion movement affects the velocities of electrons and as a result the electron temperature and the emitted radiation.

3. Thermal radiation from shock waves in real air

The main air species that can be electronically excited, and therefore radiate, are six, namely N_2, O_2, NO, N, O and N_2^+. For the four molecular species Bird (1987) has proposed a set with the states that are the more likely to be excited. Electronic excitation is considered only for these states because the electronic excitation probability of the other states is significantly lower. The number of electronic states for the atomic species is very large, so they have to be grouped into a managable set. Following Bird (1987), eight main groups of states are considered for both atomic oxygen and atomic nitrogen. For this study the same electronic states considered by Bird have been adopted for the molecular and atomic species.

In the context of a DSMC simulation we can determine if electronic excitation will take place if we know the probability of an encounter resulting in the excitation. The probability of electronic excitation resulting from a collision can be calculated with the aid of electronic excitation cross sections. The electronic excitation cross sections in general depend on the species in question, the available energy, the states between which the transition takes place and on the energy configuration of the particle (vibrational, rotational and electronic energy state).

Comprehensive measurements or calculations of the cross sections are not available for all the transitions of the species of interest. Most of the data cover transitions from the ground state only. A short review of the data available for cross sections will be presented in the following paragraphs. A more detailed review can be found in Gallis (1993).

Atomic species: The available experimental data for the electronic excitation of atoms is very limited and they can only be used to validate the theoretical approaches. Due to the grouping of atom electronic states that was done to make the simulation of atomic excitation possible the states cannot be addressed either with their spectroscopic code or with their quantum numbers. This almost excludes all the quantum mechanical approaches that could be used (Vainstain & Born approximation). In addition to these models there is a number of phenomenological formulae that have been proposed by Seaton (Green 1969), Van Regemorter and others for the cross sections. These formulae were originally proposed for atomic species but they can be used for molecular species as well. In most cases these formulae are in good agreement with experimental data. The phenomenological formulae of this type require information about the quantum states between which the excitation takes place (oscillator strength) limiting their applicability to cases where the quantum number of a state can be defined. The excitation probability of a group of states can be calculated with the excitation formula of Gryzinski (1965), which by using the principles of classical mechanics, requires only the energy of the final state as input. According to this method the excitation is reduced to a Coulomb interaction between the orbiting and the projectile electron.

The groups of states include allowed and forbidden transitions. Since in all these cases the forbidden transitions are only a small fraction of the total, the formulas for the allowed transitions can be used. There is one exception to this: the first two states of atomic nitrogen and oxygen can only be excited from the ground state via forbidden transitions. For these transitions we have a number of theories which are in very good agreement among themselves and with experimental data. Nesbet's variational method was used here to calculate the cross sections for those states (see Nesbet 1980).

In the case of atoms which are already electronically excited the excitation energy is added to the electron energy thus assuming that the excitation rate for the transition from i to j is independent of the initial state i.

Molecular species: In the case of molecular electronic excitation the theoretical prediction of the excitation cross section is more difficult due to the greater number of degrees of freedom. The simultaneous re-arrangement of the rotational and vibrational energy modes makes the theoretical evaluation of the excitation probability very difficult. Since the impact of the electron collision on the vibrational and rotational energy of a molecule is rather small the contribution of these energy modes to the electronic excitation energy was ignored. The measurements of the cross sections have been made for the ground vibrational state in all cases. With the use of the Frank-Condon factor we can define an average cross section over all the vibrational states. The Frank Condon factors for all the molecular species of interest have been given by Nicholls (1964).

For N_2 the measurements of Cartwright (1977) and for O_2 the measurements of Wakiya (1978) were used. The fifth state of O_2 where no measurements are available was approximated with Seaton's formula using the oscillator strength proposed by Allen (1965). For NO we have the oscillator strengths as calculated by Ory (Green (1969)) for four of the transitions in question. The last state was assumed to have the same oscillator strength as the fourth state. These oscillator strengths can be used with Seaton's phenomenological formula (the most frequently used) to give the cross sections for NO. For N_2^+ we have the measurement of Crandall (1974) for the second state. For the first and third state of N_2^+ no measurements are available and we have to resort to the phenomenological formula. For the first state we have the oscillator strength proposed by Allen (1965). For the third state of N_2^+ we assumed that the oscillator strength was equal to the oscillator strength for the second. In the light of the uncertainty for the order of magnitude of the measured cross sections the formula of Seaton can be considered to give the most reliable results.

4. Absorption of emitted radiation

In high altitude rarefied flows the fraction of emitted radiation that is absorbed by the flow is quite small and hence the energy transferred throughout the flow is small enough to be ignored. Absorption calculation is only made to determine the attenuation of the emitted radiation signature. In this study, Bird's approximate approach for the absorbed radiation was adopted with one difference; when a photon is emitted the probability of it being absorbed is calculated and a percentage of its energy equal to the absorption probability is absorbed instead of applying an accept-reject technique to absorb the whole energy of the photon. This way the sample of the absorbed energy is larger and the effect of absorbing rare high energy photons is weakened. In each cell the atomic species are assumed to be absorbing with their maximum cross section over all wavelengths. The absorption cross section of N and O was assumed to be 4×10^{-17} cm^2. The effect of an absorbed photon is totally ignored. This violates conservation of energy and the procedure cannot be used if significant energy is absorbed.

5. Computational results and comparison with experiment

As a test case to validate this model the conditions of a shock tube experiment (Allen 1962) were used. In this experiment the spectral radiation emission from the equilibrium and nonequilibrium region of the shock layer in air were measured. The experiment was made with simulated air (79% N_2, 21% O_2) at Mach 29 and pressure 0.1 Torr. The temperature profiles for this case are presented in Fig.1. It is worth noticing that the Rankine Hugoniot formulas predict for a non-reacting gas at the same conditions a post-shock temperature of 49,000 K. Downstream the

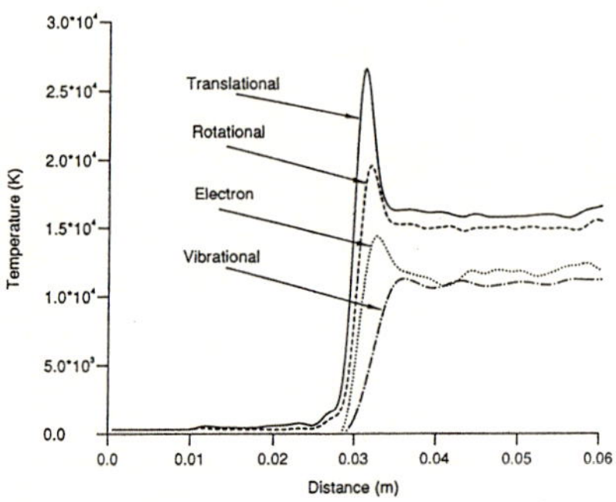

Fig. 1. Air Mach 29. Temperature profiles

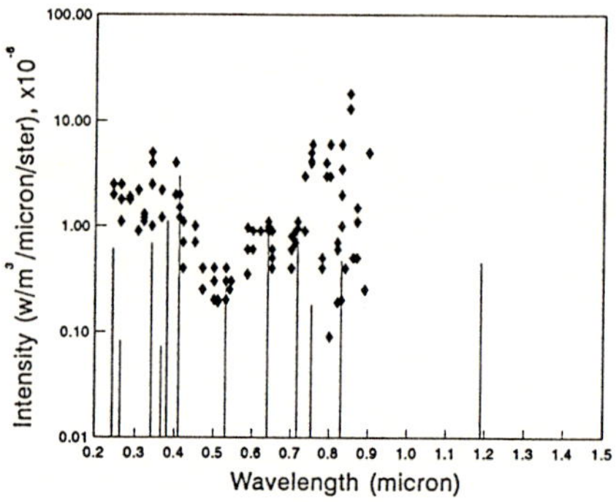

Fig. 2. Air Mach 29. Nonequilibrium spectral intensity

degree of ionisation was found to be 2.7%. The species concentration was not measured in this experiment so no direct validation of the chemical reaction modelling can be done. The prediction of the radiation spectrum could be used as an implicit validation of it.

Allen's measurements for the spectral radiation from the non-equilibrium region are presented in Fig.2 with solid markers. The predictions of the DSMC code used in this study are presented in the same figures with vertical lines. The equilibrium region was estimated to occupy the last 75% of the flowfield. Close examination of this figure shows an encouraging agreement for most frequencies with the experimental data although some of the spectral intensities due to atomic species seem to be underestimated.

The total radiation emission from the shock wave was not measured in the experiment but qualitative oscillograph records of the total emitted radiation were obtained. These oscillograms show that the radiation peaks in the non-equilibrium region and then it drops exponentially. From tables (Camm et al. 1959; Allen 1965) giving the absolute intensity of radiation at various altitudes and velocities we read that the equilibrium level of radiation for the conditions of the experiment should be 10 W/cm^3. From the oscillograms presented (Allen 1962) we conclude that the non-equilibrium radiation peak should be around 2 to 3 times the value of the equilibrium value. The total radiation intensity as calculated from the DSMC code used in this study follows the electron temperarure closely. The non-equilibrium peak value is almost 20 W/cm^3 while the equilibrium value is about 10 W/cm^3. Although some discrepancies have been observed in the spectral intensity distribution for both the equilibrium and nonequilibrium region the total radiation seems to be in good agreement with the available data both qualitatively and quantitatively. Bird for the same case predicted a peak non-equilibrium value of 0.1 W/cm^3. Judging from the uncertainty of the cross section data the prediction of the spectral intensities as a first approximation can be considered as acceptable. It should be noted that the experimental data of Allen present significant scatter in some wavelengths and therefore cannot be used as definitive data.

Table 1. *Mars-net* flight conditions

Altitude (km)	Speed (km/s)	T_{wall} (K)	T_{inf} (K)	Mach	Max.Temp. (K)	Ionisation %	Radiation (MW/m^3)
60	10	1000	247.021	26.15	17700	8.0	0.22
70	10	1000	219.585	27.75	20900	7.5	0.55
80	12	1500	198.639	35.00	44550	15.0	1.65

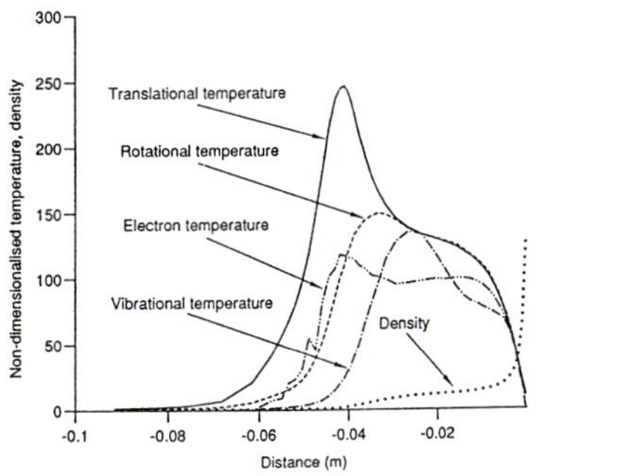

Fig. 3. Air Mach 35. Temperature and density profiles

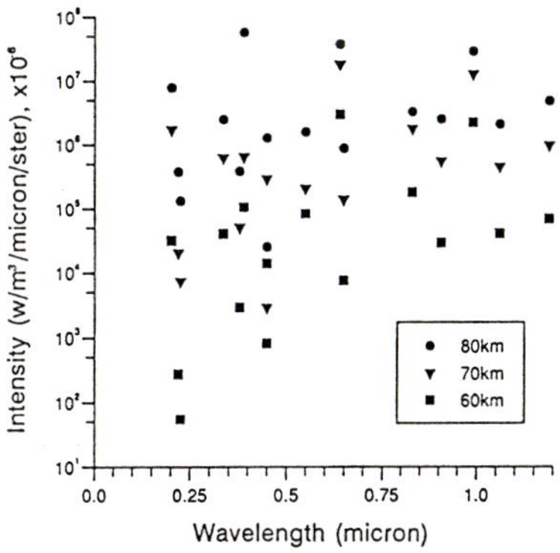

Fig. 4. Spectral intensities

As a second example of the use of the present DSMC scheme the radiation emission from the axisymmetric shock layer around a *Mars-net* reentry type of vehicle (60° degree spherically blunted cone) has been calculated. The flowfield has been studied for the case of flight at 80, 70 and 60 km. The conditions of the simulation are presented in Table 1. For the case of flight at 80 km the temperature profiles along the stagnation streamline are shown in Fig.3. The prominent characteristic of the radiation profiles is the very sharp radiation peak on the stagnation line. The radiation emission along the stagnation line rises and peaks very rapidly in the non-equilibrium area. The radiation peak coincides spatially with the temperature peak justifying its epithet as thermal radiation. The radiation peak falls very rapidly away from the stagnation streamline due to the expansion of the flow in the radial direction causing both temperature and electron density to fall. Likewise, as the temperature drops towards the body the radiation emission also drops. In Table 1 the maximum temperature on the stagnation point, the degree of ionisation and the corresponding peak radiation values are presented for all three cases. From the comparison of these data we note that the total radiation follows the temperature of the flow.

Since radiation emission is significant only from a small area of the flow and since the total energy radiated is a small fraction of the total energy its effect on the flowfield is not be expected to be large. Our assumption that the fraction of the radiated energy that is absorbed can be ignored without any significant error appears to be justified.

The spectral radiation profiles for all three cases are presented in Fig.4. For the 80 km case the contribution of N_2^+ was found to be the most significant. In the case of 70 and 60 km the significance of the atomic species is more pronounced since the percentage of ionisation drops.

6. Conclusions

Simulations of the thermal radiation under conditions of nonequilibrium have been made. To do so available measured and calculated cross sections were used. In agreement with the energy exchange and chemical reaction model employed, equilibrium conditions were not assumed anywhere in the calculation. Comparison of the computational results with the experiment indicates encouraging agreement validating therefore the argument that the assumption of equilibrium distribution of energy states for the prediction of electronic excitation under rarefied hypersonic conditions can be avoided.

Acknowledgement

This work was supported by the British Ministry of Defence (D.R.A. Farnborough).

References

Allen RA, Rose PH, Camm JC (1962) Nonequilibrium and equilibrium radiation at super satellite re-entry velocities. Avco Everett Research Laboratory, Research report 156

Allen RA (1965) Air radiation graphs: spectrally integrated fluxes including line contributions and self absorption. Avco Everett Research Laboratory, Research Note 561

Bates DR, Seaton MJ (1949) The quantal theory of continuous absorption of radiation by various atoms in their ground states.II. Further calculations in Nitrogen, Oxygen and Carbon. Monthly Not. Roy. Ast. Soc. 109: 698-704

Bird GA (1987) Non-equilibrium radiation during re-entry at 10 km/s. AIAA Thermophysics Conference, Honolulu, Hi, AIAA Paper 87-1543

Camm JC et al. (1959) Absolute intensity of non-equilibrium radiation in air and stagnation heating at high altitudes. Avco Everett Research Laboratory, Research Report 93

Cartwright DC et al. (1977) Electron impact excitation of the electronic states of N_2. Integral cross sections at incident energies from 10 to 50 eV. Physical Review A, 16 3:1041-1051

Crandall DH, Kaupilla WE (1974) Absolute cross sections for electron-impact excitation of N_2^+. Physical Review A, 9, 6: 2545-2551

Gallis MA (1993) Simulation of radiating ionised flows in the rarefied hypersonic regime. PhD Thesis, University of London, London, England

Green AES (1969) The Middle Ultraviolet: Its Science and Technology. J. Wiley & Sons

Gryzinski M (1965),Two-particle collisions. II. Coulomb collisions in the laboratory frame of reference. Physical Review 138 2A: 322- 335

Nesbet K (1980) Variational Methods in Electron-Atom Scattering Theory. Plenum Press, New York

Nicholls RW (1964) Frank-Condon factors to high vibrational quantum numbers I: N_2 and N_2^+. Journal of the National Bureau of Standards-A. Physics and Chemistry 68A, 5

Wakiya K (1978) Differential and integral cross sections for the electron impact of O_2. II. Optically forbidden transitions from the ground state. Journal of Physics B. Atomic and Molecular Physics 11, 22

Weak Shock Waves Passing over a Hemicylinder in Carbon Dioxide

H. Honma*, K. Maeno*, M. Itabashi† and N. Watanabe*
*Dept. of Mechanical Engineering, Chiba University, Yayoi-cho 1-33, Inage-ku, Chiba 263, Japan
† Taisei Kensetsu Co.

Abstract. Interferograms and their numerical simulations are presented for the dispersed irregular reflection of weak shock waves over a hemicylinder in carbon dioxide. The equilibrium Mach numbers of incident waves are 1.027, 1.045, 1.070 and 1.106. Some features of the dispersive effect on the wave pattern are discussed. The trajectory of the separation point, which is defined as a point of separation of the reflected wave from the incident wave, depends on the ratio of the shock thickness to the diameter of the cylinder. The trajectory angles of the triple point for partly dispersed waves in CO_2 are larger than the angles for shock waves in perfect gas with $\gamma = 1.4$, for the identical frozen Mach number.

Key words: Dispersed waves, Mach reflection, Triple point, Hemicylinder

1. Introduction

Weak shock waves in carbon dioxide exhibit dispersive characteristics at the shock front due to vibrational nonequilibrium of molecules. The thickness of the wave front becomes remarkably larger than the thickness of a shock wave in monatomic gas. It is measured as a few milimeter even at one atmosphere. The wave patterns and nonstationary behaviours of oblique reflection from wedges for these waves have been investigated in detail for several years (Honma and Henderson 1989; Honma et al. 1990a; Maekawa et al. 1990b; Honma et al. 1992a; Honma et al. 1992b). In this paper, as an extended work, the behaviour of dispersed waves passing over a hemicylinder is investigated in experimental and computational means. The characteristic features of dispersed irregular reflection and the effects of finite thickness of the shock front on reflection patterns are sought for, in a similar way to that in which they were investigated for oblique reflection from wedges.

Reflection and diffraction patterns of shock waves with discontinuous fronts passing over a hemicylinder have been studied by many researchers (e.g. Ben-Dor 1992). Several years ago those patterns were observed for carbon dioxide by interferometry in our laboratory (Honma et al. 1984; Honma et al. 1987), but the poorer techniques in those days prevented us to get detailed data. Combined with the numerical technique which was developed for oblique reflection studies, the study is now advanced.

2. Experimental and numerical methods

The experimental and numerical methods are essentially the same as in the previous papers on oblique reflection in carbon dioxide. A diaphragmless shock tube is used to generate weak shock waves in high pressure carbon dioxide (0.1 - 0.4 MPa). The test section is 40 mm × 40 mm square in cross section. The hemisphere model with 20 mm diameter is inserted in the lower wall of the test section. The equilibrium shock Mach number M_s is derived from the measured pressure ratio across the incident shock wave by using the shock relation for a perfect gas with $\gamma = 1.29$. The pressure is measured by a piezoelectric transducer. The shock waves are photographed with a Mach-Zehnder interferometer with 60 mm aperture. The light source is a pulsed nitrogen laser with a pulse width of 10 ns at the wavelength of 337 nm.

Shock Waves @ Marseille II
Editors: R. Brun, L. Z. Dumitrescu

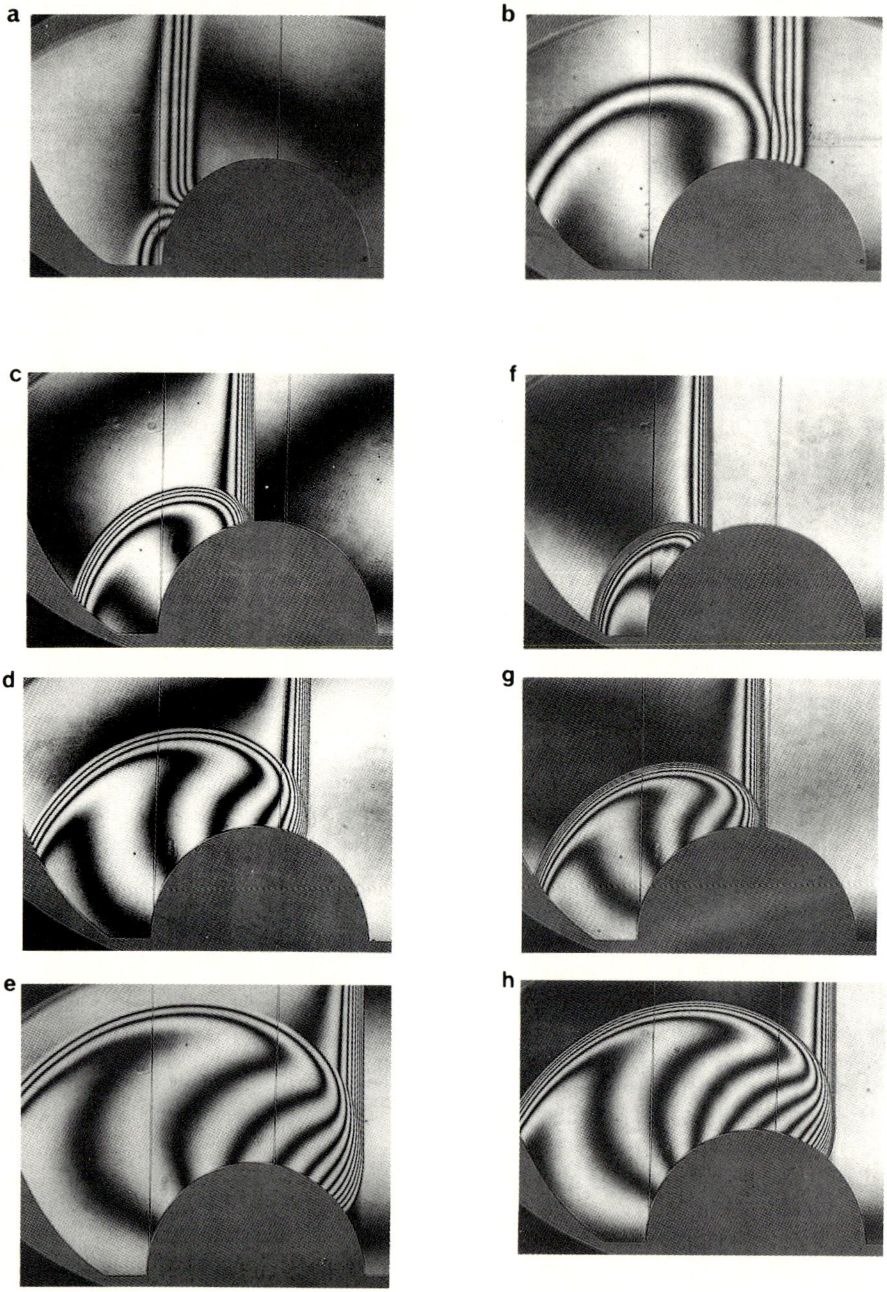

Fig. 1. Interferograms for dispersed waves passing over a hemicylinder in CO_2:
δ: shock thickness. $P_1 = 200$ kPa $d = 20$ mm
a,b: fully / $M_s = 1.027$, $\delta/d = 0.2$
c,d,e: near critical / $M_s = 1.045$, $\delta/d = 0.13$
f,g,h: partly /$M_s = 1.070$, $\delta/d = 0.1$

The two-dimensional Euler equations are used with a relaxation equation for carbon dioxide as basic equations. An operator splitting technique is applied to solve numerically the basic equations, which are transformed to the generalized coordinates fitting to the cylinder surface. A MacCormack-FCT combined method is used as a numerical scheme. As initial condition, the one-dimensional flow solution is given a-priori ahead of the hemicylinder. It is an analytical solution of Johannesen and Hodgson (1979) for a dispersed wave, and it is a discontinuous shock solution for a perfect gas with $\gamma = 1.4$. The mirror image method is used to produce a slip boundary condition at the wall.

3. Results and discussion

Fig.1 shows three cases of infinite-fringe interferograms in carbon dioxide for the equilibrium Mach numbers $M_s = 1.027$, 1.045 and 1.070, which correspond to, respectively, fully, critical and partly dispersed waves, where a "critical" dispersed wave means a shock wave with the frozen Mach number M_f close to one. In our case, $M_s = 1.042$ for $M_f = 1$. Photos a and b are for $M_s = 1.027$, c, d and e for $M_s = 1.045$, and f, g and h for $M_s = 1.070$. The initial pressure is 200 kPa. Fig.1a shows a typical example of dispersed irregular reflection corresponding to regular reflection for a discontinuous shock wave. The reflection pattern exhibits a stem-like structure near the cylinder surface due to the finite thickness of the wave front, in spite of the condition of the regular reflection. Fig.1b exhibits that the stem length rapidly increases as the incident wave passes over the top of the cylinder. In Fig.1c, a discontinuous reflected wave is formed near the top, and then it is fully dispersed in Fig.1e as the wave is diffracted along the right surface of the cylinder. In a case of a partly dispersed wave (Figs.1f - 1h), the reflection pattern of the discontinuous front has the usual shape, as a nondispersed wave. A short relaxation zone follows the discontinuous front. The thickness of the incident wave δ is defined by the distance between 10% – 90% values for the equilibrium density increment behind the shock. The nondimensional parameter δ/d is a parameter for controlling the wave pattern of dispersed irregular reflection, where d is the diameter of the cylinder.

Fig.2 shows numerical results for isopycnics, corresponding to the interferograms of Fig.1. The total number of grid points on the surface of the hemicylinder is 200. The essential characteristics of reflection patterns are reproduced by the present numerical simulation, though there are some slight discrepancies between Figs.1 and 2. The sharp reflected wave is not seen in Fig.2d due to the effect of numerical viscosity. The effect of numerical viscosity also appears on the foot of the stem shock in Fig.2h.

As an index of the reflection pattern characteristics, the separation point is defined as a point which separates the reflected wave from the incident wave. A schematic diagram of its definition is shown in Fig. 3a. The separation point S is determined by the following procedure: (1) The left boundary of isopycnics of the incident wave is extended downwards (broken line I). (2) The contour line of the same density in the stem is also extended upwards (broken line II). (3) The angle between two lines I and II is divided into two equal parts (chain line III). (4) The middle points between the left boundary of the incident wave and the outer boundary of the reflected wave are determined and connected with each other (dotted line IV). (5) The separation point S is defined as the point of intersection between lines III and IV. In Fig.3b, both measured and calculated separation points are plotted for two cases: $M_s = 1.027$ and 1.070. In both cases, the calculated trajectory of separation points is in a good agreement with the measured trajectory. It should be noted that these trajectories depend on the parameter δ/d, which is 0.2 for $M_s = 1.027$ and 0.1 for $M_s = 1.070$. The separation points go away from the cylinder as δ/d increases.

Fig.4 shows a comparison between partly dispersed waves in CO_2 and discontinuous shock waves in perfect gas, corresponding to the same frozen Mach number. The pressure or density jump across the discontinuous front of a partly dispersed wave is theoretically identical with the

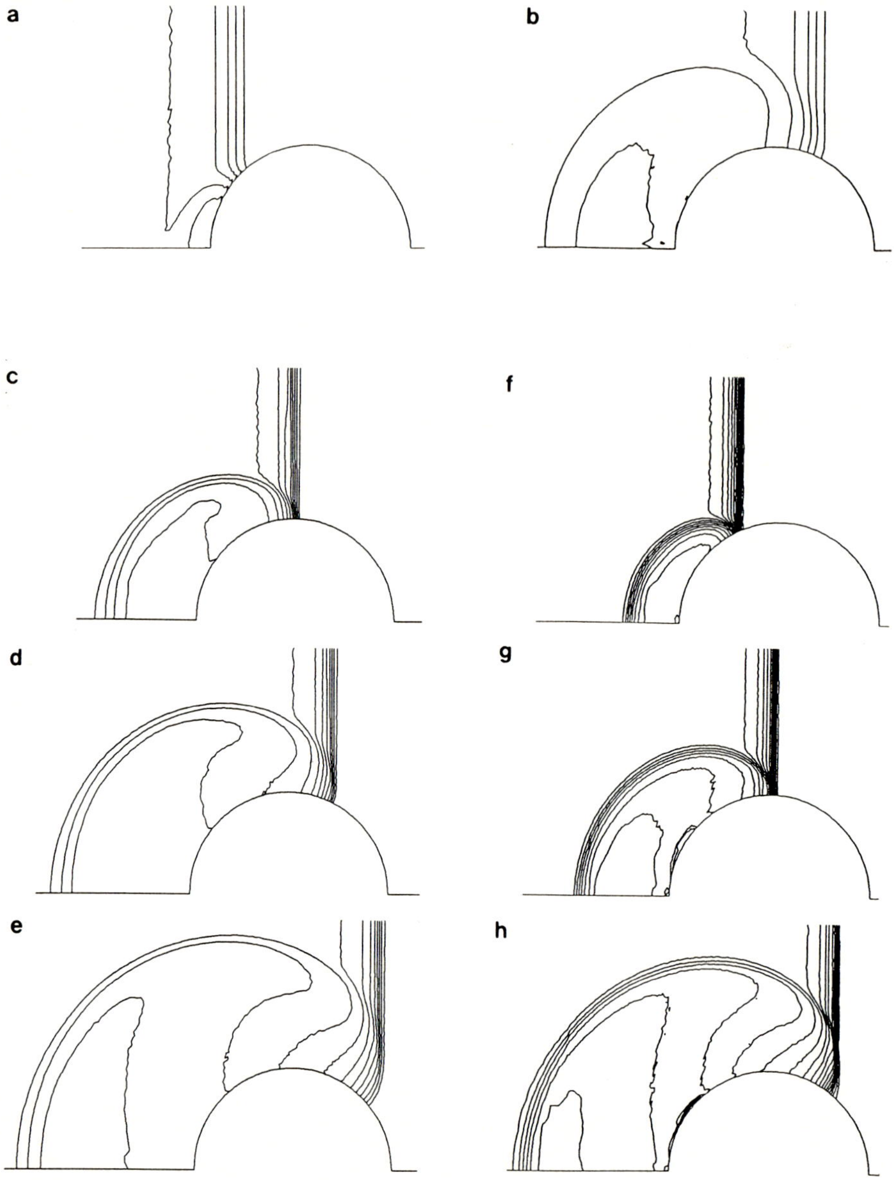

Fig. 2. Numerical simulation for dispersed waves passing over a hemicylinder in CO_2:
a,b: fully / $M_s = 1.027$, $\delta/d = 0.2$
c,d,e: near critical / $M_s = 1.045$, $\delta/d = 0.13$
f,g,h: partly /$M_s = 1.070$, $\delta/d = 0.1$

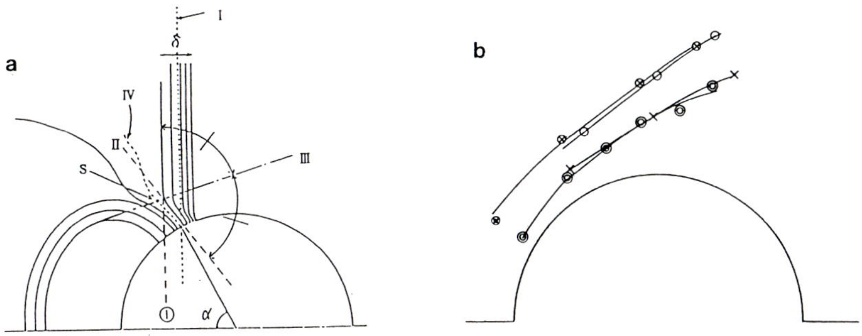

Fig. 3. Trajectory of the separation point
a: definition of the separation point (δ: shock thickness, α: position angle for incident wave)
b: trajectory of the separation point
$M_s = 1.027$, $\delta/d = 0.2$ (o: numerical; ⊗: experimental)
$M_s = 1.070$, $\delta/d = 0.1$ (⊙: numerical; ×: experimental)

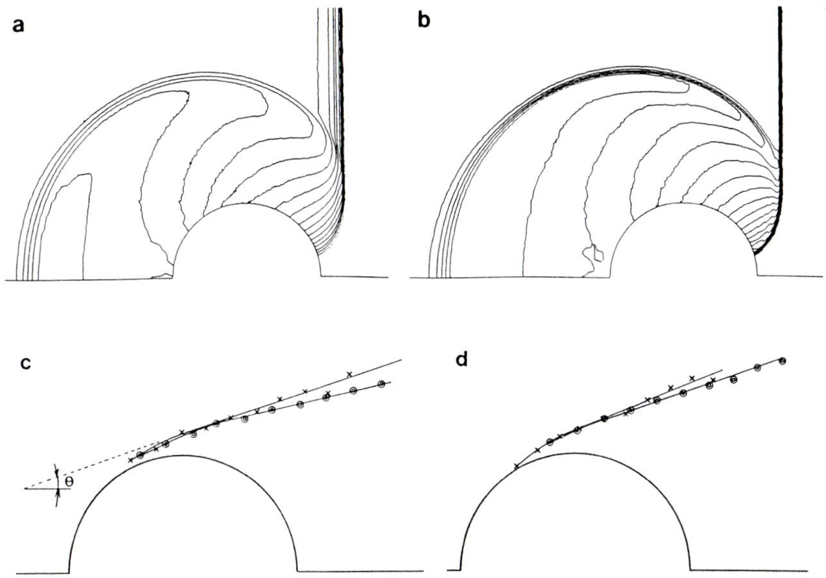

Fig. 4. Comparison between partly dispersed waves and shock waves in perfect gas with $\gamma = 1.4$
a: partly dispersed wave in CO_2 ($M_s = 1.070$)
b: shock wave in perfect gas with $\gamma = 1.4$ ($M_s = 1.026$)
c: trajectory of the triple point (× : $M_s = 1.070$, CO_2; ⊙ : $M_s = 1.026$, $\gamma = 1.4$)
d: trajectory of the triple point (× : $M_s = 1.106$, CO_2; ⊙ : $M_s = 1.061$, $\gamma = 1.4$)

jump across a shock wave in perfect gas for the same frozen Mach number. Identical profiles are expected for reflection patterns of shock waves with the same frozen Mach numbers. Two cases of numerical simulation are shown in Fig.4: $M_s = 1.070$ and 1.106, corresponding to, respectively, $M_f = 1.026$ and 1.061. Figs.4a and 4b are the isopycnics, respectively, for $M_s = 1.070$ in CO_2 and $M_f = 1.026$ in perfect gas. The wave patterns are quite similar to each other. The diffracted wave was expected to show the features of a fully dispersed wave. But in the present result no essential difference can be found between the diffracted waves in Figs.4a and 4b, mainly due to the effect of numerical viscosity. Figs.4c and 4d show the trajectories of triple points, respectively, for $M_f = 1.026$ and 1.061. The trajectory for CO_2 does not coincide with the trajectory for perfect gas, but the former exhibits a slight deviation from the latter. The deviation increases as Mach number decreases.

The trajectory angles θ, defined in Fig.4c, are 18.5° and 22.0°, respectively, for partly dispersed waves of $M_s = 1.070$ and 1.106, and 12.5° and 19.0° for shock waves of $M_f = 1.026$ and 1.061. These angles are remarkably smaller than the constant angle 33° for stronger shock waves of $M_f > 2$, which was derived from interferometer observation (Kaca, 1988) and numerical simulation (Yang et al., 1987).

4. Conclusion

The results are summarized as:

(1) The isopycnics patterns of dispersed irregular reflection from a hemicylinder in CO_2 are presented for typical cases of fully, critical and partly dispersed waves by interferograms and their numerical simulation.

(2) The trajectory of the separation point depends on the ratio δ/d, and goes away from the cylinder surface as the ratio increases.

(3) The trajectory angles of the triple point for partly dispersed waves in CO_2 are larger than the angles for shock waves in perfect gas with $\gamma = 1.4$, for the identical frozen Mach number, and remarkably smaller than the angles for stronger shock waves of $M_f > 2$.

References

Ben-Dor G (1992) Shock Wave Reflection Phenomena. Springer-Verlag, New York

Honma H, Henderson LF (1989) Irregular reflection of weak shock waves in polyatomic gases. Phys. Fluids A1:597-599

Honma H,Iimura M, Maekawa H (1990) Dispersed irregular reflection of weak shock waves in carbon dioxide. Current Topics in Shock Waves. In: Kim YW (ed) Proc. 17th Intl. Symp. Shock Waves and Shock Tubes, AIP Conference Proccedings 208, pp 222-227

Honma H, Itabashi M, Maekawa H, Usui T (1991) Nonstationary reflection of dispersed waves in carbon dioxide. In: Takayama K (ed) Proc. 18th Intl. Symp. Shock Waves, Springer-Verlag, Berlin, pp 267-272

Honma H, Maekawa H, Usui T (1992) Numerical analysis of the nonstationary oblique reflection of weak shock waves. Computers & Fluids 21:201-210

Honma H, Tomita H, Inomata K (1984) Nonstationary structures of weak shock waves in carbon dioxide with vibrational excitation. In: Oguchi H (ed) Rarefied Gas Dynamics, University of Tokyo Press, Tokyo, pp 533-540

Honma H, Yoshida H, Inomata K (1987) Reflection of weak shock waves in carbon dioxide. In: Grönig H (ed) Proc. 16th Intl. Symp. Shock Tubes and Waves, VCH, pp 581-587

Johannesen NH, Hodgson JP (1979) The physics of weak shock waves in gases. Report on Progress in Physics 42:629-676

Kaca J (1988) An interferometric investigation of the diffraction of a planar shock wave over a semicircular cylinder. UTIAS Technical note 269

Thompson JF, Warsi ZUA, Mastin CW (1985) Numerical Grid Generation. North-Holland

Yang JY, Liu Y, Lomax H (1987) Computation of shock-wave reflection by circular cylinders. AIAA J. 25, 5:683-689

Part 4: Plasmas, Astrophysics

Exact Hydrodynamic Equations and Transport Coefficients in Higher-Order Approximations for Partially-Ionized Gases

G.A. Tirskiy
Laboratory of Physics and Chemical Gasdynamics Institute of Mechanics of Moscow State University, Michurinsky Pr. 1, 119899, Moscow, Russia

Abstract. The derivation of the hydrodynamic set of equations and transport coefficients with diffusion and heat transfer in a multicomponent partially ionized gas in the presence also or the absence of an external magnetic field is considered by the methods of the kinetic theory of gases. Expressions for heat and diffusion fluxes and formulas for transport coefficients have been obtained by the Chapman-Enskog and Grad methods. The latter formulas are fairly elaborate for practical calculations. The aim of this paper is to obtain mass and heat transfer equations, and also exact formulas for transport coefficients in a simpler form, convenient for the solution of hydrodynamics problems.

Key words: Hydrodynamics equations, Multicomponent partially ionized gases, Transport equations and coefficients

1. Introduction

Hydrodynamic equations for multicomponent, multitemperature, chemically active, relaxing and emitting gas mixtures, which consist of both neutral and charged particles are the mathematical foundation for the study of hypersonic aerodynamics and heat transfer when dissipative and physico-chemical processes are present. These equations may be obtained either phenomenologically or from kinetic models (Tirskiy 1993).

The standard procedure in the Chapman - Enskog method yields the transfer equations in the form of expressions for conductive, diffusive and viscous fluxes across the boundaries via gradients of hydrodynamic variables (Navier-Stokes approximation). In this case the procedure of expansion in Sonine polynomials of the coefficients for the gradients of hydrodynamic parameters in the expression for the disturbed distribution functions yields an infinite system of algebraic equations for the determination of expansion coefficients. The solution of this system by standard methods yields complicated expressions in the form of ratios of $(N\xi + 1)$ and $(N\xi)$ - order determinants for the multicomponent diffusion $\mathcal{D}_{ij}(\xi)$ and thermal diffusion coefficients $\mathcal{D}_i^T(\xi)$, and for the heat conduction coefficients $\lambda(\xi)$ (Ferziger and Kaper 1972). Here N is the number of species in the a gas mixture and ξ is the number of terms retained in the Sonine polynomial expansion of the perturbed distribution function.

The thermal conductivity coefficient is obtained by inverting two matrices of order $N\xi$ and then order N, the elements of which are determinants of order $N\xi$ (Devoto 1966).

It is known, that for partially ionized gas mixtures four or more expansion terms need to be retained to obtain the transport coefficients with an error of about 1%. For partially ionized air one needs to compute determinants of order no less than 15 -30. Thus, the above approach is not useful for solving flow problems involving multicomponent partially ionized gases.

Shock Waves @ Marseille II
Editors: R. Brun, L. Z. Dumitrescu

2. A new approach for deriving the exact equations and transfer coefficients using higher order approximations

Using a modified approach for solving the infinite system of algebraic equations for the diffusion and heat fluxes, the exact mass and heat transfer equations with simpler expression for the transfer coefficients have been obtained at any approximation order (Kolesnikov and Tirskiy 1982, Vasil'evsky and al. 1984)

$$\mathbf{d}_i = \sum_{k=1}^{N} \frac{x_i x_k}{\mathcal{D}_{ik}(1) f_{ik}(\xi)} (\mathbf{V}_k - \mathbf{V}_j) - k_{T_i} \nabla \ln T, \tag{1}$$

$$\mathbf{q} = -\lambda(\xi) \nabla T + nkT \sum_{j=1}^{N} (k_{T_j}/\rho_j)\mathbf{J}_j + \sum_{k=1}^{N} h_k \mathbf{J}_k \tag{2}$$

where $\mathbf{d}_i = \nabla x_i + (x_i - c_i) \nabla \ln p - (c_i/p)\left(\rho \mathbf{F}_i - \sum_{k=1}^{N} \rho_k \mathbf{F}_k\right)$.

Here $\mathbf{V}_i$ is the diffusion rate, x_i is the molar concentration, c_i is the mass concentration, T is the temperature, k is the Boltzmann constant, n is the particle number density, ξ is the number of approximations in expansion of the distribution function in Sonine polynomials, p is the pressure, h_i is the partial entalpy, $\mathbf{J}_i$ is the mass diffusive flux, $\mathbf{F}_i$ is the mass force on the i-th species.

The correction function $f_{ik}(\xi)$, thermal diffusion ratio k_{Ti} and heat conduction coefficient $\lambda(\xi)$ are expressed in terms of determinants of order $N(\xi-1)$. The determinant components are just full integrals, which depend on the collision integrals of each species pair and on the mixture composition and $\mathcal{D}_{ik}(1)$ is the first-order binary diffusion coefficient, the other conventions are common. Transfer equations in the form of Eqs.1 and 2 have also been derived phenomenologically using the method of thermodynamics of irreversible processes (Kolesnichenko and Tirskiy 1976).

The above method and the method of thermodynamics of irreversible processes, together with Onsager's principle of kinetic coefficient symmetry have been used to obtain transfer equations for heat and diffusion fluxes and formulas for transfer coefficients for partially ionized gases in the presence of an external electromagnetic field also. It is assumed that the collisions of particles are two-body and elastic, and that particles do not possess internal degrees of freedom. Furthermore, an external electromagnetic field has no influence upon the collisions of particles, and there is also no effect of electron interaction with plasma oscillations on transport coefficients. The conditions under which the approximation in question is valid are known (Ferziger and Kaper 1972).

The initial set of transport equations for diffusion and partial flows in an N-component plasma in the presence of an external magnetic field takes the form (Kolesnikov 1975)

$$\mathbf{d}_i = \frac{x_i m_i}{kT} \sum_{i=1}^{N} (c_j \omega_j - \omega_j \delta_{ij}) \mathbf{b} \times \mathbf{V}_j - \sum_{j=1}^{N} \Lambda_{ij}^{oo} \mathbf{V}_j - \sum_{j=1}^{N} \sum_{p=1}^{\xi-1} \Lambda_{ij}^{op} \xi_{jp}, \tag{3}$$

$$x_i \nabla \ln T \delta_{r1} = \frac{3\sqrt{\pi}}{4} \frac{\Gamma(r+1)}{\Gamma(r+5/2)} \frac{m_i x_i \omega_i}{kT} \mathbf{b} \times \xi_{ir} + \sum_{j=1}^{N} \Lambda_{ij}^{ro} \mathbf{V}_j + \sum_{j=1}^{N} \sum_{p=1}^{\xi-1} \Lambda_{ij}^{rp} \xi_{jp} \quad (r = 1, \ldots, \xi - 1) \tag{4}$$

where $\mathbf{d}_i = \nabla x_i + (x_i - c_i) \nabla \ln p + (c_i q/n - x_i e_i)\mathbf{E}'/kT$

Here m_i is the mass, e_i is the charge, ω_i is the cyclotron frequency of the i-component, q is the space charge, $\mathbf{E}' = \mathbf{E} + \mathbf{v} \times \mathbf{B}$ where $\mathbf{E}$ is the electric field intensity, $\mathbf{B}$ is the magnetic induction vector, $(\mathbf{b} = \mathbf{B}/|\mathbf{B}|)$ and $\mathbf{v}$ is the mean mass velocity. Vectors $\boldsymbol{\xi}_{j1}$ are interrelated with reduced partial heat flows of component: $\boldsymbol{\xi}_{i1} = -\mathbf{q}_i/p_i$, where p_i is the partial pressure. The vectors $\boldsymbol{\xi}_{ip}$ $(p = 2, \ldots, \xi - 1)$ are expressed in terms of higher moments of the distribution function. The set of mass and heat transport equations (3) and (4) has been obtained on the basis of the solution of the Boltzmann equations by the Chapman-Enskog method with regard to an arbitrary number of terms in the expansion of the perturbation function in Sonine polynomials. In the absence of an external magnetic field (or $\omega_i \ll 1$) from Eqs.3 and 4 it is possible to solve Eqs.1 and 2.

We turn now to the derivation of the Stefan-Maxwell relation for diffusion flow and heat flow in a magnetic field: from Eqs.3 and 4 we can obtain the expressions

$$\begin{aligned} d_i = \sum_{k=1}^{N} \Big\{ \Big[\frac{m}{kT} c_i \omega_k (c_k - \delta_{ik}) + \varphi^{\wedge}_{ik}(\xi) \Big] \boldsymbol{b} \times \boldsymbol{V}_k + \frac{x_i x_k}{\mathcal{D}_{ik}(1) f''_{ik}(\xi)} (\boldsymbol{V}^{\parallel}_k - \boldsymbol{V}^{\parallel}_i) \\ + \frac{x_i x_k}{\mathcal{D}_{ik}(1) f^{\perp}_{ik}(\xi)} (\boldsymbol{V}^{\perp}_k - \boldsymbol{V}^{\perp}_i) \Big\} - K^{\parallel}_{Ti} \nabla^{\parallel} \ln T - K^{\perp}_{T_i} \nabla^{\perp} \ln T - K^{\wedge}_{Ti} \boldsymbol{b} \times \nabla \ln T \end{aligned} \tag{5}$$

$$\boldsymbol{q} = \boldsymbol{q}^{\parallel} + \boldsymbol{q}^{\perp} = -\lambda^{\parallel} \nabla^{\parallel} T - \lambda^{\perp} \nabla^{\perp} T - \lambda^{\wedge} \boldsymbol{b} \times \nabla T + P \sum_{j=1}^{N} \left(K^{\parallel}_{Tj} \boldsymbol{V}^{\parallel}_j + K^{\perp}_{Tj} \boldsymbol{V}^{\perp}_j + K^{\wedge}_{Tj} \boldsymbol{b} \times \boldsymbol{V}_j \right) \tag{6}$$

where the notation "⊥" denotes transport coefficients in the normal direction to the magnetic field, "||" denotes coefficients in the parallel direction to the field and the notation "∧" in the transverse one. Thermal conductivity $\lambda^{\perp}$ and $\lambda^{\wedge}$ and thermodiffusion ratios $k^{\perp}_{Tj}$ and $k^{\wedge}_{Tj}$ were to be calculated directly without preliminary determination of all other transport coefficients - diffusion, thermodiffusion and λ', in contrast to the standard Chapman - Enskog procedure (Ferziger and Kaper 1972), $\varphi^{\wedge}_{ik}(\xi) = \varphi^{\wedge}_{ki}(\xi)$ are corrections taking into consideration the effect of higher approximations on the diffusion in a transverse direction, $f^{\perp}_{ik} = f^{\perp}_{ki} = 1/(1 - \nabla^{\perp}_{ki})$, where $\nabla^{\perp}_{ki} = \varphi^{\perp}_{ki}(\xi)/\Lambda^{oo}_{ki}$ are the correction factors for binary coefficients diffusion. In the presence of a magnetic field $f^{\parallel}_{ik} \neq f^{\perp}_{ik}$, that is the anisotropy of diffusion coefficients in the longitudinal and transverse direction to the field takes place if at least the second approximation is allowed for. In Eqs.5 and 6 thermal conductivity, thermodiffusion ratios and multicomponent correction factors for binary diffusion coefficients act as transport coefficients. Such a presentation of the relationship between the fluxes and the thermodynamic forces is most expedient for solving specific problems. When an electromagnetic field is present the transfer equations are derived in the litterature by Grad's 13-moment approximation using only two members in the Sonine polynomial expansion which consequently gives less accurate transport coefficients. Thermodiffusion ratios are here given directly in terms of collision integrals and cyclotron frequencies.

3. Numerical results of calculations for transport coefficients in higher approximations

The numerical calculation of a number of transport coefficients $\lambda^{\parallel}(\xi)$, $K^{\parallel}_{Ti}(\xi)$, $f^{\parallel}_{ik}$ has been carried out for different approximations of the expansion in Sonine polynomials (Vasil'evskiy et al. 1986). The convergence in ξ has also been studied. The calculations have been made in the region of pressure and temperatures ($p = 1.013 \times 10^3$ N/m^2 to 10^7 N/m^2, $T = 2000 - 20000$ K) which embraces all the characteristic states of multicomponent air, from the neutral state to total ionization. As a result of an extensive study of the high approximations effect on the effective transport coefficients in the transport equations for chemically equilibrium air we have constructed the chart (Fig.1), which shows the necessary order of higher approximations to be taken when calculating the effective transport coefficients with errors $\leq$5%.

Curve 1 shows the boundary on the left of which (region I) the thermal conductivity and viscosity coefficient may be calculated using the first nonzero approximations (i.e. $\lambda(2)$, $\mu(1)$), while the correction ratios for all effective coefficients may be neglected (i.e. $f_{ik} = 1$, $K_{Ti} = 0$). On the right of Curve 1 (regions II, III, IV, V) the thermal conductivity must be calculated in the third approximation (the second nonzero approximation). On the right of Curve 2 (regions II, IV, V) the viscosity coefficient must be calculated in the second approximation ξ; on the

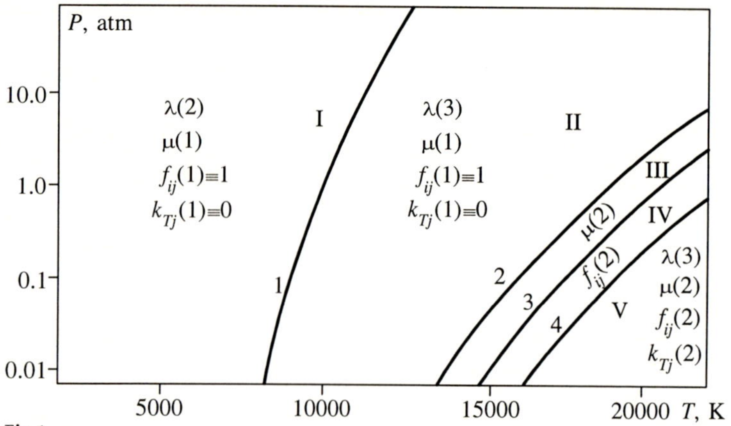

Fig. 1.

right of Curve 3 (regions IV, V) the correction coefficients become essential and only on the right of Curve 4 (region V) the thermodiffusion ratios $K_{Ti}(\xi)$ must be calculated in the second approximation for obtaining the effective coefficients. In this region the correction coefficients $f_{ik}(\xi)$ and the thermodiffusion ratios $K_{Ti}(\xi)$ may be calculated in the second approximation (first nonzero approximation)

Numerical results show that the corrections to the effective coefficients, such as the effective Schmidt number $S_{j_{eff}}$ and Lewis number $Le_{j_{eff}}$ due to the value of $f_{ij}(\xi)$ amount to 12% and for the other effective coefficients these corrections give contribution $\leq$ 5%. When calculating the effective Prandtl number σ_{eff} of highly ionized air the thermal conductivity $\lambda(\xi)$ should be computed in the third order approximation, and the second approximation is needed for the viscosity coefficient: so, determinants in the expression for the thermal conductivity and viscosity have the same order $2 \times N$. Using the lower orders of approximations leads to an error in the effective Prandtl number of up to 60%.

For the neutral components of the partially ionized air the thermodiffusion ratios $K_{T_i}(\xi)$ and corrections $fi_{ij}(\xi)$ are small ($K_{T_i} \leq 0.01, f_{ij} \leq 0.004$), and in the Stefan-Maxwell relation (Eq.1) these coefficients must only account for such pairs in which at least one of the component is an ionized particle. Thus the complete set of all the transport coefficients based on the higher approximations ($\lambda(3), \mu(2), K_{T_i}(2)$ and $f_{ij}(2)$) is needed only for highly ionized air ($x_E \sim 0.5$) corresponding to region V on Fig.1.

4. Simplified expressions for the transport coefficients of partially-ionized air

Exact formulae of the kinetic theory of gases are too complicated for practical use considering the inaccuracy of input data. The scattering of the particle interaction potentials in the region of developed ionization gives an error up to 30% for the effective equilibrium thermal conductivity and viscosity coefficients (Zluktov et al. 1990).

The approximations commonly used for neutral gases (i.e. the Buddenberg-Wilke expression for the viscosity and the Mason-Saxena equation for the translational thermal conductivity) are known to give very poor results when applied to partially ionized gases. Analysis of the formulae (Armaly and Sutton 1980, 1982) and calculation of the transfer coefficients show poor accuracy (the error is about 60% for the calculation of thermal conductivity coefficients and about 15% for viscosity coefficients for strongly ionized air). Approximate formulae for the viscosity and thermal conductivity coefficients for partially dissociated and ionized air, which give good approximation for exact solutions taking into account the high expansions, were derived (Zluktov et al. 1990).

5. Applications

1. Transfer equations in the forms (1) and (2) have been widely applied to solve various viscous multicomponent gas flow problems; the concentration gradient (Eq.1) and the temperature gradient (Eq.2) are explicitly expressed. This allows the stationary parabolized Navier-Stokes eqs. to be expressed in the normal Cauchy form. In this form the derivatives normal to the body are explicit. In such a system a higher-order numerical approximation (4th order) in the normal direction is applicable (Tirskiy 1993).
2. Using Eqs.1 and 2, hydrodynamic equations for flows of gas mixtures in local chemical and ionized equilibrium with different species-diffusive properties have been derived in a rigorous analysis (element diffusion should be included in the model, since the different diffusive properties of the constituents cause their separation) with the full set of the effective transfer coefficients for ionized air. The effect of elemental separation was detected (Vasil'evsky and Tirskiy 1991).
3. An expression for the electrical current density in a partially ionized plasma with any ionization degree, taking into account the thermo- and baro-diffusion, has been obtained. Also, an expression for the generalized Ohm's law has been derived for the case of ionized equilibrium plasmas (Sokolova and Tirskiy 1981).

References

Armaly BF, Sutton K (1980) Viscosity of multicomponent partially ionized gas mixtures. AIAA Paper N80-1495.

Armaly BF, Sutton K (1982) Thermal conductivity of partially ionized gas mixtures. AIAA Paper N82-0469.

Devoto RS (1966) Transport properties of ionized monatomic gases. Phys. Fluids 9: 1230-40.

Ferziger TH, Kaper HG (1972) Mathematical Theory of Transport Processes in Gases. Elzevier, Amsterdam - London

Kolesnikov AF (1975) Transport equations for high temperature ionized mixtures of gases in an electromagnetic field. In: Tirskiy GA (ed) Trudy Inst. Mech. MGU, Moscow 39:39 (in Russian)

Kolesnikov AF, Tirskiy GA (1982) Hydrodynamic equations of partially ionized multicomponent mixture of gases with transport coefficients in higher - order approximations. In Struminskiy VV (ed) Molecular Gasdynamics, pp. 20-44. Moscow, Nauka (in Russian)

Kolesnichenko AV, Tirskiy GA (1976) Stefan-Maxwell Relations and heat transfer in the non-ideal multicomponent continuous media. Chislen. Metody Mech. Sploshnoy Sredy 7: 106-121 (in Russian)

Sokolova IA, Tirskiy GA (1981) Ohm's law for multicomponent chemically reacting gas mixtures. In: Grigorian SS (ed) Izbran. Voprosy Sovremen. Mekh., pp. 91-109, Moscow Inst. Mech. Moscow Univ. (in Russian)

Tirskiy GA (1993) Up-to-date gasdynamic models of hypersonic aerodynamics and heat transfer with real gas properties. Ann. Rev. Fluid Mech. 25: 151-181

Vasil'evskiy SA, Sokolova IA, Tirskiy GA (1984) Exact equations and transport coefficients for multicomponent gas mixtures and partially ionized plasmas. Zh. Prikl. Mech. Tehn. Fiz. 4: 15-24 (in Russian)

Vasil'evskiy SA, Sokolova IA, Tirskiy GA (1984) The determination and calculation of effective transport coefficients for chemically equilibrium flows of partially ionized and dissociated gas mixtures. Zh. Prikl. Mech. Tehn. Fiz. 1: 68-79 (in Russian)

Vasil'evskiy SA, Tirskiy GA (1991) The effect of element diffusion in chemical equilibrium flows of multicomponent gases. In: Sedov LI (ed) Sovrem. Gasodinamich. i Fizikokhimich. Modeli Giperzvuk. Aerodinamiki i Teploobmena, , pp. 195-230. Moscow Inst. Mech. Moscow Univ. (in Russian)

Zhluktov SV, Sokolova IA, Tirskiy GA (1990) Simplified formulas for the viscosity and thermal conductivity coefficients of the multicomponent partially dissociated and ionized air. Zh. Prikl. Mech. Tehn. Fiz. 1: 41-50 (in Russian)

Reaction Kinetics Accounting for Carbon Impurities in Krypton Shock Waves

Dieter Klages and Frank Demmig
Institut für Plasmaphysik, Universität Hannover, Callinstr. 38, D 30167 Hannover, Hannover, Germany

Abstract. Impurities of the test gas can give rise to a strong increase of the electron density within the relaxation zone. Therefore, we extended our ionization relaxation model to comprise all relevant atomic reactions of the impurities. Consequently we treated the real test gas as a mixture of Krypton and impurity Carbon atoms.

In the initial stage of relaxation electron-electron and superelastic collisions are too infrequent to replenish the high energy tail of the electron velocity distribution function depleted by inelastic electron-atom collisions. Therefore the electron velocity distribution function is determined by a numerical solution of the non-stationary Boltzmann equation with Fokker-Planck collision terms. There is excellent agreement between our computed and measured electron densities. In comparison to computations under the assumption of a Maxwell distribution for the electrons our computations yield a substantial slow-down of the relaxation process and a qualitatively different time development of the electron temperature within the relaxation zone. Especially at the beginning of the relaxation the high energy tail of the electron distribution function is strongly underpopulated as compared to the Maxwellian yielding smaller excitation and ionization reaction rates. For a plasma temperature of about 9000 K and an impurity concentration of 500 parts per million there is a strong decrease of relaxation time of about 60 per cent.

Key words: Ionization relaxation, Krypton, Shock waves, Carbon impurities, EEDF

1. Introduction

The influence of impurities on rare gas shock waves has been studied by a number of authors. Usually the effect of impurities on the relaxation is roughly taken into account by adjusting the atom-atom excitation cross-section constant to measured electron densities, and thus the cross section constant will depend on the impurity concentration and the plasma temperature. Therefore, in the history of shock tube experiments with Krypton the value of the collisional excitation cross-section constant was decreased by a factor of fifty by reducing the impurity concentration to the parts per million range. A definite improvement, however, can only be achieved if the ionization relaxation model is extended to comprise all relevant atomic reactions of the impurities. Consequentley, we treated the real test gas as a mixture of Krypton and Carbon atoms.

If one studies the reaction kinetics of rare gas shock tube plasmas within the relaxation zone the question arises whether the velocity distribution of the electrons is Maxwellian or not. In our case the electron velocity distribution function deviates strongly from a Maxwellian in the high energy tail (Meyer-Prüßner and Demmig 1979, Shaw et al. 1970). Especially in the initial stage of relaxation, when the electron density and the population densities are small, electron-electron and superelastic collisions are too few to replenish the high energy tail of the distribution function which is depleted by inelastic electron-atom collisions leading to excitation and ionization.

Shock Waves @ Marseille II
Editors: R. Brun, L. Z. Dumitrescu

2. Ionization relaxation model

Since our main concern is the investigation of ionization relaxation in shock tube plasmas we confine ourselves to one-dimensional flow (Demmig 1983 b). In order to describe the influence of impurities a gas mixture is assumed and the flow is treated as unsteady in view of shock attenuation (Demmig 1983 a,b). The effect of wall boundary layers on the free flow is taken into account approximately by a mass loss term in the continuity equation.

We study the ionization relaxation in Krypton shock waves at Mach numbers M_s varying between 8 and 12, corresponding to plasma temperatures (i.e. the heavy particle temperature immediately behind the shock front) from 7500 to 13000 K. Hence, it is justified to neglect multiple ionization. Radiation processes are approximately taken into account by a loss term in the energy equation of the gas flow. In view of the experimental conditions we study impurity concentrations in the ppm range. We choose a two step model for the test gas atom Krypton as well as for the impurity Carbon atoms. The energy of the model excitation level of Krypton comprising the 5 s-states amounts to 10.18 eV (Klages and Demmig 1991). For the ionization energy of Krypton atoms we use a value of 14.0 eV. For the Carbon model it is vital to take the lower real excitation levels into consideration (Klages and Demmig 1991). This leads to a model excitation and ionization energy for Carbon atoms of 5.61 eV and 10.81 eV, respectively.

For heavy particles we assume a Maxwell distribution of velocities. In contrast to the heavy particles the velocity distribution function of the electrons deviates strongly from a Maxwellian in the high energy tail. Therefore we have to formulate a non-stationary Boltzmann equation with Fokker-Planck collision terms for the electron gas. After insertion of the collision terms for all relevant reactions we obtain the following second-order partial differential equation of parabolic type for the electron distribution function $f(\epsilon,t)$:

$$d(\epsilon,t) = \frac{Df(\epsilon,t)}{Dt} - a(\epsilon,t)\frac{\partial^2 f(\epsilon,t)}{\partial\epsilon^2} - 2b(\epsilon,t)\frac{\partial f(\epsilon,t)}{\partial\epsilon} + c(\epsilon,t)f(\epsilon,t) \tag{1}$$

This differential equation is nonlinear, since in addition to the electron energy ϵ the coefficients a, b, c and d depend on the electron distribution function and the plasma particle densities. In our case the distribution function is isotropic in velocity space because there are no external electromagnetic fields.

We obtain the electron temperature defined by the mean kinetic energy of the electron gas by solving the electron energy equation, which represents the energy balance of the electrons.

The five independent particle densities follow from a solution of the rate equation system (Klages and Demmig 1991). Under the assumption of a linear increase of the collision cross-section Q_{ij}^P above the threshold energy E_{th} and a Maxwell distribution of both particle species i and j the rate coefficient has the following well-known form:

$$S_{ij}^P = 4C_{ij}^P(2\pi\mu)^{-1/2}(k_BT_n)^{1/2}(E_{th}+2k_BT_n)\exp\left[-E_{th}/(k_BT_n)\right] \tag{2}$$

In Eq.2 C_{ij}^P is the cross-section constant of the collision process P, k_B is the Boltzmann constant, and μ and T_n are the reduced mass and the temperature of the colliding particles, respectively. In our case Eq.3 is valid for inelastic atom-atom collisions. For inelastic electron-atom collisions the rate coefficient reads:

$$S_{ej}^P = 8\pi/m_e^2\int_{E_{th}}^{+\infty}\epsilon Q_{ej}^P(\epsilon)f(\epsilon,t)d\epsilon \tag{3}$$

In contrast to Eq.2 the rate coefficient in Eq.3 has to be computed numerically. For all inelastic atom-atom and electron-atom collisions we assume a linear increasing collision cross-section except for electron- Krypton excitation collisions. The values of this cross-section are taken from Specht (1980) and Trajamar (1981). Whereas most of the cross-section constants of the Krypton reactions are known from previous studies none of the Carbon-Krypton cross-sections were known originally. The values of the mean momentum transfer cross section of elastic electron Krypton collisions are taken from Mitroy (1990), Hunter (1988) and England (1988). Since there are no literature data for elastic electron-Carbon collisions we use the corresponding values of the mean momentum transfer cross-section of elastic electron-Oxygen collisions from Thomas and Nesbet (1975).

3. Numerical methods

We utilized a Lagrangian method for the numerical treatment of the gasdynamic equations (Hoskin 1964). The stiff rate equations have been discretized according to a second-order process given by Scraton (1981). For the numerical solution of the Boltzmann Eq.1 we choose an unconditionally stable method (Keller 1972) which has no coupling between time and energy stepwidth. We solve Eq.1 under the following two assunptions:

$$f(\epsilon = \epsilon_{\text{max}}, t) = 0 \qquad \forall t$$
$$f(\epsilon = 0, t) = f_{\text{Max}} \qquad \forall t$$

For ϵ_{max} we choose an energy of 20 eV which is much larger than the mean thermal energy of about 1 eV. This limitation of the energy range has no influence on computed particle densities and temperatures since the value of the distribution function at 20 eV is already negligibly small. The assumption of a Maxwell distribution for zero energy is very well justified in our case (Fig.5). Furtermore the distribution function is normalized to unity.

4. Results

To find the values of the unknown atom-atom cross section constants we adjusted our computations to measured electron densities in the initial relaxation zone. For our adjustment procedure we chose a number of measured electron density profiles within the relaxation zone covering a relatively large range of plasma temperatures (Sect.2) and impurity concentrations (Klages and Demmig 1991, Müller et al. 1983). We derived values of the excitation cross section constants for Krypton-Krypton and Carbon-Krypton collisions varying between 4×10^{-6} m^2/J and 6×10^{-6} m^2/J and between 5×10^{-4} m^2/J and 7×10^{-4} m^2/J, respectively. The cross-section constants of the other reactions are known from previous studies (Klages and Demmig 1991).

Fig.1 shows a comparison between our model computations and experimental data points of the electron density at a shock Mach number of about 11, and an impurity concentration of 4.5 ppm. There is excellent agreement between computed and measured electron densities. The broken curve shows the development of the electron density under the assumption of a Maxwell distribution for the electrons. The substantial slow-down of the relaxation process in case of a non-Maxwell computation is due to the strongly underpopulated high energy tail of the electron distribution function yielding smaller excitation and ionization reaction rates.

The corresponding computed electron and heavy particle temperatures can be seen in Fig.2. In the initial relaxation zone the electron temperature T_e in the non-Maxwell case increases steeply due to the electron heating by ionizing atom-atom collisions. After a few microseconds the electron temperature decreases, since the energy loss of the electron gas by inelastic electron-atom collisions now exceeds the electron heating by elastic electron-heavy particle collisions and ionizing atom-atom collisions. The higher electron temperature in comparison to the Maxwell case is a consequence of the smaller excitation and ionization rates which lead to a smaller energy loss of the electron gas. In local thermodynamic equilibrium (LTE) electron and heavy particle temperatures, T_e and T_h differ slightly, since in our model we neglect radiation processes in the electron energy equation.

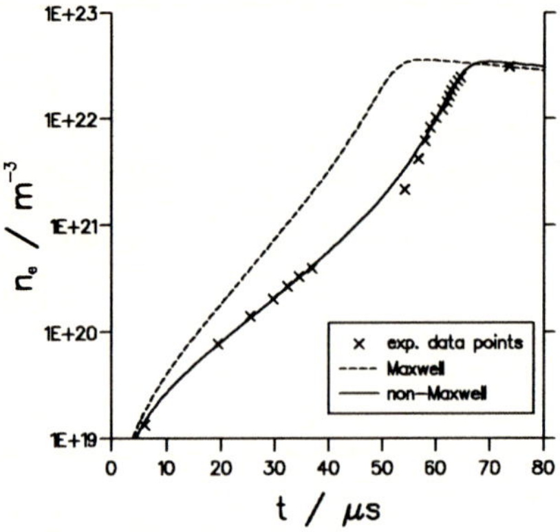

Fig. 1. Comparison between measured and computed electron densities at a fixed measurement station for an impurity concentration of 4.5 ppm and a temperature of about 10 300 K

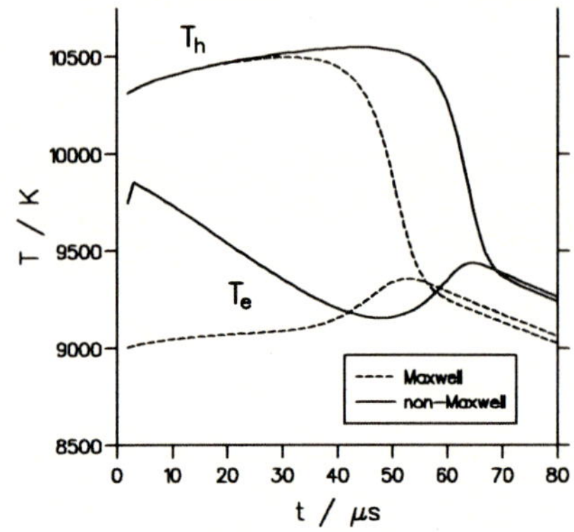

Fig. 2. Computed electron and heavy particle temperatures assuming a Maxwell and a non-Maxwell electron velocity distribution for a plasma temperature of about 10 300 K and an impurity concentration of 4.5 ppm

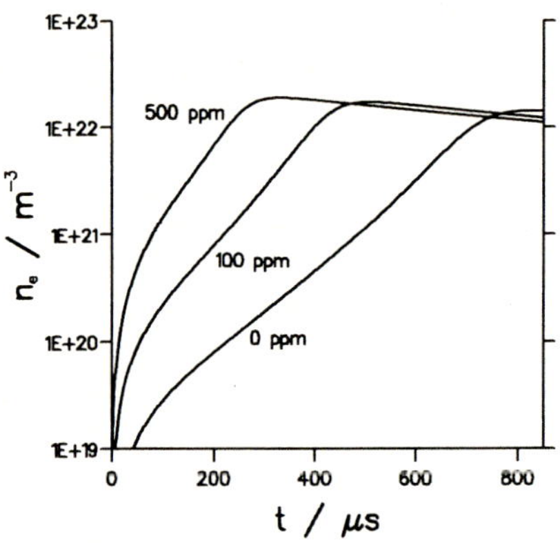

Fig. 3. Dependence of the electron density on the impurity concentration along a particle trajectory for a temperature of about 9000 K

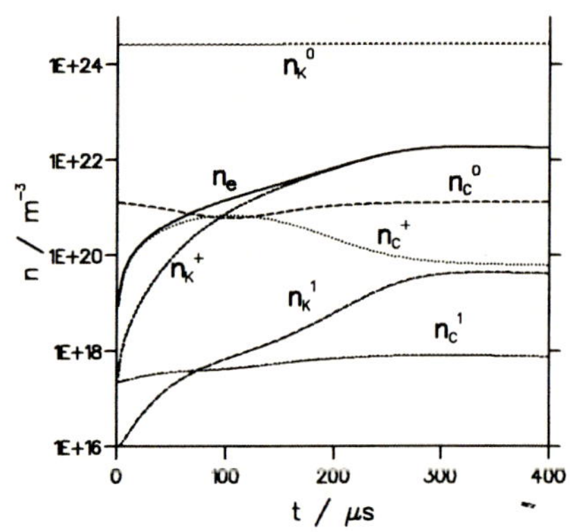

Fig. 4. Particle densities along a particle trajectory for an impurity concentration of 500 ppm and a temperature of about 9000 K

We have carried out a number of model computations with impurity concentrations varying between 0 and 500 ppm. Fig.3 shows computed electron density profiles for a temperature of about 9000 K for different impurity concentrations. Due to the already mentioned enhancement of the electron production there is a strong decrease of relaxation time with increasing impurity concentration. A comparison between the 500 ppm curve and the 0 ppm curve shows a difference in the electron density of about two orders of magnitude during the relaxation process. This result demonstrates the strong influence of Carbon impurities in the ppm range on the ionization relaxation in Krypton shock waves.

In order to study the impurity effect of Carbon in more detail all particle densities within the relaxation zone are shown in Fig.4. Carbon particles, Krypton particles and electrons are denoted by C, K, and e, respectively. The ground state, the excited state and the ground state of the singly ionized heavy particles are denoted by 0, 1, and +, repectively. In the initial relaxation zone the Carbon ion density is much higher than the Krypton ion density. This effect is caused by relatively low Carbon threshold energies in comparison to Krypton atoms leading to a very efficient excitation and ionization of Carbon atoms in Carbon-Krypton collisions. With increasing electron density the three-body electron-Carbon ion recombination rate exceeds the electron-Carbon ionization rate. Thus, the Carbon ion density decreases and at the same time ground state and excited Carbon atom densities increase to reach their LTE-values. At the end of the relaxation process electron density and Krypton ion density are nearly the same, since the Carbon impurity concentration is only in the parts per million range.

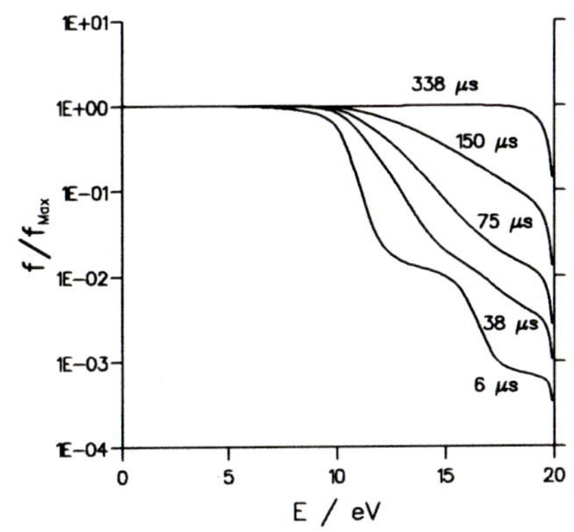

Fig. 5. Time development of the electron velocity distribution function normalized to a Maxwell distribution along a particle trajectory for an impurity concentration of 500 ppm and a temperature of about 9000 K

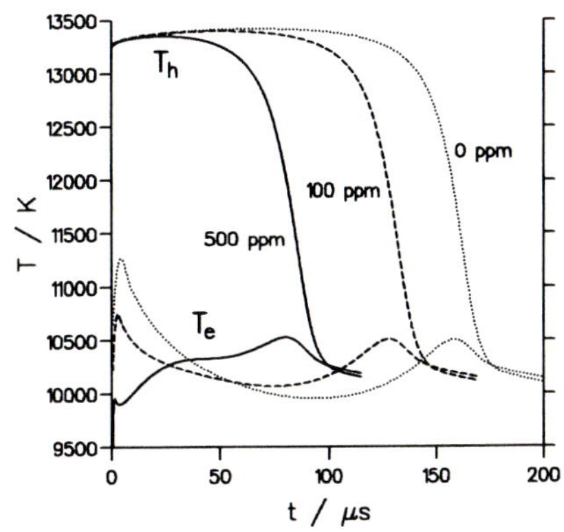

Fig. 6. Influence of the impurity concentration on the heavy particle and electron temperatures along a particle trajectory

Fig.5 shows the electron velocity distribution normalized to a Maxwell distribution as a function of electron energy at different times. At the beginning of the relaxation the high-energy tail of the distribution function is strongly underpopulated as compared to the Maxwellian. In contrast to high energies the electrons have almost a Maxwell distribution in the low-energy region. There is a partial replenishment between 12 and 17 eV since the number of superelastic electron-Carbon collisions is much higher than the inverse process of Carbon excitation collisions within this energy range. With increasing electron density and excited particle densities during the relaxation the electron-electron interaction, which attempts to establish a Maxwell distribution, becomes more and more important. In addition the increasing number of superelastic electron-Krypton collisions leads to a substantial replenishment of the high energy tail of the distribution function. In the LTE, of course, the electron gas has a Maxwell velocity distribution. The steep decrease of the distribution function for energies greater than 18 eV is due to the assumed boundary condition, that there are no electrons having energies higher than 20 eV (Sect.3).

Finally, we want to discuss heavy particle and electron temperatures for different impurity concentrations, as can be seen in Fig.6. In the initial relaxation zone the considerably lower electron temperatures for 500 ppm in comparison to the pure Krypton gas are due to a significant energy loss of the electron gas by inelastic electron-Carbon collisions. In the final stage of relaxation, where the electron density increases strongly, there are numerous elastic electron-heavy particle collisions leading to a steep decrease of the heavy particle temperature until LTE is reached.

5. Conclusions

There is an excellent agreement between our computed and measured electron densities within the relaxation zone. For Krypton shock waves at Mach numbers between 8 and 12 our computations show that Carbon impurities in the ppm range play a dominant role regarding temperatures and particle densities during the relaxation process. Furthermore there is a significant decrease of relaxation time with increasing impurity concentration. For example, an impurity concentration of 500 ppm shortens the relaxation time by about 60 %. Furthermore there is an increase of the electron density of about two orders of magnitude in comparison to the pure Krypton gas. These results demonstrate the necessity to take into account all relevant atomic reactions of Carbon impurities in rare gas shock waves.

At the beginning of the relaxation there is a strongly underpopulated high energy tail of the electron velocity distribution function as compared to the Maxwellian. This effect leads to a substantially slow-down of the relaxation process. Thus, the assumption of a Maxwell distribution function for the electrons can result in considerably erroneous particle densities and temperatures within the relaxation zone.

References

Demmig F (1983a) Models for non-equilibrium flows in real shock tubes. In: Archer RD, Milton BE (eds) Proc. 14th Intl. Symp. on Shock Tubes and Waves, Sydney, p 51

Demmig F (1983b) Investigation of ionization relaxation in shock tubes; merits and limitations. In: Archer RD, Milton BE (eds) Proc. 14th Intl. Symp. on Shock Tubes and Waves, Sydney, p 744

England JP, Elford MT (1988) Momentum transfer cross-section for electrons in Krypton derived from measurements of the drift velocity in $H_2 - Kr$ mixtures. Aust. J. Phys. 41:701

Hoskin NE (1964) Solution by characteristics of the equations of one-dimensional unsteady flow. Meth. Comput. Phys.: Adv. Res. Applics. 3:265

Hunter SR, Carter JG, Christophorou LG (1988) Low-energy electron drift and scattering in krypton and xenon. Phys. Rev. A 38:5539

Keller HB, Rolston A, Wulf HS (1972) Mathematische Methoden für Digitalrechner (in German). R. Oldenbourg Verlag, München

Klages D, Demmig F (1991) Model computations of the influence of carbon impurities on the ionization relaxation in krypton shock waves. J. Fluid Mech. 232:455

Meyer-Prüßner R, Demmig F (1979) Ionization relaxation in shock tubes taking into account a non-Maxwellian electron velocity distribution. In: Lifshitz A, Rom J (eds) Proc. 12th Intl. Symp. on Shock Tubes and Waves, Jerusalem, p197

Mitroy J (1990) The momentum transfer cross section for Krypton. Aust. J. Phys. 43:19

Müller BH, Fröbe U, Bötticher W (1983) Ionization relaxation in shock-heated Krypton at electron densities from $1 - 50 \times 10^{19} m^{-3}$. J. Phys. B: Atom. Molec. Phys. 16:4259

Scraton RE (1981) Some L-stable methods for stiff differential equations. Intl. J. Comput. Maths. B 9:81

Shaw JF, Mitchner M, Krueger CH (1970) Effects of nonelastic collisions in partially ionized gases; I. Analytical solutions and results. Phys. Fluids 13:325

Specht LT, Lawton SA, DeTemple TA (1980) Electron ionization and excitation coefficients for argon, krypton and xenon in the low E/N region. J. Appl. Phys. 51:166

Thomas LD, Nesbet RK (1975) Low-energy electron scattering by atomic oxygen. Phys. Rev. A 12:

Trajamar S, Srivastava SK, Tanaka H, Nishimura H (1981) Excitation cross sections for Krypton by electrons in the 15-100 eV impact-energy range. Phys. Rev. A 23:2167

Electron Driven Shock Waves: Antiforce Class II Waves

Mostafa Hemmati
Department of Physics, Arkansas Tech University, Russellville, Arkansas 72801 USA

Abstract. The propagation of electrical breakdown waves in a gas has been investigated by applying a set of three-component fluid equations. The basic set consists of equations of conservation of mass, momentum, and energy, plus Poisson's equation. In the fluid model, the wave front is assumed to be an electron shock wave followed by a transition region. The transition region in which the electron drift velocity and electric field go to zero is referred to as the sheath region. The sheath is very thin, being roughly a few mean free path long. Following the sheath region lies a comparatively thick region, in which the electron gas by means of further ionization cools to room temperature. This is referred to as the Quasi-Neutral Region. In a 1991 article (Proc. 18th ISSW), the solutions to different categories of breakdown waves for Quasi-Neutral Region, using the fluid model were reported. Now the boundary conditions and solutions to the Antiforce Class II Waves will be discussed.

Key words: Ionised gases, Electron-driven shock waves

1. Introduction

Breakdown waves for which the force on the electrons due to an electric field is in the opposite direction of wave propagation will be referred to as antiforce waves. Fowler (1976) revised category naming. He called those waves which move into a medium of significant electron concentration as Class II waves. These waves have a different structure than those moving into a nonionized medium (Class I waves). In the atmosphere the ionized medium is composed of positive nitrogen and negative oxygen ions with equal number densities $N_{i0}^+ = N_{i0}^-$, where N_{i0} represents the ion number density ahead of the wave. A weak electric field far ahead of the wave front (E_∞) accelerates the positive and negative ions in opposite directions with a speed v_0, generating a current ahead of the wave. This current will alter the shock conditions on electron temperature and velocity as well as Poisson's equation. The changes in the shock conditions and Poisson's equation will be discussed later in the theory. The equations of conservation of mass, momentum and energy, however, remain the same as the case of Antiforce Class I Waves (Hemmati and Fowler 1985). In dimensionless variables, they are respectively:

$$\frac{d(\nu\psi)}{d\xi} = \kappa\mu\nu, \tag{1}$$

$$\frac{d}{d\xi}\{\nu\psi(\psi-1)+\alpha\nu\theta\} = \nu\eta - \kappa\nu(\psi-1), \tag{2}$$

$$\frac{d}{d\xi}\left\{\nu\psi(\psi-1)^2+\alpha\nu\theta(5\psi-2)+\alpha\nu\psi+\alpha\eta^2-\frac{5\alpha^2\nu\theta}{\kappa}\frac{d\theta}{d\xi}\right\} = -\omega\kappa\nu\{3\alpha\theta+(\psi-1)^2\}. \tag{3}$$

The dimensionless variables, which are slightly different from those used in the proforce case, are:

$$\omega = \frac{2m}{M}, \qquad \kappa = -\frac{mVK}{eE_0}, \qquad \mu = \frac{\beta}{K}, \qquad \alpha = \frac{2e\phi}{mV^2}, \qquad n = \frac{\epsilon_0 E_0^2}{2e\phi}\nu,$$

$$T_e = \frac{2e\phi}{\kappa}\theta, \qquad E = \eta E_0, \qquad x = -\frac{mV^2}{eE_0}\xi, \qquad v = V\psi.$$

Shock Waves @ Marseille II
Editors: R. Brun, L. Z. Dumitrescu

In the above equations ν, ψ, θ, μ, κ, η,and ξ are the dimensionless electron concentration, electron velocity, electron temperature, ionization rate, elastic collision frequency, electric field, and position inside the wave respectively. The symbols m, e, n, and T_e represent electron mass, charge, number density, and temperature inside the sheath, and β, ϕ, V, M, E_0 are ionization frequency, ionization potential, wave velocity, neutral particle mass, and electric field at the wave front.

2. Theory

Assuming that in the lab frame the wave is moving in the positive x direction with a speed V_0, then in the wave frame the neutral particles will be swept into the wave front with a velocity $-V_0$. Ahead of the wave, due to the electric field E_∞ and its direction, the positive and negative ions will possess speeds $-(V_0+v_0)$, and $-(V_0-v_0)$ or the other way around. Ahead of the wave, electrons attached to the negative ions have equal number density (n_0) and velocity as those of negative ions ($n_0 = N_{i0}$). When the wave overruns the ionized medium, the ions will no longer be under the influence of the weak field E_∞, but rather of the strong electric field at the wave front E_0. This field will rip the loosely bound electrons off of the negative ions. In antiforce waves, inside the sheath, the electrons will be driven away from the wave front with a speed v, but the electron temperature and therefore electron gas pressure is assumed to be large enough to drive the wave. After making a few collisions, the ions entering the wave will acquire the same speed as neutral atoms. Assuming that the electric field ahead of the wave points in the negative x direction, the equation of conservation of current in the wave front becomes

$$env - eN_iV_0 = en_0(V_0 - v_0) - en_0(V_0 + v_0), \tag{4}$$

where n and N_i are the electron and ion number densities inside the sheath respectively. Solving Eq. (4) for N_i, one has

$$N_i = \frac{nv}{V_0} + \frac{2n_0v_0}{V_0}. \tag{5}$$

Substituting N_i in the Poisson's equation $\frac{dE}{dx} = \frac{e}{\epsilon_0}(N_i - n)$, it becomes

$$\frac{dE}{dx} = \frac{en}{\epsilon_0}\left(\frac{v}{V_0} - 1\right) + \frac{2en_0v_0}{\epsilon_0 V_0} \tag{6}$$

In terms of dimensionless variables, then Poisson's equation becomes

$$\frac{d\eta}{d\xi} = \frac{\nu}{\alpha}(1-\psi) - \frac{2\nu_0\psi_0}{\alpha}, \tag{7}$$

where ν_0 and ψ_0 are the dimensionless ion concentration and velocity in front of the wave.

The global (all particle) momentum equation can be written in the following way:

$$nmv^2 + N_iM_iV^2 + NMV^2 + nkT + (N+N_i)kT = \frac{\epsilon_0}{2}E_0^2 + C, \tag{8}$$

where $M_i = (M-m)$ is the ion mass and C is a constant to be determined by conditions ahead of the wave.

$$\begin{aligned} n_1mv_1^2 + N_{i1}(M-m)V_0^2 + MNV_0^2 + n_1kT_{e1} + (N+N_i)kT + \frac{\epsilon_0}{2}(E_0^2 - E_0^2) - n_0kT_{e0} - \\ - (N_0 + N_{i0})kT_0 - n_0m(V_0-v_0)^2 - N_{i0}(M-m)(V_0+v_0)^2 - MN_0V_0^2 = 0, \end{aligned} \tag{9}$$

where T_{e0} and T_0 are the electron and neutral temperatures ahead of the wave respectively, and the electric field at the wave onset is no longer E_∞, but becomes E_0. Also at the wave front $N = N_0$, and $N_i = N_{i0} = n_0$, because the ionization does not take place at the wave front but inside the sheath. Sanmann (1975) calculated the temperature difference term $T - T_0$, arguing that its source must be collisions of electrons with the neutrals, since the electron-ion collisions are infrequent. He concluded that even at extreme cases if the wave velocity is larger than 10^5 m/s, this term is negligible. Considering the above arguments, the terms $(N+N_i)kT$ and $(N_0+N_{io})kT_0$ cancel each other. The electron temperature ahead of the wave, T_{e0}, which is in the order of room temperature is negligible compared to the electron temperature inside the sheath (T_{e1}). Compared to the wave velocity V_0, the neutral-ion velocity v_0 ahead of the wave is very small. The terms having v_0^2, therefore, could be neglected. Taking all the above into consideration and employing the equation of conservation of particles:

$$N_0 + n_0 = N + N_{i1}, \tag{10}$$

the global momentum equation reduces to

$$n_1 kT_{e1} + n_1 m v_1^2 + MV_0^2[N_{i1} - n_0 + N - N_0] - mN_{i1}V_0^2 + 4n_0 m v_0 V_0 - 2n_0 M v_0 V_0 = 0. \tag{11}$$

Substituting for ion concentration at the wave front (N_{i1}) from Eq. (5) in the above equation reduces it to:

$$n_1 kT_{e1} = -nmv_1^2 + mn_1 v_1 V_0 + 2n_0 v_0 V_0 m\left(\frac{M}{m} - 1\right). \tag{12}$$

When we introduce dimensionless variables into Eq. (12), the initial condition on electron temperature becomes:

$$\theta_1 = \frac{\psi_1}{\alpha}(1 - \psi_1) + \frac{2\psi_0\nu_0}{\alpha\nu_1}\left(\frac{M}{m} - 1\right). \tag{13}$$

θ_1, ν_1 and ψ_1 are the electron temperature, density, and velocity at the wave front respectively.

3. Analysis

At the wave front, since the electron velocity is less than wave velocity $(\psi_1 < 1)$, its immediate consequence in the Poisson's equation becomes $\left(\frac{d\eta}{d\xi} > 0\right)$. $\frac{d\eta}{d\xi}$ remains positive until the right side of the Poisson's equation becomes negative. This implies that in traversing the interior of the wave the magnitude of the electric field η will increase until the electrons gain speeds in excess of the ion speeds. At this point the field magnitude starts decreasing until electrons and ions reach the same speed. The current entering the wave at the wave front, according to Kirchhoff's law, must be leaving the wave at the end of the wave. At the end of the wave, therefore, an electric field must be remaining to drive the current. This field magnitude at the end of the wave will depend on the current ahead of the wave. Since a conductor or a plasma in equilibrium cannot support a field, if there is no current present ahead of the wave, the field must go to zero at the end of the wave $(\eta \to 0)$. The electrons, after attaining speeds larger than the ions, gradually slow down to speeds comparable to those of ions $(\psi \to 1)$ as a result of collisions with neutral particles. To have a drastic effect on electron temperature at the wave front a current $J_0 = \nu_0\psi_0$ of the order of 10^{-7}ahead of the wave is required. This represents an ion density of the order of $10^{16} \sim 10^{17}/\text{m}^3$ ahead of the wave.

The solution to the set of electron-fluid dynamical equations, equations of conservation of mass, momentum, and energy, Eqs.1-3, and Poisson's equation, Eq.7, should meet the above-mentioned conditions at the trailing edge of the wave. In nitrogen and for α= 0.01, the set of equations were integrated through the sheath with the set of values of ν_1, ψ_1, and κ dependent on the value ofJ_0. κ is referred to as wave constant and it determines the relation between the initial applied field and the laboratory value of the wave speed. α= 0.01 represents a fast moving wave $(V \approx 3 \times 10^7\text{m/s})$.

The expected physical conditions at the trailing edge of the wave ($\psi_2 \rightarrow 1$, $\eta_2^{'} \rightarrow 0$) allows one to determine the value of ν_1 by trial and error for a given set of values of κ and ψ_1.

Solutions for the profiles of these antiforce breakdown waves for three different current values ahead of the wave are given in Figs.1-3. Figs.1 and 2 display electric field magnitude (η) and electron velocity (ψ) inside the sheath as a function of position (ξ) respectively. Fig.3 displays the electric field as a function of velocity inside the sheath. As one can see from Fig.3, Poisson's equation does not allow solutions inside the rectangle (0, 1).

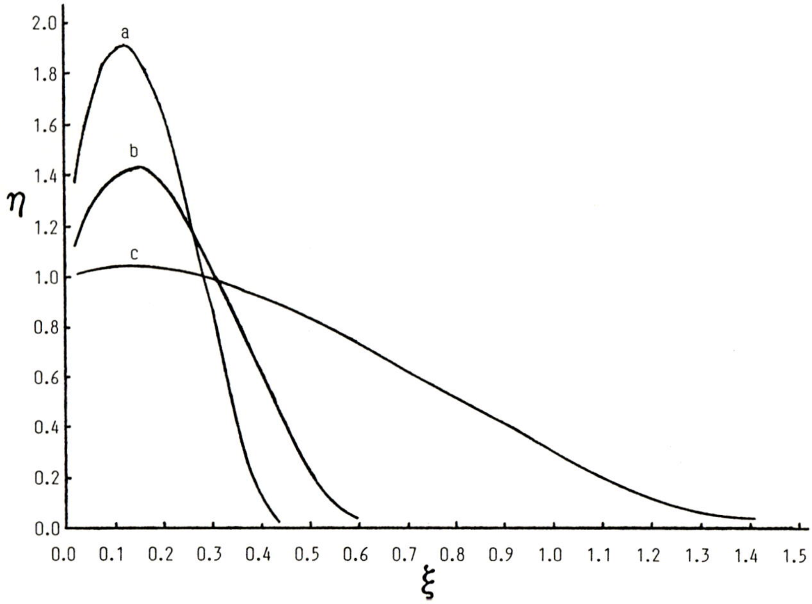

Fig. 1. Electric field η as a function of position ξ inside the sheath for: (a) $J_0 = 2 \times 10^{-6}$, (b) $J_0 = 6 \times 10^{-7}$, (c) $J_0 = 6 \times 10^{-8}$

The solutions for $\alpha = 0.01$ and three different values of J_0 were found at the following values of κ, ν_1, and ψ_1: (a) $J_0 = 2 \times 10^{-6}$, $\kappa = 1.05$, $\nu_1 = 0.694$, $\psi_1 = 0.795$; (b) $J_0 = 6 \times 10^{-7}$, $\kappa = 0.783$, $\nu_1 = 0.355$, $\psi_1 = 0.840$; (c) $J_0 = 6 \times 10^{-8}$, $\kappa = 0.455$, $\nu_1 = 0.05495$, $\psi_1 = 0.880$. The sheath thickness for the current value of 2×10^{-6} was 0.44. To connect to the physical world, the values of $\nu_1 = 0.694$ and $\xi = 0.44$ represent electron density and sheath thickness of $7.7 \times 10^{19}/\text{m}^3$ and 0.22 mm respectively. Comparing this sheath thickness with those for antiforce waves moving into a non-ionized medium (Hemmati 1985), one can see a reduction of one-half in the sheath thickness. The electron temperature at the wave front has a value of $\theta_1 = 31.88$ ($T_e = 1.85 \times 10^7$ K), and it increases away from the wave front to a value of 60.6 ($T_e = 3.51 \times 10^7$ K) at the trailing edge of the wave.

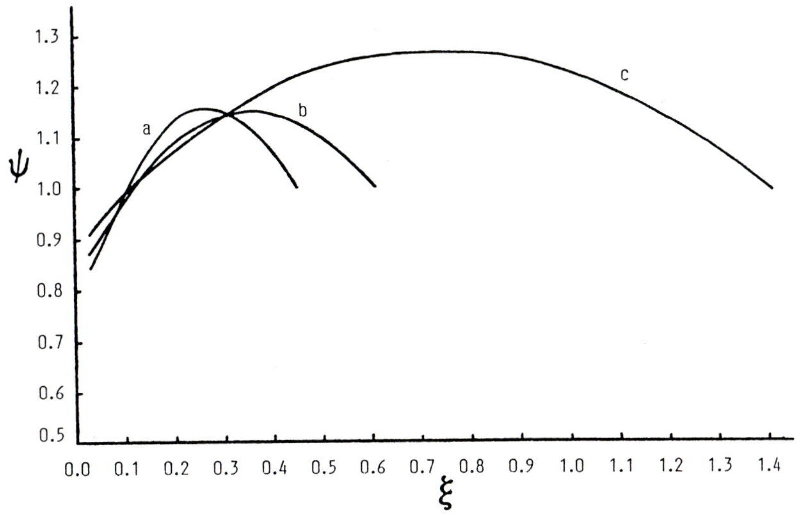

Fig. 2. Electron velocity ψ as a function of position ξ inside the sheath for: (a) $J_0 = 2 \times 10^{-6}$, (b) $J_0 = 6 \times 10^{-7}$, (c) $J_0 = 6 \times 10^{-8}$

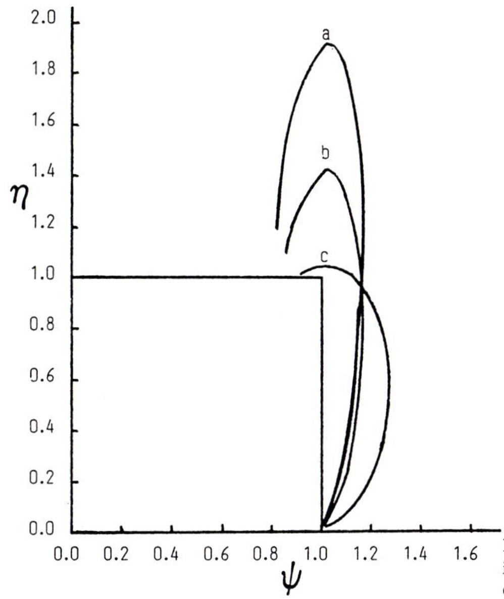

Fig. 3. Electric field η as a function of drift velocity ψ inside the sheath for: (a) $J_0 = 2 \times 10^{-6}$, (b) $J_0 = 6 \times 10^{-7}$, (c) $J_0 = 6 \times 10^{-8}$

4. Conclusions

Solutions to the antiforce breakdown waves moving into a preionized medium do exist. The initial conditions at the wave front are very sensitive to the current magnitude ahead of the wave. Larger currents produce higher electron temperatures at the wave front. The ionization rate remains constant at the beginning part of the sheath, however, it slightly increases toward the end of the sheath. Higher currents ahead of the wave produce larger thicknesses for the sheath region.

References

Fowler RG (1976) Nonlinear electron acoustic waves, Part II. Advances in Electronics and electron physics, Vol.41. Academic Press, pp 1-72

Hemmati M (1990) Exact solutions of the electron fluid-dynamical equations for the quasi-neutral region. In: Takayama K (ed) Proc. 18th Intl. Symp. on Shock Waves, Sendai, Japan, pp 1145-1149

Hemmati M, Fowler RG (1985) Electric breakdown waves: Exact solutions. Part II. Phys. Fluids 28:2744-2750

Sanmann EE, Fowler RG (1975) Structure of electron fluid dynamical plane waves. Antiforce waves. Phys. Fluids 18:1433-1438

Modelling of Shock Wave Propagation in a Developing Discharge in Nitrogen

N.Yu. Babaeva, A.Kh. Mnatsakanyan and G.V. Naidis
Institute for High Temperatures, Russian Academy of Sciences 127412, Moscow, Russia

Abstract. The results of computer simulation of the gas heating process and SW propagation in the plasma of a developing discharge are discussed. It was shown that the non-monotonous dependence of the SW velocity on the time delay between switching-on the discharge and SW generation is entirely determined by the time evolution of the gas temperature.

Key words: SW propagation in plasma, Discharge in molecular gas

1. Introduction

Shock wave (SW) propagation in a discharge plasma is accompanied by some interesting phenomena among which the observed acceleration of the SW is to be mentioned. In molecular gases this acceleration may be caused either by the density nonuniformity (thermal mechanism, evidently present in the case of monoatomic gases also) or by the rapid release of the vibrational energy due to VT relaxation behind the SW front (relaxation mechanism) (Mnatsakanyan et al. 1987).

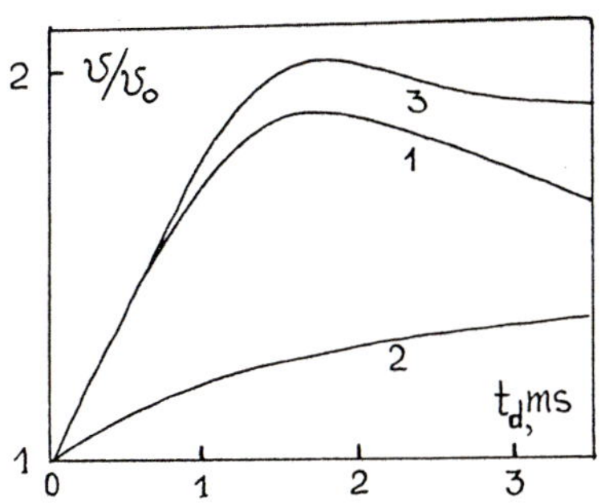

Fig. 1. SW velocity dependence on time delay t_d between switching-on the discharge and SW generation. v_0 is the SW velocity in the unperturbed gas at the corresponding point; $j = 0.3$ A/cm^2, $R = 1$ cm, $E/n = 10^{-15}$ V.cm^2. 1: experiment (Chutov et al. 1991), nitrogen, 2: experiment (Chutov et al. 1991), argon, 3: numerical calculations

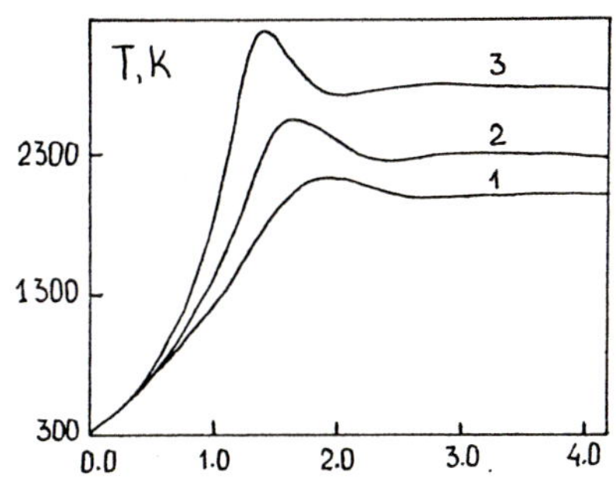

Fig. 2. Time dependence of the gas temperature in a developing discharge in nitrogen. $j = 0.3$ A/cm^2, $R = 1$ cm, $E/n = 10^{-15}$ V.cm^2. p (torr)=: 1.5 (1), 2.0 (2), 3.0 (3)

This SW acceleration was studied experimentally by Chutov et al. (1991) where the time delay t_d between switching-on the discharge and SW generation was varied and the dependence of the shock velocity V on t_d was investigated. In this experiment the shock waves were generated by an impulse discharge which created so-called blast or N-waves with a rapid fall of the pressure behind the front. The experimental results revealed that the dependence of the SW velocity V on t_d for nitrogen was nonmonotonous (Fig.1, curve 1), in contrast to the argon case (curve 2). In Chutov et al. (1991) it was suggested that this behaviour of the experimental curves may be attributed to a relaxation mechanism. In our paper the results of computer simulation of the gas heating

Shock Waves @ Marseille II
Editors: R. Brun, L. Z. Dumitrescu

process and the SW propagation in the plasma of a developing discharge for the conditions of the experiment mentioned above are discussed.

2. The gas heating process

Heating of a molecular gas is a complicated process due to the interplay of the excitation of molecule vibrations and vibrational-translational (VT) relaxation. The time evolution of the gas temperature T and vibrational energy of molecules ε_V (the number of vibrational quanta per one molecule) on the discharge axis was calculated using the balance equations which take into account the energy input, VT relaxation and diffusional transport of translational and vibrational energy to the tube walls. In the isobaric approximation this system can be written in the following form:

$$\begin{aligned} nc_P\frac{\partial T}{\partial t} &= \eta_T jE + n\omega_V\frac{\varepsilon_V - \varepsilon_V^{(0)}}{\tau_{VT}} + \frac{1}{r}\frac{\partial}{\partial r}\left(r\lambda\frac{\partial T}{\partial r}\right) \\ \frac{\partial \varepsilon_V}{\partial t} &= \frac{\eta_V jE}{n\omega_V} - \frac{\varepsilon_V - \varepsilon_V^{(0)}}{\tau_{VT}} + \frac{1}{r}\frac{\partial}{\partial r}\left(rD\frac{\partial \varepsilon_V}{\partial r}\right) \end{aligned} \tag{1}$$

Here n is the concentration of molecules, ω is the vibrational quantum, $c_P = \kappa/(\kappa-1)$, κ is the specific heat ratio, $\varepsilon_V^{(0)}$ is the equilibrium value of ε_V, D and λ are the diffusion and heat conduction coefficients, τ_{VT} is the VT-relaxation time, j is the discharge current density, E is the electric field, η_T and η_V are the fractions of energy input going into the translational and vibrational degrees of freedom.

The system (1) is to be coupled with the boundary conditions (see, for example, Gershenzon et al. 1977):

$$\begin{aligned} &\left.\frac{d\varepsilon_V}{dr}\right|_{r=0} = 0 \qquad &&\left.\frac{d\varepsilon_V}{dr}\right|_{r=R} + \frac{\bar{v}\gamma\varepsilon_V(R)}{4D} = 0 \\ &\left.\frac{dT}{dr}\right|_{r=0} = 0 \qquad &&T(R) = T_0 \end{aligned} \tag{2}$$

Here γ is the probability of quenching of the vibrationally-excited molecules in collisions with the tube walls, $\bar{v}$ is the thermal velocity of the molecules in the vicinity of the walls, R is the tube radius, T_0 is the wall temperature.

The values of η_T and η_V depend on the reduced field E/n and on ε_V. At $E/n \sim 10^{-15}$ V.cm^2 (values typical for low pressure discharges in nitrogen) the electron energy goes into vibrational and electronic excitation of the molecules in proportions ϕ_V and $\phi_e = 1 - \phi_V$, direct heating being negligibly small. In the conditions of the experiment (Chutov et al. 1991) the densities of electronically excited molecules have quasistationary values, and the energy stored in these states is small in comparison with the full energy input. The process of quenching of the electronically-excited molecules redistributes the stored energy between translational and vibrational degrees of freedom in proportions ξ_T and $\xi_V = 1 - \xi_T$, the value of ξ_T being approximately equal to 0.5 (Devyatov et al. 1991). So the values of η_T and η_V can be written as

$$\eta_V = \phi_V + \xi_V\phi_e \qquad \eta_T = 1 - \eta_V\,. \tag{3}$$

According to the calculations made by Mnatsakanyan et al. (1976) at $E/n \sim 10^{-15}$ V.cm^2, we have $\phi_V = 0.8(1 - 0.1\varepsilon_V)$.

For calculating the values of T and ε_V at the discharge tube axis an approximate method was used. It was assumed that the radial profiles of the energy sources (sums of the first two terms in the right parts of Eqs.1 for T and ε_V) have the form of Bessel functions or (which is almost the same) a parabolic form. The equation for the gas temperature can be solved analytically for two extreme cases: when heat conduction to the tube walls is neglected and when the gas heating is stationary. The difference of the radial profiles of T in these two cases is small, and we suppose that in the general case (in nonstationary conditions with heat conduction to the walls) the profile of T would be approximately the same as in the stationary case. The corresponding value of heat conduction term at the discharge axis is

$$-\left(T(0)^{\beta+1} - T_0^{\beta+1}\right) \Big/ \left[3(\beta+1)R^2/16a\right] \tag{4}$$

where $T(0)$ is the value of T at the axis, β and a are the parameters in the expression for the heat conduction coefficient $\lambda = aT^{\beta}$ (for nitrogen $\beta = 0.8$, $a = 2.6 \times 10^{-6}$ W/cm/K$^{1.8}$ (Cebeci et al. 1984)). The diffusion term in Eq.1 for ε_V can be estimated analogously. In the stationary case for an energy input having a Bessel profile and boundary conditions (2) this term is equal to $-\varepsilon_V(0)/\tau_d$, where $\tau_d = R^2/(5.8D) + R/(1.15\bar{v}\gamma)$ (Likalter et al. 1981). Note that estimations using the value $\gamma = 10^{-3}$ (Gershenzon et al. 1977) show that in the conditions of the experiment of Chutov et al. (1991) the role of vibrational energy transfer to the walls is negligible in comparison with volume VT relaxation.

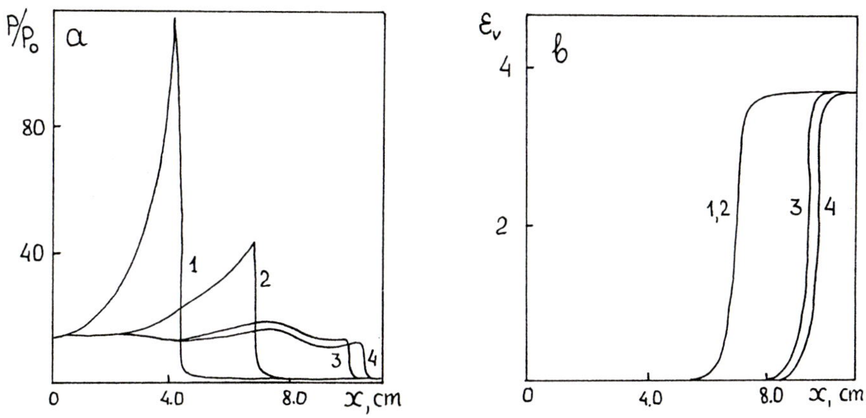

Fig. 3. Profiles of pressure (a) and vibrational energy (b) during the SW propagation through a discharge plasma. p_0 is the pressure ahead of the SW, $t_d = 2.5$ ms. Time from SW initiation, μs: 10 (1), 20 (2), 30 (3), 32 (4)

The time for gas heating in conditions of the experiment (~ 0.3 ms) is large enough and the gas pressure in the radial direction is constant. The time dependence of the pressure in the process of gas heating depends upon the relation between the discharge volume and the full volume of the system. The maximum pressure rise takes place when these volumes are equal. In this case the displacement of gas particles occurs only in radial direction.

Fig.3 shows the results of calculations of ε_V and T versus time for various initial pressures. The discharge current and reduced field are assumed to be constant in time. The values of ε_V and T depend nonmonotonously on time. Note that such a character of gas heating is typical only for molecular gases, due to their intermediate energy reservoir. Heating of atomic gases proceeds in a monotonous way (curve 2 in Fig.1).

3. Shock wave propagation through the developing plasma

The values of gas temperature and vibrational energy at the moment when the SW enters into the discharge plasma determine the acceleration of the SW. Estimations of times of VT relaxation behind the wave front (~0.1ms) show that release of vibrational energy during the time of SW propagation is small, and the relaxation mechanism of SW acceleration in the given experimental conditions does not work. This conclusion is ascertained by the direct numerical modelling of SW propagation through a discharge plasma with parameters corresponding to various values of the time delay. The model includes the system of gasdynamic equations coupled with an equation for the vibrational energy (see Mnatsakanyan et al. 1987).

The N-wave was generated by forward and backward movement of a piston, i.e. of the left boundary of the computational region. The piston's law of motion was chosen so that the N-wave intensity variation along distance in the uniform gas resembled the experimental one. The N-wave thus formed came into the discharge region with definite parameters, calculated beforehand for the given time delay t_d. The discharge boundary was prescribed by some smooth distribution of T and ε_V along the x-axis. Calculated profiles of pressure (Fig.3,a) are typical for SW propagation through a thermal nonuniformity (Mnatsakanyan et al. 1987). Profiles of vibrational energy (Fig.3,b) are displaced by the gas motion behind the SW front without changing their maximum values (the relaxation is insignificant). Thus, the dependence of the SW velocity on time delay t_d (curve 3 in Fig.1) is entirely determined by the time evolution of the gas temperature in the developing discharge.

4. Conclusion

The nonmonotonous dependence of the SW velocity on the time delay between switching-on the discharge and SW generation, as observed by Chutov et al. (1991) does not produce unambiguous evidence about the essential role of the relaxation mechanism for SW acceleration. A more definite conclusion about its role can be drawn by studying the corresponding dependence of the SW intensity which, according to Mnatsakanyan et al. (1987), differs in the cases of dominant relaxation or thermal mechanisms.

References

Berdyshev AV, Kochetov IV, Napartovich AP (1988) Simplified model of vibrational kinetics for CO_2 lasers. Khim. Fizika 7: 470 (in Russian)

Cebeci T, Bradshaw P (1984) Computational aspects of convective heat transfer. Springer, Berlin

Chutov YuI, Podolsky VN, Brayon DA (1991) Shock waves in developing discharge. In: Proc. XIX Conf. Phenom. Ionized Gases, Pisa, pp 550-551

Devyatov AM, Kuzovnikov AA, Lodinev VV, Shibkov VM (1991) On the mechanism of molecular gas heating in impulse microwave discharge. Vestnik MGU, Fiz. 32: 29 (in Russian)

Gershenzon YuM, Rozenshtein VB, Umansky SYa (1977) Heterogeneous relaxation of molecular vibrations. In: Plasma chemistry, Vol. 4, Atomizdat, Moscow, pp 61-97 (in Russian)

Likalter AA, Naidis GV (1981) Vibrational distributions in strongly excited molecular gases. In: Plasma Chemistry, vol.8, Energoizdat, Moscow, pp 156-189 (in Russian)

Mnatsakanyan AKh, Naidis GV (1976) Dependence of electron energy distribution function in nitrogen on vibrational temperature and ionization degree. Fiz. Plazmy 2: 152 (in Russian)

Mnatsakanyan AKh, Naidis GV heating (1987) Shock wave propagation through nonuniform and nonequilibrium gas regions. In: Grönig H (ed) Proc. 17th Sypm. on Shock Tubes and Waves, Aachen, pp 201-205

Shock Wave Structure in a High-Frequency Discharge Plasma

S.A. Bystrov, E.N. Fomenko, F.V. Shugaev and G.I. Shved
Moscow University, 119899 Moscow, Russia

Abstract. Shock wave structure has been investigated in a low-temperature slightly ionized plasma of a transverse HF ($f = 13.6$ MHz) gas discharge. The experiments were carried out in a shock tube ($P_0 = 650$ Pa, $M = 2 \cdots 5$). Ar, CO_2 and N_2 were chosen as test gases. The gas density was measured by a laser Schlieren technique, the intensity of IR radiation ($\lambda = 2.7\,\mu$) was used to measure the vibrational temperature of carbon dioxide. It is found that the shock has a two-step structure. The vibrational temperature of carbon dioxide in the post-shock flow exceeds its equilibrium value. A precursor has been observed ahead of the shock in CO_2. The radiation intensity ($\lambda = 0.4\,\mu$) of the precursor diminishes in the direction towards the shock front.

Key words: Shock, Plasma, Discharge

1. Introduction

The propagation of shock waves through a gas discharge is of practical interest in the development of gas lasers, plasma-chemical reactors and other similar devices. Shock waves may arise in such systems near zones of energy release. The process under consideration here involves two independent phenomena: the motion of a shock wave and a gas discharge. The structure of a shock wave in a fully ionized plasma has been investigated theoretically by Shafranov (1957). Some experimental results on the propagation of a shock wave through a slightly ionized plasma have been reported by several authors (Klimov et al. 1982, Yevt'ukin et al. 1984). In the present work the post-shock flow was very close to a one-dimensional motion.

2. Experimental apparatus and measurements

The experiments were carried out in a single-diaphragm shock tube of rectangular 41×60 mm^2 cross-section. The test chamber was dielectric. It had glass windows of 60 mm dia. Two metal plates were mounted on the upper and lower walls of the test chamber. A transverse discharge was created between the plates by means of a high-frequency generator ($f = 13.6$ MHz), with a power of 3 kW. The current density was 40 mA. The length of the discharge zone was 80 mm. The density of the post-shock flow was measured by a laser-Schlieren technique (Kiefer et al. 1980). We used a He-Ne laser with a beam diameter of 1 mm.

The shock velocity was measured with pressure transducers beyond the discharge zone and from the Schlieren signals within the discharge zone. The initial translation temperature was measured with the aid of a Fabry-Perot interferometer. The vibrational temperature of CO_2 was determined from the infrared radiation by the reference intensity technique. The electron density was measured by a microwave interferometer ($\lambda = 8$ mm). The radiation passed through a slit of 1 mm width and fell on a detector.

Shock Waves @ Marseille II
Editors: R. Brun, L. Z. Dumitrescu

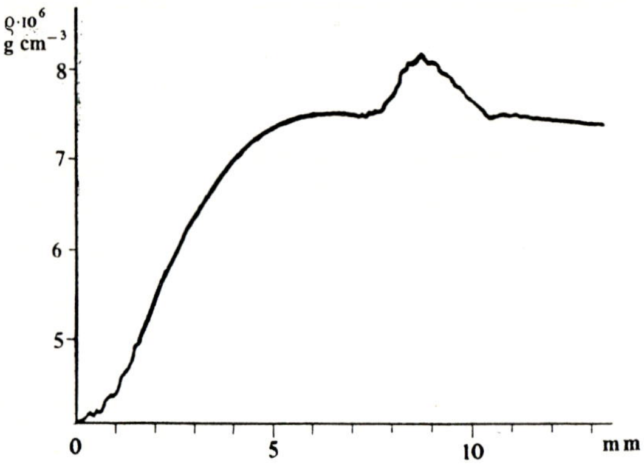

Fig. 1. N_2, $P_0 = 650$ Pa, $V_0 = 1.67$ km/s

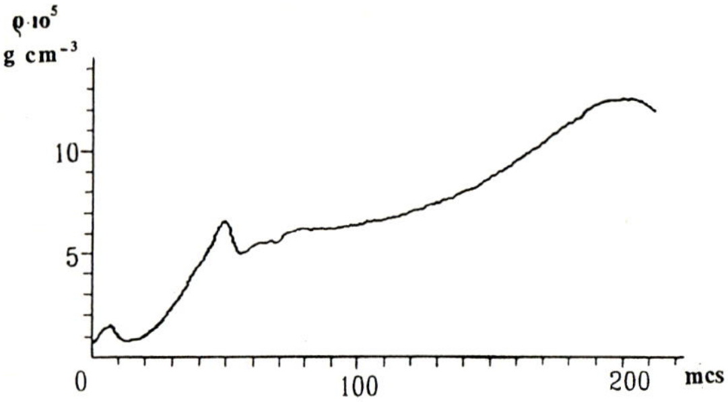

Fig. 2.

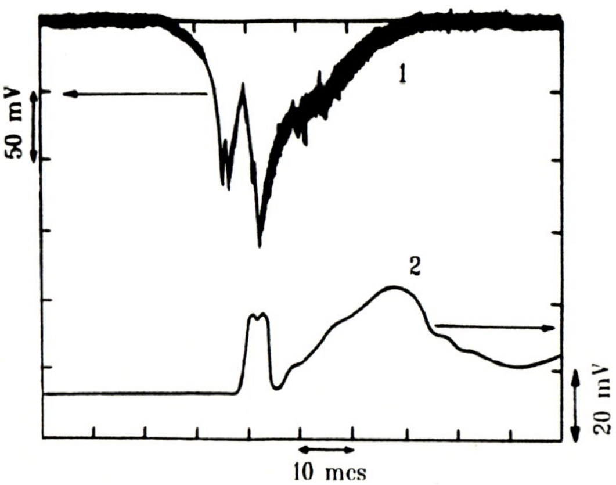

Fig. 3. 1 - precursor radiation; 2 - Schlieren signal

3. Results and discussion

The electron number density was equal to 1.3×10^{10} cm^{-3} ahead of the shock front, and 2.5×10^{11} cm^{-3} behind the shock front (CO_2 as test gas, $V_0 = 1.2$ km/s, the velocity of the shock out of the discharge zone). The initial translational temperature of the test gas was equal to 1200 K, the initial vibrational temperature of CO_2 was 2000 K. In order to obtain the density distribution we had to solve a convolution-type linear integal equation by the method of regularization developed by A. N. Tikhonov. It was found that the density distribution in the post-shock flow was steady inside the discharge zone. The shock waves, both in inert gas (Ar) and in molecular gases (CO_2, N_2) displayed a two-step structure (Fig.1). This means that the structure of the shock waves in our experiments depends mainly upon the processes of ionization and recombination, but not on the excitation of the internal degrees of freedom of the molecules. A similar structure is observed when a shock wave moves in a mixture of light and heavy gases (Raines 1991). The shock thickness in plasma is greater than in gas. Fig.2 shows that the density distribution in the post-shock flow is non-monotonous. The density first decreases and then increases. This phenomenon occurs due to the energy release behind the shock front. The magnitude of the density jump across the shock is less than the predicted value, if one does not take into account the above-mentioned energy release.

A precursor was observed ahead of the shock front in CO_2 (Fig.3). The precursor radiation ($\lambda = 0.4\ \mu$) falls in the direction towards the shock front, in contrast with the case of strong shock waves in gases (Zeldovich and Raizer 1967). The emergence of a precursor may be caused by an increase in the number of excited molecules due to the motion of surplus electron charges in front of the shock.

The radiation from the band of 0.4 μ is known to arise during electrical discharges in carbon dioxide, the excited state being $^1\Delta_u$. The molecule of carbon dioxide is considerably bent in the excited state. Transitions from vibrational levels which correpond to the minimum of potential energy of the excited state to very high vibrational levels of the ground state involve radiation from the above-mentioned band (Hertzberg 1966).

In our case the radiation arises probably because of the process of recombination. The decrease of the intensity of radiation immediately ahead of the shock front may be explained as follows: It is known that there is a broad region of elevated electron temperature ahead of the shock front in a plasma. The density of the plasma is almost constant in that region. At moderate Mach numbers the electron temperature rises slowly at first, and then more and more rapidly near the shock front. The heating of the electrons ahead of the shock must involve the decrease of recombination and consequently the decrease of the intensity of radiation (Derzhiev et al. 1986). This fact takes place when excited levels are populated during processes of recombination.

4. Conclusions

An experimental investigation of plane shock waves in a transverse HF gas discharge has been performed. The conclusions are as follows:

(1) The shock waves are steady in the gas discharge (the energy release behind the shock wave did not exceed 1 W/cm^2).

(2) Shock waves, both in inert and in molecular gases, display a two-step structure.

(3) The shock wave thickness is greater in plasma than in gas.

(4) The density distribution in the post-shock flow is non-monotonous. Namely, the density has its minimum at some distance from the shock front.

(5) A precursor was observed in CO_2. The intensity of its radiation ($\lambda = 0.4\ \mu$) decreases in the direction towards the shock front. This phenomenon is apparently due to the fact that the radiation is a recombinational one. The heating of electrons ahead of the shock front causes the decrease of recombination and, as a consequence, the decrease of the intensity of radiation.

References

Derzhiev VI, Zhidkov AG, Yakovlenko SI (1986) Ion radiation in a non-equilibrium dense plasma. Energoizdat, Moscow (in Russian)

Hertzberg G (1966) Molecular spectra and molecular structure. 3. Electronic spectra and electronic structure of polyatomic molecules. Van Nostrand, New York

Kiefer JN, Hayduc JC (1980) Rate measurements in shock waves with the laser-Schlieren technique. In: Lifshitz A, Rom J (eds) Proc. 12th Intl. Symp. on Shock Tubes and Waves, Magnes Press, Jerusalem, p 97-110

Klimov AI, Koblov AN, Mishin GI, Serov YuL, Yavor IP (1982) Propagation of shock waves through glow discharge plasma. Pis'ma Zh. Tekhn. Fiz. 8:439 (in Russian)

Raines AA (1991) Numerical solution of the Boltzmann equation for the one-dimensional problem in a binary gas mixture. In: Beylich A (ed) Proc. 17th Intl. RGD Symp., VCH, Weinheim, pp 328-331

Shafanov VD (1957) Shock wave structure in a fully ionized plasma. Zh. Exp. Theor. Fiz. 32:1453 (in Russian)

Yevtukin NV, Margolin AD, Shmeliov VM (1984) On the increase of shock wave velocity in a glow discharge plasma. Zh. Khim. Fiz. 3:1322 (in Russian)

Zeldovich YaB, Raizer YuP (1967) Physics of Shock Waves and High-Temperature Hydrodynamic Phenomena. Academic Press

Shock Waves and Flows in UV Laser-Produced Plasmas

Th. Sarnet, J.E. Montagne, G. Inglesakis and M. Autric
Institute of Fluid Mechanics - Laser-Matter Interaction Group. Parc scientifique et technologique de Luminy, 13009 Marseille, France

Abstract. An experimental study of the flow induced by a high-energy long pulse KrF excimer laser in front of metallic samples is described in this paper. The samples (aluminum and titanium alloys) have been irradiated with an energy density from 5 J/cm^2 up to 130 J/cm^2 with a pulse duration τ=300 ns and at different ambient pressures (from 5.10^{-3} to 10^5 Pa) for different background gases (air, nitrogen, argon, helium). We set up a Mach-Zehnder interferometer, using a cw argon laser, and a fast image converter camera to visualize the plasma flow. The absorption of the 5145 Å argon wavelength by the neutral titanium vapor plume has permitted to visualize simultaneously the expansion of the plasma flow, the formation of the shock wave and the neutral titanium vapor.

Key words: Excimer laser, Plasma, Hydrodynamic, Interferometry

1. Introduction

Laser-material processing such as welding, cutting, drilling, surface treatment or shock hardening involves complex phenomena including, at sufficient irradiance, the ionisation of the vapor above the material, leading to the formation of a plasma. The coupling of the laser energy to the material is strongly dependent on the characteristics of this plasma created above the irradiated surface. The dynamics of the plasma is a function of: the laser parameters (wavelength, pulse duration and shape), the incident flux level, the pressure and nature of the ambient surrounding gas and the material. This paper presents an experimental study concerning the different dynamics of a UV laser-produced plasma, using a time-resolved interferometry method and a high-speed camera.

2. Experimental set-up

The experiments have been performed using a KrF laser (KIR) built at IMFM by the Group. This laser delivers pulses of 400 ns duration with a maximum energy of about 200 Joules at λ=248 nm. A double-sided pumping scheme has been adopted using two cold-cathode diodes with "cheese grater emitters", each one powered by a six-stage, 600 kV, 36 kJ fast-rise Marx generator.

For these experiments the laser was configured with a 80 mm diameter stable resonator which produced a pulse width of 300 ns and an output energy of about 20 J on the target. This allowed the irradiation of the samples with an energy density from 5 J/cm2 to 130 J/cm^2 (i.e. 17 to 430 MW/cm^2). The laser diagnostic system consists in a calorimeter for the energy and a fast photodiode for the temporal evolution of the pulse.

The samples investigated were aluminum (AU4G/2017A) and titanium (TA6V/Ti-6Al-4V) alloys. A polishing of the surface was made in order to obtain a roughness $Ra < 0.1$ μm. The samples have been irradiated under different ambient atmospheric pressures (10^{-2} to 10^5 Pa) and with different background gases (nitrogen, argon, helium) within the experimental chamber.

A Mach-Zehnder interferometer using a cw argon laser at a wavelength of 514.5 nm has been set up (Fig.1). The fringe shift induced by both the shock wave and the plasma flow has been recorded with a fast converter camera (Imacon 790) allowing a framing rate up to 2.10^7 photographs/sec. and 1-100 ns/mm in streak configuration.

Shock Waves @ Marseille II
Editors: R. Brun, L. Z. Dumitrescu

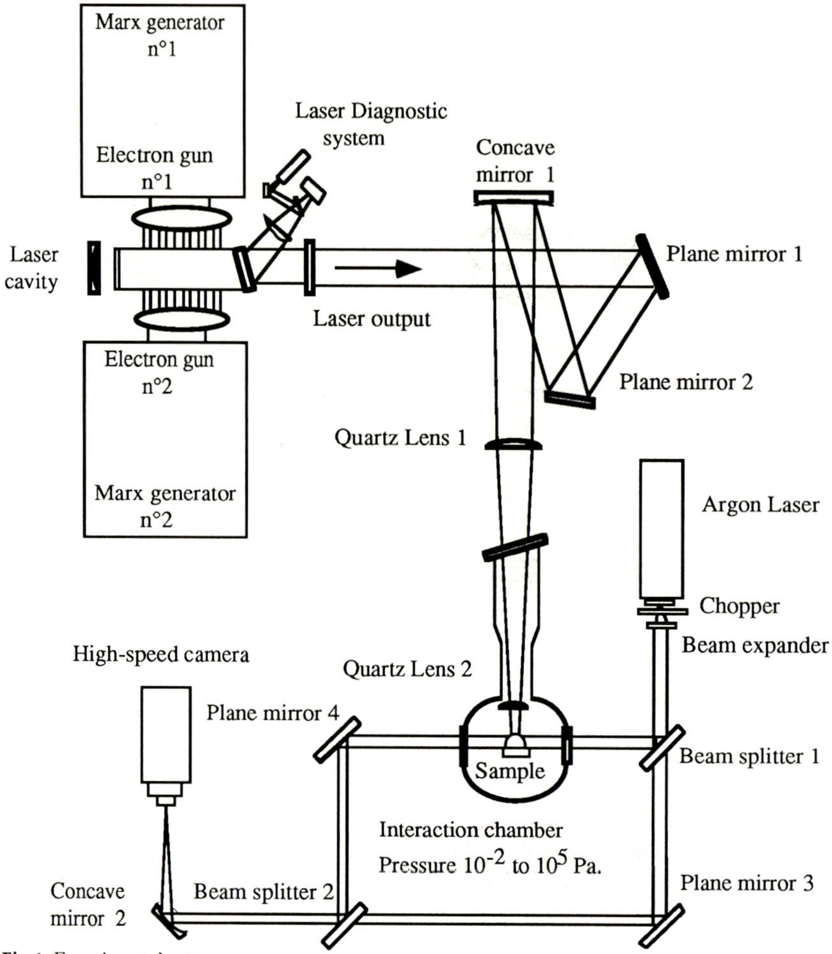

Fig. 1. Experimental set-up

3. Discussion

Both axial and radial expansion velocities of the plasma were determined from these interferograms. It has been shown that the buffer gas has an important influence on the plasma flow: the average velocity during the first microsecond can be increased from $V = 5.10^3$ m.s^{-1} in the atmospheric air to $V = 1.10^4$ m.s^{-1} if helium is the surrounding gas at the same pressure ($P = 10^5$ Pa, F=90 J/cm^2).

Different regimes of the flowfield have been put into evidence depending on the ambient gas pressure:

$10^5 > P > 10^3$ Pa

At these high pressures ($P_{atm} > P > 10^3$ Pa), laser absorption waves and spherical blast waves have been observed in all cases. However, due to the differences in the ionisation mechanism, the electron density is maximum in the air where a typical dense plasma is formed close to the sample surface. Under helium background, the electron density is lower permitting to follow the fringe shift near the surface.

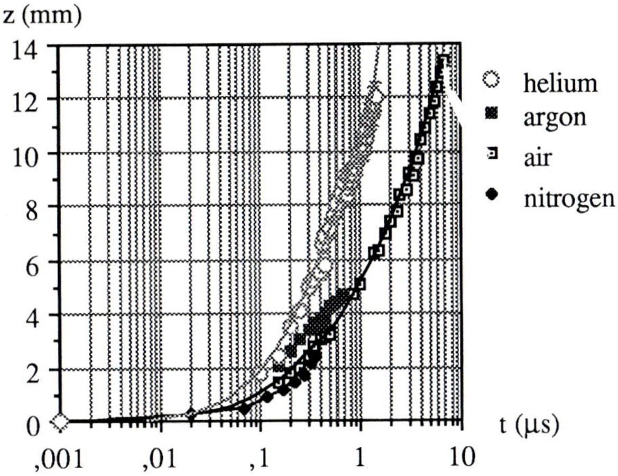

Fig. 2. Experimental axial propagation of the wave front versus time with different surrounding gases. Sample: TA6V, Pressure $P = 10^5$ Pa, $F = 90$ J/cm^2

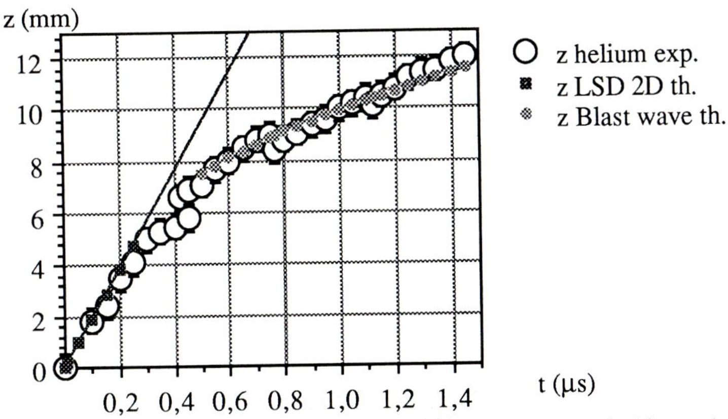

Fig. 3. Theoretical (th.) LSD regime and spherical blast wave compared with experimental (exp.) propagation. Sample: TA6V, Pressure $P = 10^5$ Pa (Helium), $F = 90$ J/cm^2

The flow is correctly described by a two-dimensional Laser Supported Detonation wave (LSD 2D) under air, helium or nitrogen. The detonation wave velocities are calculated using the relation given first by Raizer (1965):

$$V_{\mathrm{LSD2D}} = \frac{1}{2}\left[2\left(\gamma^2 - 1\right)\frac{I_0}{\rho_0}\right]^{1/3}$$

where I_0 is the incident flux, ρ_0 the ambient density in front of the wave and γ the specific heat ratio, while

$$Z_{\mathrm{Blastwave}} = \lambda_0 \left[\frac{E}{\rho_0}\right]^{0.2} t^{0.4} \qquad \text{with } \lambda_0 = \left[\frac{75(\gamma-1)(\gamma+1)^2}{16\pi(3\gamma-1)}\right]^{1/(\gamma+3)}$$

However, under argon the axial velocity of the leading edge is higher than the velocity predicted by the LSD 2D model.

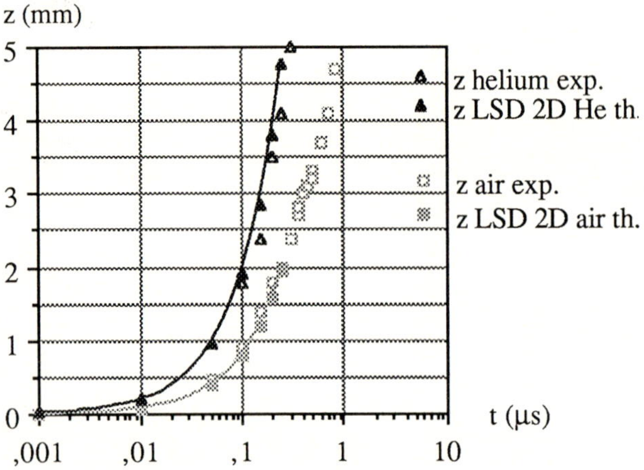

Fig. 4. Theoretical (th.) and experimental (exp.) LSD 2D propagation under helium and air. Sample: TA6V, Pressure $P = 10^5$ Pa, $F = 90$ J/cm^2

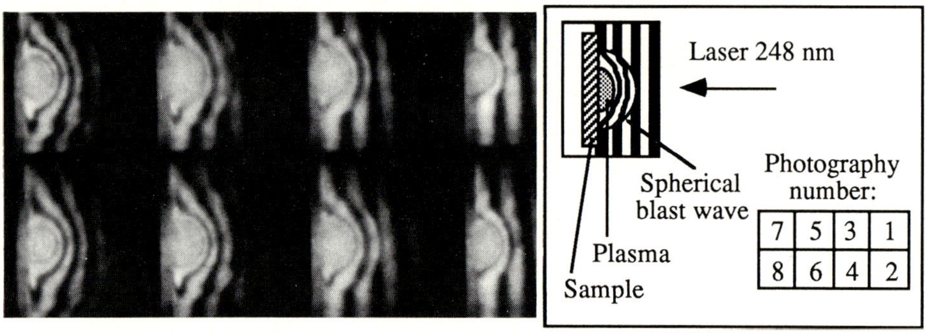

Fig. 5. Photographs showing the plasma flow and the expansion of a spherical blast wave. Interframe time: 50 ns, titanium alloy sample, $P = 10^2$ Pa, Fluence: 102 J/cm^2

$P = 10^2$ Pa

The formation of a spherical shock wave, completely transparent to the incident radiation (blast wave) has been evidenced in a rarefied atmosphere for pressures close to $P = 10^2$ Pa, with a velocity $V > 1.5.10^4$ m.s^{-1} (energy density= 90 J/cm^2).

This wave is a result of the ambient air compression due to the ejection of particles from the surface sample. These blast waves have been studied theoretically (Sedov 1959) and are usually attributed to the large amount of energy deposition in an infinitesimally small region of the gaseous medium.

It has been shown, in CO_2 laser produced plasmas, that some projections called "Aneurisms" (Stamper 1988, Ghrab 1992) can be observed, at these pressures. In our experiments, no aneurisms have been evidenced and the blast front produced can be considered as hydrodynamically stable (Fig.5).

$5.10^{-3} < P < 10^1$ Pa

At lower pressure ($5.10^{-3} < P < 10^1$ Pa), the collisional transfer of the energy from the sample to the ambient gas was not sufficient to form this spherical shock wave; therefore, material vapor

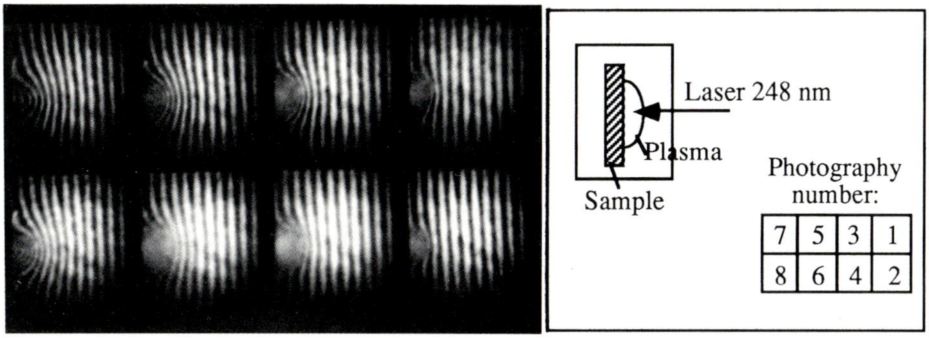

Fig. 6. Interferogram of a plasma propagation above a metallic sample under vacuum. Sample: TA6V, Pressure $P = 10^{-2}$ Pa, $F = 90$ J/cm^2. Interframe time 50 ns

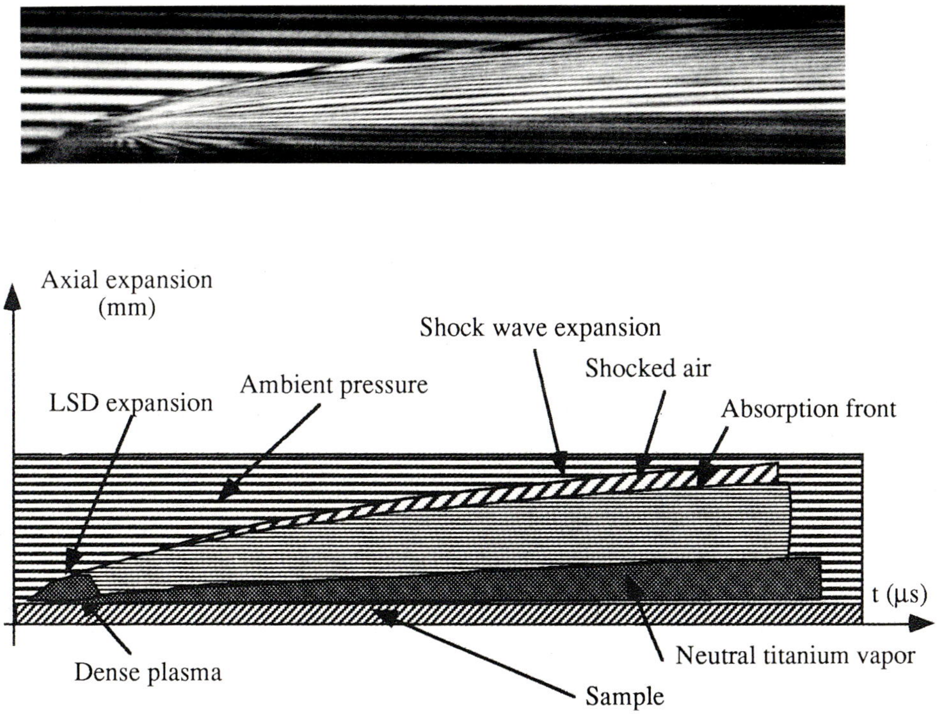

Fig. 7. Up: Streak photography of the plasma propagation. Sample: TA6V, Pressure $P = 10^5$ Pa, $F = 90$ J/cm^2, Down: illustration of the phenomena with the different regimes of expansion

expansion under vacuum becomes the main process. The fringe shift induced by the plasma flow allows to follow the density during the expansion (Fig.6). It has been shown that the illumination of a plasma with a laser tuned on a ground state neutral transition of the sample allows the visualization of the neutral plume (Gildenbach 1991, Ventzek 1992). The dynamics of the titanium neutral plume was observed by laser resonance absorption photography using the 514.5 nm wavelength of the Mach-Zehnder interferometer.

In this case, the tuning with the Ti absorption line is $\Delta\lambda = 0.01$ nm. Combining this technique with the interferometric method has permitted to visualize simultaneously the shock wave, the plasma flow and the neutral titanium vapor flow on a single shot (Fig.7).

4. Conclusions

Different hydrodynamic regimes have been evidenced by irradiating metallic samples with a long-pulse KrF laser. For high pressures ($10^5 > P > 10^3$ Pa), the Laser Supported Detonation Waves theory agrees with the experimental data. Blast waves have been observed when the pressure decreases to $P = 10^2$ Pa. Free expansion under vacuum occurs when the pressure reaches $P = 10$ Pa and below. Moreover these experiments have permitted the simultaneous visualization of LSD waves, shock waves, neutral titanium plume and plasma expansion.

All these phenomena, concerning the UV laser-matter interaction, complete the experimental and theoretical studies concerning the CO_2 laser-matter interaction field that have been carried out by the group for many years.

References

Ghrab M et. al (1992) 4th Intl. Conf. of Fluid Mechanics, Alexandria, Egypt, Vol. 3, pp 319
Gildenbach RM et al. (1991) Appl. Phys. Lett., 58: 1597
Raizer YuP et al. (1965) Heating of a gas by a powerful light pulse. Sov. Phys. JETP 21: 1009
Reilly JP et al. (1979) Modelling of momentum transfer to a surface by laser supported absorption wave. AIAA Journal, 17, 10: 1098
Sedov L (1959) Similarity and Dimensional Methods in Mechanics. Academic Press
Stamper J et al. (1988) Aneurisms in laser-driven blast waves. Phys. Fluids 31, 11: 3353-3361
Ventzek PLG et al. (1992) Appl. Phys. Lett., 72: 1696, 1992
Weyl G et al. (1981) AIAA Journal 19: 440

Influence of Shock Waves on the Beam Quality of a High Pulse Repetition Frequency XeCl Laser

O.P. Uteza, M.L. Sentis, Ph.C. Delaporte, B.M. Forestier and B.L. Fontaine
Institut de Mécanique des Fluides de Marseille, UM 34 CNRS, Marseille, France

Abstract. The influence of acoustic perturbations of an active disturbing medium on the far-field diffraction pattern is both theoretically and experimentally investigated. The theoretical results are in good agreement with experiments performed with a high repetition rate excimer laser ($\lambda = 308$ nm, PRF = 500 Hz). This theoretical approach explains the decrease of laser energy and fluence and the distortion of the beam shape in the far-field occurring in high pulse repetition frequency excimer lasers.

Key words: Excimer laser, Repetition rate, Beam quality, Far-field

1. Introduction

Rare-gas halide excimer lasers are known to be efficient sources of high power ultraviolet radiation for a wide range of scientific, industrial and medical applications (Niino et al. 1990; Taylor et al. 1990). Since the effectiveness of the laser interaction depends strongly on the fluence and the spatial distribution of the laser radiation, low beam divergence and optimum beam uniformity are required.

For high pulse repetition frequency (PRF) excimer lasers, the output characteristics, such as pulse energy and beam profile homogeneity, are very sensitive to discharge stability and uniformity. Indeed, acoustic fluctuations generated by the fast electrical energy deposition in the medium may strongly decrease laser energy and fluence in the far-field (Sentis et al. 1991; Baranov et al. 1988). Furthermore, the perturbations induce refractive index disturbances which degrade the far-field beam profile (Siegman 1986; Sutton 1969).

Calculations and experimental results have been obtained for the laser LUX (Laser Ultraviolet preionized by X rays) device with the LASERDOT 430 UV laser head.

2. Theoretical study

A complete and extensive description of the theoretical study is presented elsewhere (Uteza et al. 1993). Hereafter, we will describe the most important features.

First, the resolution of the wave equation in an optically inhomogeneous medium disturbed by a pressure field allows us to determine the amplitude of the electric field on the exit mirror. Then, the far-field diffraction pattern is obtained by taking the Fourier transform of the amplitude distribution of the electric field on the exit mirror.

The results show that the density perturbations along the integration axis (Oz) induce constant aberrations which have a minor influence on the far-field beam profile. On the other hand, it is not the case of the acoustic waves between electrodes. Indeed, they lead to amplitude and phase aberrations of the electric field which are very injurious to the beam quality. Their influence will be discussed in the experimental study. Lastly, the acoustic gradients along the flow axis (Ox) are sufficiently damped and do not degrade the beam quality in the far-field.

Shock Waves @ Marseille II
Editors: R. Brun, L. Z. Dumitrescu

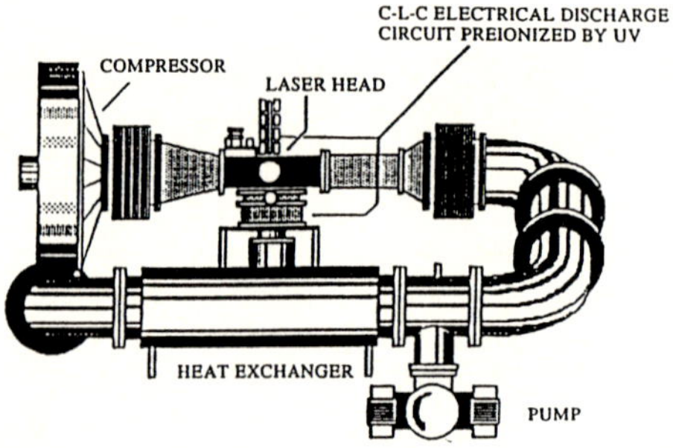

Fig. 1. LUX test-bed schematic

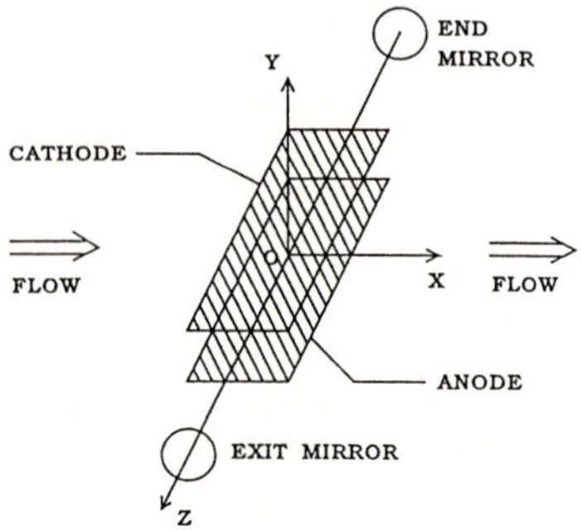

Fig. 2. Laser head schematic

3. Experimental apparatus and conditions

The XeCl excimer laser used in our experiments is depicted on Figs.1,2 and has been extensively described in previous papers (Sentis et al. 1991). A subsonic flow loop removes the gas between two successive shots and the flow speed (50 m/s) is sufficient to ensure a clearing ratio of 5 at 500 Hz. The electrical discharge circuit consists of a UV preionization of corona type and of a thyratron-switched C-L-C transfer circuit with nominal primary and secondary capacitances of 112 and 90 nF respectively for the main discharge. The electrodes are 30 cm long with a separation of 3 cm and the discharge is 2 cm wide. The mixture was made up of 99.37% Ne, 0.5% Xe, 0.13% HCl at a total pressure of 2.8 bars. The laser head uses a stable plane - parallel resonator geometry with a 40% reflecting output coupler. In single shot operation, a 300 mJ output laser beam of 50 ns duration (FWHM) and of 2×3 cm^2 cross section was obtained for an energy deposition of 50 J/l.bar.

Experimental beam diagnostics were performed by a Spiricon LBA 100 A laser beam analyser associated with a Pulnix TM 745 CCD camera. In order to reduce the laser beam divergence, two lenses are used to produce a plane-parallel beam from the laser source. The far-field conditions are then simulated by a third lens which focuses the beam.

4. Results

We realized bursts of 100 shots with flow at PRF : 50 Hz, 125 Hz, 250 Hz and 500 Hz and for an energy deposition of 50 J/l.bar. The flow evacuates the heated gas volume between two successive shots and the bursts are too short to let a strong temperature rise near the electrodes. In these conditions, thermal gradients have a minor influence on the beam characteristics compared to acoustic perturbations.

Fig.3 shows a single shot beam profile which is used as reference shot representative of a medium without perturbations. It also corresponds to the ideal case presented on the theoretical figures (Figs.5,9).

Fig.4a,b describe the evolution of means of laser energy and beam aera measured at $1/e^2$ of energy and calculated from bursts versus PRF. The comparison of these two figures shows that the fluence fall is not attributed to a beam spreading but to a laser energy decrease. So, how pressure variations induce the laser energy fall can be explained by two points:

- the first and principal reason has an electric origin. Due to working at high PRF, the transverse waves between optics create gradients parallel to the electrodes. They lead to spatial density distortion which locally affects the uniformity of the energy deposition and that could

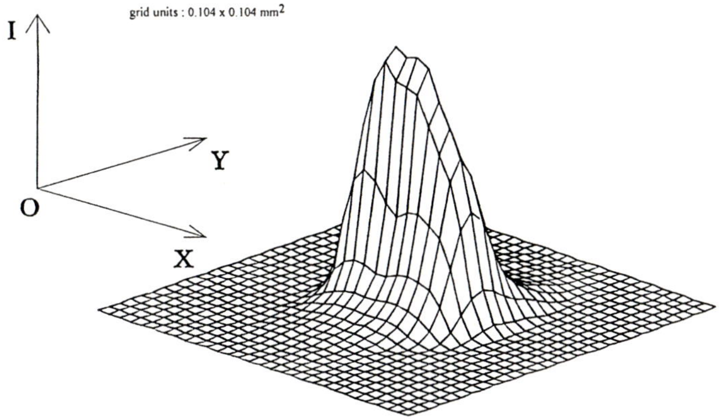

Fig. 3. Experimental far-field diffraction pattern of a single shot (reference shot)

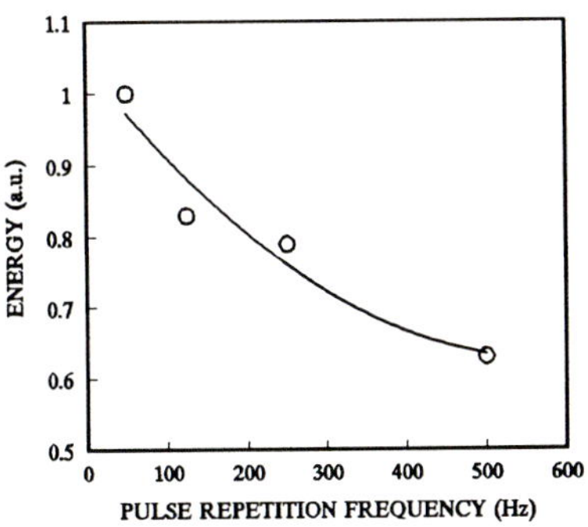

Fig. 4a. Laser mean energy versus PRF

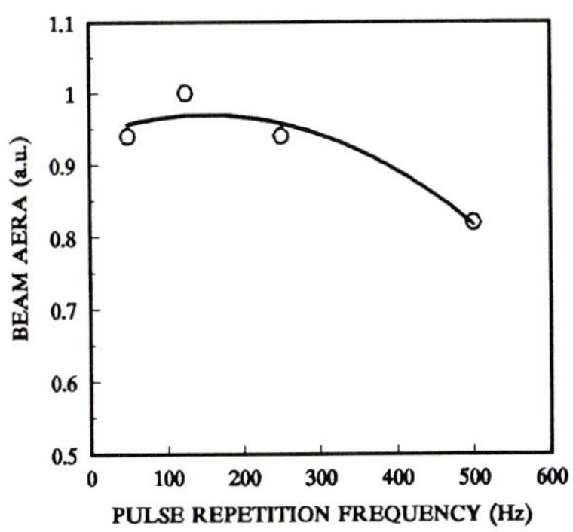

Fig. 4b. Beam area versus PRF

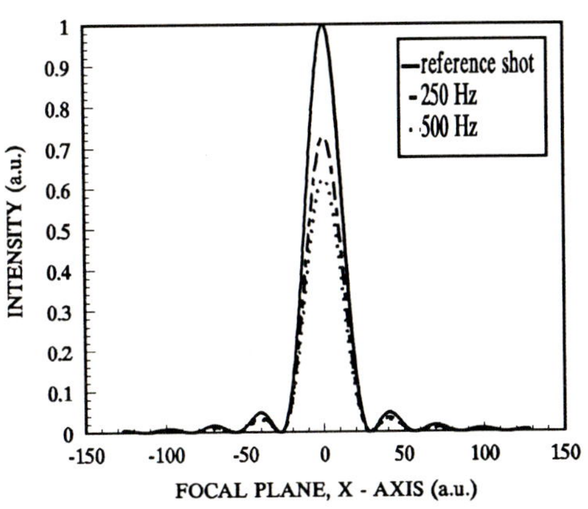

Fig. 5. Theoretical far-field diffraction pattern showing the effects of the amplitude aberrations due to the transverse waves between electrodes at 250 and 500 Hz

induce even the establishment of arcs. As the discharge instability amplifies clearly according to PRF, it is logical to record a decrease of the total output pulse laser energy with PRF increase.

- the second reason comes from an optical origin. Acoustic gradients perpendicular to the electrodes are then important. This result is confirmed by the model which shows that the acoustic gradients between optics induce no perturbations in the expression of the electric field. So, only the transverse waves between electrodes generate severe index variations and consequently aberrations which attenuate the beam, affecting the laser pulse energy in the far-field (Fig.5).

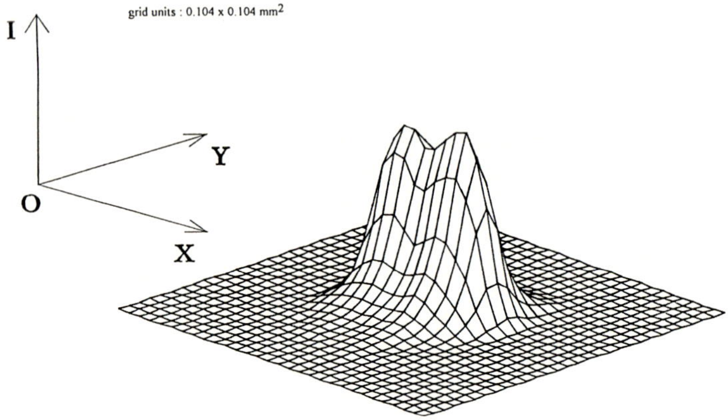

Fig. 6. Experimental far-field diffraction pattern of a representative shot of a burst at 250 Hz with flow

Careful analysis of beam profiles at high PRF also suggests that PRF increase affects the beam shape. Indeed, Figs.6,7 show the apparition of a second peak in the central part of the focused spot.

Measurements of the peak position fluctuations around its central position (Fig.8) even demonstrate that the peak location cannot be known precisely. Fig.8 confirms this phenomenon but also shows that PRF effects, energy shift and uncertainty on peak location mainly, occur essentially along the focal plane x - axis until 250 Hz. At this point, we recall that, in accordance with the properties of the Fourier transform, the axes of the focal plane are inverted. Therefore, what we observe along the x-axis of the focal plane is the result of the influence of the perturbations along the y-axis of the cavity. So, to understand why the beam shape is degraded until 250 Hz along a single dimension mainly, we must remember that three wave families beating along their respective propagation direction are created by the discharge in the cavity. But, as we have seen earlier, longitudinal waves do not produce strong perturbations and the theoretical model demonstrates that pressure gradients along the optical axis due to the transverse waves between optics have no influence on the far-field diffraction pattern. But, pressure perturbations between electrodes induce phase aberrations along Oy with the same spatial period which is equal to the inter-electrode dimension. So, the diffraction angles produced by these phase aberrations are small enough to keep the energy within the main beam angle but sufficient to induce an energy shift along the x - axis of the focal plane (Siegman 1986). It is important to underline that Fig.9, which shows theoretically the influence of phase aberrations due to the transverse waves between electrodes, confirms qualitatively this result.

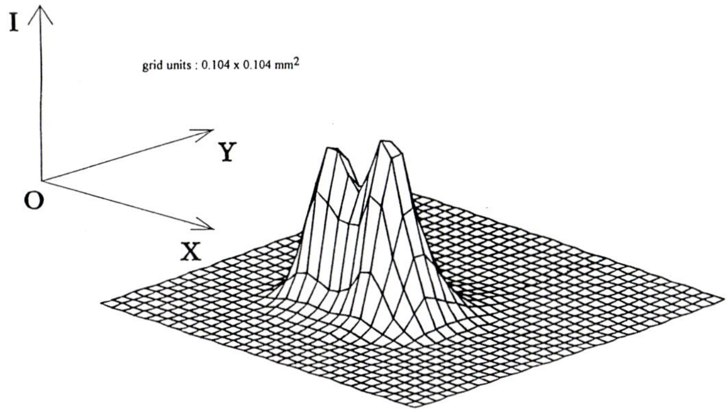

Fig. 7. Experimental far-field diffraction pattern of a representative shot of a burst at 500 Hz with flow

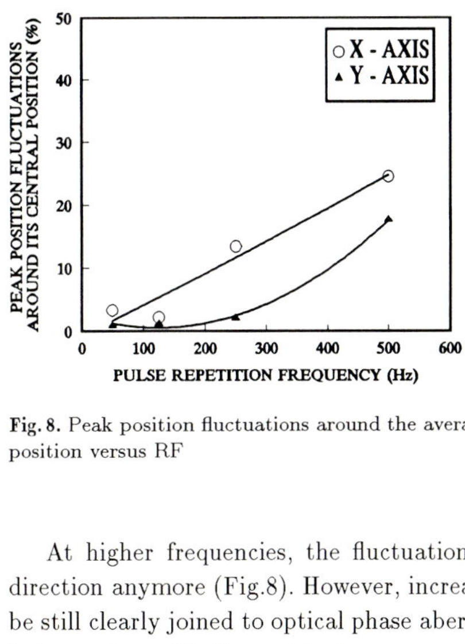

Fig. 8. Peak position fluctuations around the average position versus RF

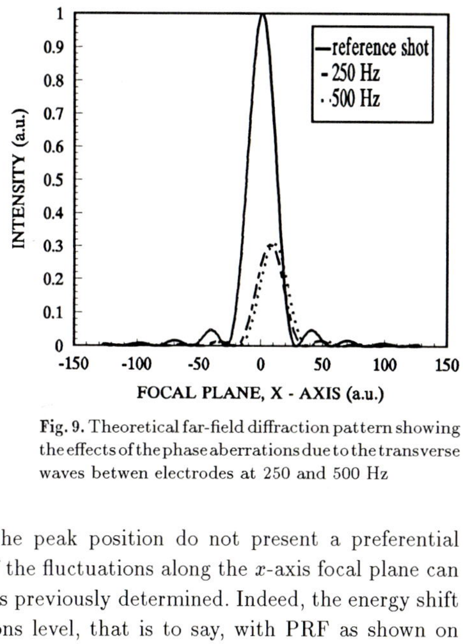

Fig. 9. Theoretical far-field diffraction pattern showing the effects of the phase aberrations due to the transverse waves betwen electrodes at 250 and 500 Hz

At higher frequencies, the fluctuations of the peak position do not present a preferential direction anymore (Fig.8). However, increases of the fluctuations along the x-axis focal plane can be still clearly joined to optical phase aberrations previously determined. Indeed, the energy shift is increasing with the residual pressure variations level, that is to say, with PRF as shown on Figs.7-9. The fluctuations along the y-axis of the focal plane can certainly be attributed to an increasing proportion of arcs. By producing strong local cylindrical pressure waves which expand in the flow axis, they create pressure gradients along a single direction, the x-axis of the cavity. As a consequence, they destroy the medium homogeneity inducing a random spatial repartition of the discharge and therefore a random peak position along the y-axis of the focal plane.

5. Conclusion

Theoretical results are well correlated with the experiments which allow to ascertain the effects of acoustic perturbations occurring in pulsed excimer lasers at high PRF on beam characteristics in the far-field.

First, the theoretical model clearly shows that acoustic gradients appearing along the optical axis produce no aberrations of the laser beam. But, they must not be neglected because they constitute the principal reason of the degradation of electrical discharge quality as soon as PRF is increased up to a few hundred Hz.

The second lesson we can draw from the theoretical study concerns the transverse waves between electrodes and the strong amplitude and phase aberrations they induce. They affect essentially the beam quality by reducing the far-field energy and fluence and seriously distorting the beam shape. This last result is due to phase aberrations and may be very injurious to the success of laser applications at high PRF. Unfortunately, this phenomenon is generally difficult to correct by simple optical elements. But, we think that the use of acoustic dampers to reduce the influence of the transverse waves between electrodes could improve efficiently the beam quality.

Acknowledgements

This work is supported by LASERDOT, French MESR and PACA region.

References

Baranov VYU, Borisov VM, Vinokhodov AYU, Kiryukhin YUB (1988) Divergence of radiation from an electric-discharge XeCl laser operating in the pulse-periodic regime. Sov. J. Quantum Electron. 18(9)

Niino H, Nakano M, Nagano S, Nitta H, Yano K, Yabe A (1990) Excimer laser ablation of polymers and carbon fiber composites. J. of Photopolymer Science and Technology 3:53-56

Sentis ML, Delaporte PhC, Forestier BM, Fontaine BL (1991) Design and characteristics of high pulse repetition rate and high average power excimer laser systems. IEEE J. of Quantum Electron. 27:2332-2339

Siegman AE (1986) Lasers. Oxford University Press, 698-743

Sutton GW (1969) Effect of turbulent fluctuations in an optically active fluid medium. AIAA J. 7:1737-1743

Taylor RS, Higginson LAJ, Leopold KE (1990) Dependence of the XeCl laser cut rate of plaque on the degree of calcification, laser fluence, and optical pulse duration. Lasers in Surgery and Medicine 10:414-419

Uteza OP, Sentis ML, Delaporte PhC, Forestier BM, Fontaine BL (1993) Laser beam quality of a high PRF XeCl laser. Optics Communications 102:523-531

Shock Wave Study in a VUV Formed-Ferrite Plasma Source

M.L. Sentis, F. Chazaud, L.H. Chinh, Ph. Delaporte and S. Moulins
Institut de Mécanique des Fluides de Marseille, UM 34 CNRS, Marseille, France

Abstract. A study of gas dynamic phenomena induced by an open discharge called "formed-ferrite plasma" (FFP) source that creates an intense radiation from the visible to the VUV region is presented. Fast side-on views of the plasma expansion were made by recording the overall visible emission from the plasma with a gated (40 ns) intensified CCD camera system. An experimental study of the plasma formation and expansion from the ferrite surface has been done as a function of the electrical energy deposition ranging from 100 J to 800 J for argon or nitrogen as background gases. We show that the propagation of the shock wave could be described accurately by the following expression : $R = kE^{\alpha}t^{n}$, where R is the distance of the shock front from the surface of the ferrite, E the deposited energy, and t the time. Comparisons with a drag model and a blast wave model are presented.

Key words: Surface discharge, Shock wave, Plasma

1. Introduction

Many industrial applications ranging from surface modification and material deposition to photolysis of molecules and laser pumping require efficient ultraviolet (UV) and vacuum ultraviolet (VUV) light sources. Commercially available VUV lamps produce rather weak intensities. This makes it important to investigate and develop new types of high power broadband radiation sources (Kogelschatz 1992).

Open discharges including exploding wire by an electrical discharge (Chase et al. 1959) and sliding arc discharges on various types of surfaces and materials (Beverly III 1986) show attractive characteristics. In this case, a maximum of the spectral distribution of photons falls generally within the UV or VUV spectral region where the strongest absorption bands of the overwhelming majority of molecules is located. Optical excitation of an active gas by the pulsed discharge radiation is feasible over a wide range of gas pressure from vacuum to several atmospheres with the possibility of long pulse operation (1 to 10 μs).

In this work, we report on gasdynamic phenomena induced by an open discharge called "formed-ferrite plasma" (FFP) source (Watanabe et al. 1987) that create an intense VUV (100 to 200 nm) radiation. This light source is of a great interest since it permits operation in a pulse-periodic regime.

The electrical energy deposition produces a highly emitting plasma column surrounded by a layer of a shock-compressed gas. These dynamic phenomena may disturb strongly the active medium in applications using a repetitive mode operation and require a good control of their influence on the active gas shot-to-shot. Thus it is necessary to characterize accurately both their formation and expansion into the surrounded volume cell.

2. Experimental apparatus and procedures

The ferrite rods (10×200 mm) are fixed by two aluminium electrodes into a vinyl chloride polymer (PVC) cell evacuated to better than 10^{-4} Torr and filled with argon or nitrogen up to 1.5 atm. The FFP source is driven by an electrical circuit which consists of an energy storage capacitor of 4 μF charged up to 27 kV and a N_2-pressurised spark gap switch. In order to initiate an electrical

Shock Waves @ Marseille II
Editors: R. Brun, L. Z. Dumitrescu

discharge on the ferrite surface a conductive channel called "formed-ferrite-filament" (FFF) has to be formed on the surface of the ferrite rod. Details on this "formation step" and modifications of the ferrite microstructure, that lead to a decrease of the material resistivity of 4 or 5 orders of magnitude, may be found elsewhere (Sentis et al. 1993).

The energy injected in the FFF and surrounding gas is measured by monitoring voltage and current waveforms with a high-voltage probe (Tektronix model P 6015) and a current transformer (Stangenes 3-0.01) respectively. Fig.1 shows the voltage (V), current (I) and deposited energy (E) waveforms for a charging voltage of 15 KV in 1 atm of argon. The voltage on the FFF remains constant and approximately equal to the charging voltage value until the ferrite impedance collapses. This interval time ($\approx$60 μs) permits the filament to be heated to very high temperatures causing thermal electrons and ion emission. Then the current becomes consistent as an intense VUV radiation of approximately 4.5 μs FWHM duration is detected. The peak corresponding to the maximum power deposition is about 30 MW. Due to impedance mismatch between the electrical circuit and the load, only 65% of the stored energy is injected ($\approx$ 285 J). Lowering the driver circuit inductance could drastically improve this percentage. Radiative characteristics of this source have been investigated in the 115 - 220 nm spectral region. (Sentis et al. 1993). Emission spectra display intense lines superimposed to a continuum due to three main mechanisms : electron-ion free-free transition, free-bound transition and electronic transitions.

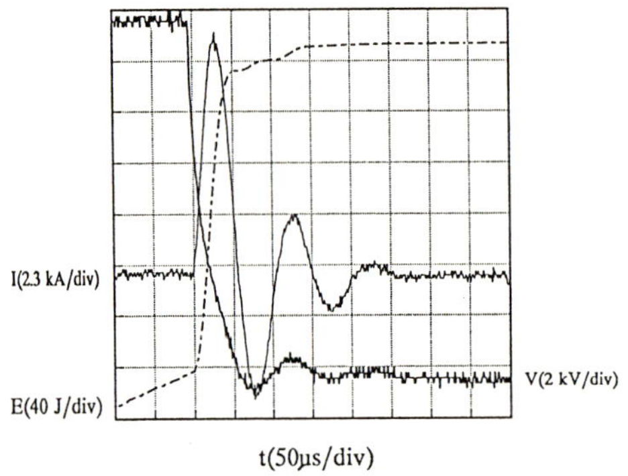

Fig. 1. Typical voltage (V), current (I) and deposited energy (E) waveforms for FFF plasma operated in 1 atm of argon with a charging voltage of 15 kV

Fast side-on views of the plasma expansion were made by recording the overall visible emission from the discharge with a gated, intensified CCD camera system (PCO model Dicam - 2). Fast (5 ns or longer) gate pulses (acting as the shutter for the camera) were supplied at variable times after the steady-state applied voltage had collapsed for "stop-action" images of the plasma expansion. For each exposure in these experiments, the gain of the microchannel plate of the detector is set just short of saturation to permit the record of both plasma and shock wave expansion. The digitized photographs are stored by computer and later analysed with image processing software.

3. Results and discussion

Fig.2 demonstrates the transverse plasma expansion initiated by an electrical discharge applied on the FFF for a charging voltage value of 19 kV in 1 atm of argon. The exposure and delay time are respectively 80 ns, 9.1 μs (a), and 13.2 μs (b). The electrical energy deposition produces a highly emitting plasma column of an approximately half-cylindrical form which expands radially. After about 4 μs, the plasma column is surrounded by a layer of a shock-compressed gas. Its visibility may be due to scattering of the plasma emission light or to emitting particles ejected from the ferrite channel and in movement with the shock compressed gas layer. A correlation with a spatial-resolved spectroscopic study would permit to bring a firm conclusion. Longitudinal photographs confirm a perfect uniformity of the plasma column along the length of the ferrite rod. The plasma axis does not correspond to the ferrite one because of the non rectilinearity of the FFF on the ferrite surface. More than one of shock fronts can be observed as the homogeneity of the FFF varies.

Position-time plots of leading edges of both plasma and shock wave emission from many ICCD images are given in Fig.3 for a fixed charging voltage value of 12 kV and a fixed argon pressure of 1 atm. The plasma remains confined in a 10 mm-width zone around the ferrite surface whereas the shock wave expands further and faster. During the first microseconds, the plasma velocity increases until 700 m/s to fall after 15 μs down to 200 m/s. After 20 μs, the emission plasma form turns into a non homogeneous plume whose intensity is getting lower. Plasma light is perceptible during more than 100 μs. A fit to the plasma expansion data is given in Fig.3 (solid line). Fits of the form $R = at^n$ suggests an exponent value of 0.5 with $a = 1.97$ mm $\mu s^{-0.5}$.

The shock front shows the same behaviour as the plasma expansion but with a velocity approximately twice the plasma one. A fit of the form $R = at^n$ suggests an exponent value of 0.75 with $a = 1.84$ mm $\mu s^{-0.75}$ (Fig.3, solid line).

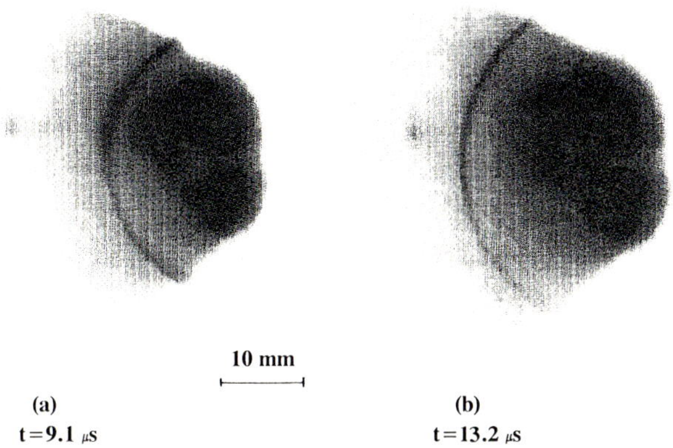

Fig. 2. Transversal plasma expansion in 1 atm argon and a charging voltage of 19 kV after a delay time of 9.1 μs (a) and 13.2 μs (b)

A blast wave model (Zel'dovich and Raizer, 1966) was developed to describe the propagation of a shock wave, caused by an explosive release of energy, E, through a background gas (density ρ_0). The propagation of the shock front follows the distance-time relation $R = \xi_0(E/{}_0)^{1/5}t^{2/5}$, where ξ_0 is a constant. This model strictly concerns instantaneous injection of energy, an assumption which is not verified in our experiment for which a part of the energy is primarily injected in the FFF and the other part in the gas surrounding the surface of the ferrite during a second time. However this model can be used here to describe the plasma expansion by functions of the form: $R = b(E/{}_0)^{\alpha}t^n$, as it will be developed further.

In Fig.4, a classical drag force model (dashed line) describes accurately the shock front propagation at early time ($0 < t < 10$ μs). The particles in motion are regarded as an ensemble which experiences a viscous force proportional to its velocity through the background gas. The equation

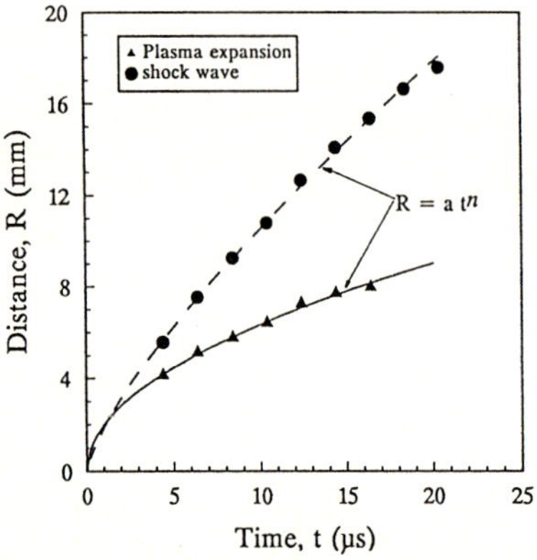

Fig. 3. $R - t$ plots of leading edges of plasma and shock wave emission for a fixed charging voltage value of 12 kV and a fixed argon pressure of 1 atm. A fit of the form $R = at^n$ is given for each set of experimental data

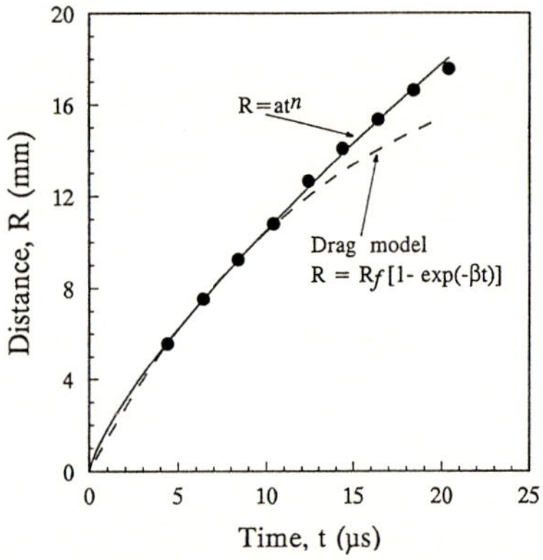

Fig. 4. $R - t$ plots of the leading edge of the shock wave emission for a fixed charging voltage value of 12 kV and a fixed argon pressure of 1 atm. A fit based on a classical drag force model is given (dashed line): $R = R_f(1 - \exp(-\beta t))$ with $R_f = 19.81$ mm and $\beta = 0.075\ \mu s^{-1}$

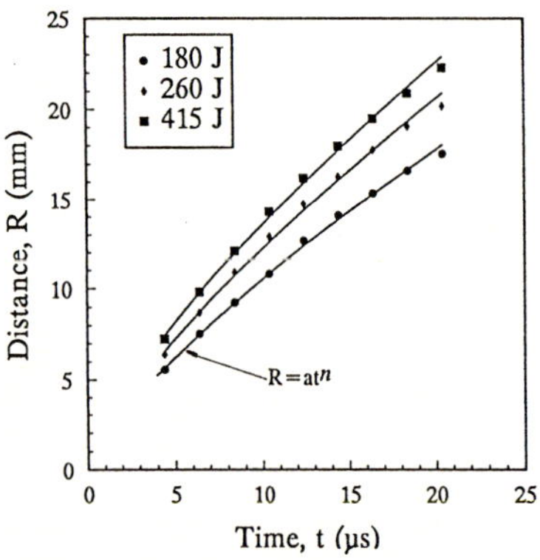

Fig. 5. $R - t$ plots of the shock front propagation for different values of the injected energy and fixed argon pressure of 1 atm. Solid lines correspond to functions of the form: $R = at^n$

of motion is $a = \beta v$, giving : $R = R_f\,[1 - \exp(-\beta t)]$, where β is the slowing coefficient. In Fig.4, the values of β and R_f are 0.075 μs^{-1} and 19.81 mm, respectively.

Fig.5 shows position-time plots of the shock front propagation for different values of the injected energy. The shock front expands faster and further as the injected energy increases. Solid lines correspond to functions of the form: $R = a\,t^n$.The value of the exponent n does not vary with E and is approximately 0.75, whereas the coefficient a increases from 1.842 to 2.52 for deposited energies of 180 J and 415 J respectively. Using the general equation of the blast wave model ($R = b(E/\rho_0)^\alpha t^n$), the coefficient a can be expressed by: $a = kE^\alpha$ where $k = 0.267$ and $a = 0.37$. Experiments have also been carried out for different values of argon pressure at a fixed charging voltage value. Lowering the density of the gas makes the propagation of both plasma and shock wave easier, in agreement with the corresponding decrease of number of collisions between the ejected particles and the gas molecules. The experimental data fit with functions of the form, $R = at^n$, where n remains constant and equal to 0.75, while the coefficient a decreases as the density ρ_0 increases in agreement with the blast wave model.

Experiments have been also produced using two ferrite rods placed face to face in the same cavity. The distance between the two rods was 42 mm, and the background gas was a mixture of argon (75%) and nitrogen (25%) introduced to prevent from arcing. Fig.6 shows a fast side-on view obtained for a charging voltage value of 19 kV in a total gas ambient pressure of 1 atm. The time exposure is 80 ns. The perfect symmetry of the photograph is the consequence of the excellent synchronization of the electrical discharge between the two ferrite rods. To our knowledge it is the first time that such a phenomenon has been observed in these experimental conditions. The two shock fronts meet after a delay time of 20 μs and still expand during a tenth of microseconds without apparently being disturbed by the other one, until being masked by the facing emission plasma. $R-t$ plots confirm this observation as it is shown on Fig.9. The t-axis is inverted for one ferrite to enhance the symmetry between the two curves and to exhibit the point corresponding to the meeting of the two shock fronts. Fits of the form $R = at^n$ give approximately the same values of parameters: $a_1 = 2.025, a_2 = 1.77$, and $n = 0.75$

In our experimental configuration and for discharges on single ferrite rod, we demonstrated that the propagation of the shock front is described by the following expression : $R = 0.267E^{0.37}t^{0.75}$. As in the blast wave model the ratio between the two exponents (n for the time and α for the deposited energy) is close to 2. A comparison of the results between experiments using one (or two) ferrite rod(s) discharges shows a good agreement and confirms this relation between R, E and t. Indeed for a charging voltage of 19 kV the total deposited energy on the surface of the two ferrite rods, deduced from the experimental waveforms of the current and voltage, is 415 J ; if we now applied the relation : $a = 0.267E^{0.35}$ with $a = a_1 = 2.025$ and $a = a_2 = 1.77$, we found that the energy deposited on the first rod, E_1, is 238 J while the energy deposited on the second rod, E_2, is 166 J. The total deposited energy determined in this way is 404 J, value very closed to the one deduced from electrical parameter measurements. The difference between the two values of the deposited energy (E_1 and E_2) is mainly due to the voltage drop generated by the resistivity of the first ferrite rod.

4. Conclusion

Useful information has been gained by performing side-on views of the plasma emitted by electrical discharges on the surface of one (or two) ferrite rod(s). The presence of a shock front has been detected which propagates in the same direction as the plasma and which has a half-cylindrical form. An experimental study of the formation and expansion of both plasma and shock wave has been done as a function of the energy injected in the FFF and the background gas pressure. A classical drag model (exponential time dependence) gives a rather good description of the expansion of the shock front during the initial stages whereas. We demonstrated that the propagation of the shock front could be described accurately by a function similar to Zel'dovich's one: $R = kE^{\alpha}t^n$. These results make the FFP source a promising candidate for many technical applications. The optical pumping of gas lasers is one of the most interesting and the characterization of the motion of plasma and shock front expansions allows to think of a direct pumping of the active medium by the FFP source. In the case of optical pumping of gas lasers and particularly excimer lasers for which the excited product can result from the photodissociation of a special molecule (e.g. XeF_2, I_2...), a bleaching wave is created in the active medium with a velocity of the order of 2000 m/s (Zuev et al. 1989). Therefore this study ensures that the stimulated emission (whose duration is about 6 μs) should not be disturbed by the shock wave expansion beyond the 1 cm-width half-cylinder around the ferrite rod. Simultaneous use of many ferrite rods providing a sufficient optical pumping power has been demonstrated.

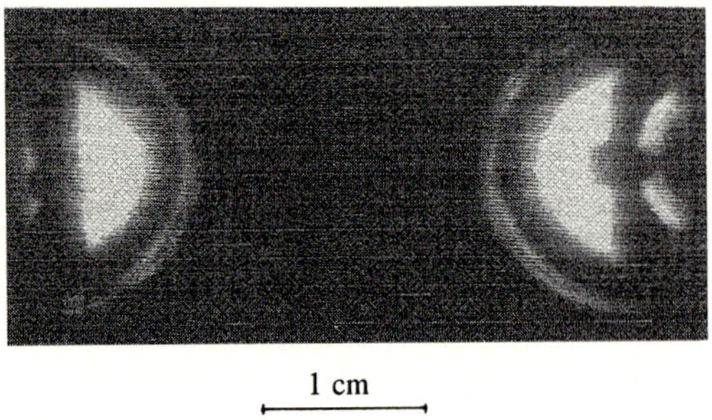

Fig. 6. Transversal plasmas and shock waves expansions view obtained for a discharge at the surface of two ferrite rods placed face to face in the same cavity

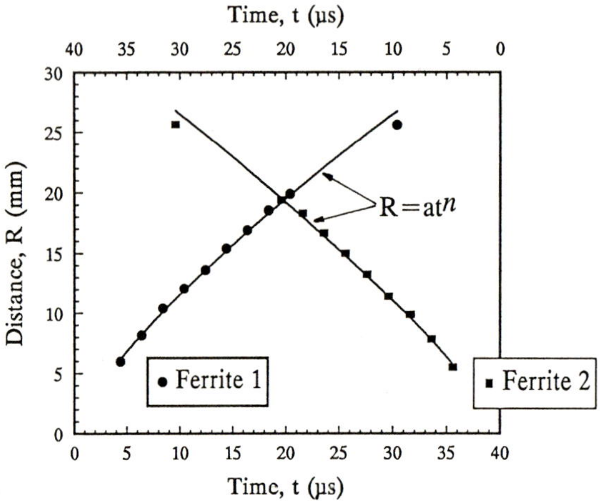

Fig. 7. $R - t$ plots of the leading edges of shock waves generated by a discharge on two ferrite rods.The background gas is a mixture of Ar (75%) and N_2 (25%) for a total pressure of 1 atm and the charging voltage value is 19 kV. The t-axis is inverted for the ferrite 2

References

Beverly III RE (1986) Electrical, gasdynamic, and radiative properties of planar surface discharges. J. Appl. Phys., 60 : 104

Chase WG, Moore HK (1959) Exploding Wires. Plenum Press, Inc., New York

Kogelschatz U (1992) Silent-discharge driven excimer UV sources and their applications. Applied Surface Science, 54 : 410

Sentis ML, Granier Ph, Chazaud F, Marine W, Gerry M, Sze RC (1993) VUV light production from a formed-ferrite plasma source. Applied Surface Science 69 : 185

Sentis ML, Chazaud F, Delaporte PhC, Marine W (1993) Emission spectra analyses of a new VUV source. To be published in App. Phys. Lett.

Watanabe K, Kashiwabara S and Fujimoto R (1987) Development of a formed-ferrite flash plasma light source for gas laser applications. Appl. Phys. Lett. 50 : 629

Zel'dovich Ya B, Raiser Yu (1966) Physics of Shock Waves and High Temperature Hydrodynamic Phenomena., Academic Press, New York, Vol.1, p 94

Zuev VS, Kashnikov GN, Kirilenko VV, Mamaev SB, Sorokin VA (1989) Photodissociation XeF laser emitting visible and ultraviolet radiation when pumped with radiation from a sectioned surface discharge. Sov. J. Quantum Electron., 19 : 748

Numerical Modelling of the Internal Ballistics of Electrothermal-Chemical Guns

Edward Figura
Fluid Gravity Engineering Ltd, Chiltlee Manor, Haslemere Road, Liphook, Hampshire. GU30 7AZ, U.K.

Abstract. We present details of numerical models which can be applied to electrothermal-chemical acceleration concepts. The idea of an ETC gun is to use a chemically reacting fluid to accelerate a projectile to high velocities with the aim of controlling the acceleration through the chemistry of the fluid. One way is to use a plasma jet to control the reactivity of the working fluid. The numerical techniques involve the solutions of a mixture of hyperbolic partial differential equations and stiff ordinary differential equations in 2D MHD and hydro-kinetic models. In this paper we present a detailed discussion of two different chemistry schemes and their incorporation into a hydrodynamic code.

Key words: ETC gun, FLIC, Finite-rate chemistry, Induction parameter model

1. Introduction

The chemical kinetics of reacting fluids usually involve a large number of species and many chemical reactions, for example, the basic hydrogen/oxygen reaction set involves 8 species and 24 reactions whilst for hydrocarbons there might be 20 species and 120 or more reactions. One approach to modelling the chemistry of such fluids is to consider all the species and reactions in detail. Each reaction can be described by a reaction rate which depends principally on an energy barrier which must be overcome before any interaction takes place. A detailed model will calculate the rate at which different reactions occur, how the species are distributed and determine how much energy has been absorbed or released in the process. Since this involves large numbers of species and reactions, this model is computationally very time-intensive.

Matters are not helped by the fact that the different reactions probably have vastly different timescales and so special numerical techniques have to be adopted in order to achieve accurate and stable solutions.

However, in a large number of cases, the kinetics of reacting flows is characterised simply by an energy release which occurs some time after reactions have been initiated. This allows us to replace the details of the chemistry scheme by an approximate two-step model which describes both the delay followed by an energy release period. Such a scheme can reproduce the essential features of reacting flows and has the main advantage of being much faster than the detailed model (from two to ten times as fast).

We shall first provide a description of the hydrodynamic code used in this work. The individual chemistry schemes are described afterwards. Finally we shall present results for an ETC gun chamber filled with a gaseous working fluid.

2. The hydrodynamic code

The hydrodynamic equations are solved in 2D using a FLIC (**FL**uid-**I**n-**C**ell) solution procedure, incorporating the van Leer flux-limiting scheme. The FLIC procedure solves the inviscid Euler equations explicitly, using a two-stage process. The first stage is a Lagrangian phase in which values of velocity and energy are calculated based on acceleration effects due to pressure gradients alone. A predictor-corrector algorithm is used to advance the solution over a timestep. During this stage, the computational grid moves with the local fluid velocity.

Shock Waves @ Marseille II
Editors: R. Brun, L. Z. Dumitrescu

The second stage involves remapping the displaced Lagrangian grid back onto the original stationary Eulerian grid. In this stage mass, momentum and energy are transported to give new values for velocity and internal energy. The velocities calculated in the first stage are used to determine the fluxes across cell boundaries. The van Leer scheme ensures positivity, monotonicity and non-linear stability. A split operator approach is used to both decouple different physical processes (e.g. the Lagrangian and Eulerian phases, chemistry etc) and to treat the 2D problem as two 1D cases (where the order of sweeps is alternated to remove biasing effects). The overall scheme is second-order accurate in both space and time.

3. Full finite-rate chemistry model

We consider I elementary reversible (or irreversible) reactions involving K chemical species. These can be represented in the general form:

$$\sum_{k=1}^{K} \nu'_{ki}\chi_{ki} \Longleftrightarrow \sum_{k=1}^{K} \nu''_{ki}\chi_{ki} \qquad (i = 1 \cdots I) \tag{1}$$

where ν'_{ki} is the stoichiometric coefficient for species k appearing as a reactant, ν''_{ki} is the coefficient for species k appearing as a product and χ_{ki} is the chemical symbol for species k.

The species conservation equations are written as:

$$\frac{dn_k}{dt} = Q_k - n_k L_k \qquad (k = 1 \cdots K) \tag{2}$$

where Q_k is the chemical creation rate, L_k is the chemical destruction rate and n_k is the number density of species k.

Often, knowledge of the individual chemical creation and destruction rates is not required and in such cases the right hand side of Eq.2 can be replaced by one variable ω_k which represents the chemical production rate of species k. This can be expressed as:

$$\omega_k = \sum_{i-1}^{K} \nu_{ki} q_i \qquad (k = 1 \cdots K) \tag{3}$$

where

$$\nu_{ki} = \nu''_{ki} - \nu'_{ki} \tag{4}$$

The rate of progress variable, q_i for the ith reaction is given by the difference of the forward rates minus the reverse rates:

$$q_i = k_{fi} \prod_{k=1}^{K} [X_k]^{\nu'_{ki}} - k_{ri} \prod_{k=1}^{K} [X_k]^{\nu''_{ki}} \tag{5}$$

where $[X_k]$ is the molar concentration of the kth species and k_{fi}, k_{ri} are the forward and reverse rate constants for the ith reaction.

It is assumed that the forward rate constants for the I reactions have the following Arrhenius temperature dependence:

$$k_{fi} = A_i T^{\beta_i} \exp\left[-E_i/RT\right] \tag{6}$$

where the pre-exponential factor A_i, the temperature exponent β_i and the activation energy E_i usually come from experiment.

The reverse rate constants are related to the forward rate constants through the equilibrium constants:

$$k_{ri} = k_{fi}/K_{ci} \tag{7}$$

Here K_{ci} is the equilibrium constant, which is written as:

$$K_{ci} = [P_{\text{atm}}/RT]^{\sum \nu_{ki}} \exp\left[\Delta S_i^0/R - \Delta H_i^0/RT\right] \tag{8}$$

where ΔS and ΔH are the changes in entropy and enthalpy due to chemical reaction.

The thermodynamic properties are taken to be in the form of polynomial fits to the specific heat at constant pressure:

$$c_{p_k}/R = \sum_{n=1}^{N} a_{n_k} T^{n-1} \tag{9}$$

Other thermodynamic properties such as entropy, enthalpy etc. are given in terms of the fits to the specific heat. The data for the polynomial coefficients is taken from the JANAF thermodynamic database.

Chemistry is introduced as the first phase of the computation. The inputs to the chemistry routine are internal energy and species mass fractions. The internal energy determines the initial temperature for the chemical reactions at the beginning of each timestep.

The chemistry phase has to be essentially Lagrangian; so we consider each cell to be a constant pressure reactor and the appropriate equations describing the evolution of temperature T and species mass fractions Y_k are:

$$\frac{dT}{dt} = -\frac{1}{\rho C_p} \sum_{k=1}^{K} h_k m_k \omega_k \tag{10}$$

$$\frac{dY_k}{dt} = m_k \omega_k / \rho \tag{11}$$

where h_k, m_k and C_p are the enthalpy, molecular weight and mass-weighted specific heat at constant pressure of species k (resp).

The chemistry equations are stiff in that the timescales for the different chemical reactions are vastly different. The chemistry routine utilises an ordinary differential equation solver which uses an implicit, back-substitution scheme for solving a system of stiff equations. This solver subdivides the timestep internally to achieve convergence and an accurate solution. The initial timestep for the routine is taken to be the hydro-timestep worked out from the Courant condition on the highest fluid velocity and sound speed in the grid.

However, if the energy release in any cell is greater than some fraction of the total energy in that cell (usually about 5 - 10%) then the timestep is reduced accordingly and the chemistry calculations performed again. This is necessary since a large and abrupt addition of energy can lead to numerical instability. The final chemistry timestep then becomes the hydro-timestep for the Lagrangian and Eulerian remap phases.

The energy released (or absorbed) as a result of the chemical reactions is the difference between the enthalpy prior to any reactions and the enthalpy after the reactions calculated at the original temperature:

$$\Delta E_{\text{kin}} = H_0(T_0) - H_1(T_0) \tag{12}$$

The released energy is added into the Lagrange phase of the hydrocode as follows: The predictor-corrector algorithm first calculates the pressure at a half-timestep ahead based on the internal energy plus $0.5\Delta E_{\text{kin}}$. This pressure is then used to compute the acceleration of each cell. At the end of the Lagrange phase, the internal energy is updated using DEkin and the pressure is recalculated.

The new temperature calculated in the chemistry routine is used to calculate new values of the specific heats, C_p and C_v. These depend on the local properties of the fluid (temperature, species mass fractions) and so are time-dependent. Hence, the fluid frozen gamma ($\gamma = C_p/C_v$) is non-stationary and non-homogeneous i.e. non-ideal. In general, the temperature, internal energy and gamma must all be self-consistent. The equation of state used is:

$$T = (\gamma - 1)\varepsilon m_w / R \tag{13}$$

where ε is the specific internal energy, m_w is the average molecular weight and R is the gas constant.

The new mass fractions feed back into the advection phase, where each species is individually advected (as a molar concentration) together with the total density, using the same algorithm. The species number densities are then normalised such that their sum is equal to the total mass density.

4. Induction parameter model

Although the full finite-rate chemistry model is able to describe all the chemical reactions from first principles, it has the drawback that it is computationally very time intensive. For performing scoping studies and for the definition of propellant characteristics it is useful to be able to use faster models for the chemistry. The induction parameter model is intended as such.

A chemically reacting flow can involve a large number of elementary reactions, each with its own timescale. In practice what happens is that there is a release of energy at any point in the flow only where that flow has been brought to the right conditions of pressure, density and temperature that allow release of energy. This means that the essential features of a reacting flow can be described by a two-step model. In this model, the detailed reaction set is replaced by the following reactions:

(1) Non-exothermic induction reaction

The rate of reaction is given by

$$\omega_\alpha \equiv \frac{d\alpha}{dt} = -\frac{1}{\tau_i} = -k_1 \rho \exp\left[-E_1/RT\right] \tag{14}$$

where τ_i is the induction time, k_1 is the effective reaction rate and E_1 is the activation energy. The induction time represents the delay between the fluid first experiencing a change in conditions (e.g. passing through a shock) and subsequently releasing energy (as a result of chemical reactions). During this induction period, no energy is released so long as $\alpha > 0$; α represents the fraction of induction time remaining during this period.

(2) Exothermic reaction

This phase starts as soon as $\alpha \leq 0$ and it describes the period of energy release. The rate of reaction is:

$$\begin{aligned} \omega_\beta \equiv \frac{d\beta}{dt} &= -k_2 p^2 \left\{\beta^2 \exp\left[-E_2/RT\right] - (1-\beta)^2 \exp\left[-(E_2+Q)/RT\right]\right\} \qquad \text{for} \quad \alpha \leq 0 \\ &= 0 \qquad \text{for} \quad \alpha > 0 \end{aligned} \tag{15}$$

where β represents the fraction of reactant remaining, k_2 is the reaction rate, E_2 is the activation energy and Q is the chemical energy of the mixture.

The continuity equation for α (and similarly for β) is written as:

$$\frac{\partial(\rho_m \alpha_m)}{\partial t} + \nabla \cdot (\rho_m \alpha_m \boldsymbol{v}) = \omega_{\alpha_m} \qquad (m = 1 \cdots MATS) \tag{16}$$

for each material m, where $MATS$ is the total number of materials present.

As with the finite-rate chemistry model, the induction parameter model is introduced as the first stage of the computation. This stage is simply the Lagrangian phase, in which Eq.14 is solved explicitly using simple forward differencing. However, since β is a sensitive function of pressure, temperature and Q, we use a 5th order, adaptive stepsize Runge-Kutta algorithm to solve Eq.15 in the Lagrangian phase. (Note that Eqs.14 and 15 use partial densities and pressures when more than one material is present).

The energy released at each timestep is computed as:

$$\Delta E_q = \sum_{m=1}^{MATS} \eta_m \Delta\beta_m Q_m \tag{17}$$

where η_m and Q_m are the partial density and chemical energy of material m in a cell and $\Delta\beta_m$ is the change in reactant mass fraction during the time step.

The parameters α and β are then transported in the advection phase, using the same advection algorithm as for density and internal energy. Also in this stage, the material partial densities are transported rather than molar concentrations (which are only used in the finite-rate chemistry model). In this model, it is assumed that the specific heat C_p for each material is independent of temperature since information about individual species is not available. Therefore, the fluid gamma is not accurate; this has to be viewed as one of the approximations inherent in the model.

5. Results and discussion: ETC chamber with Hydrogen/Oxygen Argon as the working fluid

The test problem is that of an ETC chamber with plasma inflow. We use a computational grid of 117 × 64 cells ($z - r$). The chamber is filled with an argon diluted, hydrogen and oxygen ($2H_2/O_2/7Ar$) mixture 1 atmosphere and 300 K. The inflow plasma conditions are assumed to linearly ramp up in time from some initial conditions which are based on previous work on modelling a plasma capillary tube.

We compute the first 10 μs of the problem and for the purposes of this calculation the thermodynamic properties of the plasma are taken to be those of helium (the same atomic weight is also used).

Fig.1 shows pressure contours from the finite-rate chemistry calculation at times 2, 4, 6, 8 and 10 μs. These pictures show a triple-shock Mach interaction and its propagation through the chamber. By 10 μs the shock has yet to reach the projectile whilst the contact surface will take much longer to traverse the chamber. Chemical energy is being released behind the main shock front and at the Mach reflection and this is reflected in the pressure contours. Fig.2 shows the equivalent induction parameter model results.

Each model has its advantages and disadvantages. The finite-rate chemistry model is the more accurate of the two since it considers all species and reactions in detail. However, computationally it is very time-intensive. A typical ETC simulation, e.g. as above, 10 μs on a grid of 117 × 64 cells (57 × 16 mm), requires about 60 hours computing time on an IBM RS/6000 550 workstation (41 MHz, 25.2 MFLOPS). On the other hand, the induction parameter model is less accurate but much quicker. For the above simulation, the equivalent computing time drops to about 10 hours. Which particular model will be used will depend on the hardware available, the time available (i.e. how quickly results are needed) and the type of resulting information that is required. The finite-rate chemistry model should be used where information about the distribution of species is vital and where accuracy (in terms of shock position, etc) is of the utmost importance. The model is also needed for extracting the induction parameters for different fluids.

Where one is more interested in the general behaviour of the system (e.g. will the fluid accelerate a projectile to x km/s), then the induction parameter model is the more appropriate one to

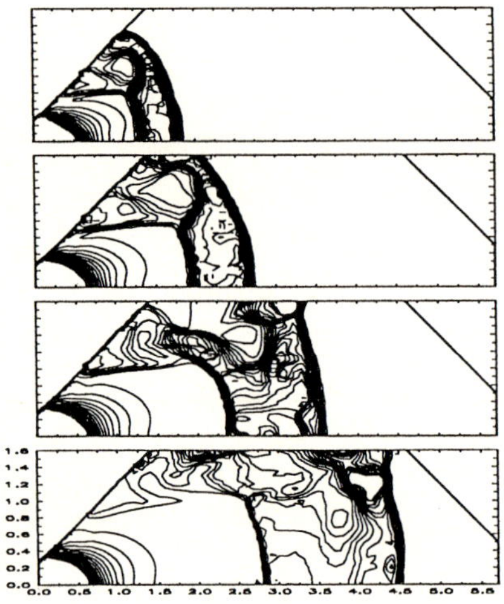

Fig. 1. Pressure contours at 4, 6, 8 & 10 μs from finite-rate chemistry model for ETC chamber with $H_2/O_2/Ar$ working fluid

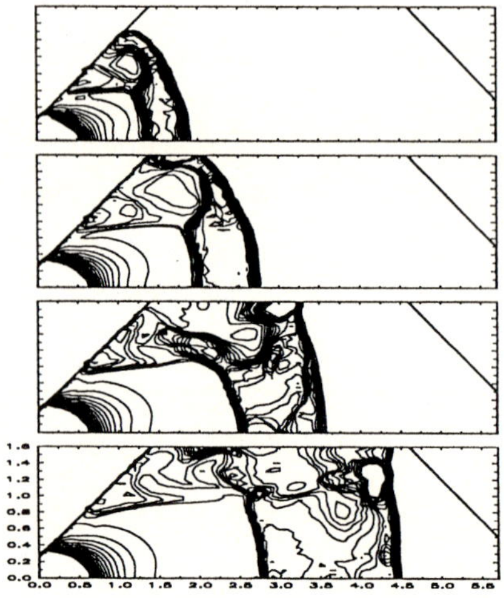

Fig. 2. Pressure contours at 4, 6, 8 & 10 μs from induction parameter chemistry model for ETC chamber with $H_2/O_2/Ar$ working fluid

use. This model is also ideal for performing scoping studies (such as trying different geometries for a particular fluid) and characterising different propellants.

6. Conclusions

In this paper, we have described two different chemistry models that can be used to numerically model a chemically reacting fluid. On the one hand there is the finite-rate chemistry model which considers all species and reactions in detail. It is very accurate but time-consuming to execute. As an alternative, we have the induction parameter chemistry model which replaces the detailed reaction set with a two-step model. It is more approximate than the first model but has the advantage that it executes much faster. We have also shown how the models are integrated into hydrodynamic scheme. The resulting hydro-kinetic code now has ability to compute the dynamics and kinetics for a variety of reacting fluid flows. Future plans include modules for thermal and magnetic diffusion and the investigation of yet different chemistry models.

Acknowledgements

We would like to acknowledge the funding of this work by Royal Ordnance, Guns and Vehicles Division (British Aerospace Defence Ltd) under Contract 31852E.

Experimental Study of a Supersonic Low-Pressure Nitrogen Plasma Jet

B.G. Cheron, L. Robin and P. Vervisch
U.R.A. 230 C.N.R.S., B.P. 118, 76130 Mont Saint Aignan, FRANCE

Abstract. A supersonic nitrogen arc plasma jet expanded in a low pressure test chamber has been analysed. Radial measurements of temperatures and densities have been performed over various sections of the jet. Nitrogen is highly dissociated in the arc chamber and the plasma is out of equilibrium. Surveys of the electron temperature and density, as well as of the Mach number have been carried out by means of electrostatic probe devices. Rotational and vibrational temperatures, and number densities of various N_2, N_2^+ and N excited levels have been drawn from emission spectroscopy. The kinetic scheme of ionic recombination of N^+ and N_2^+ is examined. A recombination path of the $N_2(C^3\Pi_u)$ electronic state via the $N_2(C''^5\Pi_u)$ state is proposed for the $N(^2D^0) + N(^4S^0)$ nitrogen recombination.

Key words: Plasma, Supersonic jet, Kinetic scheme

1. Introduction

We have carried out the experimental analysis of a highly dissociated and ionized (respectively $\alpha_{dis} = 90\%$ and $\alpha_{ion} = 10^{-1}$ to 10^{-2}) low pressure nitrogen plasma jet. Its high velocity allows a simplification of the kinetic scheme, the slow reactions being nearly frozen. The kinetic study has been performed in the subsonic region, where the radial gradients are weak.

In the vicinity of the nozzle outlet a complex structure of the jet flow has been observed. A shock structure has been detected and analysed. Along the axis, classical evolutions of species and temperatures have been observed. On the contrary, uncommon radial evolutions for ionic and molecular species have been uncovered.

2. Experimental set-up

The plasma is a D.C. arc jet. The nitrogen mass flow rate is 0.22 $g.s^{-1}$ and the static pressure is 1 mbar in the vacuum chamber. A primary arc is initiated between the tip of a cone-shaped tungsten cathode and the throat (5 mm in diameter) of a nozzle copper anode. Just behind, a ring-shaped secondary tungsten anode allows to increase the applied power from 9 kW to 14 kW. Electrodes are water cooled. A shock has been observed at $x \approx 38$ mm downstream the source outlet. In the subsonic region, the plasma spread is about 14 cm in diameter, and temperatures as well as densities profiles are nearly smooth, with, on the jet axis, $T_e \approx 5500$ K, $T_a \approx 2000$ K and $N_e \approx 6.10^{19}$ $part.m^{-3}$.

3. Experimental results

3.1. Optical measurements

Spectral scans are dominated by N-atomic lines. The $N_2(C^3\Pi_u \longrightarrow B^3\Pi_g)$ second positive system is present. The more intense molecular bands belong to the $N_2^+(B^2\Sigma_u^+ \longrightarrow X^2\Sigma_g^+)$ first negative system. In most cases, the profiles drawn from Abel's inversion present a bimodal shape, characterized by a secondary peak in the radial number density. Comparison between computed (Loftus et al. 1977) and experimental N_2^+ spectra has provided the rotational temperature T_{rot}, usually equated to the heavy particles temperature T_a. Its radial profile, shown in Fig.1, also presents a bimodal shape. Fig.2 shows a strong axial decrease, followed by an increase of T_a,

Shock Waves @ Marseille II
Editors: R. Brun, L. Z. Dumitrescu

in good agreement with a Mach compression region just behind the shock ($x > 40$ mm). Only one shock cell has been observed in the complex structure of the expanding stream. Near the plasma edge, the jet boundary is characterized by abrupt temperature and intensity decreases. The analysis of the $N_2(C^3\Pi_u \longrightarrow B^3\Pi_g)$ and $N_2^+(B^2\Sigma_u^+ \longrightarrow X^2\Sigma_g^+)$ systems in the $x \leq 20$ mm region has provided the vibrational temperature T_v, the values and profiles of which are similar to the electron ones.

3.2. Electrostatic probe measurements

Flow velocity measurements have been carried out from the perpendicular to parallel ionic current ratio $I_{i\perp}/I_{i\parallel}$ of a Langmuir probe device (Kanal 1967). The Mach number (Fig.2) reaches its maximum at $x = 38$ mm together with the T_a minimum value. Although the calibration method of this probe may be questionable (Johnson et al. 1969), the Mach number evolution agrees with the optical investigations. The radial measurements performed at various locations from the plasma source exit have shown neither a Mach disk at the conical shocks intersection, nor barrel shocks.

3.3. The $N_2(C^3\Pi_u)$ production path

The population densities strongly decrease downstream of the source outlet. The $N_2^+(B^2\Sigma_u^+)$ number density increases when crossing the shock. No population jump is really detected for excited N-atomic levels: the decrease is simply stopped in the relaxation zone on account of the deceleration. No emission signal from the $N_2(B^3\Pi_g \longrightarrow A^3\Sigma_u^+)$ system has been detected. Considering the experimental detection limit, an upper population density of the $B^3\Pi_g$ electronic state can be evaluated to $N_2(B^3\Pi_g) \leq 10^{15}$ m^{-3}. Near the nozzle outlet, the $N_2(C^3\Pi_u)$ density is about 10^{14} m^{-3}. In our situation, neither an energy pooling reaction between two $N_2(A^3\Sigma_u^+)$ molecules (Golubovskii et al. 1990), nor a collisional process (Bol'shakova et al. 1969, Golubovskii et al. 1984) between $N_2(A^3\Sigma_u^+)$ and $N_2(X^1\Sigma_g^+, v = 20)$, nor a direct electron impact is sufficiently efficient to explain the $N_2(C^3\Pi_u)$ population. It can be suggested that the formation of the $C^3\Pi_u$ state is arised, via a spin-orbit or collisional coupling with the neighbouring level $C''^5\Pi_u$, from the pairing of $N(^4S^0)$ and $N(^2D^0)$ atoms to form the N_2 molecule. Considering the numerous coincidences of the vibrational-rotational energy levels of these two states, it is possible (Partridge et al. 1988, Antonov et al. 1989) that the $C''^5\Pi_u$ state lifetime derives from a strong collisional coupling with the $C^3\Pi_u$ state. Reasonably, metastable $N(^2D^0)$ and N excited states atoms follow the same radial evolution with an off-axis peak concentration due to an important N^+ recombination rate at the jet boundary. The observed $C^3\Pi_u$ state and N evolutions are similar, which is consistent with the proposed scheme of excitation by recombination of $N(^4S^0)$ with $N(^2D^0)$ atoms. Assuming that the steady state for the $C^3\Pi_u$ level is held by the balance between the SPS radiative decay and the atomic recombination, provides a 10^{21} m^{-3}.s^{-1} recombination rate. This value is consistent with the order of magnitude of the expected one for the N-atom recombination, within our temperature range.

4. The plasma reaction processes

On the axis and close to the nozzle exit, the plasma parameters are: $T_e \approx 1.2\ 10^4$ K, $N_e \approx 2.10^{20}$ m^{-3}, $N_2^+(X) \approx 2.10^{17}$ m^{-3} and $T_a \approx 6000$ K.

4.1. The dissociation of N_2 into N by impact with heavy particles and electrons

The direct dissociation processes (Cacciatore et al. 1982):

$$N_2 + N_2 \longrightarrow 2N + N_2 \tag{1}$$

$$N + N_2 \longrightarrow 3N \tag{2}$$

$$N_2 + e^- \longrightarrow 2N + e^- \tag{3}$$

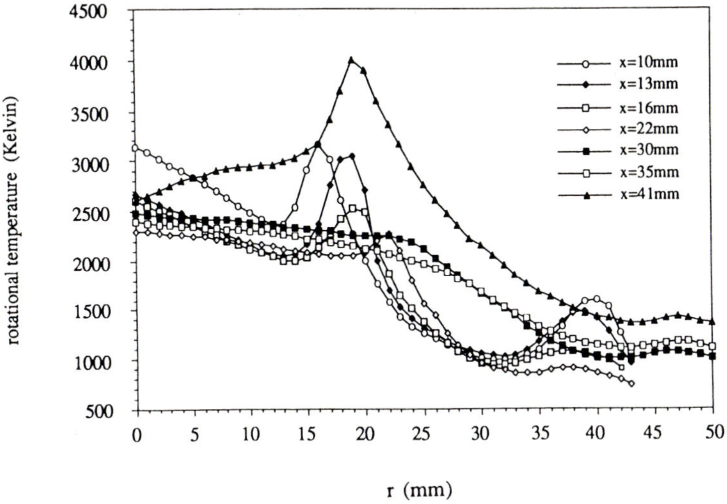

Fig. 1. Radial evolutions of the rotational temperature in the shock region

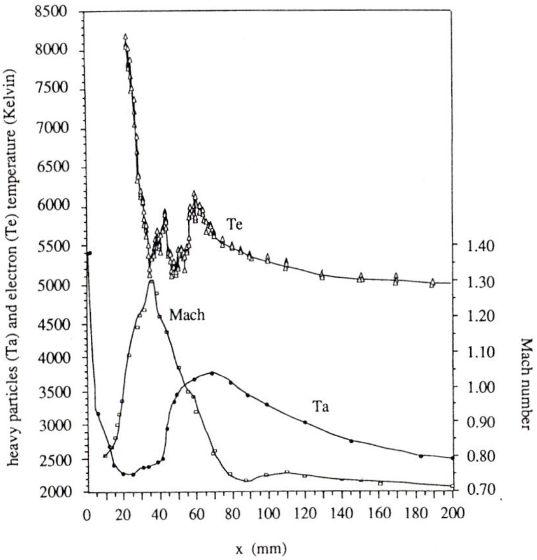

Fig. 2. Axial evolutions of the Mach number, electron and rotational temperatures

and the step by step excitation process (Morgan 1986):

$$N_2(X, v) + M \longrightarrow N_2(X, v') + M \tag{4a}$$

$$N_2(X, v' = 65) + M \longrightarrow 2N(^4S) + M \tag{4b}$$

where the collision partner M is a nitrogen atom (N) or an electron (e^-), are very long compared to the transient times. So, the dissociation of N_2 can be disregarded in our plasma jet.

4.2. The three body recombination of 2N into N_2

This process is generally described through the $A'^5\Sigma_g^+$ and $A^3\Sigma_u^+$ electronic levels:

$$N(^4S^0) + N(^4S^0) + M \longrightarrow N_2(A'^5\Sigma_g^+) + M \quad (5a)$$

$$N_2(A'^5\Sigma_g^+) + M \longrightarrow N_2(B^3\Pi_g) + M \quad (5b)$$

$$N_2(B^3\Pi_g) \longrightarrow N_2(A^3\Sigma_u^+) + \hbar\nu \quad (5c)$$

$$N_2(B^3\Pi_g) + M \longrightarrow \text{product} \quad (5d)$$

The production rate as to Park(1989) (M= e^-) remains weak in our experimental conditions. As a consequence, the recombination of atomic nitrogen is nearly frozen in the jet expansion.

4.3. The charge exchange reaction

The production of N_2^+ is usually (Young et al.1990) driven by the reaction:

$$N_2 + N^+ \xrightarrow{k_6} N_2^+ + N \quad (6)$$

From the $\sigma(E)$ curve proposed by Phelps (1991), we draw the analytic form:

$$k_6 = 2.6 \cdot 10^{-13} T_a^{0.85} \exp(-13000/T_a)(\text{cm}^3.\text{s}^{-1})$$

4.4. The dissociative recombination of N_2^+

Metastable and ground state N atoms are produced by dissociative recombination:

$$N_2^+ + e^- \xrightarrow{k_7} N(^2D^0) + N(^4S^0) \quad (7)$$

In the 1000-3000 K range, the reaction rate k_7 ($v = 0$) is well quantified (Zipf 1980), with a $T_e^{-0.37}$ temperature dependence. At high temperature it seems difficult to neglect the contribution of the $v = 0$, 1 and 2 vibrational levels. Extrapolating by:

$$k_7 = k_7(v=0)\left(1 - \exp\frac{-T_c}{T_e}\right)\left[1 + \exp\frac{-T_c}{T_e} + \exp\frac{-2T_c}{T_e}\right]$$

we obtain a good agreement with the results of Dunn and Lordi (1970) ($T_c = 3120$ K is defined as the vibrational energy spacing of the $N_2^+(X^2\Sigma_g^+)$ electronic state). A quasi stationary state is established between the N_2^+ creation and loss processes:

$$[N_2].[N^+].k_6 = [N_2^+].]N_e].k_7$$

The neutrality equation and experimental measurements lead to $N_2^+ \ll N_e \approx N^+$, providing:

$$\frac{[N_2^+]}{[N_2]} = \frac{k_6}{k_7}$$

Only the $N_2^+(B)$ number density has been measured. When N_e is large enough, production and loss of $N_2^+(B)$ are approximatively described (Orel et al. 1990) by the reactions:

$$N_2^+(B) \rightleftharpoons N_2'^+(A) + \hbar\nu \quad (8a)$$

$$N_2^+(B) + e^- \rightleftharpoons N_2^+(X) + e^- \quad (8b)$$

In the first part of the jet ($x < 100$ mm) and on its axis, T_c and N_e are high enough to suppose that the $N_2^+(B^2\Sigma_u^+)$ population is driven by electron excitation. Then, the population of this level becoming stationary, the reactions (8) provide the $N_2^+(X)$ number density, and it becomes possible to deduce the N_2 density from the (k_6/k_7) ratio. Unfortunately, the obtained evolution is totally in discrepancy with the expected evolution for an under-expanded free jet. In particular, the experimental $N_2^+(X^2\Sigma_g^+)$ number density is found strongly overpopulated upstream of the shock. This result is not changed by supposing an equilibrium at T_e between $N_2^+(B^2\Sigma_u^+)$ and $N_2^+(X^2\Sigma_g^+)$ states. In this region, the loss mechanism by dissociative recombination remains important on account of the high electron density while the production rate by charge exchange decreases dramatically on account of the threshold level of the reaction (6).

The experimental evolutions would be explained by a lowering of this threshold through the following vibrational transitions :

$$N_2(v) + N^+ \longrightarrow N_2^+(v=0) + N \tag{9}$$

Bortner (1963) proposes a k_6 expression varying as T_a^{-2}, which is a manner to lower the threshold. Retaining this law, the N_2 density along the centerline becomes coherent. In particular the pressure evolution, deduced from the relationship: $P = (1+\beta)N_2kT_a$ with $\beta = N/N_2 =$ constant, is satisfactory. Assuming $\beta = 10$ and a jet pressure equal to the chamber pressure leads to:

$$k_6 \approx 4\ 10^{-3}T_a^{-2}\exp(-13000/T_a)(\mathrm{cm^3.s^{-1}})$$

This analytic form, which is not theoretically satisfactory, clearly depends on our experimental situation. For a better description, it would be necessary to determine the reaction rate coefficient for each individual N_2 vibrational level. Elsewhere, namely at the jet boundary and downstream from the shock zone ($x > 60$ mm), the following resonant process:

$$N_2^+(X^2\Sigma_g^+) + N_2(X^1\Sigma_g^+, v \geq 12) \longrightarrow N_2^+(B^2\Sigma_u^+) + N_2(X^1\Sigma_g^+, v' \leq v) \tag{10}$$

which is important in the pink afterglows, cannot justify the measured $N_2^+(B^2\Sigma_u^+)$ number density. Then, only a recombination process is possible. A probable mechanism is the radiative association of a N^+ ion with a N-atom:

$$\begin{array}{rl} N + N^+ \longrightarrow & N_2^+(X^2\Sigma_g^+, \text{high v}) + \hbar\nu \\ (10) & \downarrow \ \rightarrow N_2^+(B^2\Sigma_u^+) \end{array} \tag{11}$$

The rate coefficient (10^{-15} $\mathrm{cm^3.s^{-1}}$) drawn from the experimental $N_2^+(B^2\Sigma_u^+)$ number densities is plausible. However some problems still remain to give grounding to this recombination pathway. The off-axis peak of $N_2^+(B^2\Sigma_u^+)$ occurs in a region where the temperature of the heavy particles decreases abruptly, the N^+ ion density varying slowly. To impute the $N_2^+(B^2\Sigma_u^+)$ number density to the reaction (10), nitrogen atoms must still remain the predominant species compared to N_2. In the considered range of temperatures (1000-1500 K), this hypothesis may be erroneous. Another possibility would be a three-body reaction with N_2 as the third body.

4.5. The recombination of N^+ ions

The N^+ density is governed by the reaction (6) and the three-body recombination reaction:

$$N^+ + 2e^- \longrightarrow N + e^- \tag{12}$$

The recombination rate can be deduced from the experimental evolution of N_e on the jet axis, far from the source outlet. The obtained value (2×10^{-23} $\mathrm{m^{-3}.s^{-1}}$) is comparable to the one (1.4×10^{-23} $\mathrm{m^{-3}.s^{-1}}$) calculated following Park's (1969) expression for the three-body recombination and to the one (7×10^{-22} $\mathrm{m^{-3}.s^{-1}}$) drawn from the experimental k_6 value. So, experimental and theoretical calculations are in good agreement.

5. Conclusions

This analysis shows that the vibrational distribution of N_2 and N_2^+ strongly influences the kinetic scheme of our supersonically expanding plasma. Reminding that there is no known technique for producing high enthalpy hypersonic flows free of vibrational excitation, these results point out the need of much more complete cross-sections data. The vibrational distribution calculation is actually limited by the lack of information concerning the cross-section for e-V and N-V processes, when the vibrational level from which the excitation starts is not the v=0 level. The dissociation rate and, as shown in this study, the charge exchange reaction rate coefficient, strongly depend on the vibrational excitation. Observation of the $N_2^+(B^2\Sigma_u^+)$ evolution in the jet has allowed to examine the kinetic scheme of N_2^+ ions, which strongly involves the N^+ ion recombination. To explain the $C^3\Pi_u$ formation, a route by $N(^4S^0)$ and $N(^2D^0)$ recombining via the $C''^5\Pi_u$ state has been proposed. This pathway can be efficient in this highly dissociated and ionized recombining plasma. This set of measurements leaves many unanswered questions and suggests a number of future experiments. In particular, inlet conditions must be known with a better accuracy for a comparison with numerical code. To that aim, measurements of fundamental N and N_2^+ number densities are currently underway.

References

Antonov EE, Popovich VI (1989) Opt. Spectrosk. (URSS) 67, 2: 172
Bol'shakova LG, Golubovskii YuB, Telezhko VM, Stoyanov DG (1990) Sov. Phys. Tech. Phys. 35, 6: 665
Bortner MH (1963) General Electric Co. Doc. R 63 SD 34 sept
Cacciatore M, Capitelli M, Gorse C (1982) J. Chem. Phys. 66,1: 41
Dunn MG, Lordi JA (1970) AIAA Journal 8,2: 339
Golubovskii YuB, Telezhko VM (1984) Teplofiz. Vys. Temp. 22: 996
Golubovskii YuB, Telezhko VM, Stoyanov DG (1990) Opt. Spektrosk. (USSR) 69, 2: 322
Johnson BH, Murphree DL (1969) AIAA J. 7, 10: 2028
Kanal M (1964) Theory of current collection of moving cylindrical probes. J. Appl. Phys. 35, 6
Loftus A, Krupenie PH (1977) J. Phys. Chem. Ref. Data 6, 1
Morgan LA (1986) J. Phys. B Atom. Mol. Phys. 19: 439
Orel AE, Rescigno TM, Lengsfield BH (1990) Phys. Rev. A 42, 9: 5292
Park C (1969) AIAA Journal 7, 8: 1653
Park C (1989) Non equilibrium hypersonic aerothermodynamics. Wiley, New York
Partridge H, Langhoff SR, Banschlicher Jr. CW (1988) J. Chem. Phys. 88, 5: 3174
Phelps AV (1991) J. Phys. Chem. Ref. Data 20, 3: 557
Young RA, Bower RD (1990) J. Chem. Phys. 92, 3: 1617
Zipf EO (1980) Geophys. Res. Letter 7, 9: 645

Shock Tube Conjugated with Nanosecond Discharge Gap for Investigating Shock Wave Propagation in Plasma

A.A. Alyapin, I.V. Filiougine, S.V. Kostiouchenko and N.N. Koudriavtsev
Moscow Institute of Physics and Technology, Moscow Region, Dolgoprydny, Russia, 141700

Abstract. A description of a shock tube set-up conjugated with a dielectric section in which an electric discharge developed in the form of a breakdown ionization wave with sublight speed is presented. This set-up allows to investigate plane shock wave propagation in an uniform, highly excited plasma and the breakdown of the gas gap by the shock wave in a wide range of the gas parameters. All disturbances existing in previous investigations of shock wave propagation in plasma with plasma production by glow or/and radio- frequency discharge such as flowing of the plasma by the gas flow etc. are eliminated by this method.

Key words: Nanosecond discharge, Shock tube

1. Introduction

The investigation of Shock Wave (SW) propagation through a plasma is of significant importance owing to such modern technical applications as hypersonic aerodynamics, supression of SW in the nozzles of jet engines, gasdynamics lasers and MHD-generators; laser-induced nuclear fission; hypersonic flights of spacecraft and aircraft in a rarefied atmosphere, etc. This phenomenon is also of serious scientific interest due to its excellent physical features. In this case gasdynamic phenomena take place in a mixture of light (electrons) and heavy (neutrals, ions) gases, with quite different mobilities and thermal conductivities. The rates of the exchange processes between these two kinds of gases are slow. This allows distinguishing the behavior of the gasdynamics parameters of the light and heavy components of the plasma in the neighborhood of the SW front (Jakes 1957, Zeldovich and Raizer 1966).

The theoretical background of SW propagation through plasma has been developed by Jakes (1957), Zeldovich and Raizer (1966), Velikovich and Liberman (1987). Only a few experimental investigations have been carried out, due to their complexity. Previously, in these experiments two different techniques have been utilized for the generation of incident plasma: glow or microwave discharge (Gorchkov et al. 1989) and laser induced optical breakdown causing bright spark and powerful UV-radiation effectively ionizing the ambient gas region (Kondrachov et al. 1987). Both techniques have serious experimental limitations. The microwave and/or glow discharge-produced plasma is quickly disturbed by the high velocity gas flow associated with the SW. For the optical breakdown technique the significant curvature of the generated SW and the two-dimensional features of the associated gas flow as well as the small dimensions of the experimental regions, seriously restrict the diagnostic possibilities.

In this work a new experimental method for investigating plane SW propagation through an uniform nonequilibrium plasma, free from the above-mentioned limitations, has been developed and realized. The method is based on our patents (Filiougine et al. 1991). Its background is the excitation of atoms and molecules by a Nanosecond Gas Discharge (NGD), highly uniform throughout the volume and instantaneous in the gasdynamic time scale (Asinovsky et al. 1983). The experimental facility consists of a Shock Tube (ST) conjugated with a dielectric discharge section for developing the NGD. This technique has been used for the investigation of some modern gasdynamics problems: plane shock wave propagation through a plasma, modelling of a big Space Shuttle-type spacecraft reentry flight, hypersonic flight of aircraft in the rarefied atmosphere, etc.

Shock Waves @ Marseille II
Editors: R. Brun, L. Z. Dumitrescu

2. Nanosecond gas discharge

It seems that this type of discharge techniques gives an unique opportunity for uniform, instantaneous gas excitation in a large volume. NGD arises after applying a high voltage pulse with a very short rise time (about several nanoseconds) and an amplitude of $10^4 - 10^6$ V across a gas gap. A discharge develops from the electrodes as an uniform Breakdown Ionization Wave (BIW), having sublight velocity and so providing instantaneous gas excitation in the time scale of the gasdynamic processes. For a gas density in the range $10^{15} - 10^{19}$ cm^3, that is interesting from a gasdynamic standpoint, this excitation may also be highly uniform throughout the volume. The main part of the discharge energy ($> 98\%$) is converted into excitation of electrons and vibration of the plasma particles.

3. Experimental facility

The main set-up layout is shown in Fig.1. A hypersonic gas flow is created behind the Incident Shock Wave (ISW), generated in the ST channel. Experiments have been carried out in a ST facility of 80 mm ID and 8.5 m length. The low pressure section (1) is connected with a discharge chamber (2) (a dielectric quartz tube) surrounded by a metallic screen (3). Its inner diameter is the same as the ST channel and its length 1.2 m. The space between the quartz tube and the metallic screen is filled with organic glass and pumped with nitrogen at a pressure up to 20 bar permitting voltage pulses of up to 10^6 V to be applied to the ring (4) and conical (5) high voltage electrodes. Special quartz windows (6) are provided for passive and active optical diagnostics of the gas in the discharge chamber.

The ST facility, described in detail previously (Asinovsky et al. 1983), enables an intensive SW with a Mach number of up to 15 and associated gas flows with Mach number up to 10 to be produced. When the SW arrives at the discharge chamber the signal from the piezoelectric transducer (7) is applied to the synchronization unit (8). This unit turns on a nanosecond voltage generator (9) with a convenient delay time and a high voltage impulse is applied to the electrodes to induce a BIW, which then propagates through the SW, which may find itself at any position in the discharge chamber, with a speed of $10^9 - 10^{10}$ cm/s (Filiougine et al. 1991), exciting the gap volume before and behind the SW front during a short time (10-20 ns). A Marx electric generator is employed to produce the impulse discharge (Doroshenko et al. 1986). It produces impulses of 100-400 kV amplitude, 30-40 ns duration, and a rise-time of less than several ns, allowing the generation of a BIW in such gases as N_2 , O_2 and air in the density range $10^{15} - 10^{19}$ cm. Typical discharge current and average impulse power are 5-20 kA and $10^8 - 10^{10}$ W respectively.

4. Set-up operation

The following parameters of NGD have been recorded: BIW velocity and time-dependent electron concentration before and behind the SW, BIW current and electrical power inserted into the discharge gap. The Doppler frequency shift of a microwave beam, reflecting form the SW front is used for continuous SW velocity recording. The optical system (insert in Fig.1) consists of the quarts windows (6), lenses (10), wide band interference filters (11), two or three steps of brightness amplifiers (12); camera (13), TV camera (14) or Multichannel Optical Analyzer (OMA) adjustable in the x and y directions (15) for getting a picture of the radiation intensity or a fast recording of its cross-section. The optical camera and TV camera are used for qualitative analyzing the radiation intensity distribution, the OMA for quantitative one-dimensional radiation intensity distribution measurements. The radiation intensity in the 2^+ positive band system of N_2, Shumann-Runge and Hertzberg band systems of O_2, NO-$\alpha, \beta, \gamma, \delta$ UV band systems, as well as in several atomic lines of O and N were monitored.

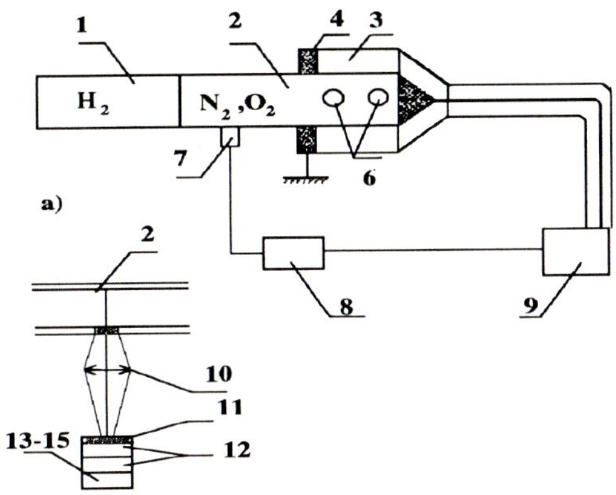

Fig. 1. Experimental set-up. Low pressure section (1), discharge chamber (2), dielectric quartz tube surrounded by metallic screen (3), ring (4) and conical (5) high voltage electrodes, quartz windows (6), piezoelectric transducer (7), synchronization unit (8) nanosecond voltage generator (9), (10) lenses, wide band interference filters (11), brightness amplifiers (12); camera (13), TV camera (14) or Multichannel Optical Analyzer (OMA) adjustable in the x and y directions (15)

5. Investigations

Several series of experiments for testing the experimental system thus developed have been performed. We explored a plane SW at $M = 5 - 10$ propagating through nitrogen, oxygen and air plasma at initial pressures of 0.005-0.1 bars. Atomic oxygen, nitrogen, nitrogen oxide molecules (in air) and electron longitudinal concentration distribution have been monitored behind the SW by microwave and optical techniques.

6. References

Asinovsky EI, Vasilyak LM, Markovetc VV (1983) Wave breakdown of the gas gaps (in Russian). Thermophysics of High Temperatures 2: 371-381; 3: 577-590

Doroshenko VM, Koudriavtsev NN, Novikov SS (1986) Determination of the time-resolved probabilities of heterogeneous recombination of atoms in shock tube experiments. Intl. J. Heat & Mass Transfer 29, 1: 1-19

Filiougine IV, Kostiouchenko SV, Koudriavtsev NN (1991) A method for shock wave generation in plasma. USSR Patent N.1669384 from 8 April 1991, (Application No. 4465946, July 26, 1987)

Gorchkov VA, Klimov AI, Fedotov AB, Shugaev FV (1989) Formation of active regions behind a shock wave in weakly ionized nonequilibrium plasma (in Russian) J. Tech. Phys. 59, 4: 135-137

Jakes JD (1957) The structure of a shock wave in a fully ionized gas. J. Fluid Mech. 3: 275-285

Kondrachov VN, Rodionov NB, Sitnikov SF, Sokolov VI (1987) Investigation of gasdynamics effects at laser spark later afterglow (in Russian). J. Tech. Phys. 56, 1: 89-97

Kovalchuk BM, Mesyats GA, Semin BN, Shpak VG (1981) High current nanosecond accelerator for investigations of high speed processes (in Russian). J. of Scientific Device and Experimental Techniques 4: 15-18

Velikovich AL, Liberman AN (1987) Physics of Shock Waves in Gases and Plasmas (in Russian). Science, Moscow

Zeldovich YaB, Raizer YuP (1966) Physics of Shock Waves and High Temperature Hydrodynamics Phenomena (in Russian). Science, Moscow

Intensive Beam Source of Fast Neutral Atoms Generated by an Electromagnetic Shock Tube

N.N. Koudriavtsev, O.A. Maziar and A.M. Sukhov
Moscow Institute of Physics and Technology, 141700, Moscow region, Dolgoprydny, Russia

Abstract. A description is made of a set-up based on an electromagnetic shock tube for producing dense beams of active atoms with velocities corresponding 1-10 ev translational energy; some illustrative results are also presented. This set-up allows to carry out investigations of fast active atom interaction with solid surfaces, taking place at vehicle orbital flights, in "TOKAMAK" controlled thermonuclear fission plants, etc.

Key words: Electromagnetic shock tube, Fast atoms, Thermonuclear fusion, Orbital flight

1. Introduction

Investigation of chemical erosion processes is of significant importance for the development of thermonuclear fusion plants such as TOKAMAK, Stellarators etc., because such processes are essential sources of impurity ions in plasma and impose restriction on the running-time of the plants (Anon. 1987). Similar phenomena take place during orbital flights of space vehicles.

In contrast to high-energy ($E > 1$ keV) pure physical ion sputtering, the mechanism of plasma-wall interaction including low-energy (1-10 ev) ions and neutral atoms (such as hydrogen, its isotopes, carbon, oxygen, metals and other impurities) is not yet clear (Post 1986). This interaction takes place in systems with complex chemical composition such as real thermonuclear reactor walls as a result of combined effects of adsorption-desorption processes, chemical reactions, energy exchange and radiative processes on the surface and in the surrounding gas phase. Only some disparate data concerning these phenomena are present in the literature (Post 1986). In particular, it is known that the influence of low-energy ions on the surface stimulates chemical interaction between the surface material and the incident beam of neutral chemically active particles, which may reveal itself as chemical erosion (Voijtsenjia et al. 1991).

Investigations of surface interaction with an incident beam containing low energy ions and neutral atoms are of significant practical interest. Experimental work in this direction is restrained now by insufficient progress in high-intensity corpuscular beam generators with strictly controlled parameters. Also modern techniques such as Time of Flight Mass-Spectrometer (TOF MS) are required for efficient control of particle-surface interaction.

2. Experimental setup

In this work a particle beam generator based on an electromagnetic shock tube is described (Fig.1). The experimental setup consists in an electromagnetic shock tube (1) designed as recommended by Ziemer (1959) and equipped with a conical skimmer (6) at the endplate, a vacuum tank (8) with high speed pumping, power supply and synchronization system. The main ideas of this design are clear from Fig.1. The main characteristics of the installation are given in Table 1.

The strong shock wave in the shock tube channel generates effective dissociation of the flow behind it. Some experimental results concerning shock wave intensity are analogous to those of Ziemer (1959). The skimmer role is to form a collisionless beam, which moves freely in the vacuum tank and interacts with the surface sample under exploration placed inside.

The translation energy of the particles depends upon their masses and lies in the interval 1-15 ev. The beam behind the skimmer contains ions and neutral atoms and molecules in definite

Shock Waves @ Marseille II
Editors: R. Brun, L. Z. Dumitrescu

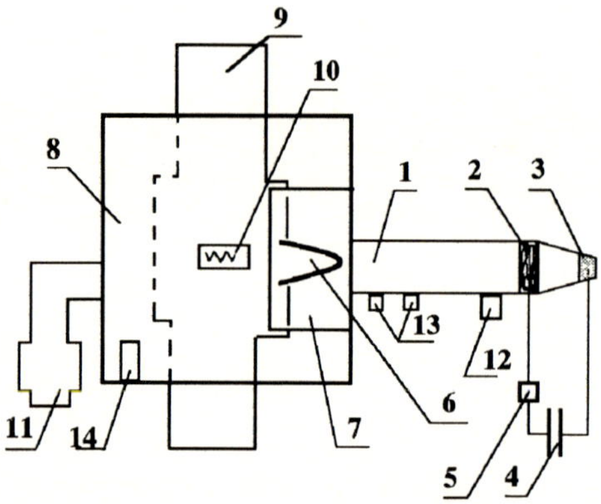

Fig. 1. Experimental set-up.
1-electomagnetic shock tube, 2,3-high voltage electrodes, 4-capacitor battery, 5-ignitron, 6-skimmer, 7-buffer volume, 8-vacuum chamber, 9-turbomolecular pump, 10-ionization density gauge, 11-forepump, 12-mechanotron, 13-piezoelectric transducer, 14-ionization pressure gauge

fractions, which can be effectively varied by controlling the shock wave Mach number, chemical composition of the gas filling up the shock tube channel and the geometrical parameters of the skimmer.

Table 1

Parameter	Value
a) Gas-dynamical (in air)	
Mach number of incident shock wave	15–40
Initial pressure in shock tube, Torr	0.02–1.0
Flow Mach number behind shock	3–4
Flow velocity behind shock, km/s	4.5–12.0
Flow temperature behind shock, K	6000-12000
Stationary flow duration, μs	30–80
Beam velocity behind skimmer, km/s	5–15
Beam density (experimental), part/cm^{-3}	10^{14}
b) Mechanical	
Shock tube ID, mm	75
Shock tube length, mm	700
Vacuum tank volume, l	40
Vacuum tank pumping speed, l/s	500
Repetition rate, Hz	up to 10
Conical skimmer half-angle	30°

The generator operates at a repetition rate up to ten pulses per second. A typical oscilloscope trace obtained in beam densities measurements by means of a special ionization gauge (Gentry and Giese 1978) is shown in Fig.2. The duration of the fronts of the signals corresponds to the flight time of molecules through the ionization region of the gauge and is about 10 μs. The amplitudes of the signals correspond to beam densities up to 3×10^{12} /cm^3. The quasistationary state duration of the particle beam produced in each impulse is of 30-80 μs.

The integrated intensity of the beam reaches 10^{18} part/cm^{-2}·s^{-1}, which is sufficient for getting a dose of more than 10^{21} part/cm^{-2} in a sufficiently short time (about a few hours). Such a dose is supposed to be characteristic for plasma-wall interaction during 6-8 months running of a thermonuclear fusion plant (Vojtsenjia et al. 1991), or one year orbital flight. Typical experimental values of the beam density vs condenser voltage are shown in Fig.3. Theoretical calculations of the

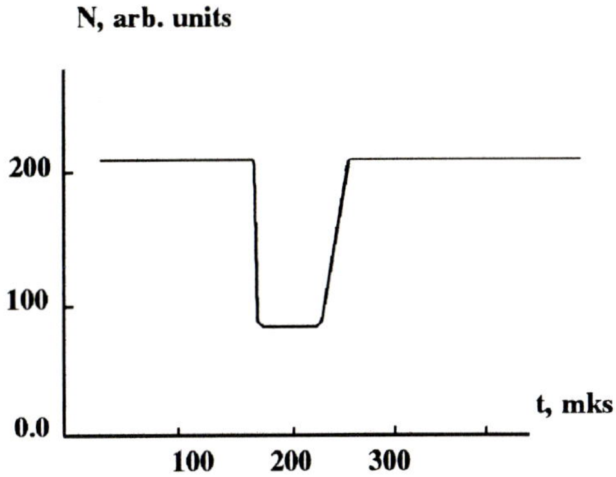

Fig. 2. Time dependence of the beam density measured by ionization gauge (10), Fig.1

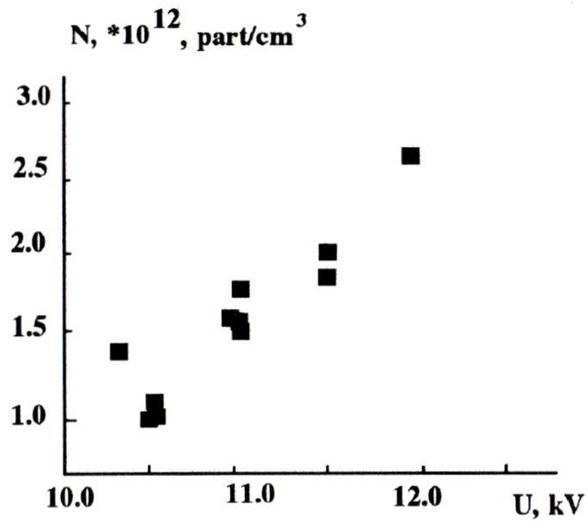

Fig. 3. Beam density of beam in air vs. capacitor battery voltage, controlling shock wave velocity. Initial pressure in the shock tube is 0.07 Torr

shock wave and beam characteristics were performed too, but the meaning of the experimental results is something greater.

3. Diagnostics

Beam parameters such as velocity and density are controlled with two ionization gauges (Gentry and Giese 1978) placed along the beam direction at a distance between them of 50 mm (Fig.1). The measurements of the beam composition and composition of the gas in the interaction region are carried out with a time-of-flight mass-spectrometer. The upper limit of the mass number recording of the mass-spectrometer is 200.

The mass-spectrometer used is supplied with a special ion source for molecular beam measurements (Pollard and Cohen 1987). The ionization region is between grids V_1 and V_2. The electron beam originates below the ionization region at the tungsten filament and is collimated to a diameter of about 7 mm at the point where it intersects the molecular beam. The grids are made from electroformed copper mesh (optical transmission $\sim$ 90%) and held parallel to one another. A cylindrical shield made of tantalum is tied to V_1 and surrounds the ion source so as to prevent the electron beam from being disturbed by stray electric fields.

Ionization should take place in a field-free region to provide a well-defined impact energy and to allow for proper space focusing. The ion extraction field must be held on long enough to sweep out all ions up to the maximum mass. These requirements are satisfied by the scheme of Pollard and Cohen (1987) in which a pulse is applied to grid V_1 to turn off the extraction field. When this pulse relaxes, the field turns back on and the ions begin their flight to the detector. During the time that the pulse to grid V is turned off the extraction field (1000 ns), a second pulse of around 600 ns duration is applied to the gate grid to admit the electrons into the ionization region. The main physical diagnostics which we intend to use for investigations of beam-surface interaction is strict spectral intensity registration of the visible and IR radiation from the beam-surface interaction region.

A secondary ion mass spectrometer is used for chemical composition measurements of the upper layer of the sample after beam exposition. Experimental data recording and processing are carried out with a personal computer. The experimental setup described here is a powerful means of investigation and will allow to obtain definite conclusions concerning the mechanism of plasma-wall interaction at high intensity and low-energy of the incident particles, typical for TOKAMAK systems and space vehicles during orbital flight.

References

Anon 51987) Proceedings of Japan-US Workshop on Plasma-Material Interaction. Tokyo, 1987.
Gentry WR, Giese CF (1978) Rev. Sci. Instrum. 49: 595
Pollard JE, Cohen RB (1987) Rev. Sci. Instrum. 58, 1: 383
Post DE (ed) (1986) Physics of Plasma-Wall Interactions in Controlled Fusion. N.Y.
Vojtsenjia VS et al. (1991) Action of Low Temperature Plasma and Electromagnetic Radiation Upon Materials, Moscow Energoatomizdat (in Russian)
Ziemer RW (1959) ARS Journal 29, 9: 642

Interaction of Strong Discontinuities in Magnetized Conducting Media

A.A. Barmin* **and E.A. Pushkar'†**
*Institute of Mechanics, Moscow University, Moscow, Russia
†General and Applied Mathematics Department, MASI (VTUZ-ZIL), Moscow, Russia

Abstract. The oblique collision of two strong discontinuities S_1 and S_1' (shock waves, rotational and/or contact discontinuities) in conducting media in the presence of an arbitrarily oriented magnetic field $\boldsymbol{H}$ is investigated. The studies are carried out in the framework of an ideal magnetohydrodynamic model. Universal methods of solving the problems under consideration are developed. Regular self-similar and non-regular Mach type interactions are studied. The numerical simulations are carried out with the entire range of all upstream parameters being covered. The solutions obtained are a generalization of the classic gasdynamic solutions (Courant and Friedrichs 1948; Chernyi 1988) for an electroconducting gas embedded in a magnetic field. A data base is created, which may be applied to analyze different shock phenomena in the presence of a magnetic field.

Key words: Magnetohydrodynamics, Shock waves, Rotational discontinuities, Interactions

1. Formulation of the problem

Let us consider the oblique collision of S_1 and S_1' propagating towards one another at an angle $\vartheta \neq 0$ through an undisturbed medium at rest. We will suppose that the conductivity is infinite so that the flow is described by the system of stationary equations of ideal magnetohydrodynamics for an inviscid non-heat-conducting gas (Kulikovskii and Lyubimov 1962)

$$\text{div}\ \rho\boldsymbol{V} = 0; \qquad \boldsymbol{V}\cdot\text{grad}(p/\rho^\gamma) = 0; \qquad \text{div}\ \boldsymbol{H} = 0$$

$$(\boldsymbol{V}\cdot\nabla)\boldsymbol{V} = -\frac{1}{\rho}\ \text{grad}\ p + \frac{1}{4\pi\rho}\ \text{rot}\ \boldsymbol{H}\times\boldsymbol{H}; \qquad \text{rot}(\boldsymbol{V}\times\boldsymbol{H}) = 0 \tag{1}$$

with the following relationships on the strong discontinuities:

$$[\rho V_n] = 0; \qquad [p + \rho V_n + H_\tau^2/8\pi] = 0; \qquad [\rho V_n \boldsymbol{V}_\tau - H_n\boldsymbol{H}_\tau/4\pi] = 0; \qquad [\boldsymbol{H}_n] = 0$$

$$[\rho(\frac{p}{(\gamma-1)\rho} + \frac{V^2}{2})V_n + pV_n + \frac{1}{4\pi}H_\tau^2 V_n - \frac{H_n}{4\pi}(\boldsymbol{H}_\tau\cdot\boldsymbol{V}_\tau)] = 0; \qquad \boldsymbol{H}_n[\boldsymbol{V}_\tau] = [\boldsymbol{H}\cdot V_n] \tag{2}$$

Here ρ is the density, p is the pressure, $\boldsymbol{V}$ is the gas velocity.

The initial values $\rho_0, p_0, \boldsymbol{H}_0, \vartheta$ and the velocities $\boldsymbol{V}_1$ and $\boldsymbol{V}_1'$ of S_1 and S_1' are given. We will consider plane-polarized flow with $\boldsymbol{V}_1, \boldsymbol{V}_1'$, and $\boldsymbol{H}_0$ orthogonal to the line B of intersection of S_1 and S_1'. In the plane of flow the line B is displayed graphically as a point B which may be imaginary if there is a shock wave moving through the undisturbed medium between S_1 and S_1' (Figs. 1a and 1b). Thus we have five dimensionless governing parameters: the Mach numbers $M_1 = |\boldsymbol{V}_1|/a_0$ and $M_1' = |V_1'|/a_0(a_0^2 = \gamma p_0/\rho_0)$, the Alfven number $N = a_A/a_0 = H_0/(4\pi\gamma p_0)^{1/2}$ (or $\beta = p_0/(H_0^2/8\pi) = 2/\gamma N^2$), ϑ, and the angle ψ between $\boldsymbol{H}_0$ and $\boldsymbol{V}_s$, where $\boldsymbol{V}_s$ is the gas velocity in the coordinate system moving with point B in which the problem is steady.

The problem formulation is self-similar as it does not contain any characteristic linear dimension. The regular solutions are constructed as two combinations of self-similar stationary solutions, diverging from the point B (Fig.1a), so that either of them consists of fast shock waves S^+ (or self-similar rarefaction waves R^+), an Alfven (rotational) discontinuity A and slow waves (S^- or R^-). There is a contact interface C separating the media between the slow waves.

There are regions of the governing parameters where regular interaction cannot occur (Barmin and Pushkar' 1990, 1991). In this case a more complicated flow occurs like the 'Mach' interaction in gas dynamics which contains a Mach shock S_M moving between S_1 and S_1' and connecting the branch points B and B' (Fig.1b). The shock S_M is characterized by its Mach number $M_M = V_M/a_0$ and the front inclination δ relative to $\boldsymbol{V}_s$. Unlike the gasdynamic case near B and B' the flow contains up to six waves, whose qualitative structure generally cannot be predicted in advance. There is no coordinate system where the flow as a whole is stationary, but locally in the vicinity of B or B' it may be steady. The flow is self-similar in the coordinates moving with the imaginary line of intersection of S_1 and S_1'(in variables $x/t, y/t$).

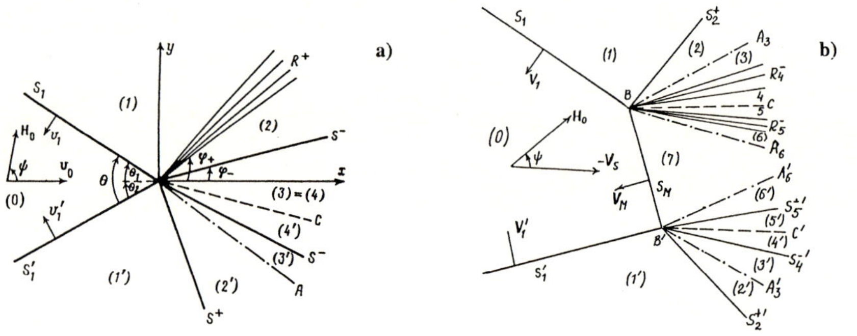

Fig. 1. Schematic of interaction between strong discontinuities S_1 and S_1': (a) regular; (b) irregular

The stationary shock waves $S^{\pm}$ are set by the angle $\varphi^{\pm}$ of inclination to the streamline and $R^{\pm}$ by the rear angle $\varphi^{\pm}$ of the rarefaction fan. The state of the medium behind the k-th wave (or k-th discontinuity) will be designated by the subscript k according to Figs.1a and 1b, the prime will denote the waves and quantities at different sides from C (in regular interaction) or in the vicinity of B'(in nonregular interaction). If the x axis is directed along the streamline ahead of $S^{\pm}$, the state behind $S^{\pm}$ is determined as follows (Pushkar' 1978):

$$[(\gamma+1)M^2\sin^2\varphi(\eta_k - \frac{\gamma-1}{\gamma+1}) - 2](\chi-\eta_k)^2 + N^2\cos^2(\varphi-\psi)\times$$
$$\times[(\chi-\eta_k)(\gamma(\eta_k-1)+1)+\eta_k(\chi-1)] = 0$$

$$\chi = \frac{N^2}{M^2}\cdot\frac{\sin^2(\varphi-\psi)}{\sin^2\varphi}; \qquad i_k = -\cos(\varphi-\psi)\cdot\frac{1-\eta_k}{\chi-\eta_k} \tag{3}$$

$$P_k = 1+\gamma M^2(1-\eta_k)\sin^2\varphi - \gamma N^2(i\cos(\varphi-\psi)+0.5i); \qquad (u_{xk}, u_{yk}) = (\frac{V_{xk}}{a_0}, \frac{V_{yk}}{a_0})$$

$$\frac{u_{xk}}{M} = 1-\sin^2\varphi(1-\eta_k) + i\frac{N^2}{M^2}\frac{\sin(\varphi-\psi)}{\tan\varphi}; \quad \frac{u_{yk}}{M} = (1-\eta_k)\sin\varphi\cos\varphi + i\frac{N^2}{M^2}\sin(\varphi-\psi)$$

$$(h_{xk}, h_{yk}) = (\frac{H_{xk}}{H_0}, \frac{H_{yk}}{H_0}); \quad h_{xk} = \cos\psi + i\cos\varphi; \quad h_{yk} = \sin\psi + i\sin\varphi; \quad \eta_k = \frac{\rho_{k-1}}{\rho_k}$$

$$P_k = \frac{p_k}{p_{k-1}}; \quad i_k = \frac{4\pi j_k}{cH_{k-1}}; \quad a_0^2 = \frac{\gamma p_{k-1}}{\rho_{k-1}}; \quad a_A^2 = \frac{H_{k-1}^2}{4\pi\rho_{k-1}}; \quad M = \frac{V_{k-1}}{a_0}; \quad N = \frac{a_A}{a_0}; \quad \psi = \psi_{k-1}$$

From the evolutionary conditions for S^+ and $S^-$the following restrictions apply on the angles of inclination φ of $S^{\pm}$ and the density ratios η_k:

$$S^+: W \geq 0; \quad \chi \leq 1; \quad \chi \leq \eta_k \leq 1; \qquad S^-: W \leq 0; \quad \chi \geq 1; \quad \eta_k \leq 1 \tag{4}$$

$$W(\varphi) = \sin^2\varphi + (\chi - N^2 - 1)\cdot M^{-2} \tag{5}$$

where Eq.5 defines the angles of inclination of the fast and slow magnetosonic characteristics.

The state behind $R^\pm$ is determined by the system of ordinary differential equations (Pushkar' 1979; Barmin and Pushkar' 1990):

$$\frac{1}{R}\cdot\frac{dR}{d\varphi} = -\frac{2}{M_\varphi}\cdot\frac{(M_\varphi^4 - N_\varphi^2)M_r + N_rN_\varphi M_\varphi}{3M_\varphi^4 - (2-\gamma)M_\varphi^2 - (\gamma-1)N_\varphi^2} \equiv Q$$

$$\frac{dM_r}{d\varphi} = (M_\varphi + \frac{N_rN_\varphi M_\varphi}{M_\varphi^2 - N_\varphi^2} - \frac{\gamma-1}{2}M_r)\cdot Q$$

$$\frac{dN_r}{d\varphi} = N_\varphi + N_r(\frac{M_\varphi^2}{M_\varphi^2 - N_\varphi^2} - 0.5\gamma)\cdot Q; \qquad \frac{dN_\varphi}{d\varphi} = -(N_r + 0.5\gamma Q N_\varphi)$$

$$M_\varphi = -0.5\cdot \mid ((N_\varphi+1)^2 + N_r^2)^{1/2} \pm ((N_\varphi-1)^2 + N_r^2)^{1/2} \mid$$

$$R = \frac{\rho_k}{\rho_{k-1}}; \qquad M_\varphi = \frac{V_\varphi}{a}; \qquad M_r = \frac{V_r}{a}; \qquad N_\varphi = \frac{H_\varphi}{a(4\pi\rho)^{1/2}} \tag{6}$$

$$N_r = \frac{H_r}{a(4\pi\rho)^{1/2}}; \qquad a^2 = C_0\gamma\rho^{\gamma-1}; \qquad \eta_k = \frac{1}{R}; \quad P_k = R^\gamma$$

$$\omega_{vk} = \varphi + \tan^{-1}\frac{M_\varphi}{M_r} - 0.5(1 - \frac{M_r}{\mid M_r \mid})\pi$$

$$(u_{xk}, u_{yk}) = R^{0.5(\gamma-1)}(M_\varphi^2 + M_r^2)^{1/2}(\cos\omega_{vk}, \sin\omega_{vk})$$

$$\omega_{Hk} = \varphi + \tan^{-1}\frac{N_\varphi}{N_r} + 0.5(1 - \frac{N_r}{\mid N_r \mid})\pi$$

$$(h_{xk}, h_{yk}) = R^{\gamma/2}N^{-1}(N_\varphi^2 + N_r^2)^{1/2}(\cos\omega_{Hk}, \sin\omega_{Hk})$$

Here subscripts r and φ designate the components of $\boldsymbol{V}$ and $\boldsymbol{H}$ in polar coordinates. In the equation for M_φ the signs $\pm$ are related to fast and slow waves, respectively. The initial value of $\varphi_0^\pm$ is determined by Eq.5, and the limiting values of $\varphi^\pm$ for R_*^+ by condition $M_\varphi = N_\varphi (V_\varphi = a_A)$, and for R_*^- by the condition of the rarefaction proceeding down to vacuum ($R = 0$) (Pushkar' 1976).

The angle of inclination φ_A of the plane-polarized rotational discontinuity is determined from equation $\chi = 1$, the medium state behind A being uniquely given by setting $P_k = 1, \eta_k = 1$, and $i = -2\cos(\varphi_A - \psi)$ in the system (3), so A has no free parameter.

The states behind either of the interacting discontinuities having a flux of mass through its surface are functions of their intensities and governing parameters and may be uniquely determined from relations (3). Thus, the spatial intersection of S_1 and S_1' leads to a break between the physical values $(\rho, p, \boldsymbol{V}, \boldsymbol{H})$ in states (1) and (1') (Figs.1a and 1b). To find the solution we must join the states on C if the interaction is regular, or on C and C' if the interaction is nonregular. On each contact discontinuity the following boundary conditions are available:

a) in the case of regular interaction:

$$P(\varphi_2^+, \varphi_4^-) - P(\varphi_2^{+\prime}, \varphi_4^{-\prime}) = 0; \boldsymbol{V}(\varphi_2^+, \varphi_4^-) - \boldsymbol{V}(\varphi_2^{+\prime}, \varphi_4^{-\prime}) = 0; \boldsymbol{H}(\varphi_2^+, \varphi_4^-) - \boldsymbol{H}(\varphi_2^{+\prime}, \varphi_4^{-\prime}) = 0 \tag{7}$$

b) on C in the case of nonregular interaction:

$$P(\varphi_2^+, \varphi_4^-) - P(\delta, M_M, \varphi_5^-) = 0; \boldsymbol{V}(\varphi_2^+, \varphi_4^-) - \boldsymbol{V}(\delta, M_M, \varphi_5^-) = 0; \boldsymbol{H}(\varphi_2^+, \varphi_4^-) - \boldsymbol{H}(\delta, M_M, \varphi_5^-) = 0 \tag{8}$$

and the same conditions for $\varphi_2^{+\prime}, \varphi_4^{-\prime}, \varphi_5^{-\prime}, \delta$, and M_M on C'.

The boundary conditions cannot be written in explicit form because the wave structure of the flow is unknown in advance and the relationships for quantities on $S^\pm$ and in $R^\pm$ are different: the finite relations (3) with restrictions (4) for $S^\pm$ and the differential equations (6) for $R^\pm$.

The local stationarity of the problem made it possible to use the generalized polars (Pushkar' 1979) proposed for investigating steady self-similar flows. The generalized polars represent two-dimensional bounded manifolds in the six-dimensional space of physical variables V_x, V_y, H_x, H_y, p, ρ, where each point is a function of two arguments φ^+, φ^-. For the numerical solution we designed a software package. This software made it possible to build up any feasible state behind any combination of waves for a given initial state, namely, for either of states (1) and (1′). In addition, it determines the kinds of waves making up the flows according to the values of $\varphi^\pm$ and selects the required system of equations for the description of their components.

The solution of the regular problem (system (7)) is reduced to seeking the absolute minimum of some functional depending on $S^\pm, A$, and $R^\pm$. From the preliminary analysis of generalized polars and their dependence on key parameters (Pushkar' 1979) a method of setting the initial approximation has been worked out which ensures convergence of the iteration process in 3-5 steps.

In the nonregular interaction it is first necessary to find the speed and the orientation of a coordinate system, in which in the vicinity of B and B' stationary solutions for branching S_1 and S_1' into systems of oblique waves exist, and to find these solutions. Such a formulation is equivalent to a problem of two-dimensional breakup of the discontinuity created by one shock wave. Thereafter, one needs to determine the overall Mach configuration joining on S_M the solutions for two branch points. The numerical method makes it possible to obtain the exact solution, all the discontinuities being separated out in explicit form.

2. Results

The solutions are constructed numerically over a wide range of the governing parameters. The effects connected with the self-consistent nature of interaction of the magnetic field and the matter are found. The wave pattern of the flow developing at the intersection of discontinuities cannot be predicted in advance. Different combinations of shock waves, rotational discontinuities and rarefaction waves may develop depending on the governing parameters, the structure of the flow being particularly sensitive to the magnetic field inclination ψ. Owing to the nonlinear nature of interaction the resulting flow downstream from the line of intersection does not have to contain the waves of the same kinds that impinge initially.

The values of the governing parameters are found at which sudden restructurings of the wave pattern occur. These sudden changes in the flow and magnetic field may be called 'catastrophes' of the flow. Several kinds of these catastrophes connected with the confluence of characteristics of different types are singled out. When one of the interacting shock waves S^+ is close to switch-on shock S_*^+ and the magnetic field passes through the normal to S^+, then the initial S_*^+ splits off into a shock wave and a rotational discontinuity of a finite intensity, the flow being completely restructured (the global catastrophe K_G). Finite changes in the pressure, velocity and magnetic field are generated throughout behind S_1 and S_1'. For example, if shocks S_1^+ and $S_1^{+\prime}$intersect when $M_1 = M_{1'} = 3, N = 2.5$, and $\vartheta = 20°$ at the change in ψ from 80° to 81° the flow $S^+AR^-CR^-S^+$ gives way the flow R^+AS$^-CS^-R^+$ (Figs.2a and 2b). Thus, the fast shocks are transformed into slow shocks. In this restructuring the pressure and the velocity on the contact interface increase by a factor of 2.5 and the induction of magnetic field decreases by a factor of 1.5.

The local catastrophes are connected with the appearance (or disappearance) of rotational discontinuities in some part of the flow: a) of small intensities when they are split off from the limiting R^+or R^-; b) of a finite intensity when the switch-on S_*^+ or switch-off S_*^- shock waves split off into a combination with A. In the latter case a system of adjacent current sheets is formed with oppositely directed currents.

In particular, the interaction of two fast shock waves is studied extensively. We singled out quasi-parallel and quasi-perpendicular types of regular interaction depending on ψ. The former

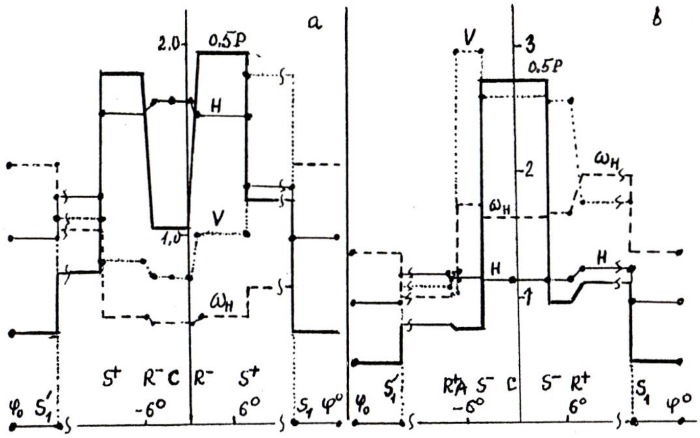

Fig. 2. Global catastrophe K_G of the flow when the angle ψ changes from 80° (case a) to 81° (case b). $M_1 = M_1' = 3$, $N = 2.5$, $\vartheta = 20°$

is characterized by a relatively small increasing strength of 'refracted' fast shocks and by a small strength of the slow waves. The rotational discontinuities are absent. The latter is characterized by restructurings of the flow at change in ψ, the strengths of all the waves (fast, Alfven and slow) are of the same order.

The two types of interaction differ in the change in the physical characteristics of flow. Figs.3-6 give the dependence of $P = p/p_0, V = V/a_0, H =| \boldsymbol{H} |/ | \boldsymbol{H}_0 |$, and $\Delta\psi = \omega_H - \psi_0$ on C as functions of ψ when the Mach numbers of S^+ and $S^{+'}$: $M_1 = M_1' = 3$ for $\vartheta = 10, 40, 60$, and 70° (curves 1-4). Figs.a, b, and c correspond to weak ($N = 0.7$), moderate ($N = 1.5$), and strong magnetic field ($N = 2$). It is clear from Fig.3 that the magnetic field has a strong influence on the resultant gasdynamic pressure, the effect being most marked for moderate field (Fig.3b, the right scale), when the ratio of the pressures in the quasi-perpendicular and quasi-parallel regimes of interaction reaches 3-4 units. In the strong field (Figs.3c-6c) the dependence $P(\psi), V(\psi), H(\psi)$, and $\Delta\psi(\psi)$ becomes disconnected as a result of the global catastrophe K_G. These relationships are also typical for weak, moderate, and strong magnetic field at other M_1, M_1', and N. In fact, when $M_1 = M_1' = M$ both in regular and nonregular interaction there is an approximate similarity for the relations $P(\psi), V(\psi), H(\psi)$, and $\Delta\psi(\psi)$ as a function of $M_A = M/N$.

Thus, in the quasi-parallel case the magnetic field smooths down the rise of pressure, decelerates the gas and grows strongly on C. In the quasi-perpendicular case the magnetic field compresses and accelerates the gas, the energy of magnetic field being transformed into thermodynamic energy and into the energy of directed flow, so that a high-velocity high-pressure stream is formed in a relatively weak magnetic field, but in the quasi-parallel case the reverse is true.

When the Alfven discontinuity impinges on the shock wave or contact discontinuity there exist directions of the magnetic field such that zones are formed where the density of the gas decreases by two to three orders of magnitude because the frozen-in magnetic field drags off the medium.

Next, the existence of regular solution and its relations with nonregular is examined. Like in the gasdynamic case two regular solutions exist (weak and strong families) if ϑ is smaller than some critical value ϑ^*. The weak and strong solutions coincide at $\vartheta = \vartheta^*$. The regular and nonregular interactions may: a) co-exist at the same governing parameters; *b*) be transformed suddenly as a result of the global catastrophe K_G so that one solution ceases to exist when the other appears; *c*) be absent. In the latter case in the formal constructed solution the point of intersection of S_1 and S_1' overtakes the Mach shock wave during a finite time. The 'parallelism' catastrophe is found to occur when $\boldsymbol{V} \parallel \boldsymbol{H}$ and the electric field near one branch point is equal

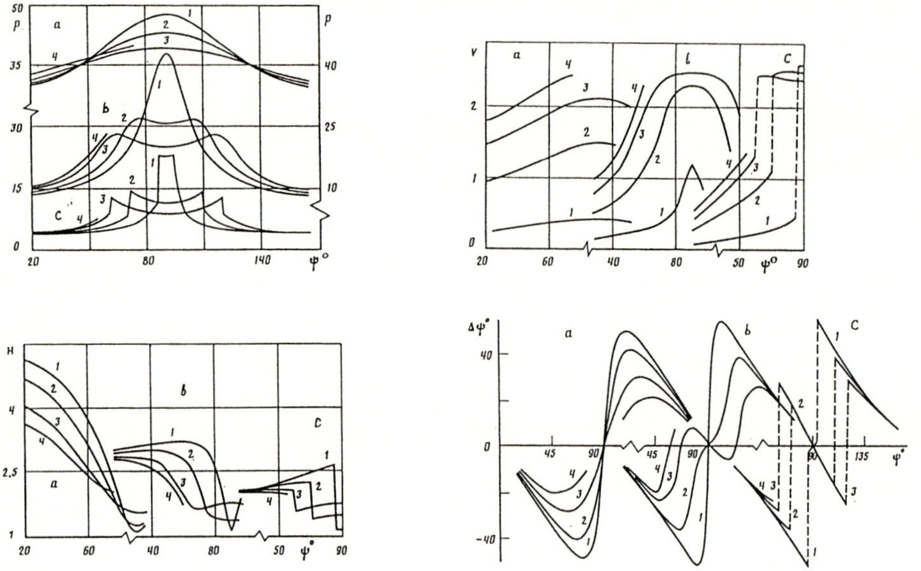

Fig. 3-6. Dependence of the physical parameters of the contact discontinuity C on the angle ψ. All quantities are related to those in the undisturbed medium ahead of S_1 and S_1'

to zero, so that the flow and the state of the medium change suddenly in the neighborhood of the contact interface.

It is found that the velocity and inclination of S_M depend essentially upon the magnetic induction vector $\boldsymbol{H}_0$. The transition to nonregular interaction brings about considerably more strong compression and acceleration of the gas by comparison with the regular interaction. The magnetic field increases more weakly. On C the pressure and velocity increase by a factor 2-5, and $|\boldsymbol{H}|$ increases only by a factor 1.5.

The features described (the sensitivity of the wave pattern to magnetic field inclination, existence of catastrophes of flow, the nature of effect of the magnetic field) have the same characteristic properties for all other kinds of interacting discontinuities.

References

Barmin AA, Pushkar' EA (1990) Oblique interaction of Alfven and contact discontinuities in magnetohydrodynamics. Izv. Akad. Nauk SSSR, Mekh. Zhidk. Gasa 1: 131

Barmin AA, Pushkar' EA (1991) Shock waves intersection in magnetohydrodynamics. Izv. Akad. Nauk SSSR, Mekh. Zhidk. Gasa 3: 132

Chernyi GG (1988) Gas Dynamics [in Russian]. Nauka, Moscow

Courant R, Friedrichs KO (1948) Hypersonic Flow and Shock Waves. Wiley-Interscience, New York

Kulikovskii AG, Lyubimov GA (1962) Magnetohydrodynamics [in Russian]. Fizmatgiz, Moscow

Pushkar' EA (1976) Simple stationary waves in an inclined magnetic field. Izv. Akad. Nauk SSSR, Mekh. Zhidk. Gasa 5: 130

Pushkar' EA (1978) Oblique magnetohydrodynamic shock waves. Izv. Akad. Nauk SSSR, Mekh. Zhidk. Gasa 4: 106

Pushkar' EA (1979) Generalized polars of plane-polarized self- similar flows in magnetohydrodynamics. Izv. Akad. Nauk SSSR, Mekh; Zhidk. Gasa 3: 111

Oblique Interaction of Strong Solar Wind Discontinuities in the Vicinity of the Terrestrial Magnetosphere

E.A. Pushkar'* and **S.A. Grib†**
*General and Applied Mathematics Department, MASI (VTUZ-ZIL), Moscow, Russia
†Central Astronomical Observatory of RAS (Pulkovo), Saint-Petersburg, Russia

Abstract. The interaction of a solar wind shock perturbation (fast shock wave or rotational discontinuity) with the bow shock wave front upstream from the magnetosphere of the Earth is considered in the ideal MHD approximation. It is shown that as the shock perturbation propagates along the bow front the character of the flow is qualitatively and quantitatively modified, being asymmetrical on the flanks. The effect of the interplanetary magnetic field orientation and the obliqueness of the arriving solar wind shock perturbation on this process is studied. It is found that there is some kind of similarity with the kinetic description of the bow front as partly quasiperpendicular and partly quasiparallel.

Key words: Solar wind shock waves, Earth's bow shock, Rotational discontinuity

1. Introduction

It is well known that strong solar wind discontinuities such as tangential ones or shock waves may cause the compression of the Earth's magnetosphere and the appearance of a so-called sudden impulse [SI^+] or sudden storm commencement [SSC] of a geomagnetic storm (Dryer 1975). These SSCs and SI^+s may trigger damped magnetic pulsations which are observed on the ground over a large area (Wedeken et al. 1986). Presently a large amount of data exist (Gosling et al. 1991) showing that the largest geomagnetic storms are associated with the passage by the Earth of nonstationary interplanetary shock waves and coronal mass ejections. The collision of the solar wind tangential discontinuity with the bow shock was studied by Neubauer (1975). In this case the vectors of the velocity and magnetic field are parallel, so the analysis was significantly simplified. The incidence of the shock on the bow front was considered by Zhuang et al. (1981) also in a simplified model in which the appearance of the Alfvén discontinuities and slow waves was neglected.

The interaction of interplanetary shock waves with the bow shock - magnetopause system was treated in the one-dimensional framework (Grib et al. 1979), for example, in the presence of an oblique interplanetary magnetic field $\boldsymbol{B}_{SW}$ in the case of 45° angle between $\boldsymbol{B}_{SW}$ and the solar wind velocity (Grib 1982). However, the two-dimensional MHD model of the interaction between the strong discontinuities is quite essential and leads to new qualitative effects both in MHD (Barmin and Pushkar' 1990, 1991) and in the magnetospheric physics (Pushkar' et al. 1991).

2. Formation

Let us consider the interaction of the plane front of a fast shock wave S_f (or rotational discontinuity A) propagating from the Sun at the velocity $\boldsymbol{V}_f$ (or $\boldsymbol{V}_A$) relative to the solar wind with the terrestrial bow shock S_b. We will assume that $\boldsymbol{B}_{SW}$ is inclined to the solar wind velocity $\boldsymbol{V}_{SW}$ at the angle ψ_B, and the normal vectors to the wave fronts, $\boldsymbol{V}_{SW}$, and $\boldsymbol{B}_{SW}$ all lie in the same plane (a plane-polarized problem) (Fig.1). The normal to S_f (or A) is inclined to $\boldsymbol{V}_{SW}$ at an angle ζ. In the coordinate system XOY moving with the Earth (the OX axis is along $\mathbf{V}_{SW}$, the OY axis is parallel to the plane of the flow), S_b is stationary and all the downstream parameters are determined by the angle α of the inclination S_b to the OX axis from the relations for the oblique MHD-shock (Pushkar' 1978).

Shock Waves @ Marseille II
Editors: R. Brun, L. Z. Dumitrescu

After contacting of the fronts S_f (or A) and S_b at the point D the arriving discontinuity moves along S_b crossing it at two points D' and D'' (Fig.1). This motion is supersonic so that the process of collision may be considered locally, taking into account the fact that the intensity S_f (or A) and the state downstream from it are constant and S_b depends parametrically on α. At the points D' and D'' each front is approximated by a plane.

The problem as a whole is determined by five dimensionless parameters: the solar wind Mach number M_{SW}, the Mach number of the travelling shock wave $M_f = V_f/a_0$, $\beta = 8\pi p/B^2_{SW}$, ψ_B and ζ. The local problem is determined by the angle α.

In a coordinate system moving with the point of intersection of the fronts for fixed α the problem is stationary and self-similar since it does not have any characteristic linear scale. Therefore, a regular solution downstream from the point of intersection can be obtained in the form of two combinations of self-similar stationary solutions: fast wave (shock S^+ or rarefaction wave R^+), plane-polarized Alfvén discontinuity A, and slow wave (S^- or R^-). There is a contact discontinuity C between these two combinations (Barmin and Pushkar' 1990, 1991). The transformed and transmitted waves moving back and forth relative to the solar wind will be designated by the subscripts b and f, respectively. The formers move through the disturbed solar wind and the latters through the undisturbed magnetosheath.

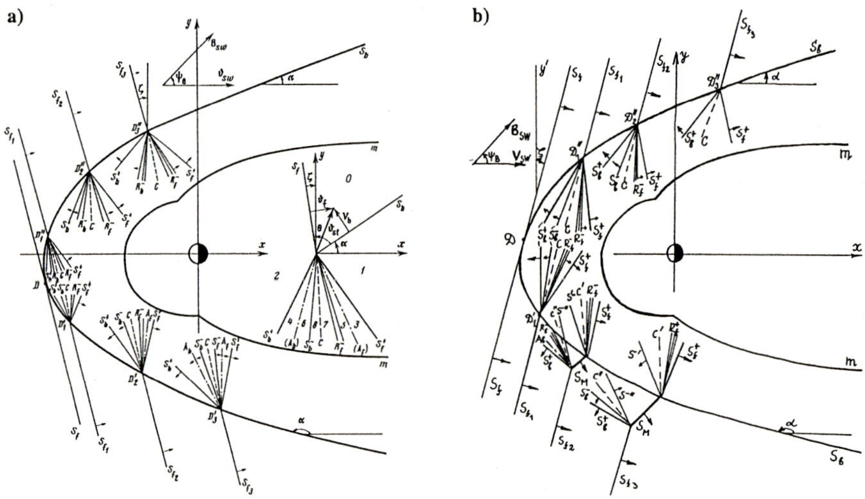

Fig. 1. Motion of the solar wind shock wave S_f over the bow shock S_b, m is the magnetopause. $S_{fi}, D_i^{'}$, and $D_i^{''}$, $i = 1, 2$, and 3 are sequential positions of S_f and points of intersection with the flanks of S_b. $M_{SW} = 3, \psi_B = 45°$, and $\beta = 1$. Case (a): $\zeta = 20°$, case (b): $\zeta = -20°$

Fast and slow waves are characterized by one free parameter and the Alfvén discontinuity in a plane-polarized case is determined only by the conditions ahead of it. The method of solving the problem consists in determining the types and intensities of the transmitted and transformed waves provided continuity of all physical values except density on the contact discontinuity C, where four independent boundary conditions are available (Barmin and Pushkar' 1990, 1991):

$$V_{Xb} = V_{Xf}; V_{Yb} = V_{Yf}; | \boldsymbol{B}_b | = | \boldsymbol{B}_f |; P_b = P_f \tag{1}$$

Boundary conditions (1) cannot be written in explicit form because the wave types in the solution are unknown in advance and the relationships connecting the flow parameters on $S^{\pm}$ and in $R^{\pm}$ are qualitatively different (Barmin and Pushkar' 1990).

Two generalized polars (Pushkar' 1979) constructed for both initial states downstream from S_f and S_b are used to find the solution. The types of the generated waves are not specified in advance. They are determined during the iteration process based on the Newtonian method (Barmin and Pushkar' 1990). The interaction of the transmitted waves on different flanks has not been considered. The solution described represents the flow in the vicinity of the bow shock.

3. Analysis

Let us consider the change in the interaction pattern as S_f moves along S_b, choosing the following dimensionless parameters: $M_{SW} = 8$, $M_f = 3$, $\beta = 1$, $\psi_B = 45°$ and $\zeta = 20°$. These values correspond to the mean values of the solar wind at 1 a.e. (Hundhauzen 1972): $V_{SW} = 390$ km·s^{-1}, $|\,\boldsymbol{B}_{SW}\,| = 6.2\gamma$, proton temperature $1.2 \cdot 10^5$ K, particle concentration 11 cm^{-3}, $V_f = 170$ km·s^{-1}. Point D (Fig.1a) corresponds to $\alpha = 110°$.

Types of waves formed in the result of the interaction are shown in Fig.1a for different points D. The quantities behind the corresponding waves divided by those in the undisturbed solar wind are given in Figs.2a-2f as functions of α. The jumps of P and $|\,\boldsymbol{B}\,|$ in S_f and S_b are a unit smaller than the ordinates of curves 1 and 2. The changes in quantities for the other waves are equal to the ordinate differences of the corresponding curves, e.g., in Fig.2a the changes in pressure are: $p_3 - p_1$ in S_f^+, $p_4 - p_3$ in S_f^- (or R_f^-), $p_4 - p_5$ in S_b^- (or R_b^-), $p_5 - p_2$ in S_b^+.

The flow in the vicinity of a head-on collision ($\alpha = 110°$) has the form $S_b^+ S_b^- C R_f^- S_f^+$ which corresponds to the results obtained by Grib (1982) for collision at the subsolar point. The types of the waves comprising the flow may be determined from Figs.2a, 2b according to the change in pressure (pressure grows in $S^{\pm}$ and drops in $R^{\pm}$) or to change in $|\,\boldsymbol{B}\,|$:$|\,\boldsymbol{B}\,|$ grows in S^+ while in S^- $|\,\boldsymbol{B}\,|$ drops, in $R^{\pm}$ vice versa. When $\alpha = 100°$ the flow is restructured into $S_b^+ R_b^- C R_f^- S_f^+$ (the point of intersection of curves 4, 5 in Figs.2a, 2 b). As S_b goes further along S_f with the reduction of the angle α, the flow is not changed qualitatively (points D''_{1-3} in Fig.1). The role of the slow waves for the change in pressure and V_x is small (of order of $5 - 10\%$ in comparison with the change in P and V_x in S^+). However, their effect is significant for the change in $|\,\boldsymbol{B}\,|$ and V_Y (up to 50% of the variations of these parameters in S^+). Thus, the slow waves should not be neglected. The characteristic non-monotonic changes in B_x, B_Y and V_Y with decreasing α are shown in Figs.2c, 2d, 2f.

On the other flank the flow has the form $S_b^+ S_b^- C R_f^- S_f^+$ up to $\alpha = 140°$, where the Alfvén discontinuity is split off from R_f^- (curve 6 in Figs.2 c - 2f), the flow becomes $S_b^+ S_b^- C R_f^- A_f S_f^+$. When this restructuring occurs the change in all quantities depends continuously on α, as the change of $\boldsymbol{B}$ in A_f and R_f^- takes place in the same direction. On further propagation of S_f the intensity of A_f (namely jumps of B_x and B_Y) grows rapidly. At $\alpha = 148°$ a slow rarefaction wave R_f^- is continuously transformed into S_f^-.

When $\alpha = 156°$, S_b^- splits up into an Alfvén discontinuity of terminal intensity and a new $S_b^{-\prime}$ (curve 7 in Figs.2c - 2e) and the interaction takes the form $S_b^+ A_b S_b^{-\prime} C S_f^- A_f S_f^+$. The intensity of $S_b^{-\prime}$ emerging from the splitting is the same as for the initial S_b^-, but the changes in $\boldsymbol{B}_\tau$ in S_b^- and $S_b^{-\prime}$ have opposite signs. The net change in $\boldsymbol{B}$ in the $A_b S_b^{-\prime}$ combination coincides with its change in S_b^- before splitting. Thus we get two current sheets in a narrow zone in which currents flow in opposite directions. As α increases the current sheets diverge. The regular solution described above does not exist for $\alpha < 27°$ and $\alpha > 166°$. At these α the flow becomes similar to Mach interaction in gas dynamics. Using the solution constructed, it is possible to determine the displacement velocity of the bow shock.

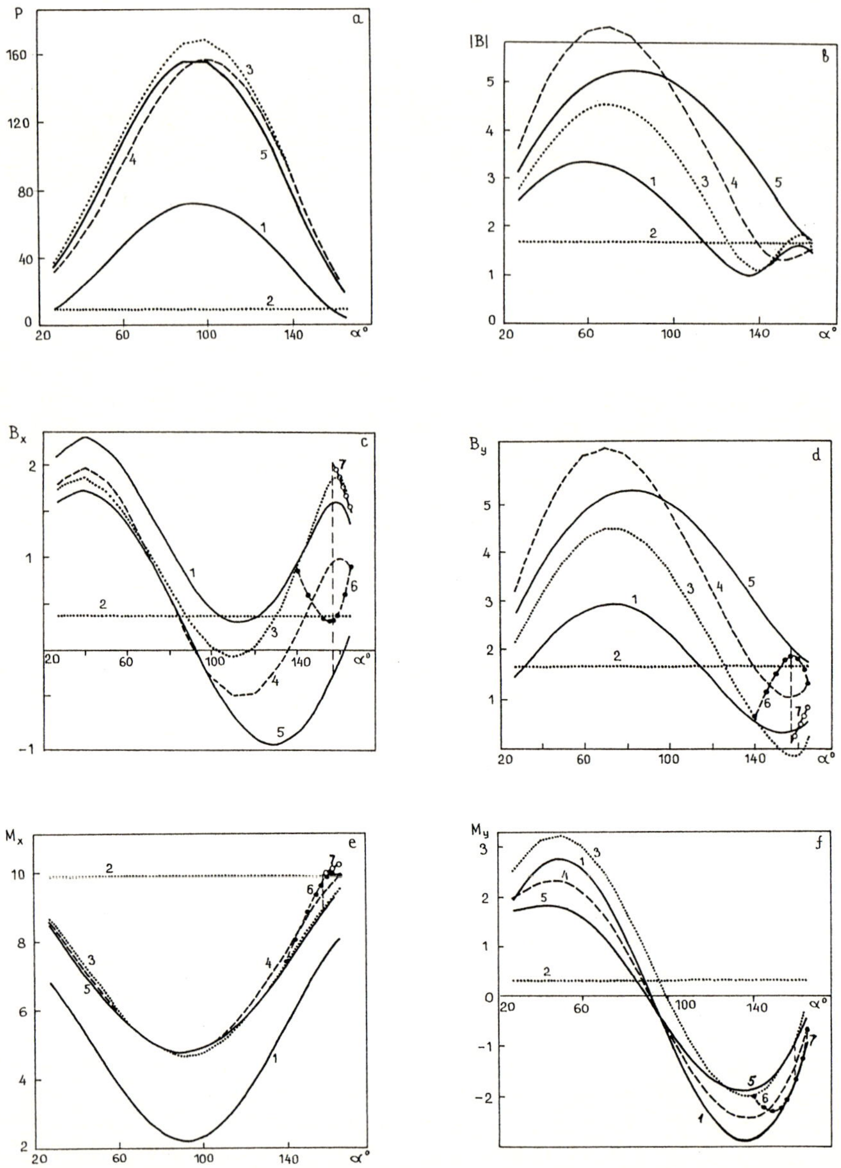

Fig. 2. Dependence of the parameters on the angle α: a) pressure p/p_0; b) $\boldsymbol{B}$ |/ | $\boldsymbol{B}_{SW}$ |; c) B_x/B_{SW}; d) B_y/B_{SW}; e) $M_x = V_x/a_0$; f) $M_y = V_y/a_0$. All quantities are related to those in the undisturbed solar wind. Curves 1 - 7 give the states: 1) behind S_b; 2) behind S_f; 3) behind S_f^+; 4) on C; 5) behind S_b^+; 6) behind A_f, 7) behind A_b

In the case when $\zeta = -20°$ (Fig.1 b), S_f impinges on the upper flank of the bow front and its normal $\boldsymbol{n}_{Sf}$ forms an angle $\delta = 65°$ with $\boldsymbol{B}_{SW}$. Near head-on collision ($\alpha = 70°$) the flow is described by $S_b^+ S_b^- CR_f^- S_f^+$, which corresponds to the case $\zeta = 20°$. As S_f moves along the upper flank of S_b the angle between the magnetic field and the velocity $\boldsymbol{V}_{st}$ of the gas in the coordinate system, in which the problem is steady, diminishes. The interaction becomes quasiparallel (Barmin and Pushkar' 1991) and the intensity of the slow waves decreases. When $\alpha = 55°$ the shock S_b^- is continuously transformed into the rarefaction wave R_b^-. At $\alpha = 25°$ the vector $\boldsymbol{V}_{st}$ is parallel to $\boldsymbol{B}_{SW}$. The intensities of both slow waves equal zero. At smaller α the flow retains the same structure $S_b^+ S_b^- C R_f^- S_f^+$ and remains regular up to the characteristic value $\alpha^* \cong 11°$.

As S_f moves in the other direction from the head-on collision point ($\alpha > 70°$) the flow $S_b^+ S_b^- C R_f^- S_f^+$ remains unchanged up to the subsolar point and beyond along the whole of lower flank, where a regular solution exists at the angles α up to about 141°. At bigger α the transition to Mach interaction takes place (Fig.1 b).

4. Conclusions

The evolution pattern of the S_f and S_b interaction presented here shows the complex character of the flow and the possibility of its nonlinear restructuring. Calculations made for $0 < \psi_B < 45°, -50° < \zeta < 50°, \beta = 0.5 - 1$ and $M_f = 2 - 3$ have shown that this evolution may take place in different ways depending on the key parameters. It is found that the most essential are ψ_B, ζ, and β. The collision causes nonlinear breakup of either of the interacting shock waves into a system of waves separated by a contact interface. These systems are transformed as the point of interaction travels along the bow shock. The transmitted fast shock wave may differ significantly from the arriving wave. The global evolution of the flow is determined by the orientation of the solar wind magnetic field and that of the arriving wave front. The wave flow structure may undergo qualitative changes.

At small angles between the normal to the arriving wave S_f and $\boldsymbol{B}_{SW}$ the Alfvén discontinuity can arise among the transmitted and transformed waves in two ways: either continuously due to splitting off from the slow rarefaction wave or jumpwise with the switch-off slow shock wave being broken up into the combination AS^-. The point of generation of the Alfvén discontinuity is determined by the angle ψ_B. When $\psi_B = 45°$ this occurs on the lower flank of the bow shock when the field is oriented upwards. Cases are possible in which with the shock propagation the Alfvén discontinuities (current sheets) are formed among the transmitted waves, which then disappear nonlinearly (jumpwise) and afterwards reappear somewhere further down the flanks.

The nonlinear restructuring of the flow determines the change in the physical quantities. The generation of the Alfvén discontinuities corresponds to a sharp change in the magnetic field and velocity, which may have a swinging effect on the magnetopause. Obviously, the adjacent current sheets, which are formed nonlinearly and generate opposite changes in the magnetic field, will be carried down to the magnetopause, possibly causing the sudden commencement of a geomagnetic storm. This may account for the fact that only some solar wind shock waves are geoeffective (Gosling et al. 1991).

From the consideration of the interaction of a rotational discontinuity with the bow shock it was found that the Alfvén discontinuity arriving with the solar wind does not penetrate through the bow shock into the magnetosheath at all inclinations: the refracted or reflected Alfvén discontinuities exist in some vicinity of head-on collision. The width of this zone in terms of the angle α is of the order of the angle of deflection of the magnetic field in the impinging A. In this process its intensity increases by two to four times. On the flanks the transmitted rotational discontinuity disappears in various ways: either "continuously" with the intensity decreasing to zero or jumpwise with the catastrophic restructuring of the flow. When the intensity of A is moderately high ($|\delta| < 35°$) the transmitted and reflected A exist at different value of α. The transmitted A can appear for long distances along the flanks.

The Alfvén discontinuities and slow waves play an important part, so that the change in the magnetic field in them may exceed the field jump in the fast waves several times. Upon the impingement of A just the slow waves have a determining effect on the magnetic field in the absence of a transmitted A_f.

A specific type of asymmetry is described in connection with the global effect of the solar wind shock wave on the magnetosphere. For example, if the shock wave impinges on the lower flank with the field oriented upwards and $\psi_B = 45°$ the Alfvén discontinuities appear only on the lower flank, while on the upper flank no such discontinuities are formed. The impingement of a rotational discontinuity results in a fast transmitted shock wave being formed on one flank and a rarefaction wave on the other. Away from head-on collision the slow waves are of opposite kinds on different flanks. Thus there is some kind of similarity with the well-known observation of the asymmetry of the kinetic plasma structure of the bow shock front (Russell and Greenstadt 1979).

Regular interaction does not occur at all points on the bow shock. When the intensity of the arriving waves is fairly high and it impinges laterally on one flank, Mach interaction always occurs on the opposite flank.

References

Barmin AA, Pushkar' EA (1990) Oblique interaction of Alfvén and contact discontinuities in magnetohydrodynamics. Izv. Akad. Nauk SSSR, Mech. Zhidk. Gaza 1: 131

Barmin AA, Pushkar' EA (1991) Shock waves intersection in magnetohydrodynamics. Izv. Akad. Nauk SSSR, Mech. Zhidk. Gaza 3: 132

Dryer M (1975) Interplanetary shocks waves: Recent developments. Space Sci. Rev. 17: 277

Gosling JT, McComas DJ, Phillips JL, Bame SJ (1991) Geomagnetic activity associated with Earth passage of interplanetary shock disturbances and coronal mass ejections. J. Geophys. Res. 96: 7831

Grib SA, Brunelli BE, Dryer M, Shen WW (1979) Interaction of interplanetary shock waves with the bow shock - magnetopause system. J. Geophys. Res. 84: 5907

Grib SA (1982) Interaction of non-perpendicular/parallel solar wind shock waves with the Earth's magnetosphere. Space Sci. Rev. 32: 43

Hundhauzen AJ (1972) Coronal expansion and solar wind. Springer, New York

Neubauer FM (1975) Nonlinear oblique interaction of interplanetary tangential discontinuities with magnetogasdynamics shocks. J. Geophys. Res. 80: 1213

Pushkar' EA (1978) On the oblique magnetohydrodynamic shock waves. Izv. Akad Nauk SSSR, Mech. Zhidk. Gasa 4:106

Pushkar' EA (1979) Generalized polars for the plane-polarized stationary self-similar flows in magnetohydrodynamics. Izv. Akad Nauk SSSR, Mech. Zhidk. Gasa 3:103

Pushkar' EA, Barmin AA, Grib SA (1991) Investigation in MHD- approximation of impingement of a solar wind shock wave on the Earth's bow shock. Geomagnetizm i Aeronomiya 31: 522

Russell CT, Greenstadt EW (1979) Initial ISEE observations of the bow shock. Nuovo Cim. 2C: 737

Wedeken U, Voelker H, Knott K, Lester M (1986) SSC-excited pulsations recorded near Noon on GEOS 2 and on the ground (CDAW 6), J. Geophys. Res. 91: 3089

Zhuang HC, Russell CT, Smith EJ, Gosling JT (1981) Three- dimensional interaction of interplanetary shock waves with the bow shock and magnetopause: A comparison of theory with ISEE observations. J. Geophys. Res. 86: 5990

Author Index (Volumes I - IV)

N

O

T

Contents - Volume I

Survey Paper

Plenary Lectures

Part 1: Hypersonic Flow and Aerospace Studies

Part 2: Combustion and Ram Accelerators

Part 3: Shock Tube Technology and Diagnostic Techniques

Part 4: Numerical Computations

Contents - Volume III

Survey Paper

Plenary Lectures

Part 1: Two-Phase Flows

Part 2: Shocks in Porous Media

Part 3: Shocks in Condensed Matter

Part 4: Industrial Applications and Environment

Part 5: Biological Aspects

Contents - Volume IV

Survey Paper

Plenary Lectures

Part 1: Shock Structure. Propagation and Focusing

Part 2: Shock Reflection and Diffraction

Part 3: Shock - Interface Interactions

Part 4: Interactions Involving Shocks

Part 5: Explosions, Blast Waves, Detonations

Shock Waves

An International Journal

This journal is aimed at publishing theoretical and experimental results on shock-wave phenomena gases, liquids, solids, and two-phase media from both the fundamental research and the applications points of view. It collates strictly refereed reviews, contributed papers, and short notes in the field, at present spread over numerous publications, in one focused journal.

ISSN 0938-1287

Subscription information 1995:

Vol. 5 (6 issues) DM 368,- (suggested list price)
plus carriage charges:
FRG DM 15,60; other countries DM 39,30

Springer

Tm.BA95.03.07

Springer-Verlag and the Environment

We at Springer-Verlag firmly believe that an international science publisher has a special obligation to the environment, and our corporate policies consistently reflect this conviction.

We also expect our business partners – paper mills, printers, packaging manufacturers, etc. – to commit themselves to using environmentally friendly materials and production processes.

The paper in this book is made from low- or no-chlorine pulp and is acid free, in conformance with international standards for paper permanency.